Prealgebra

Third Edition

Marvin L. Bittinger
Indiana University—Purdue University
at Indianapolis

David J. Ellenbogen
Community College of Vermont

ADDISON-WESLEY
An imprint of Addison Wesley Longman, Inc.

Reading, Massachusetts • Menlo Park, California • New York • Harlow, England
Don Mills, Ontario • Sydney • Mexico City • Madrid • Amsterdam

Publisher	Jason A. Jordan
Project Manager	Ruth Berry
Assistant Editor	Susan Connors Estey
Managing Editor	Ron Hampton
Production Supervisor	Kathleen A. Manley
Production Coordinator	Jane DePasquale
Design Direction	Susan Carsten
Text Designer	Rebecca Lloyd Lemna
Editorial and Production Services	Jennifer Bagdigian
Copy Editor	Martha Morong/Quadrata, Inc.
Art Editor	Janet Theurer
Photo Researcher	Naomi Kornhauser
Marketing Managers	Craig Bleyer and Laura Rogers
Illustrators	Scientific Illustrators, Gayle Hayes, and Rolin Graphics
Compositor	The Beacon Group
Cover Designer	Jeannet Leendertse
Cover Photographs	© Photodisc, 1998; © Navaswan/FPG International
Prepress Buyer	Caroline Fell
Manufacturing Coordinator	Evelyn Beaton

PHOTO CREDITS

1, Pascal Rondeau/Tony Stone Images   **4,** Copyright 1997, USA TODAY. Reprinted with permission   **69,** Pascal Rondeau/Tony Stone Images   **83,** Erlanson Productions/The Image Bank   **86,** AP/Wide World Photos   **99,** AP/Wide World Photos   **102,** Erlanson Productions/The Image Bank   **104,** John Banagan/The Image Bank   **143,** Amy Reichman/Envision   **178,** Amy Reichman/Envision   **205,** David Young-Wolff/PhotoEdit   **251,** Jeff Greenberg/Photo Researchers   **256,** Kirsten Willis   **258,** David Young-Wolff/PhotoEdit   **273,** Walter Hodges/Tony Stone Images   **320,** Gary Gay/The Image Bank   **333,** Mark Richards/PhotoEdit   **334,** Walter Hodges/Tony Stone Images   **351,** Penny Tweedie/Tony Stone Images   **353,** Steven Needham/Envision   **368,** Andrea Mohin/NYT Pictures   **370,** Steven Needham/Envision   **393,** AP/Wide World Photos   **394,** Penny Tweedie/Tony Stone Images   **405,** Mary Dyer   **417,** Andy Sacks/Tony Stone Images   **441,** Andy Sacks/Tony Stone Images   **453,** Bruce Forster/Tony Stone Images   **479,** Bruce Forster/Tony Stone Images   **481,** Reuters/Rebecca Cook/Archive Photos   **484,** Brian Spurlock   **484,** Jonathan Nourok/PhotoEdit   **511,** David Young-Wolff/PhotoEdit   **518,** Owen Franken/Stock Boston   **572,** David Young-Wolff/PhotoEdit   **581,** Dagmar Ehling/Photo Researchers   **590,** Dagmar Ehling/Photo Researchers

LIBRARY OF CONGRESS CATALOGING-IN-PUBLICATION DATA

Bittinger, Marvin L.
  Prealgebra/Marvin L. Bittinger, David J. Ellenbogen.—3rd ed.
      p.   cm.
  Includes index.
  ISBN 0-201-34024-0
  1. Mathematics. I. Ellenbogen, David. II. Title.
QA39.2.B585   1999
513'.14—dc21   99-26893
                                                       CIP

Reprinted with corrections, November 1999
Copyright © 2000 by Addison Wesley Longman, Inc. All rights reserved.
No part of this publication may be reproduced, stored in a retrieval system, or transmitted, in any form or by any means, electronic, mechanical, photocopying, recording, or otherwise, without the prior written permission of the publisher.
Printed in the United States of America.

2 3 4 5 6 7 8 9 10—VH—02010099

In Memory of Saul Ellenbogen,
Whose Love of Art and Mathematics
Served as Inspiration

Contents

1 Operations on the Whole Numbers

Pretest 2

1.1 Standard Notation 3
   Improving Your Math Study Skills: Tips for Using This Textbook 6
1.2 Addition 9
   Improving Your Math Study Skills: Getting Started in a Math Class: The First-Day Handout or Syllabus 14
1.3 Subtraction 19
1.4 Rounding and Estimating; Order 27
1.5 Multiplication and Area 33
1.6 Division 43
   Improving Your Math Study Skills: Homework 50
1.7 Solving Equations 55
1.8 Applications and Problem Solving 61
1.9 Exponential Notation and Order of Operations 73

Summary and Review Exercises 79
Test 81

2 Introduction to Integers and Algebraic Expressions 83

Pretest 84

2.1 Integers and the Number Line 85
   Improving Your Math Study Skills: Learning Resources and Time Management 90
2.2 Addition of Integers 93
2.3 Subtraction of Integers 97
   Improving Your Math Study Skills: Studying for Tests and Making the Most of Tutoring Sessions 100
2.4 Multiplication of Integers 105
2.5 Division of Integers 111
2.6 Introduction to Algebra and Expressions 115

2.7 Like Terms and Perimeter 121
2.8 Solving Equations 131

Summary and Review Exercises 137
Test 139

Cumulative Review: Chapters 1–2 141

3 Fractional Notation: Multiplication and Division 143

Pretest 144

3.1 Multiples and Divisibility 145
Improving Your Math Study Skills: Classwork Before and During Class 150
3.2 Factorizations 153
3.3 Fractions 159
3.4 Multiplication 165
3.5 Simplifying 171
3.6 Multiplying, Simplifying, and More with Area 177
3.7 Reciprocals and Division 187
3.8 Solving Equations: The Multiplication Principle 195

Summary and Review Exercises 199
Test 201

Cumulative Review: Chapters 1–3 203

4 Fractional Notation: Addition and Subtraction 205

Pretest 206

4.1 Least Common Multiples 207
4.2 Addition and Order 213
4.3 Subtraction, Equations, and Applications 223
4.4 Solving Equations: Using the Principles Together 231
4.5 Mixed Numerals 237
4.6 Addition and Subtraction Using Mixed Numerals; Applications 243
4.7 Multiplication and Division Using Mixed Numerals; Applications 253

Summary and Review Exercises 265
Test 269

Cumulative Review: Chapters 1–4 271

5 Decimal Notation 273

Pretest 274

- **5.1** Decimal Notation 275
- **5.2** Addition and Subtraction with Decimals 283
- **5.3** Multiplication with Decimals 291
- **5.4** Division with Decimals 299
- **5.5** More with Fractional Notation and Decimal Notation 309
- **5.6** Estimating 319

    Improving Your Math Study Skills: Study Tips for Trouble Spots 322

- **5.7** Solving Equations 325
- **5.8** Applications and Problem Solving 333

    Summary and Review Exercises 345

    Test 347

Cumulative Review: Chapters 1–5 349

6 Introduction to Graphing and Statistics 351

Pretest 352

- **6.1** Tables and Pictographs 353
- **6.2** Bar Graphs and Line Graphs 363

    Improving Your Math Study Skills: How Many Women Have Won the Ultimate Math Contest? 368

- **6.3** Ordered Pairs and Equations in Two Variables 375
- **6.4** Graphing Linear Equations 383
- **6.5** Means, Medians, and Modes 393
- **6.6** Predictions and Probability 399

    Summary and Review Exercises 407

    Test 411

Cumulative Review: Chapters 1–6 415

7 Ratio and Proportion 417

Pretest 418

- **7.1** Introduction to Ratios 419
- **7.2** Rates and Unit Prices 423
- **7.3** Proportions 429
- **7.4** Applications of Proportions 435
- **7.5** Geometric Applications 439

    Summary and Review Exercises 447

    Test 449

Cumulative Review: Chapters 1–7 451

8 Percent Notation 453

Pretest 454

- **8.1** Percent Notation 455
- **8.2** Solving Percent Problems Using Proportions 467
- **8.3** Solving Percent Problems Using Equations 473
- **8.4** Applications of Percent 479
- **8.5** Consumer Applications: Sales Tax, Commission, and Discount 487
- **8.6** Consumer Applications: Interest 497

Summary and Review Exercises 505
Test 507

Cumulative Review: Chapters 1–8 509

9 Geometry and Measures 511

Pretest 512

- **9.1** Systems of Linear Measurement 513
- **9.2** More with Perimeter and Area 523
- **9.3** Converting Units of Area 535
- **9.4** Angles 541
- **9.5** Square Roots and the Pythagorean Theorem 547
- **9.6** Volume and Capacity 555
- **9.7** Weight, Mass, and Temperature 565

Summary and Review Exercises 573
Test 577

Cumulative Review: Chapters 1–9 579

10 Polynomials 581

Pretest 582

- **10.1** Addition and Subtraction of Polynomials 583
- **10.2** Introduction to Multiplying and Factoring Polynomials 591

Improving Your Math Study Skills: Preparing for a Final Exam 596

- **10.3** More Multiplication of Polynomials 599
- **10.4** Integers as Exponents 603

Summary and Review Exercises 609
Test 611

Final Examination 613

Developmental Units 619

- **A** Addition 620
- **S** Subtraction 627
- **M** Multiplication 635
- **D** Division 643

Answers A-1

Index I-1

Index of Applications I-9

Preface

This text is the first in a series of texts that includes the following:

Bittinger: *Basic Mathematics,* Eighth Edition

Bittinger: *Introductory Algebra,* Eighth Edition

Bittinger: *Intermediate Algebra,* Eighth Edition

Bittinger/Beecher: *Introductory and Intermediate Algebra: A Combined Approach*

Bittinger/Ellenbogen: *Elementary Algebra: Concepts and Applications,* Fifth Edition

Bittinger/Ellenbogen: *Intermediate Algebra: Concepts and Applications,* Fifth Edition

Bittinger/Ellenbogen/Johnson: *Elementary and Intermediate Algebra: Concepts and Applications—A Combined Approach,* Second Edition

Prealgebra, Third Edition, is a significant revision of the Second Edition, particularly with respect to design, art program, pedagogy, features, and supplements package. Its unique approach, which has been developed and refined over three editions, continues to blend the following elements in order to bring students success:

- *Writing style.* The authors write in a clear, easy-to-read style that helps students progress from concepts through examples and margin exercises to section exercises.

- *Problem-solving approach.* The basis for solving problems and real-data applications is a five-step process (*Familiarize, Translate, Solve, Check,* and *State*) introduced early in the text and used consistently throughout. This problem-solving approach provides students with a consistent framework for solving applications. (See pages 61, 258, 335, and 338.)

- *Real data.* Real-data applications aid in motivating students by connecting the mathematics to their everyday lives. Extensive research was conducted to find new applications that relate mathematics to the real world.

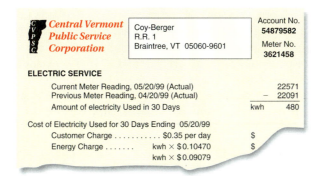

Bicycle Color Preferences

Other $\frac{4}{25}$ Blue $\frac{6}{25}$
Silver $\frac{1}{20}$
Yellow $\frac{1}{50}$ Black $\frac{23}{100}$
White $\frac{2}{25}$
Red $\frac{11}{50}$

Source: Bicycle Market Research Institute

- ***Art program.*** The art program has been expanded to improve the visualization of mathematical concepts and to enhance the real-data applications.
- ***Reviewer feedback.*** The authors solicit feedback from reviewers and students to help fulfill student and instructor needs.
- ***Accuracy.*** The manuscript is subjected to an extensive accuracy-checking process to eliminate errors.
- ***Supplements package.*** All ancillary materials are closely tied with the text and have been created by members of the author team to provide a complete and consistent package for both students and instructors.

What's New in the Third Edition?

The style, format, and approach have been strengthened in this new edition in a number of ways.

Updated Applications   Extensive research has been done to make the applications in the Third Edition even more up-to-date and realistic. A large number of the applications are new to this edition, and many are

drawn from the fields of business and economics, life and physical sciences, social sciences, and areas of general interest such as sports and health. To encourage students to understand the relevance of mathematics, many applications are enhanced by graphs and drawings similar to those found in today's newspapers and magazines. Many applications are also titled for quick and easy reference, and use real-data applications. (See pages 61, 256, 336, 354, 365, and 399.) Often applications are credited with a source line.

New Art and Design   To enhance the greater emphasis on real data and applications, we have extensively increased the number of pieces of technical and situational art (see pages 3, 333, and 341). Color has been used in a methodical and precise manner so that its use carries a consistent meaning, which enhances the readability of the text. *Prealgebra* now appears in full color, so virtually every page looks more inviting to the reader.

World Wide Web Integration   In an effort to encourage students to benefit from the resources that are available on the Addison Wesley Longman MathMax Web site (see description of the Web site later in this preface), we have added Web site icons to a number of key locations throughout the book. The Web site icons are intended to communicate to students that they will find additional related content on the Web site. For example, students can further explore the subject of the chapter opening applications (see icons on pages 83, 205, and 273) or seek additional practice exercises (see icons on pages 15, 175, and 259) on the MathMax *Prealgebra* Web site. In addition, students can visit the Web site to access InterAct Math exercises, extensive Chapter Reviews, and other helpful resources.

Preface

Collaborative Learning Features An icon located at the end of an exercise set signals the existence of a Collaborative Learning Activity correlating to that section in Irene Doo's *Collaborative Learning Activities Manual* (see pages 72, 230, and 496). Please contact your Addison Wesley Longman sales consultant for details on ordering this supplement.

Exercises The deletion of answer lines in the exercise sets has allowed us to include more exercises in the Third Edition. Exercises are paired, meaning that each even-numbered exercise is very much like the odd-numbered one that precedes it. This gives the instructor several options: If an instructor wants the student to have answers available, the odd-numbered exercises are assigned; if an instructor wants the student to practice (perhaps for a test) with no answers available, then the even-numbered exercises are assigned. In this way, each exercise set actually serves as two exercise sets. Answers to all odd-numbered exercises, with the exception of the Thinking and Writing exercises, and *all* Skill Maintenance exercises appear at the back of the text.

Skill Maintenance Exercises The Skill Maintenance exercises have been enhanced by the inclusion of 60% more exercises in this edition. These exercises review important skills and concepts from earlier chapters of the book. Section and objective codes now appear next to each Skill Maintenance exercise for easy reference (see pages 110, 212, and 318). Answers to all Skill Maintenance exercises appear at the back of the book.

Synthesis Exercises These exercises now appear in every exercise set, Summary and Review, Chapter Test, and Cumulative Review. Synthesis exercises help build critical thinking skills by challenging students to synthesize or combine learning objectives from the section being studied as well as preceding sections in the book.

Thinking and Writing Exercises Each Synthesis exercise section begins with two or three exercises that require written answers that aid in comprehension, critical thinking, and conceptualization. These exercises are designated with the icon ◆ and have been written with all students in mind—not just those looking for a challenge. Answers to the writing ex-exercises are not given at the back of the book because some instructors may collect answers to these exercises, and because correct answers will vary from student to student (see pages 152, 230, and 332). In response to user feedback, the number of Thinking and Writing Exercises has been increased by 50%.

Content We have made the following improvements to the content of *Prealgebra*.

- It is more important than ever for students to be statistically literate. To address this need, the Third Edition of *Prealgebra* expands the coverage of statistics first introduced in the Second Edition. Specifically, increased use of bar graphs and averages appears throughout the text (see pages 61, 240, 258, 304, and 393). New material on the basics of interpolation, extrapolation, and probability is also now included (see pages 399–406).
- The incorporation of algebraic concepts has always been emphasized in *Prealgebra*. In keeping with this emphasis and in response to user feedback, the Third Edition includes a new section devoted solely to solving a mix of equations that involve either addition or multiplication (see pages 131–136). This introduction is then reinforced throughout the text (see pages 195, 225, 231, and 326). New emphasis is placed on

- distinguishing between equivalent expressions and equivalent equations (see pages 131, 231, and 325).

- Geometry is a topic that has always been an important part of *Prealgebra*. In the Third Edition, coverage has been expanded to integrate more work with circles, angles, and circle graphs (see pages 337 and 545).

- 🖩 Calculator exercises and "spotlights" appeared in the Second Edition and, in response to user and reviewer feedback, have been expanded in this edition. Several new Calculator Spotlights now appear, and nearly all exercise sets now include at least one pair of synthesis exercises designed to provide practice using a scientific calculator (see pages 112, 130, 258, and 438). Answers to all Calculator Spotlight exercises appear at the back of the text.

- To impove the flow of topics in Chapter 5, material on estimation now follows rather than precedes the section titled "More with Fractional Notation and Decimal Notation."

- Beginning in Section 6.5, the word "mean" is used in place of "average."

- Chapter 10, Polynomials, now includes the use of FOIL when multiplying two binomials, as well as improved coverage of negative exponents (see pages 599 and 604).

- Approximately once per chapter, a mini-lesson, Improving Your Math Study Skills, appears. (See pages 6, 100, and 322.) These features are referenced in the Table of Contents and can be covered in their entirety at the beginning of the course or as they arise in the text. These features can also be used in conjunction with the authors' "Math Study Skills" Videotape. Please see your Addison Wesley Longman sales consultant for details on how to obtain this videotape.

Learning Aids

Interactive Worktext Approach   The pedagogy of this text is designed to provide an interactive learning experience between the student and the exposition, annotated examples, art, margin exercises, and exercise sets. This approach provides students with a clear set of learning objectives, involves them with the development of the material, and provides immediate and continual reinforcement and assessment.

Section objectives are keyed by letter not only to section subheadings, but also to exercises in the exercise sets and Summary and Review, and to the answers to the Pretest, Chapter Test, and Cumulative Review questions. This enables students to find appropriate review material easily if they need help with a particular exercise.

Throughout the text, students are directed to numerous *margin exercises,* which provide immediate reinforcement of the concepts covered in each section.

Review Material   The Third Edition of *Prealgebra* continues to provide many opportunities for students to prepare for final assessment.

Now in a two-column format, a *Summary and Review* appears at the end of each chapter and provides an extensive set of review exercises. Reference codes beside each exercise or direction line direct the student to the specific subsection being reviewed (see pages 137, 265, and 447).

Also included at the end of every chapter but Chapters 1 and 10 is a *Cumulative Review*, which reviews material from all preceding chapters. At the back of the text are answers to all Cumulative Review exercises, together with section and objective references, so that students know exactly what material to study if they need help with a review exercise (see pages 141, 349, and 451). A final examination follows Chapter 10.

For Extra Help Many valuable study aids accompany this text. Below the list of objectives found at the beginning of each section are references to appropriate videotape, tutorial software, and CD-ROM programs to make it easy for the student to find the correct support materials.

Testing The following assessment opportunities exist in the text.

The *Diagnostic Pretest*, provided at the beginning of the text, helps place students in the appropriate chapter for their skill level by identifying familiar material and specific trouble areas (see page xxi).

Chapter Pretests can then be used to place students in a specific section of the chapter, allowing them to concentrate on topics with which they have particular difficulty (see pages 84, 206, and 418).

Chapter Tests allow students to review and test comprehension of skills developed in each chapter (see pages 139, 269, and 411).

Answers to all Diagnostic Pretest, Chapter Pretest, and Chapter Test questions are found at the back of the book, along with appropriate section and objective references.

Objectives

a	Add using mixed numerals.
b	Subtract and combine using mixed numerals.
c	Solve applied problems involving addition and subtraction with mixed numerals.
d	Add and subtract using negative mixed numerals.

For Extra Help

TAPE 8   MAC WIN   CD-ROM

Supplements for the Instructor

Annotated Instructor's Edition
0-201-64602-1

The Annotated Instructor's Edition is a specially bound version of the student text with answers to all margin exercises, exercise sets, and chapter tests printed in blue near the corresponding exercises.

Instructor's Solutions Manual
by Judith A. Penna
ISBN 0-201-64603-X

This manual contains brief, worked-out solutions to all even-numbered exercises in the text's exercise sets and answers to all Thinking and Writing exercises.

Printed Test Bank/Instructor's Resource Guide
by Laurie A. Hurley
ISBN 0-201-64604-8

The test-bank section of this supplement contains the following:

- Two alternate test forms for each chapter modeled after the Chapter Tests in the text
- Two alternate test forms for each chapter, with questions presented in the same topical order as the chapter objectives
- Two alternate test forms for each chapter designed for a 50-minute class period

- Two multiple-choice test forms for each chapter
- Two cumulative review tests for each chapter, with the exception of Chapters 1 and 10
- Six alternate test forms of the final examination: two with questions organized by chapter, two with questions scrambled, and two with multiple-choice questions
- Answers for the chapter tests and final examinations

The Instructor's Resource Manual section contains the following:

- Extra practice exercise sheets and answers for several of the most challenging topics in the text
- A conversion guide from the Second Edition to the Third Edition
- A videotape index for the "Steps to Success" video series that accompanies the text
- Black-line masters of grids and number lines for transparency masters or test preparation

Collaborative Learning Activities Manual

by Irene Doo
ISBN 0-201-66198-5

The Collaborative Learning Activities Manual, written by Irene Doo of Austin Community College, features group activities that are tied graphically to sections of the textbook via an icon. This manual also provides instruction on the setup and administration of a collaborative classroom environment.

TestGen-EQ/QuizMaster-EQ CD-ROM

ISBN 0-201-64607-2

This powerful test-generation software is provided on a dual-platform Windows/Macintosh CD-ROM. TestGen-EQ's friendly graphical interface enables instructors to easily view, edit, and add questions, transfer questions to tests, and print tests in a variety of fonts and forms. Search and sort features help the instructor locate questions quickly and arrange them in a preferred order. Several question formats are available, including short-answer, true/false, multiple-choice, essay, matching, and bimodal (a bimodal question is one that can be saved in either multiple-choice or short-answer form). A built-in question editor allows the instructor to create graphs, import graphics, and insert variable numbers, text, and mathematical symbols and templates. Computerized test banks include algorithmically defined problems organized according to the textbook table of contents. Instructors can create and export practice tests as HTML for use on the World Wide Web.

Using QuizMaster-EQ, instructors can post tests and quizzes created in TestGen-EQ to a computer network so that students can take them on-line. Instructors can set preferences for how and when tests are administered. QuizMaster automatically grades the exams and allows the instructor to view or print a variety of reports for individual students, classes, or courses.

InterAct Math Plus Instructor's Package

ISBN 0-201-63555-0 (Windows), ISBN 0-201-64805-9 (Macintosh)

Used in conjunction with the InterAct Math Tutorial Software for students (0-201-64609-9), this networkable software provides instructors with full course management capabilities for tracking and reporting student use of the tutorial software. Instructors can create and administer on-line tests, summarize student results, and monitor student progress in the software.

Instructors should also review the list of supplements for students that follows.

Supplements for the Student

Student's Solutions Manual
by Judith A. Penna
ISBN 0-201-34026-7

This manual provides completely worked-out solutions with step-by-step annotations for all odd-numbered exercises except the Thinking and Writing exercises. It may be purchased by students from Addison Wesley Longman.

"Steps to Success" Videotapes
ISBN 0-201-64807-5

This videotape series features an engaging team of mathematics instructors, including your authors, presenting comprehensive coverage of each section of the text in a student-interactive format. The lecturers' presentations include examples and problems from the text and support an approach that emphasizes visualization and problem solving. A video icon at the beginning of each section references the appropriate videotape number.

InterAct Math Tutorial Software
ISBN 0-201-64609-9

The InterAct Math Tutorial Software, provided on a dual-platform Windows/Macintosh CD-ROM has been developed by professional software engineers working closely with a team of experienced developmental mathematics instructors. This software includes exercises that are linked one-to-one to the odd-numbered exercises in the text; the InterAct Math exercises require the same computational and problem-solving skills as their companion exercises in the book. Each exercise is accompanied by an example and an interactive guided solution designed to involve students in the solution process and help them identify precisely where they are having trouble. For each section of the text, the software tracks student activity and scores, which can be printed out in summary form. An InterAct Math icon at the beginning of each section identifies section coverage.

MathMax Multimedia CD-ROM for *Prealgebra*
ISBN 0-201-64806-7

This dual-platform Windows/Macintosh CD-ROM provides an interactive environment using graphics, animations, and audio narration to build on some of the unique and proven features of the MathMax series. The content of the CD is tightly and consistently integrated with the text, highlighting key concepts and referencing the *Prealgebra* numbering scheme so that students can move smoothly between the CD and other supplements. The CD includes narrated animations that provide step-by-step explanations of many of the examples in the text, and the narrations are accompanied by multiple-choice exercises. Also included are interactive chapter reviews, multimedia study skills presentations, InterAct Math exercises for every section of the text, and a glossary of key terms. A CD-ROM icon at the beginning of each section indicates section coverage.

An Addison Wesley Longman sales consultant can arrange for a demonstration of the *Prealgebra* MathMax CD-ROM.

World Wide Web Supplement (www.mathmax.com)

This on-line supplement provides additional practice and reinforcement for students through detailed chapter review material, extra practice worksheets, and expanded chapter openers. In addition, students can download a plug-in for Addison Wesley Longman's InterAct Math exercises that allows them to access tutorial problems directly through their Web browser. The site also provides teaching tips and information for instructors about all the MathMax supplements available from Addison Wesley Longman.

MathPass, Version 2.0 for Windows
ISBN 0-201-66192-6

MathPass helps students succeed in their developmental mathematics courses by creating customized study plans based on diagnostic test results from ACT, Inc.'s Computer-Adaptive Placement Assessment and Support System (COMPASS®). MathPass pinpoints topics where the student needs in-depth study or targeted review and correlates these topics with the student's textbook and related supplements (such as videos, student's solutions manuals, Web sites, tutorial software, and multimedia CD-ROMs). The MathPass Learning System provides diagnostic assessment, focused instruction, and exit placement all in one package. Contact your local Addison Wesley Longman sales consultant for more information about MathPass.

MathXL
ISBN 0-201-68154-4
(*Prealgebra* bundled with the MathXL coupon package)

MathXL is a Web-based testing and tutorial system that allows students to take practice tests similar to the chapter tests in their book. MathXL generates a personalized study plan that indicates students' strengths and pinpoints topics where they need more practice. The program then provides the practice and instruction students need to improve their skills. Test scores and practice sessions are tracked by the program so that students can monitor their progress throughout the semester. The MathXL Web site requires each student to purchase a user ID and password.

"Math Study Skills for Students" Videotape
ISBN 0-201-88039-3

Designed to help students make better use of their math study time, this videotape helps students improve retention of concepts and procedures taught in classes from basic mathematics through intermediate algebra. Through carefully crafted graphics and comprehensive on-camera explanation, author Marvin L. Bittinger helps viewers focus on study skills that are commonly overlooked.

Overcoming Math Anxiety, Second Edition
by Randy Davidson and Ellen Levitov
ISBN 0-321-06918-8

Written to help students succeed with college-level mathematics coursework, this book includes step-by-step guidelines to problem solving, note taking, and word problems. This book helps students discover the reasons behind math anxiety, learn relaxation techniques, build better math skills, and improve learning strategies for math.

Spanish Glossary
ISBN 0-321-01647-5

This pocket-sized glossary includes concise, easy-to-understand English-to-Spanish translations for key mathematical terms.

Math Tutor Center
ISBN 0-201-66352-X
(*Prealgebra* bundled with Math Tutor Center registration)

The Addison Wesley Longman Math Tutor Center provides FREE tutoring to students enrolled in courses ranging from developmental mathematics through calculus. Assisted by qualified mathematics instructors via phone, fax, or e-mail, students can receive tutoring on examples, exercises, and problems contained in their Addison Wesley Longman math textbooks. Upon request, each new book can be bundled with a free registration number that provides the student with a 6-month subscription to the service. The Math Tutor Center is open Sunday through Thursday from 3 pm to 10 pm Eastern Standard Time. Contact your local Addison Wesley Longman sales consultant for more information about the Math Tutor Center.

Acknowledgments

We would like to express our appreciation to the many people who helped make this book possible. Judy Penna deserves special thanks for her preparation of the *Student's Solution Manual*, the *Instructor's Solutions Manual*, and her careful checking of the manuscript. Laura Hurley deserves special thanks for her preparation of the *Printed Test Bank/ Instructor's Resource Manual* and her careful checking of page proofs and compilation of the answers. Sincere thanks to Irene Doo for her outstanding work on the *Collaborative Learning Activities Manual*. Heartfelt thanks also to Professor Susan McAuliffe of the University of Vermont, Tracy Psaute and Thomas Schicker of the Community College of Vermont, and Vincent Koehler for their thorough inspection of the typeset pages and answers.

Martha Morong again showed why she's the best copyeditor in the business, and Jenny Bagdigian provided superb stewardship skills in juggling many different tasks at once, while preserving the high quality that we all consider essential. Thanks also to Janet Theurer for her fine work as art editor and to Kathy Manley for her work as production supervisor and for reworking numerous pages of manuscript. A big thank you to Ruth Berry for her fine work as project manager, and we also thank Tricia Mescall for supervising the production of our videotape series. Finally, a big thank you to Jason Jordan, our most trusted and talented publisher, for making sound decisions and providing us with the freedom we need to do our jobs right.

In addition, we thank the following professors for their thoughtful reviews, suggestions, and comments.

    Kenneth Benson, *University of Illinois at Urbana-Champaign*
    Timmy G. Bremer, *Prestonburg Community College*
    Beverly R. Broomell, *Suffolk Community College*
    Debra Bryant, *Tennessee Technological University*
    Jean-Marie Magnier, *Springfield Technical Community College*
    Tom Carson, *Midlands Technical College*
    Mary Jane Cordon
    Katherine Creery, *University of Memphis*
    Carol Flakus, *Lower Columbia College*
    Linda Galloway, *Macon Sate College*
    Kay Haralson, *Austin Peay State University*
    Celeste Hernandez, *Richland College*
    Courtney Hubbell, *Nicholls State University*
    Marilyn Jacobi, *Gateway Community-Technical College*
    Nancy Johnson, *Broward Community College, North Campus*
    Steve Kahn, *Anne Arundel Community College*
    Joanne Kelly, *Palm Beach Community College*
    Theodore Lai, *Hudson County Community College*

Nancy Lehmann, *Austin Community College*
Bob Maynard, *Tidewater Community College*
Frank Miller, *Orange Coast College*
Gary Nelson, *Asheville-Buncombe Technical Community College*
David Newell, *Pierce College / Cypress College*
Patricio A. Rojas, *Albuquerque Technical Vocational Institute*
Mickey Sargent, *Ranger College*
Sally Sestini, *Cerritos College*
David Swarbrick, *St. Edwards University / Austin Community College*
Sharon Testone, *Onondaga Community College*
David E. Thielk, *Port Townsend High School*
Victor Thomas, *Holyoke Community College*
Peter Wursthorn, *Capitol Community Tech College*
Kevin Yokoyama, *College of the Redwoods*

Diagnostic Pretest

Chapter 1

1. Add: 549 + 3764.

2. S

3. Multiply:
$$\begin{array}{r} 4\,5\,3 \\ \times \;\; 3\,7 \\ \hline \end{array}$$

4.

Chapter 2

5. Evaluate $6x^2$ for $x = -5$.

6. Find the opposite, or additive inverse, of -17.

Compute and simplify.

7. $-8 + (-5)$

8. $-20 + 17$

9. $-5 - 12$

10. $-4 - (-9)$

11. $(-6)(-9)$

12. $10 \div (-5) + 3^2$

13. Find the perimeter of a 20-ft by 15-ft garden.

14. Combine like terms: $9 - 3x + 4 + 10x$.

15. Solve: $3 + t = -14$.

16. Solve: $-6x = -24$.

Chapter 3

17. Multiply and simplify: $-\dfrac{6}{7} \cdot \dfrac{21}{12}$.

18. Solve: $\dfrac{3}{5}x = -24$.

19. Find another name for $\dfrac{7}{8}$, but with 40 as the denominator.

20. A group of 5 diners shares $\dfrac{3}{4}$ of a pie equally. How much of the pie does each person receive?

Chapter 4

21. Solve: $x + \dfrac{4}{5} = \dfrac{9}{10}$.

22. Divide and simplify: $3\dfrac{1}{2} \div \dfrac{3}{4}$.

23. Find the average: $3\dfrac{1}{2}, 5, 1\dfrac{1}{4}$.

24. Solve: $\dfrac{11}{5} = \dfrac{8}{5} + \dfrac{5}{4}a$.

Chapter 5

Perform the indicated operation.

25. $3.95 + 5.026$

26. $152.7 \div (-1.5)$

27. Combine like terms: $8.2x - 9.4y - 5.7x + 2.1y$.

28. Solve: $1.5t + 4.2 = 7.8$.

Diagnostic Pretest, p. xxiii

1. [1.2b] 4313   2. [1.7b] 28   3. [1.5b] 16,761
4. [1.8a] 23; 2   5. [2.6a] 150   6. [2.1d] 17
7. [2.2a] -13 8. [2.2a] -3 9. [2.3a] -17
10. [2.3a] 5   11. [2.4a] 54   12. [2.5b] 7
13. [2.7c] 70 ft   14. [2.7b] $7x + 13$ 15. [2.8a] -17
16. [2.8b] 4   17. [3.6a] $-\dfrac{3}{2}$ 18. [3.8a] -40
19. [3.5a] $\dfrac{35}{40}$ 20. [3.7c] $\dfrac{3}{20}$ 21. [4.3b] $\dfrac{1}{10}$
22. [4.7b] $4\dfrac{2}{3}$ 23. [4.6a], [4.7b] $3\dfrac{1}{4}$ 24. [4.4a] $\dfrac{12}{25}$
25. [5.2a] 8.976   26. [5.4a] -101.8
27. [5.2d] $2.5x - 7.3y$ 28. [5.7a] 2.4

Chapter 6

29. Graph: $x + 2y = -6$.

30. Find the mean, the median, and the mode:

9, 12, 12, 14, 18, 19.

Chapter 7

31. The ratio of the number of adults to the number of children at the Wildflower Daycare Center cannot be less than 1 to 8. If there are 30 children at the center, how many adults must be present?

32. How tall is a street sign that casts a 12-ft shadow at the same time that a 5-ft woman casts an 8-ft shadow?

Chapter 8

33. Find percent notation: $\frac{5}{8}$.

34. Find decimal notation: 2.4%.

35. The price of a bicycle helmet was reduced from $45 to $36. Find the percent of decrease in price.

36. A principal of $2000 earned 4.5% simple interest for one year. How much was in the account at the end of the year?

Chapter 9

Complete.

37. 2 yd = _____ in.

38. 4 cm = _____ km

39. 64 oz = _____ lb

40. 5 mg = _____ g

41. 2 yd² = _____ ft²

42. 10 qt = _____ gal

43. Find the area and the circumference of a circle with a diameter of 12 cm. Leave answers in terms of π.

44. Each side of a cube is 5 m long. Find the volume of the cube.

Chapter 10

45. Add: $(5x^3 + 4x^2 - 9) + (3x^3 - 7x^2 - 1)$.

47. Multiply: $(2x - 1)(x + 3)$.

49. Evaluate $(2x + 3)^0$ for $x = 7$.

Diagnostic Pretest

xxiv

29. [6.4b]

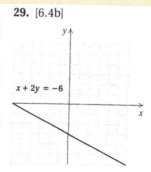

$x + 2y = -6$

30. [6.5a, b, c] 14; 13; 12 **31.** [7.4a] 4 **32.** [7.5a] 7.5 ft
33. [8.1c] 62.5% **34.** [8.1b] 0.024 **35.** [8.4b] 20%
36. [8.6a] $2090 **37.** [9.1a] 72 **38.** [9.1b] 0.00004
39. [9.7a] 4 **40.** [9.7b] 0.005 **41.** [9.3a] 18
42. [9.6b] 2.5 **43.** [9.2b] 36π cm²; 12π cm
44. [9.6a] 125 m³ **45.** [10.1a] $8x^3 - 3x^2 - 10$
46. [10.1c] $4x^4 - 8x^3 - x^2 - 2$
47. [10.3a] $2x^2 + 5x - 3$ **48.** [10.2c] $2x^3(5x^2 + 4x - 6)$
49. [10.4a] 1 **50.** [10.4b] $10a^4b^5$

1

Operations on the Whole Numbers

Introduction

In this chapter, we consider addition, subtraction, multiplication, and division of whole numbers. Then we study the solving of simple equations and apply all these skills to the solving of problems.

- **1.1** Standard Notation
- **1.2** Addition and Perimeter
- **1.3** Subtraction
- **1.4** Rounding and Estimating; Order
- **1.5** Multiplication and Area
- **1.6** Division
- **1.7** Solving Equations
- **1.8** Applications and Problem Solving
- **1.9** Exponential Notation and Order of Operations

An Application	The Mathematics
Total sales, in millions of dollars, of bicycles and related sporting supplies were $3534 in 1993, $3470 in 1994, $3435 in 1995, and $3356 in 1996. (**Source**: National Sporting Goods Association.) Find the total sales for the entire four-year period. This problem appears as Exercise 4 in Exercise Set 1.8.	We let T = the total sales, in millions of dollars. Since we are combining sales, addition can be used. We translate the problem to the equation $3534 + 3470 + 3435 + 3356 = T.$ ↑ This is how addition can occur in applications and problem solving.

 World Wide Web For more information, visit us at www.mathmax.com

Pretest: Chapter 1

1. Write a word name: 3,078,059.

2. Write in expanded notation: 6987.

3. Write in standard notation: Two billion, forty-seven million, three hundred ninety-eight thousand, five hundred eighty-nine.

4. What does the digit 6 mean in 2,967,342?

5. Round 956,449 to the nearest thousand.

6. Estimate the product 594 · 126 by first rounding the numbers to the nearest hundred.

7. Add.

$$\begin{array}{r} 7\,3\,1\,2 \\ +\,2\,9\,0\,4 \\ \hline \end{array}$$

8. Subtract.

$$\begin{array}{r} 7\,0\,1\,2 \\ -\,2\,9\,0\,4 \\ \hline \end{array}$$

9. Multiply: 359 · 64.

10. Divide: 23,149 ÷ 46.

Use either < or > for ▨ to form a true sentence.

11. 346 ▨ 364

12. 54 ▨ 45

Solve.

13. $326 \cdot 17 = m$

14. $y = 924 \div 42$

15. $19 + x = 53$

16. $34 \cdot n = 850$

Solve.

17. Anna weighs 121 lb and Kari weighs 109 lb. How much more does Anna weigh?

18. How many 12-jar cases are needed for 1512 jars of spaghetti sauce?

19. *Population.* The population of Illinois is 11,830,000. The population of Ohio is 11,151,000. (*Source*: U.S. Bureau of the Census.) What is the total population of Illinois and Ohio?

20. A rectangular lot measures 48 ft by 54 ft. A pool that is 15 ft by 20 ft is constructed on the lot. How much area is left over?

Evaluate.

21. 5^2

22. 4^3

Simplify.

23. $8^2 \div 8 \cdot 2 - (2 + 2 \cdot 7)$

24. $108 \div 9 - \{4 \cdot [18 - (5 \cdot 3)]\}$

Chapter 1 Operations on the Whole Numbers

1.1 Standard Notation

We study mathematics in order to be able to solve problems. In this chapter, we learn how to use operations on the whole numbers. We begin by studying how numbers are named.

a   From Standard Notation to Expanded Notation

To answer questions such as "How many?", "How much?", and "How tall?", we use whole numbers. The set, or collection, of **whole numbers** is

0, 1, 2, 3, 4, 5, 6, 7, 8, 9, 10, 11, 12, ...

The set goes on indefinitely. There is no largest whole number, and the smallest whole number is 0. Each whole number can be named using various notations. The set 1, 2, 3, 4, 5, ..., without 0, is called the set of **natural numbers**.

As examples, we use data from the bar graph shown here.

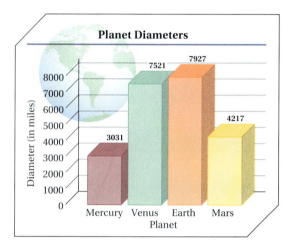

Note that the diameter of Mars is 4217 miles (mi). **Standard notation** for this number is 4217. We find **expanded notation** for 4217 as follows:

4217 = 4 thousands + 2 hundreds + 1 ten + 7 ones.

Example 1   Write expanded notation for 3031 mi, the diameter of Mercury.

3031 = 3 thousands + 0 hundreds + 3 tens + 1 one

Example 2   Write expanded notation for 54,567.

54,567 = 5 ten thousands + 4 thousands
          + 5 hundreds + 6 tens + 7 ones

Do Exercises 1 and 2 (in the margin at the right).

Objectives

a Convert from standard notation to expanded notation.

b Convert from expanded notation to standard notation.

c Write a word name for a number given standard notation.

d Write standard notation for a number given a word name.

e Given a standard notation like 278,342, tell what 8 means, what 3 means, and so on; identify the hundreds digit, the thousands digit, and so on.

For Extra Help

TAPE 1   MAC   CD-ROM
                WIN

Write in expanded notation.

1. 1805

2. 36,223

Answers on page A-1

Write in expanded notation.
3. 3210

4. 2009

5. 5700

Write in standard notation.
6. 5 thousands + 6 hundreds + 8 tens + 9 ones

7. 8 ten thousands + 7 thousands + 1 hundred + 2 tens + 8 ones

8. 9 thousands + 3 ones

Write a word name.
9. 57

10. 29

11. 88

rs on page A-1

Operations on the Whole Numbers

Example 3 Write expanded notation for 3400.

   3400 = 3 thousands + 4 hundreds + 0 tens + 0 ones, or
           3 thousands + 4 hundreds

Do Exercises 3–5.

b From Expanded Notation to Standard Notation

Example 4 Write standard notation for 2 thousands + 5 hundreds + 7 tens + 5 ones.

   Standard notation is 2575.

Example 5 Write standard notation for 9 ten thousands + 6 thousands + 7 hundreds + 1 ten + 8 ones.

   Standard notation is 96,718.

Example 6 Write standard notation for 2 thousands + 3 tens.

   Standard notation is 2030.

Do Exercises 6–8.

c Word Names

"Three," "two hundred one," and "forty-two" are **word names** for numbers. When we write word names for two-digit numbers like 42, 76, and 91, we use hyphens. For example, U.S. Olympic softball pitcher Michelle Granger can pitch a softball at a speed of 72 miles per hour (mph). A word name for 72 is "seventy-two."

Examples Write a word name.

7. 43   Forty-three   8. 91   Ninety-one

Do Exercises 9–11.

For large numbers, digits are separated into groups of three, called **periods**. Each period has a name: *ones, thousands, millions, billions,* and so on. When we write or read a large number, we start at the left with the largest period. The number named in the period is followed by the name of the period; then a comma is written and the next period is named. Recently, the U.S. national debt was $5,103,040,000,000. We can use a **place-value** chart to illustrate how to use periods to read the number 5,103,040,000,000.

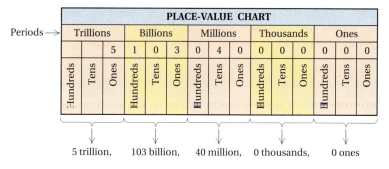

The U.S. debt was five trillion, one hundred three billion, forty million dollars.

Example 9 Write a word name for 46,605,314,732.

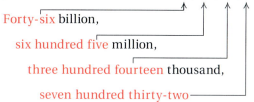

The word "and" *should not* appear in word names for whole numbers. Although we commonly hear such expressions as "two hundred *and* one," the use of "and" is not, strictly speaking, correct in word names for whole numbers. In decimal notation, we will find it appropriate to use "and" for the decimal point. For example, 317.4 is read as "three hundred seventeen *and* four tenths."

Do Exercises 12–15.

d | From Word Names to Standard Notation

Example 10 Write in standard notation.

Five hundred six million,
three hundred forty-five thousand,
two hundred twelve

Standard notation is 506,345,212.

Do Exercise 16.

e | Digits

A **digit** is a number 0, 1, 2, 3, 4, 5, 6, 7, 8, or 9 that names a place-value location.

Examples What does the digit 8 mean in each case?

11. 27**8**,342   8 thousands
12. **8**72,342   8 hundred thousands
13. ,343,399,223   8 billions

Do Exercises 17–20.

Write a word name.
12. 204

13. 79,204

14. 1,879,204

15. 22,301,879,204

16. Write in standard notation.
    Two hundred thirteen million, one hundred five thousand, three hundred twenty-nine

What does the digit 2 mean in each case?
17. 526,555

18. 265,789

19. 42,789,654

20. 24,789,654

Answers on page A-1

1.1 Standard Notation

Golf Balls. On an average day, Americans buy 486,574 golf balls. In 486,574, what digit tells the number of:

21. Thousands?

22. Ten thousands?

23. Ones?

24. Hundreds?

Example 14 *Dunkin Donuts.* On an average day, about 2,739,626 Dunkin Donuts are served in the United States. In 2,739,626, what digit tells the number of:

a) Hundred thousands?   7

b) Thousands?   9

Do Exercises 21–24.

Improving Your Math Study Skills

Tips for Using This Textbook

Throughout this textbook, you will find a feature called "Improving Your Math Study Skills." Some students find it helpful to read all of these early in the course. Each topic title is listed in the table of contents beginning on p. v.

One of the most important ways to improve your math study skills is to learn the proper use of the textbook. Here we highlight a few points that we consider most helpful.

- **Be sure to note the special symbols a , b , c , and so on, that correspond to the objectives you are to be able to perform.** They appear in many places throughout the text. The first time you see them is in the margin at the beginning of each section. The second time is in the subheadings of each section, and the third time is in the exercise set. You will also find them next to the skill maintenance exercises in each exercise set and in the review exercises at the end of each chapter, as well as in the answers to the chapter tests, pretests, and cumulative reviews. These objective symbols allow you to refer to the appropriate section whenever you need to review a topic.

- **Note the symbols in the margin under the list of objectives at the beginning of each section.** These refer to the many distinctive study aids that accompany the book.

- **Read and study each step of each example.** The examples include important side comments that explain each step. These carefully chosen examples and comments prepare you for success in the exercise set.

- **Stop and do the margin exercises as you study a section.** When our students come to us troubled about how they are doing in the course, the first question we ask is "Are you doing the margin exercises when directed to do so?" This is one of the most effective ways to enhance your ability to learn mathematics from this text. Don't deprive yourself of its benefits!

- **When you study the book, don't mark the points that you think are important, but mark the points you do not understand!** This book includes many design features that highlight important points. Use your efforts to mark where you are having trouble. Then when you go to class, a math lab, or a tutoring session, you will be prepared to ask questions that pertain to your difficulties rather than spend time going over what you already understand.

- **If you are having trouble, consider using the *Student's Solutions Manual*, which contains worked-out solutions to the odd-numbered exercises in the exercise sets.**

- **Try to keep one section ahead of your syllabus.** If you study ahead of your lectures, you can concentrate on what is being explained in them, rather than try to write everything down. You can then take notes only of special points or of questions related to what is happening in class.

Answers on page A-1

Exercise Set 1.1

Always review the objectives before doing an exercise set. See page 3. Note how the objectives are keyed to the exercises.

a Write in expanded notation.

1. 5742
2. 3897
3. 27,342
4. 93,986

5. 5609
6. 9990
7. 2300
8. 7020

b Write in standard notation.

9. 2 thousands + 4 hundreds + 7 tens + 5 ones
10. 7 thousands + 9 hundreds + 8 tens + 3 ones

11. 6 ten thousands + 8 thousands + 9 hundreds + 3 tens + 9 ones
12. 1 ten thousand + 8 thousands + 4 hundreds + 6 tens + 1 one

13. 7 thousands + 3 hundreds + 0 tens + 4 ones
14. 8 thousands + 0 hundreds + 2 tens + 0 ones

15. 1 thousand + 9 ones
16. 2 thousands + 4 hundreds + 5 tens

c Write a word name.

17. 85
18. 48
19. 88,000
20. 45,987

21. 123,765
22. 111,013
23. 7,754,211,577
24. 43,550,651,808

Write a word name for the number in (...)

25. *NBA Salaries.* In a recent year, the (...) the National Basketball Association (...)

27. *Population.* The population of Sout(...) 1,583,141,000.

Exercise Set 1.1, p. 7

1. 5 thousands + 7 hundreds + 4 tens + 2 ones
3. 2 ten thousands + 7 thousands + 3 hundreds + 4 tens + 2 ones
5. 5 thousands + 6 hundreds + 9 ones
7. 2 thousands + 3 hundreds
9. 2475
11. 68,939
13. 7304
15. 1009
17. Eighty-five
19. Eighty-eight thousand
21. One hundred twenty-three thousand, seven hundred sixty-five
23. Seven billion, seven hundred fifty-four million, two hundred eleven thousand, five hundred seventy-seven
25. One million, eight hundred sixty-seven thousand
27. One billion, five hundred eighty-three million, one hundred forty-one thousand
29. 2,233,812
31. 8,000,000,000
33. 9,460,000,000,000
35. 2,974,600
37. 5 thousands
39. 5 hundreds
41. 3
43. 0
45. ◆
47. All 9's as digits. Answers may vary. For an 8-digit readout, it would be 99,999,999. This number has three periods.

(...acific Ocean is (...) game sponsored by (...)s of winning the (...) 467,322,388 to 1.

d Write in standard notation.

29. Two million, two hundred thirty-three thousand, eight hundred twelve

30. Three hundred fifty-four thousand, seven hundred two

31. Eight billion

32. Seven hundred million

Write standard notation for the number in each sentence.

33. *Light Distance.* Light travels nine trillion, four hundred sixty billion kilometers in one year.

34. *Pluto.* The distance from the sun to Pluto is three billion, six hundred sixty-four million miles.

35. *Area of Greenland.* The area of Greenland is two million, nine hundred seventy-four thousand, six hundred square kilometers.

36. *Memory Space.* On computer hard drives, one gigabyte is actually one billion, seventy-three million, seven hundred forty-one thousand, eight hundred twenty-four bytes of memory.

e What does the digit 5 mean in each case?

37. 235,888

38. 253,888

39. 488,526

40. 500,346

In 89,302, what digit tells the number of:

41. Hundreds?

42. Thousands?

43. Tens?

44. Ones?

Synthesis

Exercises designated as *Synthesis exercises* will challenge you to combine two or more objectives at once. The icon ◆ denotes synthesis exercises that are writing exercises. Writing exercises are meant to be answered in one or more complete sentences and are usually less challenging than other synthesis exercises. Because answers to writing exercises may vary, they are not listed at the back of the book. Exercises marked with a 🖩 are meant to be solved using a calculator.

45. ◆ Write an English sentence in which the number 260,000,000 is used.

46. ◆ Explain why we use commas when writing large numbers.

47. 🖩 What is the largest number that you can name on your calculator? How many digits does that number have? How many periods?

48. How many whole numbers between 100 and 400 contain the digit 2 in their standard notation?

1.2 Addition and Perimeter

a Addition and the Real World

Addition of whole numbers corresponds to combining or putting things together. Let's look at various situations in which addition applies.

We combine two sets.

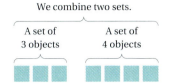

This is the resulting set.

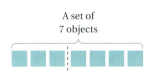

The addition that corresponds to the figure above is

3 + 4 = 7.

The number of objects in a set can be found by counting. We count and find that the two sets have 3 members and 4 members, respectively. After combining, we count and find that there are 7 objects. We say that the **sum** of 3 and 4 is 7. The numbers added are called **addends**.

This is read "3 plus 4 equals 7."

Example 1 Write an addition sentence that corresponds to this situation.

Kelly has $3 and earns $10 more. How much money does she have?

An addition that corresponds is $3 + $10 = $13.

Do Exercises 1 and 2.

Addition also corresponds to combining distances or lengths.

Example 2 Write an addition sentence that corresponds to this situation.

A car is driven 44 mi from San Francisco to San Jose. It is then driven 42 mi from San Jose to Oakland. How far is it from San Francisco to Oakland along the same route?

44 mi + 42 mi = 86 mi

Do Exercises 3 and 4.

Objectives

a Write an addition sentence that corresponds to a situation.

b Add whole numbers.

For Extra Help

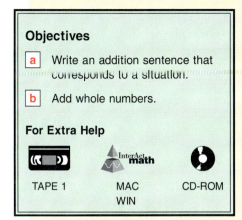

TAPE 1   MAC WIN   CD-ROM

Write an addition sentence that corresponds to each situation.

1. John has 8 music CD-ROMs in his backpack. Then he buys 2 educational CD-ROMs at the bookstore. How many CD-ROMs does John have in all?

2. Sue earns $45 in overtime pay on Thursday and $33 on Friday. How much overtime pay does she earn altogether on the two days?

Write an addition sentence that corresponds to each situation.

3. A car is driven 100 mi from Austin to Waco. It is then driven 93 mi from Waco to Dallas. How far is it from Austin to Dallas along the same route?

4. A coaxial cable 5 feet (ft) long is connected to a cable 7 ft long. How long is the resulting cable?

Answers on page A-1

Write an addition sentence that corresponds to each situation.

5. Find the perimeter of (distance around) the figure.

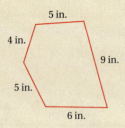

6. Find the perimeter of (distance around) the figure.

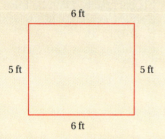

Write an addition sentence that corresponds to each situation.

7. The front parking lot of Sparks Electronics contains 30,000 square feet (sq ft) of parking space. The back lot contains 40,000 sq ft. What is the total area of the two parking lots?

8. You own a small rug that contains 8 square yards (sq yd) of fabric. You buy another rug that contains 9 sq yd. What is the area of the floor covered by both rugs?

Answers on page A-1

Chapter 1   Operations on the Whole Numbers

When we find the sum of the distances around an object, we are finding its **perimeter.**

Example 3 Write an addition sentence that corresponds to this situation.

A computer salesperson travels the following route to visit various electronics stores. How long is the route?

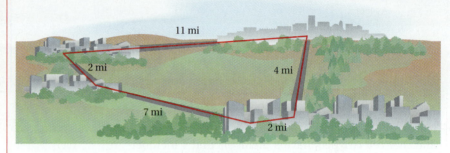

2 mi + 7 mi + 2 mi + 4 mi + 11 mi = 26 mi

Do Exercises 5 and 6.

Addition also corresponds to combining areas.

Example 4 Write an addition sentence that corresponds to this situation.

The area of a standard large index card is 40 square inches (sq in.). The area of a standard small index card is 15 sq in. Altogether, what is the total area of a large and a small card?

The area of the large index card is 40 sq in.	The area of the small index card is 15 sq in.	The total area of the two cards is 55 sq in.
40 sq in. +	15 sq in. =	55 sq in.

Do Exercises 7 and 8.

Addition corresponds to combining volumes as well.

Example 5 Write an addition sentence that corresponds to this situation.

Two trucks haul dirt to a construction site. One hauls 5 cubic yards (cu yd) and the other hauls 7 cu yd. Altogether, how many cubic yards of dirt have they hauled to the site?

5 cu yd + 7 cu yd = 12 cu yd

Do Exercises 9 and 10.

b Addition of Whole Numbers

To add numbers, we add the ones digits first, then the tens, then the hundreds, and so on.

Example 6   Add: 7312 + 2504.

Place values are lined up in columns.

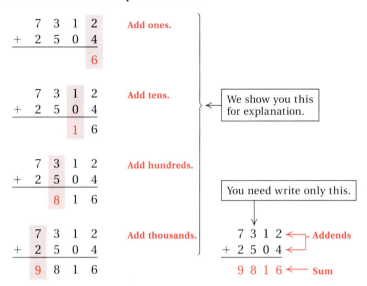

Do Exercise 11.

Example 7   Add: 6878 + 4995.

```
      1
   6 8 7 8    Add ones. We get 13 ones, or 1 ten + 3 ones.
 + 4 9 9 5    Write 3 in the ones column and 1 above the tens.
         3    This is called carrying, or regrouping.


     1 1
   6 8 7 8    Add tens. We get 17 tens, or 1 hundred + 7 tens.
 + 4 9 9 5    Write 7 in the tens column and 1 above the hundreds.
       7 3


   1 1 1
   6 8 7 8    Add hundreds. We get 18 hundreds, or 1 thousand +
 + 4 9 9 5    8 hundreds.
     8 7 3    Write 8 in the hundreds column and 1 above the thousands.


   1 1 1
   6 8 7 8    Add thousands. We get 11 thousands.
 + 4 9 9 5
 1 1 8 7 3
```

Do Exercises 12 and 13.

Write an addition sentence that corresponds to each situation.

9. Two trucks haul sand to a construction site to use in a driveway. One hauls 6 cu yd and the other hauls 8 cu yd. Altogether, how many cubic yards of sand are they hauling to the site?

10. A football fan drives to all college football games using a motor home. On one trip the fan buys 80 gallons (gal) of gasoline and on another, 56 gal. How many gallons were bought in all?

11. Add.

$$\begin{array}{r} 6\,2\,0\,3 \\ +\,3\,5\,4\,2 \\ \hline \end{array}$$

Add.

12. $\begin{array}{r} 7\,9\,6\,8 \\ +\,5\,4\,9\,7 \\ \hline \end{array}$

13. $\begin{array}{r} 9\,8\,0\,4 \\ +\,6\,3\,7\,8 \\ \hline \end{array}$

Answers on page A-1

1.2 Addition and Perimeter

Add from the top.

14. 9
     9
     4
   + 5

15. 8
     6
     9
     7
   + 4

16. Add from the bottom.
        9
        9
        4
      + 5

Answers on page A-1

How do we do an addition of three numbers, like $2 + 3 + 6$? We do so by adding 3 and 6, and then 2. We can show this with parentheses:

$2 + (3 + 6) = 2 + 9 = 11.$ **Parentheses tell what to do first.**

We could also add 2 and 3, and then 6:

$(2 + 3) + 6 = 5 + 6 = 11.$

Either way we get 11. It does not matter how we group the numbers. This illustrates the **associative law of addition**, $a + (b + c) = (a + b) + c$. We can also add whole numbers in any order. That is, $2 + 3 = 3 + 2$. This illustrates the **commutative law of addition**, $a + b = b + a$. Together the commutative and associative laws tell us that to add more than two numbers, we can use any order and grouping we wish.

Example 8 Add from the top.

  8
  9
  7
+ 6

We first add 8 and 9, getting 17; then 17 and 7, getting 24; then 24 and 6, getting 30.

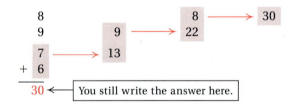

Do Exercises 14 and 15.

Example 9 Add from the bottom.

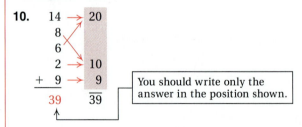

Do Exercise 16.

Sometimes it is easier to look for pairs of numbers whose sums are 10 or 20 or 30, and so on.

Examples Add.

10. 14 → 20
      8
      6
      2 → 10
    + 9 → 9
     ──   ──
     39   39

You should write only the answer in the position shown.

Chapter 1   Operations on the Whole Numbers

11. 23 + 19 + 7 + 21 + 4 = 74

30 + 40 + 4
74

Do Exercises 17–19.

Example 12   Add: 2391 + 3276 + 8789 + 1498.

```
      2
    2 3 9 1
    3 2 7 6
    8 7 8 9
  + 1 4 9 8
          4
```
Add ones: We get 24, so we have 2 tens + 4 ones. Write 4 in the ones column and 2 above the tens.

```
    3 2
    2 3 9 1
    3 2 7 6
    8 7 8 9
  + 1 4 9 8
        5 4
```
Add tens: We get 35 tens, so we have 30 tens + 5 tens, or 3 hundreds + 5 tens. Write 5 in the tens column and 3 above the hundreds.

```
  1 3 2
    2 3 9 1
    3 2 7 6
    8 7 8 9
  + 1 4 9 8
      9 5 4
```
Add hundreds: We get 19 hundreds, or 1 thousand + 9 hundreds. Write 9 in the hundreds column and 1 above the thousands.

```
  1 3 2
    2 3 9 1
    3 2 7 6
    8 7 8 9
  + 1 4 9 8
  1 5 9 5 4
```
Add thousands: We get 15 thousands.

Do Exercise 20.

Margin Exercises, Section 1.2, pp. 9–13

1. 8 + 2 = 10   **2.** $45 + $33 = $78   **3.** 100 mi + 93 mi = 193 mi   **4.** 5 ft + 7 ft = 12 ft   **5.** 4 in. + 5 in. + 9 in. + 6 in. + 5 in. = 29 in.   **6.** 5 ft + 6 ft + 5 ft + 6 ft = 22 ft   **7.** 30,000 sq ft + 40,000 sq ft = 70,000 sq ft   **8.** 8 sq yd + 9 sq yd = 17 sq yd   **9.** 6 cu yd + 8 cu yd = 14 cu yd   **10.** 80 gal + 56 gal = 136 gal   **11.** 9745   **12.** 13,465   **13.** 16,182   **14.** 27   **15.** 34   **16.** 27   **17.** 38   **18.** 47   **19.** 61   **20.** 27,474

Add. Look for pairs of numbers whose sums are 10, 20, 30, and so on.

17.
```
    1 5
      7
      5
      3
  +   8
```

18. 6 + 12 + 14 + 8 + 7

19. 27 + 8 + 13 + 2 + 11

20. Add.
```
    1 9 3 2
    6 7 2 3
    9 8 7 8
  + 8 9 4 1
```

To the instructor and the student: This section presented a review of addition of whole numbers. Students who are successful should go on to Section 1.3. Those who have trouble should study developmental unit A near the back of this text and then repeat Section 1.2.

Answers on page A-1

1.2   Addition and Perimeter

Improving Your Math Study Skills

Getting Started in a Math Class: The First-Day Handout or Syllabus

There are many ways in which to improve your math study skills. We have already considered some tips on using this book (see Section 1.1). We now consider some more general tips.

- **Textbook.** On the first day of class, most instructors distribute a handout that lists the textbook and other materials needed in the course. If possible, call the instructor or the department office before the term begins to find out which textbook you will be using and visit the bookstore to pick it up. This way, you can be ready to begin studying as soon as class starts. Delay in obtaining a copy of the textbook may cause you to fall behind in your homework.

- **Attendance.** The handout may also describe the attendance policy for your class. Some instructors take attendance at every class, while others use different methods to track students' attendance. Regardless of the policy, you should plan to attend class every time. Missing even one class can cause you to fall behind. If attendance counts toward your course grade, find out if there is a way to make up for missed days. In general, missing a class is not catastrophic if you put in the effort to catch up by studying the material on your own.

   If you do miss a class, call the instructor as soon as possible to find out what material was covered and what was assigned for the next class. If you have a study partner, call this person; ask if you can make a copy of his or her notes and find out what the homework assignment was. It is a good idea to meet with your instructor in person to clarify any concepts that you do not understand. This way, when you do return to class, you will be able to follow along with the rest of the group.

- **Homework.** The first-day handout may also detail how homework is handled. Find out when, and how often, homework will be assigned, whether homework is collected or graded, and whether there will be quizzes on the homework material. If the homework will be graded, find out what part of the final grade it will determine. Also, ask what the policy is for late homework and the format in which homework should be submitted. If you do miss a homework deadline, be sure to do the assigned homework anyway, as this is the best way to learn the material.

- **Grading.** The handout may also provide information on how your grade will be calculated at the end of the term. Typically, there will be tests during the term and a final exam at the end of the term. Frequently, homework is counted as part of the grade calculation, as are the quizzes. Find out how many tests will be given, if there is an option for make-up tests, or if any test grades will be dropped at the end of the term.

   Some instructors keep the class grades on a computer. If this is the case, find out if you can receive current grade reports throughout the term. This will help you focus on what is needed to obtain the desired grade in the course. Although a good grade should not be your only goal in this class, most students find it motivational to know what their grade is at any time during the term.

- **Get to know your classmates.** It can be a big help in a math class to get to know your fellow students. You might consider forming a study group. To do so, simply exchange phone numbers and schedules with other group members so that you can coordinate study time for homework or tests.

- **Get to know your instructor.** It can, of course, help immensely to get to know your instructor. Trivial though it may seem, get basic information like his or her name, how he or she can be contacted outside of class, and where the office is.

   Learn about your instructor's teaching style. Does he or she use an overhead projector or the board? Will there be frequent in-class questions? Try to adjust to your instructor's style.

Exercise Set 1.2, p. 15

1. $7 + 8 = 15$ **3.** 500 acres + 300 acres = 800 acres
5. 114 mi **7.** 52 in. **9.** 1300 ft **11.** 387
13. 5188 **15.** 164 **17.** 100 **19.** 900 **21.** 1010
23. 8503 **25.** 5266 **27.** 4466 **29.** 8310 **31.** 6608
33. 16,784 **35.** 34,432 **37.** 101,310 **39.** 100,101
41. 28 **43.** 26 **45.** 67 **47.** 230 **49.** 130
51. 1349 **53.** 36,926 **55.** 18,424 **57.** 2320
59. 31,685 **61.** 11,679 **63.** 22,654 **65.** 12,765,097
67. 7992 **68.** Nine hundred twenty-four million, six hundred thousand **69.** 8 ten thousands
70. 23,000,000 **71.** ◆ **73.** 56,055,667
75. $1 + 99 = 100, 2 + 98 = 100, \ldots, 49 + 51 = 100$. Then $49 \cdot 100 = 4900$ and $4900 + 50 + 100 = 5050$.

Exercise Set 1.2

a Write an addition sentence that corresponds to each situation.

1. Isabel receives 7 e-mail messages on Tuesday and 8 on Wednesday. How many e-mail messages did she receive altogether on the two days?

2. At a construction site, there are two gasoline containers to be used by earth-moving vehicles. One contains 400 gal and the other 200 gal. How many gallons do both contain altogether?

3. A builder buys two parcels of land to build a housing development. One contains 500 acres and the other 300 acres. What is the total number of acres purchased?

4. During March and April, Deron earns extra money doing income taxes part time. In March he earned $220, and in April he earned $340. How much extra did he earn altogether in March and April?

Find the perimeter of (distance around) each figure.

5.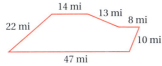

6.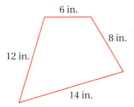

7.

8.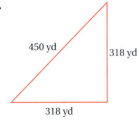

9.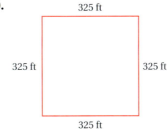

10.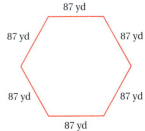

b Add.

11. 364
    + 23

12. 1521
    + 348

13. 1706
    +3482

14. 7503
    +2683

15. 8 6
    + 7 8

16. 7 3
    + 6 9

17. 9 9
    +   1

18. 9 9 9
    +   1 1

19. 789 + 111

20. 839 + 386

21. 909 + 101

22. 707 + 909

23. 8113 + 390

24. 271 + 3338

25. 356 + 4910

26. 280 + 34,702

27. 3870 + 92 + 7 + 497

28. 10,120 + 12,989 + 5738

29. 5 0 9 3
    + 3 2 1 7

30. 3 6 5 4
    + 2 7 0 0

31. 4 8 2 5
    + 1 7 8 3

32. 6 7 7 5
    + 1 4 3 2

33. 9 9 9 9
    + 6 7 8 5

34. 4 5,8 7 9
    + 2 1,7 8 6

35. 2 3,4 4 3
    + 1 0,9 8 9

36. 6 7,6 5 4
    + 9 8,7 8 6

37. 7 7,5 4 3
    + 2 3,7 6 7

38. 4 4,6 5 4
    +   4,7 6 5

39. 9 9,9 9 9
    +   1 0 2

40. 1 2 7,5 5 6
    +   6 8,7 6 6

Chapter 1 Operations on the Whole Numbers

Add from the top. Then check by adding from the bottom.

41. 7
     9
     4
 + 8

42. 4
     3
     9
     1
 + 8

43. 8
     6
     2
     3
 + 7

44. 9
     4
     7
     8
 + 7

Add. Look for pairs of numbers whose sums are 10, 20, 30, and so on.

45.   7
   1 8
     3
   3 7
 +   2

46. 2 3
   1 6
   1 1
   1 8
 + 1 9

47. 4 5
   2 5
   3 6
   4 4
 + 8 0

48. 3 8
   2 7
   3 2
   1 4
 + 7 6

Add.

49. 2 3
   6 2
 + 4 5

50. 4 3
   1 1
 + 3 7

51. 4 5 1
     3 6
 + 8 6 2

52.   3 1
   7 5 3
 + 9 2 4

53.   2,6 0 3
  2 8,2 1 4
 + 6,1 0 9

54. 9 3,2 4 9
    1,2 6 8
 + 7 4,8 2 3

55. 1 2,0 7 0
   2,9 5 4
 +   3,4 0 0

56. 4 2,4 8 7
  8 3,1 4 1
 + 3 6,7 1 2

57. 3 2 7
   4 2 8
   5 6 9
   7 8 7
 + 2 0 9

58. 9 8 9
   5 6 6
   8 3 4
   9 2 0
 + 7 0 3

59. 4 8 3 5
    7 2 9
   9 2 0 4
   8 9 8 6
 + 7 9 3 1

60.   5,9 4 6
       8 3 4
    1 2,9 5 6
  9 2 8,3 4 2
    3 4,9 0 1
 + 5 6,0 0 0

Exercise Set 1.2

61. $\begin{array}{r} 2\,0\,3\,7 \\ 4\,9\,2\,3 \\ 3\,4\,7\,1 \\ +1\,2\,4\,8 \end{array}$
62. $\begin{array}{r} 4\,5\,6\,7 \\ 1\,0\,2\,3 \\ 4\,8\,2\,1 \\ +3\,6\,8\,3 \end{array}$
63. $\begin{array}{r} 3\,4\,2\,0 \\ 8\,7\,1\,9 \\ 4\,3\,1\,2 \\ +6\,2\,0\,3 \end{array}$
64. $\begin{array}{r} 2\,0\,0\,3 \\ 1\,4\,9 \\ 5\,8 \\ +3\,4\,2\,6 \end{array}$

65. $\begin{array}{r} 5{,}6\,7\,8{,}9\,8\,7 \\ 1{,}4\,0\,9{,}3\,1\,2 \\ 8\,9\,8{,}8\,8\,8 \\ +4{,}7\,7\,7{,}9\,1\,0 \end{array}$
66. $\begin{array}{r} 7\,8{,}8\,9\,9{,}3\,1\,1 \\ 6{,}7\,8\,4{,}1\,7\,0 \\ 1\,1{,}5\,4\,1{,}9\,1\,3 \\ +\phantom{00{,}}1\,0\,0{,}8\,1\,7 \end{array}$

Skill Maintenance

The exercises that follow begin an important feature called *skill maintenance exercises*. These exercises provide an ongoing review of any preceding objective in the book. You will see them in virtually every exercise set. It has been found that this kind of extensive review can significantly improve your performance on a final examination.

67. Write standard notation for 7 thousands + 9 hundreds + 9 tens + 2 ones.   [1.1b]

68. Write a word name for the number in the following sentence: [1.1c]

Recently, the National Basketball Association's gross revenue was $924,600,000 (**Source**: *Wall Street Journal*).

69. What does the digit 8 mean in 486,205?   [1.1e]

70. Write in standard notation: [1.1d]

Twenty-three million.

Synthesis

71. ◆ Describe a situation that corresponds to the addition 80 sq ft + 140 sq ft. (See Examples 2–5.)

72. ◆ Explain in your own words what the associative law of addition means.

Add.

73. 5,987,943 + 328,959 + 49,738,765

74. 39,487,981 + 8,709,486 + 989,765

75. A fast way to add all the numbers from 1 to 10 inclusive is to pair 1 with 9, 2 with 8, and so on. Use a similar approach to add all the numbers from 1 to 100 inclusive.

1.3 Subtraction

a | Subtraction and the Real World: Take Away

Subtraction of whole numbers corresponds to two kinds of situations. The first one is called "take away."

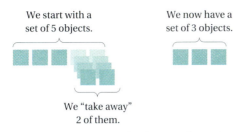

The subtraction that corresponds to the figure above is as follows.

$$5 - 2 = 3 \quad \text{This is read "5 minus 2 equals 3."}$$

A **subtrahend** is the number being subtracted. A **difference** is the result of subtracting one number from another. That is, it is the result of subtracting the subtrahend from the **minuend**.

Examples Write a subtraction sentence that corresponds to each situation.

1. Juan goes to a music store and chooses 10 CDs to take to the listening station. He rejects 7 of them, but buys the rest. How many CDs did Juan buy?

$$10 - 7 = 3$$

2. Kaitlin has $300 and spends $85 for office supplies. How much money is left?

$$\$300 - \$85 = \$215$$

Do Exercises 1 and 2.

Objectives

a Write a subtraction sentence that corresponds to a situation involving "take away."

b Given a subtraction sentence, write a related addition sentence; and given an addition sentence, write two related subtraction sentences.

c Write a subtraction sentence that corresponds to a situation involving "how much more?".

d Subtract whole numbers.

For Extra Help

TAPE 1   MAC WIN   CD-ROM

Write a subtraction sentence that corresponds to each situation.

1. A contractor removes 5 cu yd of sand from a pile containing 67 cu yd. How many cubic yards of sand are left in the pile?

2. Sparks Electronics owns a field next door that has an area of 20,000 sq ft. Deciding they need more room for parking, the owners have 12,000 sq ft paved. How many square feet of field are left unpaved?

Answers on page A-2

1.3 Subtraction

19

Write a related addition sentence.

3. $7 - 5 = 2$

4. $17 - 8 = 9$

Write two related subtraction sentences.

5. $5 + 8 = 13$

6. $11 + 3 = 14$

Answers on page A-2

Chapter 1 Operations on the Whole Numbers

20

b Related Sentences

Subtraction is defined in terms of addition. For example, $5 - 2$ is that number which when added to 2 gives 5. Thus for the subtraction sentence

$5 - 2 = 3$,   Taking away 2 from 5 gives 3.

there is a *related addition* sentence

$5 = 3 + 2$.   Putting back the 2 gives 5 again.

In fact, we know answers to subtractions are correct only because of the related addition, which provides a handy way to check a subtraction.

Example 3 Write a related addition sentence: $8 - 5 = 3$.

$8 - 5 = 3$

This number gets added (after 3).

$8 = 3 + 5$

By the commutative law of addition, there is also another addition sentence:

$8 = 5 + 3$.

The related addition sentence is $8 = 3 + 5$.

Do Exercises 3 and 4.

Example 4 Write two related subtraction sentences: $4 + 3 = 7$.

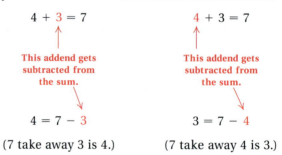

$4 + 3 = 7$ $4 + 3 = 7$

This addend gets subtracted from the sum.

$4 = 7 - 3$ $3 = 7 - 4$

(7 take away 3 is 4.)   (7 take away 4 is 3.)

The related subtraction sentences are $4 = 7 - 3$ and $3 = 7 - 4$.

Do Exercises 5 and 6.

c How Much More?

The second kind of situation to which subtraction corresponds is called "how much more?". We need the concept of a missing addend for "how-much-more" problems. From the related sentences, we see that finding a *missing addend* is the same as finding a *difference*.

Missing addend   Difference

$12 = 3 + \blacksquare$ $12 - 3 = \blacksquare$

Examples Write a subtraction sentence that corresponds to each situation.

5. A student has $47 and wants to buy a graphing calculator that costs $89. How much more is needed to buy the calculator?

To find the subtraction sentence, we first consider addition.

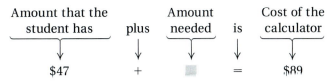

Now we write a related subtraction sentence:

47 + ■ = 89

■ = 89 − 47.   **The addend 47 gets subtracted.**

6. Cathy is reading *True Success: A New Philosophy of Excellence,* by Tom Morris, as part of her philosophy class. It contains 288 pages. She has read 126 pages. How many more pages must she read?

Now we write a related subtraction sentence:

126 + ■ = 288

■ = 288 − 126.   **126 gets subtracted.**

Do Exercises 7 and 8.

d Subtraction of Whole Numbers

To subtract numbers, we subtract the ones digits first, then the tens, then the hundreds, and so on.

Example 7 Subtract: 9768 − 4320.

```
  9 7 6 8
− 4 3 2 0
          8     Subtract ones.

  9 7 6 8
− 4 3 2 0
      4 8       Subtract tens.

  9 7 6 8
− 4 3 2 0
    4 4 8       Subtract hundreds.

  9 7 6 8
− 4 3 2 0
  5 4 4 8       Subtract thousands.

  9 7 6 8
− 4 3 2 0
  5 4 4 8       You should write only this.
```

This is for explanation.

Do Exercise 9.

Write an addition sentence and a related subtraction sentence corresponding to each situation. You need not carry out the subtraction.

7. It is 348 mi from Miami to Jacksonville. Alice has driven 67 mi from Miami to West Palm Beach on the way to Jacksonville. How much farther does she have to drive to get to Jacksonville?

8. A bricklayer estimates that it will take 1200 bricks to complete the side of a building but has only 800 bricks on the job site. How many more bricks will be needed?

9. Subtract.

    7 8 9 3
  − 4 0 9 2

Answers on page A-2

Subtract. Check by adding.

10. 8686
       −2358

11. 7145
       −2398

Subtract.

12. 70
       −14

13. 503
       −298

Subtract.

14. 7007
       −6349

15. 6000
       −3149

16. 9035
       −7489

Sometimes we need to borrow.

Example 8 Subtract: 6246 − 1879.

$$\begin{array}{r} \overset{3\ 16}{6\,2\,\cancel{4}\,\cancel{6}} \\ -1\,8\,7\,9 \\ \hline 7 \end{array}$$

We cannot subtract 9 ones from 6 ones, but we can subtract 9 ones from 16 ones. We borrow 1 ten to get 16 ones.

$$\begin{array}{r} \overset{\ \ 13}{1\ \cancel{3}\ 16} \\ 6\,\cancel{2}\,\cancel{4}\,\cancel{6} \\ -1\,8\,7\,9 \\ \hline 6\,7 \end{array}$$

We cannot subtract 7 tens from 3 tens, but we can subtract 7 tens from 13 tens. We borrow 1 hundred to get 13 tens.

$$\begin{array}{r} \overset{11\ 13}{5\ \cancel{1}\ \cancel{3}\ 16} \\ \cancel{6}\,\cancel{2}\,\cancel{4}\,\cancel{6} \\ -1\,8\,7\,9 \\ \hline 4\,3\,6\,7 \end{array}$$

We cannot subtract 8 hundreds from 1 hundred, but we can subtract 8 hundreds from 11 hundreds. We borrow 1 thousand to get 11 hundreds.

We can always check the answer by adding it to the number being subtracted.

This is what you should write.

$$\begin{array}{r} \overset{11\ 13}{5\ \cancel{1}\ \cancel{3}\ 16} \\ \cancel{6}\,\cancel{2}\,\cancel{4}\,\cancel{6} \\ -1\,8\,7\,9 \\ \hline 4\,3\,6\,7 \end{array}$$

Check:

$$\begin{array}{r} \overset{1\ 1\ 1}{\ } \\ 4\,3\,6\,7 \\ +1\,8\,7\,9 \\ \hline 6\,2\,4\,6 \end{array}$$

This answer checks because this is the top number in the subtraction.

Do Exercises 10 and 11.

Example 9 Subtract: 902 − 477.

$$\begin{array}{r} \overset{8\ 9\ 12}{9\,\cancel{0}\,\cancel{2}} \\ -4\,7\,7 \\ \hline 4\,2\,5 \end{array}$$

We cannot subtract 7 ones from 2 ones. We have 9 hundreds, or 90 tens. We borrow 1 ten to get 12 ones. We then have 89 tens.

Do Exercises 12 and 13.

Example 10 Subtract: 8003 − 3667.

$$\begin{array}{r} \overset{7\ 9\ 9\ 13}{8\,\cancel{0}\,\cancel{0}\,\cancel{3}} \\ -3\,6\,6\,7 \\ \hline 4\,3\,3\,6 \end{array}$$

We have 8 thousands, or 800 tens. We borrow 1 ten to get 13 ones. We then have 799 tens.

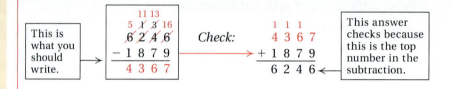

11. Subtract: 6000 − 3762.

$$\begin{array}{r} \overset{5\ 9\ 9\ 10}{\cancel{6}\,\cancel{0}\,\cancel{0}\,\cancel{0}} \\ -3\,7\,6\,2 \\ \hline 2\,2\,3\,8 \end{array}$$

12. Subtract: 6024 − 2968.

$$\begin{array}{r} \overset{\ \ \ \ 11}{5\ 9\ \cancel{1}\ 14} \\ \cancel{6}\,\cancel{0}\,\cancel{2}\,\cancel{4} \\ -2\,9\,6\,8 \\ \hline 3\,0\,5\,6 \end{array}$$

Do Exercises 14–16.

Margin Exercises, Section 1.3, pp. 19–22

1. 67 cu yd − 5 cu yd = 62 cu yd   **2.** 20,000 sq ft − 12,000 sq ft = 8,000 sq ft   **3.** 7 = 2 + 5, or 7 = 5 + 2   **4.** 17 = 9 + 8, or 17 = 8 + 9   **5.** 5 = 13 − 8; 8 = 13 − 5   **6.** 11 = 14 − 3; 3 = 14 − 11   **7.** 67 + ▨ = 348; ▨ = 348 − 67   **8.** 800 + ▨ = 1200; ▨ = 1200 − 800   **9.** 3801   **10.** 6328   **11.** 4747   **12.** 56   **13.** 205   **14.** 658   **15.** 2851   **16.** 1546

Exercise Set 1.3

a Write a subtraction sentence that corresponds to each situation. You need not carry out the subtraction.

1. Jeanne has $1260 in her college checking account. She spends $450 for her food bill at the dining hall. How much is left in her account?

2. *Frozen Yogurt.* A dispenser at a frozen yogurt store contains 126 ounces (oz) of strawberry yogurt. A 13-oz cup is sold to a customer. How much is left in the dispenser?

3. A host pours 5 oz of salsa from a jar containing 16 oz. How many ounces are left?

4. *Chocolate Cake.* One slice of chocolate cake with fudge frosting contains 564 calories (cal). One cup of hot cocoa made with skim milk contains 188 calories. How many more calories are in the cake than in the cocoa?

b Write a related addition sentence.

5. $7 - 4 = 3$
6. $12 - 5 = 7$
7. $13 - 8 = 5$
8. $9 - 9 = 0$

9. $23 - 9 = 14$
10. $20 - 8 = 12$
11. $43 - 16 = 27$
12. $51 - 18 = 33$

Write two related subtraction sentences.

13. $6 + 9 = 15$
14. $7 + 9 = 16$
15. $8 + 7 = 15$
16. $8 + 0 = 8$

17. $17 + 6 = 23$
18. $11 + 8 = 19$
19. 2

c Write an addition sentence and a related subtraction sentence ... out the subtraction.

21. *Kangaroos.* There are 32 million kangaroos in Australia and 17 million people. How many more kangaroos are there than people?

22.

Exercise Set 1.3, p. 23

1. $1260 - $450 =
3. 16 oz - 5 oz =
5. $7 = 3 + 4$, or $7 = 4 + 3$
7. $13 = 5 + 8$, or $13 = 8 + 5$
9. $23 = 14 + 9$, or $23 = 9 + 14$
11. $43 = 27 + 16$, or $43 = 16 + 27$
13. $6 = 15 - 9$; $9 = 15 - 6$
15. $8 = 15 - 7$; $7 = 15 - 8$
17. $17 = 23 - 6$; $6 = 23 - 17$
19. $23 = 32 - 9$; $9 = 32 - 23$
21. $17 + \_\_ = 32$; $\_\_ = 32 - 17$
23. $10 + \_\_ = 23$; $\_\_ = 23 - 10$
25. 12   27. 44
29. 533   31. 1126   33. 39   35. 298   37. 226
39. 234   41. 5382   43. 1493   45. 2187   47. 3831
49. 7748   51. 33,794   53. 2168   55. 43,028
57. 56   59. 36   61. 84   63. 454   65. 771
67. 2191   69. 3749   71. 7019   73. 5745   75. 95,974
77. 9989   79. 83,818   81. 4206   83. 10,305
85. 7 ten thousands   86. Six million, three hundred seventy-five thousand, six hundred two   87. 29,708
88. 22,692   89. ◆   91. 2,829,177   93. 3; 4

23. A set of drapes requires 23 yards (yd) of material. The decorator has 10 yd of material in stock. How much more must be ordered?

24. Marv needs to bowl a score of 223 in order to beat his opponent. His score with one frame to go is 195. How many pins does Marv need in the last frame to beat his opponent?

d Subtract.

25. 16
    − 4

26. 86
    − 13

27. 65
    − 21

28. 87
    − 34

29. 866
    − 333

30. 526
    − 323

31. 4547
    − 3421

32. 6875
    − 2111

33. 86 − 47

34. 73 − 28

35. 625 − 327

36. 726 − 509

37. 835 − 609

38. 953 − 246

39. 981 − 747

40. 887 − 698

41. 7769
    − 2387

42. 6431
    − 2896

43. 3982
    − 2489

44. 7650
    − 1765

45. 5046
    − 2859

46. 6308
    − 2679

47. 7640
    − 3809

48. 8003
    − 599

49. $12{,}647$
    $-4{,}899$

50. $16{,}222$
    $-5{,}888$

51. $46{,}771$
    $-12{,}977$

52. $95{,}654$
    $-48{,}985$

53. $10{,}002 - 7834$

54. $23{,}048 - 17{,}592$

55. $90{,}237 - 47{,}209$

56. $84{,}703 - 298$

57. 80
    -24

58. 40
    -37

59. 90
    -54

60. 90
    -78

61. 140
    -56

62. 470
    -188

63. 690
    -236

64. 803
    -418

65. 903
    -132

66. 6408
    -258

67. 2300
    -109

68. 3506
    -1293

69. 6808
    -3059

70. 7840
    -3027

71. 8092
    -1073

72. 6007
    -1589

73. 5843 − 98 **74.** 10,002 − 398 **75.** 101,734 − 5760 **76.** 15,017 − 7809

77. 10,008 − 19 **78.** 21,043 − 8909 **79.** 83,907 − 89 **80.** 311,568 − 19,394

81.   7 0 0 0
       − 2 7 9 4

82.   8 0 0 1
       − 6 5 4 3

83.   4 8,0 0 0
       − 3 7,6 9 5

84.   1 7,0 4 3
       − 1 1,5 9 8

Skill Maintenance

85. What does the digit 7 mean in 6,375,602? [1.1e]

86. Write a word name for 6,375,602. [1.1c]

87. Write standard notation for 2 ten thousands + 9 thousands + 7 hundreds + 8 ones. [1.1b]

88. Add: 9807 + 12,885. [1.2b]

Synthesis

89. ◈ Is subtraction commutative (is there a commutative law of subtraction)? Why or why not?

90. ◈ Describe a situation that corresponds to the subtraction $20 − $17. (See Examples 2 and 5.)

Subtract.

91. ▦ 3,928,124 − 1,098,947

92. ▦ 21,431,206 − 9,724,837

93. Fill in the missing digits to make the equation true:
9,▨48,621 − 2,097,▨81 = 7,251,140.

Chapter 1 Operations on the Whole Numbers

1.4 Rounding and Estimating; Order

a Rounding

We round numbers in various situations if we do not need an exact answer. For example, we might round to check if an answer to a problem is reasonable or to check a calculation done by hand or on a calculator. We might also round to see if we are being charged the correct amount in a store.

To understand how to round, we first look at some examples using number lines, even though this is not the way we normally do rounding.

Example 1 Round 47 to the nearest ten.

Here is a part of a number line; 47 is between 40 and 50.

Since 47 is closer to 50 than it is to 40, we round up to 50.

Example 2 Round 42 to the nearest ten.

42 is between 40 and 50.

Since 42 is closer to 40 than it is to 50, we round down to 40.

Do Exercises 1–4.

Example 3 Round 45 to the nearest ten.

45 is halfway between 40 and 50.

We could round 45 down to 40 or up to 50. We agree to round up to 50.

> When a number is halfway between rounding numbers, round up.

Do Exercises 5–7.

Here is a rule for rounding.

To round to a certain place:
a) Locate the digit in that place.
b) Consider the next digit to the right.
c) If the digit to the right is 5 or higher, round up; if the digit to the right is 4 or lower, round down.
d) Change all digits to the right of the rounding location to zeros.

Objectives

a) Round to the nearest ten, hundred, or thousand.
b) Estimate sums and differences by rounding.
c) Use < or > for ▮ to write a true sentence in a situation like 6 ▮ 10.

For Extra Help

TAPE 1   MAC WIN   CD-ROM

Round to the nearest ten.

1. 37

2. 52

3. 73

4. 98

Round to the nearest ten.

5. 35

6. 75

7. 85

Answers on page A-2

Round to the nearest ten.

8. 137

9. 473

10. 235

11. 295

Round to the nearest hundred.

12. 641

13. 759

14. 750

15. 9325

Round to the nearest thousand.

16. 7896

17. 8459

18. 19,843

19. 68,500

Answers on page A-2

Chapter 1   Operations on the Whole Numbers

Example 4   Round 6485 to the nearest ten.

a) Locate the digit in the tens place.

   6 4 **8** 5
       ↑

b) Consider the next digit to the right.

   6 4 8 **5**
         ↑

c) Since that digit is 5 or higher, round 8 tens up to 9 tens.

d) Change all digits to the right of the tens digit to zeros.

   6 4 9 0 ← This is the answer.

Example 5   Round 6485 to the nearest hundred.

a) Locate the digit in the hundreds place.

   6 **4** 8 5
     ↑

b) Consider the next digit to the right.

   6 4 **8** 5
       ↑

c) Since that digit is 5 or higher, round 4 hundreds up to 5 hundreds.

d) Change all digits to the right of hundreds to zeros.

   6 5 0 0 ← This is the answer.

Example 6   Round 6485 to the nearest thousand.

a) Locate the digit in the thousands place.

   6 4 8 5
   ↑

b) Consider the next digit to the right.

   6 **4** 8 5
     ↑

c) Since that digit is 4 or lower, round down, meaning that 6 thousands stays as 6 thousands.

d) Change all digits to the right of thousands to zeros.

   6 0 0 0 ← This is the answer.

CAUTION!   7000 is not a correct answer to Example 6. It is incorrect to round from the ones digit over, as follows:

   6485,   6490,   6500,   7000.

Do Exercises 8–19.

There are many methods of rounding. For example, in computer applications, the rounding of 8563 to the nearest hundred might be done using a different rule called **truncating**, meaning that we simply change all digits to the right of the rounding location to zeros. Thus, 8563 would round to 8500, which is not the same answer that we would get using the rule discussed in this section.

b Estimating

Estimating is used to simplify a problem so that it can then be solved easily or mentally. Rounding is used when estimating. There are many ways to estimate.

Example 7 Michelle earned $21,791 as a consultant and $17,239 as an instructor in a recent year. Estimate Michelle's yearly earnings.

There are many ways to get an answer, but there is no one perfect answer based on how the problem is worded. Let's consider two methods.

METHOD 1. Round each number to the nearest thousand and then add.

```
  2 1,7 9 1        2 2,0 0 0
+ 1 7,2 3 9      + 1 7,0 0 0
                 $ 3 9,0 0 0  ← Estimated answer
```

METHOD 2. We might use a less formal approach, depending on how specific we want the answer to be. We note that both amounts are close to $20,000, and so the total is close to $40,000. In some contexts, such as retirement planning, this might be sufficient.

The point to be made is that estimating can be done in many ways and can have many answers, even though in the problems that follow we ask you to round in a specific way.

Example 8 Estimate this sum by first rounding to the nearest ten:

78 + 49 + 31 + 85.

We round each addend to the nearest ten. Then we add.

```
  7 8        8 0
  4 9        5 0
  3 1        3 0
+ 8 5      + 9 0
            2 5 0  ← Estimated answer
```

Do Exercise 20.

Example 9 Estimate this sum by first rounding to the nearest hundred:

850 + 674 + 986 + 839.

We have

```
  8 5 0        9 0 0
  6 7 4        7 0 0
  9 8 6      1 0 0 0
+ 8 3 9      +  8 0 0
             3 4 0 0
```

Do Exercise 21.

20. Estimate the sum by first rounding to the nearest ten. Show your work.

```
  7 4
  2 3
  3 5
+ 6 6
```

21. Estimate the sum by first rounding to the nearest hundred. Show your work.

```
  6 5 0
  6 8 5
  2 3 8
+ 1 6 8
```

Margin Exercises, Section 1.4, pp. 27–30

1. 40   2. 50   3. 70   4. 100   5. 40   6. 80   7. 90
8. 140   9. 470   10. 240   11. 300   12. 600
13. 800   14. 800   15. 9300   16. 8000   17. 8000
18. 20,000   19. 69,000   20. 200   21. 1800
22. 2600   23. 11,000   24. <   25. >   26. >
27. <   28. <   29. >

Answers on page A-2

1.4 Rounding and Estimating; Order

22. Estimate the difference by first rounding to the nearest hundred. Show your work.

$$\begin{array}{r}9285\\-6739\end{array}$$

23. Estimate the difference by first rounding to the nearest thousand. Show your work.

$$\begin{array}{r}23{,}278\\-11{,}698\end{array}$$

Use < or > for ▨ to form a true sentence. Draw a number line if necessary.

24. 8 ▨ 12

25. 12 ▨ 8

26. 76 ▨ 64

27. 64 ▨ 76

28. 217 ▨ 345

29. 345 ▨ 217

Answers on page A-2

Example 10 Estimate the difference by first rounding to the nearest thousand: 9324 − 2849.

We have

$$\begin{array}{r}9324\\-2849\end{array}\qquad\begin{array}{r}9000\\-3000\\\hline 6000\end{array}$$

Do Exercises 22 and 23.

The sentence 7 − 5 = 2 says that 7 − 5 is the same as 2. Later we will use the symbol ≈ when rounding. This symbol means **"is approximately equal to."** Thus, when 687 is rounded to the nearest ten, we may write

687 ≈ 690.

c Order

We know that 2 is not the same as 5. This can be written 2 ≠ 5. In fact, 2 is less than 5. We can see this order on a number line: 2 is to the left of 5. The number 0 is the smallest whole number.

> For any whole numbers *a* and *b*:
> 1. *a* < *b* (read "*a* is less than *b*") is true when *a* is to the left of *b* on a number line.
> 2. *a* > *b* (read "*a* is greater than *b*") is true when *a* is to the right of *b* on a number line.
>
> We call < and > **inequality symbols.**

Example 11 Use < or > for ▨ to form a true sentence: 7 ▨ 11.

Since 7 is to the left of 11, we write 7 < 11.

Example 12 Use < or > for ▨ to form a true sentence: 92 ▨ 87.

Since 92 is to the right of 87, we write 92 > 87.

A sentence like 8 + 5 = 13 is called an **equation**. A sentence like 7 < 11 is called an **inequality**. The sentence 7 < 11 is a true inequality. The sentence 23 > 69 is a false inequality.

Do Exercises 24–29.

Exercise Set 1.4

a Round to the nearest ten.

1. 48
2. 17
3. 67
4. 99

5. 731
6. 532
7. 895
8. 798

Round to the nearest hundred.

9. 146
10. 874
11. 957
12. 650

13. 9079
14. 4645
15. 32,850
16. 198,402

Round to the nearest thousand.

17. 5876
18. 4500
19. 7500
20. 2001

21. 45,340
22. 735,562
23. 373,405
24. 6,713,855

b Estimate each sum or difference by first rounding each addend to the nearest ten. Show your work.

25. 7 8
    + 9 7

26. 6 2
     9 7
     4 6
    + 8 8

27. 8 0 7 4
    − 2 3 4 7

28. 6 7 3
    −   2 8

Estimate each sum by first rounding each addend to the nearest ten. Do any of the given sums seem to be incorrect when compared to the estimate? Which ones?

29. 4 5
     7 7
     2 5
    + 5 6
     3 4 3

30. 4 1
     2 1
     5 5
    + 6 0
     1 7 7

31. 6 2 2
     7 8
     8 1
    + 1 1 1
     9 3 2

32. 8 3 6
     3 7 4
     7 9 4
    + 9 3 8
     3 9 4 7

Estimate each sum or difference by first rounding to the nearest hundred. Show your work.

33. 7 3 4 8
    + 9 2 4 7

34. 5 6 8
     4 7 2
     9 3 8
    + 4 0 2

35. 6 8 5 2
    − 1 7 4 8

36. 9 4 3 8
    − 2 7 8 7

Exercise Set 1.4, p. 31

1. 50 **3.** 70 **5.** 730 **7.** 900 **9.** 100 **11.** 1000
13. 9100 **15.** 32,900 **17.** 6000 **19.** 8000
21. 45,000 **23.** 373,000 **25.** 180 **27.** 5720
29. 220; incorrect **31.** 890; incorrect **33.** 16,500
35. 5200 **37.** 1600 **39.** 1500 **41.** 31,000
43. 69,000 **45.** < **47.** > **49.** < **51.** > **53.** >
55. > **57.** 86,754 **58.** 13,589 **59.** 48,824
60. 4415 **61.** ◆ **63.** 30,411 **65.** 69,594

Estimate each sum by first rounding the addends to the nearest hundred. Do any of the given sums seem to be incorrect when compared to the estimate? Which ones?

37. 216 84 745 +595 1640	**38.** 481 702 623 +1043 1849	**39.** 750 428 63 +205 1446	**40.** 326 275 758 +943 2302

Estimate each sum or difference by first rounding to the nearest thousand. Show your work.

41. 9643 4821 8943 +7004	**42.** 7648 9348 7842 +2222	**43.** 92,149 −22,555	**44.** 84,890 −11,110

c Use < or > for ▆ to form a true sentence. Draw a number line if necessary.

45. 0 ▆ 17 **46.** 32 ▆ 0 **47.** 34 ▆ 12 **48.** 28 ▆ 18

49. 1000 ▆ 1001 **50.** 77 ▆ 117 **51.** 133 ▆ 132 **52.** 999 ▆ 997

53. 460 ▆ 17 **54.** 345 ▆ 456 **55.** 37 ▆ 11 **56.** 12 ▆ 32

Skill Maintenance

Add. [1.2b]

		Subtract. [1.3d]	
57. 67,789 +18,965	**58.** 9002 +4587	**59.** 67,789 −18,965	**60.** 9002 −4587

Synthesis

61. ◆ When rounding 748 to the nearest hundred, a student rounds to 750 and then to 800. What mistake is the student making?

62. ◆ Explain how estimating and rounding can be useful when shopping for groceries.

63.–66. 🖩 Use a calculator to find the sums or differences in Exercises 41–44. Since you can still make errors on a calculator—say, by pressing the wrong buttons—you can check your answers by estimating.

Chapter 1   Operations on the Whole Numbers

1.5 Multiplication and Area

a | Multiplication and the Real World

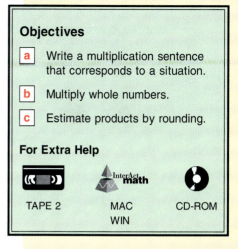

Objectives

a. Write a multiplication sentence that corresponds to a situation.
b. Multiply whole numbers.
c. Estimate products by rounding.

For Extra Help

TAPE 2   MAC WIN   CD-ROM

Multiplication of whole numbers corresponds to two kinds of situations.

Repeated Addition

The multiplication 3×5 corresponds to this repeated addition:

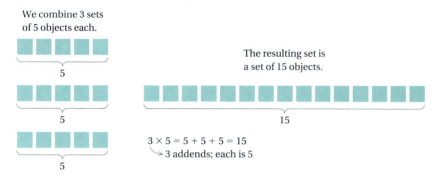

We combine 3 sets of 5 objects each.

The resulting set is a set of 15 objects.

$3 \times 5 = 5 + 5 + 5 = 15$
3 addends; each is 5

We say that the *product* of 3 and 5 is 15. The numbers 3 and 5 are called *factors*.

$3 \times 5 = 15$ This is read "3 times 5 equals 15."

Factors   Product

The numbers that we multiply can be called **factors**. The result of the multiplication is a number called a **product**.

Rectangular Arrays

The multiplication 3×5 corresponds to this rectangular array:

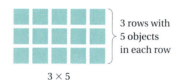

3 rows with 5 objects in each row

3×5

When you write a multiplication sentence corresponding to a real-world situation, you should think of either a rectangular array or repeated addition. In some cases, it may help to think both ways.

We have used an "×" to denote multiplication. A dot " · ", as in $3 \cdot 5 = 15$, is also commonly used. (Use of the dot is attributed to the German mathematician Gottfried Wilhelm von Leibniz in 1698.) Parentheses are also used to denote multiplication—for example, $(3)(5) = 15$, or $3(5) = 15$.

1.5   Multiplication and Area

33

Write a multiplication sentence that corresponds to each situation.

1. Marv practices for the U.S. Open bowling tournament. He bowls 8 games each day for 7 days. How many games does he bowl altogether for practice?

2. A lab technician pours 75 milliliters (mL) of acid into each of 10 beakers. How much acid is poured in all?

3. *Checkerboard.* A checkerboard consists of 8 rows with 8 squares in each row. How many squares in all are there on a checkerboard?

Answers on page A-2

Chapter 1 Operations on the Whole Numbers

34

Examples Write a multiplication sentence that corresponds to each situation.

1. It is known that Americans drink 24 million gal of soft drinks per day (*per day* means *each day*). What quantity of soft drinks is consumed every 5 days?

We draw a picture in which 🥫 = 1 million gallons or we can simply visualize the situation. Repeated addition fits best in this case.

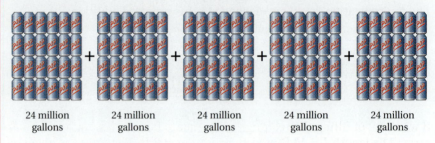

| 24 million gallons | 24 million gallons | 24 million gallons | 24 million gallons | 24 million gallons |

5 · 24 million gallons = 120 million gallons

2. One side of a building has 6 floors with 7 windows on each floor. How many windows are there on that side of the building?

We have a rectangular array and can easily draw a sketch.

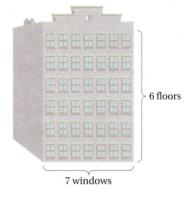

6 · 7 = 42

Do Exercises 1–3.

Area

The area of a rectangular region is often considered to be the number of square units needed to fill it. Here is a rectangle 4 cm (centimeters) long and 3 cm wide. It takes 12 square centimeters (sq cm) to fill it.

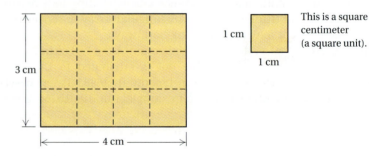

This is a square centimeter (a square unit).

In this case, we have a rectangular array. The number of square units is 3 · 4, or 12. The total area is 12 sq cm.

Once exponents are introduced in Section 1.9, we will often abbreviate square centimeters as cm^2, square feet as ft^2, and so on.

Example 3 Write a multiplication sentence that corresponds to this situation.

A rectangular room is 10 ft long and 8 ft wide. Find its area.

We draw a picture.

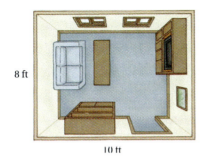

If we think of filling the rectangle with square feet, we have a rectangular array. The length $l = 10$ ft, and the width $w = 8$ ft. The area A is given by the formula

$$A = l \cdot w = 10 \cdot 8 = 80 \text{ sq ft.}$$

Do Exercise 4.

b Multiplication of Whole Numbers

Let's find the product

$$\begin{array}{r} 5\,4 \\ \times\,3\,2 \\ \hline \end{array}$$

To do this, we multiply 54 by 2, then 54 by 30, and then add.

$$\begin{array}{r} 5\,4 \\ \times\,2 \\ \hline 1\,0\,8 \end{array} \qquad \begin{array}{r} 1 \\ 5\,4 \\ \times\,3\,0 \\ \hline 1\,6\,2\,0 \end{array}$$

Since we are going to add the results, let's write the work this way.

$$\begin{array}{r} 5\,4 \\ \times\,3\,2 \\ \hline 1\,0\,8 \quad \text{Multiplying 54 by 2} \\ 1\,6\,2\,0 \quad \text{Multiplying 54 by 30} \\ \hline 1\,7\,2\,8 \quad \text{Adding to obtain the product} \end{array}$$

The fact that we can do this is based on a property called the **distributive law.** It says that to multiply a number by a sum, $a \cdot (b + c)$, we can multiply each part by a and then add like this: $(a \cdot b) + (a \cdot c)$. Thus, $a \cdot (b + c) = (a \cdot b) + (a \cdot c)$. Applied to the example above, the distributive law gives us

$$\begin{aligned} 54 \cdot 32 = 54 \cdot (30 + 2) &= \underbrace{(54 \cdot 30)} + \underbrace{(54 \cdot 2)} \\ &= 1620 + 108 \\ &= 1728. \end{aligned}$$

4. What is the area of this pool table?

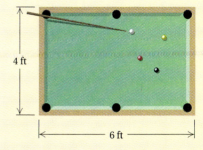

Margin Exercises, Section 1.5, pp. 34–38

1. $8 \cdot 7 = 56$ **2.** $10 \cdot 75 = 750$ mL
3. $8 \cdot 8 = 64$ **4.** $4 \cdot 6 = 24$ sq ft   **5.** 1035
6. 3024   **7.** 46,252   **8.** 205,065   **9.** 144,432
10. 287,232   **11.** 14,075,720   **12.** 391,760
13. 17,345,600   **14.** 56,200   **15.** 562,000
16. (a) 1081; **(b)** 1081; **(c)** same   **17.** 40   **18.** 15
19. 210,000; 160,000

Answer on page A-2

Multiply.

5. 4 5
     × 2 3

6. 48 × 63

Multiply.

7. 7 4 6
     ×   6 2

8. 245 × 837

Answers on page A-2

Chapter 1 Operations on the Whole Numbers

Example 4 Multiply: 43 × 57.

$$\begin{array}{r} \overset{2}{5}\,7 \\ \times\ 4\,3 \\ \hline 1\,7\,1 \end{array}$$ Multiplying 57 by 3

$$\begin{array}{r} \overset{2}{\underset{}{\,}}\\ \overset{2}{5}\,7 \\ \times\ 4\,3 \\ \hline 1\,7\,1 \\ 2\,2\,8\,0 \end{array}$$ Multiplying 57 by 40 (We write a 0 and then multiply 57 by 4.)

You may have learned that such a 0 does not have to be written. You may omit it if you wish. If you do omit it, remember, when multiplying by tens, to put the answer in the tens place.

$$\begin{array}{r} 2 \\ 2 \\ 5\,7 \\ \times\ 4\,3 \\ \hline 1\,7\,1 \\ 2\,2\,8\,0 \\ \hline 2\,4\,5\,1 \end{array}$$ Adding to obtain the product

Do Exercises 5 and 6.

Example 5 Multiply: 457 × 683.

$$\begin{array}{r} \overset{5}{\,}\overset{2}{\,} \\ 6\,8\,3 \\ \times\ 4\,5\,7 \\ \hline 4\,7\,8\,1 \end{array}$$ Multiplying 683 by 7

$$\begin{array}{r} \overset{4}{\,}\overset{1}{\,} \\ \overset{5}{\,}\overset{2}{\,} \\ 6\,8\,3 \\ \times\ 4\,5\,7 \\ \hline 4\,7\,8\,1 \\ 3\,4\,1\,5\,0 \end{array}$$ Multiplying 683 by 50

$$\begin{array}{r} \overset{3}{\,}\overset{1}{\,} \\ \overset{4}{\,}\overset{1}{\,}\overset{}{\,} \\ \overset{5}{\,}\overset{2}{\,} \\ 6\,8\,3 \\ \times\ 4\,5\,7 \\ \hline 4\,7\,8\,1 \\ 3\,4\,1\,5\,0 \\ 2\,7\,3\,2\,0\,0 \\ \hline 3\,1\,2{,}1\,3\,1 \end{array}$$

Multiplying 683 by 400

Adding

Do Exercises 7 and 8.

Zeros in Multiplication

Example 6 Multiply: 306 × 274.

Note that 306 = 3 hundreds + 6 ones.

```
      2 7 4
    × 3 0 6
    ───────
      1 6 4 4    Multiplying 274 by 6
    8 2 2 0 0    Multiplying 274 by 3 hundreds (We write 00
    ───────      and then multiply 274 by 3.)
    8 3,8 4 4    Adding
```

Do Exercises 9–11.

Example 7 Multiply: 360 × 274.

Note that 360 = 3 hundreds + 6 tens.

```
      2 7 4
    × 3 6 0
    ───────
      1 6 4 4 0  ← Multiplying by 6 tens (We write 0
    8 2 2 0 0 ←    and then multiply 274 by 6.)
    ───────      Multiplying by 3 hundreds (We write 00
    9 8,6 4 0    and then multiply 274 by 3.)
                 Adding
```

Do Exercises 12–15.

Note the following.

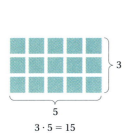

3 · 5 = 15

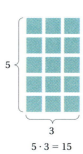

5 · 3 = 15

If we rotate the array on the left, we get the array on the right. The answers match. This illustrates the **commutative law of multiplication.** It says that we can multiply two numbers in any order, $a \cdot b = b \cdot a$, and still get the same answer.

Do Exercise 16.

Multiply.

9.   4 7 2   10. 408 × 704
     × 3 0 6

11.   2 3 4 4
     × 6 0 0 5

Multiply.

12.   4 7 2   13.   2 3 4 4
     × 8 3 0   × 7 4 0 0

14. 100 × 562

15. 1000 × 562

16. a) Find 23 · 47.

    b) Find 47 · 23.

    c) Compare your answers from parts (a) and (b).

Answers on page A-2

1.5 Multiplication and Area

37

Multiply.

17. $5 \cdot 2 \cdot 4$

18. $5 \cdot 1 \cdot 3$

19. Estimate the product twice: first by rounding to the nearest ten and then by rounding to the nearest hundred. Show your work.

$$\begin{array}{r} 8\ 3\ 7 \\ \times\ 2\ 4\ 5 \\ \hline \end{array}$$

To the instructor and the student: This section presented a review of multiplication of whole numbers. Students who are successful should go on to Section 1.6. Those who have trouble should study developmental unit M near the back of this text and then repeat Section 1.5.

Answers on page A-2

Chapter 1 Operations on the Whole Numbers

38

To multiply three or more numbers, we usually group them so that we multiply two at a time. Consider $2 \cdot (3 \cdot 4)$ and $(2 \cdot 3) \cdot 4$. The parentheses tell what to do first:

$$2 \cdot (3 \cdot 4) = 2 \cdot (12) = 24.$$ We multiply 3 and 4, then that result and 2.

We can also multiply 2 and 3, then that result and 4:

$$(2 \cdot 3) \cdot 4 = (6) \cdot 4 = 24.$$

Either way we get 24. It does not matter how we group the numbers. This illustrates that **multiplication is associative**: $a \cdot (b \cdot c) = (a \cdot b) \cdot c$. Together the commutative and associative laws tell us that to multiply more than two numbers, we can use any order and grouping we wish.

CAUTION! Do not confuse the associative law with the distributive law. To multiply $2 \cdot (3 \cdot 4)$, each number is used twice. To multiply $2 \cdot (3 + 4)$, the 2 is used twice: $2 \cdot 3 + 2 \cdot 4$.

Do Exercises 17 and 18.

c Rounding and Estimating

Example 8 Estimate the following product twice: first by rounding to the nearest ten and then by rounding to the nearest hundred: 683×457.

Nearest ten	Nearest hundred	Exact
$6\ 8\ 0$	$7\ 0\ 0$	$6\ 8\ 3$
$\times\ 4\ 6\ 0$	$\times\ 5\ 0\ 0$	$\times\ 4\ 5\ 7$
$4\ 0\ 8\ 0\ 0$	$3\ 5\ 0\ 0\ 0\ 0$	$4\ 7\ 8\ 1$
$2\ 7\ 2\ 0\ 0\ 0$		$3\ 4\ 1\ 5\ 0$
$3\ 1\ 2\ 8\ 0\ 0$		$2\ 7\ 3\ 2\ 0\ 0$
		$3\ 1\ 2\ 1\ 3\ 1$

Note in Example 8 that the estimate, having been rounded to the nearest ten, is

312,800.

The estimate, having been rounded to the nearest hundred, is

350,000.

Note how the estimates compare to the exact answer,

312,131.

Do Exercise 19.

Exercise Set 1.5

a Write a multiplication sentence that corresponds to each situation.

1. The *Los Angeles Sunday Times* crossword puzzle is arranged rectangularly with squares in 21 rows and 21 columns. How many squares does the puzzle have altogether?

2. *Pixels.* A computer screen consists of small rectangular dots called *pixels*. How many pixels are there on a screen that has 600 rows with 800 pixels in each row?

3. A new soft drink beverage carton contains 8 cans, each of which holds 12 oz. How many ounces are there in the carton?

4. There are 7 days in a week. How many days are there in 18 weeks?

Find the area of each region.

5. 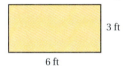 3 ft, 6 ft

6. 7 mi, 7 mi

7. 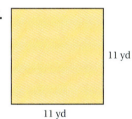 11 yd, 11 yd

8.  16 cm, 9 cm

9. 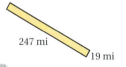 3 mm, 48 mm

10. 247 mi, 19 mi

Exercise Set 1.5, p. 39

1. 21 · 21 = 441   **3.** 8 · 12 oz = 96 oz   **5.** 18 sq ft
7. 121 sq yd   **9.** 144 sq mm   **11.** 870   **13.** 2,340,000
15. 520   **17.** 564   **19.** 65,200   **21.** 4,371,000
23. 1527   **25.** 64,603   **27.** 4770   **29.** 3995
31. 46,080   **33.** 14,652   **35.** 207,672   **37.** 798,408
39. 166,260   **41.** 11,794,332   **43.** 20,723,872
45. 362,128   **47.** 20,064,048   **49.** 25,236,000
51. 302,220   **53.** 49,101,136   **55.** 30,525
57. 298,738   **59.** 50 · 70 = 3500   **61.** 30 · 30 = 900
63. 900 · 300 = 270,000   **65.** 400 · 200 = 80,000
67. 6000 · 5000 = 30,000,000
69. 8000 · 6000 = 48,000,000   **71.** 4370   **72.** 3109
73. 2350; 2300; 2000   **75.** ◆

b Multiply.

11. 87   12. 100   13.   2340   14.   800
    ×10   × 96   ×1000   × 70

15. 65   16. 87   17.   94   18.   76
    × 8   × 4   × 6   × 9

19. 652   20. 652   21. 4371   22. 4371
    ×100   × 10   ×1000   × 100

23. 3 · 509   24. 7 · 806   25. 7(9229)   26. 4(7867)

27. 90(53)   28. 60(78)   29. (47)(85)   30. (34)(87)

31. 640   32. 666   33. 444   34. 509
    × 72   × 66   × 33   × 88

Chapter 1 Operations on the Whole Numbers

35. 509
   ×408

36. 432
   ×375

37. 853
   ×936

38. 346
   ×650

39. 489
   ×340

40. 7080
   × 160

41. 4378
   ×2694

42. 8007
   × 480

43. 6428
   ×3224

44. 8928
   ×3172

45. 3482
   × 104

46. 6408
   ×6064

47. 5006
   ×4008

48. 6789
   ×2330

49. 5608
   ×4500

50. 4560
   ×7890

51. 876
   ×345

52. 355
   ×299

53. 7889
   ×6224

54. 6501
   ×3449

55. 555
   × 55

56. 888
   × 88

57. 734
   ×407

58. 5080
   × 302

Exercise Set 1.5

41

c Estimate each product by first rounding to the nearest ten. Show your work.

59. 45
      × 67

60. 51
      × 78

61. 34
      × 29

62. 63
      × 54

Estimate each product by first rounding to the nearest hundred. Show your work.

63. 876
      × 345

64. 355
      × 299

65. 432
      × 199

66. 789
      × 434

Estimate each product by first rounding to the nearest thousand. Show your work.

67. 5608
      × 4576

68. 2344
      × 6123

69. 7888
      × 6224

70. 6501
      × 3449

Skill Maintenance

71. Add. [1.2b]

        20
       850
     +3500

72. Subtract. [1.3d]

      6003
     −2894

73. Round 2345 to the nearest ten, to the nearest hundred, and to the nearest thousand. [1.4a]

Synthesis

74. ◆ Explain in your own words what it means to say that multiplication is commutative.

75. ◆ Describe a situation that corresponds to the multiplication 4 · $150. (See Examples 1 and 2.)

76. 🖩 An 18-story office building is box-shaped. Each floor measures 172 ft by 84 ft with a 20-ft by 35-ft rectangular area lost to an elevator and a stairwell. How much area is available as office space?

1.6 Division

a Division and the Real World

Division of whole numbers corresponds to two kinds of situations. In the first, consider the division 20 ÷ 5, read "20 divided by 5." We can think of 20 objects arranged in a rectangular array. We ask "How many rows, each with 5 objects, are there?"

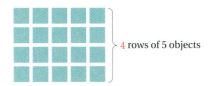

4 rows of 5 objects

Since there are 4 rows of 5 objects each, we have

20 ÷ 5 = 4.   **This is read "20 divided by 5 equals 4."**

In the second situation, we can ask, "If we make 5 rows, how many objects will there be in each row?"

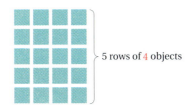

5 rows of 4 objects

Since there are 4 objects in each of the 5 rows, we have

20 ÷ 5 = 4.

We say that the **dividend** is 20, the **divisor** is 5, and the **quotient** is 4.

Dividend   Divisor   Quotient

The *dividend* is what we are dividing into. The result of the division is the *quotient*.

We also write a division such as 20 ÷ 5 as

20/5   or $\frac{20}{5}$ or   .

Example 1 Write a division sentence that corresponds to this situation.

A parent gives $24 to 3 children, with each child getting the same amount. How much does each child get?

We think of an array with 3 rows. Each row will go to a child. How many dollars will be in each row?

3 rows with 8 in each row

24 ÷ 3 = 8

Objectives

a Write a division sentence that corresponds to a situation.

b Given a division sentence, write a related multiplication sentence; and given a multiplication sentence, write two related division sentences.

c Divide whole numbers.

For Extra Help

TAPE 2   MAC WIN   CD-ROM

1.6 Division

43

Write a division sentence that corresponds to each situation. You need not carry out the division.

1. There are 112 students in a college band, and they are marching with 14 in each row. How many rows are there?

2. A college band is in a rectangular array. There are 112 students in the band, and they are marching in 8 rows. How many students are there in each row?

Example 2 Write a division sentence that corresponds to this situation. You need not carry out the division.

How many mailboxes that cost $45 each can be purchased for $495?

We think of an array with 45 one-dollar bills in each row. The money in each row will buy a mailbox. How many rows will there be?

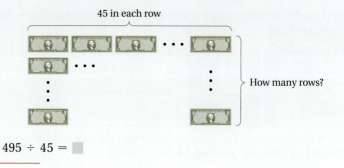

$495 \div 45 = \square$

> Whenever we have a rectangular array, we know the following:
> (The total number) ÷ (The number of rows) = (The number in each row).
>
> Also:
> (The total number) ÷ (The number in each row) = (The number of rows).

Do Exercises 1 and 2.

b Related Sentences

By looking at rectangular arrays, we can see how multiplication and division are related. The following array shows that $4 \cdot 5 = 20$.

$4 \cdot 5 = 20$

The array also shows the following:

$20 \div 5 = 4$ and $20 \div 4 = 5$.

Division is actually defined in terms of multiplication. For example, $20 \div 5$ is defined to be the number that when multiplied by 5 gives 20. Thus, for every division sentence, there is a related multiplication sentence.

$20 \div 5 = 4$ **Division sentence**

$20 = 4 \cdot 5$ **Related multiplication sentence**

> To get the related multiplication sentence, we use
> Dividend = Quotient · Divisor.

Answers on page A-2

Chapter 1 Operations on the Whole Numbers

Example 3 Write a related multiplication sentence: $12 \div 6 = 2$.

We have

$12 \div 6 = 2$ Division sentence

$12 = 2 \cdot 6$.   Related multiplication sentence

The related multiplication sentence is $12 = 2 \cdot 6$.

> By the commutative law of multiplication, there is also another multiplication sentence: $12 = 6 \cdot 2$.

Do Exercises 3 and 4.

For every multiplication sentence, we can write related divisions, as we can see from the preceding array.

Example 4 Write two related division sentences: $7 \cdot 8 = 56$.

We have

$7 \cdot 8 = 56$ $7 \cdot 8 = 56$

This factor becomes a divisor.   This factor becomes a divisor.

$7 = 56 \div 8$. $8 = 56 \div 7$.

The related division sentences are $7 = 56 \div 8$ and $8 = 56 \div 7$.

Do Exercises 5 and 6.

c Division of Whole Numbers

Multiplication can be thought of as repeated addition. Division can be thought of as repeated subtraction. Compare.

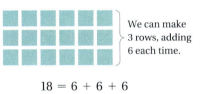
We can make 3 rows, adding 6 each time.

$18 = 6 + 6 + 6$

3 additions

$= 3 \cdot 6$

If we take away 6 objects at a time, we can do so 3 times.

$18 - 6 - 6 - 6 = 0$

3 subtractions

$18 \div 6 = 3$

Margin Exercises, Section 1.6, pp. 44–49

1. $112 \div 14 =$ 2. $112 \div 8 =$ 3. $15 = 5 \cdot 3$, or $15 = 3 \cdot 5$ 4. $72 = 9 \cdot 8$, or $72 = 8 \cdot 9$
5. $6 = 12 \div 2$; $2 = 12 \div 6$ 6. $6 = 42 \div 7$; $7 = 42 \div 6$ 7. 6; $6 \cdot 9 = 54$ 8. 6 R 7; $6 \cdot 9 = 54$, $54 + 7 = 61$ 9. 4 R 5; $4 \cdot 12 = 48$, $48 + 5 = 53$
10. 6 R 13; $6 \cdot 24 = 144$, $144 + 13 = 157$ 11. 59 R 3
12. 1475 R 5   13. 1015   14. 134   15. 63 R 12
16. 807 R 4   17. 1088   18. 360 R 4   19. 800 R 47

Write a related multiplication sentence.

3. $15 \div 3 = 5$

4. $72 \div 8 = 9$

Write two related division sentences.

5. $6 \cdot 2 = 12$

6. $7 \cdot 6 = 42$

Answers on page A-2

Divide by repeated subtraction. Then check.

7. $54 \div 9$

8. $61 \div 9$

9. $53 \div 12$

10. $157 \div 24$

Answers on page A-2

To divide by repeated subtraction, we keep track of the number of times we subtract.

Example 5 Divide by repeated subtraction: $20 \div 4$.

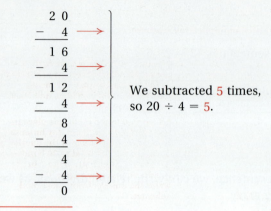

We subtracted 5 times, so $20 \div 4 = 5$.

Example 6 Divide by repeated subtraction: $23 \div 5$.

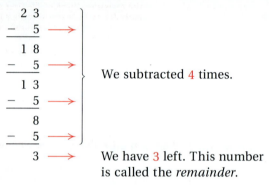

We subtracted 4 times.

We have 3 left. This number is called the *remainder*.

We write

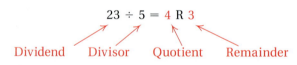

$$23 \div 5 = 4 \text{ R } 3$$

Dividend   Divisor   Quotient   Remainder

CHECKING DIVISIONS. To check a division, we multiply. Suppose we divide 98 by 2 and get 49:

$98 \div 2 = 49$.

To check, we think of the related multiplication sentence $49 \cdot 2 = \square$. We multiply 49 by 2 and see if we get 98.

If there is a remainder, we add it after multiplying.

Example 7 Check the division in Example 6.

We found that $23 \div 5 = 4$ R 3. To check, we multiply 5 by 4. This gives us 20. Then we add 3 to get 23. The dividend is 23, so the answer checks.

Do Exercises 7–10.

When we use the process of long division, we are doing repeated subtraction, even though we are going about it in a different way.

To divide, we start from the digit of highest place value in the dividend and work down to the lowest place value through the remainders. At each step we ask if there are multiples of the divisor in the quotient.

Example 8 Divide and check: 3642 ÷ 5.

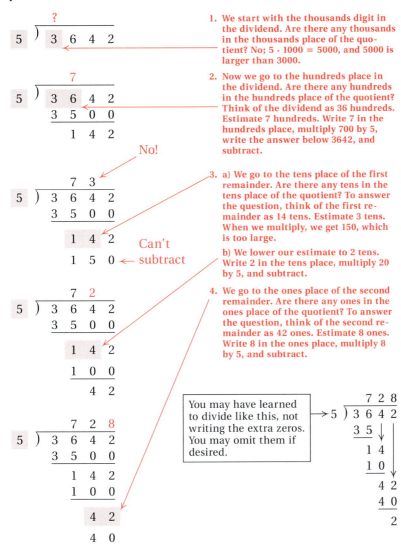

1. We start with the thousands digit in the dividend. Are there any thousands in the thousands place of the quotient? No; 5 · 1000 = 5000, and 5000 is larger than 3000.

2. Now we go to the hundreds place in the dividend. Are there any hundreds in the hundreds place of the quotient? Think of the dividend as 36 hundreds. Estimate 7 hundreds. Write 7 in the hundreds place, multiply 700 by 5, write the answer below 3642, and subtract.

3. a) We go to the tens place of the first remainder. Are there any tens in the tens place of the quotient? To answer the question, think of the first remainder as 14 tens. Estimate 3 tens. When we multiply, we get 150, which is too large.

   b) We lower our estimate to 2 tens. Write 2 in the tens place, multiply 20 by 5, and subtract.

4. We go to the ones place of the second remainder. Are there any ones in the ones place of the quotient? To answer the question, think of the second remainder as 42 ones. Estimate 8 ones. Write 8 in the ones place, multiply 8 by 5, and subtract.

You may have learned to divide like this, not writing the extra zeros. You may omit them if desired.

The answer is 728 R 2. To check, we multiply the quotient 728 by the divisor 5. This gives us 3640. Then we add 2 to get 3642. The dividend is 3642, so the answer checks.

Do Exercises 11–13.

We can summarize our division procedure as follows.

To do division of whole numbers:
a) Estimate.
b) Multiply.
c) Subtract.

Divide and check.

11. 4) 2 3 9

12. 6) 8 8 5 5

13. 5) 5 0 7 5

Answers on page A-2

1.6 Division

47

Divide.

14. 4 5) 6 0 3 0

15. 5 2) 3 2 8 8

Sometimes rounding the divisor helps us find estimates.

Example 9 Divide: $8904 \div 42$.

We mentally round 42 to 40.

$$
\begin{array}{r}
2 \\
42{\overline{\smash{\big)}\,8904}} \\
\underline{8400} \\
504
\end{array}
$$
← *Think*: 89 hundreds ÷ 40. Estimate 2 hundreds, but write $2 \times 42 = 84$.

$$
\begin{array}{r}
21 \\
42{\overline{\smash{\big)}\,8904}} \\
\underline{8400} \\
504 \\
\underline{420} \\
84
\end{array}
$$
← *Think*: 50 tens ÷ 40. Estimate 1 ten, but write $1 \times 42 = 42$.

$$
\begin{array}{r}
212 \\
42{\overline{\smash{\big)}\,8904}} \\
\underline{8400} \\
504 \\
\underline{420} \\
84 \\
\underline{84} \\
0
\end{array}
$$
← *Think*: 84 ones ÷ 40. Estimate 2 ones, but write $2 \times 42 = 84$.

> **CAUTION!** Be careful to keep the digits lined up correctly.

The answer is 212. *Remember*: If after estimating and multiplying you get a number that is larger than the divisor, you cannot subtract, so lower your estimate.

Do Exercises 14 and 15.

Calculator Spotlight

Calculators usually provide division answers in decimal notation. (Decimal notation will be considered in Chapter 5.) There are calculators that give quotients and remainders directly. One such calculator, the *TI Math Explorer,* uses a special division key [INT÷]. An I indicator is displayed when [INT÷] is pressed. Then a [Q] and an [R] in brackets indicate the quotient and the remainder.

Example Divide: $3642 \div 5$.

Press	Display
3642 [INT÷]	I 3642
5 =	728 2
	[Q] [R]

The quotient is 728 and the remainder is 2.

Exercises

Use a calculator that finds quotients and remainders to do the following divisions.

1. $8855 \div 6$
2. $9724 \div 27$
3. $44{,}847 \div 56$
4. $6030 \div 45$

Answers on page A-2

Calculator Spotlight, p. 48

1. 1475 R 5
2. 360 R 4
3. 800 R 47
4. 134

Zeros in Quotients

Example 10 Divide: $6341 \div 7$.

```
        9
     ┌─────────
  7  ) 6 3 4 1
       6 3 0 0
       ───────
             4 1
```
← *Think*: 63 hundreds ÷ 7. Estimate 9 hundreds.

```
        9 0
     ┌─────────
  7  ) 6 3 4 1
       6 3 0 0
       ───────
             4 1
```
← *Think*: 4 tens ÷ 7. There are no tens in the quotient (other than the tens in 900). We write a 0 to show this.

```
        9 0 5
     ┌─────────
  7  ) 6 3 4 1
       6 3 0 0
       ───────
             4 1
             3 5
             ───
               6
```
← *Think*: 41 ones ÷ 7. Estimate 5 ones.

The answer is 905 R 6.

Do Exercises 16 and 17.

Example 11 Divide: $8889 \div 37$.

We mentally round 37 to 40.

```
           2
        ┌─────────
   3 7  ) 8 8 8 9
          7 4 0 0
          ───────
          1 4 8 9
```
← *Think*: 37 ≈ 40; 88 hundreds ÷ 40. Estimate 2 hundreds, but write 2 × 37 = 74.

```
           2 4
        ┌─────────
   3 7  ) 8 8 8 9
          7 4 0 0
          ───────
          1 4 8 9
          1 4 8 0
          ───────
                9
```
← *Think*: 148 tens ÷ 40. Estimate 4 tens, but write 4 × 37 = 148.

```
           2 4 0
        ┌─────────
   3 7  ) 8 8 8 9
          7 4 0 0
          ───────
          1 4 8 9
          1 4 8 0
          ───────
                9
```
← *Think*: 9 ones ÷ 40. There are no ones in the quotient.

The answer is 240 R 9.

Do Exercises 18 and 19.

Divide.

16. $6 \overline{) 4846}$

17. $7 \overline{) 7616}$

Divide.

18. $27 \overline{) 9724}$

19. $56 \overline{) 44{,}847}$

To the instructor and the student: This section presented a review of division of whole numbers. Students who are successful should go on to Section 1.7. Those who have trouble should study developmental unit D near the back of this text and then repeat Section 1.6.

Answers on page A-2

1.6 Division

Improving Your Math Study Skills

Homework

Before Doing Your Homework

- **Setting.** Consider doing your homework as soon as possible after class, before you forget what you learned in the lecture. Research has shown that after 24 hours, most people forget about half of what is in their short-term memory. To avoid this "automatic" forgetting, you need to transfer the knowledge into long-term memory. The best way to do this with math concepts is to perform practice exercises repeatedly. This is the "drill-and-practice" part of learning math that comes when you do your homework. It cannot be overlooked if you want to succeed in your study of math.

   Try to set a specific time for your homework. Then choose a location that is quiet and free from interruptions. Some students find it helpful to listen to music when doing homework. Research has shown that classical music creates the best atmosphere for studying: Give it a try!

- **Reading.** Before you begin doing the homework exercises, you should reread the assigned material in the textbook. You may also want to look over your class notes again and rework some of the examples given in class.

   You should not read a math textbook as you would a novel or history textbook. Math texts are not meant to be read passively. Be sure to stop and do the margin exercises when directed. Also be sure to reread any paragraphs as you see the need.

While Doing Your Homework

- **Study groups.** For some students, forming a study group can be helpful. Many times, two heads are better than one. Also, it is true that "to teach is to learn again." Thus, when you explain a concept to your classmate, you often gain a deeper understanding of the concept yourself. If you do study in a group, resist the temptation to waste time by socializing.

   If you work regularly with someone, be careful not to become dependent on that person. Work on your own some of the time so that you do not rely heavily on others and are able to learn even when they are not available.

- **Notebook.** When doing your homework, consider using notebook paper in a spiral or three-ring binder. You want to be able to go over your homework when studying for a test. Therefore, you need to be able to easily access any problem in your homework notebook. Write legibly in your notebook so you can check over your work. Label each section and each exercise clearly, and show all steps. Your clear writing will also be appreciated by your instructor should your homework be collected. Most tutors and instructors can be more helpful if they can see and understand all the steps in your work.

   When you are finished with your homework, check the answers to the odd-numbered exercises at the back of the book or in the *Student's Solutions Manual* and make corrections. If you do not understand why an answer is wrong, draw a star by it so you can ask questions in class or during your instructor's office hours.

After Doing Your Homework

- **Review.** If you complete your homework several days before the next class, review your work every day. This will keep the material fresh in your mind. You should also review the work immediately before the next class so that you can ask questions as needed.

Exercise Set 1.6

a Write a division sentence that corresponds to each situation. You need not carry out the division.

1. *Canyonlands.* The trail boss for a trip into Canyonlands National Park divides 760 pounds (lb) of equipment among 4 mules. How many pounds does each mule carry?

2. *Surf Expo.* In a swimwear showing at Surf Expo, a trade show for retailers of beach supplies, each swimsuit test takes 8 minutes (min). If the show runs for 240 min, how many tests can be scheduled?

3. A lab technician pours 455 mL of sulfuric acid into 5 beakers, putting the same amount in each. How much acid is in each beaker?

4. A computer screen is made up of a rectangular array of pixels. There are 480,000 pixels in all, with 800 pixels in each row. How many rows are there on the screen?

b Write a related multiplication sentence.

5. $18 \div 3 = 6$

6. $72 \div 9 = 8$

7. $22 \div 22 = 1$

8. $32 \div 1 = 32$

9. $54 \div 6 = 9$

10. $40 \div 8 = 5$

11. $37 \div 1 = 37$

12. $28 \div 28 = 1$

Write two related division sentences.

13. $9 \times 5 = 45$

14. $2 \cdot 7 = 14$

15. $37 \cdot 1 = 37$

16. $4 \cdot 12 = 48$

17. $8 \times 8 = 64$

18. $9 \cdot 7 = 63$

19. $11 \cdot 6 = 66$

20. $1 \cdot 43 = 43$

c Divide.

21. 277 ÷ 5

22. 699 ÷ 3

23. 864 ÷ 8

24. 869 ÷ 8

25. 4) 1 2 2 8

26. 3) 2 1 2 4

27. 6) 4 5 2 1

28. 9) 9 1 1 0

29. 297 ÷ 4

30. 389 ÷ 2

31. 738 ÷ 8

32. 881 ÷ 6

33. 5) 8 5 1 5

34. 3) 6 0 2 7

35. 9) 8 8 8 8

36. 8) 4 1 3 9

37. 127,000 ÷ 10

38. 127,000 ÷ 100

39. 127,000 ÷ 1000

40. 4260 ÷ 10

41. 7 0) 3 6 9 2

42. $20\overline{)5798}$ **43.** $30\overline{)875}$ **44.** $40\overline{)987}$

45. 852 ÷ 21 **46.** 942 ÷ 23 **47.** $85\overline{)7672}$

48. $54\overline{)2729}$ **49.** $111\overline{)3219}$ **50.** $102\overline{)5612}$

51. $8\overline{)843}$ **52.** $7\overline{)749}$ **53.** $5\overline{)8047}$

54. $9\overline{)7273}$ **55.** $5\overline{)5036}$ **56.** $7\overline{)7074}$

57. 1058 ÷ 46 **58.** 7242 ÷ 24 **59.** 3425 ÷ 32

60. $48\overline{)4899}$ **61.** $24\overline{)8880}$ **62.** $36\overline{)7563}$

Exercise Set 1.6

63. 28) 17,067 **64.** 36) 28,929 **65.** 80) 24,320

66. 90) 88,560 **67.** 285) 999,999 **68.** 306) 888,888

69. 456) 3,679,920 **70.** 803) 5,622,60_

Exercise Set 1.6, p. 51

1. 760 ÷ 4 = ▨   **3.** 455 ÷ 5 = ▨   **5.** 18 = 3 · 6, or 18 = 6 · 3   **7.** 22 = 22 · 1, or 22 = 1 · 22   **9.** 54 = 6 · 9, or 54 = 9 · 6   **11.** 37 = 1 · 37, or 37 = 37 · 1   **13.** 9 = 45 ÷ 5; 5 = 45 ÷ 9   **15.** 37 = 37 ÷ 1; 1 = 37 ÷ 37   **17.** 8 = 64 ÷ 8   **19.** 11 = 66 ÷ 6; 6 = 66 ÷ 11   **21.** 55 R 2   **23.** 108   **25.** 307   **27.** 753 R 3   **29.** 74 R 1   **31.** 92 R 2   **33.** 1703   **35.** 987 R 5   **37.** 12,700   **39.** 127   **41.** 52 R 52   **43.** 29 R 5   **45.** 40 R 12   **47.** 90 R 22   **49.** 29   **51.** 105 R 3   **53.** 1609 R 2   **55.** 1007 R 1   **57.** 23   **59.** 107 R 1   **61.** 370   **63.** 609 R 15   **65.** 304   **67.** 3508 R 219   **69.** 8070   **71.** 7 thousands + 8 hundreds + 8 tens + 2 ones   **72.** >   **73.** 21 = 16 + 5, or 21 = 5 + 16   **74.** 56 = 14 + 42, or 56 = 42 + 14   **75.** 47 = 56 − 9; 9 = 56 − 47   **76.** 350 = 414 − 64; 64 = 414 − 350   **77.** ◆   **79.** 30

Skill Maintenance

71. Write expanded notation for 7882.   [1.1a]

72. Use < or > for ▨ to write a true sentence: [1.4c]
   888 ▨ 788.

Write a related addition sentence.   [1.3b]

73. 21 − 16 = 5

74. 56 − 14 = 42

Write two related subtraction sentences.   [1.3b]

75. 47 + 9 = 56

76. 350 + 64 = 414

Synthesis

77. ◆ Describe a situation that corresponds to the division 1180 ÷ 295. (See Examples 1 and 2.)

78. ◆ Is division associative? Why or why not?

79. A group of 1231 college students is going to take buses for a field trip. Each bus can hold only 42 students. How many buses are needed?

80. 🖩 Fill in the missing digits to make the equation true:
   34,584,132 ÷ 76▨ = 4▨,386.

Chapter 1   Operations on the Whole Numbers

1.7 Solving Equations

a Solutions by Trial

Let's find a number that we can put in the blank to make this sentence true:

$9 = 3 + \_\_$.

We are asking "9 is 3 plus what number?" The answer is 6.

$9 = 3 + 6$

Do Exercises 1 and 2.

A sentence with $=$ is called an **equation**. A **solution** of an equation is a number that makes the sentence true. Thus, 6 is a solution of

$9 = 3 + \_\_$ because $9 = 3 + 6$ is true.

However, 7 is not a solution of

$9 = 3 + \_\_$ because $9 = 3 + 7$ is false.

Do Exercises 3 and 4.

We can use a letter instead of a blank. For example,

$9 = 3 + x$.

We call x a **variable** because it can be replaced by a variety of numbers.

> A **solution** is a replacement for the variable that makes the equation true. When we find all the solutions, we say that we have **solved** the equation.

Example 1 Solve $x + 12 = 27$ by trial.

We replace x with several numbers.

    If we replace x with 13, we get a false equation: $13 + 12 = 27$.
    If we replace x with 14, we get a false equation: $14 + 12 = 27$.
    If we replace x with 15, we get a true equation: $15 + 12 = 27$.

No other replacement makes the equation true, so the solution is 15.

Examples Solve.

2. $7 + n = 22$
(7 plus what number is 22?)
The solution is 15.

3. $8 \cdot 23 = y$
(8 times 23 is what?)
The solution is 184.

Note, as in Example 3, that when the variable is alone on one side of the equation, the other side shows us what calculations to do in order to find the solution.

Do Exercises 5–8.

Objectives

a Solve simple equations by trial.

b Solve equations like
$t + 28 = 54$, $28 \cdot x = 168$, and
$98 \div 2 = y$.

For Extra Help

TAPE 2   MAC WIN   CD-ROM

Find a number that makes the sentence true.

1. $8 = 1 + \_\_$

2. $\_\_ + 2 = 7$

3. Determine whether 7 is a solution of $\_\_ + 5 = 9$.

4. Determine whether 4 is a solution of $\_\_ + 5 = 9$.

Solve by trial.

5. $n + 3 = 8$

6. $x - 2 = 8$

7. $45 \div 9 = y$

8. $10 + t = 32$

Answers on page A-3

1.7 Solving Equations

55

Solve.

9. $346 \times 65 = y$

10. $x = 2347 + 6675$

11. $4560 \div 8 = t$

12. $x = 6007 - 2346$

Solve.

13. $x + 9 = 17$

14. $77 = m + 32$

Answers on page A-3

Chapter 1   Operations on the Whole Numbers

---

b Solving Equations

We now begin to develop more efficient ways to solve certain equations. When an equation has a variable alone on one side, it is easy to see the solution or to compute it. For example, the solution of

$$x = 12$$

is 12. When a calculation is on one side and the variable is alone on the other, we can find the solution by carrying out the calculation.

Example 4 Solve: $x = 245 \times 34$.

To solve the equation, we carry out the calculation.

$$\begin{array}{r} 245 \\ \times\ 34 \\ \hline 980 \\ 7350 \\ \hline 8330 \end{array}$$

The solution is 8330.

Do Exercises 9–12.

Look at the equation

$$x + 12 = 27.$$

We can get x alone on one side of the equation by writing a related subtraction sentence:

$x = 27 - 12$ 12 is subtracted to find the related subtraction sentence.

$x = 15.$ Doing the subtraction

It is useful in our later study of algebra to think of this as "subtracting 12 on both sides." Thus,

$x + 12 - 12 = 27 - 12$ Subtracting 12 on both sides

$x + 0 = 15$ Carrying out the subtraction

$x = 15.$

▶ To solve $x + a = b$, subtract a on both sides.

Note that $x = 15$ is easier to solve than $x + 12 = 27$. This is because we see easily that when x is replaced with 15, we get a true equation: $15 = 15$. The solution of $x = 15$ is 15, which is also the solution of $x + 12 = 27$. When two equations have the same solution(s), the equations are said to be **equivalent**. Thus, $x = 15$ and $x + 12 = 27$ are equivalent.

Example 5 Solve: $t + 28 = 54$.

We have

$t + 28 = 54$

$t + 28 - 28 = 54 - 28$ Subtracting 28 on both sides

$t + 0 = 26$

$t = 26.$

The solution is 26.

Do Exercises 13 and 14.

Example 6 Solve: $182 = 65 + n$.

We have

$$182 = 65 + n$$
$$182 - 65 = 65 + n - 65 \quad \text{Subtracting 65 on both sides}$$
$$117 = 0 + n \quad \text{65 plus } n \text{ minus 65 is } 0 + n.$$
$$117 = n.$$

The solution is 117.

Do Exercise 15.

Example 7 Solve: $7381 + x = 8067$.

We have

$$7381 + x = 8067$$
$$7381 + x - 7381 = 8067 - 7381 \quad \text{Subtracting 7381 on both sides}$$
$$x = 686.$$

The solution is 686.

Do Exercises 16 and 17.

We now learn to solve equations like $8 \cdot n = 96$. Look at

$$8 \cdot n = 96.$$

We can get n alone by writing a related division sentence:

$$n = 96 \div 8 = \frac{96}{8} \quad \text{96 is divided by 8.}$$
$$= 12. \quad \text{Doing the division}$$

Note that $n = 12$ is equivalent to $8 \cdot n = 96$ but easier to solve.

It is useful in our later study of algebra to think of "dividing by the same number on both sides to form an equivalent equation." Thus,

$$\frac{8 \cdot n}{8} = \frac{96}{8} \quad \text{Dividing by 8 on both sides}$$
$$n = 12. \quad \text{8 times } n \text{ divided by 8 is } n.$$

> To solve $a \cdot x = b$, divide by a on both sides.

15. Solve: $155 = t + 78$.

Solve.
16. $4566 + x = 7877$

17. $8172 = h + 2058$

Margin Exercises, Section 1.7, pp. 55–58

1. 7   2. 5   3. No   4. Yes   5. 5   6. 10   7. 5
8. 22   9. 22,490   10. 9022   11. 570   12. 3661
13. 8   14. 45   15. 77   16. 3311   17. 6114   18. 8
19. 16   20. 644   21. 96   22. 94

Answers on page A-3

1.7 Solving Equations

57

Solve.

18. $8 \cdot x = 64$

19. $144 = 9 \cdot n$

20. Solve: $5152 = 8 \cdot t$.

21. Solve: $18 \cdot y = 1728$.

22. Solve: $n \cdot 48 = 4512$.

Example 8 Solve: $10 \cdot x = 240$.

We have

$$10 \cdot x = 240$$

$$\frac{10 \cdot x}{10} = \frac{240}{10} \quad \text{Dividing by 10 on both sides}$$

$$x = 24.$$

The solution is 24.

Do Exercises 18 and 19.

Example 9 Solve: $5202 = 9 \cdot t$.

We have

$$5202 = 9 \cdot t$$

$$\frac{5202}{9} = \frac{9 \cdot t}{9} \quad \text{Dividing by 9 on both sides}$$

$$578 = t.$$

The solution is 578.

Do Exercise 20.

Example 10 Solve: $14 \cdot y = 1092$.

We have

$$14 \cdot y = 1092$$

$$\frac{14 \cdot y}{14} = \frac{1092}{14} \quad \text{Dividing by 14 on both sides}$$

$$y = 78.$$

The solution is 78.

Do Exercise 21.

Example 11 Solve: $n \cdot 56 = 4648$.

We have

$$n \cdot 56 = 4648$$

$$\frac{n \cdot 56}{56} = \frac{4648}{56} \quad \text{Dividing by 56 on both sides}$$

$$n = 83.$$

The solution is 83.

Do Exercise 22.

Answers on page A-3

Exercise Set 1.7

a Solve by trial.

1. $x + 0 = 14$
2. $x - 7 = 18$
3. $y \cdot 17 = 0$
4. $56 \div m = 7$

b Solve.

5. $13 + x = 42$
6. $15 + t = 22$
7. $12 = 12 + m$
8. $16 = t + 16$

9. $3 \cdot x = 24$
10. $6 \cdot x = 42$
11. $112 = n \cdot 8$
12. $162 = 9 \cdot m$

13. $45 \times 23 = x$
14. $23 \times 78 = y$
15. $t = 125 \div 5$
16. $w = 256 \div 16$

17. $p = 908 - 458$
18. $9007 - 5667 = m$
19. $x = 12{,}345 + 78{,}555$
20. $5678 + 9034 = t$

21. $3 \cdot m = 96$
22. $4 \cdot y = 96$
23. $715 = 5 \cdot z$
24. $741 = 3 \cdot t$

25. $10 + x = 89$
26. $20 + x = 57$
27. $61 = 16 + y$
28. $53 = 17 + w$

29. $6 \cdot p = 1944$
30. $4 \cdot w = 3404$
31. $5 \cdot x = 3715$
32. $9 \cdot x = 1269$

33. $47 + n = 84$
34. $56 + p = 92$
35. $x + 78 = 144$
36. $z + 67 = 133$

Exercise Set 1.7, p. 59

1. 14 3. 0 5. 29 7. 0 9. 8 11. 14 13. 1035
15. 25 17. 450 19. 90,900 21. 32 23. 143
25. 79 27. 45 29. 324 31. 743 33. 37 35. 66
37. 15 39. 48 41. 175 43. 335 45. 104
47. 45 49. 4056 51. 17,603 53. 18,252 55. 205
57. $7 = 15 - 8$; $8 = 15 - 7$ 58. $6 = 48 \div 8$; $8 = 48 \div 6$ 59. < 60. > 61. 142 R 5
62. 334 R 11 63. ◆ 65. 347

37. $165 = 11 \cdot n$ **38.** $660 = 12 \cdot n$ **39.** $624 = t \cdot 13$ **40.** $784 = y \cdot 16$

41. $x + 214 = 389$ **42.** $x + 221 = 333$ **43.** $567 + x = 902$ **44.** $438 + x = 807$

45. $18 \cdot x = 1872$ **46.** $19 \cdot x = 6080$ **47.** $40 \cdot x = 1800$ **48.** $20 \cdot x = 1500$

49. $2344 + y = 6400$ **50.** $9281 = 8322 + t$ **51.** $8322 + 9281 = x$ **52.** $9281 - 8322 = y$

53. $234 \times 78 = y$ **54.** $10{,}534 \div 458 = q$ **55.** $58 \cdot m = 11{,}890$ **56.** $233 \cdot x = 22{,}135$

Skill Maintenance

57. Write two related subtraction sentences: $7 + 8 = 15$.   [1.3b]

58. Write two related division sentences: $6 \cdot 8 = 48$.   [1.6b]

Use > or < for ▨ to write a true sentence.   [1.4c]

59. 123 ▨ 789

60. 342 ▨ 339

Divide.   [1.6c]

61. $1283 \div 9$

62. $17 \overline{)5689}$

Synthesis

63. ◆ Describe a procedure that can be used to convert any equation of the form $a + b = c$ to a related subtraction equation.

64. ◆ Describe a procedure that can be used to convert any equation of the form $a \cdot b = c$ to a related division equation.

Solve.

65. $23{,}465 \cdot x = 8{,}142{,}355$

66. $48{,}916 \cdot x = 14{,}332{,}388$

1.8 Applications and Problem Solving

a Applications and problem solving are the main uses of mathematics. To solve a problem using the operations on the whole numbers, we first look at the situation. We try to translate the problem to an equation. Then we solve the equation. We check to see if the solution of the equation is a solution of the original problem. Thus we are using the following five-step strategy.

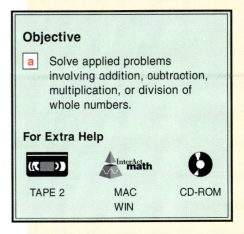

Objective

a Solve applied problems involving addition, subtraction, multiplication, or division of whole numbers.

For Extra Help

TAPE 2   MAC WIN   CD-ROM

> **FIVE STEPS FOR PROBLEM SOLVING**
>
> 1. *Familiarize* yourself with the situation. If it is described in words, as in a textbook, *read carefully*. In any case, think about the situation. Draw a picture whenever it makes sense to do so. Choose a letter, or *variable*, to represent the unknown quantity to be solved for.
> 2. *Translate* the problem to an equation.
> 3. *Solve* the equation.
> 4. *Check* the answer in the original wording of the problem.
> 5. *State* the answer to the problem clearly with appropriate units.

Example 1 *Minivan Sales.* Recently, sales of minivans have stabilized. The bar graph at right shows the number of Chrysler Town & Country LXi minivans sold in recent years. Find the total number of minivans sold during those years.

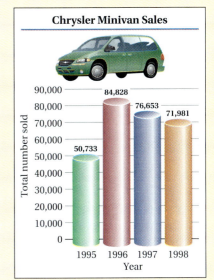

Chrysler Minivan Sales

(Bar graph: 1995: 50,733; 1996: 84,828; 1997: 76,653; 1998: 71,981)

Source: Chrysler Corporation

1. **Familiarize.** We can make a drawing or at least visualize the situation.

50,733	+	84,828	+	76,653	+	71,981	=	n
in 1995		in 1996		in 1997		in 1998		Total sold

   Since we are combining objects, addition can be used. First we define the unknown. We let n = the total number of minivans sold.

2. **Translate.** We translate to an equation:

   $50{,}733 + 84{,}828 + 76{,}653 + 71{,}981 = n.$

3. **Solve.** We solve the equation by carrying out the addition.

   ```
     1 3 1 1
     5 0,7 3 3
     8 4,8 2 8
     7 6,6 5 3
   + 7 1,9 8 1
   ─────────
     2 8 4,1 9 5
   ```

   Thus, $284{,}195 = n$, or $n = 284{,}195$.

4. **Check.** We check 284,195 in the original problem. There are many ways in which this can be done. For example, we can repeat the calculation. (We leave this to the student.) Another way is to check the reasonableness of the answer by noting that it is larger than the sales in any of the individual years. We can also estimate by rounding. Here, we round to the nearest thousand:

   $50{,}733 + 84{,}828 + 76{,}653 + 71{,}981$
   $\approx 51{,}000 + 85{,}000 + 77{,}000 + 72{,}000$
   $= 285{,}000.$

1. *Teacher needs in 2005.* The data in the table show the estimated number of new jobs for teachers in the year 2005. The reason is an expected boom in the number of youngsters under the age of 18. Find the total number of new jobs available for teachers in 2005.

Type of Teacher	Number of New Jobs
Secondary	386,000
Aide	364,000
Childcare worker	248,000
Elementary	220,000
Special education	206,000

Source: Bureau of Labor Statistics

2. *Checking Account.* You have $756 in your checking account. You write a check for $387 to pay for a sound system for your campus apartment. How much is left in your checking account?

Answers on page A-3

Chapter 1 Operations on the Whole Numbers

62

Since $284{,}195 \approx 285{,}000$, we have a partial check. If we had an estimate like 236,000 or 580,000, we might be suspicious that our calculated answer is incorrect. Since our estimated answer is close to our calculation, we are further convinced that our answer checks.

5. **State.** The total number of Town and Country minivans sold during these years is 284,195.

Do Exercise 1.

Example 2 *Hard-Drive Space.* The hard drive on your computer has 572 megabytes (MB) of storage space available. You install a software package called Microsoft® Office, which uses 84 MB of space. How much storage space do you have left after the installation?

1. **Familiarize.** We first make a drawing or at least visualize the situation. We let $M =$ the amount of space left.

572 MB   84 MB

2. **Translate.** We see that this is a "take-away" situation. We translate to an equation.

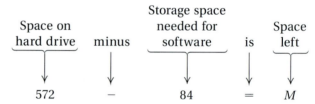

3. **Solve.** This sentence tells us what to do. We subtract.

$$\begin{array}{r} \overset{16}{\overset{4\,612}{\cancel{5}\,\cancel{7}\,\cancel{2}}} \\ -8\,4 \\ \hline 4\,8\,8 \end{array}$$

Thus, $488 = M$, or $M = 488$.

4. **Check.** We check our answer of 488 MB by repeating the calculation. We note that the answer should be less than the original amount of memory, 572 MB, which it is. We can also add the difference, 488, to the subtrahend, 84: $84 + 488 = 572$. We can also estimate:

$$572 - 84 \approx 600 - 100 = 500 \approx 488.$$

5. **State.** There are 488 MB of memory left.

Do Exercise 2.

In the real world, problems may not be stated in written words. You must still become familiar with the situation before you can solve the problem.

Example 3 *Travel Distance.* Vicki is driving from Indianapolis to Salt Lake City to work during summer vacation. The distance from Indianapolis to Salt Lake City is 1634 mi. She travels 1154 mi to Denver. How much farther must she travel?

1. **Familiarize.** We first make a drawing or at least visualize the situation. We let x = the remaining distance to Salt Lake City.

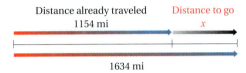

2. **Translate.** We see that this is a "how-much-more" situation. We translate to an equation.

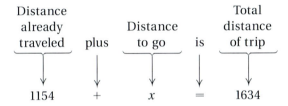

3. **Solve.** We solve the equation.

$$1154 + x = 1634$$
$$1154 + x - 1154 = 1634 - 1154 \quad \text{Subtracting 1154 on both sides}$$
$$x = 480$$

```
    5 13
  1 6̸ 3̸ 4
-  1 1 5 4
  ─────────
      4 8 0
```

4. **Check.** We check our answer of 480 mi in the original problem. This number should be less than the total distance, 1634 mi, which it is. We can add the difference, 480, to the subtrahend, 1154: $1154 + 480 = 1634$. We can also estimate:

$$1634 - 1154 \approx 1600 - 1200$$
$$= 400 \approx 480.$$

The answer, 480 mi, checks.

5. **State.** Vicki must travel 480 mi farther to Salt Lake City.

Do Exercise 3.

---

3. *Calculator Purchase.* Bernardo has $76. He wants to purchase a graphing calculator for $94. How much more does he need?

Answer on page A-3

4. *Total Cost of Laptop Computers.* What is the total cost of 12 laptop computers with CD-ROM drive and printer if each one costs $3249?

Example 4 *Total Cost of VCRs.* What is the total cost of 5 four-head VCRs if each one costs $289?

1. **Familiarize.** We first make a drawing or at least visualize the situation. We let n = the cost of 5 VCRs. Repeated addition works well here.

2. **Translate.** We translate to an equation.

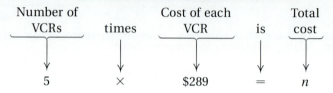

3. **Solve.** This sentence tells us what to do. We multiply.

$$\begin{array}{r} \overset{4\ 4}{2\ 8\ 9} \\ \times\quad\ 5 \\ \hline 1\ 4\ 4\ 5 \end{array}$$

Thus, $n = 1445$.

4. **Check.** We have an answer that is much larger than the cost of any individual VCR, which is reasonable. We can repeat our calculation. We can also check by estimating:

$5 \times 289 \approx 5 \times 300 = 1500 \approx 1445.$

The answer checks.

5. **State.** The total cost of 5 VCRs is $1445.

Do Exercise 4.

Answer on page A-3

Example 5 *Bed Sheets.* The dimensions of a sheet for a king-size bed are 108 in. by 102 in. What is the area of the sheet? (The dimension labels on sheets list width × length.)

1. **Familiarize.** We first make a drawing. We let A = the area.

2. **Translate.** Using a formula for area, we have

   A = length · width = $l \cdot w$ = 102 · 108.

3. **Solve.** We carry out the multiplication.

   ```
       1 0 8
   ×   1 0 2
       2 1 6
   1 0 8 0 0
   1 1 0 1 6
   ```

   Thus, A = 11,016.

4. **Check.** We repeat our calculation. We also note that the answer is larger than both the length and the width, which it should be. (This would not be the case if we were using numbers smaller than 1.) The answer checks.

5. **State.** The area of a king-size bed sheet is 11,016 sq in.

Do Exercise 5.

Example 6 *Diet Cola Packaging.* Diet Cola has become popular in the quest to control weight. A bottling company produces 2203 cans of cola. How many 8-can packages can be filled? How many cans will be left over?

1. **Familiarize.** We first draw a picture. We let n = the number of 8-can packages to be filled.

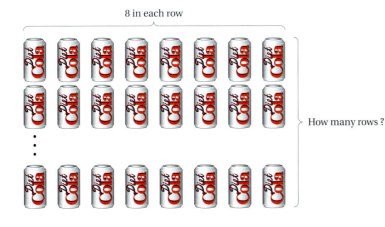

5. *Bed Sheets.* The dimensions of a sheet for a queen-size bed are 90 in. by 102 in. What is the area of the sheet?

Answer on page A-3

5. *Bed Sheets.*

1.8 Applications and Problem Solving

65

6. *Diet Cola Packaging.* The bottling company also uses 6-can packages. How many 6-can packages can be filled with 2269 cans of cola? How many cans will be left over?

2. Translate. We can translate to an equation as follows.

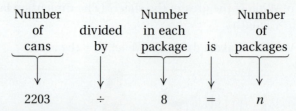

$$2203 \div 8 = n$$

3. Solve. We solve the equation by carrying out the division.

```
        2 7 5
    8 ) 2 2 0 3
        1 6 0 0
          6 0 3
          5 6 0
            4 3
            4 0
               3
```

4. Check. We can check by multiplying the number of packages by 8 and adding the remainder, 3:

$$8 \cdot 275 = 2200, \quad 2200 + 3 = 2203.$$

5. State. Thus, 275 8-can packages can be filled. There will be 3 cans left over.

Do Exercise 6.

Example 7 *Automobile Mileage.* The Chrysler Town & Country LXi minivan featured in Example 1 gets 18 miles to the gallon (mpg) in city driving. How many gallons will it use in 4932 mi of city driving?

1. Familiarize. We first make a drawing. It is often helpful to be descriptive about how you define a variable. In this example, we let g = the number of gallons (g comes from "gallons").

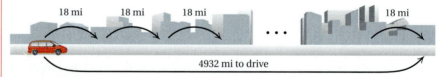

2. Translate. Repeated addition applies here. Thus the following multiplication corresponds to the situation.

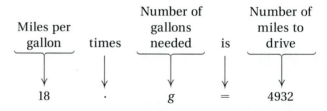

$$18 \cdot g = 4932$$

Answer on page A-3

3. **Solve.** To solve the equation, we divide by 18 on both sides.

$$18 \cdot g = 4932$$
$$\frac{18 \cdot g}{18} = \frac{4932}{18}$$
$$g = 274$$

```
      2 7 4
18 ) 4 9 3 2
     3 6 0 0
     1 3 3 2
     1 2 6 0
         7 2
         7 2
          0
```

4. **Check.** To check, we multiply 274 by 18: $18 \cdot 274 = 4932$.
5. **State.** The minivan will use 274 gal.

Do Exercise 7.

Multistep Problems

Sometimes we must use more than one operation to solve a problem, as in the following example.

Example 8 *Weight Loss.* Many Americans exercise for weight control. It is known that one must burn off about 3500 calories in order to lose one pound. The chart shown here details how many calories are burned by certain activities. How long would an individual have to run at a brisk pace in order to lose one pound?

1. **Familiarize.** We can first make a chart.

ONE POUND			
3500 calories			
100 cal 8 min	100 cal 8 min	... ...	100 cal 8 min

2. **Translate.** Repeated addition applies here. Thus the following multiplication corresponds to the situation. We must find out how many 100's there are in 3500. We let $x =$ the number of 100's in 3500.

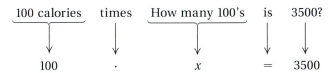

7. *Automobile Mileage.* The Chrysler Town & Country LXi minivan gets 24 miles to the gallon (mpg) in highway driving. How many gallons will it use in 888 mi of highway driving?

Answer on page A-3

7. *Automobile Mileage.*

1.8 Applications and Problem Solving

8. **Weight Loss.** Using the chart for Example 8, determine how long an individual must swim in order to lose one pound.

9. **Bones in the Hands and Feet.** There are 27 bones in each human hand and 26 bones in each human foot. How many bones are there in all in the hands and feet?

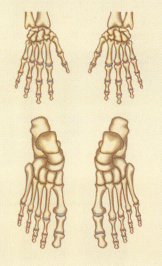

Answers on page A-3

Chapter 1 Operations on the Whole Numbers

68

3. **Solve.** To solve the equation, we divide by 100 on both sides.

$$100 \cdot x = 3500$$

$$\frac{100 \cdot x}{100} = \frac{3500}{100}$$

$$x = 35$$

```
         3 5
100 ) 3 5 0 0
      3 0 0 0
          5 0 0
          5 0 0
              0
```

We know that running for 8 min will burn off 100 calories. To do this 35 times will burn off one pound, so you must run for 35 times 8 minutes in order to burn off one pound. We let t = the time it takes to run off one pound.

$$35 \times 8 = t$$
$$280 = t$$

```
    3 5
  × 8
  2 8 0
```

4. **Check.** Suppose you run for 280 min. Every 8 minutes you burn 100 calories. If we divide 280 by 8, we get 35, and 35 times 100 is 3500, the number of calories it takes to lose one pound.

5. **State.** It will take 280 min, or 4 hr, 40 min, of running to lose one pound.

Do Exercises 8 and 9.

As you consider the following exercises, here are some words and phrases that may be helpful to look for when you are translating problems to equations.

Addition:	sum, total, increase, altogether, plus
Subtraction:	difference, minus, how much more?, how many more?, decrease, deducted, how many left?
Multiplication:	given rows and columns, how many in all?, product, total from a repeated addition, area, of
Division:	how many in each row?, how many rows?, how many pieces?, how many parts in a whole?, quotient, divisible

Margin Exercises, Section 1.8, pp. 62–68

1. 1,424,000 2. $369 3. $18 4. $38,988
5. 9180 sq in. 6. 378 packages; 1 can left over
7. 37 gal 8. 70 min, or 1 hr, 10 min 9. 106

Exercise Set 1.8

a Solve.

1. During the first four months of a recent year, Campus Depot Business Machine Company reported the following sales:
   January $3572
   February $2718
   March $2809
   April $3177
   What were the total sales over this time period?

2. A family travels the following miles during a five-day trip:
   Monday   568
   Tuesday   376
   Wednesday   424
   Thursday   150
   Friday   224
   How many miles did they travel altogether?

Bicycle Sales. The bar graph below shows the total sales, in millions of dollars, for bicycles and related supplies in recent years. Use this graph for Exercises 3–6.

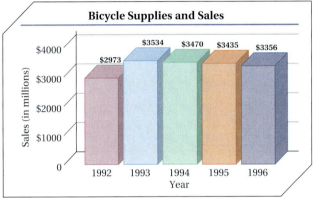

3. What were the total sales for 1993 and 1994?

4. What were the total sales for 1993 through 1996?

5. How much more were the sales in 1993 than in 1994?

6. How much more were the sales in 1994 than in 1992?

7. *Longest Rivers.* The longest river in the world is the Nile, which has a length of 4145 mi. It is 138 mi longer than the next longest river, which is the Amazon in South America. How long is the Amazon?

8. *Largest Lakes.* The largest lake in the world is the Caspian Sea, which has an area of 317,000 square kilometers (sq km). The Caspian is 288,900 sq km larger than the second largest lake, which is Lake Superior. What is the area of Lake Superior?

9. *Sheet Perimeter.* The dimensions of a sheet for a queen-size bed are 90 in. by 102 in. What is the perimeter of the sheet?

10. Sh...

Exercise Set 1.8, p. 69

1. $12,276   3. $7,004,000,000   5. $64,000,000
7. 4007 mi   9. 384 in.   11. 4500   13. 7280
15. $247   17. 54 weeks; 1 episode left over
19. 168 hr   21. $400   23. (a) 4700 sq ft; (b) 288 ft
25. 44   27. 56 cartons; 11 books left over
29. 1600 mi; 27 in.   31. 18   33. 22   35. $704
37. 525 min, or 8 hr, 45 min   39. 3000 sq in.
41. 234,600   42. 235,000   43. 22,000   44. 16,000
45. 320,000   46. 720,000   47. ◆   49. 792,000 mi; 1,386,000 mi

www.mathmax.com

11. *Paper Quantity.* A ream of paper contains 500 sheets. How many sheets are in 9 reams?

12. *Reading Rate.* Cindy's reading rate is 205 words per minute. How many words can she read in 30 min?

13. *Elvis Impersonators.* When Elvis Presley died in 1977, there were already 48 professional Elvis impersonators (**Source**: *Chance Magazine* 9, no. 1, Winter 1996). In 1995, there were 7328. How many more were there in 1995?

14. *LAV Vehicle.* A fully-loaded U.S. Light Armed Vehicle 25 (LAV-25) weighs 3930 lb more than its empty curb weight. The loaded LAV-25 weighs 28,400 lb. (**Source**: *Car & Driver* 42, no. 1, July 1996: 153–155) What is its curb weight?

15. Dana borrows $5928 for a used car. The loan is to be paid off in 24 equal monthly payments. How much is each payment (excluding interest)?

16. A family borrows $4824 to build a sunroom on the back of their house. The loan is to be paid off in equal monthly payments of $134 (excluding interest). How many months will it take to pay off the loan?

17. *Cheers Episodes.* *Cheers* is the longest-running comedy in the history of television, with 271 episodes created. A local station picks up the syndicated reruns. If the station runs 5 episodes per week, how many full weeks will pass before it must start over with past episodes? How many episodes will be left for the last week?

18. A lab technician separates a vial containing 70 cubic centimeters (cc) of blood into test tubes, each of which contains 3 cc of blood. How many test tubes can be filled? How much blood is left over?

19. There are 24 hours (hr) in a day and 7 days in a week. How many hours are there in a week?

20. There are 60 min in an hour and 24 hr in a day. How many minutes are there in a day?

21. You have $568 in your checking account. You write checks for $46, $87, and $129. Then you deposit $94 back in the account upon the return of some books. How much is left in your account?

22. The balance in your checking account is $749. You write checks for $34 and $65. Then you make a deposit of $123 from your paycheck. What is your new balance?

23. *NBA Court.* The standard basketball court used by college and NBA players has dimensions of 50 ft by 94 ft (*Source*: National Basketball Association).

a) What is its area?
b) What is its perimeter?

24. *High School Court.* The standard basketball court used by high school players has dimensions of 50 ft by 84 ft.

a) What is its area? What is its perimeter?
b) How much larger is the area of an NBA court than a high school court? (See Exercise 23.)

25. Sixteen-ounce bottles of catsup are generally shipped in cartons containing 12 bottles each. How many cartons are needed to ship 528 bottles of catsup?

26. Copies of this book are generally shipped from the warehouse in cartons containing 24 books each. How many cartons are needed to ship 840 books?

27. Copies of this book are generally shipped from the warehouse in cartons containing 24 books each. How many cartons are completely filled when shipping 1355 books? How many books are left over?

28. Sixteen-ounce bottles of catsup are generally shipped in cartons containing 12 bottles each. How many cartons are completely filled when shipping 1033 bottles of catsup? How many bottles are left over?

29. *Map Drawing.* A map has a scale of 64 mi to the inch. How far apart *in reality* are two cities that are 25 in. apart on the map? How far apart *on the map* are two cities that, in reality, are 1728 mi apart?

30. *Map Drawing.* A map has a scale of 25 mi to the inch. How far apart *on the map* are two cities that, in reality, are 2200 mi apart? How far apart *in reality* are two cities that are 13 in. apart on the map?

31. A carpenter drills 216 holes in a rectangular array in a pegboard. There are 12 holes in each row. How many rows are there?

32. Lou works as a CPA. He arranges 504 entries on a spreadsheet in a rectangular array that has 36 rows. How many entries are in each row?

33. Elaine buys 5 video games at $44 each and pays for them with $10 bills. How many $10 bills does it take?

34. Lowell buys 5 video games at $44 each and pays for them with $20 bills. How many $20 bills does it take?

35. Before going back to college, David buys 4 shirts at $59 each and 6 pairs of pants at $78 each. What is the total cost of this clothing?

36. Ann buys office supplies at Office Depot. One day she buys 8 boxes of paper at $24 each and 16 pens at $3 each. How much does she spend?

37. *Weight Loss.* Use the information from the chart on page 67. How long must you do aerobic exercises in order to lose one pound?

38. *Weight Loss.* Use the information from the chart on page 67. How long must you bicycle at 9 mph in order to lose one pound?

39. *Index Cards.* Index cards of dimension 3 in. by 5 in. are normally shipped in packages containing 100 cards each. How much writing area is available if one uses the front and back sides of a package of these cards?

40. An office for adjunct instructors at a community college has 6 bookshelves, each of which is 3 ft long. The office is moved to a new location that has dimensions of 16 ft by 21 ft. Is it possible for the bookshelves to be put side by side on the 16-ft wall?

Skill Maintenance

Round 234,562 to the nearest: [1.4a]

41. Hundred.

42. Thousand.

Estimate each computation by rounding to the nearest thousand. [1.4b]

43. 2783 + 4602 + 5797 + 8111

44. 28,430 − 11,977

Estimate each product by rounding to the nearest hundred. [1.5c]

45. 787 · 363

46. 887 · 799

Synthesis

47. ◆ Of the five problem-solving steps listed at the beginning of this section, which is the most difficult for you? Why?

48. ◆ Write a problem for a classmate to solve. Design the problem so that the solution is "The driver still has 329 mi to travel."

49. *Speed of Light.* Light travels about 186,000 miles per second (mi/sec) in a vacuum as in outer space. In ice it travels about 142,000 mi/sec, and in glass it travels about 109,000 mi/sec. In 18 sec, how many more miles will light travel in a vacuum than in ice? in glass?

50. Carney Community College has 1200 students. Each professor teaches 4 classes and each student takes 5 classes. There are 30 students and 1 teacher in each classroom. How many professors are there at Carney Community College?

Make a budget for a road trip to your favorite destination.

1.9 Exponential Notation and Order of Operations

a | Exponential Notation

Consider the product $3 \cdot 3 \cdot 3 \cdot 3$. Such products occur often enough that mathematicians have found it convenient to create a shorter notation, called **exponential notation**, explained as follows.

$3 \cdot 3 \cdot 3 \cdot 3$ (4 factors) is shortened to 3^4 ← exponent, base

We read 3^4 as "three to the fourth power," 5^3 as "five to the third power," or "five cubed," and 5^2 as "five squared." The latter comes from the fact that a square of side s has area A given by $A = s^2$.

$A = s^2$

Example 1 Write exponential notation for $10 \cdot 10 \cdot 10 \cdot 10 \cdot 10$.

Exponential notation is 10^5. 5 is the *exponent*. 10 is the *base*.

Example 2 Write exponential notation for $2 \cdot 2 \cdot 2$.

Exponential notation is 2^3.

Do Exercises 1–4.

b | Evaluating Exponential Notation

We evaluate exponential notation by rewriting it as a product and computing the product.

Example 3 Evaluate: 10^3.

$10^3 = 10 \cdot 10 \cdot 10 = 1000$

Example 4 Evaluate: 5^4.

$5^4 = 5 \cdot 5 \cdot 5 \cdot 5 = 625$

Caution! 5^4 does not mean $5 \cdot 4$.

Do Exercises 5–8.

Objectives

a | Write exponential notation for products such as $4 \cdot 4 \cdot 4$.
b | Evaluate exponential notation.
c | Simplify expressions using the rules for order of operations.
d | Remove parentheses within parentheses.

For Extra Help

TAPE 2 | MAC WIN | CD-ROM

Write in exponential notation.

1. $5 \cdot 5 \cdot 5 \cdot 5$ 5^4

2. $5 \cdot 5 \cdot 5 \cdot 5 \cdot 5$ 5^5

3. $10 \cdot 10$ 10^2

4. $10 \cdot 10 \cdot 10 \cdot 10$ 10^4

Evaluate.

5. 10^4 10,000

6. 10^2 100

7. 8^3 512

8. 2^5 32

Answers on page A-3

Margin Exercises, Section 1.9, pp. 73–76

1. 5^4 2. 5^5 3. 10^2 4. 10^4 5. 10,000   6. 100
7. 512   8. 32   9. 51   10. 30   11. 584   12. 84
13. 4; 1   14. 52; 52   15. 29   16. 1880   17. 253
18. 93   19. 1880   20. 305   21. 93   22. 87 in.
23. 46   24. 4

Simplify.

9. $93 - 14 \cdot 3$ 51

10. $104 \div 4 + 4$ 30

11. $25 \cdot 26 - (56 + 10)$ 584

12. $75 \div 5 + (83 - 14)$ 84

Simplify and compare.   4; 1
13. $64 \div (32 \div 2)$ and
    $(64 \div 32) \div 2$

14. $(28 + 13) + 11$ and   52; 52
    $28 + (13 + 11)$

Answers on page A-3

Chapter 1   Operations on the Whole Numbers

c Simplifying Expressions

Suppose we have a calculation like the following:

$$3 + 4 \cdot 8.$$

How do we find the answer? Do we add 3 to 4 and then multiply by 8, or do we multiply 4 by 8 and then add 3? In the first case, the answer is 56. In the second, the answer is 35. We agree to compute as in the second case. Consider the calculation

$$7 \cdot 14 - (12 + 18).$$

What do the parentheses mean? To deal with these questions, we must make some agreement regarding the order in which we perform operations. The rules are as follows.

> **RULES FOR ORDER OF OPERATIONS**
>
> 1. Do all calculations within parentheses ( ), brackets [ ], or braces { } before operations outside.
> 2. Evaluate all exponential expressions.
> 3. Do all multiplications and divisions in order from left to right.
> 4. Do all additions and subtractions in order from left to right.

It is worth noting that these are the rules that a computer uses to do computations. In order to program a computer, you must know these rules.

Example 5   Simplify: $16 \div 8 \times 2$.

There are no parentheses or exponents, so we start with the third step.

$16 \div 8 \times 2 = 2 \times 2$ Doing all multiplications and divisions in order from left to right
$ = 4$

Example 6   Simplify: $7 \cdot 14 - (12 + 18)$.

$7 \cdot 14 - (12 + 18) = 7 \cdot 14 - 30$ Carrying out operations inside parentheses
$ = 98 - 30$ Doing all multiplications and divisions
$ = 68$ Doing all additions and subtractions

Do Exercises 9–12.

Example 7   Simplify and compare: $23 - (10 - 9)$ and $(23 - 10) - 9$.

We have

$23 - (10 - 9) = 23 - 1 = 22;$
$(23 - 10) - 9 = 13 - 9 = 4.$

We can see that $23 - (10 - 9)$ and $(23 - 10) - 9$ represent different numbers. Thus subtraction is not associative.

Do Exercises 13 and 14.

The Future Of B.S.D Begins With Me Today

Example 8 Simplify: $7 \cdot 2 - (12 + 0) \div 3 - (5 - 2)$.

$7 \cdot 2 - (12 + 0) \div 3 - (5 - 2) = 7 \cdot 2 - 12 \div 3 - 3$ Carrying out operations inside parentheses

$= 14 - 4 - 3$ Doing all multiplications and divisions in order from left to right

$= 7$ Doing all additions and subtractions in order from left to right

Do Exercise 15.

Example 9 Simplify: $15 \div 3 \cdot 2 \div (10 - 8)$.

$15 \div 3 \cdot 2 \div (10 - 8) = 15 \div 3 \cdot 2 \div 2$ Carrying out operations inside parentheses

$= 5 \cdot 2 \div 2$ Doing all multiplications and divisions in order from left to right

$= 10 \div 2$

$= 5$

Do Exercises 16–18.

Example 10 Simplify: $4^2 \div (10 - 9 + 1)^3 \cdot 3 - 5$.

$4^2 \div (10 - 9 + 1)^3 \cdot 3 - 5$

$= 4^2 \div (1 + 1)^3 \cdot 3 - 5$ Carrying out operations inside parentheses

$= 4^2 \div 2^3 \cdot 3 - 5$ Adding inside parentheses

$= 16 \div 8 \cdot 3 - 5$ Evaluating exponential expressions

$= 2 \cdot 3 - 5$

$= 6 - 5$ Doing all multiplications and divisions in order from left to right

$= 1$

Do Exercises 19–21.

Calculator Spotlight

Calculators often have an $\boxed{x^y}$, $\boxed{a^x}$, or $\boxed{\wedge}$ key for raising a base to a power. To find 3^5 with such a key, we press $\boxed{3}$ $\boxed{x^y}$ $\boxed{5}$ $\boxed{=}$. The result is 243.

1. Find 4^5.   **2.** Find 7^9.   **3.** Find 2^{20}.

To determine whether a calculator is programmed to follow the rules for order of operations, press $\boxed{3}$ $\boxed{+}$ $\boxed{4}$ $\boxed{\times}$ $\boxed{2}$ $\boxed{=}$. If the result is 11, that particular calculator follows the rules. If the result is 14, the calculator performs operations as they are entered. To compensate for the latter case, we would press $\boxed{4}$ $\boxed{\times}$ $\boxed{2}$ $\boxed{=}$ $\boxed{+}$ $\boxed{3}$ $\boxed{=}$.

4. Find $84 - 5 \cdot 7$.   **5.** Find $80 + 50 \div 10$.

When a calculator has parentheses, $\boxed{(}$ and $\boxed{)}$, expressions like $5(4 + 3)$ can be found without first entering the addition. We simply press $\boxed{5}$ $\boxed{\times}$ $\boxed{(}$ $\boxed{4}$ $\boxed{+}$ $\boxed{3}$ $\boxed{)}$ $\boxed{=}$. The result is 35.

6. Find $9(7 + 8)$.   **7.** Find $8[4 + 3(7 - 1)]$.

Calculator Spotlight, p. 75

1. 1024   **2.** 40,353,607   **3.** 1,048,576   **4.** 49   **5.** 85
6. 135   **7.** 176

---

15. Simplify:

$9 \times 4 - (20 + 4) \div 8 - (6 - 2)$.

29

Simplify.

16. $5 \cdot 5 \cdot 5 + 26 \cdot 71 - (16 + 25 \cdot 3)$

1880

17. $30 \div 5 \cdot 2 + 10 \cdot 20 + 8 \cdot 8 - 23$

253

18. $95 - 2 \cdot 2 \cdot 2 \cdot 5 \div (24 - 4)$

93

Simplify.

19. $5^3 + 26 \cdot 71 - (16 + 25 \cdot 3)$

1880

20. $(1 + 3)^3 + 10 \cdot 20 + 8^2 - 23$

305

21. $95 - 2^3 \cdot 5 \div (24 - 4)$

93

Answers on page A-3

1.9 Exponential Notation and Order of Operations

22. NBA Tall Men. The heights, in inches, of several of the tallest players in the NBA are given in the bar graph below. Find the average height of these players.

87 IN

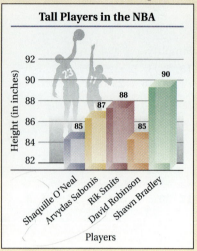

Source: NBA

Simplify.

23. $9 \times 5 + \{6 \div [14 - (5 + 3)]\}$

46

24. $[18 - (2 + 7) \div 3] - (31 - 10 \times 2)$

4

Answers on page A-3

To find the **average** of a set of numbers, we first add the numbers and then divide by the number of addends.

Example 11 *Average Height of Waterfalls.* The heights of the four tallest waterfalls in the world are given in the bar graph at right. Find the average height of all four.

Source: World Almanac

The average is given by

$(3212 + 2425 + 2149 + 2014) \div 4.$

To find the average, we carry out the computation using the rules for order of operations:

$(3212 + 2425 + 2149 + 2014) \div 4 = 9800 \div 4$
$= 2450.$

Thus the average height of the four tallest waterfalls is 2450 ft.

Do Exercise 22.

d Parentheses Within Parentheses

When parentheses occur within parentheses, we can make them different shapes, such as [ ] (also called "brackets") and { } (also called "braces"). All of these have the same meaning. When parentheses occur within parentheses, computations in the innermost pair are to be done first.

Example 12 Simplify: $16 \div 2 + \{40 - [13 - (4 + 2)]\}$.

$16 \div 2 + \{40 - [13 - (4 + 2)]\}$
$= 16 \div 2 + \{40 - [13 - 6]\}$ Doing the calculations in the innermost parentheses first
$= 16 \div 2 + \{40 - 7\}$ Again, doing the calculations in the innermost parentheses
$= 16 \div 2 + 33$
$= 8 + 33$ Doing all multiplications and divisions in order from left to right
$= 41$ Doing all additions and subtractions in order from left to right

Example 13 Simplify: $[25 - (4 + 3) \times 3] \div (11 - 7)$.

$[25 - (4 + 3) \times 3] \div (11 - 7) = [25 - 7 \times 3] \div (11 - 7)$
$= [25 - 21] \div (11 - 7)$
$= 4 \div 4$
$= 1$

Do Exercises 23 and 24.

Exercise Set 1.9

a Write in exponential notation.

1. $3 \cdot 3 \cdot 3 \cdot 3$
2. $2 \cdot 2 \cdot 2 \cdot 2 \cdot 2$
3. $5 \cdot 5$
4. $13 \cdot 13 \cdot 13$
5. $7 \cdot 7 \cdot 7 \cdot 7 \cdot 7$
6. $10 \cdot 10$
7. $10 \cdot 10 \cdot 10$
8. $1 \cdot 1 \cdot 1 \cdot 1$

b Evaluate.

9. 7^2
10. 5^3
11. 9^3
12. 10^2
13. 12^4
14. 10^5
15. 11^2
16. 6^3

c Simplify.

17. $12 + (6 + 4)$
18. $(12 + 6) + 18$
19. $52 - (40 - 8)$
20. $(52 - 40) - 8$
21. $1000 \div (100 \div 10)$
22. $(1000 \div 100) \div 10$
23. $(256 \div 64) \div 4$
24. $256 \div (64 \div 4)$
25. $(2 + 5)^2$
26. $2^2 + 5^2$
27. $(11 - 8)^2 - (18 - 16)^2$
28. $(32 - 27)^3 + (19 + 1)^3$
29. $16 \cdot 24 + 50$
30. $23 + 18 \cdot 20$
31. $83 - 7 \cdot 6$
32. $10 \cdot 7 - 4$
33. $10 \cdot 10 - 3 \cdot 4$
34. $90 - 5 \cdot 5 \cdot 2$
35. $4^3 \div 8 - 4$
36. $8^2 - 8 \cdot 2$
37. $17 \cdot 20 - (17 + 20)$
38. $1000 \div 25 - (15 + 5)$
39. $6 \cdot 10 - 4 \cdot 10$
40. $3 \cdot 8 + 5 \cdot 8$
41. $300 \div 5 + 10$
42. $144 \div 4 - 2$
43. $3 \cdot (2 + 8)^2 - 5 \cdot (4 - 3)^2$
44. $7 \cdot (10 - 3)^2 - 2 \cdot (3 + 1)^2$
45. $4^2 + 8^2 \div 2^2$
46. $6^2 - 3^4 \div 3^3$

47. $10^3 - 10 \cdot 6 - (4 + 5 \cdot 6)$

48. $7^2 + 20 \cdot 4 - (28 + 9 \cdot 2)$

49. $6 \cdot 11 - (7 + 3) \div 5 - (6 - 4)$

50. $8 \times 9 - (12 - 8) \div 4 - (10 - 7)$

51. $120 - 3^3 \cdot 4 \div (5 \cdot 6 - 6 \cdot 4)$

52. $80 - 2^4 \cdot 15 \div (7 \cdot 5 - 45 \div 3)$

53. Find the average of $64, $97, and $121.

54. Find the average of four test grades of 86, 92, 80, and 78.

d Simplify.

55. $8 \times 13 + \{42 \div [18 - (6 + 5)]\}$

56. $72 \div 6 - \{2 \times [9 - (4 \times 2)]\}$

57. $[14 - (3 + 5) \div 2] - [18 \div (8 - 2)]$

58. $[92 \times (6 - 4) \div 8] + [7 \times (8 - 3)]$

59. $(82 - 14) \times [(10 + 45 \div 5) - (6 \cdot 6 - 5 \cdot 5)]$

60. $(18 \div 2) \cdot \{[(9 \cdot 9 - 1) \div 2] - [5 \cdot 20 - (7 \cdot 9 - 2)]\}$

61. $4 \times \{(200 - 50 \div 5) - [(35 \div 7) \cdot (35 \div 7) - 4 \times 3]\}$

62. $\{[18 - 2 \cdot 6] - [40 \div (17 - 9)]\} + \{48 - 13 \times 3 + [(50 - 7 \cdot 5) + 2]\}$

Skill Maintenance

Solve. [1.7b]

63. $x + 341 = 793$

64. $7 \cdot x = 91$

Solve. [1.8a]

65. *Colorado.* The state of Colorado is roughly the shape of a rectangle that is 270 mi by 380 mi. What is its area?

66. On a long four-day trip, a family bought the following amounts of gasoline for their motor home:
    23 gallons,   24 gallons,
    26 gallons,   25 gallons.
    How much gasoline did they buy in all?

Synthesis

67. ◆ The expression $9 - (4 \times 2)$ contains parentheses. Are they necessary? Why or why not?

68. ◆ The expression $(3 \cdot 4)^2$ contains parentheses. Are they necessary? Why or why not?

Simplify.

69. $15(23 - 4 \cdot 2)^3 \div (3 \cdot 25)$

70. $(19 - 2^4)^5 - (141 \div 47)^2$

Each of the equations in Exercises 71–73 is incorrect. First find the correct answer. Then place as many parentheses as needed in the original equation in order to make the incorrect answer correct.

71. $1 + 5 \cdot 4 + 3 = 36$

72. $12 \div 4 + 2 \cdot 3 - 2 = 2$

73. $12 \div 4 + 2 \cdot 3 - 2 = 4$

74. Use the symbols $+$, $-$, $\times$, $\div$, and ( ) and one occurrence each of 1, 2, 3, 4, 5, 6, 7, 8, and 9 to represent 100.

Use the order of operations as a group to simplify expressions.

Summary and Review Exercises: Chapter 1

The review exercises that follow are for practice. Answers are given at the back of the book. If you miss an exercise, restudy the objective indicated in blue next to the exercise or direction line that precedes it.

Write in expanded notation. [1.1a]

1. 2793
2. 56,078

Write in standard notation. [1.1b]

3. 8 thousands + 6 hundreds + 6 tens + 9 ones

4. 9 ten thousands + 8 hundreds + 4 tens + 4 ones

Write a word name. [1.1c]

5. 67,819
6. 2,781,427

Write in standard notation. [1.1d]

7. Four hundred seventy-six thousand, five hundred eighty-eight

8. *San Francisco International.* The total number of passengers passing through San Francisco International Airport in a recent year was thirty-six million, two hundred sixty thousand, sixty-four.

9. What does the digit 8 mean in 4,678,952? [1.1e]

10. In 13,768,940, what digit tells the number of millions? [1.1e]

11. Write an addition sentence that corresponds to the situation. [1.2a]

    Toni has $406 in her checking account. She is paid $78 for a part-time job and deposits that in her checking account. How much is then in the account?

12. Find the perimeter. [1.2a]

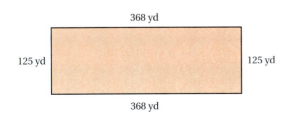

368 yd
125 yd   125 yd
368 yd

Add. [1.2b]

13. 7304 + 6968
14. 27,609 + 38,415

15. 2743 + 4125 + 6274 + 8956

16. 9 1,4 2 6
    +   7,4 9 5

Write a subtraction sentence that corresponds to each situation. [1.3a], [1.3c]

17. By exercising daily, you lose 12 lb in one month. If you weighed 151 lb at the beginning of the month, what is your weight now?

18. Natasha has $196 and wants to buy a fax machine for $340. How much more does she need?

19. Write a related addition sentence: [1.3b]
    $10 - 6 = 4$.

20. Write two related subtraction sentences: [1.3b]
    $8 + 3 = 11$.

Subtract. [1.3d]

21. 8045 − 2897
22. 8465 − 7312

23. 6003 − 3729
24. 3 7,4 0 5
    − 1 9,6 4 8

Round 345,759 to the nearest: [1.4a]

25. Hundred.
26. Ten.

27. Thousand.

Estimate each sum, difference, or product by first rounding to the nearest hundred. Show your work. [1.4b], [1.5c]

28. 41,348 + 19,749
29. 38,652 − 24,549

30. 396 · 748

Use < or > for ▓ to form a true sentence. [1.4c]

31. 67 ▓ 56
32. 1 ▓ 23

33. Write a multiplication sentence that corresponds to the situation. [1.5a]

A farmer plants apple trees in a rectangular array. He plants 15 rows with 32 trees in each row. How many apple trees does he have altogether?

34. Find the area of the rectangle in Exercise 12.

Multiply. [1.5b]

35. $700 \cdot 600$ **36.** $7846 \cdot 800$

37. $726 \cdot 698$ **38.** $587 \cdot 47$

39. $8\,3\,0\,5$
$\times6\,4\,2$

Write a division sentence that corresponds to each situation. You need not carry out the division. [1.6a]

40. A cheese factory made 176 lb of Monterey Jack cheese. The cheese was placed in 4-lb boxes. How many boxes were filled?

41. A beverage company packed 222 cans of soda into 6-can cartons. How many cartons did they fill?

42. Write a related multiplication sentence: [1.6b]
$56 \div 8 = 7$.

43. Write two related division sentences: [1.6b]
$13 \cdot 4 = 52$.

Divide. [1.6c]

44. $63 \div 5$ **45.** $80 \div 16$

46. $7\overline{)6\,3\,9\,4}$ **47.** $3073 \div 8$

48. $6\,0\overline{)2\,8\,6}$ **49.** $4266 \div 79$

50. $3\,8\overline{)1\,7{,}1\,7\,6}$ **51.** $1\,4\overline{)7\,0{,}1\,1\,2}$

52. $52{,}668 \div 12$

Solve. [1.7b]

53. $46 \cdot n = 368$ **54.** $47 + x = 92$

55. $x = 782 - 236$

Solve. [1.8a]

56. An apartment builder bought 3 electric ranges at $299 each and 4 dishwashers at $379 each. What was the total cost?

57. *Lincoln-Head Pennies.* In 1909, the first Lincoln-head pennies were minted. Seventy-three years later, these pennies were first minted with a decreased copper content. In what year was the copper content reduced?

58. A family budgets $4950 for food and clothing and $3585 for entertainment. The yearly income of the family was $28,283. How much of this income remained after these two allotments?

59. A chemist has 2753 mL of alcohol. How many 20-mL beakers can be filled? How much will be left over?

60. Write in exponential notation: $4 \cdot 4 \cdot 4$. [1.9a]

Evaluate. [1.9b]

61. 10^4 **62.** 6^2

Simplify. [1.9c, d]

63. $8 \cdot 6 + 17$

64. $10 \cdot 24 - (18 + 2) \div 4 - (9 - 7)$

65. $7 + (4 + 3)^2$

66. $7 + 4^2 + 3^2$

67. $(80 \div 16) \times [(20 - 56 \div 8) + (8 \cdot 8 - 5 \cdot 5)]$

68. Find the average of 157, 170, and 168.

Synthesis

69. ◆ Write a problem for a classmate to solve. Design the problem so that the solution is "Each of the 144 bottles will contain 8 oz of hot sauce." [1.8a]

70. ◆ Is subtraction associative? Why or why not? [1.2b], [1.3d]

71. ■ Determine the missing digit d. [1.5b]

$9\,d$
$\times\,\,\,d\,2$
$\overline{8\,0\,3\,6}$

72. ■ Determine the missing digits a and b. [1.6c]

$9\,a\,1$
$2\,b\,1\,\overline{)\,2\,3\,6{,}4\,2\,1}$

73. A mining company estimates that a crew must tunnel 2100 ft into a mountain to reach a deposit of copper ore. Each day the crew tunnels about 500 ft. Each night about 200 ft of loose rocks roll back into the tunnel. How many days will it take the mining company to reach the copper deposit? [1.8a]

Test: Chapter 1

1. Write in expanded notation: 8843.

2. Write a word name: 38,403,277.

3. In the number 546,789, which digit tells the number of hundred thousands?

Add.

4.
```
  6 8 1 1
+ 3 1 7 8
```

5.
```
  4 5,8 8 9
+ 1 7,9 0 2
```

6.
```
   1 2
    8
    3
    7
+   4
```

7.
```
  6 2 0 3
+ 4 3 1 2
```

Subtract.

8.
```
  7 9 8 3
- 4 3 5 3
```

9.
```
  2 9 7 4
- 1 9 3 5
```

10.
```
  8 9 0 7
- 2 0 5 9
```

11.
```
  2 3,0 6 7
- 1 7,8 9 2
```

Multiply.

12.
```
  4 5 6 8
×       9
```

13.
```
  8 8 7 6
×   6 0 0
```

14.
```
    6 5
  × 3 7
```

15.
```
    6 7 8
  × 7 8 8
```

Divide.

16. 15 ÷ 4

17. 420 ÷ 6

18. 8 9) 8 6 3 3

19. 4 4) 3 5,4 2 8

Solve.

20. *James Dean.* James Dean was 24 yr old when he died. He was born in 1931. In what year did he die?

21. A beverage company produces 739 cans of soda. How many 8-can packages can be filled? How many cans will be left over?

22. *Area of New England.* Listed below are the areas, in square miles, of the New England states (**Source:** U.S. Bureau of the Census). What is the total area of New England?

Maine	30,865
Massachusetts	7,838
New Hampshire	8,969
Vermont	9,249
Connecticut	4,845
Rhode Island	1,045

23. A rectangular lot measures 200 m by 600 m. What is the area of the lot? What is the perimeter of the lot?

24. A sack of oranges weighs 27 lb. A sack of apples weighs 32 lb. Find the total weight of 16 bags of oranges and 43 bags of apples.

25. A box contains 5000 staples. How many staplers can be filled from the box if each stapler holds 250 staples?

Answers

1. _____
2. _____
3. _____
4. _____
5. _____
6. _____
7. _____
8. _____
9. _____
10. _____
11. _____
12. _____
13. _____
14. _____
15. _____
16. _____
17. _____
18. _____
19. _____
20. _____
21. _____
22. _____
23. _____
24. _____
25. _____

Answers

26. _____

27. _____

28. _____

29. _____

30. _____

31. _____

32. _____

33. _____

34. _____

35. _____

36. _____

37. _____

38. _____

39. _____

40. _____

41. _____

42. _____

43. _____

44. _____

45. _____

46. _____

47. _____

48. _____

49. _____

Solve.

26. $28 + x = 74$ **27.** $169 \div 13 = n$ **28.** $38 \cdot y = 532$

Round 34,578 to the nearest:

29. Thousand. **30.** Ten. **31.** Hundred.

Estimate each sum, difference, or product by first rounding to the nearest hundred. Show your work.

32. $2\,3{,}6\,4\,9$
$+\,5\,4{,}7\,4\,6$

33. $5\,4{,}7\,5\,1$
$-\,2\,3{,}6\,4\,9$

34. $8\,2\,4$
$\times\,4\,8\,9$

Use < or > for ▓ to form a true sentence.

35. 34 ▓ 17 **36.** 117 ▓ 157

37. Write in exponential notation: $12 \cdot 12 \cdot 12 \cdot 12$.

Evaluate.

38. 7^3 **39.** 2^3

Simplify.

40. $(10 - 2)^2$ **41.** $10^2 - 2^2$ **42.** $(25 - 15) \div 5$

43. 8 × ... ... $2^4 + 24 \div 12$

45. Fi...

Synth...

46. A... ...29 a month to repay ...n. If she has already ...he 10-yr loan, how ...s remain?

48. Je... ...d the single digit a... ...hich av... $a \div 3 \times 25 - 7^2 = 339$.

Test: Chapter 1, p. 81

1. [1.1a] 8 thousands + 8 hundreds + 4 tens + 3 ones
2. [1.1c] Thirty-eight million, four hundred three thousand, two hundred seventy-seven **3.** [1.1e] 5
4. [1.2b] 9989 **5.** [1.2b] 63,791 **6.** [1.2b] 34
7. [1.2b] 10,515 **8.** [1.3d] 3630 **9.** [1.3d] 1039
10. [1.3d] 6848 **11.** [1.3d] 5175 **12.** [1.5b] 41,112
13. [1.5b] 5,325,600 **14.** [1.5b] 2405
15. [1.5b] 534,264 **16.** [1.6c] 3 R 3 **17.** [1.6c] 70
18. [1.6c] 97 **19.** [1.6c] 805 R 8 **20.** [1.8a] 1955
21. [1.8a] 92 packages, 3 cans left over
22. [1.8a] 62,811 mi² **23.** [1.8a] 120,000 m²; 1600 m
24. [1.8a] 1808 lb **25.** [1.8a] 20 **26.** [1.7b] 46
27. [1.7b] 13 **28.** [1.7b] 14 **29.** [1.4a] 35,000
30. [1.4a] 34,580 **31.** [1.4a] 34,600
32. [1.4b] 23,600 + 54,700 = 78,300
33. [1.4b] 54,800 − 23,600 = 31,200
34. [1.5c] 800 · 500 = 400,000 **35.** [1.4c] >
36. [1.4c] < **37.** [1.9a] 12⁴ **38.** [1.9b] 343
39. [1.9b] 8 **40.** [1.9c] 64 **41.** [1.9c] 96
42. [1.9c] 2 **43.** [1.9d] 216 **44.** [1.9c] 18
45. [1.9c] 92 **46.** [1.5a], [1.8a] 336 in² **47.** [1.8a] 80
48. [1.9c] 83 **49.** [1.9c] 9

Chapter 1 Operations on the Whole Numbers

2

Introduction to Integers and Algebraic Expressions

Introduction

This chapter is actually our first look at algebra. We introduce numbers called *integers* that represent an extension of the set of whole numbers. We will learn to add, subtract, multiply, and divide integers.

2.1 Integers and the Number Line
2.2 Addition of Integers
2.3 Subtraction of Integers
2.4 Multiplication of Integers
2.5 Division of Integers
2.6 Introduction to Algebra and Expressions
2.7 Like Terms and Perimeter
2.8 Solving Equations

An Application

Midway through a movie, Lisa resets the counter on her VCR to 0. She then fast-forwards the tape 12 min and rewinds it 19 min. What does the counter now read?

This problem appears as Exercise 69 in Exercise Set 2.3.

The Mathematics

If we let t be the final counter reading, we have

$$12 + (-19) = t.$$

This is addition of integers.

World Wide Web   For more information, visit us at www.mathmax.com

Pretest: Chapter 2

1. Tell which integers correspond to the following situation: Bill lost $35 in Atlantic City and Janet won $67 in Las Vegas.

Use either < or > for ▮ to form a true sentence.

2. −7 ▮ 0

3. −2 ▮ −17

4. −12 ▮ 7

Find the absolute value.

5. $|73|$

6. $|-57|$

7. Find $-x$ when x is -32.

8. Find the opposite, or additive inverse, of -17.

Compute and simplify.

9. $-6 + (-12)$

10. $15 + (-9)$

11. $-19 + 10$

12. $-13 + 13$

13. $-7 - 9$

14. $-8 - (-10)$

15. $7 - (-11)$

16. $3 - 9$

17. $-37 \cdot 0$

18. $-8 \cdot (-6)$

19. $(-4)^3$

20. $45 \div (-9)$

21. $(-33) \div (-3)$

22. $\dfrac{-400}{5}$

23. $\dfrac{0}{-7}$

24. $10 \div 2 \cdot 5 - 5^2$

25. Evaluate $7a - b$ for $a = -2$ and $b = -1$.

Multiply.

26. $3(x + 5)$

27. $7(2x - 3y - 1)$

Combine like terms.

28. $7a + 8a$

29. $-9x + 7 - x + 8$

30. Find the perimeter of a 12-ft by 20-ft deck.

Solve.

31. $-3x = 54$

32. $-5 + t = 17$

33. $x \cdot 8 = 48$

Chapter 2 Integers and Algebraic Expressions

2.1 Integers and the Number Line

In this section, we extend the set of whole numbers to form the set of *integers*.

Integers

To create the set of integers, we begin with the set of whole numbers, 0, 1, 2, 3, and so on. For each number 1, 2, 3, and so on, we obtain a new number to the left of zero on the number line:

For the number 1, there will be an *opposite* number -1 (negative 1).

For the number 2, there will be an *opposite* number -2 (negative 2).

For the number 3, there will be an *opposite* number -3 (negative 3), and so on.

The **integers** consist of the whole numbers and these new numbers. We illustrate them on a number line as follows.

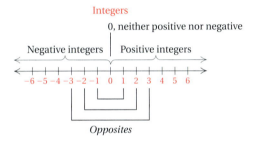

The integers to the left of zero are called **negative integers** and those to the right of zero are called **positive integers.** Zero is neither positive nor negative.

> The **integers**: ..., $-5, -4, -3, -2, -1, 0, 1, 2, 3, 4, 5, ...$

Objectives

a. Tell which integers correspond to a real-world situation.

b. Form a true sentence using $<$ or $>$.

c. Find the absolute value of any integer.

d. Find the opposite of any integer.

For Extra Help

TAPE 3   MAC WIN   CD-ROM

a   Integers and the Real World

Integers correspond to many real-world problems and situations. The following examples will help you get ready to translate problem situations to mathematical language.

Example 1 Tell which integer corresponds to this situation: The coldest temperature ever recorded in Vermont is 55° below zero Fahrenheit (F).

55° below zero is $-55°$.

Tell which integers correspond to each situation.

1. The halfback gained 8 yd on first down. The quarterback was sacked for a 5-yd loss on second down.

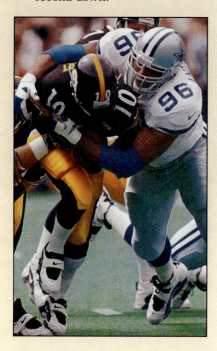

2. The highest temperature ever recorded in the United States was 134° in Death Valley on July 10, 1913. The coldest temperature ever recorded in the United States was 80° below zero in Prospect Creek, Alaska, in January 1971.

3. At 10 sec before liftoff, ignition occurs. At 148 sec after liftoff, the first stage is detached from the rocket.

4. Jacob owes $137 to the bookstore. Fortunately, he has $289 in a savings account.

Answers on page A-4

Chapter 2   Integers and Algebraic Expressions

Example 2   Tell which integer corresponds to this situation: Death Valley is 280 feet below sea level.

The integer −280 corresponds to the situation. The elevation is −280 ft.

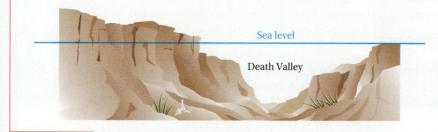

Example 3   Tell which integers correspond to this situation: Elaine rewound the videotape in her VCR 17 min and then fast-forwarded it 25 min.

The integers −17 and 25 correspond to the situation. The integer −17 corresponds to the rewinding and 25 corresponds to the fast-forwarding.

Some common uses for negative integers are as follows:

Time:	Before an event;
Temperature:	Degrees below zero;
Money:	Amount lost, spent, owed, or withdrawn;
Elevation:	Depth below sea level;
Travel:	Motion in the backward (reverse) direction.

Do Exercises 1–4.

Calculator Spotlight

To enter a negative number on most calculators, we use the $\boxed{+/-}$ key. This key gives the opposite of whatever number is currently displayed. Thus, to enter −27, we press $\boxed{2}\ \boxed{7}\ \boxed{+/-}$. Some graphing calculators have a $\boxed{(-)}$ key. To enter −27 on such a graphing calculator, we simply press $\boxed{(-)}\ \boxed{2}\ \boxed{7}$.

Exercises

Press the appropriate keys so that your calculator displays each of the following numbers.

1. −9   2. −57   3. −1996

b Order on the Number Line

Numbers are named in order on the number line, with numbers increasing as we move to the right. For any two numbers on the line, the one to the left is *less than* the one to the right.

Since the symbol $<$ means "is less than," the sentence $-5 < 9$ means "-5 is less than 9." The symbol $>$ means "is greater than," so the sentence $-4 > -8$ means "-4 is greater than -8."

Examples Use either $<$ or $>$ for ▮ to form a true sentence.

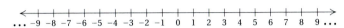

4. -9 ▮ 2 Since -9 is to the left of 2, we have $-9 < 2$.
5. 7 ▮ -13 Since 7 is to the right of -13, we have $7 > -13$.
6. -19 ▮ -6 Since -19 is to the left of -6, we have $-19 < -6$.

Do Exercises 5–8.

c Absolute Value

From the number line, we see that some integers, like 5 and -5, are the same distance from zero.

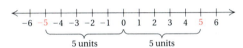

How far is 5 from 0? How far is -5 from 0? Since distance is never negative (it is "nonnegative," that is, either positive or zero), it follows that 5 is 5 units from 0 and -5 is also 5 units from 0.

> The **absolute value** of a number is its distance from zero on a number line. We use the symbol $|x|$ to represent the absolute value of a number x.

Examples Find the absolute value.

7. $|-3|$ The distance of -3 from 0 is 3, so $|-3| = 3$.
8. $|25|$ The distance of 25 from 0 is 25, so $|25| = 25$.
9. $|0|$ The distance of 0 from 0 is 0, so $|0| = 0$.
10. $|-17|$ The distance of -17 from 0 is 17, so $|-17| = 17$.
11. $|9|$ The distance of 9 from 0 is 9, so $|9| = 9$.

> To find a number's absolute value:
> 1. If a number is positive or zero, use the number itself.
> 2. If a number is negative, make the number positive.

Do Exercises 9–12.

Use either $<$ or $>$ for ▮ to form a true sentence.

5. 15 ▮ 7

6. 12 ▮ -3

7. -13 ▮ -3

8. -4 ▮ -20

Find the absolute value.

9. $|18|$

10. $|-9|$

11. $|-29|$

12. $|52|$

Answers on page A-4

In each case draw a number line, if necessary.

13. Find $-x$ when x is 1.

14. Find $-x$ when x is -2.

15. Evaluate $-x$ when x is 0.

---

d | Opposites, or Additive Inverses

The set of integers is represented below on a number line.

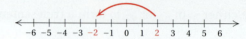

Given a number on one side of 0, we can get a number on the other side by *reflecting* the number across zero. For example, the *reflection* of 2 is -2.

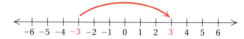

We can read -2 as "negative 2," "the opposite of 2," or "the additive inverse of 2." We read $-x$ as "the opposite of x."

> The **opposite** of a number x is written $-x$.

Example 12 If x is -3, find $-x$.

To find the opposite of x when x is -3, we reflect -3 to the other side of 0.

We have $-(-3) = 3$. The opposite of -3 is 3.

Example 13 Find $-x$ when x is 0.

When we try to reflect 0 "to the other side of 0," we go nowhere:

$-x = 0$ when x is 0.

In Examples 12 and 13, the variable was replaced with a number. When this occurs, we say that we are **evaluating** the expression.

Example 14 Evaluate $-x$ when x is 4.

To find the opposite of x when x is 4, we reflect 4 to the other side of 0.

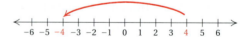

We have $-(4) = -4$. The opposite of 4 is -4.

Do Exercises 13–15.

Answers on page A-4

A negative number is sometimes said to have a "negative sign." A positive number is said to have a "positive sign." Replacing a number with its opposite, or additive inverse, is sometimes called *changing the sign.*

Examples Change the sign. (Find the opposite, or additive inverse.)

15. -6 $-(-6) = 6$
16. -10 $-(-10) = 10$
17. 0 $-(0) = 0$
18. 14 $-(14) = -14$

Do Exercises 16–19.

Note that when we change a number's sign twice, we return to the original number.

Example 19 If x is 2, find $-(-x)$.

We replace x with 2 and find $-(-2)$.

We see from the figure that $-(-2) = 2$.

Example 20 Evaluate $-(-x)$ for $x = -4$.

We replace x with -4 and find $-(-(-4))$.

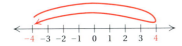

Reflecting -4 to the other side of 0 and then back again gives us -4. Thus, $-(-(-4)) = -(\ 4\ ) = -4$.

Do Exercises 20–23.

Change the sign. (Find the opposite, or additive inverse.)

16. -4

17. -13

18. 28

19. 0

20. If x is 7, find $-(-x)$.

21. If x is 1, find $-(-x)$.

22. Evaluate $-(-x)$ for $x = -6$.

23. Evaluate $-(-x)$ for $x = -2$.

To the student and the instructor: Recall that the Skill Maintenance Exercises, which occur at the end of most exercise sets, review any skill that has been studied before in the text. Often skill maintenance exercises are chosen to provide preparation for the next section in the text.

Answers on page A-4

2.1 Integers and the Number Line

Improving Your Math Study Skills

Learning Resources and Time Management

Two important topics to consider in enhancing your math study skills are learning resources and time management.

Learning Resources

- **Textbook supplements.** Are you aware of all the supplements that exist for this textbook? Many details are given in the Preface. Now that you are more familiar with the book, let's discuss them.

  1. The *Student's Solutions Manual* contains worked-out solutions to the odd-numbered exercises in the exercise sets. Consider obtaining a copy if you are having trouble. It should be your first choice if you can make an additional purchase.

  2. An extensive set of *videotapes* supplement this text. These may be available to you on your campus at a learning center or math lab. Check with your instructor.

  3. *Tutorial software* also accompanies the text. If not available in the campus learning center or lab, this software can be ordered by calling 1-800-322-1377.

  4. The Math Tutor Center is a free tutoring resource for students possessing a valid registration number. This service is available Sunday through Thursday from 3 PM to 10 PM eastern standard time. To receive help by phone, FAX, or Email with any odd-numbered exercise, simply dial 1-888-777-0463.

- **The Internet.** Our on-line World Wide Web supplement provides additional practice resources. If you have internet access, you can reach this site through the address:

  http://www.mathmax.com

  It contains many helpful ideas as well as many links to other resources for learning mathematics.

- **Your college or university.** Your own college or university probably has resources to enhance your math learning.

  1. For example, there may be a learning lab or tutoring center for drop-in tutoring.
  2. There may be special lab classes or group tutoring sessions tailored for the specific course you are taking.
  3. Perhaps there is a bulletin board or network where you can locate the names of experienced private tutors.
  4. Often classmates interested in forming a study group can be found.

- **Your instructor.** Although it may seem obvious, you should consider an often overlooked resource: your instructor. Find out your instructor's office hours and make it a point to visit when you need additional help.

Time Management

- **Juggling time.** Have reasonable expectations about the time you need to study math. Unreasonable expectations may lead to lower grades and increased frustrations. Working 40 hours per week and taking 12 hours of credit is equivalent to working two full-time jobs. Can you handle such a load? As a rule of thumb, your ratio of work hours to credit load should be about 40/3, 30/6, 20/9, 10/12, and 5/14. Budget about 2–3 hours of homework and studying per hour of class.

- **Daily schedule.** Make an hour-by-hour schedule of your typical week. Include work, college, home, personal, sleep, study, and leisure times. Be realistic about the amount of time needed for sleep and home duties. If possible, try to schedule study time for when you are most alert.

Other study tips appear on pages 50, 100, 150, 322, 368, and 596.

Exercise Set 2.1

a Tell which integers correspond to each situation.

1. Hewlett-Packard stock recently dropped 2 points.

2. Redbank, Montana, once recorded a temperature of 70° below zero.

3. The Dead Sea, between Jordan and Israel, is 1286 ft below sea level, whereas Mt. Everest is 29,028 ft above sea level.

4. The space shuttle stood ready, 3 sec before liftoff. Solid fuel rockets were released 128 sec after liftoff.

5. Terry deposited $850 in a savings account. Two weeks later, she withdrew $432.

6. Ben & Jerry's stock recently rose 3 points after having dropped 1 point.

b Use either < or > for ☐ to form a true sentence.

7. 7 ☐ 0
8. 9 ☐ 0
9. −9 ☐ 5
10. 8 ☐ −8

11. −6 ☐ 6
12. 0 ☐ −7
13. −8 ☐ −5
14. −5 ☐ −3

15. −5 ☐ −11
16. −3 ☐ −4
17. −6 ☐ −5
18. −10 ☐ −14

c Find the absolute value.

19. $|23|$
20. $|11|$
21. $|0|$
22. $|-4|$
23. $|-24|$

24. $|-36|$
25. $|53|$
26. $|54|$
27. $|-8|$
28. $|-79|$

d Find $-x$ when x is each of the following.

29. −8
30. −6
31. −7
32. 6
33. 0

34. −15
35. −19
36. 50
37. 42
38. −73

Change the sign. (Find the opposite, or additive inverse.)

39. −8
40. −7
41. 7
42. 10
43. −29

44. −14
45. −22
46. 0
47. 1
48. −53

Evaluate $-(-x)$ when x is each of the following.

49. 3 **50.** -7 **51.** -8 **52.** 1 **53.** 2

54. 19 **55.** 0 **56.** -2 **57.** -34 **58.** -23

Skill Maintenance

59. Add: $327 + 498$. [1.2b]

60. Evaluate: 5^3. [1.9b]

61. Multiply: $209 \cdot 34$. [1.5b]

62. Solve: $300 \cdot x = 1200$. [1.7b]

63. Evaluate: 9^2. [1.9b]

64. Multiply: $31 \cdot 50$. [1.5b]

Synthesis

65. ◆ Explain in your own words why $-(-x)$ is x.

66. ◆ Does $-x$ always represent a negative number? Why or why not?

67. ◆ Does $|x|$ always represent a positive number? Why or why not?

68. 🖩 List the keystrokes needed to find the opposite of the sum of 972 and 589 on your calculator.

69. 🖩 List the keystrokes needed to find the opposite of the product of 327 and 83 on your calculator.

Simplify.

70. $-|3|$ **71.** $-|-8|$ **72.** $-|-2|$ **73.** $-|7|$

Solve. Consider only integer replacements.

74. $|x| = 7$

75. $|x| < 2$

76. Simplify $-(-x)$, $-(-(-x))$, and $-(-(-(-x)))$.

77. List these integers in order from least to greatest.
2^{10}, -5, $|-6|$, 4, $|3|$, -100, 0, 2^7, 7^2, 10^2

2.2 Addition of Integers

a | Addition

To explain addition of integers, we can use the number line.

> To do the addition $a + b$, we start at a, and then move according to b.
>
> a) If b is positive, we move to the right.
> b) If b is negative, we move to the left.
> c) If b is 0, we stay at a.

Example 1 Add: $2 + (-5)$.

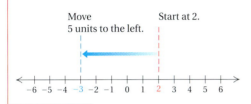

$2 + (-5) = -3$

Example 2 Add: $-1 + (-3)$.

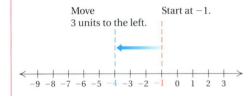

$-1 + (-3) = -4$

Example 3 Add: $-4 + 9$.

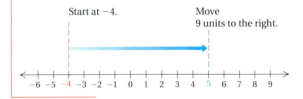

$-4 + 9 = 5$

Do Exercises 1–7.

You may have noticed a pattern in Example 2 and Margin Exercises 2 and 6. When two negative integers are added, the result is negative.

> To add two negative integers, add their absolute values and change the sign (making the answer negative).

Examples Add.

4. $-5 + (-7) = -12$ *Think*: Add the absolute values: $5 + 7 = 12$. Make the answer negative, -12.

5. $-8 + (-2) = -10$

Do Exercises 8–11.

Objective

a | Add integers without using a number line.

For Extra Help

TAPE 3   MAC WIN   CD-ROM

Add, using a number line.

1. $3 + (-4)$

2. $-3 + (-5)$

3. $-3 + 7$

4. $-5 + 5$

Write an addition sentence.

5.

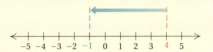

6.

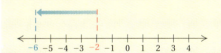

7.

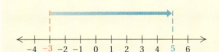

Add. Do not use a number line except as a check.

8. $-5 + (-6)$

9. $-9 + (-3)$

10. $-20 + (-14)$

11. $-11 + (-11)$

Answers on page A-4

2.2 Addition of Integers

93

Add. Do not use a number line except as a check.

12. $-4 + 6$

13. $-7 + 3$

14. $5 + (-7)$

15. $10 + (-7)$

16. $5 + (-5)$

17. $-6 + 6$

18. $-10 + 10$

19. $89 + (-89)$

Add.

20. $(-15) + (-37) + 25 + 42 + (-59) + (-14)$

21. $42 + (-81) + (-28) + 24 + 18 + (-31)$

22. $-35 + 17 + 14 + (-27) + 31 + (-12)$

Answers on page A-4

When we add a positive integer and a negative integer, as in Examples 1 and 3, the answer is negative or positive, depending on which number has the greater absolute value.

> To add a positive integer and a negative integer, find the difference of their absolute values.
> a) If the negative integer has the greater absolute value, the answer is negative.
> b) If the positive integer has the greater absolute value, the answer is positive.

Examples Add.

6. $3 + (-5) = -2$ *Think*: The absolute values are 3 and 5. The difference is 2. Since the negative number has the larger absolute value, the answer is *negative*, -2.

7. $11 + (-8) = 3$ *Think*: The absolute values are 11 and 8. The difference is 3. The positive number has the larger absolute value, so the answer is *positive*, 3.

8. $1 + (-5) = -4$
9. $-7 + 4 = -3$
10. $7 + (-3) = 4$
11. $-6 + 10 = 4$

We call $-a$ the *additive inverse* of a, because adding any number to its additive inverse always gives 0, the *additive identity*:

$$-8 + 8 = 0, \quad 14 + (-14) = 0, \quad \text{and} \quad 0 + 0 = 0.$$

> For any integer a,
> $$a + (-a) = -a + a = 0.$$
> (The sum of any number and its opposite is 0.)

Do Exercises 12–19.

Suppose we wish to add several numbers, positive and negative:

$$15 + (-2) + 7 + 14 + (-5) + (-12).$$

Because of the commutative and associative laws for addition, we can group the positive numbers together and the negative numbers together and add them separately. Then we add the two results.

Example 12 Add: $15 + (-2) + 7 + 14 + (-5) + (-12)$.

First add the positive numbers: $15 + 7 + 14 = 36$.
Then add the negative numbers: $-2 + (-5) + (-12) = -19$.
Finally, add the results: $36 + (-19) = 17$.

We can also add in any other order we wish, say, from left to right:

$$\begin{aligned}
15 + (-2) + 7 + 14 + (-5) + (-12) &= 13 + 7 + 14 + (-5) + (-12) \\
&= 20 + 14 + (-5) + (-12) \\
&= 34 + (-5) + (-12) \\
&= 29 + (-12) \\
&= 17.
\end{aligned}$$

Do Exercises 20–22.

Exercise Set 2.2

a Add, using a number line.

1. $-7 + 2$
2. $2 + (-5)$
3. $-9 + 5$
4. $8 + (-3)$
5. $-3 + 9$
6. $5 + (-5)$
7. $-7 + 7$
8. $-8 + (-5)$
9. $-3 + (-1)$
10. $-2 + (-9)$

Add. Use a number line only as a check.

11. $-4 + (-11)$
12. $-3 + (-7)$
13. $-6 + (-5)$
14. $-10 + (-14)$
15. $9 + (-9)$
16. $10 + (-10)$
17. $-2 + 2$
18. $-3 + 3$
19. $0 + 8$
20. $7 + 0$
21. $0 + (-8)$
22. $-7 + 0$
23. $-25 + 0$
24. $-43 + 0$
25. $0 + (-27)$
26. $0 + (-19)$
27. $17 + (-17)$
28. $-13 + 13$
29. $-25 + 25$
30. $11 + (-11)$
31. $8 + (-5)$
32. $-7 + 8$
33. $-4 + (-5)$
34. $0 + (-3)$
35. $0 + (-5)$
36. $10 + (-12)$
37. $14 + (-5)$
38. $-3 + 14$
39. $-11 + 8$
40. $0 + (-34)$
41. $-19 + 19$
42. $-10 + 3$
43. $-17 + 7$
44. $-15 + 5$
45. $-17 + (-7)$
46. $-15 + (-5)$
47. $11 + (-16)$
48. $-7 + 15$
49. $-15 + (-6)$
50. $-8 + 8$
51. $11 + (-9)$
52. $-14 + (-19)$
53. $-20 + (-6)$
54. $19 + (-19)$
55. $-15 + (-7) + 1$
56. $23 + (-5) + 4$
57. $30 + (-10) + 5$
58. $40 + (-8) + 5$

59. $-23 + (-9) + 15$

60. $-25 + 25 + (-9)$

61. $40 + (-40) + 6$

62. $63 + (-18) + 12$

63. $85 + (-65) + (-12)$

64. $-35 + (-63) + (-27) + (-14) + (-59)$

65. $-24 + (-37) + (-19) + (-45) + (-35)$

66. $75 + (-14) + (-17) + (-5)$

67. $28 + (-44) + 17 + 31 + (-94)$

68. $27 + (-54) + (-32) + 65 + 46$

Skill Maintenance

69. Write in expanded notation: 39,417. [1.1a]

70. Round to the nearest hundred: 746. [1.4a]

71. Round to the nearest thousand: 32,831. [1.4a]

72. Multiply: $42 \cdot 56$. [1.5b]

73. Divide: $288 \div 9$. [1.6c]

74. Round to the nearest ten: 3496. [1.4a]

Synthesis

75. ◆ Why was the concept of absolute value introduced in Section 2.1?

76. ◆ Explain in your own words why the sum of two negative numbers is always negative.

77. ◆ A student states that -93 is "bigger than" -47. What mistake is the student making?

Add.

78. $-|27| + (-|-13|)$

79. $|-32| + (-|15|)$

80. ▦ $-3496 + (-2987)$

81. ▦ $497 + (-3028)$

82. For what numbers x is $x + (-7)$ positive?

83. For what numbers x is $-7 + x$ negative?

Tell whether each sum is positive, negative, or zero.

84. If n is positive and m is negative, then $-n + m$ is _____.

85. If $n = m$ and n is negative, then $-n + (-m)$ is _____.

86. If n is negative and m is less than n, then $n + m$ is _____.

87. If n is positive and m is greater than n, then $n + m$ is _____.

Add integers using a variety of methods.

2.3 Subtraction of Integers

a Subtraction

We now consider subtraction of integers. Subtraction is defined as follows.

> The difference $a - b$ is the number that when added to b gives a.

For example, $45 - 17 = 28$ because $28 + 17 = 45$. Let's consider an example in which the answer is a negative number.

Example 1 Subtract: $5 - 8$.

Think: $5 - 8$ is the number that when added to 8 gives 5. What number can we add to 8 to get 5? The number must be negative. The number is -3:

$$5 - 8 = -3.$$

That is, $5 - 8 = -3$ because $5 = -3 + 8$.

Do Exercises 1–3.

The definition above does *not* provide the most efficient way to do subtraction. From that definition, however, a faster way can be developed. Look for a pattern in the following table.

Subtractions	Adding an Opposite
$5 - 8 = -3$	$5 + (-8) = -3$
$-6 - 4 = -10$	$-6 + (-4) = -10$
$-7 - (-10) = 3$	$-7 + 10 = 3$
$-7 - (-2) = -5$	$-7 + 2 = -5$

Do Exercises 4–7.

Perhaps you have noticed that we can subtract by adding the opposite of the number being subtracted. This can always be done.

> To subtract, add the opposite, or additive inverse, of the number being subtracted:
> $$a - b = a + (-b).$$

This is the method generally used for quick subtraction of integers.

Objectives

a Subtract integers and simplify combinations of additions and subtractions.

b Solve applied problems involving addition and subtraction of integers.

For Extra Help

TAPE 3   MAC   CD-ROM
             WIN

Subtract.
1. $-6 - 4$
   Think: What number can be added to 4 to get -6?

2. $-7 - (-10)$
   Think: What number can be added to -10 to get -7?

3. $-7 - (-2)$
   Think: What number can be added to -2 to get -7?

Complete the addition and compare with the subtraction.

4. $4 - 6 = -2$;
   $4 + (-6) = $ _____

5. $-3 - 8 = -11$;
   $-3 + (-8) = $ _____

6. $-5 - (-9) = 4$;
   $-5 + 9 = $ _____

7. $-5 - (-3) = -2$;
   $-5 + 3 = $ _____

Answers on page A-4

Equate each subtraction with a corresponding addition. Then write the equation in words.

8. $3 - 10$

9. $13 - 5$

10. $-12 - (-9)$

11. $-12 - 10$

12. $-14 - (-14)$

Subtract.

13. $2 - 8$

14. $-6 - 10$

15. $14 - 9$

16. $-8 - (-11)$

17. $-8 - (-2)$

18. $5 - (-8)$

Answers on page A-4

Examples Equate each subtraction with a corresponding addition. Then write the equation in words.

2. $3 - 7$;
$3 - 7 = 3 + (-7)$ Adding the opposite of 7

Three minus seven is three plus negative seven.

3. $-14 - (-23)$;
$-14 - (-23) = -14 + 23$ Adding the opposite of -23

Negative fourteen minus negative twenty-three is negative fourteen plus twenty-three.

4. $-12 - 30$;
$-12 - 30 = -12 + (-30)$ Adding the opposite of 30

Negative twelve minus thirty is negative twelve plus negative thirty.

5. $-20 - (-17)$;
$-20 - (-17) = -20 + 17$ Adding the opposite of -17

Negative twenty minus negative seventeen is negative twenty plus seventeen.

Do Exercises 8–12.

Once the subtraction has been rewritten as addition, we add as in Section 2.2.

Examples Subtract.

6. $2 - 6 = 2 + (-6)$ The opposite of 6 is -6. We change the subtraction to addition and add the opposite. Instead of subtracting 6, we add -6.
$= -4$

7. $4 - (-9) = 4 + 9$ The opposite of -9 is 9. We change the subtraction to addition and add the opposite. Instead of subtracting -9, we add 9.
$= 13$

8. $-3 - 8 = -3 + (-8)$ We change the subtraction to addition and add the opposite. Instead of subtracting 8, we add -8.
$= -11$

9. $10 - 7 = 10 + (-7)$ We change the subtraction to addition and add the opposite. Instead of subtracting 7, we add -7.
$= 3$

10. $-4 - (-9) = -4 + 9$ Instead of subtracting -9, we add 9.
$= 5$ To check, note that $5 + (-9) = -4$.

11. $-5 - (-3) = -5 + 3$ Instead of subtracting -3, we add 3.
$= -2$ Check: $-2 + (-3) = -5$.

Do Exercises 13–18.

Chapter 2 Integers and Algebraic Expressions

When several additions and subtractions occur together, we can make them all additions. The commutative law for addition can then be used.

Example 12 Simplify: $-3 - (-5) - 9 + 4 - (-6)$.

$$
\begin{aligned}
-3 - (-5) - 9 + 4 - (-6) &= -3 + 5 + (-9) + 4 + 6 &&\text{Adding opposites} \\
&= -3 + (-9) + 5 + 4 + 6 &&\text{Using a commutative law} \\
&= -12 + 15 \\
&= 3
\end{aligned}
$$

Do Exercises 19 and 20.

b Applications and Problem Solving

Let's now see how we can use addition and subtraction of integers to solve applied problems.

Example 13 *Home-Run Differential.* In baseball the difference between the number of home runs hit by a team's players and the number given up by its pitchers is called the *home-run differential*, that is,

$$\text{Home run differential} = \begin{pmatrix}\text{Number of home} \\ \text{runs hits}\end{pmatrix} - \begin{pmatrix}\text{Number of home} \\ \text{runs allowed}\end{pmatrix}.$$

Teams strive for a positive home-run differential.

a) In a recent year, Atlanta hit 215 home runs and gave up 117. Find its home-run differential.

b) In a recent year, San Francisco hit 161 home runs and gave up 171. Find its home-run differential.

We solve as follows.

a) We subtract 117 from 215 to find the home-run differential for Atlanta:

Home-run differential = $215 - 117 = 98$.

b) We subtract 171 from 161 to find the home-run differential for San Francisco:

Home-run differential = $161 - 171 = -10$.

Do Exercises 21 and 22.

Simplify.

19. $-6 - (-2) - (-4) - 12 + 3$

20. $9 - (-6) + 7 - 11 - 14 - (-20)$

21. *Home-Run Differential.* Complete the following table to find the home-run (HR) differentials (Diff) for all the National League baseball teams.

National League			
	HRs hit	HRs allowed	Diff.
Atlanta	215	117	98
St. Louis	223	151	
Chicago	212	180	
San Diego	167	139	
Los Angeles	159		24
Houston	166		19
Colorado	183	174	
Montreal	147	156	
San Francisco	161	171	
New York	136	152	
Arizona	159		−29
Cincinnati	138		−32
Milwaukee	152	188	
Pittsburgh	107	147	
Philadelphia	126		−62
Florida	114	182	

22. *Temperature Extremes.* In Churchill, Manitoba, Canada, the average daily low temperature in January is −31° Celsius (C). The average daily low temperature in Key West, Florida, is 19°C. How much higher is the average daily low temperature in Key West, Florida?

Answers on page A-4

Improving Your Math Study Skills

Studying for Tests and Making the Most of Tutoring Sessions

This math study skill feature focuses on the very important task of test preparation.

Test-Taking Tips

- **Make up your own test questions as you study.** You have probably become accustomed by now to the section and objective codes that appear throughout the book. After you have done your homework for a particular objective, write one or two questions on your own that you think might be on a test. This allows you to carry out a task similar to what a teacher does in preparing an exam. You will be amazed at the insight this will provide.

- **Do an overall review of the chapter focusing on the objectives and the examples.** This should be accompanied by a thorough review of any class notes you may have taken.

- **Do the review exercises at the end of the chapter.** Check your answers at the back of the book. If you have trouble with an exercise, use the objective symbol as a guide to go back for further study of that objective. These review exercises are very much like a sample test.

- **Do the chapter test at the end of the chapter.** This is like taking a second sample test. Check the answers and objective symbols at the back of the book.

- **Ask your instructor or former students for old exams.** Working such exams can be very helpful and allows you to see what your instructor thinks is important.

- **When taking a test, read each question carefully and try to do all the questions the first time through, but pace yourself.** Answer all the questions, and mark those to recheck if you have time at the end. Very often, your first hunch will be correct.

- **Try to write your test in a neat and orderly manner.** Very often, instructors try to award partial credit when grading an exam. If your test paper is sloppy and disorderly, it is difficult to verify the partial credit. Doing your work neatly can ease such a task for the instructor. Try using a pencil with soft lead or an erasable pen to make your writing darker and therefore more readable.

- **What about the student who says, "I could do the work at home, but on the test I made silly mistakes"?** Yes, all of us, including instructors, make silly computational mistakes in class, on homework, and on tests. But your instructor, if he or she has taught for some time, is probably aware that 90% of students who make such comments in truth do not have sufficient depth of knowledge of the subject matter. Silly mistakes often are a sign that the student has not mastered the material. There is no way we can make that analysis for you. It will have to be unraveled by some careful soul searching on your part or by a conference with your instructor.

Making the Most of Tutoring and Help Sessions

Often students find that a tutoring session would be helpful. The following comments may help you to make the most of such sessions.

- **Work on the topics before you go to the help or tutoring session. Do not go to such sessions viewing yourself as an empty cup and the tutor as a magician who will pour in the learning.** The primary source of your ability to learn is within you. We have seen many students over the years go to help or tutoring sessions with no advanced preparation. When students do this they waste time and, in many cases, money. Go to class, study the textbook, and mark trouble spots. Then use the help and tutoring sessions to work on these trouble spots efficiently.

- **Do not be afraid to ask questions in these sessions!** The more you talk to your tutor, the more the tutor can help you with your difficulties.

- **Try being a "tutor" yourself.** Explaining a topic to someone else—a classmate, your instructor—is often the best way to master that topic.

Exercise Set 2.3

a Subtract.

1. $4 - 7$
2. $3 - 8$
3. $0 - 8$
4. $0 - 9$

5. $-8 - (-4)$
6. $-6 - (-8)$
7. $-11 - (-11)$
8. $-6 - (-6)$

9. $12 - 17$
10. $14 - 19$
11. $20 - 27$
12. $30 - 4$

13. $-9 - (-5)$
14. $-7 - (-9)$
15. $-40 - (-40)$
16. $-9 - (-9)$

17. $7 - 7$
18. $9 - 9$
19. $7 - (-7)$
20. $4 - (-4)$

21. $8 - (-3)$
22. $-7 - 4$
23. $-6 - 8$
24. $6 - (-10)$

25. $-4 - (-9)$
26. $-14 - 2$
27. $1 - 8$
28. $2 - 8$

29. $-6 - (-5)$
30. $-4 - (-3)$
31. $8 - (-10)$
32. $5 - (-6)$

33. $0 - 10$
34. $0 - 18$
35. $-5 - (-2)$
36. $-3 - (-1)$

37. $-7 - 14$
38. $-9 - 16$
39. $0 - (-5)$
40. $0 - (-1)$

41. $-8 - 0$
42. $-9 - 0$
43. $7 - (-5)$
44. $7 - (-4)$

45. 2 − 25 **46.** 18 − 63 **47.** −42 − 26 **48.** −18 − 63

49. −71 − 2 **50.** −49 − 3 **51.** 24 − (−92) **52.** 48 − (−73)

53. −50 − (−50) **54.** −70 − (−70) **55.** −30 − (−85) **56.** −25 − (−15)

Simplify.

57. 7 − (−5) + 4 − (−3) **58.** −5 − (−8) + 3 − (−7) **59.** −31 + (−28) − (−14) − 17

60. −43 − (−19) − (−21) + 25 **61.** −34 − 28 + (−33) − 44 **62.** 39 + (−88) − 29 − (−83)

63. −93 − (−84) − 41 − (−56) **64.** 84 + (−99) + 44 − (−18) − 43 **65.** −5 − (−30) + 30 + 40 − (−12)

66. 14 − (−50) + 20 − (−32) **67.** 132 − (−21) + 45 − (−21) **68.** 81 − (−20) − 14 − (−50) + 53

 Solve.

69. Midway through a movie, Lisa resets the counter on her VCR to 0. She then fast-forwards the tape 12 min and rewinds it 19 min. What does the counter now read?

70. Laura has a charge of $476.89 on her credit card, but she then returns a sweater that cost $128.95. How much does she now owe on her credit card?

71. Jan is $120 in debt. How much money does Jan need to earn in order to raise her total assets to $350?

72. Through exercise, Rod went from 8 lb above his "ideal" body weight to 9 lb below it. How many pounds did Rod lose?

73. *Offshore Oil.* In 1993, the elevation of the world's deepest offshore oil well was −2860 ft. In 1998, the deepest well was 360 ft deeper. (**Source:** *New York Times*, 12/7/94, p. D1.) What was the elevation of the deepest well in 1998?

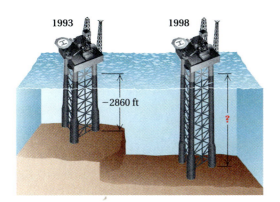

74. *Oceanography.* The deepest point in the Pacific Ocean is the Marianas Trench, with a depth of 11,033 m. The deepest point in the Atlantic Ocean is the Puerto Rico Trench, with a depth of 8648 m. What is the difference in the elevation of the two trenches?

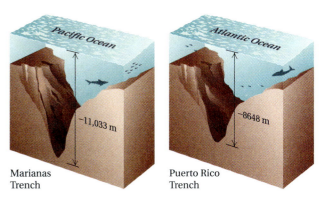

Skill Maintenance

Evaluate.

75. 4^3 [1.9b]

76. 1^7 [1.9b]

77. How many 12-oz cans of soda can be filled with 96 oz of soda? [1.8a]

78. A case of soda contains 24 bottles. If each bottle contains 12 oz, how many ounces of soda are in the case? [1.8a]

Simplify.

79. $5 + 4^2 + 2 \cdot 7$ [1.9c]

80. $45 \div (2^2 + 11)$ [1.9c]

Synthesis

81. ◆ Explain why the commutative law was used in Example 12.

82. ◆ If a negative number is subtracted from a positive number, will the result always be positive? Why or why not?

83. ◆ Write a problem for a classmate to solve. Design the problem so that the solution is "The temperature dropped to −9°F."

Exercise Set 2.3

Subtract.

84. 📱 123,907 − 433,789

85. 📱 23,011 − (−60,432)

Tell whether each statement is true or false for all integers a and b. If false, show why.

86. $a - 0 = 0 - a$

87. $0 - a = a$

88. If $a \neq b$, then $a - b \neq 0$.

89. If $a = -b$, then $a + b = 0$.

90. If $a + b = 0$, then a and b are opposites.

91. If $a - b = 0$, then $a = -b$.

92. Doreen is a stockbroker. She kept track of the changes in the stock market over a period of 5 weeks. By how many points (pts) had the market risen or fallen over this time?

Week 1	Week 2	Week 3	Week 4	Week 5
Down 13 pts	Down 16 pts	Up 36 pts	Down 11 pts	Up 19 pts

93. *Blackjack Counting System.* The casino game of blackjack makes use of many card-counting systems to give players a winning edge if the count becomes negative. One such system is called *High–Low*, first developed by Harvey Dubner in 1963. Each card counts as −1, 0, or 1 as follows:

  2, 3, 4, 5, 6   count as +1;
  7, 8, 9   count as 0;
  10, J, Q, K, A   count as −1.

  (*Source*: Patterson, Jerry L., *Casino Gambling*. New York: Perigee, 1982)

  a) Find the final count on the sequence of cards

    K, A, 2, 4, 5, 10, J, 8, Q, K, 5.

  b) Does the player have a winning edge?

Chapter 2 Integers and Algebraic Expressions

Subtract integers using tiles.

2.4 Multiplication of Integers

a Multiplication

Multiplication of integers is very much like multiplication of whole numbers. The only difference is that we must determine whether the answer is positive or negative.

Multiplication of a Positive Integer and a Negative Integer

To see how to multiply a positive integer and a negative integer, consider the pattern of the following.

This number decreases by 1 each time.   This number decreases by 5 each time.

$$4 \cdot 5 = 20$$
$$3 \cdot 5 = 15$$
$$2 \cdot 5 = 10$$
$$1 \cdot 5 = 5$$
$$0 \cdot 5 = 0$$
$$-1 \cdot 5 = -5$$
$$-2 \cdot 5 = -10$$
$$-3 \cdot 5 = -15$$

Do Exercise 1.

According to this pattern, it looks as though the product of a negative integer and a positive integer is negative. This leads to the first part of the rule for multiplying integers.

> To multiply a positive integer and a negative integer, multiply their absolute values and make the answer negative.

Examples Multiply.

1. $8(-5) = -40$
2. $50(-1) = -50$
3. $-7 \cdot 6 = -42$

Do Exercises 2–4.

Multiplication of Two Negative Integers

How do we multiply two negative integers? Again we look for a pattern.

This number decreases by 1 each time.   This number increases by 5 each time.

$$4 \cdot (-5) = -20$$
$$3 \cdot (-5) = -15$$
$$2 \cdot (-5) = -10$$
$$1 \cdot (-5) = -5$$
$$0 \cdot (-5) = 0$$
$$-1 \cdot (-5) = 5$$
$$-2 \cdot (-5) = 10$$
$$-3 \cdot (-5) = 15$$

Do Exercise 5.

Objective

a Multiply integers.

b Find products of three or more integers and simplify powers of integers.

For Extra Help

TAPE 3   MAC WIN   CD-ROM

1. Complete, as in the example.

$$4 \cdot 10 = 40$$
$$3 \cdot 10 = 30$$
$$2 \cdot 10 =$$
$$1 \cdot 10 =$$
$$0 \cdot 10 =$$
$$-1 \cdot 10 =$$
$$-2 \cdot 10 =$$
$$-3 \cdot 10 =$$

Multiply.

2. $-3 \cdot 6$

3. $20 \cdot (-5)$

4. $9(-1)$

5. Complete, as in the example.

$$3 \cdot (-10) = -30$$
$$2 \cdot (-10) = -20$$
$$1 \cdot (-10) =$$
$$0 \cdot (-10) =$$
$$-1 \cdot (-10) =$$
$$-2 \cdot (-10) =$$
$$-3 \cdot (-10) =$$

Answers on page A-5

Multiply.

6. $(-3)(-4)$

7. $-16(-2)$

8. $(-1)(-7)$

Multiply.

9. $0 \cdot (-5)$

10. $-23 \cdot 0$

Answers on page A-5

Chapter 2 Integers and Algebraic Expressions

According to the pattern, the product of two negative integers is positive. This leads to the second part of the rule for multiplying integers.

> To multiply two negative integers, multiply their absolute values. The answer is positive.

Examples Multiply.

4. $(-2)(-4) = 8$
5. $(-10)(-7) = 70$
6. $(-9)(-1) = 9$

Do Exercises 6–8.

The following is another way to state the rules for multiplication.

> To multiply two integers:
> a) Multiply the absolute values.
> b) If the signs are the same, the answer is positive.
> c) If the signs are different, the answer is negative.

Multiplication by Zero

No matter how many times 0 is added to itself, the answer is 0. This leads to the following result.

> For any integer a,
> $$a \cdot 0 = 0.$$
> (The product of 0 and any integer is 0.)

Examples Multiply.

7. $-19 \cdot 0 = 0$
8. $0(-7) = 0$

Do Exercises 9 and 10.

b Multiplication of More Than Two Integers

Because of the commutative and the associative laws, to multiply three or more integers, we can group as we please.

Examples Multiply.

9. a) $-8 \cdot 2(-3) = -16(-3)$ Multiplying the first two numbers
$= 48$ Multiplying the results

b) $-8 \cdot 2(-3) = 24 \cdot 2$ Multiplying the negatives
$= 48$ The result is the same as above.

10. $7(-1)(-4)(-2) = (-7)8$ Multiplying the first two numbers and the last two numbers
$= -56$

11. a) $-5 \cdot (-2) \cdot (-3) \cdot (-6) = 10 \cdot 18$ Each pair of negatives gives a positive product.
$= 180$

b) $-5 \cdot (-2) \cdot (-3) \cdot (-6) \cdot (-1) = 10 \cdot 18 \cdot (-1)$ Making use of Example 11(a)
$= -180$

We can see the following pattern in the results of Examples 9–11.

> The product of an even number of negative integers is positive.
> The product of an odd number of negative integers is negative.

Do Exercises 11–13.

Powers of Integers

The result of raising a negative number to a power is positive or negative, depending on the exponent.

Examples Simplify.

12. $(-7)^2 = (-7)(-7) = 49$ The result is positive.

13. $(-4)^3 = (-4)(-4)(-4)$
$= 16(-4)$
$= -64$ The result is negative.

14. $(-3)^4 = (-3)(-3)(-3)(-3)$
$= 9 \cdot 9$
$= 81$ The result is positive.

15. $(-2)^5 = (-2)(-2)(-2)(-2)(-2)$
$= 4 \cdot 4 \cdot (-2)$
$= 16(-2)$
$= -32$ The result is negative.

Perhaps you noted the following.

> When a negative number is raised to an even exponent, the result is positive.
> When a negative number is raised to an odd exponent, the result is negative.

Do Exercises 14–16.

We have seen that when an integer is multiplied by -1, the result is the opposite of that integer. That is, $-1 \cdot a = -a$, for any integer a.

Multiply.
11. $-2 \cdot (-5) \cdot (-4) \cdot (-3)$

12. $(-4)(-5)(-2)(-3)(-1)$

13. $(-1)(-1)(-2)(-3)(-1)(-1)$

Simplify.
14. $(-2)^3$

15. $(-9)^2$

16. $(-1)^7$

Answers on page A-5

17. Simplify: -5^2.

Example 16 Simplify: -7^2.

We first note that -7^2 lacks parentheses so the base is 7, not -7. Thus we regard -7^2 as $-1 \cdot 7^2$:

$-7^2 = -1 \cdot 7^2$
$\quad\ \ = -1 \cdot 7 \cdot 7$ **The rules for order of operations tell us to square first.**
$\quad\ \ = -1 \cdot 49$
$\quad\ \ = -49$.

Compare Examples 12 and 16 and note that $(-7)^2 \neq -7^2$. In fact, the expressions $(-7)^2$ and -7^2 are not read the same way: $(-7)^2$ is read "negative seven squared," whereas -7^2 is read "the opposite of seven squared."

Do Exercises 17 and 18.

18. Write $(-8)^2$ and -8^2 in words.

Calculator Spotlight

When using a calculator to calculate numbers like $(-39)^4$, it is important to use the correct sequence of keystrokes. On most scientific calculators, the appropriate keystrokes are

$\boxed{3}\ \boxed{9}\ \boxed{+/-}\ \boxed{x^y}\ \boxed{4}\ \boxed{=}$.

Note that in this instance the $\boxed{+/-}$ key is used to change a number's sign. To calculate -39^4, the following keystrokes are needed:

$\boxed{1}\ \boxed{+/-}\ \boxed{\times}\ \boxed{3}\ \boxed{9}\ \boxed{x^y}\ \boxed{4}\ \boxed{=}$,

or

$\boxed{3}\ \boxed{9}\ \boxed{x^y}\ \boxed{4}\ \boxed{=}\ \boxed{+/-}$.

On many graphing calculators, $(-39)^4$ is found by pressing

$\boxed{(}\ \boxed{(-)}\ \boxed{3}\ \boxed{9}\ \boxed{)}\ \boxed{\wedge}\ \boxed{4}\ \boxed{\text{ENTER}}$

and -39^4 is found by pressing

$\boxed{(-)}\ \boxed{3}\ \boxed{9}\ \boxed{\wedge}\ \boxed{4}\ \boxed{\text{ENTER}}$.

You can either experiment or consult a user's manual if you are unsure of the proper keystrokes for your calculator.

Exercises

Use a calculator to determine each of the following.

1. $(-23)^6$ **2.** $(-17)^5$
3. $(-104)^3$ **4.** $(-4)^{10}$
5. 9^6 **6.** -7^6
7. -6^5 **8.** -3^9

Answers on page A-5

Chapter 2 Integers and Algebraic Expressions

Exercise Set 2.4

a Multiply.

1. $-3 \cdot 7$
2. $-8 \cdot 2$
3. $-9 \cdot 2$
4. $-7 \cdot 6$

5. $8 \cdot (-6)$
6. $8 \cdot (-3)$
7. $-10 \cdot 3$
8. $-9 \cdot 8$

9. $-2 \cdot (-5)$
10. $-8 \cdot (-2)$
11. $-9 \cdot (-2)$
12. $(-8)(-9)$

13. $-7 \cdot (-6)$
14. $-8 \cdot (-3)$
15. $-10(-3)$
16. $-9(-8)$

17. $12(-10)$
18. $15(-8)$
19. $-6(-50)$
20. $-25(-8)$

21. $(-72)(-1)$
22. $41(-3)$
23. $(-20)17$
24. $(-1)43$

25. $-23 \cdot 0$
26. $-17 \cdot 0$
27. $0(-14)$
28. $0(-38)$

b Multiply.

29. $3 \cdot (-8) \cdot 4$
30. $(-3) \cdot (-4) \cdot (-5)$
31. $7(-4)(-3)5$

32. $9(-2)(-6)7$
33. $-2(-5)(-7)$
34. $(-2)(-5)(-3)(-5)$

35. $(-5)(-2)(-3)(-1)$
36. $-6(-5)(-9)$
37. $(-15)(-29)0 \cdot 8$

38. $19(-7)(-8)0 \cdot 6$
39. $(-7)(-1)(7)(-6)$
40. $(-5)6(-4)5$

Simplify.

41. $(-5)^2$ **42.** $(-8)^2$ **43.** $(-5)^3$ **44.** $(-2)^4$

45. $(-10)^4$ **46.** $(-1)^5$ **47.** -2^4 **48.** $(-2)^6$

49. $(-3)^5$ **50.** -10^4 **51.** $(-1)^{12}$ **52.** $(-1)^{13}$

53. -3^6 **54.** -2^6 **55.** -5^3 **56.** -2^5

Write each of the following expressions in words.

57. -7^4 **58.** $(-6)^8$ **59.** $(-9)^6$ **60.** -5^4

Skill Maintenance

61. Round 532,451 to the nearest hundred. [1.4a]

62. Write standard notation for sixty million. [1.1d]

63. Divide: $2880 \div 36$. [1.6c]

64. Multiply: 75×34. [1.5b]

65. A rectangular rug measures 5 ft by 8 ft. What is the area of the rug? [1.8a]

66. How many 12-egg cartons can be filled with 2880 eggs? [1.8a]

Synthesis

67. ◈ Explain in your own words why $(-8)^{12}$ is positive.

68. ◈ Explain in your own words why $-7(-1)^3$ is the opposite of -7.

69. ◈ Which number is larger, $(-3)^{79}$ or $(-5)^{79}$? Why?

Simplify.

70. $(-3)^5(-1)^{379}$ **71.** $(-2)^3 \cdot [(-1)^{29}]^{46}$ **72.** $(-2)^6(-1^8)$ **73.** $-5^2(-1)^{29}$

74. $|(-2)^5 + 3^2| - (3-7)^2$

75. $|-12(-3)^2 - 5^3 - 6^2 - (-5)^2|$

76. ▦ $-935(238 - 243)^3$

77. ▦ $(-17)^4(129 - 133)^5$

78. Jo wrote seven checks for $13 each. If she had a balance of $68 in her account, what was her balance after writing the checks?

79. After diving 95 m below the surface, a diver rises at a rate of 7 meters per minute for 9 min. What is the diver's new elevation?

80. What must be true of m and n if $[(-5)^m]^n$ is to be (a) negative? (b) positive?

81. What must be true of m and n if $-mn$ is to be (a) positive? (b) zero? (c) negative?

2.5 Division of Integers

We now consider division of integers. The definition of division results in rules for division that are the same as those for multiplication.

a Division of Integers

Objectives

a. Divide integers.
b. Use the rules for order of operations with integers.

For Extra Help

TAPE 4   MAC WIN   CD-ROM

> The quotient $\frac{a}{b}$ (or $a \div b$) is the number, if there is one, that when multiplied by b gives a.

Let's use the definition to divide integers.

Examples Divide, if possible. Check each answer.

1. $14 \div (-7) = -2$ *Think*: What number multiplied by -7 gives 14? The number is -2. *Check*: $(-2)(-7) = 14$.

2. $\dfrac{-32}{-4} = 8$ *Think*: What number multiplied by -4 gives -32? The number is 8. *Check*: $8(-4) = -32$.

3. $\dfrac{-21}{7} = -3$ *Think*: What number multiplied by 7 gives -21? The number is -3. *Check*: $(-3) \cdot 7 = -21$.

4. $0 \div (-5) = 0$ *Think*: What number multiplied by -5 gives 0? The number is 0. *Check*: $0(-5) = 0$.

The rules for division are the same as those for multiplication. We state them together.

> To multiply or divide two integers:
> a) Multiply or divide the absolute values.
> b) If the signs are the same, the answer is positive.
> c) If the signs are different, the answer is negative.

Do Exercises 1–6.

In Example 4, we divided *into* 0. Consider now division of a number *by* 0, as in $9 \div 0$. The expression $9 \div 0$ represents the number that when multiplied by 0 gives 9. But any number times 0 gives 0, not 9. For this reason, we say that $9 \div 0$ is **undefined**. This result is generalized as follows.

> Division by zero is undefined: $a \div 0$, or $\dfrac{a}{0}$, is undefined for all integers a.

Example 5 Divide, if possible: $-17 \div 0$.

$\dfrac{-17}{0}$ is undefined. *Think*: What number multiplied by 0 gives -17? There is no such number because the product of 0 and *any* number is 0.

Do Exercises 7–9.

Divide.

1. $6 \div (-3)$

   Think: What number multiplied by -3 gives 6?

2. $\dfrac{-15}{-3}$

   Think: What number multiplied by -3 gives -15?

3. $-24 \div 8$

   Think: What number multiplied by 8 gives -24?

4. $\dfrac{0}{-4}$

5. $\dfrac{30}{-5}$

6. $\dfrac{-45}{9}$

Divide, if possible.

7. $26 \div 0$

8. $0 \div (-12)$

9. $-52 \div 0$

Answers on page A-5

Simplify.

10. $5 - (-7)(-3)^2$

11. $(-2) \cdot |3 - 2^2| + 5$

12. $\dfrac{(-5)(-9)}{1 - 2 \cdot 2}$

b Order of Operations

When several operations are to be done in a calculation or a problem, we apply the same rules that were used in Section 1.9. We repeat them here for review, now including absolute-value symbols.

> **RULES FOR ORDER OF OPERATIONS**
>
> 1. Do all calculations within parentheses, brackets, braces, or absolute-value symbols. Simplify, if possible, above and below any fraction bars.
> 2. Evaluate all exponential expressions.
> 3. Do all multiplications and divisions in order from left to right.
> 4. Do all additions and subtractions in order from left to right.

Examples Simplify.

6. $17 - 10 \div 2 \cdot 4$

There are no parentheses or powers so we begin with the third rule.

$17 - 10 \div 2 \cdot 4 = 17 - 5 \cdot 4$
$= 17 - 20$
$= -3$

Carrying out all multiplications and divisions in order from left to right

7. $|(-2)^3 \div 4| - 5(-2)$

We first simplify within the absolute-value symbols.

$|(-2)^3 \div 4| - 5(-2) = |-8 \div 4| - 5(-2)$ $(-2)^3 = (-2)(-2)(-2) = -8$

$\qquad\qquad\qquad\qquad = |-2| - 5(-2)$ Dividing

$\qquad\qquad\qquad\qquad = 2 - 5(-2)$ Finding the absolute value of -2

$\qquad\qquad\qquad\qquad = 2 - (-10)$ Multiplying

$\qquad\qquad\qquad\qquad = 12$ Subtracting by adding the opposite of -10

A fraction bar is a grouping symbol. It separates any calculations in the numerator from those in the denominator.

Example 8 Simplify: $\dfrac{5 - (-3)^2}{-2}$.

$\dfrac{5 - (-3)^2}{-2} = \dfrac{5 - 9}{-2}$ Simplifying: $(-3)^2 = (-3)(-3) = 9$

$\qquad\qquad = \dfrac{-4}{-2}$ Subtracting

$\qquad\qquad = 2$ Dividing

Do Exercises 10–12.

Calculator Spotlight

Most calculators now provide grouping symbols. Such keys may appear as (and) or [(... and ...)] . Grouping symbols can be useful when we are simplifying expressions written in fractional form. For example, to simplify an expression like

$$\dfrac{38 + 142}{2 - 47},$$

we press (3 8 + 1 4 2) ÷ (2 − 4 7) = .

Failure to include grouping symbols in the above keystrokes would mean that we are simplifying a different expression:

$$38 + \dfrac{142}{2} - 47.$$

Exercises

Use a calculator with grouping symbols to simplify each of the following.

1. $\dfrac{38 - 178}{5 + 30}$ **2.** $\dfrac{311 - 17^2}{2 - 13}$

3. $785 - \dfrac{285 - 5^4}{17 + 3 \cdot 51}$

Answers on page A-5

Exercise Set 2.5

a Divide, if possible. Check each answer.

1. $28 \div (-4)$
2. $\dfrac{35}{-7}$
3. $\dfrac{28}{-2}$
4. $26 \div (-13)$

5. $\dfrac{18}{-2}$
6. $-22 \div (-2)$
7. $\dfrac{-48}{-12}$
8. $-63 \div (-9)$

9. $\dfrac{-72}{8}$
10. $\dfrac{-50}{25}$
11. $-100 \div (-50)$
12. $\dfrac{-400}{8}$

13. $-344 \div 8$
14. $\dfrac{-128}{8}$
15. $\dfrac{200}{-25}$
16. $-651 \div (-31)$

17. $\dfrac{-56}{0}$
18. $\dfrac{0}{-5}$
19. $\dfrac{88}{-11}$
20. $\dfrac{-145}{-5}$

21. $-\dfrac{276}{12}$
22. $-\dfrac{217}{7}$
23. $\dfrac{0}{-2}$
24. $\dfrac{-13}{0}$

25. $\dfrac{19}{-1}$
26. $\dfrac{-17}{1}$
27. $-41 \div 1$
28. $23 \div (-1)$

b Simplify, if possible.

29. $8 - 2 \cdot 3 - 9$
30. $8 - (2 \cdot 3 - 9)$
31. $8 - 2(3 - 9)$
32. $(8 - 2)(3 - 9)$

33. $16 \cdot (-24) + 50$
34. $10 \cdot 20 - 15 \cdot 24$
35. $40 - 3^2 - 2^3$
36. $2^4 + 2^2 - 10$

37. $4 \cdot (6 + 8)/(4 + 3)$
38. $4^3 + 10 \cdot 20 + 8^2 - 23$
39. $4 \cdot 5 - 2 \cdot 6 + 4$
40. $5^3 + 4 \cdot 9 - (8 + 9 \cdot 3)$

41. $\dfrac{9^2 - 1}{1 - 3^2}$
42. $\dfrac{100 - 6^2}{(-5)^2 - 3^2}$
43. $8(-7) + 6(-5)$
44. $10(-5) \div 1(-1)$

45. $20 \div 5(-3) + 3$
46. $14 \div 2(-6) + 7$
47. $8 \div 2 \cdot 0 \div 6$
48. $9 \cdot 0 \div 5 \cdot 4$

49. $4 \cdot 5^2 \div 10$
50. $(2 - 5)^2 \div (-9)$
51. $(3 - 8)^2 \div (-1)$
52. $3 - 3^2$

53. $12 - 20^3$ **54.** $20 + 4^3 \div (-8)$ **55.** $2 \times 10^3 - 5000$ **56.** $-7(3^4) + 18$

57. $6[9 - (3 - 4)]$ **58.** $8[(6 - 13) - 11]$ **59.** $-1000 \div (-100) \div 10$ **60.** $256 + (-32) \div (-4)$

61. $8 - |7 - 9| \cdot 3$ **62.** $|8 - 7 - 9| \cdot 2 + 1$ **63.** $9 - |7 - 3^2|$ **64.** $9 - |5 - 7|^3$

65. $\dfrac{(-5)^3 + 17}{10(2 - 6) - 2(5 + 2)}$ **66.** $\dfrac{(3 - 5)^2 - (7 - 13)}{(2 - 5)3 + 2 \cdot 4}$ **67.** $\dfrac{2 \cdot 4^3 - 4 \cdot 32}{19^3 - 17^4}$ **68.** $\dfrac{-16 \cdot 28 \div 2^2}{5 \cdot 25 - 5^3}$

Skill Maintenance

69. Fabrikant Fine Diamonds ran a 4-in. by 7-in. advertisement in *The New York Times*. Find the area of the ad.   [1.8a]

70. A classroom contains 7 rows of chairs with 6 chairs in each row. How many chairs are there in the classroom?   [1.8a]

71. A Ford Windstar gets 25 miles per gallon (mpg). How many gallons will it take to travel 350 mi?   [1.8a]

72. A Honda Passport gets 20 mpg. How many gallons will it take to travel 340 mi?   [1.8a]

73. A 7-oz bag of tortilla chips contains 1050 calories. How many calories are in a 1-oz serving?   [1.8a]

74. A 7-oz bag of tortilla chips contains 8 grams (g) of fat. How many grams of fat are in a carton containing 12 bags of chips?   [1.8a]

Synthesis

75. ◆ Explain in your own words why $23 \div 0$ is undefined.

76. ◆ Explain how multiplication can be used to justify why the quotient of two negative integers is a positive integer.

77. ◆ Explain how multiplication can be used to justify why a negative integer divided by a positive integer is a negative integer.

Simplify.

78. $\dfrac{(25 - 4^2)^3}{17^2 - 16^2} \cdot ((-6)^2 - 6^2)$ **79.** $\dfrac{(7 - 8)^{37}}{7^2 - 8^2} \cdot (98 - 7^2 \cdot 2)$ **80.** 🖩 $\dfrac{19 - 17^2}{13^2 - 34}$ **81.** 🖩 $\dfrac{195 + (-15)^3}{195 - 7 \cdot 5^2}$

Determine the sign of each expression if m is negative and n is positive.

82. $\dfrac{-n}{m}$ **83.** $\dfrac{-n}{-m}$ **84.** $-\left(\dfrac{-n}{m}\right)$ **85.** $-\left(\dfrac{n}{-m}\right)$ **86.** $-\left(\dfrac{-n}{-m}\right)$

Use the order of operations as a group to simplify expressions.

2.6 Introduction to Algebra and Expressions

a Algebraic Expressions

In arithmetic, we work with expressions such as

$$37 + 86, \quad 7 \times 8, \quad 19 - 7, \quad \text{and} \quad \frac{3}{8}.$$

In algebra, we use both numbers and variables and work with *algebraic expressions* such as

$$x + 86, \quad 7 \times t, \quad 19 - y, \quad \text{and} \quad \frac{a}{b}.$$

Expressions like these should be familiar from the equation and problem solving that we have already done.

When a letter can stand for various numbers, we call the letter a **variable**. A number or a letter that stands for just one number is called a **constant**. Let c = the speed of light. Then c is a constant. Let a = your age in years. Then a is a variable since the value of a changes every second.

An **algebraic expression** consists of variables, numerals, and operation signs. When we replace a variable with a number, we say that we are **substituting** for the variable. This process is called **evaluating the expression**.

Example 1   Evaluate $x + y$ for $x = 37$ and $y = 29$.

We substitute 37 for x and 29 for y and carry out the addition:

$$x + y = 37 + 29 = 66.$$

The number 66 is called the **value** of the expression.

Algebraic expressions involving multiplication can be written in several ways. For example, "8 times a" can be written as $8 \times a$, $8 \cdot a$, $8(a)$, or simply $8a$. Two letters written together without an operation symbol, such as ab, also indicates multiplication.

Example 2   Evaluate $3y$ for $y = -14$.

$3y = 3(-14) = -42$ Parentheses are required here.

Do Exercises 1–3.

Algebraic expressions involving division can also be written in several ways. For example, "8 divided by t" can be written as $8 \div t$, $8/t$, or $\dfrac{8}{t}$.

Example 3   Evaluate $\dfrac{a}{b}$ and $\dfrac{-a}{-b}$ for $a = 35$ and $b = 7$.

We substitute 35 for a and 7 for b:

$$\frac{a}{b} = \frac{35}{7} = 5; \qquad \frac{-a}{-b} = \frac{-35}{-7} = 5.$$

Note that $\dfrac{-a}{-b} = \dfrac{a}{b}$, as the rules for division would lead us to expect.

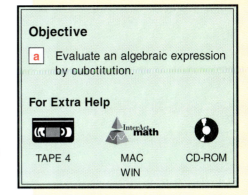

Objective

a Evaluate an algebraic expression by substitution.

For Extra Help

TAPE 4   MAC WIN   CD-ROM

1. Evaluate $a + b$ for $a = 38$ and $b = 26$.

2. Evaluate $x - y$ for $x = 57$ and $y = 29$.

3. Evaluate $4t$ for $t = -15$.

Answers on page A-5

For each number, find two equal expressions with negative signs in different places.

4. $\dfrac{-7}{x}$

5. $-\dfrac{m}{n}$

6. $\dfrac{r}{-4}$

7. Evaluate $\dfrac{a}{-b}, \dfrac{-a}{b},$ and $-\dfrac{a}{b}$ for $a = 28$ and $b = 4$.

8. Find the Fahrenheit temperature that corresponds to 10 degrees Celsius (see Example 5).

9. Evaluate $3x^2$ for $x = 4$ and $x = -4$.

10. Evaluate a^4 for $a = 3$ and $a = -3$.

11. Evaluate $(-x)^2$ and $-x^2$ for $x = 3$.

12. Evaluate $(-x)^2$ and $-x^2$ for $x = 2$.

13. Evaluate x^5 for $x = 2$ and $x = -2$.

Answers on page A-5

Example 4 Evaluate $-\dfrac{a}{b}, \dfrac{-a}{b},$ and $\dfrac{a}{-b}$ for $a = 15$ and $b = 3$.

We substitute 15 for a and 3 for b:

$$-\dfrac{a}{b} = -\dfrac{15}{3} = -5; \quad \dfrac{-a}{b} = \dfrac{-15}{3} = -5; \quad \dfrac{a}{-b} = \dfrac{15}{-3} = -5.$$

Note that $-\dfrac{a}{b}, \dfrac{-a}{b},$ and $\dfrac{a}{-b}$ all represent the same number.

Do Exercises 4–7.

Example 5 Evaluate $\dfrac{9C}{5} + 32$ for $C = 20$.

This expression can be used to find the Fahrenheit temperature that corresponds to 20 degrees Celsius:

$$\dfrac{9C}{5} + 32 = \dfrac{9 \cdot 20}{5} + 32 = \dfrac{180}{5} + 32 = 36 + 32 = 68.$$

Do Exercise 8.

Example 6 Evaluate $5x^2$ for $x = 3$ and $x = -3$.

The rules for order of operations specify that the replacement for x be squared first. That result is then multiplied by 5:

$$5x^2 = 5(3)^2 = 5(9) = 45;$$
$$5x^2 = 5(-3)^2 = 5(9) = 45.$$

Example 6 shows that when opposites are raised to an even power, the results are the same.

Do Exercises 9 and 10.

Example 7 Evaluate $(-x)^2$ and $-x^2$ for $x = 7$.

We have

$$(-x)^2 = (-7)^2 = (-7)(-7) = 49. \quad \text{Substitute 7 for } x. \text{ Then evaluate the power.}$$

To evaluate $-x^2$, recall from Section 2.4 that taking the opposite of a number is the same as multiplying that number by -1. Thus, $-x^2$ can be rewritten as $-1 \cdot x^2$. When x is replaced with 7, we will need to square first and then take the opposite or multiply by -1:

$$-x^2 = -1 \cdot x^2$$
$$= -1 \cdot 7^2$$
$$= -1 \cdot 49$$
$$= -49.$$

These steps emphasize that we find -7^2 by first squaring 7 and then finding the opposite.

CAUTION! Example 7 shows that
$$(-x)^2 \neq -x^2.$$

Do Exercises 11—13.

Chapter 2 Integers and Algebraic Expressions

Exercise Set 2.6

a Evaluate.

1. $7t$, for $t = 2$
   (The cost, in cents, of using a microwave for 2 hr)

2. $40t$, for $t = 2$
   (The cost, in cents, of using an electric oven for 2 hr)

3. $\dfrac{x}{y}$, for $x = 9$ and $y = -3$

4. $\dfrac{m}{n}$, for $m = 14$ and $n = 2$

5. $\dfrac{3p}{q}$, for $p = 2$ and $q = 6$

6. $\dfrac{5y}{z}$, for $y = 15$ and $z = -25$

7. $\dfrac{x + y}{5}$, for $x = -10$ and $y = 20$

8. $\dfrac{p - q}{2}$, for $p = 16$ and $q = -2$

9. $3 + 5 \cdot x$, for $x = 2$

10. $9 - 2 \cdot x$, for $x = 3$

11. $2l + 2w$, for $l = 3$ and $w = 4$
    (The perimeter, in feet, of a 3-ft by 4-ft rectangle)

12. $3(a + b)$, for $a = 2$ and $b = 4$

13. $2(l + w)$, for $l = 3$ and $w = 4$
    (The perimeter, in feet, of a 3-ft by 4-ft rectangle)

14. $3a + 3b$, for $a = 2$ and $b = 4$

15. $7a - 7b$, for $a = 5$ and $b = 2$

16. $4x - 4y$, for $x = 6$ and $y = 1$

17. $7(a - b)$, for $a = 5$ and $b = 2$

18. $4(x - y)$, for $x = 6$ and $y = 1$

19. $16t^2$, for $t = 5$
    (The distance, in feet, that an object falls in 5 sec)

20. $\dfrac{49t^2}{10}$, for $t = 10$
    (The distance, in meters, that an object falls in 10 sec)

Exercise Set 2.6

117

21. $9m - m^2$, for $m = -4$

22. $7n - n^2$, for $n = -5$

23. $a + (b - a)^2$, for $a = 6$ and $b = 4$

24. $(x + y)^2 - y$, for $x = 2$ and $y = 3$

25. $a + b - a^2$, for $a = 6$ and $b = 4$

26. $x + y^2 - y$, for $x = 2$ and $y = 3$

27. $\dfrac{n^2 - n}{2}$, for $n = 9$

(For determining the number of handshakes possible among 9 people)

28. $\dfrac{5(F - 32)}{9}$, for $F = 50$

(For converting 50 degrees Fahrenheit to degrees Celsius)

For each expression, write two equal expressions with negative signs in different places.

29. $-\dfrac{3}{a}$

30. $\dfrac{7}{-x}$

31. $\dfrac{-n}{b}$

32. $-\dfrac{3}{r}$

33. $\dfrac{9}{-p}$

34. $\dfrac{-u}{5}$

35. $\dfrac{-14}{w}$

36. $\dfrac{-23}{m}$

Evaluate $\dfrac{-a}{b}$, $\dfrac{a}{-b}$, and $-\dfrac{a}{b}$ for the given values.

37. $a = 35$, $b = 7$

38. $a = 40$, $b = 5$

39. $a = 81$, $b = 3$

40. $a = 56$, $b = 7$

Evaluate.

41. $(-3x)^2$ and $-3x^2$, for $x = 7$

42. $(-2x)^2$ and $-2x^2$, for $x = 3$

43. $5x^2$, for $x = 2$ and $x = -2$

44. $2x^2$, for $x = 5$ and $x = -5$

45. x^3, for $x = 6$ and $x = -6$

46. x^6, for $x = 2$ and $x = -2$

47. x^6, for $x = 1$ and $x = -1$

48. x^5, for $x = 3$ and $x = -3$

49. a^7, for $a = 2$ and $a = -2$

50. a^7, for $a = 1$ and $a = -1$

51. $-m^2 + m$, for $m = -4$

52. $-n^3 - n$, for $n = 5$

53. $a - 3a^3$, for $a = -5$

54. $x^3 - 5x$, for $x = -3$

55. $x^2 + 5x \div 2$, for $x = -6$

56. $6x \div x^2 - 2x$, for $x = 3$

57. $m^3 - m^2$, for $m = 5$

58. $a^6 - a$, for $a = -2$

Exercise Set 2.6

Skill Maintenance

59. Write a word name for 23,043,921. [1.1c]

60. Multiply: $17 \cdot 53$. [1.5b]

61. Estimate by rounding to the nearest ten. Show your work. [1.4b]

$$\begin{array}{r} 5\,2\,8\,3 \\ -\,2\,4\,7\,5 \\ \hline \end{array}$$

62. Divide: $2982 \div 3$. [1.6c]

63. On January 6, it snowed 9 in., and on January 7, it snowed 8 in. How much did it snow altogether? [1.8a]

64. On March 9, it snowed 12 in., but on March 10, the sun melted 7 in. How much snow remained? [1.8a]

Synthesis

65. ◆ Under what condition(s) will the expression ax^2 be nonnegative? Explain.

66. ◆ A student evaluates $a + a^2$ for $a = 5$ and gets 100 as the result. What mistake did the student probably make?

67. ◆ Does $-\dfrac{a}{b}$ always represent a negative number? Why or why not?

Evaluate.

68. 🖩 $a - b^3 + 17a$, for $a = 19$ and $b = -16$

69. 🖩 $x^2 - 23y + y^3$, for $x = 18$ and $y = -21$

70. $a^{1996} - a^{1997}$, for $a = -1$

71. $x^{1492} - x^{1493}$, for $x = -1$

72. $(m^3 - mn)^m$, for $m = 4$ and $n = 6$

73. $5a^{3a-4}$, for $a = 2$

Classify each statement as true or false. If false, write an example showing why.

74. For any choice of x, $x^2 = (-x)^2$.

75. For any choice of x, $x^3 = -x^3$.

76. For any choice of x, $x^6 + x^4 = (-x)^6 + (-x)^4$.

77. For any choice of x, $(-3x)^2 = 9x^2$.

2.7 Like Terms and Perimeter

Two of the most important topics in algebra are solving equations and forming *equivalent expressions*. In this section, we see how the *distributive law* can be used to form equivalent expressions.

a | Equivalent Expressions and the Distributive Law

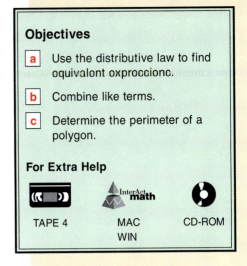

Objectives

a Use the distributive law to find equivalent expressions.

b Combine like terms.

c Determine the perimeter of a polygon.

For Extra Help

TAPE 4   MAC WIN   CD-ROM

It is useful to know when two algebraic expressions will represent the same number.

Example 1 Evaluate $1 \cdot x$ for $x = 5$ and $x = -8$ and compare the results to x.

We substitute 5 for x:

$1 \cdot x = 1 \cdot 5 = 5.$

Next, we substitute -8 for x:

$1 \cdot x = 1 \cdot (-8) = -8.$

We see that $1 \cdot x$ and x represent the same number.

Do Exercises 1 and 2.

Example 1 and Margin Exercise 1 illustrate that the expressions $1 \cdot x$ and x represent the same number for any replacement of x. Because of this, $1 \cdot x$ and x are said to be **equivalent expressions.**

> Two expressions that have the same value for all allowable replacements are called **equivalent**.

In Section 2.6, we saw that the expressions $\dfrac{-a}{-b}$ and $\dfrac{a}{b}$ are equivalent but that the expressions $(-x)^2$ and $-x^2$ are *not* equivalent.

An important concept, known as the **distributive law,** is useful for finding equivalent algebraic expressions. The distributive law involves two operations: multiplication and either addition or subtraction.

To review how the distributive law works, consider the following:

```
   4 5
 ×   7
 ─────
   3 5   ← This is 7 · 5.
 2 8 0   ← This is 7 · 40.
 ─────
 3 1 5   ← This is the sum 7 · 40 + 7 · 5.
```

To carry out the multiplication, we actually added two products. That is,

$7 \cdot 45 = 7(40 + 5) = 7 \cdot 40 + 7 \cdot 5.$

Complete each table by evaluating each expression for the given values.

1.

	$1 \cdot x$	x
$x = 3$		
$x = -6$		
$x = 0$		

2.

	$-1 \cdot x$	$-x$
$x = 2$		
$x = -6$		
$x = 0$		

Answers on page A-5

Multiply.

3. $4(a + b)$

4. $5(x + y + z)$

Multiply.

5. $3(x - y)$

6. $2(a - b + c)$

Answers on page A-5

Chapter 2   Integers and Algebraic Expressions

The distributive law says that if we want to multiply a sum of several numbers by a number, we can either add within the grouping symbols and then multiply, or multiply each of the terms separately and then add.

> **THE DISTRIBUTIVE LAW**
> For any numbers a, b, and c,
> $$a(b + c) = ab + ac.$$

In the statement of the distributive law, we know that in an expression such as $ab + ac$, the multiplications are to be done first according to the rules for order of operations. This means that in $a(b + c)$, we cannot omit the parentheses. If we did we would have $ab + c$, which means $(ab) + c$.

To see that $a(b + c)$ and $ab + ac$ are equivalent, note that

$$3(4 + 2) = 3 \cdot 4 + 3 \cdot 2$$
$$3 \cdot 6 \;\; = \;\; 12 + 6$$
$$18 = 18.$$

To see that $a(b + c) \neq ab + c$, note that

$$3(4 + 2) \neq 3 \cdot 4 + 2$$
$$3 \cdot 6 \;\; \neq \;\; 12 + 2$$
$$18 \neq 14.$$

Example 2   Multiply: $2(l + w)$.

We use the distributive law to write an equivalent expression:

$$2(l + w) = 2 \cdot l + 2 \cdot w$$
$$= 2l + 2w. \quad \text{Try to go directly to this step.}$$

Exercises 11 and 13 in Section 2.6 can serve as a check that $2(l + w)$ and $2l + 2w$ are equivalent.

Do Exercises 3 and 4.

Since subtraction can be regarded as addition of the opposite, it follows that the distributive law holds in cases involving subtraction.

Example 3   Multiply: $7(a - b)$.

$$7(a - b) = 7 \cdot a - 7 \cdot b$$
$$= 7a - 7b \quad \text{Try to go directly to this step.}$$

Exercises 15 and 17 in Section 2.6 can serve as a check that $7(a - b)$ and $7a - 7b$ are equivalent.

Do Exercises 5 and 6.

In more complicated problems, it is sometimes helpful to write one or more middle steps.

Example 4 Multiply: **(a)** $9(x - 5)$; **(b)** $8(a + 2b - 7)$; **(c)** $-4(x - 2y + 3z)$.

a) $9(x - 5) = 9x - 9(5)$ Using the distributive law
$= 9x - 45$

b) $8(a + 2b - 7) = 8 \cdot a + 8 \cdot 2b - 8 \cdot 7$ Using the distributive law
$= 8a + 16b - 56$

c) $-4(x - 2y + 3z) = -4 \cdot x - (-4)(2y) + (-4)(3z)$ Using the distributive law
$= -4x - (-4 \cdot 2)y + (-4 \cdot 3)z$ Using an associative law (twice)
$= -4x - (-8y) + (-12z)$
$= -4x + 8y - 12z$ Try to go directly to this step.

Do Exercises 7–10.

b Combining Like Terms

A **term** is a number, a variable, or a product of numbers and/or variables, or a quotient of numbers and/or variables. Terms are separated by addition signs. If there are subtraction signs, we can find an equivalent expression that uses addition signs.

Example 5 What are the terms of $3x - 4y + \dfrac{2}{z}$?

$3x - 4y + \dfrac{2}{z} = 3x + (-4y) + \dfrac{2}{z}$ Separating parts with + signs

The terms are $3x$, $-4y$, and $\dfrac{2}{z}$.

Example 6 What are the terms of $5xy + 3x^2 - 8$?

$5xy + 3x^2 - 8 = 5xy + 3x^2 + (-8)$ Separating parts with + signs

The terms are $5xy$, $3x^2$, and -8.

Do Exercises 11 and 12.

Multiply.
7. $3(x - 5)$

8. $5(x - y + 4)$

9. $-2(x - 3)$

10. $-5(x - 2y + 4z)$

What are the terms of each expression?
11. $5x - 4y + 3$

12. $-4y - 2x + \dfrac{x}{y}$

Answers on page A-5

Identify the like terms.

13. $9a^3 + 4ab + a^3 + 3ab + 7$

Terms in which the variable factors are exactly the same, such as $9x$ and $-4x$, are called **like**, or **similar, terms**. For example, $3y^2$ and $7y^2$ are like terms, whereas $5x$ and $6x^2$ are not.

Examples Identify the like terms.

7. $7x + 5x^2 + 2x + 8 + 5x^3 + 1$

$7x$ and $2x$ are like terms; 8 and 1 are like terms.

8. $5ab + a^3 - a^2b - 2ab + 7a^3$

$5ab$ and $-2ab$ are like terms; a^3 and $7a^3$ are like terms.

14. $3xy - 5x^2 + y^2 - 4xy + y$

Do Exercises 13 and 14.

When an algebraic expression contains like terms, an equivalent expression can be formed by **combining**, or **collecting, like terms**. To combine like terms, we rely on the distributive law even though that step is often not written out.

Example 9 Combine like terms: **(a)** $5x + 3x$; **(b)** $6mn - 7mn$; **(c)** $7y - 5 - 3y + 8$; **(d)** $2a^5 + 9ab + 3 + a^5 - 7 - 4ab$.

a) $5x + 3x = (5 + 3)x$ Using the distributive law (in "reverse")
$\qquad\quad\; = 8x$ We usually go directly to this step.

Combine like terms.

15. $4a + 7a$

b) $6mn - 7mn = (6 - 7)mn$ Try to do this mentally.
$\qquad\qquad\;\;\, = -1mn$, or simply $-mn$

c) $7y - 5 - 3y + 8 = 7y + (-5) + (-3y) + 8$ Rewriting as addition
$\qquad\qquad\qquad\;\; = 7y + (-3y) + (-5) + 8$ Using a commutative law
$\qquad\qquad\qquad\;\; = 4y + 3$ Try to go directly to this step.

d) $2a^5 + 9ab + 3 + a^5 - 7 - 4ab$
$= 2a^5 + 9ab + 3 + a^5 + (-7) + (-4ab)$
$= 2a^5 + a^5 + 9ab + (-4ab) + 3 + (-7)$ Rearranging terms
$= 3a^5 + 5ab + (-4)$ Think of a^5 as $1a^5$.
$= 3a^5 + 5ab - 4$

16. $5x^2 - 9 + 2x^2 + 3$

As you gain more experience, you will perform several of these steps mentally.

Do Exercises 15–17.

17. $4m - 2n^2 + 3 + n^2 + m - 7$

Answers on page A-5

c Perimeter

> A **polygon** is a closed geometric figure with three or more sides. The **perimeter** of a polygon is the distance around it, or the sum of the lengths of its sides.

Example 10 Find the perimeter of this polygon.

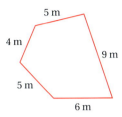

We add the lengths of the sides. Since all the units are the same, we are effectively combining like terms.

Perimeter = 6 m + 5 m + 4 m + 5 m + 9 m

= (6 + 5 + 4 + 5 + 9) m   **Using the distributive law**

= 29 m   **Try to go directly to this step.**

Do Exercises 18 and 19.

A **rectangle** is a polygon with four sides and four 90° angles. Opposite sides of a rectangle have the same measure. The symbol ⌐ indicates a 90° angle.

Example 11 Find the perimeter of a rectangle that is 3 cm by 4 cm.

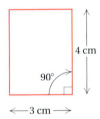

Perimeter = 3 cm + 3 cm + 4 cm + 4 cm

= (3 + 3 + 4 + 4) cm

= 14 cm

Do Exercise 20.

Note that the perimeter of the rectangle in Example 11 is 2 · 3 cm + 2 · 4 cm, or equivalently 2(3 cm + 4 cm). This can be generalized, as follows.

> The **perimeter P of a rectangle** of length l and width w is given by
> $P = 2l + 2w,$ or $P = 2 \cdot (l + w).$
>
>

Find the perimeter of the polygon.

18.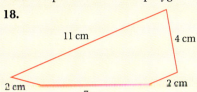

19.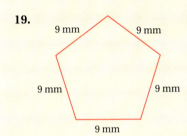

20. Find the perimeter of a rectangle that is 2 cm by 4 cm.

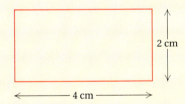

Answers on page A-5

21. Find the perimeter of a 4-ft by 8-ft sheet of plywood.

Example 12 Find the perimeter of a rectangular table that is 4 ft by 6 ft.

$P = 2l + 2w$ **We could also use $P = 2(l + w)$.**

$ = 2 \cdot 6 \text{ ft} + 2 \cdot 4 \text{ ft}$

$ = (2 \cdot 6) \text{ ft} + (2 \cdot 4) \text{ ft}$ **Try to do this mentally.**

$ = 12 \text{ ft} + 8 \text{ ft}$

$ = 20 \text{ ft}$

The perimeter of the table is 20 ft.

Do Exercise 21.

A **square** is a rectangle in which all sides have the same length.

Example 13 Find the perimeter of a square with sides of length 9 mm.

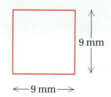

$P = 9 \text{ mm} + 9 \text{ mm} + 9 \text{ mm} + 9 \text{ mm}$

$ = (9 + 9 + 9 + 9) \text{ mm}$

$ = 36 \text{ mm}$

22. Find the perimeter of a square with sides of length 10 km.

Do Exercise 22.

> The **perimeter P of a square** is four times s, the length of a side:
> $P = s + s + s + s$
> $ = 4s$.

Example 14 Find the perimeter of a square garden with sides of length 12 ft.

$P = 4s$

$ = 4 \cdot 12 \text{ ft}$

$ = 48 \text{ ft}$

The perimeter of the garden is 48 ft.

23. Find the perimeter of a square sandbox with sides of length 6 ft.

Do Exercise 23.

Answers on page A-5

Exercise Set 2.7

a Multiply.

1. $5(a + b)$
2. $7(x + y)$
3. $4(x + 1)$

4. $6(a + 1)$
5. $2(b + 5)$
6. $4(x + 3)$

7. $7(1 - t)$
8. $4(1 - y)$
9. $6(5x + 2)$

10. $9(6m + 7)$
11. $8(x + 7 + 6y)$
12. $4(5x + 8 + 3p)$

13. $-7(y - 2)$
14. $-9(y - 7)$
15. $-9(-5x - 6y + 8)$

16. $-7(-2x - 5y + 9)$
17. $-4(x - 3y - 2z)$
18. $8(2x - 5y - 8z)$

19. $8(a - 3b + c)$
20. $-6(a + 2b - c)$
21. $4(x - 3y - 7z)$

22. $5(9x - y + 8z)$
23. $5(4a - 5b + c - 2d)$
24. $7(9a - 4b + 3c - d)$

b Combine like terms.

25. $7a + 12a$
26. $12x + 2x$

27. $10a - a$
28. $-16x + x$

29. $2x + 6z + 9x$
30. $3a - 5b + 7a$

31. $27a + 70 - 40a - 8$
32. $42x - 6 - 4x + 2$

33. $23 + 5t + 7y - t - y - 27$

34. $45 - 90d - 87 - 9d + 3 + 7d$

35. $5x - 12x$

36. $9t - 17t$

37. $y - 17y$

38. $3m - 9m + 4$

39. $-8 + 11a - 5b + 6a - 7b + 7$

40. $8x - 5x + 6 + 3y - 2y - 4$

41. $8x + 3y - 2x$

42. $8y - 3z + 4y$

43. $11x + 2y - 4x - y$

44. $13a + 9b - 2a - 4b$

45. $a + 3b + 5a - 2 + b$

46. $x + 7y + 5 - 2y + 3x$

47. $6x^3 + 2x - 5x^3 + 7x$

48. $9a^2 - 4a + a - 3a^2$

49. $3a^2 + 7a^3 - a^2 + 5 + a^3$

50. $x^3 - 5x^2 + 2x^3 - 3x^2 + 4$

51. $9xy + 4y^2 - 2xy + 2y^2 - 1$

52. $7a^3 + 4ab - 5 - 7ab + 8$

53. $8a^2b - 3ab^2 - 4a^2b + 2ab$

54. $9x^3y + 4xy^3 - 6xy^3 + 3xy$

55. $3x^4 - 2y^4 + 8x^4y^4 - 7x^4 + 8y^4$

56. $3a^6 - 9b^4 + 2a^6b^4 - 7a^6 - 2b^4$

c Find the perimeter of each polygon.

57.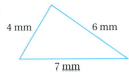

58.

59.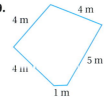

60.

61.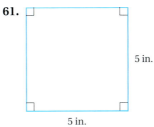

62.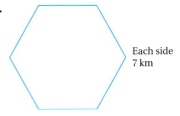

Soccer Field. A soccer field contains many rectangles. Use the diagram of a regulation soccer field (at right) to find the perimeter of each of the following rectangles.

63. The perimeter of the smallest possible regulation field

64. The perimeter of the largest possible regulation field

65. The perimeter of each penalty area

66. The perimeter of each goal area

Exercise Set 2.7

67. Find the perimeter of a 12-ft by 14-ft bedroom.

68. Find the perimeter of a 3-ft by 7-ft screen door.

69. Find the perimeter of a checkerboard that is 14 in. on each side.

70. Find the perimeter of a square skylight that is 2 m on each side.

71. Find the perimeter of a square frame that is 75 cm on each side.

72. Find the perimeter of a square garden that is 12 yd on each side.

73. Find the perimeter of a 12-ft by 20-ft deck.

74. Find the perimeter of a 40-ft by 35-ft backyard.

Skill Maintenance

75. A box of Grand Union Corn Flakes contains 510 grams (g) of corn flakes. A serving of corn flakes weighs 30 g. How many servings are in one box? [1.8a]

76. Estimate the difference by rounding to the nearest ten. [1.4b]

$$\begin{array}{r} 7\,0\,4 \\ -\,4\,8\,6 \end{array}$$

77. Multiply: $6 \cdot 529$. [1.5b]

78. Write a word name: 3,534,512. [1.1c]

79. Divide: $3549 \div 5$. [1.6c]

80. Subtract: $3000 - 2189$. [1.3d]

Synthesis

81. ◆ Explain in your own words what it means for two algebraic expressions to be equivalent.

82. ◆ Can the formula for the perimeter of a rectangle be used to find the perimeter of a square? Why or why not?

83. ◆ Why do you think the distributive law is introduced *before*, rather than after, the material on combining like terms.

Replace the blanks with $+$, $-$, $\times$, or $\div$ to make each statement true.

84. ▦ -32 ▮ $(88$ ▮ $29) = -1888$

85. ▦ 59 ▮ 17 ▮ 59 ▮ $8 = 1475$

86. ▦ 27 ▮ 18 ▮ 21 ▮ $18 = 864$

Simplify. (Multiply and then combine like terms.)

87. $5(x + 3) + 2(x - 7)$

88. $3(a - 7) + 7(a + 4)$

89. $2(3 - 4a) + 5(a - 7)$

90. $7(2 - 5x) + 3(x - 8)$

91. $-5(2 + 3x + 4y) + 7(2x - y)$

92. $3(4 - 2x) + 5(9x - 3y + 1)$

93. In order to save energy, Andrea plans to run a bead of caulk sealant completely around each of 3 doors and 13 windows. Each door measures 3 ft by 7 ft, each window measures 3 ft by 4 ft, and each sealant cartridge covers 56 ft. If each cartridge costs $5.95, how much will it cost Andrea to seal the windows and doors?

94. Eric is attaching lace trim to small tablecloths that are 5 ft by 5 ft, and to large tablecloths that are 7 ft by 7 ft. If the lace costs $1.95 per yard, how much will the trim cost for 6 small tablecloths and 6 large tablecloths?

2.8 Solving Equations

In Section 1.7, we learned to solve certain equations by writing a "related equation." We now formalize this approach in a manner that will be used throughout this book.

a | The Addition Principle

In Section 1.7, we learned to solve an equation like $x + 12 = 27$ by writing the related subtraction, $x = 27 - 12$, or $x = 15$. We can easily see that the solution of $x = 15$ is 15: Replacing x with 15, we get

$15 = 15$, which is true.

You should check that the solution of $x + 12 = 27$ is also 15. Because their solutions are identical, $x = 15$ and $x + 12 = 27$ are said to be **equivalent equations**.

Objectives

a | Use the addition principle to solve equations.

b | Use the division principle to solve equations.

c | Decide which principle should be used to solve an equation.

For Extra Help

TAPE 4   MAC WIN   CD-ROM

> Equations with the same solutions are called **equivalent equations**.

Note the difference between equivalent *expressions* and equivalent *equations*:

$6a$ and $4a + 2a$ are equivalent expressions because, for any replacement of a, both expressions represent the same number.

$3x = 15$ and $4x = 20$ are equivalent equations because any solution of one equation is also a solution of the other equation.

There are principles that enable us to begin with one equation and create an equivalent equation similar to $x = 15$, for which the solution is obvious. One such principle, the *addition principle*, is stated below.

Suppose that a and b stand for the same number and some number c is added to a. We get the same result if we add c to b, because a and b represent the same number.

> **THE ADDITION PRINCIPLE**
> For any numbers a, b, and c,
> $a = b$ is equivalent to $a + c = b + c$.

Solve.

1. $x - 5 = 19$

2. $x - 9 = -12$

Solve.

3. $42 = x + 17$

4. $a + 8 = -6$

Answers on page A-6

Example 1 Solve: $x - 7 = -2$.

We have

$$x - 7 = -2$$
$$x - 7 + 7 = -2 + 7 \quad \text{Using the addition principle: adding 7 on both sides}$$
$$x + 0 = 5 \quad \text{Adding 7 "undoes" the subtraction of 7.}$$
$$x = 5.$$

The solution appears to be 5. To be sure, we check in the original equation.

Check:
$$\begin{array}{c|c} x - 7 = -2 \\ \hline 5 - 7 \,?\, -2 \\ -2 \,|\, \quad \text{TRUE} \end{array}$$

The solution is 5.

Do Exercises 1 and 2.

Recall from Section 2.3 that subtraction can be regarded as adding the opposite of the number being subtracted. Because of this, the addition principle allows us to subtract the same number on both sides of the equation.

Example 2 Solve: $23 = t + 7$.

We have

$$23 = t + 7$$
$$23 - 7 = t + 7 - 7 \quad \text{Using the addition principle to add } -7 \text{ or to subtract 7 on both sides}$$
$$16 = t + 0 \quad \text{Subtracting 7 "undoes" the addition of 7.}$$
$$16 = t.$$

The solution is 16. The check is left to the student.

To visualize the addition principle, think of a jeweler's balance. When both sides of the balance hold equal amounts of weight, the balance is level. If weight is added or removed, equally, on both sides, the balance remains level.

Do Exercises 3 and 4.

b The Division Principle

In Section 1.7, we found that $8n = 96$ could be solved by dividing by 8 on both sides:

$$8 \cdot n = 96$$
$$\frac{8 \cdot n}{8} = \frac{96}{8} \qquad \text{Dividing by 8 on both sides}$$
$$n = 12. \qquad \text{8 times } n \text{, divided by 8, is } n. \; 96 \div 8 \text{ is } 12.$$

You can check that $8 \cdot n = 96$ and $n = 12$ are equivalent. We can divide an equation by any nonzero number in order to find an equivalent equation.

> **THE DIVISION PRINCIPLE**
> For any numbers a, b, and c ($c \neq 0$),
>
> $a = b$ is equivalent to $\dfrac{a}{c} = \dfrac{b}{c}$.

In Chapter 3, after we have discussed multiplication of fractions, we will use an equivalent form of this principle: the multiplication principle.

Example 3 Solve: $9x = 63$.

We have

$$9x = 63$$
$$\frac{9x}{9} = \frac{63}{9} \qquad \text{Using the division principle to divide by 9 on both sides}$$
$$x = 7.$$

CHECK:
$$\begin{array}{c} 9x = 63 \\ \hline 9 \cdot 7 \; ? \; 63 \\ 63 \; | \qquad \text{TRUE} \end{array}$$

The solution is 7.

Do Exercises 5 and 6.

Solve.

5. $7x = 42$

6. $-24 = 3t$

Answers on page A-6

Solve.

7. $63 = -7n$

8. $-6x = 72$

Example 4 Solve: $48 = -8n$.

It is important to distinguish between a negative sign, as we have in $-8n$, and a minus sign, as we had in Example 1. To undo multiplication by -8, we use the division principle:

$$48 = -8n$$
$$\frac{48}{-8} = \frac{-8n}{-8} \quad \text{Dividing by } -8 \text{ on both sides}$$
$$-6 = n.$$

CHECK:
$$\frac{48 = -8n}{48 \;?\; -8(-6)}$$
$$| \; 48 \quad \text{TRUE}$$

The solution is -6.

Do Exercises 7 and 8.

c | Selecting the Correct Approach

It is important for you to be able to determine which principle should be used to solve a particular equation. In future chapters, you will need to use both principles together as well as individually.

Examples Solve.

Solve.

9. $-2x = -52$

5. $39 = -3 + t$

To undo addition of -3, we subtract -3 or simply add 3 on both sides:

$$3 + 39 = 3 + (-3) + t \quad \text{Using the addition principle}$$
$$42 = 0 + t$$
$$42 = t.$$

CHECK:
$$\frac{39 = -3 + t}{39 \;?\; -3 + 42}$$
$$| \; 39 \quad \text{TRUE}$$

10. $-2 + x = -52$

The solution is 42.

6. $39 = -3t$

To undo multiplication of -3, we divide by -3 on both sides:

$$39 = -3t$$
$$\frac{39}{-3} = \frac{-3t}{-3} \quad \text{Using the division principle}$$
$$-13 = t.$$

11. $x \cdot 7 = -28$

CHECK:
$$\frac{39 = -3t}{39 \;?\; -3(-13)}$$
$$| \; 39 \quad \text{TRUE}$$

The solution is -13.

Answers on page A-6

Do Exercises 9–11.

Exercise Set 2.8

a Solve.

1. $x - 7 = 5$
2. $x - 2 = 13$
3. $x - 6 = -9$
4. $x - 5 = -12$

5. $t + 5 = 13$
6. $a + 7 = 25$
7. $x + 9 = -3$
8. $x + 8 = -6$

9. $17 = n - 6$
10. $24 = t - 7$
11. $-9 = x + 3$
12. $-12 = x + 5$

13. $-9 + t = 8$
14. $-5 + a = 13$
15. $3 = 17 + x$
16. $9 = 14 + t$

b Solve.

17. $3x = 24$
18. $6x = 30$
19. $-8t = 32$
20. $-6t = 42$

21. $-5n = -65$
22. $-7n = -35$
23. $64 = 8x$
24. $72 = 3x$

25. $81 = -3t$
26. $55 = -5t$
27. $-x = 83$
28. $-x = 56$

29. $n(-6) = -42$
30. $n(-4) = -48$
31. $-x = -475$
32. $-x = -390$

c Solve.

33. $t - 7 = -2$
34. $3t = 48$
35. $6x = -90$
36. $x + 9 = -15$

37. $8 + x = 43$
38. $-40 = 10x$
39. $18 = x - 27$
40. $-33 = x(-11)$

41. $35 = -5t$ **42.** $9 + t = -19$ **43.** $19x = -171$ **44.** $-36 = x - 12$

45. $19 + x = -171$ **46.** $-36 = x(-12)$ **47.** $-38 = t + 43$ **48.** $-135 = -9t$

Skill Maintenance

Simplify. [1.9c]

49. $5 + 3 \cdot 2^3$ **50.** $(9 - 7)^4 - 3^2$ **51.** $12 \div 3 \cdot 2$

52. $27 \div 3(2 + 1)$ **53.** $15 - 3 \cdot 2 + 7$ **54.** $30 - 4^2 \div 8 \cdot 2$

Synthesis

55. ◆ Explain in your own words the difference between equivalent equations and equivalent expressions.

56. ◆ Debra decides to solve $x - 9 = -5$ by adding 5 on both sides of the equation. Is there anything wrong with her doing this? Why or why not?

57. ◆ James decides to solve $-3x = 15$ by adding 3 on both sides of the equation. Is there anything wrong with his doing this? Why or why not?

Solve.

58. $2x - 7x = -40$ **59.** $9 + x - 5 = 23$ **60.** $17 - 3^2 = 4 + t - 5^2$

61. $(-9)^2 = 2^3 t + (3 \cdot 6 + 1)t$ **62.** $(-7)^2 - 5 = t + 4^3$ **63.** 🖩 $(-17)^3 = 15^3 x$

64. 🖩 $x - (19)^3 = -18^3$ **65.** 🖩 $23^2 = x + 22^2$ **66.** $3x - 5 = 13$

67. $4x + 3 = 31$ **68.** $6 - 2x = -12$ **69.** $8 + 3x = -22$

Summary and Review: Chapter 2

Important Properties and Formulas

For any integers a, b, and c: $a + (-a) = 0$; $a - b = a + (-b)$;
$a \cdot 0 = 0$; $a(b + c) = ab + ac$
Perimeter of a Rectangle: $P = 2l + 2w$, or $P = 2(l + w)$
Perimeter of a Square: $P = 4s$

Review Exercises

1. Tell which integers correspond to this situation:

   Bonnie has $527 in her savings account and Roger is $53 in debt.

Use either < or > for ⬚ to form a true statement.

2. 0 ⬚ −5
3. −7 ⬚ 6
4. −4 ⬚ −19

Find the absolute value.

5. $|-39|$
6. $|12|$
7. $|0|$

8. Find $-x$ when $x = -53$.

9. Find $-(-x)$ when $x = 29$.

Compute and simplify.

10. $-14 + 5$
11. $-5 + (-6)$
12. $14 + (-8)$
13. $0 + (-24)$
14. $15 - 24$
15. $9 - (-14)$
16. $-8 - (-7)$
17. $-3 - (-10)$
18. $-3 + 7 + (-8)$
19. $8 - (-9) - 7 + 2$
20. $-23 \cdot (-4)$
21. $7(-12)$
22. $2(-4)(-5)(-1)$
23. $15 \div (-5)$
24. $\dfrac{-55}{11}$
25. $\dfrac{0}{7}$
26. $7 \div 1^2 \cdot (-3) - 4$
27. $(-3)|4 - 3^2| - 5$

28. Evaluate $3a + b$ for $a = 4$ and $b = -5$.

29. Evaluate $\dfrac{-x}{y}$, $\dfrac{x}{-y}$, and $-\dfrac{x}{y}$ for $x = 20$ and $y = 5$.

Multiply.

30. $4(5x + 9)$

31. $3(2a - 4b + 5)$

Combine like terms.

32. $5a + 13a$

33. $-7x + 13x$

34. $9m + 14 - 12m - 8$

35. Find the perimeter of a frame that is 8 in. by 10 in.

36. Find the perimeter of a square pane of glass that is 25 cm on each side.

Solve.

37. $x - 9 = -17$ **38.** $-4t = 36$ **39.** $x \cdot 7 = -42$

Skill Maintenance

40. Write a word name for 386,451.

41. Estimate by rounding to the nearest ten. Show your work.

$$\begin{array}{r} 7296 \\ -2741 \end{array}$$

42. Estimate by rounding to the nearest hundred. Show your work.

$$2481 - 1729$$

43. In 1998, Mark McGwire hit 70 home runs and Sammy Sosa hit 66. How many did they hit altogether that year?

Multiply.

44. $3 \cdot 8495$

45. $\begin{array}{r} 734 \\ \times 29 \end{array}$

Synthesis

46. ◆ A classmate insists on reading $-x$ as "negative x." When asked why, the response is "because $-x$ is negative." What mistake is this student making?

47. ◆ Are $(a - b)^2$ and $(b - a)^2$ equivalent for all choices of a and b? Why or why not? Experiment with different replacements for a and b.

Simplify.

48. 🖩 $87 \div 3 \cdot 29^3 - (-6)^6 + 1957$

49. 🖩 $1969 + (-8)^5 - 17 \cdot 15^3$

50. 🖩 $\dfrac{113 - 17^3}{15 + 8^3 - 507}$

51. For what values of x is $|x| > x$?

52. For what values of x will $8 + x^3$ be negative?

Test: Chapter 2

1. Tell which integers correspond to this situation: The Tee Shop sold 542 fewer tee-shirts than expected in January and 307 more than expected in February.

2. Use either < or > for ▓ to form a true statement.
   -14 ▓ -21

3. Find the absolute value: $|-429|$.

4. Find $-(-x)$ when $x = -19$.

Compute and simplify.

5. $6 + (-17)$

6. $-9 + (-12)$

7. $-8 + 17$

8. $0 - 12$

9. $7 - 22$

10. $-5 - 19$

11. $-8 - (-27)$

12. $17 - (-3) - 5 + 9$

13. $(-4)^3$

14. $13(-10)$

15. $-9 \cdot 0$

16. $-72 \div (-9)$

17. $\dfrac{-56}{7}$

18. $8 \div 2 \cdot 2 - 3^2$

19. $29 - (3 - 5)^2$

Answers

1. _____
2. _____
3. _____
4. _____
5. _____
6. _____
7. _____
8. _____
9. _____
10. _____
11. _____
12. _____
13. _____
14. _____
15. _____
16. _____
17. _____
18. _____
19. _____

Test: Chapter 2

Answers

20. _____

21. _____

22. _____

23. _____

24. _____

25. _____

26. _____

27. _____

28. _____

29. _____

30. _____

31. _____

32. _____

33. _____

34. _____

---

20. On January 9, the temperature dropped from −2°F to −15°F. How many degrees did it drop?

21. Evaluate $\dfrac{a-b}{6}$ for $a = -8$ and $b = 10$.

22. Multiply: $7(2x + 3y - 1)$.

23. Combine like terms: $9x - 14 - 5x - 3$.

Solve.

24. $-7x = -35$

25. $a + 7 = -5$

Skill Maintenance

26. Write a word name for 2,308,451.

27. Estimate by rounding to the nearest ten. Show your work.
$$\begin{array}{r} 3\,2\,0\,4 \\ -\,1\,9\,1\,5 \end{array}$$

28. Estimate the difference by rounding to the nearest hundred. Show your work.
$9247 - 2879$

29. Maurice shoveled 9 driveways while Phyllis shoveled 12. How many driveways did they shovel between the two of them?

Multiply.

30. $8 \cdot 706$

31. $\begin{array}{r} 3\,0\,2 \\ \times\ \ \ 6\,8 \end{array}$

Synthesis

32. A carpenter plans to attach trim around a doorway and along the base of all walls in a 12-ft by 14-ft room. If the doorway is 3 ft by 7 ft, how many feet of trim are needed? (Only three sides of a doorway get trim.)

Simplify.

33. $9 - 5[x + 2(3 - 4x)] + 14$

34. $15x + 3(2x - 7) - 9(4 + 5x)$

Chapter 2 Integers and Algebraic Expressions

Cumulative Review: Chapters 1–2

1. Write standard notation for the number in the following sentence: The earth travels five hundred eighty-four million, seventeen thousand, eight hundred miles around the sun.

2. Write a word name for 5,380,621.

Add.

3. $14{,}862$
    $+2{,}935$

4. 7989
    789
    $+79$

Subtract.

5. 5376
    -430

6. 2004
    -579

Multiply.

7. 621
    $\times27$

8. 2505
    $\times 3300$

9. $31 \cdot (-8)$

10. $-12(-6)$

Divide.

11. $19 \overline{)4580}$

12. $62 \overline{)3844}$

13. $0 \div (-32)$

14. $60 \div (-12)$

15. Round 427,931 to the nearest thousand.

16. Round 5309 to the nearest hundred.

Estimate each sum or product by rounding to the nearest hundred. Show your work.

17. $749{,}559$
     $+301{,}362$

18. 749
     $\times 531$

19. Use < or > for ■ to form a true sentence: -26 ■ 19.

20. Find the absolute value: $|-279|$.

Cumulative Review: Chapters 1–2

Simplify.

21. $35 - 25 \div 5 + 2 \times 3$

22. $\{17 - [8 - (5 - 2 \times 2)]\} \div (3 + 12 \div 6)$

23. $10 \div 1(-5) - 6^2$

24. 5^3

25. Evaluate $\dfrac{x+y}{5}$ for $x = 11$ and $y = 4$.

26. Evaluate $7x^2$ for $x = -2$.

Multiply.

27. $-2(x + 5)$

28. $6(3x - 2y + 4)$

Simplify.

29. $-12 + (-14)$

30. $-17 - 14$

31. $23 - 38$

32. $-12 - (-25)$

Solve.

33. $x + 8 = 35$

34. $-12t = 36$

35. $-6 + x = -9$

36. $384 \div 16 = n$

Solve.

37. The Barnes & Noble Bookstore in lower Manhattan is the world's largest, covering 154,250 sq ft. Although W. & G. Foyle Ltd. of London has the most titles in the world, it covers "only" 75,825 sq ft. (*Source*: *The Guinness Book of Records,* 1998). How much larger is the Barnes & Noble store?

38. Four of the largest hotels in the United States are in Las Vegas. One has 3174 rooms, the second has 2920 rooms, the third has 2832 rooms, and the fourth has 5005 rooms. What is the total number of rooms in these four hotels?

39. Amanda is offered a part-time job paying $3900 a year. How much is each weekly paycheck?

40. Eastside Appliance sells a refrigerator for $600 and $30 tax with no delivery charge. Westside Appliance sells the same model for $560 and $28 tax plus a $25 delivery charge. Which is the better buy?

41. Combine like terms: $-9 + 10x - 5 + 13x$.

Synthesis

42. A soft drink distributor has 142 loose cans of cola. The distributor wishes to form as many 24-can cases as possible and then, with any remaining cans, as many six-packs as possible. How many cases will be filled? How many six-packs? How many loose cans will remain?

43. Simplify: $a - \{3a - [4a - (2a - 4a)]\}$.

3

Fractional Notation: Multiplication and Division

Introduction

Multiplication and division using fractional notation are examined in this chapter. To aid our study, the chapter begins with rules for divisibility and factorizations. After multiplication and division are discussed, those skills are used to solve equations and problems.

3.1 Multiples and Divisibility
3.2 Factorizations
3.3 Fractions
3.4 Multiplication
3.5 Simplifying
3.6 Multiplying, Simplifying, and More with Area
3.7 Reciprocals and Division
3.8 Solving Equations: The Multiplication Principle

An Application	The Mathematics
For their annual pancake breakfast, the Colchester Boy Scouts need $\frac{2}{3}$ cup of Bisquick® per person. If 135 people are expected, how much Bisquick do they need? This problem appears as Example 5 in Section 3.6.	We let $n =$ the number of cups of Bisquick needed. The problem then translates to $$n = 135 \cdot \frac{2}{3}.$$ ↑ Multiplication using fractional notation occurs often in problem solving.

For more information, visit us at www.mathmax.com

Pretest: Chapter 3

1. Determine whether 165 is divisible by 3. Do not use long division.

2. Determine whether 1645 is divisible by 5. Do not use long division.

3. Determine whether 67 is prime, composite, or neither.

4. Find the prime factorization of 280.

Simplify.

5. $\dfrac{75}{75}$

6. $\dfrac{7x}{1}$

7. $\dfrac{0}{50}$

8. $\dfrac{-8}{32}$

9. $\dfrac{10a}{35a}$

10. Find an equivalent expression for $\dfrac{3}{7}$ with a denominator of 28.

Multiply and simplify.

11. $\dfrac{1}{3} \cdot \dfrac{18}{5}$

12. $\dfrac{5}{6} \cdot (-24)$

13. $\dfrac{2a}{5} \cdot \dfrac{25}{8}$

Find the reciprocal.

14. $\dfrac{7}{8}$

15. 11

Divide and simplify.

16. $15 \div \dfrac{5}{8}$

17. $\dfrac{2}{3} \div \left(-\dfrac{8}{9}\right)$

18. $\dfrac{14}{9} \div (7x)$

Solve.

19. $-\dfrac{8}{7} = 4x$

20. $\dfrac{7}{10} \cdot x = 21$

21. $\dfrac{5}{12} = \dfrac{2}{3} \cdot a$

Solve.

22. Heather earns $72 for working a full day. How much will she earn for working $\frac{3}{4}$ of a day?

23. A piece of twine $\frac{5}{8}$ m long is to be cut into 15 pieces of the same length. What is the length of each piece?

24. A triangular sign with a base of 3 ft and a height of 4 ft is cut from a 4-ft by 8-ft sheet of plywood. Find the area of the leftover plywood.

Chapter 3 Fractions:
Multiplication & Division

3.1 Multiples and Divisibility

In this chapter, we begin our work with fractions. Certain skills make this work easier. For example, in order to simplify fractions like

$$\frac{15}{40},$$

it will be useful to learn about *multiples* and *divisibility*.

a Multiples

A **multiple** of a number is a product of it and some integer. For example, some multiples of 2 are:

  2 (because 2 = 1 · 2);
  4 (because 4 = 2 · 2);
  6 (because 6 = 3 · 2);
  8 (because 8 = 4 · 2);
  10 (because 10 = 5 · 2).

We can also find multiples of 2 by counting by twos: 2, 4, 6, 8, and so on.

Example 1 Show that each of the numbers 3, 6, 9, and 15 is a multiple of 3.

We show that each of 3, 6, 9, and 15 can be expressed as a product of 3 and some integer:

3 = 1 · 3;   6 = 2 · 3;   9 = 3 · 3;   15 = 5 · 3.

Do Exercises 1 and 2.

Example 2 Multiply by 1, 2, 3, and so on, to find ten multiples of 7.

1 · 7 = 7	6 · 7 = 42
2 · 7 = 14	7 · 7 = 49
3 · 7 = 21	8 · 7 = 56
4 · 7 = 28	9 · 7 = 63
5 · 7 = 35	10 · 7 = 70

Do Exercise 3.

> A number b is said to be **divisible** by another number a if b is a multiple of a.

Thus,

  6 is divisible by 2 because 6 is a multiple of 2 (6 = 3 · 2);
  27 is divisible by 3 because 27 is a multiple of 3 (27 = 9 · 3);
  100 is divisible by 25 because 100 is a multiple of 25 (100 = 4 · 25).

> A number b is divisible by another number a if division of b by a results in a remainder of zero. We sometimes say that a divides b "evenly."

Objectives

a Find some multiples of a number, and determine whether a number is divisible by another number.

b Test to see if a number is divisible by 2, 3, 5, 6, 9, or 10.

For Extra Help

TAPE 5   MAC   CD-ROM
          WIN

1. Show that each of the numbers 5, 45, and 100 is a multiple of 5.

2. Show that each of the numbers 10, 60, and 110 is a multiple of 10.

3. Multiply by 1, 2, 3, and so on, to find ten multiples of 5.

Answers on page A-7

4. Determine whether 16 is divisible by 2.

Example 3 Determine whether 24 is divisible by 3.

We divide 24 by 3:

$$\begin{array}{r} 8 \\ 3\overline{)24} \\ \underline{24} \\ 0 \end{array}$$

The remainder of 0 indicates that 24 is divisible by 3.

Example 4 Determine whether 98 is divisible by 4.

We divide 98 by 4:

$$\begin{array}{r} 24 \\ 4\overline{)98} \\ \underline{8} \\ 18 \\ \underline{16} \\ 2 \end{array}$$ ← Not 0!

Since the remainder is not 0 we know that 98 is *not* divisible by 4.

Do Exercises 4–6.

5. Determine whether 125 is divisible by 5.

6. Determine whether 125 is divisible by 6.

Calculator Spotlight

Rather than list remainders, most calculators display quotients using decimal notation. Although decimal notation is not studied until Chapter 5, it is still possible for us to now check for divisibility using a calculator.

To see if a number, like 551, is divisible by another number, like 19, we simply press [5][5][1][÷][1][9][=]. If the resulting quotient contains no digits to the right of the decimal point, the first number is divisible by the second. Thus, since 551 ÷ 19 = 29, we know that 551 is divisible by 19. On the other hand, since 551 ÷ 20 = 27.55, we know that 551 is *not* divisible by 20.

Exercises

For each pair of numbers, determine whether the first number is divisible by the second number.

1. 731, 17
2. 1502, 79
3. 1053, 36
4. 4183, 47

Answers on page A-7

Chapter 3 Fractions: Multiplication & Division

b Tests for Divisibility

We now learn quick ways of checking for divisiblity by 2, 3, 5, 6, 9, or 10 without actually performing long division.

Divisibility by 2

You may already know the test for divisibility by 2.

> A number is divisible by 2 (is *even*) if it has a ones digit of 0, 2, 4, 6, or 8 (that is, it has an even ones digit).

To see why this test works, consider 354, which is

$$3 \text{ hundreds} + 5 \text{ tens} + 4 \text{ ones}.$$

Hundreds and tens are both multiples of 2. If the ones digit is a multiple of 2, then the entire number is a multiple of 2.

Examples Determine whether each of the following numbers is divisible by 2.

5. 355 *is not* divisible by 2; 5 is not even.
6. 4786 *is* divisible by 2; 6 is even.
7. 8990 *is* divisible by 2; 0 is even.
8. 4261 *is not* divisible by 2; 1 is not even.

Do Exercises 7–10.

Divisibility by 3

> A number is divisible by 3 if the sum of its digits is divisible by 3.

An explanation of why this test works is outlined in Exercise 52.

Examples Determine whether each of the following numbers is divisible by 3.

9. 18 $1 + 8 = 9$
10. 93 $9 + 3 = 12$
11. 201 $2 + 0 + 1 = 3$

All are divisible by 3 because the sums of their digits are divisible by 3.

12. 256 $2 + 5 + 6 = 13$ The sum is not divisible by 3, so 256 is not divisible by 3.

Do Exercises 11–14.

Divisibility by 6

A number divisible by 6 is a multiple of 6. But $6 = 2 \cdot 3$, so the number is also a multiple of 2 and 3. Thus a number is divisible by 6 if it is divisible by both 2 and 3.

> A number is divisible by 6 if its ones digit is even (0, 2, 4, 6, or 8) and the sum of its digits is divisible by 3.

Determine whether each of the following numbers is divisible by 2.

7. 84

8. 59

9. 998

10. 2225

Determine whether each of the following numbers is divisible by 3.

11. 111

12. 1111

13. 309

14. 17,216

Answers on page A-7

Determine whether each of the following numbers is divisible by 6.

15. 420

16. 106

17. 321

18. 444

Determine whether each of the following numbers is divisible by 9.

19. 16

20. 117

21. 930

22. 29,223

Answers on page A-7

Examples Determine whether each of the following numbers is divisible by 6.

13. 720

Because 720 is even, it is divisible by 2. Since $7 + 2 + 0 = 9$ and 9 is divisible by 3, we know that 720 is also divisible by 3. Since 720 is divisible by both 2 and 3, we know that 720 *is* divisible by 6.

720      $7 + 2 + 0 = 9$
↑   ↑
Even   Divisible by 3

14. 531

Because 531 is not divisible by 2, we know that 531 *is not* divisible by 6.

531
↑
Not even

15. 478

Because the sum of its digits is not divisible by 3, we know that 478 is not divisible by 3. Since 478 is not divisible by 3, we know that 478 *is not* divisible by 6.

$4 + 7 + 8 = 19$
↑
Not divisible by 3

Do Exercises 15–18.

Divisibility by 9

The test for divisibility by 9 is similar to the test for divisibility by 3. An explanation of why it works is outlined in Exercise 52.

 A number is divisible by 9 if the sum of its digits is divisible by 9.

Example 16 The number 6984 *is* divisible by 9 because

$6 + 9 + 8 + 4 = 27$

and 27 is divisible by 9.

Example 17 The number 322 *is not* divisible by 9 because

$3 + 2 + 2 = 7$

and 7 is not divisible by 9.

Do Exercises 19–22.

Divisibility by 10

> A number is divisible by 10 if its ones digit is 0.

We know that this test works because the product of 10 and *any* number has a ones digit of 0.

Examples Determine whether each of the following numbers is divisible by 10.

18. 3440 *is* divisible by 10 because its ones digit is 0.
19. 3447 *is not* divisible by 10 because its ones digit is not 0.

Do Exercises 23–26.

Divisibility by 5

> A number is divisible by 5 if its ones digit is 0 or 5.

Examples Determine whether each of the following numbers is divisible by 5.

20. 220 *is* divisible by 5 because its ones digit is 0.
21. 475 *is* divisible by 5 because its ones digit is 5.
22. 6514 *is not* divisible by 5 because its ones digit is neither 0 nor 5.

Do Exercises 27–30.

To see why the test for 5 works, consider 7830:

$$7830 = 10 \cdot 783 = 5 \cdot 2 \cdot 783.$$

Since 7830 is a multiple of 10 and 10 is a multiple of 5, it follows that 7830 is divisible by 5.
  Next, consider 6325:

$$6325 = 632 \text{ tens} + 5 \text{ ones}.$$

Tens are multiples of 5, and the ones digit, 5, is as well. Thus, 6325 is a multiple of 5. Only if the ones digit is 0 or 5, as in 7830 or 6325, will the entire number be divisible by 5.

Divisibility by 4, 7, and 8

Although tests exist for divisibility by 4, 7, and 8, they are often more difficult to perform than the actual long division.

Determine whether each of the following numbers is divisible by 10.

23. 305

24. 300

25. 847

26. 8760

Determine whether each of the following numbers is divisible by 5.

27. 5780

28. 3427

29. 34,678

30. 7775

Answers on page A-7

3.1 Multiples and Divisibility

149

Improving Your Math Study Skills

Classwork: Before and During Class

Before Class

Textbook

- Check your syllabus (or ask your instructor) to find out which sections will be covered during the next class. Then be sure to read, or at least skim, these sections *before* class. Although you may not understand all the concepts, you will at least familiarize yourself with the material. This will help you to understand the next lesson.

- This book makes use of color, shading, and design elements to highlight important concepts, so you do not need to highlight these. Instead, it is more productive for you to note trouble spots with either a highlighter or Post-It™ notes. Then use these marked points as possible questions for clarification by your instructor at the appropriate time. Be sure to always have a pencil or erasable pen in hand when reading this book.

Homework

- Review the previous day's homework just before class. This will refresh your memory on the concepts covered in the last class, and again provide you with possible questions to ask your instructor.

During Class

Class Seating

- If possible, choose a seat at the front of the class. In most classes, the more serious students tend to sit up front so you will probably be able to concentrate better if you do the same. You should also avoid sitting next to noisy or distracting students.

- If your instructor uses an overhead projector, select a seat that will give you an unobstructed view of the screen.

Taking Notes

- This textbook has been written and laid out so that it represents a quality set of notes at the same time that it teaches. Thus you might not need to take many notes in class. Just watch, listen, and ask yourself questions as the class moves along, rather than continually taking notes.

  However, if you still feel more comfortable taking your own notes, consider using the following two-column method. Divide your page in half vertically so that you have two columns side by side. Write down what is on the board or screen in the left column; then, in the right column, write clarifying comments or questions.

- If you have any difficulty keeping up with the instructor, use abbreviations to speed up your note-taking. Consider standard abbreviations like "Ex" for "Example," "≈" for "approximately equal to," or "∴" for "therefore." Create your own abbreviations as well.

- Another shortcut for note-taking is to write only the beginning of a word, leaving space for the rest. Be sure you write enough of the word to know what it means later on!

- Some students find it helpful to follow lectures with their textbooks open so that they can see exactly which portion of the book is being discussed. These students write their notes directly in the margins of their book.

- Whatever approach you use for note-taking, be sure to review the notes that you write while they are still fresh in your mind. Sometimes you may need to insert corrections if you wrote too quickly or copied something incorrectly.

Other study tips appear on pages 50, 90, 100, 322, 368, and 596.

Exercise Set 3.1

a Multiply by 1, 2, 3, and so on, to find ten multiples of each number.

1. 6
2. 14
3. 20
4. 50
5. 3
6. 7

7. 13
8. 17
9. 10
10. 4
11. 9
12. 11

13. Determine whether 26 is divisible by 7.

14. Determine whether 29 is divisible by 9.

15. Determine whether 1880 is divisible by 8.

16. Determine whether 4227 is divisible by 3.

17. Determine whether 106 is divisible by 4.

18. Determine whether 196 is divisible by 16.

19. Determine whether 4227 is divisible by 9.

20. Determine whether 200 is divisible by 25.

21. Determine whether 8650 is divisible by 16.

22. Determine whether 4143 is divisible by 7.

b To answer Exercises 23–28, consider the following numbers. Use the tests for divisibility.

46	300	85	256
224	36	711	8064
19	45,270	13,251	1867
555	4444	254,765	21,568

23. Which of the above are divisible by 2?

24. Which of the above are divisible by 3?

25. Which of the above are divisible by 10?

26. Which of the above are divisible by 5?

27. Which of the above are divisible by 6?

28. Which of the above are divisible by 9?

To answer Exercises 29–34, consider the following numbers.

56	200	75	35
324	42	812	402
784	501	2345	111,111
55,555	3009	2001	1005

29. Which of the above are divisible by 3?

30. Which of the above are divisible by 2?

31. Which of the above are divisible by 5?

32. Which of the above are divisible by 10?

33. Which of the above are divisible by 9?

34. Which of the above are divisible by 6?

Skill Maintenance

Solve.

35. $16 \cdot t = 848$ [1.7b], [2.8b]

36. $m + 9 = 14$ [1.7b], [2.8a]

37. $23 + x = 15$ [1.7b], [2.8a]

38. $24 \cdot m = -576$ [1.7b], [2.8b]

39. Find the total cost of 12 shirts at $37 each and 4 pairs of pants at $59 each.   [1.8a]

40. Add: $-34 + 76$.   [2.2a]

Synthesis

41. ◆ Describe a test that could be used to determine whether a number is divisible by 25.

42. ◆ Is every counting number a multiple of 1? Why or why not?

43. ◆ Describe a manner in which Exercises 23, 24, and 26 can be used to answer Exercises 25 and 27.

44. ◆ Describe a test for determining whether a number is divisible by 30.

45. ▦ Find the largest five-digit number that is divisible by 47.

46. ▦ Find the largest six-digit number that is divisible by 53.

Find the smallest number that is simultaneously a multiple of the given numbers.

47. 2, 3, and 5

48. 3, 5, and 7

49. 4, 6, and 10

50. 6, 10, and 14

51. A passenger in a taxicab asks for the driver's company number. The driver says abruptly, "Sure–it's the smallest multiple of 11 that, when divided by 2, 3, 4, 5, or 6, has a remainder of 1." What is the number?

52. ◆ To help see why the tests for division by 3 and 9 work, note that any four-digit number $abcd$ can be rewritten as $1000 \cdot a + 100 \cdot b + 10 \cdot c + d$, or $999a + 99b + 9c + a + b + c + d$.

a) Explain why $999a + 99b + 9c$ is divisible by both 9 and 3 for all choices of a, b, c, and d.
b) Explain why the four-digit number $abcd$ is divisible by 9 if $a + b + c + d$ is divisible by 9 and is divisible by 3 if $a + b + c + d$ is divisible by 3.

Use the divisibility rules and properties of numbers to discover an unknown number.

3.2 Factorizations

In Section 3.1, we saw that both 28 and 35 are multiples of 7. Another way of saying this is to state that 7 is a *factor* of both 28 and 35. When a number is expressed as a product of two or more factors, we say that we have *factored* the original number. Thus the word "factor" can be used as either a noun or a verb. Being able to factor is an important skill for our study of fractions.

a | Factoring Numbers

Looking at the equation 3 · 4 = 12, we see that 3 and 4 are *factors* of 12. Since 12 = 12 · 1, we know that 12 and 1 are also factors of 12.

 A number c is a **factor** of the number a if a is divisible by c.

A **factorization** of a number expresses that number as a product of natural numbers.

For example, each of the following gives a factorization of 12.

12 = 4 · 3 ⟵ This factorization shows that 4 and 3 are factors of 12.
12 = 12 · 1 ⟵ This factorization shows that 12 and 1 are factors of 12.
12 = 6 · 2 ⟵ This factorization shows that 6 and 2 are factors of 12.
12 = 2 · 3 · 2 ⟵ This factorization shows that 2 and 3 are factors of 12.

This shows that 1, 2, 3, 4, 6, and 12 are all factors of 12. Note that since $n = n \cdot 1$, every number has a factorization, and every number has at least itself and 1 as factors.

Example 1 Find all the factors of 24.

To help us get started, we can use some of the tests for divisibility. For example, since 24 is even, we know that 2 is a factor. Since the sum of the digits in 24 is 6 and 6 is divisible by 3, we know that 3 is a factor. We can use trial and error to determine that 4 is also a factor, but that 5 is not. A list of factorizations can then be used to make a complete list of factors.

Factorizations: 1 · 24; 2 · 12; 3 · 8; 4 · 6;
Factors: 1, 2, 3, 4, 6, 8, 12, 24

Note that, apart from the number itself, no factor can be more than half the size of the number of which it is a factor.

Do Exercises 1–4.

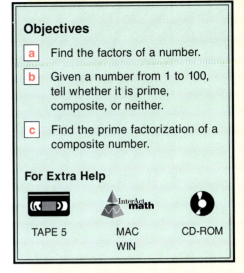

Objectives

a | Find the factors of a number.
b | Given a number from 1 to 100, tell whether it is prime, composite, or neither.
c | Find the prime factorization of a composite number.

For Extra Help

TAPE 5   MAC WIN   CD-ROM

Find all the factors of each number listed. (*Hint*: Find some factorizations of the number.)

1. 6

2. 8

3. 10

4. 32

Answers on page A-7

3.2 Factorizations

153

5. Tell whether each number is prime, composite, or neither.

1, 4, 6, 8, 13, 19, 41

b Prime and Composite Numbers

> A natural number that has exactly two different factors, itself and 1, is called a **prime number**.

Example 2 Tell whether the numbers 2, 3, 5, 7, and 11 are prime.

The number 2 is prime. It has only the factors 1 and 2.

The number 5 is prime. It has only the factors 1 and 5.

The numbers 3, 7, and 11 are also prime.

Example 3 Tell whether the numbers 4, 6, 8, 10, 63, and 1 are prime.

The number 4 is not prime. It has the factors 1, 2, and 4.

The numbers 6, 8, 10, and 63 are not prime. Each has factors other than itself and 1. For instance, 2 is a factor of 6, 8, and 10, and 7 is a factor of 63.

The number 1 is not prime. It does not have two *different* factors.

> A natural number, other than 1, that is not prime is called a **composite number**.

In other words, if a number has at least one factor other than itself and 1, it is composite. Thus, from Examples 2 and 3, we see that

2, 3, 5, 7, and 11 are prime;

4, 6, 8, 10, and 63 are composite;

and   1 is neither prime nor composite.

Do Exercise 5.

Below is a list of the prime numbers 2 to 157. The ability to recognize primes will save you time as you progress through this text.

A List of Primes (from 2 to 157)
2, 3, 5, 7, 11, 13, 17, 19, 23, 29, 31, 37, 41, 43, 47, 53, 59, 61, 67, 71, 73, 79, 83, 89, 97, 101, 103, 107, 109, 113, 127, 131, 137, 139, 149, 151, 157

Mathematicians continue to search for bigger and bigger primes. Prime numbers can be very useful when encoding messages and programming computers.

Answer on page A-7

Chapter 3 Fractions: Multiplication & Division

It can be useful to note that when two different prime numbers are factors of a number, the product of those primes will also be a factor. For instance, in Example 1, since 2 and 3 are both prime factors of 24, their product, 6, is also a factor.

c Prime Factorizations

To express a composite number as a product of primes is to find a **prime factorization** of the number. To do this, we consider the primes

$$2, 3, 5, 7, 11, 13, 17, 19, 23, \text{ and so on,}$$

and determine whether a given number is divisible by any of them.

Example 4 Find the prime factorization of 39.

a) We check for divisibility by the first prime, 2. Since 39 is not even, 2 is not a factor of 39.

b) Since the sum of the digits in 39 is 12 and 12 is divisible by 3, we know that 39 is divisible by 3. We then perform the division.

$$\begin{array}{r} 13 \\ 3\overline{)39} \end{array} \quad R = 0 \quad \text{A remainder of 0 confirms that 3 is a factor of 39.}$$

Because 13 is a prime, we are finished. The prime factorization is

$$39 = 3 \cdot 13.$$

Example 5 Find the prime factorization of 76.

a) Since 76 is even, it must have the first prime, 2, as a factor.

$$\begin{array}{r} 38 \\ 2\overline{)76} \end{array} \quad \text{We can write } 76 = 2 \cdot 38.$$

b) Because 38 is also even, we see that 76 contains a second factor of 2.

$$\begin{array}{r} 19 \\ 2\overline{)38} \end{array} \quad \text{Note that } 38 = 2 \cdot 19, \text{ so } 76 = 2 \cdot 2 \cdot 19.$$

Because 19 is prime, the complete factorization is

$$76 = 2 \cdot 2 \cdot 19. \quad \text{All factors are prime.}$$

We abbreviate our procedure as follows.

$$\begin{array}{r} 19 \\ 2\overline{)38} \\ 2\overline{)76} \end{array} \quad \longleftarrow \text{ We begin here.}$$

$$76 = 2 \cdot 2 \cdot 19$$

A factorization like $2 \cdot 2 \cdot 19$ can be written as $2^2 \cdot 19$ or $2 \cdot 19 \cdot 2$ or $19 \cdot 2 \cdot 2$ or $19 \cdot 2^2$. In any case, the prime factors are the same. For this reason, we agree that any of these may be considered "the" prime factorization of 76.

> Each composite number has just one (unique) prime factorization.

Find the prime factorization of each number.

6. 6

7. 12

8. 45

9. 98

10. 126

11. 144

Example 6 Find the prime factorization of 72.

$$\begin{array}{r} 3 \\ 3\overline{)9} \\ 2\overline{)18} \\ 2\overline{)36} \\ 2\overline{)72} \end{array} \leftarrow \text{Begin here.}$$

$$72 = 2 \cdot 2 \cdot 2 \cdot 3 \cdot 3$$

Another way to find a prime factorization is by using a **factor tree** as follows:

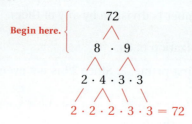

Begin here.

$2 \cdot 2 \cdot 2 \cdot 3 \cdot 3 = 72$

Had we begun with $2 \cdot 36$, $3 \cdot 24$, $4 \cdot 18$, or $6 \cdot 12$, the same prime factorization would result.

Example 7 Find the prime factorization of 189.

We can use a string of successive divisions.

$$\begin{array}{r} 7 \\ 3\overline{)21} \\ 3\overline{)63} \\ 3\overline{)189} \end{array}$$ Since 189 is odd, 2 will not be a factor. We begin with 3.

$$189 = 3 \cdot 3 \cdot 3 \cdot 7$$

We can also use a factor tree.

```
      189
      /\
     3 · 63
        /\
       3 · 7 · 9
              /\
3 · 7 · 3 · 3 = 189
```

Example 8 Find the prime factorization of 65.

We can use a string of successive divisions.

$$\begin{array}{r} 13 \\ 5\overline{)65} \end{array}$$ 65 is not divisible by 2 or 3 but *is* divisible by 5.

$$65 = 5 \cdot 13$$

We can also use a factor tree.

```
    65
    /\
5 · 13 = 65
```

Do Exercises 6–11.

Answers on page A-7

Exercise Set 3.2

a Find all the factors of each number.

1. 18 **2.** 16 **3.** 54 **4.** 48

5. 4 **6.** 9 **7.** 7 **8.** 11

9. 1 **10.** 3 **11.** 98 **12.** 100

13. 42 **14.** 105 **15.** 385 **16.** 110

17. 36 **18.** 196 **19.** 225 **20.** 441

b State whether each number is prime, composite, or neither.

21. 17 **22.** 24 **23.** 22 **24.** 31

25. 48 **26.** 43 **27.** 31 **28.** 54

29. 1 **30.** 2 **31.** 9 **32.** 19

33. 47 **34.** 27 **35.** 29 **36.** 49

c Find the prime factorization of each number.

37. 16 **38.** 8 **39.** 14 **40.** 15

41. 22 **42.** 32 **43.** 25 **44.** 40

45. 62 **46.** 169 **47.** 140 **48.** 50

49. 100   **50.** 110   **51.** 35   **52.** 70

53. 78   **54.** 86   **55.** 77   **56.** 99

57. 112   **58.** 142   **59.** 300   **60.** 175

Skill Maintenance

Multiply.

61. $-2 \cdot 13$ [2.4a]   **62.** $(-8)(-32)$ [2.4a]

Add.

63. $-17 + 25$ [2.2a]   **64.** $-9 + (-14)$ [2.2a]

Divide.

65. $0 \div 22$ [2.5a]   **66.** $22 \div 22$ [1.6c]

Synthesis

67. ◆ Explain a method for constructing a composite number that contains exactly two factors other than itself and 1.

68. ◆ Are the divisibility tests of Section 3.1 useful for finding prime factorizations? Why or why not?

69. ◆ If a and b are both factors of c, does it follow that $a \cdot b$ is also a factor of c? Why or why not?

Find the prime factorization of each number.

70. 🖩 473,073,361   **71.** 🖩 28,502,923   **72.** 7800

73. 2520   **74.** 2772   **75.** 1998

76. Describe an arrangement of 54 objects that corresponds to the factorization $54 = 6 \times 9$.

77. Describe an arrangement of 24 objects that corresponds to the factorization $24 = 2 \cdot 3 \cdot 4$.

78. Two numbers are **relatively prime** if there is no prime number that is a factor of both numbers. For example, 10 and 21 are relatively prime but 15 and 18 are not. List five pairs of composite numbers that are relatively prime.

Find all the prime numbers less than 100, using the Sieve of Eratosthenes.

3.3 Fractions

The study of arithmetic begins with the set of whole numbers

0, 1, 2, 3, 4, 5, 6, 7, 8, 9, 10, 11, and so on.

The need soon arises for fractional parts of numbers such as halves, thirds, fourths, and so on. Here are some examples:

$\frac{1}{25}$ of the parking spaces in a commercial area in Indiana must be marked for the handicapped.

For $\frac{9}{10}$ of the people in the United States, English is the primary language.

$\frac{1}{11}$ of all women develop breast cancer at some point in their life.

$\frac{43}{200}$ of the world's population is in China.

a Identifying Numerators and Denominators

The following are some additional examples of fractions:

$$\frac{1}{2}, \quad \frac{13}{41}, \quad \frac{-8}{5}, \quad \frac{x}{y}, \quad -\frac{4}{25}, \quad \frac{2a}{7b}.$$

This way of writing number names is called **fractional notation**. The top number is called the **numerator** and the bottom number is called the **denominator**.

Example 1 Identify the numerator and the denominator.

$\frac{7}{8}$ ← Numerator
    ← Denominator

Do Exercises 1–3.

b Fractions and the Real World

Example 2 What part is shaded?

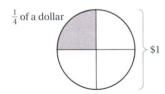

When an object is divided into 4 parts of the same size, each of these parts is $\frac{1}{4}$ of the object. Thus, $\frac{1}{4}$ (*one-fourth* or *one-quarter*) of a dollar is shaded.

Do Exercises 4–7.

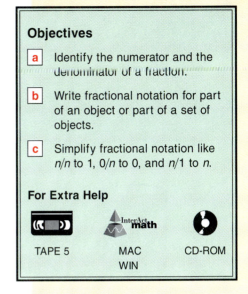

Objectives

a Identify the numerator and the denominator of a fraction.

b Write fractional notation for part of an object or part of a set of objects.

c Simplify fractional notation like n/n to 1, $0/n$ to 0, and $n/1$ to n.

For Extra Help

TAPE 5   MAC   CD-ROM
          WIN

Identify the numerator and the denominator of each fraction.

1. $\frac{5}{7}$ 2. $\frac{5a}{7b}$ 3. $\frac{-22}{3}$

What part is shaded?

4.

5.

6.

7.

Answers on page A-7

3.3 Fractions

159

What amount is shaded?

8.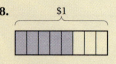

9.

10.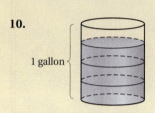

11.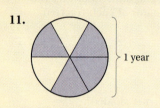

What amount is shaded?

12.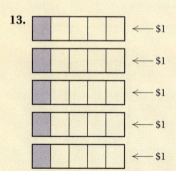

13.

14.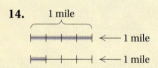

Answers on page A-7

The diagram on the preceding page is an example of a *circle graph*, or *pie chart*. Circle graphs are often used to illustrate the relationships of fractional parts of a whole. The following graph shows color preferences of bicycles as determined by the Bicycle Market Research Institute.

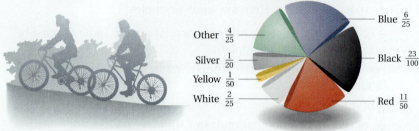

Example 3 What amount is shaded?

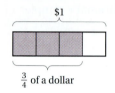

The object is divided into 4 parts of the same size, and 3 of them are shaded. This is $3 \cdot \frac{1}{4}$, or $\frac{3}{4}$. Thus, $\frac{3}{4}$ (*three-fourths* or *three-quarters*) of a dollar is shaded.

Do Exercises 8–11.

The fraction $\frac{3}{4}$ corresponds to another situation. We take 3 objects, divide them into fourths, and take $\frac{1}{4}$ of the entire amount (which appears now as $\frac{12}{4}$). This is $\frac{1}{4} \cdot 3$, or $\frac{3}{4}$, or $3 \div 4$.

Example 4 What amount is shaded?

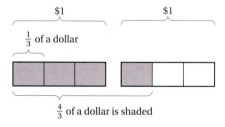

The objects are divided into 3 equally sized parts each, and 4 of these parts are shaded. We have more than one whole object. In this case, it is $4 \cdot \frac{1}{3}$, or $\frac{4}{3}$ of a dollar.

Do Exercises 12–15 on this page and the next.

Fractional notation also corresponds to situations involving part of a set.

Example 5 What part of this set, or collection, of workers is women?

4 carpenters   3 electricians

There are 7 workers, and 3 are women. We say that three-sevenths, or $\frac{3}{7}$, of the workers are women.

Do Exercises 16–18.

c Some Fractional Notation for Integers

Fractional Notation for 1

The number 1 corresponds to situations like the following.

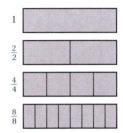

If we divide an object into n parts and take n of them, we get all of the object (1 whole object). Since a negative divided by a negative is a positive, the following is stated for *all* nonzero integers.

> $\frac{n}{n} = 1$, for any integer n that is not 0.

Example 6 Simplify: a) $\frac{5}{5}$; b) $\frac{-9}{-9}$; c) $\frac{17x}{17x}$ (assume $x \neq 0$).

a) $\frac{5}{5} = 1$ b) $\frac{-9}{-9} = 1$ c) $\frac{17x}{17x} = 1$

Do Exercises 19–24.

15.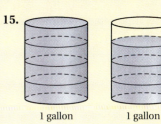

   1 gallon   1 gallon

16. Referring to Example 5, what part of the set of workers has dark hair?

17. What part of this set of shapes is shaded?

18. What part of this set were elected United States president? are recording stars?

   Abraham Lincoln
   Whitney Houston
   Garth Brooks
   Bill Clinton
   Sheryl Crow
   Gloria Estefan

Simplify. Assume that $a \neq 0$.

19. $\frac{7}{7}$ 20. $\frac{a}{a}$

21. $\frac{-34}{-34}$ 22. $\frac{1}{1}$

23. $\frac{-2347}{-2347}$ 24. $\frac{54a}{54a}$

Answers on page A-7

3.3 Fractions

161

Simplify, if possible. Assume that $x \neq 0$.

25. $\dfrac{0}{2}$ **26.** $\dfrac{0}{-8}$

27. $\dfrac{0}{7x}$ **28.** $\dfrac{4-4}{236}$

29. $\dfrac{7}{0}$ **30.** $\dfrac{-4}{0}$

Simplify.

31. $\dfrac{8}{1}$ **32.** $\dfrac{-10}{1}$

33. $\dfrac{-346}{1}$ **34.** $\dfrac{24-1}{23}$

Answers on page A-7

Chapter 3 Fractions: Multiplication & Division

Fractional Notation for 0

Consider $\frac{0}{4}$. This corresponds to dividing an object into 4 parts and taking none of them. We get 0. This result also extends to all nonzero integers.

▶ $\dfrac{0}{n} = 0$, for any integer n that is not 0.

Example 7 Simplify: a) $\dfrac{0}{9}$; b) $\dfrac{0}{1}$; c) $\dfrac{0}{5a}$ (assume $a \neq 0$); d) $\dfrac{0}{-23}$.

a) $\dfrac{0}{9} = 0$ b) $\dfrac{0}{1} = 0$

c) $\dfrac{0}{5a} = 0$ d) $\dfrac{0}{-23} = 0$

Fractional notation with a denominator of 0, such as $n/0$, is meaningless because we cannot speak of an object as divided into *zero* parts. (If it is not divided at all, then we say that it is undivided and remains in one part.)

▶ $\dfrac{n}{0}$ is not defined.

(When asked to simplify $\dfrac{n}{0}$, we write *undefined*.)

Do Exercises 25–30.

Other Integers

Consider $\frac{4}{1}$. This corresponds to taking 4 objects and dividing each into 1 part. (We do not divide them.) We have 4 objects.

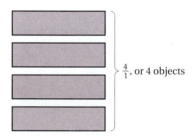

$\frac{4}{1}$, or 4 objects

▶ Any integer divided by 1 is the original integer. That is,

$\dfrac{n}{1} = n$, for any integer n.

Example 8 Simplify: a) $\dfrac{2}{1}$; b) $\dfrac{-9}{1}$; c) $\dfrac{3x}{1}$.

a) $\dfrac{2}{1} = 2$ b) $\dfrac{-9}{1} = -9$ c) $\dfrac{3x}{1} = 3x$

Do Exercises 31–34.

Exercise Set 3.3

a Identify the numerator and the denominator of each fraction.

1. $\dfrac{3}{4}$
2. $\dfrac{-9}{10}$
3. $\dfrac{7}{-9}$
4. $\dfrac{15}{8}$
5. $\dfrac{2x}{3z}$
6. $\dfrac{9a}{2b}$

b For each figure, what amount is shaded?

7. 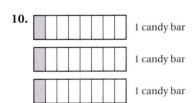 $1
8. 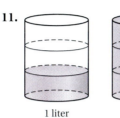 $1
9. 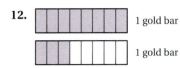 1 mile / 1 mile

10. 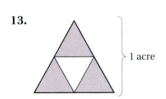 1 candy bar / 1 candy bar / 1 candy bar
11. 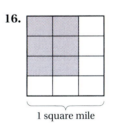 1 liter   1 liter
12. 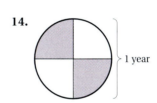 1 gold bar / 1 gold bar

13. 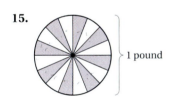 1 acre
14.  1 year

15. 1 pound
16. 1 square mile

What fractional part of each set is shaded?

17.
18.
19.
20.

c Simplify, if possible. Assume that all variables are nonzero.

21. $\dfrac{0}{17}$

22. $\dfrac{19}{19}$

23. $\dfrac{15}{1}$

24. $\dfrac{10}{1}$

25. $\dfrac{20}{20}$

26. $\dfrac{-20}{1}$

27. $\dfrac{-14}{-14}$

28. $\dfrac{4a}{1}$

29. $\dfrac{0}{-234}$

30. $\dfrac{37a}{37a}$

31. $\dfrac{3n}{3n}$

32. $\dfrac{0}{-1}$

33. $\dfrac{9x}{9x}$

34. $\dfrac{-12a}{1}$

35. $\dfrac{-63}{1}$

36. $\dfrac{-3x}{-3x}$

37. $\dfrac{0}{2a}$

38. $\dfrac{0}{8}$

39. $\dfrac{52}{0}$

40. $\dfrac{8-8}{1247}$

41. $\dfrac{7n}{1}$

42. $\dfrac{247}{0}$

43. $\dfrac{6}{7-7}$

44. $\dfrac{15}{9-9}$

Skill Maintenance

Multiply.

45. $-7(30)$ [2.4a]

46. $23 \cdot (-14)$ [2.4a]

47. $(-71)(-12)0$ [2.4b]

48. $32(-29)0$ [2.4b]

49. Recently, the average annual income of people living in Connecticut was $30,303 per person. In Mississippi, the average annual income was $16,531. How much more do people in Connecticut make, on average, than those living in Mississippi? [1.8a]

50. Sandy can type 62 words per minute. How long will it take Sandy to type 12,462 words? [1.8a]

Synthesis

51. ◆ Explain in your own words why $n/1 = n$, for any integer n.

52. ◆ Explain in your own words why $0/n = 0$, for any nonzero integer n.

53. ◆ Explain in your own words why $n/n = 1$, for any nonzero integer n.

54. The surface of Earth is 3 parts water and 1 part land. What fractional part of Earth is water? land?

55. The year 1999 began on a Friday. What fractional part of 1999 were Mondays?

56. Rayona earned $2700 one summer. During the following semester, she spent $1200 for tuition, $540 for rent, and $360 for food. The rest went for miscellaneous expenses. What part of the income went for tuition? rent? food? miscellaneous expenses?

57. A couple had 3 sons, each of whom had 3 daughters. If each daughter gave birth to 3 sons, what fractional part of the couple's descendants is female?

3.4 Multiplication

a **Multiplication by an Integer**

We can find $3 \cdot \frac{1}{4}$ by thinking of repeated addition. We add three $\frac{1}{4}$'s.

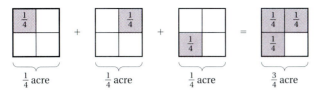

We see that $3 \cdot \frac{1}{4} = \frac{1}{4} + \frac{1}{4} + \frac{1}{4} = \frac{3}{4}$.

Do Exercises 1 and 2.

To multiply a fraction by an integer,

a) multiply the top number (the numerator) by the integer and

b) keep the same denominator.

$$6 \cdot \frac{4}{5} = \frac{6 \cdot 4}{5} = \frac{24}{5}$$

Examples Multiply.

1. $5 \times \frac{3}{8} = \frac{5 \times 3}{8} = \frac{15}{8}$

   Skip this step when you feel comfortable doing so.

2. $\frac{2}{5} \cdot 13 = \frac{2 \cdot 13}{5} = \frac{26}{5}$

3. $-10 \cdot \frac{1}{3} = \frac{-10}{3}$, or $-\frac{10}{3}$ Recall that $\frac{-a}{b} = -\frac{a}{b}$.

4. $a \cdot \frac{4}{7} = \frac{4a}{7}$ Recall that $a \cdot 4 = 4 \cdot a$.

Do Exercises 3–6.

b **Multiplication Using Fractional Notation**

We find a product such as $\frac{9}{7} \cdot \frac{3}{4}$ as follows. An explanation appears on the next page.

To multiply a fraction by a fraction,

a) multiply the numerators and

b) multiply the denominators.

$$\frac{9}{7} \cdot \frac{3}{4} = \frac{9 \cdot 3}{7 \cdot 4} = \frac{27}{28}$$

Objectives

a Multiply an integer and a fraction.

b Multiply using fractional notation.

c Solve problems involving multiplication of fractions.

For Extra Help

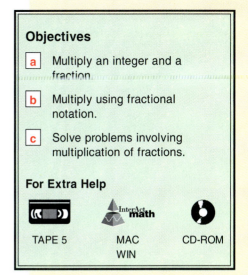

TAPE 5   MAC WIN   CD-ROM

1. Find $2 \cdot \frac{1}{3}$.

2. Find $5 \cdot \frac{1}{8}$.

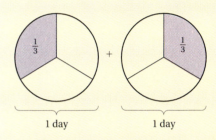

Multiply.

3. $5 \times \frac{2}{3}$

4. $(-11) \times \frac{3}{8}$

5. $23 \cdot \frac{2}{5}$

6. $x \cdot \frac{4}{9}$

Answers on page A-7

3.4 Multiplication

165

Multiply.

7. $\dfrac{3}{8} \cdot \dfrac{5}{7}$

8. $\dfrac{4}{3} \times \dfrac{8}{5}$

9. $\left(-\dfrac{3}{10}\right)\left(-\dfrac{1}{10}\right)$

10. $(-7)\dfrac{a}{b}$

11. Draw diagrams like those in the text to show how the multiplication $\dfrac{4}{5} \cdot \dfrac{1}{3}$ corresponds to a real-world situation.

Answers on page A-7

Chapter 3 Fractions: Multiplication & Division

166

Examples Multiply.

5. $\dfrac{5}{6} \times \dfrac{7}{4} = \underbrace{\dfrac{5 \times 7}{6 \times 4}}_{} = \dfrac{35}{24}$

Skip this step when you feel comfortable doing so.

6. $\dfrac{3}{5} \cdot \dfrac{7}{8} = \dfrac{3 \cdot 7}{5 \cdot 8} = \dfrac{21}{40}$

7. $\dfrac{4}{x} \cdot \dfrac{y}{9} = \dfrac{4y}{9x}$

8. $(-6)\left(-\dfrac{4}{5}\right) = \dfrac{-6}{1} \cdot \dfrac{-4}{5} = \dfrac{24}{5}$ Recall that $\dfrac{n}{1} = n$.

Do Exercises 7–10.

Unless one of the factors is a whole number, multiplication of fractions is hard to imagine as repeated addition. To see how multiplication of fractions corresponds to situations in the real world, consider the expressions

$$\dfrac{2}{5} \cdot \dfrac{3}{4} \quad \text{and} \quad \dfrac{2}{3} \text{ of } \dfrac{3}{4}.$$

Imagine some object and take $\dfrac{3}{4}$ of it. We divide it into 4 parts and take 3 of them. That is shown in the shading below.

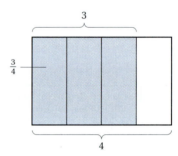

Next, we take $\dfrac{2}{5}$ of the result. We divide the shaded part into 5 parts and take 2 of them. That is shown below as the heavily shaded region.

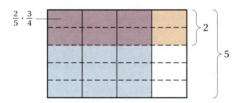

The entire object has been divided into 20 parts, of which 6 have been shaded heavily:

$$\dfrac{2}{5} \cdot \dfrac{3}{4} = \dfrac{2 \cdot 3}{5 \cdot 4} = \dfrac{6}{20}.$$

The figure above shows a rectangular array inside a rectangular array. The number of pieces in the entire array is $5 \cdot 4$ (the product of the denominators). The number of pieces shaded heavily is $2 \cdot 3$ (the product of the numerators). For the answer, we take 6 pieces out of a set of 20 to get $\dfrac{6}{20}$. We have shown that $\dfrac{2}{5}$ *of* $\dfrac{3}{4}$ corresponds to the multiplication $\dfrac{2}{5} \cdot \dfrac{3}{4}$.

Do Exercise 11.

c Applications and Problem Solving

Most problems that can be solved by multiplying fractions can be thought of in terms of rectangular arrays.

Example 9 A rancher owns a square mile of land. He gives $\frac{4}{5}$ of it to his daughter and she gives $\frac{2}{3}$ of her share to her son. How much land goes to the daughter's son?

1. **Familiarize.** We first make a drawing to help solve the problem. The land may not be square. It could be in a shape like A or B below, or it could even be in more than one piece. But to think about the problem, we can visualize a square, as shown by shape C.

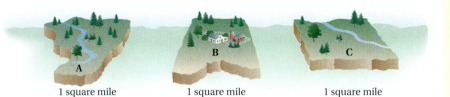

The daughter gets $\frac{4}{5}$ of the land. We shade $\frac{4}{5}$.

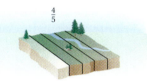

Her son gets $\frac{2}{3}$ of her part. We "raise" that.

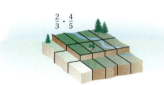

2. **Translate.** We let $n =$ the part of the land that goes to the daughter's son. We are taking "two-thirds of four-fifths." The word "of" corresponds to multiplication. Thus the following multiplication sentence corresponds to the situation:

$$\frac{2}{3} \cdot \frac{4}{5} = n.$$

3. **Solve.** The number sentence tells us what to do. We have

$$\frac{2}{3} \cdot \frac{4}{5} = n, \quad \text{or} \quad \frac{8}{15} = n.$$

4. **Check.** We can check this in the figure above, where we see that 8 of 15 equally sized parts will go to the daughter's son.

5. **State.** The daughter's son gets $\frac{8}{15}$ of a square mile of land.

Do Exercise 12.

12. A seaside hotel uses $\frac{3}{4}$ of its extra land for recreational purposes. Of that, $\frac{1}{2}$ is used for swimming pools. What part of the extra land is used for swimming pools?

Answer on page A-7

13. The length of a button on a fax machine is $\frac{9}{10}$ cm. The width is $\frac{7}{10}$ cm. What is the area?

14. Of the students at Overton Community College, $\frac{1}{8}$ participate in sports and $\frac{3}{5}$ of these play soccer. What fractional part of the student body plays soccer?

We have seen that the area of a rectangular region is found by multiplying length by width. That is true whether length and width are whole numbers or not. Remember, the area of a rectangular region is given by the formula

$$A = l \cdot w \quad (Area = length \cdot width).$$

Example 10 The length of a rectangular key on a calculator is $\frac{7}{10}$ cm. The width is $\frac{3}{10}$ cm. What is the area?

1. **Familiarize.** Recall that area is length times width. We make a drawing, letting $A =$ the area of the calculator key.

2. **Translate.** Next, we translate.

   Area is Length times Width
   $A = \frac{7}{10} \times \frac{3}{10}$

3. **Solve.** The sentence tells us what to do. We multiply:

   $$\frac{7}{10} \cdot \frac{3}{10} = \frac{7 \cdot 3}{10 \cdot 10} = \frac{21}{100}.$$

4. **Check.** To check, we can repeat the calculation or draw a grid, as in Example 9. This is left to the student.

5. **State.** The area of the key is $\frac{21}{100}$ cm².

Do Exercise 13.

Example 11 A cornbread recipe calls for $\frac{3}{4}$ cup of cornmeal. A chef is making $\frac{1}{2}$ of the recipe. How much cornmeal will the chef need?

1. **Familiarize.** We make a drawing or at least visualize the situation. We let $n =$ the number of cups of cornmeal the chef will need.

2. **Translate.** The multiplication sentence $\frac{1}{2} \cdot \frac{3}{4} = n$ corresponds to the situation.

3. **Solve.** We carry out the multiplication:

   $$\frac{1}{2} \cdot \frac{3}{4} = \frac{1 \cdot 3}{2 \cdot 4} = \frac{3}{8}.$$

4. **Check.** To check, we can determine what fractional part of the drawing has been heavily shaded. This is left to the student.

5. **State.** The chef will need $\frac{3}{8}$ cup of cornmeal.

Do Exercise 14.

Answers on page A-7

Exercise Set 3.4

a Multiply.

1. $3 \cdot \dfrac{1}{7}$
2. $2 \cdot \dfrac{1}{5}$
3. $(-5) \times \dfrac{1}{6}$
4. $(-4) \times \dfrac{1}{7}$
5. $\dfrac{2}{3} \cdot 7$

6. $\dfrac{2}{5} \cdot 6$
7. $(-1)\dfrac{7}{9}$
8. $(-1)\dfrac{4}{11}$
9. $\dfrac{2}{5} \cdot x$
10. $\dfrac{3}{8} \cdot y$

11. $\dfrac{2}{5}(-3)$
12. $\dfrac{3}{5}(-4)$
13. $a \cdot \dfrac{3}{4}$
14. $b \cdot \dfrac{2}{5}$
15. $17 \times \dfrac{m}{6}$

16. $\dfrac{n}{7} \cdot 40$
17. $-3 \cdot \dfrac{-2}{5}$
18. $-4 \cdot \dfrac{-5}{7}$
19. $-\dfrac{2}{7}(-x)$
20. $-\dfrac{3}{4}(-a)$

b Multiply.

21. $\dfrac{1}{2} \cdot \dfrac{1}{5}$
22. $\dfrac{1}{4} \cdot \dfrac{1}{3}$
23. $\left(-\dfrac{1}{4}\right) \times \dfrac{1}{10}$
24. $\left(-\dfrac{1}{3}\right) \times \dfrac{1}{10}$
25. $\dfrac{2}{3} \times \dfrac{1}{5}$

26. $\dfrac{3}{5} \times \dfrac{1}{5}$
27. $\dfrac{2}{y} \cdot \dfrac{x}{5}$
28. $\left(-\dfrac{3}{4}\right)\left(-\dfrac{3}{5}\right)$
29. $\left(-\dfrac{3}{4}\right)\left(-\dfrac{3}{4}\right)$
30. $\dfrac{3}{b} \cdot \dfrac{a}{7}$

31. $\dfrac{2}{3} \cdot \dfrac{7}{13}$
32. $\dfrac{3}{11} \cdot \dfrac{4}{5}$
33. $\dfrac{1}{10}\left(\dfrac{-3}{5}\right)$
34. $\dfrac{3}{10}\left(\dfrac{-7}{5}\right)$
35. $\dfrac{7}{8} \cdot \dfrac{a}{8}$

36. $\dfrac{4}{5} \cdot \dfrac{4}{b}$
37. $\dfrac{1}{y} \cdot \dfrac{1}{100}$
38. $\dfrac{x}{10} \cdot \dfrac{7}{100}$
39. $\dfrac{-14}{15} \cdot \dfrac{13}{19}$
40. $\dfrac{-12}{13} \cdot \dfrac{12}{13}$

c Solve.

41. A rectangular table top measures $\frac{4}{5}$ m long by $\frac{3}{5}$ m wide. What is its area?

42. If each slice of pie is $\frac{1}{6}$ of a pie, how much of the pie is $\frac{1}{2}$ of a slice?

43. *Forestry.* A chain saw holds $\frac{1}{5}$ gal of fuel. Chain-saw fuel is $\frac{1}{16}$ two-cycle oil and $\frac{15}{16}$ unleaded gasoline. How much two-cycle oil is in a freshly filled chain saw?

44. *Football.* One of 39 high school football players plays college football. One of 39 college players plays professional football. What fraction of the number of high school players plays professional football?

45. *Cooking.* A recipe for a batch of granola calls for $\frac{2}{3}$ cup of molasses. How much molasses is needed to make $\frac{3}{4}$ of a batch?

46. *Sewing.* It takes $\frac{2}{3}$ yd of ribbon to make a decorative bow. How much ribbon is needed for 4 bows?

47. *Municipal Waste.* Of every 3 tons of municipal waste, 2 tons is dumped in landfills. Of the municipal waste that goes into landfills, $\frac{1}{10}$ is yard trimmings. What fractional part of municipal waste is trimmings that are landfilled?

48. *Muncipal Waste.* Of every 3 tons of municipal waste, 1 ton is paper and paperboard. If $\frac{2}{3}$ of all municipal waste is landfilled, what fractional part of municipal waste is paper and paperboard that is landfilled?

Skill Maintenance

Simplify.

49. $5 - 3^2$ [2.5b]

50. $(5 - 3)^2$ [1.9c]

51. $8 \cdot 12 - (7 + 13)$ [1.9c]

52. $8 \cdot 12 - 7 + 13$ [1.9c]

53. What does the digit 6 mean in 4,678,952?   [1.1e]

54. What does the digit 4 mean in 4,678,952?   [1.1e]

Synthesis

55. ◆ Following Example 8, we explained, using words and pictures, why $\frac{2}{5} \cdot \frac{3}{4}$ equals $\frac{6}{20}$. Present a similar explanation of why $\frac{2}{3} \cdot \frac{4}{7}$ equals $\frac{8}{21}$.

56. ◆ Write a problem for a classmate to solve. Design the problem so that the solution is "About $\frac{1}{30}$ of the students are lefthanded women."

57. ◆ Is mulitplication of fractions commutative? Why or why not?

Multiply. Write each answer using fractional notation.

58. 🖩 $\frac{341}{517} \cdot \frac{209}{349}$

59. 🖩 $\left(-\frac{57}{61}\right)^3$

60. $\left(\frac{2}{5}\right)^3 \left(-\frac{7}{9}\right)$

61. $\left(-\frac{1}{2}\right)^5 \left(\frac{3}{5}\right)$

62. $\left(-\frac{3}{4}\right)^2 \left(-\frac{5}{7}\right)^2$

63. Evaluate $-\frac{2}{3}xy$ for $x = \frac{2}{5}$ and $y = -\frac{1}{7}$.

64. Evaluate $-\frac{3}{4}ab$ for $a = \frac{2}{5}$ and $b = \frac{7}{5}$.

3.5 Simplifying

a Multiplying by 1

Recall the following:

$$1 = \frac{1}{1} = \frac{2}{2} = \frac{3}{3} = \frac{4}{4} = \frac{-13}{-13} = \frac{45}{45} = \frac{100}{100} = \frac{n}{n}.$$

Any nonzero number divided by itself is 1.

> When we multiply a number by 1, we get the same number.
>
> $$\frac{3}{5} = \frac{3}{5} \cdot 1 = \frac{3}{5} \cdot \frac{4}{4} = \frac{12}{20}$$

Since $\frac{3}{5} \cdot 1 = \frac{12}{20}$, we know that $\frac{3}{5}$ and $\frac{12}{20}$ are two names for the same number. This means that $\frac{3}{5}$ and $\frac{12}{20}$ are *equivalent* (see Section 2.7).

Do Exercises 1–4.

Suppose we want to rename $\frac{2}{3}$, using a denominator of 15. We can multiply by 1 to find a number equivalent to $\frac{2}{3}$:

$$\frac{2}{3} = \frac{2}{3} \cdot \frac{5}{5} = \frac{2 \cdot 5}{3 \cdot 5} = \frac{10}{15}.$$

We chose $\frac{5}{5}$ for 1 because $15 \div 3$ is 5.

Example 1 Find a number equivalent to $\frac{1}{4}$ with a denominator of 24.

Since $24 \div 4 = 6$, we multiply by 1, using $\frac{6}{6}$:

$$\frac{1}{4} = \frac{1}{4} \cdot \frac{6}{6} = \frac{1 \cdot 6}{4 \cdot 6} = \frac{6}{24}.$$

Example 2 Find a number equivalent to $\frac{2}{5}$ with a denominator of -35.

Since $-35 \div 5 = -7$, we multiply by 1, using $\frac{-7}{-7}$:

$$\frac{2}{5} = \frac{2}{5}\left(\frac{-7}{-7}\right) = \frac{2(-7)}{5(-7)} = \frac{-14}{-35}.$$

Example 3 Find an expression equivalent to $\frac{9}{8}$ with a denominator of $8a$.

Since $8a \div 8 = a$, we multiply by 1, using $\frac{a}{a}$:

$$\frac{9}{8} \cdot \frac{a}{a} = \frac{9a}{8a}.$$

Do Exercises 5–9.

Objectives

a Multiply by 1 to find an equivalent expression using a different denominator.

b Simplify fractional notation.

For Extra Help

TAPE 6   MAC WIN   CD-ROM

Multiply.

1. $\frac{1}{2} \cdot \frac{8}{8}$

2. $\frac{3}{5} \cdot \frac{x}{x}$

3. $-\frac{13}{25} \cdot \frac{4}{4}$

4. $\frac{8}{3}\left(\frac{-2}{-2}\right)$

Find an equivalent expression for each number, but with the denominator indicated. Use multiplication by 1.

5. $\frac{4}{3} = \frac{?}{9}$

6. $\frac{3}{4} = \frac{?}{-24}$

7. $\frac{9}{10} = \frac{?}{10x}$

8. $\frac{3}{15} = \frac{?}{45}$

9. $\frac{-8}{7} = \frac{?}{49}$

Answers on page A-8

3.5 Simplifying

Simplify.

10. $\dfrac{6}{14}$

11. $\dfrac{-10}{12}$

12. $\dfrac{40}{8}$

13. $\dfrac{4a}{3a}$

14. $-\dfrac{50}{30}$

Answers on page A-8

b Simplifying

All of the following are names for three-fourths:

$$\frac{3}{4}, \frac{-6}{-8}, \frac{9}{12}, \frac{12}{16}, \frac{-15}{-20}.$$

We say that $\frac{3}{4}$ is **simplest** because it has the smallest positive denominator. Note that 3 and 4 have no factor in common other than 1.

To simplify, we reverse the process of multiplying by 1. This is accomplished by removing any factors other than 1 and −1 that the numerator and the denominator have in common.

$$\frac{12}{18} = \frac{2 \cdot 6}{3 \cdot 6} \quad \leftarrow \text{Factoring the numerator}$$
$$\phantom{\frac{12}{18}} \quad \leftarrow \text{Factoring the denominator}$$
$$= \frac{2}{3} \cdot \frac{6}{6} \quad \text{Factoring the fraction}$$
$$= \frac{2}{3} \cdot 1 \quad \frac{6}{6} = 1$$
$$= \frac{2}{3} \quad \text{Removing the factor 1: } \frac{2}{3} \cdot 1 = \frac{2}{3}$$

Examples Simplify.

4. $\dfrac{-8}{20} = \dfrac{-2 \cdot 4}{5 \cdot 4} = \dfrac{-2}{5} \cdot \dfrac{4}{4} = \dfrac{-2}{5}$ Removing a factor equal to 1: $\dfrac{4}{4} = 1$

5. $\dfrac{2}{6} = \dfrac{1 \cdot 2}{3 \cdot 2} = \dfrac{1}{3} \cdot \dfrac{2}{2} = \dfrac{1}{3}$ Writing 1 allows for pairing of factors in the numerator and the denominator.

6. $\dfrac{30}{6} = \dfrac{5 \cdot 6}{1 \cdot 6} = \dfrac{5}{1} \cdot \dfrac{6}{6} = \dfrac{5}{1} = 5$ ← We could also simplify $\dfrac{30}{6}$ by doing the division 30 ÷ 6. That is, $\dfrac{30}{6} = 30 \div 6 = 5$.

7. $-\dfrac{15}{10} = -\dfrac{3 \cdot 5}{2 \cdot 5}$
$$= -\dfrac{3}{2} \cdot \dfrac{5}{5}$$
$$= -\dfrac{3}{2}$$
Removing a factor equal to 1: $\dfrac{5}{5} = 1$

8. $\dfrac{4x}{15x} = \dfrac{4 \cdot x}{15 \cdot x}$ (Assume that $x \neq 0$.)
$$= \dfrac{4}{15} \cdot \dfrac{x}{x}$$
$$= \dfrac{4}{15}$$
Removing a factor equal to 1: $\dfrac{x}{x} = 1$

Note that $\dfrac{4}{15}$ is considered simplified—the numbers 4 and 15 have no factors in common.

Do Exercises 10–14.

The tests for divisibility are also helpful when simplifying.

Example 9 Simplify: $\dfrac{105}{135}$.

Since both 105 and 135 end in 5, we know that 5 is a factor of both the numerator and the denominator:

$$\dfrac{105}{135} = \dfrac{21 \cdot 5}{27 \cdot 5} = \dfrac{21}{27} \cdot \dfrac{5}{5} = \dfrac{21}{27}.$$

A fraction is not "simplified" if common factors of the numerator and the denominator remain. Because 21 and 27 are both divisible by 3, we must simplify further:

$$\dfrac{105}{135} = \dfrac{21}{27} = \dfrac{7 \cdot 3}{9 \cdot 3} = \dfrac{7}{9} \cdot \dfrac{3}{3} = \dfrac{7}{9}.$$

Example 10 Simplify: $\dfrac{90}{84}$.

Since 90 and 84 are both even, we know that 2 is a common factor:

$$\dfrac{90}{84} = \dfrac{2 \cdot 45}{2 \cdot 42}$$

$$= \dfrac{2}{2} \cdot \dfrac{45}{42} = \dfrac{45}{42}. \quad \text{Removing a factor equal to 1: } \dfrac{2}{2} = 1$$

Before stating that $\dfrac{45}{42}$ represents simplified form, we must check to see whether 45 and 42 share a common factor. Since the sum of the digits in 45 is 9 and 9 is divisible by 3, we know that 45 is divisible by 3. Similarly, it can be shown that 42 is divisible by 3. Thus, 3 is a common factor and we can simplify further:

$$\dfrac{45}{42} = \dfrac{3 \cdot 15}{3 \cdot 14}$$

$$= \dfrac{3}{3} \cdot \dfrac{15}{14} = \dfrac{15}{14}. \quad \text{Removing a factor equal to 1: } \dfrac{3}{3} = 1$$

Thus $\dfrac{90}{84}$ simplifies to $\dfrac{15}{14}$.

Do Exercises 15–18.

Simplify.

15. $\dfrac{35}{40}$

16. $\dfrac{801}{702}$

17. $\dfrac{-24}{21}$

18. Simplify each fraction in this circle graph.

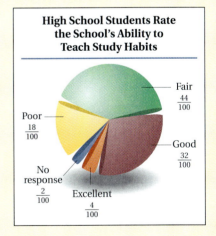

High School Students Rate the School's Ability to Teach Study Habits

Fair $\dfrac{44}{100}$
Poor $\dfrac{18}{100}$
Good $\dfrac{32}{100}$
No response $\dfrac{2}{100}$
Excellent $\dfrac{4}{100}$

Answers on page A-8

3.5 Simplifying

173

Calculator Spotlight

 Fraction calculators are equipped with a key, often labeled $a^b/_c$, that allows for simplification with fractional notation. To simplify

$$\frac{208}{256}$$

with such a fraction calculator, the following keystrokes can be used.

| 2 | 0 | 8 | $a^b/_c$ |
| 2 | 5 | 6 | = |

The display that appears

| 13 ⌐ 16. |

represents simplified fractional notation $\frac{13}{16}$.

Exercises

Use a fraction calculator to simplify each of the following.

1. $\frac{84}{90}$ 2. $\frac{35}{40}$ 3. $\frac{690}{835}$ 4. $\frac{42}{150}$

CANCELING   Canceling is a shortcut that you may have used for removing a factor that equals 1 when working with fractional notation. With *great* concern, we mention it as a possibility for speeding up your work. Canceling may be done only when removing common factors in numerators and denominators. Each common factor allows us to remove a factor equal to 1 in a product.

Our concern is that canceling be done with care and understanding. In effect, slashes are used to indicate factors equal to 1 that have been removed. For instance, Example 10 might have been done faster as follows:

$$\frac{90}{84} = \frac{2 \cdot 45}{2 \cdot 42} \quad \text{Factoring the numerator and the denominator}$$

$$= \frac{\cancel{2} \cdot 45}{\cancel{2} \cdot 42} \quad \text{When a factor equal to 1 is noted, it is "canceled" as shown: } \frac{2}{2} = 1.$$

$$= \frac{45}{42} = \frac{\cancel{3} \cdot 15}{\cancel{3} \cdot 14} = \frac{15}{14}.$$

CAUTION!   The difficulty with canceling is that it is often applied incorrectly in situations like the following:

$$\frac{\cancel{2} + 3}{\cancel{2}} = 3; \quad \frac{\cancel{4} + 1}{\cancel{4} + 2} = \frac{1}{2}; \quad \frac{\cancel{15}}{\cancel{54}} = \frac{1}{4}.$$

Wrong!   Wrong!   Wrong!

The correct answers are

$$\frac{2+3}{2} = \frac{5}{2}; \quad \frac{4+1}{4+2} = \frac{5}{6}; \quad \frac{15}{54} = \frac{\cancel{3} \cdot 5}{\cancel{3} \cdot 18} = \frac{5}{18}.$$

In each of the incorrect cancellations, the numbers canceled did not form a factor equal to 1. Factors are parts of products. For example, in $2 \cdot 3$, the numbers 2 and 3 are factors, but in $2 + 3$, the numbers 2 and 3 are terms, not factors.

- If you cannot factor, do not cancel! If in doubt, do not cancel!
- Only factors can be canceled and factors are never separated by $+$ or $-$ signs.

Exercise Set 3.5

a Find an equivalent expression for the given number, with the denominator indicated. Use multiplication by 1.

1. $\dfrac{1}{2} = \dfrac{?}{10}$
2. $\dfrac{1}{8} = \dfrac{?}{12}$
3. $\dfrac{3}{4} = \dfrac{?}{-48}$
4. $\dfrac{2}{9} = \dfrac{?}{-18}$

5. $\dfrac{9}{10} = \dfrac{?}{30}$
6. $\dfrac{3}{8} = \dfrac{?}{48}$
7. $\dfrac{11}{5} = \dfrac{?}{5t}$
8. $\dfrac{5}{3} = \dfrac{?}{3a}$

9. $\dfrac{5}{12} = \dfrac{?}{48}$
10. $\dfrac{7}{8} = \dfrac{?}{56}$
11. $-\dfrac{17}{18} = -\dfrac{?}{54}$
12. $-\dfrac{11}{16} = -\dfrac{?}{256}$

13. $\dfrac{2}{-5} = \dfrac{?}{-25}$
14. $\dfrac{7}{-8} = \dfrac{?}{-32}$
15. $\dfrac{-7}{22} = \dfrac{?}{132}$
16. $\dfrac{-10}{21} = \dfrac{?}{126}$

17. $\dfrac{5}{8} = \dfrac{?}{8x}$
18. $\dfrac{2}{7} = \dfrac{?}{7a}$
19. $\dfrac{7}{11} = \dfrac{?}{11m}$
20. $\dfrac{4}{3} = \dfrac{?}{3n}$

21. $\dfrac{4}{9} = \dfrac{?}{9ab}$
22. $\dfrac{8}{11} = \dfrac{?}{11xy}$
23. $\dfrac{4}{9} = \dfrac{?}{27b}$
24. $\dfrac{8}{11} = \dfrac{?}{55y}$

b Simplify.

25. $\dfrac{2}{4}$
26. $\dfrac{3}{6}$
27. $-\dfrac{6}{9}$
28. $\dfrac{-9}{12}$
29. $\dfrac{10}{25}$

30. $\dfrac{8}{10}$
31. $\dfrac{24}{-8}$
32. $\dfrac{36}{-4}$
33. $\dfrac{27}{36}$
34. $\dfrac{30}{40}$

35. $-\dfrac{24}{14}$
36. $-\dfrac{16}{10}$
37. $\dfrac{16n}{48n}$
38. $\dfrac{150a}{25a}$
39. $\dfrac{-17}{51}$

40. $\dfrac{-425}{525}$
41. $\dfrac{420}{480}$
42. $\dfrac{180}{240}$
43. $\dfrac{136}{153}$
44. $\dfrac{117}{91}$

45. $\dfrac{3ab}{8ab}$ 46. $\dfrac{6xy}{7xy}$ 47. $\dfrac{9xy}{6x}$ 48. $\dfrac{10ab}{15a}$

Skill Maintenance

49. A soccer field is 90 yd long and 40 yd wide. What is its area? [1.8a]

50. Yardbird Landscaping buys 13 maple saplings and 17 oak saplings for a project. A maple costs $23 and an oak costs $37. How much is spent altogether for the saplings? [1.8a]

Subtract.

51. $34 - 39$ [2.3a] 52. $50 - 68$ [2.3a] 53. $803 - 617$ [1.3d] 54. $8344 - 5607$ [1.3d]

Solve.

55. $30 \cdot x = -150$ [1.7b], [2.5a], [2.8b]

56. $5280 = 1760 + t$ [1.7b], [2.8a]

Synthesis

57. ◆ Explain in your own words when it *is* possible to "cancel" and when it *is not* possible to "cancel."

58. ◆ Can fractional notation be simplified if the numerator and the denominator are two different prime numbers? Why or why not?

59. ◆ Why is multiplication of fractions (Section 3.4) discussed before simplification of fractions (Section 3.5)?

Simplify. Use the list of prime numbers on p. 154.

60. $\dfrac{221}{247}$

61. $\dfrac{209ab}{247ac}$

62. $-\dfrac{253x}{143y}$

63. $-\dfrac{187a}{289b}$

64. 🖩 $\dfrac{2603}{2831}$

65. 🖩 $\dfrac{3473}{3197}$

66. Sociologists have found that 4 of 10 people are shy. Write fractional notation for the part of the population that is shy; the part that is not shy. Simplify.

67. Sociologists estimate that 3 of 20 people are left-handed. In a crowd of 460 people, how many would you expect to be left-handed?

68. The circle graph below shows how long shoppers stay when visiting a mall. What portion of shoppers stay for 0–2 hr?

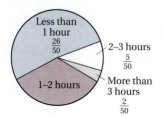

69. A new Chevrolet Prizm costs $14,400. Pam will pay $\tfrac{1}{2}$ of the cost, Sam will pay $\tfrac{1}{4}$ of the cost, Jan will pay $\tfrac{1}{6}$ the cost, and Nan will pay the rest.
   a) How much will Nan pay?
   b) What fractional part will Nan pay?

Use fraction bars to represent equivalent fractions.

3.6 Multiplying, Simplifying, and More with Area

a Simplifying When Multiplying

Objectives

a Multiply and simplify using fractional notation.

b Solve applied problems involving multiplication.

For Extra Help

TAPE 6   MAC WIN   CD-ROM

We usually want a simplified answer when we multiply. To make such simplifying easier, it is generally best not to calculate the products in the numerator and the denominator until we have first factored and simplified. Consider

$$\frac{5}{6} \cdot \frac{14}{15}.$$

We proceed as follows:

$$\frac{5}{6} \cdot \frac{14}{15} = \frac{5 \cdot 14}{6 \cdot 15}$$ We do not yet carry out the multiplication. Note that 2 is a factor of 6 and 14. Also, note that 5 is a factor of 5 and 15.

$$= \frac{5 \cdot 2 \cdot 7}{2 \cdot 3 \cdot 5 \cdot 3}$$ Factoring and identifying common factors

$$= \frac{5 \cdot 2}{5 \cdot 2} \cdot \frac{7}{3 \cdot 3}$$ Factoring the fraction

$$= 1 \cdot \frac{7}{3 \cdot 3}$$

$$= \frac{7}{3 \cdot 3}$$ Removing a factor equal to 1: $\frac{5 \cdot 2}{5 \cdot 2} = 1$

$$= \frac{7}{9}.$$

To multiply and simplify:

a) Write the products in the numerator and the denominator, but do not calculate the products.

b) Identify any common factors of the numerator and the denominator.

c) Factor the fraction to remove any factors that equal 1.

d) Calculate the remaining products.

Examples   Multiply and simplify.

1. $\dfrac{2}{3} \cdot \dfrac{5}{4} = \dfrac{2 \cdot 5}{3 \cdot 4}$ Note that 2 is a common factor of 2 and 4.

$$= \frac{2 \cdot 5}{3 \cdot 2 \cdot 2}$$ Try to go directly to this step.

$$= \frac{2}{2} \cdot \frac{5}{3 \cdot 2}$$

$$= 1 \cdot \frac{5}{3 \cdot 2} = \frac{5}{6}$$ Removing a factor equal to 1: $\frac{2}{2} = 1$

2. $\dfrac{6}{7} \cdot \dfrac{-5}{3} = \dfrac{3 \cdot 2 \cdot (-5)}{7 \cdot 3}$ Note that 3 is a common factor of 6 and 3.

$$= \frac{3}{3} \cdot \frac{2(-5)}{7} = \frac{-10}{7}, \text{ or } -\frac{10}{7}$$ Removing a factor equal to 1: $\frac{3}{3} = 1$

Multiply and simplify.

1. $\dfrac{2}{3} \cdot \dfrac{7}{8}$

2. $\dfrac{4}{5} \cdot \dfrac{-5}{12}$

3. $16 \cdot \dfrac{3}{8}$

4. $\dfrac{5}{2x} \cdot 6$

3. $\dfrac{10}{21} \cdot \dfrac{14a}{15} = \dfrac{5 \cdot 2 \cdot 7 \cdot 2a}{7 \cdot 3 \cdot 5 \cdot 3}$ Note that 5 is a common factor of 10 and 15.
Note that 7 is a common factor of 21 and 14a.

$= \dfrac{5 \cdot 7}{5 \cdot 7} \cdot \dfrac{2 \cdot 2a}{3 \cdot 3}$

$= \dfrac{4a}{9}$ Removing a factor equal to 1: $\dfrac{5 \cdot 7}{5 \cdot 7} = 1$

4. $40 \cdot \dfrac{7}{8} = \dfrac{8 \cdot 5 \cdot 7}{8 \cdot 1}$ Note that 8 is a common factor of 40 and 8.

$= \dfrac{8}{8} \cdot \dfrac{5 \cdot 7}{1} = 35$ Removing a factor equal to 1: $\dfrac{8}{8} = 1$

CAUTION! Canceling can be used as follows for these examples.

1. $\dfrac{2}{3} \cdot \dfrac{5}{4} = \dfrac{\cancel{2} \cdot 5}{3 \cdot \cancel{2} \cdot 2} = \dfrac{5}{6}$ Removing a factor equal to 1: $\dfrac{2}{2} = 1$

2. $\dfrac{6}{7} \cdot \dfrac{-5}{3} = \dfrac{\cancel{3} \cdot 2(-5)}{7 \cdot \cancel{3}} = \dfrac{-10}{7}$ Removing a factor equal to 1: $\dfrac{3}{3} = 1$

3. $\dfrac{10}{21} \cdot \dfrac{14a}{15} = \dfrac{\cancel{5} \cdot 2 \cdot \cancel{7} \cdot 2a}{\cancel{7} \cdot 3 \cdot \cancel{5} \cdot 3} = \dfrac{4a}{9}$ Removing a factor equal to 1: $\dfrac{5 \cdot 7}{5 \cdot 7} = 1$

4. $40 \cdot \dfrac{7}{8} = \dfrac{\cancel{8} \cdot 5 \cdot 7}{\cancel{8} \cdot 1} = 35$ Removing a factor equal to 1: $\dfrac{8}{8} = 1$

Remember, if you can't factor, you can't cancel!

Do Exercises 1–4.

b Solving Problems

Example 5 For their annual pancake breakfast, the Colchester Boy Scouts need $\frac{2}{3}$ cup of Bisquick® per person. If at most 135 guests are expected, how much Bisquick do the scouts need?

1. **Familiarize.** We first make a drawing or at least visualize the situation. Repeated addition will work here.

$\frac{2}{3}$ cup per guest — 135 guests

We let n = the number of cups of Bisquick needed.

2. **Translate.** The problem translates to the following equation:

$n = 135 \cdot \dfrac{2}{3}.$

Answers on page A-8

3. **Solve.** To solve the equation, we carry out the multiplication:

$$n = 135 \cdot \frac{2}{3} = \frac{135 \cdot 2}{3}$$ Multiplying

$$= \frac{3 \cdot 45 \cdot 2}{3 \cdot 1}$$ Note that 135 is divisible by 3.

$$= \frac{3}{3} \cdot \frac{45 \cdot 2}{1}$$ Removing the factor $\frac{3}{3}$

$$= 90.$$ Simplifying

4. **Check.** We could repeat the calculation but this check is left to the student. We can also think about the reasonableness of the answer. Since each guest requires less than 1 cup, it makes sense that 135 guests requires fewer than 135 cups. This provides a partial check of the answer.

5. **State.** The scouts will need 90 cups of Bisquick.

Do Exercise 5.

5. Yardbird Landscaping uses $\frac{2}{5}$ lb of peat moss for a rosebush. How much will be needed for 25 rosebushes?

Area

Multiplication of fractions can arise in geometry problems involving the area of a triangle. Consider a triangle with a base of length b and a height of h, as shown.

A rectangle can be formed by splitting and inverting a copy of this triangle:

The rectangle's area, $b \cdot h$, is exactly twice the area of the triangle. We have the following result.

> The **area A of a triangle** is half the length of the base b times the height h:
>
> $$A = \frac{1}{2} \cdot b \cdot h.$$

Example 6 Find the area of this triangle.

$$A = \frac{1}{2} \cdot b \cdot h$$

$$= \frac{1}{2} \cdot 9 \text{ yd} \cdot 6 \text{ yd}$$

$$= \frac{9 \cdot 6}{2} \text{ yd}^2$$

$$= 27 \text{ yd}^2$$

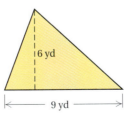

Answers on page A-8

Find the area.

6.
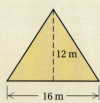

Example 7 Find the area of this triangle.

$A = \frac{1}{2} \cdot b \cdot h$

$= \frac{1}{2} \cdot \frac{10}{3} \text{ cm} \cdot 4 \text{ cm}$

$= \frac{1 \cdot 10 \cdot 4}{2 \cdot 3} \text{ cm}^2$

$= \frac{1 \cdot 2 \cdot 5 \cdot 4}{2 \cdot 3} \text{ cm}^2$ Removing a factor equal to 1: $\frac{2}{2} = 1$

$= \frac{20}{3} \text{ cm}^2$

Do Exercises 6 and 7.

Example 8 Find the area of this kite.

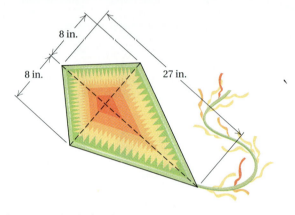

1. **Familiarize.** We look for figures with areas we can calculate using area formulas that we already know. We let $K =$ the kite's area.
2. **Translate.** The kite consists of two triangles, each with a base of 27 in. and a height of 8 in. We can apply the formula $A = \frac{1}{2} \cdot b \cdot h$ for the area of a triangle and then multiply by 2.

   Kite's area is twice Area of long triangle
   $K \quad = \quad 2 \quad \cdot \quad \frac{1}{2}(27 \text{ in.}) \cdot (8 \text{ in.})$

3. **Solve.** We have

   $K = 2 \cdot \frac{1}{2} \cdot (27 \text{ in.}) \cdot (8 \text{ in.})$

   $= 1 \cdot 27 \text{ in} \cdot 8 \text{ in.} = 216 \text{ in}^2.$

4. **Check.** We can check by repeating the calculations.
5. **State.** The area of the kite is 216 in².

Do Exercise 8.

7.
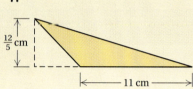

8. Find the area.
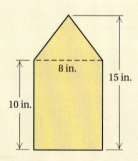

Answers on page A-8

Exercise Set 3.6

a Multiply. Don't forget to simplify.

1. $\dfrac{3}{8} \cdot \dfrac{5}{3}$
2. $\dfrac{4}{5} \cdot \dfrac{1}{4}$
3. $\dfrac{7}{8} \cdot \dfrac{-1}{7}$
4. $\dfrac{5}{6} \cdot \dfrac{-1}{5}$

5. $\dfrac{1}{8} \cdot \dfrac{6}{5}$
6. $\dfrac{2}{5} \cdot \dfrac{1}{12}$
7. $\dfrac{1}{6} \cdot \dfrac{4}{3}$
8. $\dfrac{3}{6} \cdot \dfrac{1}{6}$

9. $\dfrac{12}{-5} \cdot \dfrac{9}{8}$
10. $\dfrac{16}{-15} \cdot \dfrac{5}{4}$
11. $\dfrac{5x}{9} \cdot \dfrac{7}{5}$
12. $\dfrac{25}{4a} \cdot \dfrac{4}{3}$

13. $\dfrac{1}{4} \cdot 8$
14. $\dfrac{1}{6} \cdot 12$
15. $15 \cdot \dfrac{1}{3}$
16. $14 \cdot \dfrac{1}{2}$

17. $-12 \cdot \dfrac{3}{4}$
18. $-18 \cdot \dfrac{5}{6}$
19. $\dfrac{3}{8} \cdot 8a$
20. $\dfrac{2}{9} \cdot 9x$

21. $\left(-\dfrac{3}{7}\right)\left(-\dfrac{7}{3}\right)$
22. $\left(-\dfrac{2}{9}\right)\left(-\dfrac{9}{2}\right)$
23. $\dfrac{a}{b} \cdot \dfrac{b}{a}$
24. $\dfrac{n}{m} \cdot \dfrac{m}{n}$

25. $\dfrac{1}{27} \cdot 360a$
26. $\dfrac{1}{28} \cdot 105n$
27. $176\left(\dfrac{1}{-6}\right)$
28. $135\left(\dfrac{1}{-10}\right)$

29. $7x \cdot \dfrac{1}{7x}$

30. $5a \cdot \dfrac{1}{5a}$

31. $\dfrac{2x}{9} \cdot \dfrac{27}{2x}$

32. $\dfrac{10a}{3} \cdot \dfrac{3}{5a}$

33. $\dfrac{7}{10} \cdot \dfrac{34}{150}$

34. $\dfrac{8}{10} \cdot \dfrac{45}{100}$

35. $\dfrac{36}{85} \cdot \dfrac{25}{-99}$

36. $\dfrac{-70}{45} \cdot \dfrac{50}{49}$

37. $\dfrac{-98}{99} \cdot \dfrac{27a}{175a}$

38. $\dfrac{70}{-49} \cdot \dfrac{63}{300x}$

39. $\dfrac{110}{33} \cdot \dfrac{-24}{25}$

40. $\dfrac{-19}{130} \cdot \dfrac{65}{38x}$

41. $\left(-\dfrac{11}{24}\right)\dfrac{3}{5}$

42. $\left(-\dfrac{15}{22}\right)\dfrac{4}{7}$

43. $\dfrac{10a}{21} \cdot \dfrac{3}{4a}$

44. $\dfrac{17}{18x} \cdot \dfrac{3x}{5}$

b Solve.

45. Anna receives $56 for working a full day doing inventory at a hardware store. How much will she receive for working $\frac{3}{4}$ of the day?

46. After Jack completes 60 hr of teacher training in college, he can earn $88 for working a full day as a substitute teacher. How much will he receive for working $\frac{3}{4}$ of a day?

47. *Food Preparation.* How much salmon is needed to serve 30 people if each person gets $\frac{2}{5}$ lb?

48. *Mailing Lists.* Business people have determined that $\frac{1}{4}$ of the addresses on a mailing list will change in one year. A business has a mailing list of 2500 people. After one year, how many addresses on that list will be incorrect?

49. *Sociology.* Sociologists have determined that $\frac{2}{5}$ of the people in the world are shy. A sales manager is interviewing 650 people for a new sales position. How many of these people might be shy?

50. *Food Preparation.* Francesca's Sandwich Shop sells subs by the foot. If one serving is $\frac{2}{3}$ ft long, how many feet are needed to serve 30 people?

51. A recipe for pie crust calls for $\frac{2}{3}$ cup of flour. A chef is making $\frac{1}{2}$ of the recipe. How much flour should the chef use?

52. Of the students in the entering class, $\frac{2}{5}$ have cameras; $\frac{1}{4}$ of these students also join the college photography club. What fraction of the students in the entering class join the photography club?

53. A house worth $124,000 was assessed at $\frac{3}{4}$ of its value. What is the assessed value of the house?

54. Roxanne's tuition was $2800. A loan was obtained for $\frac{3}{4}$ of the tuition. How much was the loan?

55. *Map Scaling.* On a map, 1 in. represents 240 mi. How much does $\frac{2}{3}$ in. represent?

56. *Map Scaling.* On a map, 1 in. represents 120 mi. How much does $\frac{3}{4}$ in. represent?

Exercise Set 3.6

57. *Household Budgets.* Vic has an annual income of $27,000. Of this, $\frac{1}{4}$ is spent for food, $\frac{1}{5}$ for housing, $\frac{1}{10}$ for clothing, $\frac{1}{9}$ for savings, $\frac{1}{4}$ for taxes, and the rest for other expenses. How much is spent for each?

58. *Household Budgets.* Heidi has an annual income of $25,200. Of this, $\frac{1}{4}$ is spent for food, $\frac{1}{5}$ for housing, $\frac{1}{10}$ for clothing, $\frac{1}{9}$ for savings, $\frac{1}{4}$ for taxes, and the rest for other expenses. How much is spent for each?

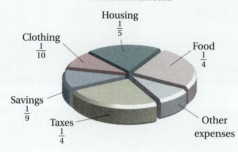

Find the area.

59.

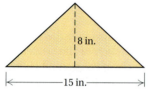

60.

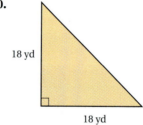

61.

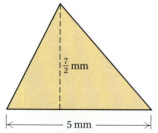

62.

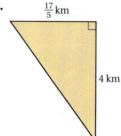

Chapter 3 Fractions: Multiplication & Division

63.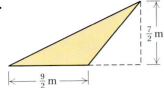

64.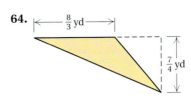

65.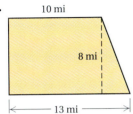

66.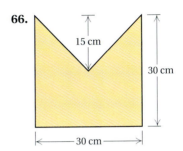

67. *Construction.* Find the total area of the sides and ends of the building.

68. *Sailing* A rectangular piece of sailcloth is 36 ft by 24 ft. A triangular sail with a height of 28 ft and a base of 16 ft is cut from the sailcloth. How much area is left over?

Skill Maintenance

Solve.

69. $48 \cdot t = 1680$ [1.7b], [2.8b]

70. $456 + x = 9002$ [1.7b], [2.8a]

71. $747 = x + 270$ [1.7b], [2.8a]

72. $280 = 4 \cdot t$ [1.7b], [2.8b]

Add.

73. $(-39) + (-72)$ [2.2a]

74. $-59 + 37$ [2.2a]

Exercise Set 3.6

Synthesis

75. ◆ When multiplying using fractional notation, we form products in the numerator and the denominator, but do not automatically calculate the products. Why?

76. ◆ If a fraction's numerator and denominator have no factors (other than 1) in common, can the fraction be simplified? Why or why not?

77. ◆ Is the product of two fractions always a fraction? Why or why not?

Simplify. Use the list of prime numbers on p. 154.

78. $\dfrac{201}{535} \cdot \dfrac{4601}{6499}$

79. $\dfrac{5767}{3763} \cdot \dfrac{159}{395}$

80. $\dfrac{667}{899} \cdot \dfrac{558}{621}$

81. *Painting.* A painter needs to determine the surface area of an octagonal steeple. Find the total area, if the dimensions are as shown below.

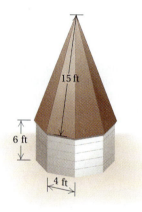

82. *Manufacturing.* A specially shaped candy box is triangular at each end, as shown below. Find the surface area of the box.

83. Of the students entering a college, $\frac{7}{8}$ have completed high school and $\frac{2}{3}$ are older than 20. If $\frac{1}{7}$ of all students are left-handed, what fraction of students entering the college are left-handed high school graduates over the age of 20?

84. Refer to the information in Exercise 83. If 480 students are entering the college, how many of them are left-handed high school graduates 20 years old or younger?

85. Refer to Exercise 83. What fraction of students entering the college did not graduate high school, are 20 years old or younger, and are left-handed?

3.7 Reciprocals and Division

a Reciprocals

Look at these products:

$$8 \cdot \frac{1}{8} = \frac{8}{8} = 1; \qquad \frac{-2}{3} \cdot \frac{3}{-2} = \frac{-6}{-6} = 1.$$

> If the product of two numbers is 1, we say that they are **reciprocals** of each other. To find a number's reciprocal, interchange the numerator and the denominator.
>
> The numbers $\frac{3}{4}$ and $\frac{4}{3}$ are reciprocals of each other.

Objectives

a Find the reciprocal of a number.

b Divide and simplify using fractional notation.

c Solve problems involving division.

For Extra Help

TAPE 6   MAC WIN   CD-ROM

Examples Find the reciprocal.

1. The reciprocal of $\frac{4}{5}$ is $\frac{5}{4}$.   Note that $\frac{4}{5} \cdot \frac{5}{4} = \frac{20}{20} = 1.$

2. The reciprocal of $\frac{a}{b}$ is $\frac{b}{a}$.   Note that $\frac{a}{b} \cdot \frac{b}{a} = \frac{ab}{ba} = 1.$

3. The reciprocal of 8 is $\frac{1}{8}$.   Think of 8 as $\frac{8}{1}$: $\frac{8}{1} \cdot \frac{1}{8} = \frac{8}{8} = 1.$

4. The reciprocal of $\frac{1}{3}$ is 3.   Note that $\frac{1}{3} \cdot 3 = \frac{3}{3} = 1.$

5. The reciprocal of $-\frac{5}{9}$ is $-\frac{9}{5}$.   Negative numbers have negative reprocals: $\left(-\frac{5}{9}\right)\left(-\frac{9}{5}\right) = \frac{45}{45} = 1.$

Find the reciprocal.

1. $\frac{2}{5}$

2. $\frac{-6}{x}$

3. 9

4. $\frac{1}{5}$

5. $-\frac{3}{10}$

Do Exercises 1–5.

Does 0 have a reciprocal? If it did, it would have to be a number x such that

$$0 \cdot x = 1.$$

But 0 times any number is 0. Thus, 0 has no reciprocal.

b Division

Recall that $a \div b$ is the number that when multiplied by b gives a. Consider the division $\frac{3}{4} \div \frac{1}{8}$. This asks how many $\frac{1}{8}$'s are in $\frac{3}{4}$. We can answer this by looking at the figure below.

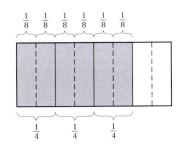

Answers on page A-8

3.7 Reciprocals and Division

187

We see that there are six $\frac{1}{8}$'s in $\frac{3}{4}$. Thus,

$$\frac{3}{4} \div \frac{1}{8} = 6.$$

We can check this by multiplying:

$$6 \cdot \frac{1}{8} = \frac{6}{8} = \frac{3}{4}.$$

Here is a faster way to divide. An explanation of why it works appears on the next page.

To divide a fraction, multiply by its reciprocal:

Multiply by the reciprocal of the divisor.

$$\frac{a}{b} \div \frac{c}{d} = \frac{a}{b} \cdot \frac{d}{c}.$$

Recall that when two numbers with unlike signs are multiplied or divided, the result is negative. When both numbers have the same sign, the result is positive.

Examples Divide and simplify.

6. $\dfrac{5}{6} \div \dfrac{2}{3} = \dfrac{5}{6} \cdot \dfrac{3}{2}$ Multiplying by the reciprocal of the divisor

$\phantom{\dfrac{5}{6} \div \dfrac{2}{3}} = \dfrac{5 \cdot 3}{3 \cdot 2 \cdot 2}$ Factoring and identifying a common factor

$\phantom{\dfrac{5}{6} \div \dfrac{2}{3}} = \dfrac{3}{3} \cdot \dfrac{5}{2 \cdot 2}$ Removing a factor equal to 1: $\dfrac{3}{3} = 1$

$\phantom{\dfrac{5}{6} \div \dfrac{2}{3}} = \dfrac{5}{4}$

7. $\dfrac{-3}{5} \div \dfrac{1}{2} = \dfrac{-3}{5} \cdot 2$ The reciprocal of $\dfrac{1}{2}$ is 2.

$\phantom{\dfrac{-3}{5} \div \dfrac{1}{2}} = \dfrac{-3 \cdot 2}{5} = \dfrac{-6}{5}$

8. $\dfrac{2a}{5} \div 7 = \dfrac{2a}{5} \cdot \dfrac{1}{7}$ The reciprocal of 7 is $\dfrac{1}{7}$.

$\phantom{\dfrac{2a}{5} \div 7} = \dfrac{2a \cdot 1}{5 \cdot 7} = \dfrac{2a}{35}$

9. $\dfrac{7}{10} \div \left(-\dfrac{14}{15}\right) = \dfrac{7}{10} \cdot \left(-\dfrac{15}{14}\right)$ Multiplying by the reciprocal of the divisor

$\phantom{\dfrac{7}{10} \div \left(-\dfrac{14}{15}\right)} = \dfrac{7 \cdot 5(-3)}{2 \cdot 5 \cdot 7 \cdot 2}$ Factoring and identifying common factors

$\phantom{\dfrac{7}{10} \div \left(-\dfrac{14}{15}\right)} = \dfrac{7 \cdot 5}{7 \cdot 5} \cdot \dfrac{-3}{4}$ Removing a factor equal to 1: $\dfrac{7 \cdot 5}{7 \cdot 5} = 1$

$\phantom{\dfrac{7}{10} \div \left(-\dfrac{14}{15}\right)} = -\dfrac{3}{4}$

CAUTION! Canceling can be used as follows for Examples 6 and 9.

6. $\dfrac{5}{6} \div \dfrac{2}{3} = \dfrac{5}{6} \cdot \dfrac{3}{2} = \dfrac{5 \cdot 3}{6 \cdot 2} = \dfrac{5 \cdot \cancel{3}}{\cancel{3} \cdot 2 \cdot 2} = \dfrac{5}{2 \cdot 2} = \dfrac{5}{4}$ Removing a factor equal to 1: $\dfrac{3}{3} = 1$

9. $\dfrac{7}{10} \div \left(-\dfrac{14}{15}\right) = \dfrac{7}{10} \cdot \left(-\dfrac{15}{14}\right) = \dfrac{\cancel{7} \cdot \cancel{5}(-3)}{2 \cdot \cancel{5} \cdot \cancel{7} \cdot 2} = \dfrac{-3}{4},\ \text{or}\ -\dfrac{3}{4}$ Removing a factor equal to 1: $\dfrac{7 \cdot 5}{7 \cdot 5} = 1$

Remember, if you can't factor, you can't cancel!

Do Exercises 6–10.

Why do we multiply by a reciprocal when dividing? To see this, let's consider $\dfrac{2}{3} \div \dfrac{7}{5}$. We will multiply by 1. The name for 1 that we will use is $(5/7)/(5/7)$; it comes from the reciprocal of $\dfrac{7}{5}$.

$\dfrac{2}{3} \div \dfrac{7}{5} = \dfrac{\frac{2}{3}}{\frac{7}{5}}$ Writing fractional notation for the division

$= \dfrac{\frac{2}{3}}{\frac{7}{5}} \cdot 1$ Multiplying by 1

$= \dfrac{\frac{2}{3}}{\frac{7}{5}} \cdot \dfrac{\frac{5}{7}}{\frac{5}{7}}$ Multiplying by 1; $\dfrac{5}{7}$ is the reciprocal of $\dfrac{7}{5}$ and $\dfrac{\frac{5}{7}}{\frac{5}{7}} = 1$

$= \dfrac{\frac{2}{3} \cdot \frac{5}{7}}{\frac{7}{5} \cdot \frac{5}{7}}$ Multiplying the numerators and the denominators

$= \dfrac{\frac{2}{3} \cdot \frac{5}{7}}{1} = \dfrac{2}{3} \cdot \dfrac{5}{7} = \dfrac{10}{21}$ After we multiplied, we got 1 for the denominator. The numerator (in color) shows the multiplication by the reciprocal.

Do Exercise 11.

Divide and simplify.

6. $\dfrac{6}{7} \div \dfrac{3}{4}$

7. $\left(-\dfrac{2}{3}\right) \div \dfrac{1}{4}$

8. $\dfrac{4}{5} \div 8$

9. $60 \div \dfrac{3a}{5}$

10. $\dfrac{3}{5} \div \dfrac{-3}{5}$

11. Divide by multiplying by 1:

$\dfrac{\frac{4}{5}}{\frac{6}{7}}.$

Answers on page A-8

3.7 Reciprocals and Division

12. Each loop in a spring uses $\frac{3}{8}$ in. of wire. How many loops can be made from 120 in. of wire?

13. For a party, Jana made an 8-foot submarine sandwich. If one serving is $\frac{2}{3}$ ft, how many servings does Jana's sub contain?

c Solving Problems

Example 10 *Chemistry.* In a chemistry experiment, Lita needs to fill as many test tubes as possible with $\frac{3}{5}$ g of salt each. If she begins with 51 g of salt, how many test tubes can she fill?

1. **Familiarize.** We first make a drawing or at least visualize the situation. Repeated subtraction, or division, will work here.

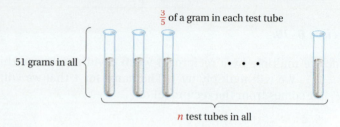

We let $n =$ the number of test tubes that can be filled.

2. **Translate.** The problem can be translated to the following equation:
$$n = 51 \div \frac{3}{5}.$$

3. **Solve.** To solve the equation, we carry out the division:

$$n = 51 \div \frac{3}{5}$$
$$= 51 \cdot \frac{5}{3} \qquad \text{Multiplying by the reciprocal}$$
$$= \frac{51 \cdot 5}{1 \cdot 3}$$
$$= \frac{3 \cdot 17 \cdot 5}{1 \cdot 3}$$
$$= \frac{3}{3} \cdot \frac{17 \cdot 5}{1} \qquad \text{Identifying a factor equal to 1}$$
$$= 85. \qquad \text{Simplifying}$$

4. **Check.** If each of 85 test tubes contains $\frac{3}{5}$ g of salt, a total of
$$85 \cdot \frac{3}{5} = \frac{85 \cdot 3}{5} = \frac{5 \cdot 17 \cdot 3}{5} = 17 \cdot 3,$$

or 51 g of salt is used. Since the problem states that Lita begins with 51 g, our answer checks.

5. **State.** Lita can fill 85 test tubes with salt.

Do Exercises 12 and 13.

Answers on page A-8

Exercise Set 3.7

a Find the reciprocal.

1. $\dfrac{7}{3}$

2. $\dfrac{6}{5}$

3. 4

4. 7

5. $\dfrac{1}{6}$

6. $\dfrac{1}{4}$

7. $-\dfrac{10}{3}$

8. $-\dfrac{12}{5}$

9. $\dfrac{2}{21}$

10. $\dfrac{3}{28}$

11. $\dfrac{-3n}{m}$

12. $\dfrac{8t}{-7r}$

13. $\dfrac{7}{-15}$

14. $\dfrac{-6}{25}$

15. $7m$

16. $5n$

b Divide. Don't forget to simplify when possible.

17. $\dfrac{3}{5} \div \dfrac{3}{4}$

18. $\dfrac{2}{3} \div \dfrac{3}{4}$

19. $\dfrac{7}{6} \div \dfrac{5}{-3}$

20. $\dfrac{5}{3} \div \dfrac{4}{-9}$

21. $\dfrac{4}{3} \div \dfrac{1}{3}$

22. $\dfrac{10}{9} \div \dfrac{1}{2}$

23. $\left(-\dfrac{1}{3}\right) \div \dfrac{1}{6}$

24. $\left(-\dfrac{1}{4}\right) \div \dfrac{1}{5}$

25. $\dfrac{3}{8} \div 24$

26. $\dfrac{5}{6} \div 45$

27. $\dfrac{12}{7} \div (4x)$

28. $\dfrac{18}{5} \div (2y)$

29. $(-12) \div \dfrac{3}{2}$

30. $(-24) \div \dfrac{3}{8}$

31. $28 \div \dfrac{4}{5a}$

32. $40 \div \dfrac{2}{3m}$

33. $\left(-\dfrac{5}{8}\right) \div \left(-\dfrac{5}{8}\right)$

34. $\left(-\dfrac{2}{5}\right) \div \left(-\dfrac{2}{5}\right)$

35. $\dfrac{-8}{15} \div \dfrac{4}{5}$

36. $\dfrac{6}{-13} \div \dfrac{3}{26}$

37. $\dfrac{77}{64} \div \dfrac{49}{18}$

38. $\dfrac{81}{42} \div \dfrac{33}{56}$

39. $120a \div \dfrac{45}{14}$

40. $360n \div \dfrac{27n}{8}$

Chapter 3 Fractions:
Multiplication & Division

c Solve.

41. Benny uses $\frac{5}{4}$ g of toothpaste each time he brushes his teeth. How many times will Benny be able to brush his teeth with a 110-g tube of toothpaste?

42. Joy uses $\frac{1}{2}$ yd of dental floss each day. How long will a 45-yd container of dental floss last for Joy?

43. *Town Planning.* The Milton road crew repaves $\frac{1}{12}$ mi of road each day. How long will it take the crew to repave a $\frac{3}{4}$-mi stretch of road?

44. *Expenditures.* An airguard unit has $9 million to spend on new helicopters. Each helicopter costs $\frac{3}{4}$ million. How many helicopters can be bought?

45. *Packaging.* Tina's Market prepackages Swiss cheese in $\frac{3}{4}$-lb packages. How many packages can be made from a 15-lb slab of cheese?

46. *Meal Planning.* Ian purchased 6 lb of cold cuts for a luncheon. If Ian is to allow $\frac{3}{8}$ lb per person, how many people can attend the luncheon?

47. *Gardening.* The Bingham community garden is to be split into 16 equally sized plots. If the garden occupies $\frac{3}{4}$ acre of land, how large will each plot be?

48. *Art Supplies.* The Ferristown School District purchased $\frac{3}{4}$ T (ton) of clay. The clay is to be shared equally among the district's 6 art departments. How much will each art department receive?

49. A piece of coaxial cable $\frac{3}{5}$ m long is to be cut into 6 pieces of the same length. What will be the length of each piece?

50. A piece of speaker wire $\frac{4}{5}$ m long is to be cut into eight pieces of the same length. What will be the length of each piece?

51. *Sewing.* A pair of basketball shorts requires $\frac{3}{4}$ yd of nylon. How many pairs of shorts can be made from 24 yd of the fabric?

52. *Sewing.* A child's shirt requires $\frac{5}{6}$ yd of cotton fabric. How many shirts can be made from 25 yd of the fabric?

53. *Knitting.* Brianna is knitting a sweater in which each stitch is $\frac{3}{8}$ in. long. How many stitches will Brianna need for a row that is 12 in. long?

54. *Knitting.* Gene is knitting a pair of socks in which each stitch is $\frac{5}{32}$ in. long. How many stitches will Gene need for a row that is 10 in. long?

Skill Maintenance

Multiply.

55. $(-17)(-30)$ [2.4a]

56. $(73)(-4)$ [2.4a]

Evaluate each of the following.

57. x^3, for $x = 3$ and $x = -3$ [2.6a]

58. $5x^2$, for $x = 4$ and $x = -4$ [2.6a]

59. $3x^2$, for $x = 7$ and $x = -7$ [2.6a]

60. x^3, for $x = 7$ and $x = -7$ [2.6a]

Synthesis

61. ◆ A student incorrectly insists that $\frac{2}{5} \div \frac{3}{4}$ is $\frac{15}{8}$. What mistake is the student probably making?

62. ◆ Write a problem for a classmate to solve. Devise the problem so that the solution requires the classmate to divide by a fraction. Arrange for the solution to be "The contents of the barrel will fill 40 bags with $\frac{3}{4}$ lb in each bag."

63. ◆ Without performing the division, explain why $5 \div \frac{1}{7}$ is a bigger number than $5 \div \frac{2}{3}$.

Simplify.

64. $\left(\frac{9}{10} \div \frac{2}{5} \div \frac{3}{8}\right)^2$

65. $\dfrac{\left(-\frac{3}{7}\right)^2 \div \frac{12}{5}}{\left(\frac{-2}{9}\right)\left(\frac{9}{2}\right)}$

66. $\left(\frac{14}{15} \div \frac{49}{65} \cdot \frac{77}{260}\right)^2$

67. $\left(\frac{10}{9}\right)^2 \div \frac{35}{27} \cdot \frac{49}{44}$

Simplify. Use the list of prime numbers on p. 154.

68. 🖩 $\dfrac{711}{1957} \div \dfrac{10{,}033}{13{,}081}$

69. 🖩 $\dfrac{8633}{7387} \div \dfrac{485}{581}$

3.8 Solving Equations: The Multiplication Principle

In Sections 1.7 and 2.8, we learned to solve an equation involving multiplication by dividing on both sides. With fractional notation, we can solve the same type of equation by using multiplication.

a The Multiplication Principle

We have seen that to divide by a fraction, we multiply by the reciprocal of that fraction. This suggests that we restate the division principle in its more common form—the multiplication principle.

Objectives

a Use the multiplication principle to solve equations.

For Extra Help

TAPE 6   MAC WIN   CD-ROM

> **THE MULTIPLICATION PRINCIPLE**
> For any numbers a, b, and c, with $c \neq 0$,
> $a = b$ is equivalent to $a \cdot c = b \cdot c$.

Example 1 Solve: $\frac{3}{4}x = 15$.

We can multiply by any nonzero number on both sides to produce an equivalent equation. Since we are looking for an equation of the form $1x = \square$, we multiply by the reciprocal of $\frac{3}{4}$ on both sides.

$$\frac{3}{4}x = 15$$

$$\frac{4}{3} \cdot \frac{3}{4}x = \frac{4}{3} \cdot 15 \qquad \text{Using the multiplication principle; note that } \frac{4}{3} \text{ is the reciprocal of } \frac{3}{4}.$$

$$\left(\frac{4}{3} \cdot \frac{3}{4}\right)x = \frac{4 \cdot 15}{3} \qquad \text{Using an associative law; try to do this mentally.}$$

$$1x = 20 \qquad \text{Multiplying; note that } \frac{4 \cdot 15}{3} = \frac{4 \cdot \cancel{3} \cdot 5}{\cancel{3}}.$$

$$x = 20 \qquad \text{Remember that } 1x \text{ is } x.$$

To confirm that 20 is the solution, we perform a check.

CHECK:
$$\begin{array}{c|c} \frac{3}{4}x = 15 \\ \hline \frac{3}{4} \cdot 20 \;?\; 15 \\ \frac{3 \cdot \cancel{4} \cdot 5}{\cancel{4}} \\ 3 \cdot 5 \;\bigg|\; 15 \quad \text{TRUE} \end{array}$$

Removing a factor equal to 1: $\frac{4}{4} = 1$

The solution is 20.

Note that using the multiplication principle to multiply by $\frac{4}{3}$ on both sides is the same as using the division principle to divide by $\frac{3}{4}$ on both sides.

Do Exercises 1 and 2.

Solve.

1. $\frac{2}{3}x = 10$

2. $\frac{2}{7}a = -8$

Answers on page A-8

Solve.

3. $-\dfrac{9}{8} = 4x$

4. $-\dfrac{6}{7}a = \dfrac{9}{14}$

In an expression like $\frac{3}{4}x$, the constant factor—in this case, $\frac{3}{4}$—is called the **coefficient**. In Example 1, we multiplied on both sides by $\frac{4}{3}$, the reciprocal of the coefficient of x.

Example 2 Solve: $5a = -\dfrac{7}{3}$.

We have

$$5a = -\dfrac{7}{3}$$

$$\dfrac{1}{5} \cdot 5a = \dfrac{1}{5} \cdot \left(-\dfrac{7}{3}\right) \quad \text{Multiplying by } \tfrac{1}{5}, \text{ the reciprocal of 5, on both sides}$$

$$1a = -\dfrac{1 \cdot 7}{5 \cdot 3}$$

$$a = -\dfrac{7}{15}.$$

CHECK:

$$\begin{array}{c|c} 5a = -\dfrac{7}{3} \\ \hline 5\left(-\dfrac{7}{15}\right) \;?\; -\dfrac{7}{3} \\ -\dfrac{5 \cdot 7}{5 \cdot 3} \\ -\dfrac{7}{3} & -\dfrac{7}{3} \quad \text{TRUE} \end{array}$$

The solution is $-\dfrac{7}{15}$.

Example 3 Solve: $\dfrac{10}{3} = -\dfrac{4}{9}x$.

We have

$$\dfrac{10}{3} = -\dfrac{4}{9}x$$

$$-\dfrac{9}{4} \cdot \dfrac{10}{3} = -\dfrac{9}{4} \cdot \left(-\dfrac{4}{9}\right)x \quad \text{The reciprocal of } -\tfrac{4}{9} \text{ is } -\tfrac{9}{4}.$$

$$-\dfrac{3 \cdot 3 \cdot 2 \cdot 5}{2 \cdot 2 \cdot 3} = x$$

$$-\dfrac{15}{2} = x. \quad \text{Removing a factor equal to 1: } \dfrac{3 \cdot 2}{2 \cdot 3} = 1$$

We leave the check to the student. The solution is $-\dfrac{15}{2}$.

Do Exercises 3 and 4.

Answers on page A-8

Exercise Set 3.8

a Use the multiplication principle to solve each equation. Don't forget to check!

1. $\dfrac{8}{5}x = 18$
2. $\dfrac{4}{3}x = 20$
3. $\dfrac{7}{3}a = 21$
4. $\dfrac{4}{5}a = 24$

5. $\dfrac{3}{7}x = -18$
6. $\dfrac{3}{8}x = -21$
7. $6a = \dfrac{12}{17}$
8. $3a = \dfrac{15}{14}$

9. $\dfrac{3}{5}x = \dfrac{2}{7}$
10. $\dfrac{3}{7}x = \dfrac{1}{4}$
11. $\dfrac{3}{2}t = -\dfrac{8}{7}$
12. $\dfrac{4}{3}t = -\dfrac{5}{2}$

13. $\dfrac{4}{5} = -10a$
14. $\dfrac{6}{5} = -12a$
15. $\dfrac{9}{5}x = \dfrac{3}{10}$
16. $\dfrac{10}{3}x = \dfrac{8}{15}$

17. $-\dfrac{3}{10}x = 8$
18. $-\dfrac{2}{11}x = 5$
19. $a \cdot \dfrac{9}{7} = -\dfrac{3}{14}$
20. $a \cdot \dfrac{9}{4} = -\dfrac{3}{10}$

21. $-x = \dfrac{9}{13}$
22. $-x = \dfrac{7}{11}$
23. $-x = -\dfrac{27}{31}$
24. $-x = -\dfrac{35}{39}$

25. $7t = 5$
26. $-6t = 1$
27. $-24 = -10a$
28. $-18 = -20a$

29. $-\dfrac{15}{7} = \dfrac{3}{2}t$

30. $-\dfrac{14}{9} = \dfrac{10}{3}t$

31. $x \cdot \dfrac{5}{16} = \dfrac{15}{14}$

32. $x \cdot \dfrac{4}{15} = \dfrac{12}{25}$

33. $-\dfrac{3}{20}x = -\dfrac{21}{10}$

34. $-\dfrac{7}{25}x = -\dfrac{21}{10}$

35. $-\dfrac{25}{17} = -\dfrac{35}{34}a$

36. $-\dfrac{49}{45} = -\dfrac{28}{27}a$

Skill Maintenance

Simplify.

37. $36 \div (-3)^2 \times (7-2)$ [2.5b]

38. $(-37 - 12 + 1) \div (-2)^3$ [2.5b]

Form an equivalent expression by combining like terms.

39. $13x + 4x$ [2.7b]

40. $9a - 5a$ [2.7b]

41. $2a + 3 + 5a$ [2.7b]

42. $3x - 7 + x$ [2.7b]

Synthesis

43. ◆ Does the multiplication principle enable us to solve any equations that could not have been solved with the division principle? Why or why not?

44. ◆ Example 1 was solved by multiplying by $\frac{4}{3}$ on both sides of the equation. Could we have divided by $\frac{3}{4}$ on both sides instead? Why or why not?

45. ◆ Can the multiplication principle be used to solve equations like $7x = 63$? Why or why not?

Solve.

46. $2x - 7x = -\dfrac{10}{9}$

47. $\left(-\dfrac{4}{7}\right)^2 = \left(\dfrac{2^3 - 9}{3}\right)^3 x$

Solve using the five-step problem-solving approach.

48. After driving 180 km, $\frac{5}{8}$ of a trip is completed. How long is the total trip? How many kilometers are left to drive?

49. After driving 240 km, $\frac{3}{5}$ of a trip is completed. How long is the total trip? How many kilometers are left to drive?

50. A package of coffee beans weighed $\frac{21}{32}$ lb when it was $\frac{3}{4}$ full. How much could the package hold when completely filled?

51. After swimming $\frac{2}{7}$ mi, Katie had swum $\frac{3}{4}$ of the race. How long a race was Katie competing in?

52. A brick of Swiss cheese is 14 in. long. How many slices will it yield if half of the brick is cut by a slicer set for $\frac{3}{32}$-in. slices and half is cut by a slicer set for $\frac{3}{64}$-in. slices?

Summary and Review: Chapter 3

Important Properties and Formulas

$\frac{0}{n} = 0$, for $n \neq 0$; $\frac{n}{0}$ is undefined; $\frac{n}{1} = n$; $\frac{n}{n} = 1$, for $n \neq 1$

Area of a Rectangle: $A = l \cdot w$

Area of a Triangle: $A = \frac{1}{2} \cdot b \cdot h$

The Multiplication Principle: For $c \neq 0$, $a = b$ is equivalent to $a \cdot c = b \cdot c$.

Review Exercises

1. Determine whether 4232 is divisible by 6. Do not use long division.

2. Determine whether 784 is divisible by 5. Do not use long division.

3. Determine whether 4347 is divisible by 9. Do not use long division.

Find the prime factorization of each number.
4. 70   5. 72   6. 150

7. Determine whether 37 is prime, composite, or neither.

8. Identify the numerator and the denominator of $\frac{9}{7}$.

What part is shaded?

9.    10.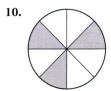

Simplify, if possible. Assume that all variables are nonzero.

11. $\frac{0}{4}$ 12. $\frac{23}{23}$

13. $\frac{48}{1}$ 14. $\frac{7x}{7x}$

15. $-\frac{10}{15}$ 16. $\frac{7}{28}$

17. $\frac{-21}{21}$ 18. $\frac{9m}{12m}$

19. $\frac{12}{30}$ 20. $\frac{-27}{0}$

21. $\frac{9n}{1}$ 22. $\frac{-9}{-27}$

Find an equivalent expression for the given number, but with the denominator indicated. Use multiplication by 1.

23. $\frac{5}{7} = \frac{?}{21}$ 24. $\frac{-6}{11} = \frac{?}{55}$

Find the reciprocal of each number.

25. $\frac{5}{9}$ 26. -7

27. $\frac{1}{8}$ 28. $\frac{3x}{5y}$

Perform the indicated operation and, if possible, simplify.

29. $\dfrac{2}{7} \cdot \dfrac{3}{5}$

30. $\dfrac{4}{x} \cdot \dfrac{y}{9}$

31. $\dfrac{3}{4} \cdot \dfrac{8}{9}$

32. $-\dfrac{5}{7} \cdot \dfrac{1}{10}$

33. $\dfrac{3a}{10} \cdot \dfrac{2}{15a}$

34. $\dfrac{4a}{7} \cdot \dfrac{7}{4a}$

35. $6 \div \dfrac{5}{3}$

36. $\dfrac{3}{14} \div \dfrac{6}{7}$

37. $180 \div \dfrac{3}{5}$

38. $-\dfrac{5}{36} \div \left(-\dfrac{25}{12}\right)$

39. $14 \div \dfrac{7}{2a}$

40. $-\dfrac{23}{25} \div \dfrac{23}{25}$

Solve.

41. The Mulligans have driven $\frac{4}{5}$ of a 275-mi trip. How far have they driven?

42. A recipe calls for $\frac{3}{4}$ cup of sugar. In making $\frac{1}{2}$ of this recipe, how much sugar should be used?

43. The Winchester swim team has 4 swimmers in a $\frac{2}{3}$-mi relay race. How far will each person swim?

44. How many $\frac{2}{3}$-cup cereal bowls can be filled from 12 cups of cornflakes?

Find the area.

45.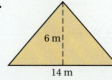

46.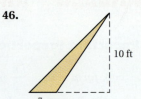

Solve.

47. $\dfrac{2}{3}x = 160$

48. $\dfrac{3}{8} = -\dfrac{5}{4}t$

Skill Maintenance

49. Solve: $17 \cdot x = 408$.

50. Simplify: $20 \div 2 \cdot 2 - 3^2$.

51. Add: $(-798) + 812$.

52. Multiply: $-3 \cdot (-9)$.

Synthesis

53. ◆ Write in your own words a series of steps that can be used when simplifying fractional notation.

54. ◆ A student claims that $\frac{20}{80}$ simplifies to $\frac{2}{8}$. Is the student correct? Why or why not?

55. Use a calculator and the list of prime numbers on p. 154 to find simplified fractional notation for the solution of
$$\dfrac{1751}{267}x = \dfrac{3193}{2759}.$$

56. Simplify: $\dfrac{15x}{14z} \cdot \dfrac{17yz}{35xy} \div \left(-\dfrac{3}{7}\right)^2$.

57. What digit(s) could be inserted in the ones place to make 574__ divisible by 6?

Test: Chapter 3

1. Determine whether 4782 is divisible by 3. Do not use long division.

2. Determine whether 5478 is divisible by 5. Do not use long division.

Find the prime factorization of each number.

3. 36

4. 60

5. Determine whether 49 is prime, composite, or neither.

6. Identify the numerator and the denominator of $\frac{4}{9}$.

7. What part is shaded?

Simplify, if possible. Assume that all variables are nonzero.

8. $\frac{26}{1}$

9. $\frac{-12}{-12}$

10. $\frac{0}{16}$

11. $\frac{-8}{24}$

12. $\frac{5x}{45x}$

13. $\frac{2}{28}$

14. Find an equivalent expression for $\frac{3}{8}$ with a denominator of 40.

Find the reciprocal.

15. $\frac{x}{27}$

16. -9

Perform the indicated operation. Simplify, if possible.

17. $\frac{5}{9} \cdot \frac{7}{2}$

18. $\frac{7}{11} \div \frac{3}{4}$

19. $3 \cdot \frac{x}{8}$

Answers

1. _____
2. _____
3. _____
4. _____
5. _____
6. _____
7. _____
8. _____
9. _____
10. _____
11. _____
12. _____
13. _____
14. _____
15. _____
16. _____
17. _____
18. _____
19. _____

20. $28 \div \left(-\dfrac{6}{7}\right)$

21. $\dfrac{4a}{13} \cdot \dfrac{9b}{30ab}$

Solve.

22. A $\tfrac{3}{4}$-lb slab of cheese is shared equally by 5 people. How much does each person receive?

23. Monroe weighs $\tfrac{5}{7}$ of his dad's weight. If his dad weighs 175 lb, how much does Monroe weigh?

24. $\dfrac{7}{8} \cdot x = 56$

25. $\dfrac{7}{10} = \dfrac{-2}{5} \cdot t$

26. Find the area.

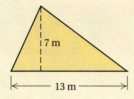

Skill Maintenance

27. Simplify: $3^2 + 2(1 + 3)^2$.

28. Solve: $47 \cdot t = 4747$.

29. Add: $(-93) + (-74)$.

30. Simplify: $(-9)(-7)$.

Synthesis

31. A recipe for a batch of buttermilk pancakes calls for $\tfrac{3}{4}$ teaspoon (tsp) of salt. Jacqueline plans to cut the amount of salt in half for each of 5 batches of pancakes. How much salt will she need altogether?

32. Grandma Phyllis left $\tfrac{2}{3}$ of her $\tfrac{7}{8}$-acre tree farm to Karl. Karl gave $\tfrac{1}{4}$ of his share to his oldest daughter, Irene. How much land did Irene receive?

33. Simplify: $\left(-\dfrac{3}{8}\right)^2 \div \dfrac{6}{7} \cdot \dfrac{2}{9} \div (-5)$.

34. Solve: $\dfrac{33}{38} \cdot \dfrac{34}{55} = \dfrac{17}{35} \cdot \dfrac{15}{19} x$.

Cumulative Review: Chapters 1–3

1. Write a word name: 2,056,783.

Add.

2. $2,739$
   $+8,243$

3. $-29 + (-14)$

4. $-45 + 12$

Subtract.

5. $4,324$
   $-2,195$

6. $17 - 40$

7. $-12 - (-4)$

Multiply and simplify.

8. 735
   $\times23$

9. $-52 \cdot 6$

10. $\dfrac{6}{7} \cdot (-35x)$

11. $\dfrac{2}{9} \cdot \dfrac{21}{10}$

Divide and simplify.

12. $13 \overline{)3058}$

13. $-85 \div 5$

14. $-16 \div \dfrac{4}{7}$

15. $\dfrac{3}{7} \div \dfrac{9}{14}$

16. Round 4514 to the nearest ten.

17. Estimate the product by rounding to the nearest hundred. Show your work.
    921
    $\times 453$

18. Find the absolute value: $|879|$.

19. Simplify: $10^2 \div 5(-2) - 8(2 - 8)$.

20. Determine whether 98 is prime, composite, or neither.

21. Evaluate $a - b^2$ for $a = -5$ and $b = 4$.

Solve.

22. $a + 24 = 49$

23. $7x = 63$

24. $\dfrac{2}{9} \cdot a = -10$

25. A 1996 van that gets 25 miles per gallon is traded in toward a 1999 truck that gets 17 miles per gallon. How many more miles per gallon did the older vehicle get?

26. A 64-oz soda is poured into 8 glasses. How much will each glass hold if the soda is poured out evenly?

Combine like terms.

27. $9 - 5x - 13 + 7x$

28. $-12x + 7y + 15x$

Simplify, if possible.

29. $\dfrac{97}{97}$

30. $\dfrac{59}{1}$

31. $\dfrac{0}{72}$

32. $\dfrac{-10}{54}$

Find the reciprocal.

33. $\dfrac{2}{5}$

34. 17

35. Find an equivalent expression for $\dfrac{3}{10}$ with a denominator of 70. Use multiplying by 1.

36. A babysitter earns \$60 for working a full day. How much is earned for working $\dfrac{3}{5}$ of a day?

37. How many $\dfrac{3}{4}$-lb servings can be made from a 9-lb roast?

38. Tony has jogged $\dfrac{2}{3}$ of a course that is $\dfrac{9}{10}$ of a mile long. How far has Tony gone?

Synthesis

39. Evaluate $\dfrac{ab}{c}$ for $a = -\dfrac{2}{5}$, $b = \dfrac{10}{13}$, and $c = \dfrac{26}{27}$.

40. Evaluate $-|xy|^2$ for $x = -\dfrac{3}{5}$ and $y = \dfrac{1}{2}$.

41. Wayne and Patty each earn \$85 a day, while Janet earns \$90 a day. They decide to pool their earnings from three days and spend $\dfrac{2}{5}$ of that on entertainment and save the rest. How much will Wayne, Patty, and Janet end up saving?

Chapter 3 Fractions: Multiplication & Division

4

Fractional Notation: Addition and Subtraction

Introduction

In this chapter, we consider addition and subtraction using fractional notation. We then examine addition, subtraction, multiplication, and division using mixed numerals. These operations are then applied to the solution of equations and problems.

4.1 Least Common Multiples
4.2 Addition and Order
4.3 Subtraction, Equations, and Applications
4.4 Solving Equations: Using the Principles Together
4.5 Mixed Numerals
4.6 Addition and Subtraction Using Mixed Numerals; Applications
4.7 Multiplication and Division Using Mixed Numerals; Applications

An Application

Melody has had three children. Their birth weights were $7\frac{1}{2}$ lb, $7\frac{3}{4}$ lb, and $6\frac{3}{4}$ lb. What was the average weight of her babies?

This problem appears as Example 12 in Section 4.7.

The Mathematics

We let w = the average weight, in pounds, of the babies. The problem then translates to

$$w = \frac{7\frac{1}{2} + 7\frac{3}{4} + 6\frac{3}{4}}{3}.$$

↑
Operations using mixed numerals occur often in problem solving.

World Wide Web   For more information, visit us at www.mathmax.com

Pretest: Chapter 4

1. Find the least common multiple of 25 and 15.

2. Use < or > for ■ to write a true sentence.
$$\frac{7}{9} \; \blacksquare \; \frac{4}{5}.$$

Perform the indicated operation and, if possible, simplify.

3. $\dfrac{-5}{8} + \dfrac{3}{8}$

4. $\dfrac{5}{6} + \dfrac{-7}{9} + \dfrac{1}{15}$

5. $\dfrac{2}{5} - \dfrac{1}{7}$

6. Convert to fractional notation: $5\dfrac{3}{8}$.

7. Convert to a mixed numeral: $\dfrac{13}{4}$.

8. Divide. Write a mixed numeral for the answer.
$$1\,2 \,\overline{)\,4\;7\;8\;9\,}$$

9. Subtract. Write a mixed numeral for the answer.
$$\begin{array}{r} 14\dfrac{1}{5} \\ -\;7\dfrac{5}{6} \\ \hline \end{array}$$

Solve.

10. $\dfrac{2}{3} + x = \dfrac{8}{9}$

11. $14 = \dfrac{2}{3}x + 20$

Perform the indicated operations. Write a mixed numeral for each answer.

12. $(-5)\left(-3\dfrac{8}{11}\right)$

13. $6\dfrac{1}{3} \cdot 5\dfrac{3}{4}$

14. $45 \div \left(-5\dfrac{5}{6}\right)$

15. $4\dfrac{5}{12} \div 3\dfrac{1}{4}$

16. Evaluate $xy \div z$ for $x = 3\dfrac{2}{5}$, $y = 6$, and $z = 2\dfrac{1}{3}$.

Solve.

17. The Colburn Inn bought 60 lb of onions and used $21\dfrac{3}{4}$ lb. How many pounds were left?

18. Water weighs $62\dfrac{1}{2}$ lb per cubic foot. How many cubic feet does $265\dfrac{5}{8}$ lb of water occupy?

19. On a trip, Janet averaged $315\dfrac{1}{2}$ miles per day for 5 days. How far did she travel altogether in those 5 days?

20. Uri is baking a birthday cake that requires $3\dfrac{3}{4}$ cups of flour and a batch of biscotti that requires $2\dfrac{1}{2}$ cups. How much flour is required altogether?

Chapter 4   Fractions: Addition & Subtraction

4.1 Least Common Multiples

In this chapter, we study addition and subtraction using fractional notation. Suppose we want to add $\frac{2}{3}$ and $\frac{1}{2}$. To do so, we use the least common multiple of the denominators: $\frac{2}{3} + \frac{1}{2} = \frac{4}{6} + \frac{3}{6}$. Then we add the numerators and keep the common denominator, 6. Before we do this, though, we study finding the **least common denominator (LCD)**, or **least common multiple (LCM)**, of the denominators.

a Finding Least Common Multiples

> The **least common multiple**, or LCM, of two natural numbers is the smallest number that is a multiple of both.

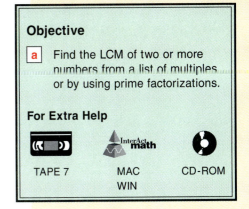

Objective

a Find the LCM of two or more numbers from a list of multiples or by using prime factorizations.

For Extra Help

TAPE 7   MAC WIN   CD-ROM

Example 1 Find the LCM of 20 and 30.

a) First list some multiples of 20 by multiplying 20 by 1, 2, 3, and so on:

   20, 40, 60, 80, 100, 120, 140, 160, 180, 200, 220, 240, ...

b) Then list some multiples of 30 by multiplying 30 by 1, 2, 3, and so on:

   30, 60, 90, 120, 150, 180, 210, 240, ...

c) Now list the numbers *common* to both lists, the common multiples:

   60, 120, 180, 240, ...

d) These are the common multiples of 20 and 30. The *least* of these common multiples is 60. Thus the LCM of 20 and 30 is 60.

Do Exercises 1 and 2.

Next we develop two methods that are more efficient for finding LCMs. You may choose to learn either method (consult with your instructor), or both, but if you are going on to study algebra, you should definitely learn Method 2.

Method 1: Finding LCMs Using One List of Multiples

Method 1. To find the LCM of a set of numbers (say, 9 and 12), determine whether the largest number is a multiple of the other(s):

1. If it is, it is the LCM.

   (Since 12 is not a multiple of 9, the LCM is not 12.)

2. If the largest number *is not* a multiple of the other(s), check consecutive multiples of the largest number until you find one that *is* a multiple of the other number(s). That number is the LCM.

   (2 · 12 = 24, but 24 is not a multiple of 9.)

   (3 · 12 = 36, and 36 *is* a multiple of 9, so the LCM of 9 and 12 is 36.)

1. By examining lists of multiples, find the LCM of 9 and 15.

2. By examining lists of multiples, find the LCM of 8 and 10.

Answer on page A-9

4.1 Least Common Multiples

207

Find the LCM.

3. 6, 9

4. 6, 8

Find the LCM.

5. 5, 10

6. 20, 40, 50

Answers on page A-9

Chapter 4 Fractions: Addition & Subtraction

Example 2 Find the LCM of 12 and 15.

1. 15 is the larger number, but it is not a multiple of 12.
2. Check multiples of 15:

$2 \cdot 15 = 30$,   Not a multiple of 12
$3 \cdot 15 = 45$,   Not a multiple of 12
$4 \cdot 15 = 60$.   A multiple of 12

The LCM = 60.

Example 3 Find the LCM of 4 and 14.

1. 14 is the larger number, but it is not a multiple of 4.
2. Check multiples of 14:

$2 \cdot 14 = 28$.   A multiple of 4

The LCM = 28.

Do Exercises 3 and 4.

Example 4 Find the LCM of 8 and 32.

1. 32 is the larger number and 32 is a multiple of 8, so it is the LCM.

The LCM = 32.

Example 5 Find the LCM of 10, 100, and 250.

1. 250 is the largest number, but it is not a multiple of 100.
2. Check multiples of 250:

$2 \cdot 250 = 500$.   A multiple of 10 and 100

The LCM = 500.

Do Exercises 5 and 6.

Method 2: Finding LCMs Using Factorizations

A second method for finding LCMs uses prime factorizations. Consider again 20 and 30. Their prime factorizations are

$20 = 2 \cdot 2 \cdot 5$ and $30 = 2 \cdot 3 \cdot 5$.

The least common multiple must include the factors of each number, so it must include each prime factor the greatest number of times that it appears in either of the factorizations. To find the LCM for 20 and 30, we select one factorization, say,

$2 \cdot 2 \cdot 5$,

and note that because it lacks the factor 3, it does not contain the entire factorization of 30. If we multiply $2 \cdot 2 \cdot 5$ by 3, every prime factor occurs just often enough to contain both 20 and 30 as factors.

LCM = $2 \cdot 2 \cdot 5 \cdot 3$

— 20 is a factor of the LCM.
— 30 is a factor of the LCM.

Note that each prime factor is used the greatest number of times that it occurs in either of the individual factorizations.

Method 2. To find the LCM of a set of numbers (say, 9 and 12):

1. Write the prime factorization of each number.

   $(9 = 3 \cdot 3; \quad 12 = 2 \cdot 2 \cdot 3)$

2. Select one of the factorizations and see whether it contains the other(s).

   $(2 \cdot 2 \cdot 3 \text{ does not contain } 3 \cdot 3.)$

   a) If it does, it represents the LCM.

   b) If it does not, multiply that factorization by those prime factors of the other number(s) that it lacks. The final product is the LCM.

   $(2 \cdot 2 \cdot 3 \cdot 3 \text{ is the LCM.})$

3. As a check, make sure that the LCM includes each factor the greatest number of times that it occurs in any one factorization.

Example 6 Find the LCM of 18 and 21.

1. We begin by writing the prime factorization of each number:

   $18 = 2 \cdot 3 \cdot 3 \quad \text{and} \quad 21 = 3 \cdot 7.$

2. a) We inspect the factorization $2 \cdot 3 \cdot 3$ and note that it does not contain the other factorization, $3 \cdot 7$.

   b) To find the LCM of 18 and 21, we multiply $2 \cdot 3 \cdot 3$ by the factor of 21 that it lacks, 7:

   $\text{LCM} = 2 \cdot 3 \cdot 3 \cdot 7.$

   — 18 is a factor.
   — 21 is a factor.

3. The greatest number of times that 2 occurs as a factor of 18 or 21 is **one** time; the greatest number of times that 3 occurs as a factor of 18 or 21 is **two** times; and the greatest number of times that 7 occurs as a factor of 18 or 21 is **one** time. To check, note that the LCM has exactly **one** 2, **two** 3's, and **one** 7. The LCM is $2 \cdot 3 \cdot 3 \cdot 7$, or 126.

Example 7 Find the LCM of 24 and 36.

1. We begin by writing the prime factorization of each number:

   $24 = 2 \cdot 2 \cdot 2 \cdot 3 \quad \text{and} \quad 36 = 2 \cdot 2 \cdot 3 \cdot 3.$

2. a) The factorization $2 \cdot 2 \cdot 2 \cdot 3$ does not contain the factorization $2 \cdot 2 \cdot 3 \cdot 3$. Nor does $2 \cdot 2 \cdot 3 \cdot 3$ contain $2 \cdot 2 \cdot 2 \cdot 3$.

   b) To find the LCM of 24 and 36, we multiply the factorization of 24, $2 \cdot 2 \cdot 2 \cdot 3$, by any prime factors of 36 that are lacking. In this case, a second factor of 3 is needed. We have

   $\text{LCM} = 2 \cdot 2 \cdot 2 \cdot 3 \cdot 3.$

   — 24 is a factor.
   — 36 is a factor.

3. To check, note that 2 and 3 appear in the LCM the greatest number of times that each appears as a factor of 24 or 36. The LCM is

   $2 \cdot 2 \cdot 2 \cdot 3 \cdot 3$, or 72.

Do Exercises 7 and 8.

Use prime factorizations to find the LCM.

7. 8, 10

8. 18, 40

Answers on page A-9

4.1 Least Common Multiples

Find the LCM.

9. 3, 18

10. 12, 24

11. 24, 35, 45

Example 8 Find the LCM of 7 and 21.

1. Because 7 is prime, we think of $7 = 7$ as a "factorization" in order to carry out our procedure:

   $7 = 7$ and $21 = 3 \cdot 7$.

2. One factorization, $3 \cdot 7$, contains the other. Thus the LCM is $3 \cdot 7$, or 21.

Example 9 Find the LCM of 27, 90, and 84.

1. We first find the prime factorization of each number:

   $27 = 3 \cdot 3 \cdot 3$, $90 = 2 \cdot 3 \cdot 3 \cdot 5$, and $84 = 2 \cdot 2 \cdot 3 \cdot 7$.

2. a) No one factorization contains the other two.

   b) We begin with the factorization of 90, $2 \cdot 3 \cdot 3 \cdot 5$ (we could have used any factorization listed). Since 27 contains a third factor of 3, we multiply by another factor of 3:

   $2 \cdot 3 \cdot 3 \cdot 5 \cdot 3$.
           90 is a factor.
           27 is a factor.

   Next, we multiply $2 \cdot 3 \cdot 3 \cdot 5 \cdot 3$ by the factors of 84 still missing, $2 \cdot 7$:

   $2 \cdot 3 \cdot 3 \cdot 5 \cdot 3 \cdot 2 \cdot 7$.
           90 and 27 are factors.
           84 is a factor.

   The LCM is $2 \cdot 3 \cdot 3 \cdot 5 \cdot 3 \cdot 2 \cdot 7$, or 3780.

3. The check is left to the student.

Do Exercises 9–11.

Exponential notation is often helpful when writing least common multiples. Let's reconsider Example 7:

$24 = 2 \cdot 2 \cdot 2 \cdot 3 = 2^3 \cdot 3^1$;

$36 = 2 \cdot 2 \cdot 3 \cdot 3 = 2^2 \cdot 3^2$;

$\text{LCM} = 2 \cdot 2 \cdot 2 \cdot 3 \cdot 3 = 2^3 \cdot 3^2$, or 72.

Note that in the factorizations of 24 and 36, the largest power of 2 is 2^3 and the largest power of 3 is 3^2. These powers are used to create the LCM, $2^3 \cdot 3^2$, or 72.

Find the LCM.

12. xy, yz

13. $5a^2$, $a^3 b$

Example 10 Find the LCM of $7a^2 b$ and ab^2.

1. We have the following factorizations:

   $7a^2 b = 7 \cdot a \cdot a \cdot b$ and $ab^2 = a \cdot b \cdot b$.

2. a) No one factorization contains the other.

   b) Consider the factorization of $7a^2 b$, $7 \cdot a \cdot a \cdot b$. Since ab^2 contains a second factor of b, we multiply by another factor of b:

   $7 \cdot a \cdot a \cdot b \cdot b$.
           $7a^2 b$ is a factor.
           ab^2 is a factor.

   The LCM is $7 \cdot a \cdot a \cdot b \cdot b$, or $7a^2 b^2$.

3. The check is left to the student.

Do Exercises 12 and 13.

Answers on page A-9

Exercise Set 4.1

a Find the LCM of each set of numbers.

1. 2, 4   **2.** 3, 15   **3.** 10, 25   **4.** 10, 15   **5.** 20, 40

6. 8, 12   **7.** 18, 27   **8.** 9, 11   **9.** 30, 50   **10.** 8, 36

11. 30, 40   **12.** 21, 27   **13.** 18, 24   **14.** 12, 18   **15.** 60, 70

16. 35, 45   **17.** 16, 36   **18.** 18, 20   **19.** 32, 36   **20.** 36, 48

21. 2, 3, 5   **22.** 5, 18, 3   **23.** 3, 5, 7   **24.** 6, 12, 18   **25.** 24, 36, 12

26. 8, 16, 22   **27.** 5, 12, 15   **28.** 12, 18, 40   **29.** 9, 12, 6   **30.** 8, 16, 12

31. 180, 100, 450   **32.** 18, 30, 50, 48   **33.** 75, 100   **34.** 81, 90   **35.** ab, bc

36. $7x$, xy **37.** $3x$, $9x^2$ **38.** $10x^4$, $5x^3$ **39.** $4x^3$, x^2y **40.** $6ab^2$, a^3b

Applications of LCMs: Planet Orbits. Earth, Jupiter, Saturn, and Uranus all revolve around the sun. Earth takes 1 yr, Jupiter 12 yr, Saturn 30 yr, and Uranus 84 yr to make a complete revolution. On a certain night, you look at those three distant planets and wonder how many years it will take before they have the same position again. To determine this, you find the LCM of 12, 30, and 84. It will be that number of years.

41. How often will Jupiter and Saturn appear in the same direction in the night sky as seen from Earth?

42. How often will Jupiter, Saturn, and Uranus appear in the same direction in the night sky as seen from Earth?

Skill Maintenance

Perform the indicated operation and, if possible, simplify.

43. $-38 + 52$ [2.2a]

44. $-18 \div \left(\dfrac{2}{3}\right)$ [3.7b]

45. $23 \cdot 345$ [1.5b]

46. $\dfrac{4}{5} \cdot \dfrac{10}{12}$ [3.6a]

47. $\dfrac{4}{5} \div \left(-\dfrac{7}{10}\right)$ [3.7b]

48. $382 - 549$ [2.3a]

Synthesis

49. ◆ Under what conditions is the LCM of two composite numbers simply the product of the two numbers?

50. ◆ Is the LCM of two prime numbers always their product? Why or why not?

51. ◆ Is the LCM of two numbers always at least twice as large as the larger of the two numbers? Why or why not?

Use a calculator and the multiples method to find the LCM of each pair of numbers.

52. 288, 324

53. 2700, 7800

54. Use Example 9 to help find the LCM of 27, 90, 84, 210, 108, and 50.

55. Use Examples 6 and 7 to help find the LCM of 18, 21, 24, 36, 63, 56, and 20.

56. The exhibits at a flea market are either 6 ft long or 8 ft long. How long is the shortest aisle that can accommodate exhibits of either length with no space left over? (*Note:* each aisle will be filled with either all 6-ft exhibits or all 8-ft exhibits.)

57. Consider 8 and 12. Determine whether each of the following is the LCM of 8 and 12. Tell why or why not.
    a) $2 \cdot 2 \cdot 3 \cdot 3$
    b) $2 \cdot 2 \cdot 3$
    c) $2 \cdot 3 \cdot 3$
    d) $2 \cdot 2 \cdot 2 \cdot 3$

58. Find three different pairs of numbers for which 56 is the LCM. Do not use 56 itself in any one of the pairs.

59. Find three different pairs of numbers for which 54 is the LCM. Do not use 54 itself in any one of the pairs.

Find the least common multiple of two or more numbers using shaped markers.

4.2 Addition and Order

a Like Denominators

Addition using fractional notation corresponds to combining or putting like things together, just as when we combined like terms in Section 2.7. For example,

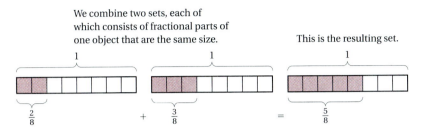

2 eighths + 3 eighths = 5 eighths,

or $2 \cdot \dfrac{1}{8} + 3 \cdot \dfrac{1}{8} = 5 \cdot \dfrac{1}{8}$,

or $\dfrac{2}{8} + \dfrac{3}{8} = \dfrac{5}{8}$.

Do Exercise 1.

> To add when denominators are the same,
> a) add the numerators,
> b) keep the denominator, and
> c) simplify, if possible.
>
> $\dfrac{2}{6} + \dfrac{5}{6} = \dfrac{2+5}{6} = \dfrac{7}{6}$

Examples Add and, if possible, simplify.

1. $\dfrac{2}{4} + \dfrac{1}{4} = \dfrac{2+1}{4} = \dfrac{3}{4}$ **No simplifying is possible.**

2. $\dfrac{3}{12} + \dfrac{5}{12} = \dfrac{3+5}{12} = \dfrac{8}{12}$ **Adding numerators; the denominator remains unchanged.**

   $= \dfrac{4}{4} \cdot \dfrac{2}{3} = \dfrac{2}{3}$ **Simplifying by removing a factor equal to 1: $\dfrac{4}{4} = 1$**

3. $\dfrac{-11}{6} + \dfrac{3}{6} = \dfrac{-11+3}{6} = \dfrac{-8}{6}$

   $= \dfrac{2}{2} \cdot \dfrac{-4}{3} = \dfrac{-4}{3}$, or $-\dfrac{4}{3}$ **Removing a factor equal to 1: $\dfrac{2}{2} = 1$**

4. $-\dfrac{2}{a} + \left(-\dfrac{3}{a}\right) = \dfrac{-2}{a} + \dfrac{-3}{a}$ **Recall that $-\dfrac{m}{n} = \dfrac{-m}{n}$. We generally try to avoid negative signs in the denominator.**

   $= \dfrac{-2 + (-3)}{a} = \dfrac{-5}{a}$, or $-\dfrac{5}{a}$

Do Exercises 2–5.

Objectives

a Add using fractional notation when denominators are the same.

b Add using fractional notation when denominators are different.

c Use $<$ or $>$ to form a true statement using fractional notation.

d Solve problems involving addition with fractional notation.

For Extra Help

TAPE 7   MAC WIN   CD-ROM

1. Find $\dfrac{1}{5} + \dfrac{3}{5}$.

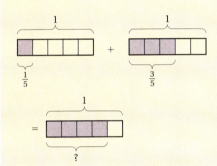

Add and, if possible, simplify.

2. $\dfrac{1}{3} + \dfrac{2}{3}$

3. $\dfrac{5}{12} + \dfrac{1}{12}$

4. $\dfrac{-9}{16} + \dfrac{3}{16}$

5. $\dfrac{3}{x} + \dfrac{-7}{x}$

Answers on page A-9

4.2 Addition and Order

213

Simplify by combining like terms.

6. $\frac{3}{10}a + \frac{1}{10}a$

In some cases, we need to add fractions when combining like terms.

Example 5 Simplify by combining like terms: $\frac{2}{7}x + \frac{3}{7}x$.

$$\frac{2}{7}x + \frac{3}{7}x = \left(\frac{2}{7} + \frac{3}{7}\right)x \quad \text{Try to do this step mentally.}$$
$$= \frac{5}{7}x.$$

Do Exercises 6 and 7.

b Addition Using the Least Common Denominator

At the beginning of this section, we visualized the addition $\frac{2}{8} + \frac{3}{8}$. Consider now the addition $\frac{1}{2} + \frac{1}{3}$.

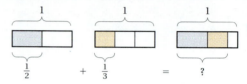

By rewriting $\frac{1}{2}$ as $\frac{1}{2} \cdot \frac{3}{3} = \frac{3}{6}$ and $\frac{1}{3}$ as $\frac{1}{3} \cdot \frac{2}{2} = \frac{2}{6}$, we can determine the sum.

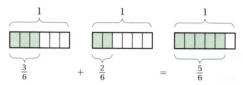

Thus, when denominators differ, before adding we must multiply by 1 to get a common denominator. There is always more than one common denominator that can be used. Consider the addition $\frac{3}{4} + \frac{1}{6}$:

A. $\frac{3}{4} + \frac{1}{6} = \frac{3}{4} \cdot 1 + \frac{1}{6} \cdot 1$

$= \frac{3}{4} \cdot \frac{6}{6} + \frac{1}{6} \cdot \frac{4}{4}$ Here 24 is the common denominator.

$= \frac{18}{24} + \frac{4}{24}$

$= \frac{22}{24} = \frac{11}{12}$;

B. $\frac{3}{4} + \frac{1}{6} = \frac{3}{4} \cdot 1 + \frac{1}{6} \cdot 1$

$= \frac{3}{4} \cdot \frac{3}{3} + \frac{1}{6} \cdot \frac{2}{2}$ Here 12 is the common denominator.

$= \frac{9}{12} + \frac{2}{12}$

$= \frac{11}{12}$

We had to simplify at the end of (A), but not in (B). In (B), we used the *least* common multiple of the denominators, 12. That number is called the **least common denominator, or LCD**.

7. $\frac{2}{19} + \frac{3}{19}x + \frac{5}{19} + \frac{7}{19}x$

To add when denominators are different:

a) Find the least common multiple of the denominators. That number is the least common denominator, LCD.

b) Multiply by 1, writing 1 in the form of n/n, to find an equivalent sum in which the LCD appears.

c) Add and, if possible, simplify.

Answers on page A-9

Example 6 Add: $\dfrac{1}{8} + \dfrac{3}{4}$.

a) Since 4 is a factor of 8, the LCM of 4 and 8 is 8. Thus the LCD is 8.

b) We need to find a fraction equivalent to $\dfrac{3}{4}$ with a denominator of 8:

$$\dfrac{1}{8} + \dfrac{3}{4} = \dfrac{1}{8} + \dfrac{3}{4} \cdot \dfrac{?}{2}.\quad \text{Think: } 4 \times \blacksquare = 8. \text{ The answer is 2,}$$
$$\text{so we multiply by 1, using } \tfrac{2}{2}.$$

c) We add: $\dfrac{1}{8} + \dfrac{6}{8} = \dfrac{7}{8}.\quad \dfrac{7}{8}$ cannot be simplified.

Do Exercise 8.

In Examples 7–10, we follow the same steps without spelling them out.

Example 7 Add: $\dfrac{5}{6} + \dfrac{1}{9}$.

The LCD is 18. $\quad 6 = 2 \cdot 3$ and $9 = 3 \cdot 3$, so the LCM of 6 and 9 is $2 \cdot 3 \cdot 3$, or 18.

$$\dfrac{5}{6} + \dfrac{1}{9} = \dfrac{5}{6} \cdot 1 + \dfrac{1}{9} \cdot 1$$

$$= \dfrac{5}{6} \cdot \dfrac{3}{3} + \dfrac{1}{9} \cdot \dfrac{2}{2}\quad \begin{array}{l}\text{Think: } 9 \times \blacksquare = 18.\\ \text{The answer is 2, so}\\ \text{we multiply by 1, using } \tfrac{2}{2}.\\[4pt] \text{Think: } 6 \times \blacksquare = 18.\\ \text{The answer is 3, so}\\ \text{we multiply by 1, using } \tfrac{3}{3}.\end{array}$$

$$= \dfrac{15}{18} + \dfrac{2}{18}$$

$$= \dfrac{17}{18}$$

Do Exercise 9.

Example 8 Add: $\dfrac{3}{-5} + \dfrac{11}{10}$.

$$\dfrac{3}{-5} + \dfrac{11}{10} = \dfrac{-3}{5} + \dfrac{11}{10}\quad \text{Recall that } \dfrac{m}{-n} = \dfrac{-m}{n}. \text{ The LCD is 10.}$$

$$= \dfrac{-3}{5} \cdot \dfrac{2}{2} + \dfrac{11}{10}$$

$$= \dfrac{-6}{10} + \dfrac{11}{10}$$

$$= \dfrac{5}{10}$$
$$= \dfrac{1}{2}\quad \left.\begin{array}{l}\text{We may still have to simplify,}\\ \text{but simplifying is almost always}\\ \text{easier if the LCD has been used.}\end{array}\right.$$

---

8. Add using the least common denominator.

$$\dfrac{2}{3} + \dfrac{1}{6}$$

9. Add: $\dfrac{3}{8} + \dfrac{5}{6}$.

Answers on page A-9

4.2 Addition and Order

215

Add.

10. $\dfrac{1}{-6} + \dfrac{7}{18}$

11. $7 + \dfrac{3}{5}$

Add.

12. $\dfrac{4}{10} + \dfrac{1}{100} + \dfrac{3}{1000}$

13. $\dfrac{7}{10} + \dfrac{-2}{21} + \dfrac{1}{7}$

Use < or > for ■ to form a true sentence.

14. $\dfrac{3}{8}$ ■ $\dfrac{5}{8}$

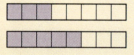

15. $\dfrac{7}{10}$ ■ $\dfrac{6}{10}$

16. $\dfrac{-2}{9}$ ■ $\dfrac{-5}{9}$

Answers on page A-9

Chapter 4 Fractions: Addition & Subtraction

216

Example 9 Add: $\dfrac{5}{8} + 2$.

$\dfrac{5}{8} + 2 = \dfrac{5}{8} + \dfrac{2}{1}$ **Rewriting 2 in fractional notation**

$= \dfrac{5}{8} + \dfrac{2}{1} \cdot \dfrac{8}{8}$ **The LCD is 8.**

$= \dfrac{5}{8} + \dfrac{16}{8}$

$= \dfrac{21}{8}$

Do Exercises 10 and 11.

Example 10 Add: $\dfrac{9}{70} + \dfrac{11}{21} + \dfrac{-6}{15}$.

We need to determine the LCM of 70, 21, and 15:

$70 = 2 \cdot 5 \cdot 7,$
$21 = 3 \cdot 7,$ The LCM is $2 \cdot 3 \cdot 5 \cdot 7$, or 210.
$15 = 3 \cdot 5$

$\dfrac{9}{70} + \dfrac{11}{21} + \dfrac{-6}{15} = \dfrac{9}{70} \cdot \dfrac{3}{3} + \dfrac{11}{21} \cdot \dfrac{2 \cdot 5}{2 \cdot 5} + \dfrac{-6}{15} \cdot \dfrac{7 \cdot 2}{7 \cdot 2}$

$= \dfrac{9 \cdot 3}{70 \cdot 3} + \dfrac{11 \cdot 10}{21 \cdot 10} + \dfrac{-6 \cdot 14}{15 \cdot 14}$

$= \dfrac{27}{210} + \dfrac{110}{210} + \dfrac{-84}{210}$

$= \dfrac{137 + (-84)}{210}$

$= \dfrac{53}{210}.$ **Since 53 is prime and not a factor of 210, we cannot simplify.**

In each case, we multiply by 1 to obtain the LCD. To form 1, look at the prime factorization of the LCD and use the factor(s) missing from each denominator.

Do Exercises 12 and 13.

c | Order

Common denominators are also important for determining the larger of two fractions. When two fractions share a common denominator, the larger number can be found by comparing numerators. For example, 4 is greater than 3, so $\tfrac{4}{5}$ is greater than $\tfrac{3}{5}$.

$\dfrac{4}{5} > \dfrac{3}{5}$

Similarly, because -6 is less than -2, we have

$\dfrac{-6}{7} < \dfrac{-2}{7},$ or $-\dfrac{6}{7} < -\dfrac{2}{7}.$

Do Exercises 14–16.

Example 11 Use $<$ or $>$ for ▓ to form a true sentence:

$$\frac{5}{8} \; \square \; \frac{2}{3}.$$

You can confirm that the LCD is 24. We multiply by 1 to make the denominators the same:

$$\frac{5}{8} \cdot \frac{3}{3} = \frac{15}{24}; \quad \frac{2}{3} \cdot \frac{8}{8} = \frac{16}{24}.$$

Since $15 < 16$, it follows that $\frac{15}{24} < \frac{16}{24}$. Thus,

$$\frac{5}{8} < \frac{2}{3}.$$

Example 12 Use $<$ or $>$ for ▓ to form a true sentence:

$$-\frac{89}{100} \; \square \; -\frac{9}{10}.$$

The LCD is 100.

$$-\frac{9}{10} \cdot \frac{10}{10} = \frac{-90}{100} \quad \text{We multiply by } \tfrac{10}{10} \text{ to get the LCD.}$$

Since $-89 > -90$, it follows that $-\frac{89}{100} > -\frac{90}{100}$, so

$$-\frac{89}{100} > -\frac{9}{10}.$$

Do Exercises 17–19.

d Solving Problems

Example 13 *Baking.* A recipe for fudge brownies calls for $\frac{1}{4}$ cup of oil and $\frac{2}{3}$ cup of milk. How many cups of liquid ingredients are in the recipe?

1. **Familiarize.** We first make a drawing and let $n =$ the total number of cups of liquid ingredients.

$\frac{1}{4}$ cup $\frac{2}{3}$ cup n cups

2. **Translate.** The problem can be translated to an equation as follows:

Amount of oil	plus	Amount of milk	is	Amount of liquid
$\frac{1}{4}$	$+$	$\frac{2}{3}$	$=$	n

Use $<$ or $>$ for ▓ to form a true sentence.

17. $\frac{2}{3} \; \square \; \frac{3}{4}$

18. $\frac{-3}{4} \; \square \; \frac{-8}{12}$

19. $\frac{5}{6} \; \square \; \frac{7}{8}$

Answers on page A-9

20. Maureen bought $\frac{1}{2}$ lb of peanuts and $\frac{3}{5}$ lb of cashews. How many pounds of nuts were bought altogether?

3. **Solve.** To solve the equation, we carry out the addition:

$$\frac{1}{4} + \frac{2}{3} = n \quad \text{The LCD is 12.}$$

$$\frac{1}{4} \cdot \frac{3}{3} + \frac{2}{3} \cdot \frac{4}{4} = n \quad \text{Multiplying by 1}$$

$$\frac{3}{12} + \frac{8}{12} = n$$

$$\frac{11}{12} = n.$$

4. **Check.** As a partial check, we note that the sum is larger than either of the individual amounts, as expected. We can also check by repeating the calculations.

5. **State.** The recipe calls for $\frac{11}{12}$ cup of liquid ingredients.

Do Exercise 20.

Calculator Spotlight

 Many calculators are equipped with a key, often labeled $\boxed{a^b/_c}$, that allows for computations with fractional notation. To calculate

$$\frac{2}{3} + \frac{4}{5}$$

with such a calculator, the following keystrokes can be used (note that the key $\boxed{a^b/_c}$ usually doubles as the $\boxed{d/c}$ key):

$\boxed{2}$ $\boxed{a^b/_c}$ $\boxed{3}$ $\boxed{+}$ $\boxed{4}$ $\boxed{a^b/_c}$ $\boxed{5}$ $\boxed{=}$ $\boxed{\text{Shift}}$ $\boxed{d/c}$.

The display that appears,

$\boxed{\quad 22 \lrcorner 15 \quad}$,

represents the fraction $\frac{22}{15}$.

Note that we used the keystrokes $\boxed{\text{Shift}}$ $\boxed{d/c}$ to convert from a mixed numeral (see Section 4.5) to fractional notation.

Graphing calculators can also perform computations with fractional notation. To do the above addition on a graphing calculator, we use the $\boxed{\text{MATH}}$ key as follows:

$\boxed{2}$ $\boxed{\div}$ $\boxed{3}$ $\boxed{+}$ $\boxed{4}$ $\boxed{\div}$ $\boxed{5}$ $\boxed{\text{MATH}}$ $\boxed{1}$ $\boxed{\text{ENTER}}$.

CAUTION! Although it is possible to add on a calculator using fractional notation, it is still very important for you to understand how such addition is performed longhand. For this reason, your instructor may disallow the use of calculators on this chapter's test.

Exercises

Calculate.

1. $\frac{3}{8} + \frac{1}{4}$
2. $\frac{5}{12} + \frac{7}{10}$
3. $\frac{15}{7} + \frac{1}{3}$
4. $\frac{19}{20} + \frac{17}{35}$
5. $\frac{29}{30} + \frac{18}{25}$
6. $\frac{17}{23} + \frac{13}{29}$

Answers on page A-9

Exercise Set 4.2

a, **b** Add and, if possible, simplify.

1. $\dfrac{4}{9} + \dfrac{5}{9}$

2. $\dfrac{1}{4} + \dfrac{1}{4}$

3. $\dfrac{1}{8} + \dfrac{5}{8}$

4. $\dfrac{7}{8} + \dfrac{1}{8}$

5. $\dfrac{7}{10} + \dfrac{3}{-10}$

6. $\dfrac{1}{-6} + \dfrac{5}{6}$

7. $\dfrac{9}{a} + \dfrac{4}{a}$

8. $\dfrac{4}{t} + \dfrac{3}{t}$

9. $\dfrac{-5}{11} + \dfrac{3}{11}$

10. $\dfrac{7}{12} + \dfrac{-5}{12}$

11. $\dfrac{2}{9}x + \dfrac{5}{9}x$

12. $\dfrac{3}{11}a + \dfrac{2}{11}a$

13. $\dfrac{5}{32} + \dfrac{3}{32}t + \dfrac{7}{32} + \dfrac{13}{32}t$

14. $\dfrac{3}{25}x + \dfrac{7}{25} + \dfrac{12}{25}x + \dfrac{-2}{25}$

15. $-\dfrac{3}{x} + \left(-\dfrac{7}{x}\right)$

16. $-\dfrac{9}{a} + \dfrac{5}{a}$

17. $\dfrac{1}{8} + \dfrac{1}{6}$

18. $\dfrac{1}{9} + \dfrac{1}{6}$

19. $\dfrac{-4}{5} + \dfrac{7}{10}$

20. $\dfrac{-3}{4} + \dfrac{1}{12}$

21. $\dfrac{5}{12} + \dfrac{3}{8}$

22. $\dfrac{7}{8} + \dfrac{1}{16}$

23. $\dfrac{3}{20} + 4$

24. $\dfrac{2}{15} + 3$

25. $\dfrac{5}{-8} + \dfrac{5}{6}$

26. $\dfrac{5}{-6} + \dfrac{7}{9}$

27. $\dfrac{3}{10}x + \dfrac{7}{100}x$

28. $\dfrac{9}{20}a + \dfrac{3}{40}a$

29. $\dfrac{5}{12} + \dfrac{4}{15}$

30. $\dfrac{3}{16} + \dfrac{1}{12}$

31. $\dfrac{9}{10} + \dfrac{-99}{100}$

32. $\dfrac{3}{10} + \dfrac{-27}{100}$

33. $5 + \dfrac{7}{12}$

34. $7 + \dfrac{3}{8}$

35. $-5t + \dfrac{2}{7}t$

36. $-4x + \dfrac{3}{5}x$

37. $-\dfrac{5}{12} + \dfrac{7}{-24}$

38. $-\dfrac{1}{18} + \dfrac{5}{-12}$

39. $\dfrac{4}{10} + \dfrac{3}{100} + \dfrac{7}{1000}$

40. $\dfrac{7}{10} + \dfrac{2}{100} + \dfrac{9}{1000}$

41. $\dfrac{3}{10} + \dfrac{5}{12} + \dfrac{8}{15}$

42. $\dfrac{1}{2} + \dfrac{3}{8} + \dfrac{1}{4}$

43. $\dfrac{5}{6} + \dfrac{25}{52} + \dfrac{7}{4}$

44. $\dfrac{15}{24} + \dfrac{7}{36} + \dfrac{91}{48}$

45. $\dfrac{2}{9} + \dfrac{7}{10} + \dfrac{-4}{15}$

46. $\dfrac{5}{12} + \dfrac{-3}{8} + \dfrac{1}{10}$

Chapter 4   Fractions: Addition & Subtraction

c Use < or > for ▓ to form a true sentence.

47. $\dfrac{5}{8}$ ▓ $\dfrac{6}{8}$

48. $\dfrac{7}{9}$ ▓ $\dfrac{5}{9}$

49. $\dfrac{2}{3}$ ▓ $\dfrac{5}{6}$

50. $\dfrac{11}{18}$ ▓ $\dfrac{5}{9}$

51. $\dfrac{-2}{3}$ ▓ $\dfrac{-5}{7}$

52. $\dfrac{-3}{5}$ ▓ $\dfrac{-4}{7}$

53. $\dfrac{11}{15}$ ▓ $\dfrac{7}{10}$

54. $\dfrac{5}{14}$ ▓ $\dfrac{8}{21}$

55. $\dfrac{3}{4} - \dfrac{1}{5}$

56. $\dfrac{3}{8} - \dfrac{13}{16}$

57. $\dfrac{-7}{20}$ ▓ $\dfrac{-6}{15}$

58. $\dfrac{-7}{12}$ ▓ $\dfrac{-9}{16}$

Arrange each grouping of fractions from smallest to largest.

59. $\dfrac{3}{10}, \dfrac{5}{12}, \dfrac{4}{15}$

60. $\dfrac{5}{6}, \dfrac{19}{21}, \dfrac{11}{14}$

d Solve.

61. Rose bought $\tfrac{1}{3}$ lb of orange pekoe tea and $\tfrac{1}{2}$ lb of English cinnamon tea. How many pounds of tea were bought altogether?

62. Mitch bought $\tfrac{1}{4}$ lb of gumdrops and $\tfrac{1}{2}$ lb of caramels. How many pounds of candy were bought altogether?

63. Ruwanda walked $\tfrac{3}{8}$ mi to Juan's dormitory, and then $\tfrac{3}{4}$ mi to class. How far did Ruwanda walk?

64. Ola walked $\tfrac{7}{8}$ mi to the student union, and then $\tfrac{2}{5}$ mi to class. How far did Ola walk?

65. *Baking.* A recipe for muffins calls for $\tfrac{1}{2}$ qt (quart) of buttermilk, $\tfrac{1}{3}$ qt of skim milk, and $\tfrac{1}{16}$ qt of oil. How many quarts of liquid ingredients does the recipe call for?

66. *Baking.* A recipe for bread calls for $\tfrac{2}{3}$ cup of water, $\tfrac{1}{4}$ cup of milk, and $\tfrac{1}{8}$ cup of oil. How many cups of liquid ingredients does the recipe call for?

Exercise Set 4.2

67. *Masonry.* A cubic meter of concrete mix contains 420 kg of cement, 150 kg of stone, and 120 kg of sand. What is the total weight of the cubic meter of concrete mix? What fractional part is cement? stone? sand? Add these amounts. What is the result?

68. *Bartending.* A recipe for cherry punch calls for $\frac{1}{5}$ L of ginger ale and $\frac{3}{5}$ L of black cherry soda. How much liquid is needed? If the recipe is doubled, how much liquid is needed? If the recipe is halved, how much liquid is needed?

69. A park ranger hikes $\frac{3}{5}$ mi to a lookout, another $\frac{3}{10}$ mi to an osprey's nest, and finally, $\frac{3}{4}$ mi to a campsite. How far did the ranger hike?

70. A triathlete runs $\frac{7}{8}$ mi, canoes $\frac{1}{3}$ mi, and swims $\frac{1}{6}$ mi. How many miles does the triathlete cover?

71. A tile $\frac{5}{8}$ in. thick is glued to a board $\frac{7}{8}$ in. thick. The glue is $\frac{3}{32}$ in. thick. How thick is the result?

72. A baker used $\frac{1}{2}$ lb of flour for rolls, $\frac{1}{4}$ lb for donuts, and $\frac{1}{3}$ lb for cookies. How much flour was used?

Skill Maintenance

Subtract. [2.3a]

73. $-7 - 6$

74. $-5 - (-9)$

75. $9 - 17$

76. $-8 - 23$

Evaluate. [2.6a]

77. $\frac{x - y}{3}$, for $x = 7$ and $y = -3$

78. $3(x + y)$ and $3x + 3y$, for $x = 5$ and $y = 9$

Synthesis

79. ◆ Suppose that a classmate believes, incorrectly, that $\frac{2}{5} + \frac{4}{5} = \frac{6}{10}$. How could you convince the classmate that he or she is mistaken?

80. ◆ Explain how pictures could be used to convince someone that $\frac{5}{7}$ is larger than $\frac{13}{21}$.

81. ◆ To add numbers with different denominators, a student consistently uses the product of the denominators as a common denominator. Is this correct? Why or why not?

Add and, if possible, simplify.

82. $\frac{3}{10}t + \frac{2}{7} + \frac{2}{15}t + \frac{3}{5}$

83. $\frac{2}{9} + \frac{4}{21}x + \frac{4}{15} + \frac{3}{14}x$

84. $5t^2 + \frac{6}{a}t + 2t^2 + \frac{3}{a}t$

Use <, >, or = for ■ to form a true sentence.

85. ■ $\frac{12}{97} + \frac{67}{137}$ ■ $\frac{8144}{13,289}$

86. ■ $\frac{37}{157} + \frac{19}{107}$ ■ $\frac{6941}{16,799}$

87. A guitarist's band is booked for Friday and Saturday night at a local club. The guitarist is part of a trio on Friday and part of a quintet on Saturday. Thus the guitarist is paid one-third of one-half the weekend's pay for Friday and one-fifth of one-half the weekend's pay for Saturday. What fractional part of the band's pay did the guitarist receive for the weekend's work? If the band was paid $1200, how much did the guitarist receive?

Arrange sockets and drill bits in fractional sizes from smallest to largest.

4.3 Subtraction, Equations, and Applications

a | Subtraction

Like Denominators

We can consider the difference $\frac{4}{8} - \frac{3}{8}$ as we did before, as either "take away" or "how much more." Let's consider "take away."

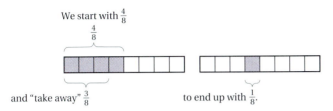

We start with 4 eighths and take away 3 eighths:

4 eighths − 3 eighths = 1 eighth,

or $\quad 4 \cdot \frac{1}{8} - 3 \cdot \frac{1}{8} = \frac{1}{8}, \quad$ or $\quad \frac{4}{8} - \frac{3}{8} = \frac{1}{8}.$

> To subtract when denominators are the same,
> a) subtract the numerators,
> b) keep the denominator, and
> c) simplify, if possible.
>
> $\frac{7}{10} - \frac{4}{10} = \frac{7-4}{10} = \frac{3}{10}$

Examples Subtract and simplify.

1. $\frac{8}{13} - \frac{3}{13} = \frac{8-3}{13} = \frac{5}{13}$

2. $\frac{3}{35} - \frac{13}{35} = \frac{3-13}{35} = \frac{-10}{35} = \frac{5}{5} \cdot \frac{-2}{7} = \frac{-2}{7},$ or $-\frac{2}{7} \quad$ Removing a factor equal to 1: $\frac{5}{5} = 1$

3. $\frac{13}{2a} - \frac{5}{2a} = \frac{13-5}{2a} = \frac{8}{2a} = \frac{2}{2} \cdot \frac{4}{a} = \frac{4}{a} \quad$ Removing a factor equal to 1: $\frac{2}{2} = 1$

Do Exercises 1–3.

Different Denominators

> To subtract when denominators are different:
> a) Find the least common multiple of the denominators. That number is the least common denominator, LCD.
> b) Multiply by 1, writing 1 in the form of n/n, to find an equivalent subtraction in which the LCD appears.
> c) Subtract and, if possible, simplify.

Objectives

a | Subtract using fractional notation.

b | Solve equations of the type $x + a = b$ and $a + x = b$, where a and b may be fractions.

c | Solve applied problems involving subtraction with fractional notation.

For Extra Help

TAPE 7 | MAC WIN | CD-ROM

Subtract and simplify.

1. $\frac{7}{8} - \frac{3}{8}$

2. $\frac{9}{5a} - \frac{6}{5a}$

3. $\frac{8}{10} - \frac{13}{10}$

Answers on page A-10

4. Subtract: $\dfrac{3}{4} - \dfrac{2}{3}$.

Subtract.

5. $\dfrac{5}{6} - \dfrac{2}{3}$

6. $\dfrac{2}{5} - \dfrac{7}{10}$

7. $\dfrac{2}{3} - \dfrac{5}{6}$

8. $\dfrac{11}{28} - \dfrac{5}{16}$

9. Simplify: $\dfrac{9}{10}x - \dfrac{3}{5}x$.

Answers on page A-10

Chapter 4 Fractions: Addition & Subtraction

Example 4 Subtract: $\dfrac{2}{5} - \dfrac{3}{8}$.

a) The LCM of 5 and 8 is 40, so the LCD is 40.

b) We need to find numbers equivalent to $\dfrac{2}{5}$ and $\dfrac{3}{8}$ with denominators of 40:

$$\dfrac{2}{5} - \dfrac{3}{8} = \dfrac{2}{5} \cdot \dfrac{8}{8} - \dfrac{3}{8} \cdot \dfrac{5}{5}$$

Think: $8 \times \blacksquare = 40$. The answer is 5, so we multiply by 1, using $\dfrac{5}{5}$.

Think: $5 \times \blacksquare = 40$. The answer is 8, so we multiply by 1, using $\dfrac{8}{8}$.

c) We subtract: $\dfrac{16}{40} - \dfrac{15}{40} = \dfrac{16 - 15}{40} = \dfrac{1}{40}$.

Do Exercise 4.

Example 5 Subtract: $\dfrac{7}{12} - \dfrac{5}{6}$.

Since 6 is a factor of 12, the LCM of 6 and 12 is 12. The LCD is 12.

$$\dfrac{7}{12} - \dfrac{5}{6} = \dfrac{7}{12} - \dfrac{5}{6} \cdot \dfrac{2}{2}$$

Think: $6 \times \blacksquare = 12$. The answer is 2, so we multiply by 1, using $\dfrac{2}{2}$.

$$= \dfrac{7}{12} - \dfrac{10}{12}$$

$$= \dfrac{7 - 10}{12} = \dfrac{-3}{12}$$ If we prefer, we can add the opposite: $7 + (-10)$.

$$= \dfrac{3}{3} \cdot \dfrac{-1}{4} = \dfrac{-1}{4}, \text{ or } -\dfrac{1}{4}$$ Simplifying by removing a factor equal to 1: $\dfrac{3}{3} = 1$

Example 6 Subtract: $\dfrac{17}{24} - \dfrac{4}{15}$.

We need to find the LCM of 24 and 15:

$\left.\begin{array}{l} 24 = 2 \cdot 2 \cdot 2 \cdot 3, \\ 15 = 3 \cdot 5 \end{array}\right\}$ The LCM is $2 \cdot 2 \cdot 2 \cdot 3 \cdot 5$, or 120.

$$\dfrac{17}{24} - \dfrac{4}{15} = \dfrac{17}{24} \cdot \dfrac{5}{5} - \dfrac{4}{15} \cdot \dfrac{8}{8}$$ Multiplying by 1 to obtain the LCD. To form 1, look at the prime factorization of the LCM and use the factors that each denominator lacks.

$$= \dfrac{85}{120} - \dfrac{32}{120} = \dfrac{85 - 32}{120} = \dfrac{53}{120}.$$

Do Exercises 5–8.

Example 7 Simplify by combining like terms: $\dfrac{7}{8}x - \dfrac{3}{4}x$.

$$\dfrac{7}{8}x - \dfrac{3}{4}x = \left(\dfrac{7}{8} - \dfrac{3}{4}\right)x$$ Try to do this step mentally.

$$= \left(\dfrac{7}{8} - \dfrac{6}{8}\right)x = \dfrac{1}{8}x$$ Multiplying $\dfrac{3}{4}$ by $\dfrac{2}{2}$ and subtracting

Do Exercise 9.

b Solving Equations

In Section 2.8, we introduced the addition principle as one way to form equivalent equations. We can use that principle here to solve equations containing fractions.

Example 8 Solve: $x - \dfrac{1}{3} = \dfrac{6}{7}$.

$$x - \dfrac{1}{3} = \dfrac{6}{7}$$

$$x - \dfrac{1}{3} + \dfrac{1}{3} = \dfrac{6}{7} + \dfrac{1}{3}$$ Using the addition principle: adding $\frac{1}{3}$ on both sides

$$x + 0 = \dfrac{6}{7} + \dfrac{1}{3}$$ Adding $\frac{1}{3}$ "undoes" the subtraction of $\frac{1}{3}$.

$$x = \dfrac{6}{7} \cdot \dfrac{3}{3} + \dfrac{1}{3} \cdot \dfrac{7}{7}$$ Multiplying by 1 to obtain the LCD, 21

$$x = \dfrac{18}{21} + \dfrac{7}{21} = \dfrac{25}{21}$$

CHECK:

$$\dfrac{x - \dfrac{1}{3} = \dfrac{6}{7}}{\begin{array}{c|c} \dfrac{25}{21} - \dfrac{1}{3} \;?\; \dfrac{6}{7} \\ \dfrac{25}{21} - \dfrac{1}{3} \cdot \dfrac{7}{7} \\ \dfrac{25}{21} - \dfrac{7}{21} \\ \dfrac{18}{21} \\ \dfrac{6 \cdot \cancel{3}}{7 \cdot \cancel{3}} & \dfrac{6}{7} \end{array}}$$ TRUE

Recall that since subtraction can be regarded as adding the opposite of the number being subtracted, the addition principle allows us to subtract the same number on both sides of an equation.

Example 9 Solve: $x + \dfrac{1}{4} = \dfrac{3}{5}$.

$$x + \dfrac{1}{4} - \dfrac{1}{4} = \dfrac{3}{5} - \dfrac{1}{4}$$ Using the addition principle to add $-\frac{1}{4}$ or to subtract $\frac{1}{4}$ on both sides

$$x + 0 = \dfrac{3}{5} \cdot \dfrac{4}{4} - \dfrac{1}{4} \cdot \dfrac{5}{5}$$ The LCD is 20. We multiply by 1 to get the LCD.

$$x = \dfrac{12}{20} - \dfrac{5}{20} = \dfrac{7}{20}$$

The solution is $\dfrac{7}{20}$. We leave the check to the student.

Do Exercises 10–12.

Solve.

10. Solve: $x - \dfrac{2}{5} = \dfrac{1}{5}$.

11. $x + \dfrac{2}{3} = \dfrac{5}{6}$

12. $\dfrac{3}{5} + t = -\dfrac{7}{8}$

Answers on page A-10

4.3 Subtraction, Equations, and Applications

13. A $\frac{4}{5}$-cup bottle of salad dressing consists of olive oil and vinegar. The bottle contains $\frac{2}{3}$ cup of oil. How much vinegar is in the bottle?

c **Solving Problems**

Example 10 *Woodworking.* Celeste is replacing a $\frac{3}{4}$-in. thick shelf in her bookcase. If her replacement board is $\frac{15}{16}$ in. thick, how much should it be planed down before the repair can be completed?

1. **Familiarize.** We first make a drawing or at least visualize the situation.

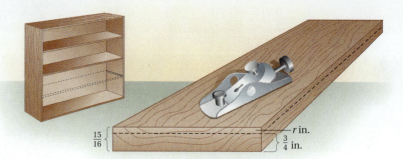

If necessary, we consult a reference book to learn that to plane a piece of wood is to use a tool for removing thin amounts of wood while producing a smooth surface. We let $r =$ the amount of wood, in inches, that needs to be removed.

2. **Translate.** The problem translates to a "take-away" situation.

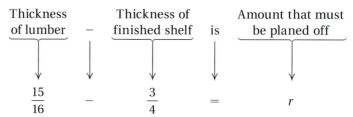

3. **Solve.** To solve the equation, we carry out the subtraction:

$$\frac{15}{16} - \frac{3}{4} = r \quad \text{The LCD is 16.}$$

$$\frac{15}{16} - \frac{3}{4} \cdot \frac{4}{4} = r \quad \text{Multiplying by 1}$$

$$\frac{15}{16} - \frac{12}{16} = r$$

$$\frac{3}{16} = r.$$

4. **Check.** We can check by adding $\frac{3}{16}$ to $\frac{3}{4}$:

$$\frac{3}{16} + \frac{3}{4} = \frac{3}{16} + \frac{12}{16} = \frac{15}{16}.$$

Since $\frac{15}{16}$ in. is the board's original thickness, our answer checks.

5. **State.** Celeste should plane $\frac{3}{16}$ in. from the replacement board.

Do Exercise 13.

Answer on page A-10

Exercise Set 4.3

a Subtract and, if possible, simplify.

1. $\dfrac{5}{6} - \dfrac{1}{6}$

2. $\dfrac{7}{5} - \dfrac{2}{5}$

3. $\dfrac{11}{16} - \dfrac{15}{16}$

4. $\dfrac{5}{12} - \dfrac{7}{12}$

5. $\dfrac{7}{a} - \dfrac{3}{a}$

6. $\dfrac{4}{t} - \dfrac{9}{t}$

7. $\dfrac{10}{3t} - \dfrac{4}{3t}$

8. $\dfrac{9}{2a} - \dfrac{5}{2a}$

9. $\dfrac{3}{5a} - \dfrac{7}{5a}$

10. $\dfrac{3}{7t} - \dfrac{9}{7t}$

11. $\dfrac{7}{8} - \dfrac{1}{16}$

12. $\dfrac{4}{3} - \dfrac{5}{6}$

13. $\dfrac{7}{15} - \dfrac{4}{5}$

14. $\dfrac{3}{4} - \dfrac{3}{28}$

15. $\dfrac{3}{4} - \dfrac{1}{20}$

16. $\dfrac{3}{4} - \dfrac{4}{16}$

17. $\dfrac{2}{15} - \dfrac{5}{12}$

18. $\dfrac{11}{16} - \dfrac{9}{10}$

19. $\dfrac{6}{10} - \dfrac{7}{100}$

20. $\dfrac{9}{10} - \dfrac{3}{100}$

21. $\dfrac{7}{15} - \dfrac{3}{25}$

22. $\dfrac{18}{25} - \dfrac{4}{35}$

23. $\dfrac{69}{100} - \dfrac{9}{10}$

24. $\dfrac{42}{100} - \dfrac{11}{20}$

25. $\dfrac{2}{3} - \dfrac{1}{8}$ **26.** $\dfrac{3}{4} - \dfrac{1}{2}$ **27.** $\dfrac{3}{5} - \dfrac{1}{2}$ **28.** $\dfrac{5}{6} - \dfrac{2}{3}$

29. $\dfrac{11}{18} - \dfrac{7}{24}$ **30.** $\dfrac{-7}{25} - \dfrac{2}{15}$ **31.** $\dfrac{13}{90} - \dfrac{17}{120}$ **32.** $\dfrac{8}{25} - \dfrac{29}{150}$

33. $\dfrac{2}{3}x - \dfrac{4}{9}x$ **34.** $\dfrac{7}{4}x - \dfrac{5}{12}x$ **35.** $\dfrac{3}{5}a - \dfrac{3}{4}a$ **36.** $\dfrac{4}{7}a - \dfrac{1}{3}a$

b Solve.

37. $x - \dfrac{5}{9} = \dfrac{2}{9}$ **38.** $x - \dfrac{3}{11} = \dfrac{7}{11}$ **39.** $a + \dfrac{2}{11} = \dfrac{8}{11}$ **40.** $a + \dfrac{4}{15} = \dfrac{13}{15}$

41. $x + \dfrac{2}{3} = \dfrac{7}{9}$ **42.** $x + \dfrac{1}{2} = \dfrac{7}{8}$ **43.** $a - \dfrac{3}{8} = \dfrac{3}{4}$ **44.** $x - \dfrac{3}{10} = \dfrac{2}{5}$

45. $\dfrac{2}{3} + x = \dfrac{4}{5}$

46. $\dfrac{4}{5} + x = \dfrac{6}{7}$

47. $\dfrac{3}{8} + a = \dfrac{1}{12}$

48. $\dfrac{5}{6} + a = \dfrac{2}{9}$

49. $n - \dfrac{1}{10} = -\dfrac{1}{30}$

50. $n - \dfrac{3}{4} = -\dfrac{5}{12}$

51. $x + \dfrac{3}{4} = -\dfrac{1}{2}$

52. $x + \dfrac{5}{6} = -\dfrac{11}{12}$

c Solve.

53. Monica spent $\tfrac{3}{4}$ hr listening to tapes of Beethoven and Brahms. She spent $\tfrac{1}{3}$ hr listening to Beethoven. How many hours were spent listening to Brahms?

54. From a $\tfrac{4}{5}$-lb wheel of cheese, a $\tfrac{1}{4}$-lb piece was served. How much cheese remained on the wheel?

55. *Exercise.* As part of an exercise program, Hugo is to walk $\tfrac{7}{8}$ mi each day. He has already walked $\tfrac{1}{3}$ mi. How much farther should Hugo walk?

56. *Fitness.* As part of a fitness program, Deb swims $\tfrac{1}{2}$ mi every day. She has already swum $\tfrac{1}{5}$ mi. How much farther should Deb swim?

57. *Tire Tread.* A new long-life tire has a tread depth of $\tfrac{3}{8}$ in. instead of the more typical $\tfrac{11}{32}$ in. (*Source*: *Popular Science*). How much deeper is the new tread depth?

58. *Furniture Cleaner.* A $\tfrac{3}{4}$-cup mixture of lemon juice and olive oil makes an excellent cleaner for wood furniture. If the mixture contains $\tfrac{1}{3}$ cup of lemon juice, how much olive oil is in the cleaner?

Exercise Set 4.3

Skill Maintenance

Divide and simplify. [3.7b]

59. $\dfrac{9}{10} \div \dfrac{3}{5}$

60. $\dfrac{3}{7} \div \dfrac{9}{4}$

61. $(-7) \div \dfrac{1}{3}$

62. $8 \div \left(-\dfrac{1}{4}\right)$

63. A small box of cornflakes weighs $\dfrac{3}{4}$ lb. How much do 8 small boxes of cornflakes weigh? [3.6b]

64. A batch of fudge requires $\dfrac{3}{4}$ cup of sugar. How much sugar is needed to make 12 batches? [3.6b]

Synthesis

65. ◆ If a negative fraction is subtracted from another negative fraction, is the result always negative? Why or why not?

66. ◆ Victor incorrectly writes $\dfrac{8}{5} - \dfrac{8}{2} = \dfrac{8}{3}$. How could you convince Victor that his subtraction is incorrect?

67. ◆ Without performing the actual computation, explain how you can tell that $\dfrac{3}{7} - \dfrac{5}{9}$ is negative.

Solve.

68. ▦ $x + \dfrac{16}{323} = \dfrac{10}{187}$

69. ▦ $x + \dfrac{7}{253} = \dfrac{12}{299}$

70. A mountain climber, beginning at sea level, climbs $\dfrac{3}{5}$ km, descends $\dfrac{1}{4}$ km, climbs $\dfrac{1}{3}$ km, and then descends $\dfrac{1}{7}$ km. At what elevation does the climber finish?

Simplify.

71. $\dfrac{2}{5} - \dfrac{1}{6}(-3)^2$

72. $\dfrac{7}{8} - \dfrac{1}{10}\left(-\dfrac{5}{6}\right)^2$

73. $-4 \times \dfrac{3}{7} - \dfrac{1}{7} \times \dfrac{4}{5}$

74. $\left(\dfrac{5}{6}\right)^2 + \left(\dfrac{3}{4}\right)^2$

75. Mazzi's meat slicer cut 8 slices of turkey and 3 slices of Vermont cheddar. If each turkey slice was $\dfrac{1}{16}$-in. thick and each cheddar slice was $\dfrac{5}{32}$-in. thick, how tall was the pile of cold cuts?

76. As part of a rehabilitation program, an athlete must swim and then walk a total of $\dfrac{9}{10}$ km each day. If one lap in the swimming pool is $\dfrac{3}{80}$ km, how far must the athlete walk after swimming 10 laps?

77. The Fullerton estate was left to four children. One received $\dfrac{1}{4}$ of the estate, one received $\dfrac{3}{8}$, and the twins split the rest. What fractional piece did each twin receive?

78. Mark Romano owns $\dfrac{7}{12}$ of Romano-Chrenka Chevrolet and Lisa Romano owns $\dfrac{1}{6}$. If Paul and Ella Chrenka own the remaining share of the dealership equally, what fractional piece does Paul own?

4.4 Solving Equations: Using the Principles Together

The equations that we solved in Sections 2.8, 3.8, and 4.3 required the use of either the addition principle or the multiplication principle. In this section, we will learn to solve equations that require the use of *both* principles. As review, let's restate both principles.

> **THE ADDITION PRINCIPLE**
>
> For any numbers a, b, and c,
>
> $a = b$ is equivalent to $a + c = b + c$.

> **THE MULTIPLICATION PRINCIPLE**
>
> For any numbers a, b, and c, with $c \neq 0$,
>
> $a = b$ is equivalent to $a \cdot c = b \cdot c$.

a Using the Principles Together

Suppose we want to determine whether 6 is the solution of $5x - 8 = 27$. To check, we replace x with 6 and simplify.

CHECK:
$$5x - 8 = 27$$
$$5 \cdot 6 - 8 \;?\; 27$$
$$30 - 8$$
$$22 \;|\; 27 \quad \text{FALSE}$$

This shows that 6 is *not* the solution.

Do Exercises 1 and 2.

In the check above, note that the rules for order of operations dictate that we multiply (or divide) before we subtract (or add).

When solving an equation, we find an equivalent equation in which the variable is isolated on one side. Until now, this has been accomplished by "undoing" the operation present:

  To solve $x - 7 = -2$, we added 7 on both sides.
  To solve $9x = 63$, we divided by 9 on both sides.

To solve an equation like $5x - 8 = 27$, we need to "undo" both subtraction *and* multiplication in order to isolate x. Because the subtraction is performed *after* the multiplication in $5x - 8$, we will use addition *before* division when solving.

Objectives

a Solve equations that require use of both the addition principle and the multiplication principle.

For Extra Help

TAPE 7   MAC WIN   CD-ROM

1. Determine whether -9 is the solution of $7x + 8 = -55$.

2. Determine whether -6 is the solution of $4x + 3 = -25$.

Answers on page A-10

3. Solve: $2x - 9 = 43$

Example 1 Solve: $5x - 8 = 27$.

We first isolate $5x$ by adding 8 on both sides:

$5x - 8 = 27$
$5x - 8 + 8 = 27 + 8$ **Using the addition principle**
$5x + 0 = 35$ **Try to do this step mentally.**
$5x = 35$.

Next, we isolate x by dividing by 5 (or multiplying by $\frac{1}{5}$) on both sides:

$5x = 35$
$\dfrac{5x}{5} = \dfrac{35}{5}$ **Using the division principle or the multiplication principle (multiplying by $\frac{1}{5}$ on both sides)**
$1x = 7$ **Try to do this step mentally.**
$x = 7$.

CHECK:
$\begin{array}{c|c} 5x - 8 = 27 \\ \hline 5 \cdot 7 - 8 \;?\; 27 \\ 35 - 8 \\ 27 \;|\; 27 \quad \text{TRUE} \end{array}$

The solution is 7.

Note that we subtracted *last* in the check and used the addition principle *first* when solving.

Do Exercise 3.

4. Solve: $-3x + 2 = 47$

Example 2 Solve: $38 = 9x + 2$.

We first isolate $9x$ by subtracting 2 on both sides:

$38 = 9x + 2$
$38 - 2 = 9x + 2 - 2$ **Subtracting 2 (or adding −2) on both sides**
$36 = 9x + 0$ **Try to do this step mentally.**
$36 = 9x$.

Now that we have isolated $9x$ on one side of the equation, we can divide by 9 to isolate x:

$36 = 9x$
$\dfrac{36}{9} = \dfrac{9x}{9}$ **Dividing by 9 (or multiplying by $\frac{1}{9}$ on both sides)**
$4 = x$.   **Simplifying**

CHECK:
$\begin{array}{c|c} 38 = 9x + 2 \\ \hline 38 \;?\; 9 \cdot 4 + 2 \\ 36 + 2 \\ 38 \;|\; 38 \quad \text{TRUE} \end{array}$

The solution is 4.

Do Exercise 4.

Answers on page A-10

Example 3 Solve: $20 = 6 - \frac{2}{3}x$.

Our plan is to first use the addition principle to isolate $-\frac{2}{3}x$ and to then use the multiplication principle to isolate x.

$$20 = 6 - \frac{2}{3}x$$

$$20 - 6 = 6 - \frac{2}{3}x - 6 \quad \text{\color{red}Subtracting 6 (or adding −6) on both sides}$$

$$14 = -\frac{2}{3}x$$

$$\left(-\frac{3}{2}\right)14 = \left(-\frac{3}{2}\right)\left(-\frac{2}{3}x\right) \quad \text{\color{red}Multiplying by } -\frac{3}{2} \text{ (or dividing by } -\frac{2}{3}\text{)} \text{ on both sides}$$

$$-\frac{3 \cdot 14}{2} = 1x$$

$$-\frac{3 \cdot 7 \cdot \cancel{2}}{\cancel{2}} = 1x \quad \text{\color{red}Removing a factor equal to 1: } \frac{2}{2} = 1$$

$$-21 = x.$$

CHECK:
$$\begin{array}{c|c} 20 = 6 - \frac{2}{3}x & \\ \hline 20 \; ? \; 6 - \frac{2}{3}(-21) & \\ & 6 + \frac{42}{3} \\ 20 & 6 + 14 \end{array} \quad \text{TRUE}$$

The solution is -21.

Do Exercise 5.

The same steps are used to solve the next example on page 234.

5. Solve: $9 - \frac{3}{4}x = 21$.

Answer on page A-10

4.4 Solving Equations: Using the Principles Together

6. Solve: $3 + \dfrac{14}{5}t = -\dfrac{21}{5}$.

Example 4 Solve: $5 + \dfrac{9}{2}t = -\dfrac{7}{2}$.

$$5 + \dfrac{9}{2}t = -\dfrac{7}{2}$$

$$5 + \dfrac{9}{2}t - 5 = -\dfrac{7}{2} - 5 \quad \text{Subtracting 5 on both sides}$$

$$\dfrac{9}{2}t = -\dfrac{7}{2} - \dfrac{10}{2} \quad \text{Writing 5 as } \dfrac{10}{2} \text{ to use the LCD}$$

$$\dfrac{9}{2}t = -\dfrac{17}{2}$$

$$\dfrac{2}{9} \cdot \dfrac{9}{2}t = \dfrac{2}{9}\left(-\dfrac{17}{2}\right) \quad \text{Multiplying by } \dfrac{2}{9} \text{ on both sides}$$

$$1t = -\dfrac{\cancel{2} \cdot 17}{9 \cdot \cancel{2}} \quad \text{Removing a factor equal to 1: } \dfrac{2}{2} = 1$$

$$t = -\dfrac{17}{9}$$

CHECK:

$$5 + \dfrac{9}{2}t = -\dfrac{7}{2}$$

$$5 + \dfrac{\cancel{9}}{2}\left(-\dfrac{17}{\cancel{9}}\right) \;?\; -\dfrac{7}{2} \quad \text{Removing a factor equal to 1: } \dfrac{9}{9} = 1$$

$$5 + \left(-\dfrac{17}{2}\right)$$

$$\dfrac{10}{2} + \left(\dfrac{-17}{2}\right)$$

$$\dfrac{10 - 17}{2}$$

$$\dfrac{-7}{2} \;\bigg|\; -\dfrac{7}{2} \quad \text{TRUE}$$

The solution is $-\dfrac{17}{9}$.

Do Exercise 6.

Answer on page A-10

Exercise Set 4.4

a Solve using the addition principle and/or the multiplication principle. Don't forget to check!

1. $6x - 4 = 14$

2. $7x - 6 = 22$

3. $3a + 8 = 23$

4. $19 = 2x - 7$

5. $4a + 9 = 37$

6. $2a - 9 = -7$

7. $31 = 3x - 5$

8. $5x + 7 = -8$

9. $-5t + 4 = 39$

10. $-8t + 7 = 39$

11. $3x + 4 = -11$

12. $3a - 7 = -1$

13. $\dfrac{4}{5}x = 20$

14. $\dfrac{3}{4}x = 18$

15. $\dfrac{3}{2}x - 3 = 12$

16. $\dfrac{7}{3}x - 1 = 6$

17. $\dfrac{3}{5}t - 4 = 8$

18. $6 - \dfrac{2}{9}t = -4$

19. $x + \dfrac{7}{3} = \dfrac{19}{6}$

20. $-\dfrac{41}{10} + x = \dfrac{7}{2}$

21. $7 = a + \dfrac{14}{5}$

22. $9 = a + \dfrac{47}{10}$

23. $\dfrac{2}{5}t - 1 = \dfrac{7}{5}$

24. $-\dfrac{53}{4} = \dfrac{3}{2}a + 2$

25. $\dfrac{39}{8} = \dfrac{11}{4} + \dfrac{1}{2}x$

26. $\dfrac{17}{2} = \dfrac{2}{7}t - \dfrac{3}{2}$

27. $\dfrac{13}{3}s + \dfrac{11}{2} = \dfrac{35}{4}$

28. $\dfrac{11}{5}t + \dfrac{36}{5} = \dfrac{7}{2}$

Skill Maintenance

Solve. [2.3b]

29. Jeremy withdraws $200 from an ATM (automated teller machine), makes a $90 deposit, and then withdraws another $40. How much has Jeremy's account balance changed?

30. Animal Instinct, a pet supply store, makes a profit of $850 on Friday, and $375 on Saturday, but suffers a loss of $45 on Sunday. Find the total profit or loss for the three days.

Divide and simplify. [3.7b]

31. $\dfrac{10}{7} \div 2m$

32. $45n \div \dfrac{9}{4}$

Multiply. [2.7a]

33. $3(a + b)$

34. $7(m - 3)$

Synthesis

35. ◆ Describe a procedure that a classmate could use to solve the equation $ax + b = c$ for x.

36. ◆ Nathan begins solving the equation $-\dfrac{2}{3}x + 7 = -9$ by adding 9 on both sides. Is this a wise thing to do? Why or why not?

37. ◆ Lorin begins solving the equation $\dfrac{2}{3}x + 1 = \dfrac{5}{6}$ by multiplying on both sides by 12. Is this a wise thing to do? Why or why not?

Solve.

38. 🖩 $\dfrac{1081}{3599}x - \dfrac{17}{61} = \dfrac{19}{59}$

39. 🖩 $\dfrac{553}{2451}a - \dfrac{13}{57} = \dfrac{29}{43}$

40. $-\dfrac{a}{5} + \dfrac{31}{4} = \dfrac{16}{3}$

41. $\dfrac{47}{5} - \dfrac{a}{4} = \dfrac{44}{7}$

42. $\dfrac{49}{8} + \dfrac{2x}{9} = 4$

43. The perimeter of the figure shown is 15 cm. Solve for x.

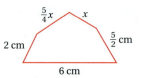

4.5 Mixed Numerals

a | What Is a Mixed Numeral?

A symbol like $2\frac{3}{4}$ is called a **mixed numeral**.

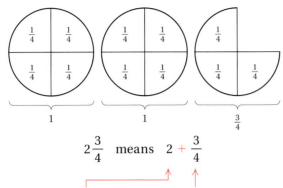

$$2\frac{3}{4} \quad \text{means} \quad 2 + \frac{3}{4}$$

This is a whole number.   This is a fraction less than 1.

Examples Convert to a mixed numeral.

1. $7 + \frac{2}{5} = 7\frac{2}{5}$

2. $4 + \frac{3}{10} = 4\frac{3}{10}$

Do Exercises 1–3.

The notation $2\frac{3}{4}$ has a plus sign left out. To aid in understanding, we sometimes write the missing plus sign. Similarly, the notation $-5\frac{2}{3}$ has a minus sign left out since $-5\frac{2}{3} = -\left(5 + \frac{2}{3}\right) = -5 - \frac{2}{3}$.

Mixed numbers can be displayed easily on a number line, as shown here.

Examples Convert to fractional notation.

3. $2\frac{3}{4} = 2 + \frac{3}{4}$ **Inserting the missing plus sign**

$= \frac{2}{1} + \frac{3}{4}$ $2 = \frac{2}{1}$

$= \frac{2}{1} \cdot \frac{4}{4} + \frac{3}{4}$ **Finding a common denominator**

$= \frac{8}{4} + \frac{3}{4}$

$= \frac{11}{4}$ **Adding**

4. $4\frac{3}{10} = 4 + \frac{3}{10} = \frac{4}{1} + \frac{3}{10} = \frac{4}{1} \cdot \frac{10}{10} + \frac{3}{10} = \frac{40}{10} + \frac{3}{10} = \frac{43}{10}$

Do Exercises 4 and 5.

Objectives

a Convert from mixed numerals to fractional notation.

b Convert from fractional notation to mixed numerals.

c Divide, writing a mixed numeral for the quotient.

For Extra Help

TAPE 8   MAC WIN   CD-ROM

---

1. $1 + \frac{2}{3} = \boxed{}$ Convert to a mixed numeral.

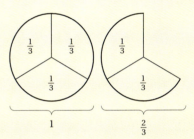

Convert to a mixed numeral.

2. $8 + \frac{3}{4}$

3. $12 + \frac{2}{3}$

Convert to fractional notation.

4. $4\frac{2}{5}$

5. $6\frac{1}{10}$

Answers on page A-10

4.5 Mixed Numerals

Convert to fractional notation. Use the faster method.

6. $4\frac{5}{6}$

7. $9\frac{1}{4}$

8. $20\frac{2}{3}$

Convert to fractional notation.

9. $-6\frac{2}{5}$

10. $-8\frac{3}{7}$

Answers on page A-10

Using Example 4, we can develop a faster way to convert.

> To convert from a mixed numeral like $4\frac{3}{10}$ to fractional notation:
> (a) Multiply: $4 \cdot 10 = 40$.
> (b) Add: $40 + 3 = 43$.
> (c) Keep the denominator.
>
> $4\frac{3}{10} = \frac{43}{10}$

Examples Convert to fractional notation.

5. $6\frac{2}{3} = \frac{20}{3}$ $6 \cdot 3 = 18, 18 + 2 = 20;$ keep the denominator

6. $8\frac{2}{9} = \frac{74}{9}$ $9 \cdot 8 = 72; 72 + 2 = 74;$ keep the denominator

7. $10\frac{7}{8} = \frac{87}{8}$ $8 \cdot 10 = 80; 80 + 7 = 87;$ keep the denominator

Do Exercises 6–8.

To find the opposite of the number in Example 5, we can write either $-6\frac{2}{3}$ or $-\frac{20}{3}$. Thus, to convert a negative mixed numeral to fractional notation, we remove the negative sign for purposes of computation and then include it in the answer.

Examples Convert to fractional notation.

8. $-5\frac{1}{3} = -\frac{16}{3}$ $3 \cdot 5 = 15; 15 + 1 = 16;$ include the negative sign

9. $-7\frac{5}{6} = -\frac{47}{6}$ $6 \cdot 7 = 42; 42 + 5 = 47$

Do Exercises 9 and 10.

b   Writing Mixed Numerals

We can find a mixed numeral for $\frac{5}{3}$ as follows:

$$\frac{5}{3} = \frac{3}{3} + \frac{2}{3} = 1 + \frac{2}{3} = 1\frac{2}{3}.$$

Fractional symbols like $\frac{5}{3}$ also indicate division:

$$\begin{array}{r} 1\frac{2}{3} \\ 3\overline{)5} \\ \underline{3} \\ 2 \end{array}$$ ← Now divide 2 by 3: $2 \div 3 = \frac{2}{3}$

Thus, $\frac{5}{3} = 1\frac{2}{3}$.

Chapter 4 Fractions: Addition & Subtraction

We can also visualize $\frac{5}{3}$ as one-third of 5 objects, as shown below.

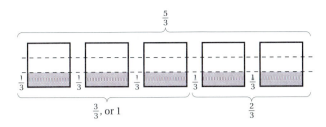

To convert from fractional notation to a mixed numeral, divide.

$\frac{13}{5}$; $\begin{array}{r} 2 \\ 5{\overline{\smash{)}13}} \\ \underline{10} \\ 3 \end{array}$ → The quotient

→ The remainder

$2\frac{3}{5}$; $\frac{13}{5} = 2\frac{3}{5}$

Examples Convert to a mixed numeral.

10. $\frac{8}{5}$ $\begin{array}{r} 1 \\ 5{\overline{\smash{)}8}} \\ \underline{5} \\ 3 \end{array}$ $\frac{8}{5} = 1\frac{3}{5}$

A fraction larger than 1, such as $\frac{8}{5}$, is sometimes referred to as an "improper" fraction. We have intentionally avoided such terminology. The use of such notation as $\frac{8}{5}$, $\frac{69}{10}$, and so on, is quite proper and very common in algebra.

11. $\frac{69}{10}$ $\begin{array}{r} 6 \\ 10{\overline{\smash{)}69}} \\ \underline{60} \\ 9 \end{array}$ $\frac{69}{10} = 6\frac{9}{10}$

12. $\frac{122}{8}$ $\begin{array}{r} 15 \\ 8{\overline{\smash{)}122}} \\ \underline{80} \\ 42 \\ \underline{40} \\ 2 \end{array}$ $\frac{122}{8} = 15\frac{2}{8} = 15\frac{1}{4}$

Whenever possible, simplify the fractional part of the numeral.

Do Exercises 11–13.

The same procedure also works with negative numbers. Of course, the result will be a negative mixed numeral.

Example 13 Convert $\frac{-9}{4}$ to a mixed numeral.

Since $\begin{array}{r} 2 \\ 4{\overline{\smash{)}9}} \\ \underline{8} \\ 1 \end{array}$, we have $\frac{9}{4} = 2\frac{1}{4}$.

Thus, $\frac{-9}{4} = -2\frac{1}{4}$.

Do Exercises 14 and 15.

Convert to a mixed numeral.

11. $\frac{7}{3}$

12. $\frac{11}{10}$

13. $\frac{110}{6}$

Convert to a mixed numeral.

14. $\frac{-12}{5}$

15. $-\frac{134}{12}$

Answers on page A-10

4.5 Mixed Numerals

239

16. Divide. Write a mixed numeral for the answer.

$$6 \overline{)4846}$$

c Finding Quotients and Averages

It is quite common when performing long division to express the quotient as a mixed numeral. As in Examples 10–13, the remainder becomes the numerator of the fractional part of the mixed numeral.

Example 14 Divide. Write a mixed numeral for the quotient.

$$7 \overline{)6341}$$

We first divide as usual.

$$\begin{array}{r} 905 \\ 7\overline{)6341} \\ \underline{6300} \\ 41 \\ \underline{35} \\ 6 \end{array}$$

The answer is 905 R 6, or, as a mixed numeral $905\frac{6}{7}$. Using fractional notation, we write $\frac{6341}{7} = 905\frac{6}{7}$.

Do Exercise 16.

Example 15 *Charities.* The American Institute of Philanthropy monitors how much charitable organizations spend in order to raise $100. (*Source:* AIP 1996–97 Watchdog Report. American Institute of Philanthropy, St. Louis, MO 63108). Five of the best organizations in this respect are listed below. How much did they spend, on average, to raise $100?

African–American Institute	$3
Asia Foundation	$4
International Rescue Committee	$8
Make-A-Wish Foundation	$7
National Hispanic Scholarship Fund	$4

Recall from Section 1.9 that to find the *average* of a set of values, we add the values and divide that sum by the number of values being added.

$$\text{Average spent} = \frac{3+4+8+7+4}{5} = \frac{26}{5} = 5\frac{1}{5}$$

On average, these groups spent $5\frac{1}{5}$ to raise $100.

Do Exercise 17.

17. Over the last 4 yr, Roland Thompson's raspberry patch has yielded 48, 35, 65, and 75 qt of berries. Find the average yield for the four years.

Calculator Spotlight

Exercises

If your calculator has the capability of finding whole-number quotients and remainders (see Section 1.6), use it to find mixed numerals for the answers to each of the following divisions.

1. $6 \overline{)8857}$
2. $9 \overline{)6088}$
3. $56 \overline{)44{,}851}$
4. $18 \overline{)234{,}567}$
5. $11 \overline{)567{,}895}$
6. $32 \overline{)234{,}567}$
7. $45 \overline{)6033}$
8. $213 \overline{)567{,}988}$
9. $112 \overline{)400{,}003}$
10. $908 \overline{)11{,}234}$

Answers on page A-10

Exercise Set 4.5

a Convert to fractional notation.

1. $3\frac{2}{5}$
2. $5\frac{2}{3}$
3. $6\frac{1}{4}$
4. $8\frac{1}{2}$
5. $20\frac{1}{8}$

6. $-10\frac{1}{3}$
7. $5\frac{1}{10}$
8. $8\frac{1}{10}$
9. $20\frac{3}{5}$
10. $30\frac{4}{5}$

11. $-9\frac{5}{6}$
12. $-8\frac{7}{8}$
13. $6\frac{9}{10}$
14. $1\frac{3}{5}$
15. $-12\frac{3}{4}$

16. $-15\frac{2}{3}$
17. $5\frac{7}{10}$
18. $7\frac{3}{100}$
19. $-5\frac{7}{100}$
20. $-6\frac{4}{15}$

b Convert to a mixed numeral.

21. $\frac{14}{3}$
22. $\frac{19}{8}$
23. $\frac{-27}{6}$
24. $\frac{30}{9}$
25. $\frac{57}{10}$
26. $\frac{-89}{10}$

27. $\frac{53}{7}$
28. $\frac{65}{8}$
29. $\frac{45}{6}$
30. $\frac{-50}{8}$
31. $\frac{46}{4}$
32. $\frac{39}{9}$

33. $\frac{-12}{8}$
34. $\frac{757}{100}$
35. $\frac{28}{6}$
36. $-\frac{345}{8}$
37. $-\frac{223}{4}$
38. $\frac{467}{100}$

c Divide. Write a mixed numeral for the answer.

39. $8 \overline{)8\ 6\ 9}$
40. $3 \overline{)2\ 1\ 2\ 6}$
41. $7 \overline{)6\ 3\ 4\ 5}$
42. $9 \overline{)9\ 1\ 1\ 0}$
43. $2\ 1 \overline{)8\ 5\ 2}$

44. $8\ 5 \overline{)7\ 6\ 7\ 2}$
45. $-302 \div 15$
46. $-475 \div 13$
47. $471 \div (-21)$
48. $542 \div (-25)$

Nutrition. For Exercises 49–52, consider the list at right of the 20 least fatty fast foods. (*Source:* The Consumer Bible by Mark Green. New York: Workman, 1995, p. 25).

Chain	Food	Fat
Boston Market	Fruit salad side dish	0 g
Boston Market	Steamed vegetables	0 g
Hardee's	Mashed potatoes and gravy side	0 g
KFC	Garden rice side dish	1 g
KFC	Mashed potatoes with gravy side dish	1 g
KFC	Green beans side dish	1 g
Arby's	Chicken Noodle Soup	2 g
Jack-in-the-Box	Chicken Teriyaki Bowl	2 g
KFC	Baked Beans side dish	2 g
KFC	Mean Greens side dish	2 g
Boston Market	Chicken Soup	3 g
Church's	Potatoes & Gravy	3 g
Jack-in-the-Box	Beef Teriyaki Bowl	3 g
KFC	Red Beans & Rice	3 g
Popeye's	Corn on the cob	3 g
Arby's	Mixed Vegetable Soup	4 g
Boston Market	Chicken breast sandwich, no mayo or mustard	4 g
Boston Market	White meat chicken quarter, no skin or wing	4 g
Dairy Queen/Brazier	BBQ Beef Sandwich	4 g
KFC	Vegetable Medley Salad	4 g

49. What is the average number of grams (g) of fat in the foods from Boston Market?

50. What is the average number of grams of fat in the foods from Arby's and Jack-in-the-Box taken together?

51. What is the average number of grams of fat for the entire list?

52. What is the average number of grams of fat for the last 10 items on the list?

Skill Maintenance

Multiply and simplify. [3.6a]

53. $\dfrac{7}{9} \cdot \dfrac{24}{21}$

54. $\dfrac{6}{5} \cdot 15$

55. $\dfrac{5}{12} \cdot (-6)$

56. $\dfrac{7}{10} \cdot \dfrac{5}{14}$

Synthesis

57. ◆ Describe in your own words a method for rewriting a mixed numeral as a fraction.

58. ◆ Describe in your own words a method for rewriting a fraction as a mixed numeral.

59. ◆ Are the numbers $2\tfrac{1}{3}$ and $2 \cdot \tfrac{1}{3}$ equal? Why or why not?

Write a mixed numeral.

60. ▦ $\dfrac{128{,}236}{541}$

61. ▦ $\dfrac{103{,}676}{349}$

62. $\dfrac{56}{7} + \dfrac{2}{3}$

63. $\dfrac{72}{12} + \dfrac{5}{6}$

64. $\dfrac{12}{5} + \dfrac{19}{15}$

65. There are $\dfrac{366}{7}$ weeks in a leap year.

66. There are $\dfrac{365}{7}$ weeks in a year.

67. *Athletics.* At a track and field meet, the hammer that is thrown has a wire length ranging from 3 ft, $10\tfrac{1}{4}$ in. to 3 ft, $11\tfrac{3}{4}$ in., a $4\tfrac{1}{8}$-in. grip, and a 16-lb ball with a diameter of from $4\tfrac{3}{8}$ in. to $5\tfrac{1}{8}$ in. Give specifications for the wire length and diameter of an "average" hammer.

4.6 Addition and Subtraction Using Mixed Numerals; Applications

a Addition

To find the sum $1\frac{5}{8} + 3\frac{1}{8}$, we first add the fractions. Then we add the whole numbers.

$$
\begin{array}{r}
1\frac{5}{8} = \\
+\,3\frac{1}{8} = \\
\hline
\frac{6}{8}
\end{array}
\qquad
\begin{array}{r}
1\frac{5}{8} \\
+\,3\frac{1}{8} \\
\hline
4\frac{6}{8} = 4\frac{3}{4}
\end{array}
$$

↑ Add the fractions.   ↑ Add the whole numbers.   ← Simplify the fractional part of the result when possible.

Objectives

a Add using mixed numerals.

b Subtract and combine like terms using mixed numerals.

c Solve applied problems involving addition and subtraction with mixed numerals.

d Add and subtract using negative mixed numerals.

For Extra Help

TAPE 8   MAC WIN   CD-ROM

Do Exercise 1.

Example 1 Add: $5\frac{2}{3} + 3\frac{5}{6}$. Write a mixed numeral for the answer.

We first rewrite $\frac{2}{3}$, using the LCD, 6. Then we add.

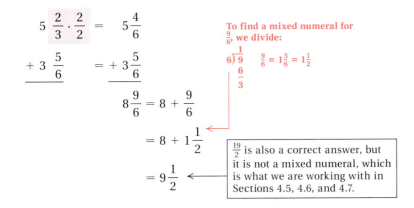

To find a mixed numeral for $\frac{9}{6}$, we divide:

$\frac{19}{2}$ is also a correct answer, but it is not a mixed numeral, which is what we are working with in Sections 4.5, 4.6, and 4.7.

1. Add.

$$
\begin{array}{r}
2\frac{3}{10} \\
+\,5\frac{1}{10} \\
\hline
\end{array}
$$

2. Add.

$$
\begin{array}{r}
8\frac{2}{5} \\
+\,3\frac{7}{10} \\
\hline
\end{array}
$$

Do Exercise 2.

Example 2 Add: $10\frac{5}{6} + 7\frac{3}{8}$.

The LCD is 24.

$$
\begin{array}{r}
10\,\frac{5}{6}\cdot\frac{4}{4} = \quad 10\frac{20}{24} \\
+\,7\,\frac{3}{8}\cdot\frac{3}{3} = +\,7\frac{9}{24} \\
\hline
17\frac{29}{24} = 18\frac{5}{24}
\end{array}
$$

The fractional part of a mixed numeral should always be less than 1.

3. Add.

$$
\begin{array}{r}
9\frac{3}{4} \\
+\,3\frac{5}{6} \\
\hline
\end{array}
$$

Do Exercise 3.

Answers on page A-10

4.6 Addition and Subtraction Using Mixed Numerals; Applications

243

Subtract.

4. $10\dfrac{7}{8}$
    $-\ 9\dfrac{3}{8}$

5. $8\dfrac{2}{3}$
    $-\ 5\dfrac{1}{2}$

6. Subtract.

$5\dfrac{1}{12}$
$-\ 1\dfrac{3}{4}$

7. Subtract.

5
$-\ 1\dfrac{1}{3}$

Answers on page A-10

Chapter 4 Fractions: Addition & Subtraction

b Subtraction

Example 3 Subtract: $7\dfrac{3}{4} - 2\dfrac{1}{4}$.

$$7\dfrac{3}{4} = \quad 7\dfrac{3}{4}$$
$$-\ 2\dfrac{1}{4} = -\ 2\dfrac{1}{4}$$
$$\overline{\dfrac{2}{4}} \quad\ \ \overline{5\dfrac{2}{4}} = 5\dfrac{1}{2}$$

↑ Subtract the fractions.
↑ Subtract the whole numbers.
↑ Simplifying

Example 4 Subtract: $9\dfrac{4}{5} - 3\dfrac{1}{2}$.

The LCD is 10.

$$9\dfrac{4}{5}\cdot\dfrac{2}{2} = \quad 9\dfrac{8}{10}$$
$$-\ 3\dfrac{1}{2}\cdot\dfrac{5}{5} = -\ 3\dfrac{5}{10}$$
$$\overline{6\dfrac{3}{10}}$$

Do Exercises 4 and 5.

Example 5 Subtract: $7\dfrac{1}{6} - 2\dfrac{1}{4}$.

The LCD is 12.

$$7\dfrac{1}{6}\cdot\dfrac{2}{2} = \quad 7\dfrac{2}{12}$$
$$-\ 2\dfrac{1}{4}\cdot\dfrac{3}{3} = -\ 2\dfrac{3}{12}$$

To subtract $2\dfrac{3}{12}$ from $7\dfrac{2}{12}$ we borrow 1, or $\dfrac{12}{12}$, from 7:
$7\dfrac{2}{12} = 6 + 1 + \dfrac{2}{12} = 6 + \dfrac{12}{12} + \dfrac{2}{12} = 6\dfrac{14}{12}$.

Once $7\dfrac{1}{6}$ has been written as $6\dfrac{14}{12}$, we can subtract as we did in Examples 3 and 4:

$$7\dfrac{2}{12} = \quad 6\dfrac{14}{12}$$
$$-\ 2\dfrac{3}{12} = -\ 2\dfrac{3}{12}$$
$$\overline{4\dfrac{11}{12}}.$$

Do Exercise 6.

Example 6 Subtract: $12 - 9\dfrac{3}{8}$.

$$12 \quad = \quad 11\dfrac{8}{8}$$
$$-\ 9\dfrac{3}{8} = -\ 9\dfrac{3}{8}$$
$$\overline{2\dfrac{5}{8}}$$

← $12 = 11 + 1 = 11 + \dfrac{8}{8} = 11\dfrac{8}{8}$

Do Exercise 7.

To combine like terms, we use the distributive law and add or subtract as above.

Example 7 Combine like terms: **(a)** $9\frac{3}{4}x - 4\frac{1}{2}x$; **(b)** $4\frac{5}{6}t + 2\frac{7}{9}t$.

a) $9\frac{3}{4}x - 4\frac{1}{2}x = \left(9\frac{3}{4} - 4\frac{1}{2}\right)x$ Using the distributive law. This is often done mentally.

$= \left(9\frac{3}{4} - 4\frac{2}{4}\right)x$ The LCD is 4.

$= 5\frac{1}{4}x$ Subtracting

b) $4\frac{5}{6}t + 2\frac{7}{9}t = \left(4\frac{5}{6} + 2\frac{7}{9}\right)t$ This step is often performed mentally.

$= \left(4\frac{15}{18} + 2\frac{14}{18}\right)t$ The LCD is 18.

$= 6\frac{29}{18}t = 7\frac{11}{18}t$

Do Exercises 8–10.

c Solving Problems

Example 8 *Intel Stock.* One day, the stock of Intel Corporation opened at $100\frac{3}{8}$ and then rose $4\frac{3}{4}$. Find the price of the stock at the end of the day.

1. **Familiarize.** We first make a drawing or at least visualize the situation. We let p = the price, in dollars, at the end of the day.

Note that $100\frac{3}{8}$ is close to $100 and that $4\frac{3}{4}$ is close to $5, so we expect the answer to be close to $100 + 5$, or $105.

2. **Translate.** From the work above, we see that the price at the end of the day is the opening price plus the amount of the rise. Thus,

$$p = 100\frac{3}{8} + 4\frac{3}{4}.$$

3. **Solve.** The equation tells us what to do. We add, using the LCD, 8:

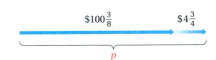

$$104\frac{9}{8} = 105\frac{1}{8}.$$

Thus, $p = \$105\frac{1}{8}$.

Combine like terms.

8. $7\frac{1}{6}t + 5\frac{2}{3}t$

9. $7\frac{11}{12}x - 5\frac{2}{3}x$

10. $4\frac{3}{10}x + 9\frac{11}{15}x$

Answers on page A-10

4.6 Addition and Subtraction Using Mixed Numerals; Applications

245

11. Executive Car Care sold two pieces of synthetic leather. One piece was $3\frac{1}{4}$ yd long and the other was $3\frac{5}{6}$ yd long. What was the total length sold?

12. *Knives.* The Damascus blade of a pearl-handled folding knife is $3\frac{3}{4}$ in. long. The same blade in an ATS-34 is $4\frac{1}{8}$ in. long (***Source:*** *Blade Magazine* 23, no. 10, October 1996: 26–27). How many inches longer is the ATS-34 blade?

Answers on page A-10

4. Check. We check by repeating the calculation or by noting that $105\frac{1}{8}$ is close to $105, as predicted in the *Familiarize* step.

5. State. The price of the stock was $105\frac{1}{8}$ at the end of the day.

Do Exercise 11.

Example 9 *NCAA Football Goalposts.* Recently, in college football, the distance between goalposts was reduced from $23\frac{1}{3}$ ft to $18\frac{1}{2}$ ft. By how much was it reduced?

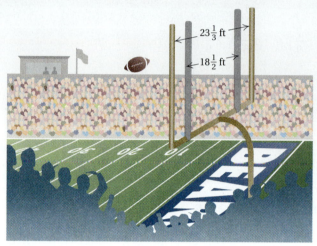

Source: NCAA

1. Familiarize. We let $d =$ the size of the reduction in feet and make a drawing to illustrate the situation.

2. Translate. We translate as follows.

$$\underbrace{\text{Former distance}}_{23\frac{1}{3}} - \underbrace{\text{New distance}}_{18\frac{1}{2}} = \underbrace{\text{Amount of reduction}}_{d}$$

3. Solve. To solve the equation, we carry out the subtraction. The LCD is 6.

$$23\frac{1}{3} = 23\frac{1}{3} \cdot \frac{2}{2} = 23\frac{2}{6} = 22\frac{8}{6}$$
$$-18\frac{1}{2} = -18\frac{1}{2} \cdot \frac{3}{3} = -18\frac{3}{6} = -18\frac{3}{6}$$
$$\overline{\phantom{-18\frac{1}{2} = -18\frac{1}{2} \cdot \frac{3}{3} = -18\frac{3}{6} = }\ 4\frac{5}{6}}$$

Thus, $d = 4\frac{5}{6}$ ft.

4. Check. To check, we add the reduction to the new distance:

$$18\frac{1}{2} + 4\frac{5}{6} = 18\frac{3}{6} + 4\frac{5}{6} = 22\frac{8}{6} = 23\frac{2}{6} = 23\frac{1}{3}.$$

This checks.

5. State. The reduction in the goalpost distance was $4\frac{5}{6}$ ft.

Do Exercise 12.

d Negative Mixed Numerals

Consider the numbers $5\frac{3}{4}$ and $-5\frac{3}{4}$ on a number line.

Note that just as $5\frac{3}{4}$ means $5 + \frac{3}{4}$, we can regard $-5\frac{3}{4}$ as $-5 - \frac{3}{4}$.

To subtract a larger number from a smaller number, we must modify the approach of Examples 3–6. To see why, consider the subtraction $4 - 4\frac{1}{2}$. We know that if we have \$4 and make a \$$4\frac{1}{2}$ purchase, we will owe half a dollar. Thus,

$$4 - 4\frac{1}{2} = -\frac{1}{2}.$$

The following is *not* correct:

$$\left.\begin{array}{rl} 4 &= 3\frac{2}{2} \\ -4\frac{1}{2} &= -4\frac{1}{2} \\ \hline & -1\frac{1}{2} \end{array}\right\} \text{Wrong!}$$

The correct answer, $-\frac{1}{2}$, can be obtained by rewriting the subtraction as addition (see Section 2.3):

$$4 - 4\frac{1}{2} = 4 + \left(-4\frac{1}{2}\right).$$

Because $-4\frac{1}{2}$ has the greater absolute value, the answer will be negative. The difference in absolute values is $4\frac{1}{2} - 4 = \frac{1}{2}$, so

$$4 - 4\frac{1}{2} = -\frac{1}{2}.$$

Do Exercise 13.

Example 10 Subtract: $3\frac{2}{7} - 4\frac{2}{5}$.

Since $4\frac{2}{5}$ is greater than $3\frac{2}{7}$, the answer will be negative. We can also see this by rewriting the subtraction as $3\frac{2}{7} + \left(-4\frac{2}{5}\right)$. The difference in absolute values is

$$\begin{array}{rl} 4\frac{2}{5} = & 4\frac{2}{5} \cdot \frac{7}{7} = 4\frac{14}{35} \\ -3\frac{2}{7} = & -3\frac{2}{7} \cdot \frac{5}{5} = -3\frac{10}{35} \\ \hline & 1\frac{4}{35}. \end{array}$$

Because $-4\frac{2}{5}$ has the larger absolute value, we make the answer negative.

Thus, $3\frac{2}{7} - 4\frac{2}{5} = -1\frac{4}{35}.$

---

13. Subtract: $7 - 7\frac{3}{4}$.

Answer on page A-10

Subtract.

14. $1\frac{3}{4} - 5\frac{1}{2}$

15. $-7\frac{1}{3} - \left(-5\frac{1}{2}\right)$

16. Subtract: $-7\frac{1}{10} - 6\frac{2}{15}$.

Answers on page A-10

Chapter 4 Fractions: Addition & Subtraction

Example 11 Subtract: $-6\frac{4}{5} - \left(-9\frac{3}{10}\right)$.

We rewrite the subtraction as addition:

$$-6\frac{4}{5} - \left(-9\frac{3}{10}\right) = -6\frac{4}{5} + 9\frac{3}{10}.$$ *Instead of subtracting, we add the opposite.*

Since $9\frac{3}{10}$ has the greater absolute value, the answer will be positive. The difference in absolute values is

$$
\begin{array}{rcrcrcr}
9\frac{3}{10} &=& 9\frac{3}{10} &=& 9\frac{3}{10} &=& 8\frac{13}{10} \\
-6\frac{4}{5} &=& -6\frac{4}{5} \cdot \frac{2}{2} &=& -6\frac{8}{10} &=& -6\frac{8}{10} \\
\hline
& & & & & & 2\frac{5}{10} = 2\frac{1}{2}.
\end{array}
$$

Thus, $-6\frac{4}{5} - \left(-9\frac{3}{10}\right) = 2\frac{1}{2}$.

Do Exercises 14 and 15.

In Section 2.2, we saw that to add two negative numbers we add absolute values and make the answer negative. The same approach is used with mixed numerals.

Example 12 Subtract: $-4\frac{1}{6} - 5\frac{2}{9}$.

We rewrite the subtraction as addition:

$$
\begin{aligned}
-4\frac{1}{6} - 5\frac{2}{9} &= -4\frac{1}{6} + \left(-5\frac{2}{9}\right) \\
&= -\left(4\frac{1}{6} + 5\frac{2}{9}\right) \quad \text{The LCD is 18.} \\
&= -\left(4\frac{3}{18} + 5\frac{4}{18}\right) \quad \frac{1}{6} \cdot \frac{3}{3} = \frac{3}{18};\ \frac{2}{9} \cdot \frac{2}{2} = \frac{4}{18} \\
&= -9\frac{7}{18}.
\end{aligned}
$$

Thus, $-4\frac{1}{6} - 5\frac{2}{9} = -9\frac{7}{18}$.

Do Exercise 16.

Exercise Set 4.6

a Perform the indicated operation. Write a mixed numeral for each answer.

1. $5\frac{7}{8}$
   $+ 3\frac{5}{8}$

2. $4\frac{5}{6}$
   $+ 3\frac{5}{6}$

3. $1\frac{1}{4}$
   $+ 1\frac{2}{3}$

4. $4\frac{1}{3}$
   $+ 5\frac{2}{9}$

5. $7\frac{3}{4}$
   $+ 5\frac{5}{6}$

6. $4\frac{3}{8}$
   $+ 6\frac{5}{12}$

7. $3\frac{2}{5}$
   $+ 8\frac{7}{10}$

8. $5\frac{1}{2}$
   $+ 3\frac{7}{10}$

9. $6\frac{3}{8}$
   $+ 10\frac{5}{6}$

10. $\frac{5}{8}$
    $+ 1\frac{5}{6}$

11. $12\frac{4}{5}$
    $+ 8\frac{7}{10}$

12. $15\frac{5}{8}$
    $+ 11\frac{3}{4}$

13. $14\frac{5}{8}$
    $+ 13\frac{1}{4}$

14. $16\frac{1}{4}$
    $+ 15\frac{7}{8}$

15. $4\frac{1}{5}$
    $- 2\frac{3}{5}$

16. $5\frac{1}{8}$
    $- 2\frac{3}{8}$

17. $6\frac{3}{5}$
    $- 2\frac{1}{2}$

18. $7\frac{2}{3}$
    $- 6\frac{1}{2}$

19. $34\frac{1}{3}$
    $- 12\frac{5}{8}$

20. $23\frac{5}{16}$
    $- 16\frac{3}{4}$

21. 21
    $- 8\frac{3}{4}$

22. 42
    $- 3\frac{7}{8}$

23. 34
    $- 18\frac{5}{8}$

24. 23
    $- 19\frac{3}{4}$

25. $21\frac{1}{6}$
    $- 13\frac{3}{4}$

26. $42\frac{1}{10}$
    $- 23\frac{7}{12}$

27. $25\frac{1}{9}$
    $- 13\frac{5}{6}$

28. $23\frac{5}{16}$
    $- 14\frac{7}{12}$

b Combine like terms.

29. $5\frac{3}{14}t + 3\frac{2}{21}t$

30. $9\frac{1}{2}x + 5\frac{3}{4}x$

31. $9\frac{1}{2}x - 7\frac{3}{8}x$

32. $7\frac{3}{4}x - 2\frac{3}{8}x$

33. $3\frac{7}{8}t + 4\frac{9}{10}t$

34. $5\frac{3}{8}x + 6\frac{2}{7}x$

35. $37\frac{5}{9}t - 25\frac{4}{5}t$

36. $23\frac{1}{6}t - 19\frac{2}{5}t$

37. $2\frac{5}{6}x + 3\frac{1}{3}x$

38. $7\frac{3}{20}t + 1\frac{2}{15}t$

39. $4\frac{3}{11}x + 5\frac{2}{3}x$

40. $4\frac{11}{12}t + 5\frac{7}{10}t$

c Solve.

41. *Fishing.* Candy caught two trout. One weighed $1\frac{1}{2}$ lb and the other weighed $2\frac{3}{4}$ lb. What was the total weight of the fish?

42. *Shopping.* Hubert purchased two packages of cheese weighing $1\frac{1}{3}$ lb and $4\frac{3}{5}$ lb. What was the total weight of the cheese?

43. *Heights.* Rocky is $187\frac{1}{10}$ cm tall and his daughter is $180\frac{3}{4}$ cm tall. How much taller is Rocky?

44. *Heights.* Aunt Louise is $168\frac{1}{4}$ cm tall and her son is $150\frac{7}{10}$ cm tall. How much taller is Aunt Louise?

45. *Plumbing.* Janet uses pipes of lengths $10\frac{5}{16}$ ft and $8\frac{3}{4}$ ft in the installation of a sink. How much pipe was used?

46. *Writing Supplies.* The standard pencil is $6\frac{7}{8}$ in. wood and $\frac{1}{2}$ in. eraser (*Source*: Eberhard Faber American). What is the total length of the standard pencil?

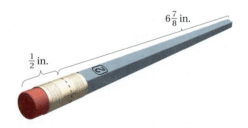

47. *Writing Supplies.* A standard sheet of paper is $8\frac{1}{2}$ in. by 11 in. What is the total distance around (perimeter of) the paper?

48. *Book Size.* One standard book size is $8\frac{1}{2}$ in. by $9\frac{3}{4}$ in. What is the total distance around (perimeter of) the front cover of such a book?

49. *Toys "R" Us Stock.* During a recent year, the price of one share of stock in Toys "R" Us varied between a low of 20\frac{1}{2}$ and a high of 37\frac{5}{8}$ (**Source:** Toys "R" Us annual report). What was the difference between the high and the low?

50. *Coca-Cola Stock.* On a recent day, the stock of Coca-Cola opened at 86\frac{1}{8}$ and closed at 84\frac{9}{16}$. How much did the price of the stock drop that day?

51. *Carpentry.* When cutting wood with a saw, a carpenter must take into account the thickness of the saw blade. Suppose that from a piece of wood 36 in. long, a carpenter cuts a $15\frac{3}{4}$-in. length with a saw blade that is $\frac{1}{8}$ in. in thickness. How long is the piece that remains?

52. *Painting.* When redecorating, a painter used $1\frac{3}{4}$ gal of paint for the living room and $1\frac{1}{3}$ gal for the family room. How much paint was used in all?

53. Sue, an interior designer, worked $10\frac{1}{2}$ hr over a three-day period. If Sue worked $2\frac{1}{2}$ hr on the first day and $4\frac{1}{5}$ hr on the second, how many hours did Sue work on the third day?

54. A DC-10 flew 640 mi on a nonstop flight. On the return flight, it landed after having flown $320\frac{3}{10}$ mi. How far was the plane from its original point of departure?

Find the perimeter of (distance around) each figure.

55.

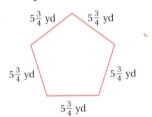

56.

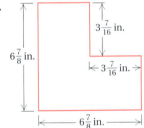

Find the length d in each figure.

57.

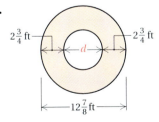

58.

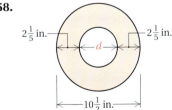

Exercise Set 4.6

59. Find the smallest length of a bolt that will pass through a piece of tubing with an outside diameter of $\frac{1}{2}$ in., a washer $\frac{1}{16}$ in. thick, a piece of tubing with a $\frac{3}{4}$-in. outside diameter, another washer $\frac{1}{16}$ in. thick, and a nut $\frac{3}{16}$ in. thick.

60. The front of the stage at the Lagrange Town Hall is $6\frac{1}{2}$ yd long. If renovation work succeeds in adding $2\frac{3}{4}$ yd in length, how long is the renovated stage?

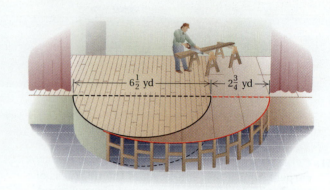

d Subtract.

61. $8\frac{3}{5} - 9\frac{2}{5}$

62. $4\frac{5}{7} - 8\frac{3}{7}$

63. $3\frac{1}{2} - 6\frac{3}{4}$

64. $5\frac{1}{2} - 7\frac{3}{4}$

65. $3\frac{4}{5} - 7\frac{2}{3}$

66. $2\frac{3}{7} - 5\frac{1}{2}$

67. $-3\frac{1}{5} - 4\frac{2}{5}$

68. $-5\frac{3}{8} - 4\frac{1}{8}$

69. $-4\frac{2}{5} - 6\frac{3}{7}$

70. $-2\frac{3}{4} - 3\frac{3}{8}$

71. $-6\frac{1}{9} - \left(-4\frac{2}{9}\right)$

72. $-2\frac{3}{5} - \left(-1\frac{1}{5}\right)$

Skill Maintenance

Perform the indicated operation and, if possible, simplify.

73. $\frac{12}{25} \div \frac{24}{5}$ [3.7b]

74. $\left(-\frac{15}{9}\right)\left(\frac{18}{39}\right)$ [3.6a]

75. $\left(-\frac{3}{4}\right)\left(-\frac{32}{33}\right)$ [3.6a]

76. $-\frac{49}{54} \div \frac{7}{6}$ [3.7b]

Synthesis

77. ◆ Explain how "borrowing" is used in this section.

78. ◆ Is the sum of two mixed numerals always a mixed numeral? Why or why not?

79. ◆ Write a problem for a classmate to solve. Design the problem so the solution is "The larger package holds $4\frac{1}{2}$ oz more than the smaller package."

Calculate each of the following. Write the result as a mixed numeral.

80. 🖩 $3289\frac{1047}{1189} + 5278\frac{32}{41}$

81. 🖩 $5798\frac{17}{53} - 3909\frac{1957}{2279}$

82. 🖩 $4230\frac{19}{73} - 5848\frac{17}{29}$

83. A post for a pier is 29 ft long. Half of the post extends above the water's surface and $8\frac{3}{4}$ ft of the post is buried in mud. How deep is the water at that location?

Solve.

84. $35\frac{2}{3} + n = 46\frac{1}{4}$

85. $42\frac{7}{9} = x - 13\frac{2}{5}$

86. $-15\frac{7}{8} = 12\frac{1}{2} + t$

4.7 Multiplication and Division Using Mixed Numerals; Applications

Whereas addition and subtraction of mixed numerals are usually performed by leaving the numbers as mixed numerals, multiplication and division are usually performed by first converting the numbers to fractional notation.

a Multiplication

> To multiply using mixed numerals, first convert to fractional notation. Then multiply with fractional notation and, if appropriate, rewrite the answer as a mixed numeral.

Objectives

a. Multiply using mixed numerals.
b. Divide using mixed numerals.
c. Evaluate expressions using mixed numerals.
d. Solve problems involving multiplication and division with mixed numerals.

For Extra Help

TAPE 8   MAC WIN   CD-ROM

Example 1 Multiply: $6 \cdot 2\frac{1}{2}$.

$$6 \cdot 2\frac{1}{2} = \frac{6}{1} \cdot \frac{5}{2} = \frac{6 \cdot 5}{1 \cdot 2} = \frac{2 \cdot 3 \cdot 5}{2 \cdot 1} = 15 \quad \text{Removing a factor equal to 1: } \frac{2}{2} = 1$$

Here we write fractional notation.

Do Exercise 1.

Example 2 Multiply: $3\frac{1}{2} \cdot \frac{3}{4}$.

$$3\frac{1}{2} \cdot \frac{3}{4} = \frac{7}{2} \cdot \frac{3}{4} = \frac{21}{8} = 2\frac{5}{8} \quad \text{Although fractional notation is needed, } common \text{ denominators are not required.}$$

Do Exercise 2.

Example 3 Multiply: $-8 \cdot 4\frac{2}{3}$.

$$-8 \cdot 4\frac{2}{3} = -\frac{8}{1} \cdot \frac{14}{3} = -\frac{112}{3} = -37\frac{1}{3}$$

Do Exercise 3.

Example 4 Multiply: $2\frac{1}{4} \cdot 3\frac{2}{5}$.

$$2\frac{1}{4} \cdot 3\frac{2}{5} = \frac{9}{4} \cdot \frac{17}{5} = \frac{153}{20} = 7\frac{13}{20}$$

CAUTION! $2\frac{1}{4} \cdot 3\frac{2}{5} \neq 6\frac{2}{20}$. A common error is to multiply the whole numbers and then the fractions. The correct answer, $7\frac{13}{20}$, is found after converting first to fractional notation.

Do Exercise 4.

1. Multiply: $6 \cdot 3\frac{1}{3}$.

2. Multiply: $2\frac{1}{2} \cdot \frac{3}{4}$.

3. Multiply: $-2 \cdot 6\frac{2}{5}$.

4. Multiply: $3\frac{1}{3} \cdot 2\frac{1}{2}$.

Answers on page A-10

5. Divide: $84 \div 5\frac{1}{4}$.

b Division

The division $1\frac{1}{2} \div \frac{1}{6}$ is shown here.

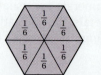

We see that $\frac{1}{6}$ goes into $1\frac{1}{2}$ nine times.

$$1\frac{1}{2} \div \frac{1}{6} = \frac{3}{2} \div \frac{1}{6}$$
$$= \frac{3}{2} \cdot 6 = \frac{3 \cdot 6}{2} = \frac{3 \cdot 3 \cdot 2}{2 \cdot 1} = \frac{3 \cdot 3}{1} \cdot \frac{2}{2} = \frac{3 \cdot 3}{1} \cdot 1 = 9$$

> To divide using mixed numerals, first write fractional notation. Then divide (multiply by the reciprocal of the divisor) and, if appropriate, rewrite the answer as a mixed numeral.

Divide.

6. $2\frac{1}{4} \div 1\frac{1}{5}$

Example 5   Divide: $32 \div 3\frac{1}{5}$.

$$32 \div 3\frac{1}{5} = \frac{32}{1} \div \frac{16}{5} \quad \text{Converting to fractional notation}$$
$$= \frac{32}{1} \cdot \frac{5}{16} = \frac{32 \cdot 5}{1 \cdot 16} = \frac{2 \cdot 16 \cdot 5}{1 \cdot 16} = 10 \quad \text{Removing a factor equal to 1: } \frac{16}{16} = 1$$

↑ Remember to multiply by the reciprocal of the divisor.

CAUTION!   The reciprocal of $3\frac{1}{5}$ is neither $5\frac{1}{3}$ nor $3\frac{5}{1}$!

Do Exercise 5.

Example 6   Divide: $2\frac{1}{3} \div 1\frac{3}{4}$.

$$2\frac{1}{3} \div 1\frac{3}{4} = \frac{7}{3} \div \frac{7}{4} = \frac{7}{3} \cdot \frac{4}{7} = \frac{7 \cdot 4}{7 \cdot 3} = \frac{4}{3} = 1\frac{1}{3} \quad \text{Removing a factor equal to 1: } \frac{7}{7} = 1$$

7. $1\frac{3}{4} \div \left(-2\frac{1}{2}\right)$

Example 7   Divide: $-1\frac{3}{5} \div \left(-3\frac{1}{3}\right)$.

$$-1\frac{3}{5} \div \left(-3\frac{1}{3}\right) = -\frac{8}{5} \div \left(-\frac{10}{3}\right) = \frac{8}{5} \cdot \frac{3}{10} \quad \text{The product or quotient of two negatives is positive.}$$
$$= \frac{2 \cdot 4 \cdot 3}{5 \cdot 2 \cdot 5} = \frac{12}{25} \quad \text{Removing a factor equal to 1: } \frac{2}{2} = 1$$

Do Exercises 6 and 7.

Answers on page A-10

c Evaluating Expressions

Mixed numerals can appear in algebraic expressions just as the integers of Section 2.6 did.

Example 8 A train traveling r miles per hour for t hours travels a total of rt miles. (*Remember*: Distance = Rate · Time.)

a) Find the distance traveled by a 60-mph train in $2\frac{3}{4}$ hr.
b) Find the distance traveled if the speed of the train is $26\frac{1}{2}$ mph and the time is $2\frac{2}{3}$ hr.

a) We evaluate rt for $r = 60$ and $t = 2\frac{3}{4}$:

$$rt = 60 \cdot 2\frac{3}{4}$$
$$= \frac{60}{1} \cdot \frac{11}{4}$$
$$= \frac{15 \cdot 4 \cdot 11}{1 \cdot 4} = 165.\quad \text{Removing a factor equal to 1: } \frac{4}{4} = 1$$

In $2\frac{3}{4}$ hr, a 60-mph train travels 165 mi.

b) We evaluate rt for $r = 26\frac{1}{2}$ and $t = 2\frac{2}{3}$:

$$rt = 26\frac{1}{2} \cdot 2\frac{2}{3}$$
$$= \frac{53}{2} \cdot \frac{8}{3} = \frac{53 \cdot 2 \cdot 4}{2 \cdot 3}\quad \text{Removing a factor equal to 1: } \frac{2}{2} = 1$$
$$= \frac{212}{3} = 70\frac{2}{3}.$$

In $2\frac{2}{3}$ hr, a $26\frac{1}{2}$-mph train travels $70\frac{2}{3}$ mi.

Example 9 Evaluate $x + yz$ for $x = 7\frac{1}{3}$, $y = \frac{1}{3}$, and $z = 5$.

We substitute and follow the rules for order of operations:

$$x + yz = 7\frac{1}{3} + \frac{1}{3} \cdot 5 \quad \text{The dot indicates the multiplication, } yz.$$
$$= 7\frac{1}{3} + \frac{1}{3} \cdot \frac{5}{1} \quad \text{Multiply first; then add.}$$
$$= 7\frac{1}{3} + \frac{5}{3}$$
$$= 7\frac{1}{3} + 1\frac{2}{3} \quad \text{Adding mixed numerals}$$
$$= 8\frac{3}{3} = 9.$$

Do Exercises 8–10.

Evaluate.

8. rt, for $r = 78$ and $t = 2\frac{1}{4}$

9. $7xy$, for $x = 9\frac{2}{5}$ and $y = 2\frac{3}{7}$

10. $x - y \div z$, for $x = 5\frac{7}{8}$, $y = \frac{1}{4}$, and $z = 2$

Answers on page A-10

11. Kyle's pickup truck travels on an interstate highway at 65 mph for $3\frac{1}{2}$ hr. How far does it travel?

12. Holly's minivan travels 302 mi on $15\frac{1}{10}$ gal of gas. How many miles per gallon did it get?

Answers on page A-10

d Applications and Problem Solving

Example 10   *Cassette Tape Music.* The tape in an audio cassette is played at a rate of $1\frac{7}{8}$ in. per second. A recording has 30 in. of damaged tape. How many seconds of music have been lost?

1. **Familiarize.** We can make a drawing.

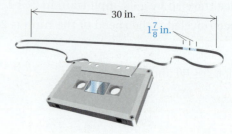

Since each $1\frac{7}{8}$ in. of tape represents 1 sec of lost music, the question can be regarded as asking how many times 30 can be divided by $1\frac{7}{8}$. We let t = the number of seconds of music lost.

2. **Translate.** The situation corresponds to a division sentence:

$$t = 30 \div 1\frac{7}{8}.$$

3. **Solve.** To solve the equation, we perform the division:

$$t = 30 \div 1\frac{7}{8}$$

$$= \frac{30}{1} \div \frac{15}{8} \qquad \text{Rewriting in fractional notation}$$

$$= \frac{30}{1} \cdot \frac{8}{15}$$

$$= \frac{15 \cdot 2 \cdot 8}{1 \cdot 15} \qquad \text{Removing a factor equal to 1: } \frac{15}{15} = 1$$

$$= 16.$$

4. **Check.** We check by multiplying. If 16 sec of music were lost, then

$$16 \cdot 1\frac{7}{8} = \frac{16}{1} \cdot \frac{15}{8}$$

$$= \frac{8 \cdot 2 \cdot 15}{1 \cdot 8} = 30 \text{ in.} \qquad \text{Removing a factor equal to 1: } \frac{8}{8} = 1$$

of tape were destroyed. Our answer checks. A quick, partial, check uses approximations:

$$16 \cdot 1\frac{7}{8} \approx 16 \cdot 2 = 32 \approx 30. \qquad \text{The symbol } \approx \text{ means "approximately equal to."}$$

5. **State.** The cassette has lost 16 sec of music.

Do Exercises 11 and 12.

Example 11 *Home Furnishings.* An L-shaped room consists of a rectangle that is $8\frac{1}{2}$ ft by 11 ft adjacent to one that is $6\frac{1}{2}$ by $7\frac{1}{2}$ ft. What is the total area of a carpet that covers the floor?

1. **Familiarize.** We make a drawing of the situation. We let $a =$ the total floor area.

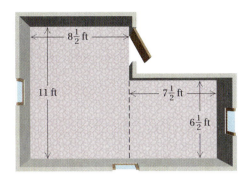

2. **Translate.** The total area is the sum of the areas of the two rectangles. This gives us the following equation:

$$a = 8\frac{1}{2} \cdot 11 + 7\frac{1}{2} \cdot 6\frac{1}{2}.$$

3. **Solve.** This is a multistep problem. We perform each multiplication and then add. This follows the rules for order of operations:

$$a = 8\frac{1}{2} \cdot 11 + 7\frac{1}{2} \cdot 6\frac{1}{2}$$

$$= \frac{17}{2} \cdot \frac{11}{1} + \frac{15}{2} \cdot \frac{13}{2} \quad \text{Rewriting in fractional notation}$$

$$= \frac{17 \cdot 11}{2 \cdot 1} + \frac{15 \cdot 13}{2 \cdot 2}$$

$$= \frac{187}{2} + \frac{195}{4}$$

$$= 93\frac{1}{2} + 48\frac{3}{4}$$

$$= 93\frac{2}{4} + 48\frac{3}{4} \quad \text{For addition, a common denominator is needed.}$$

$$= 141\frac{5}{4}$$

$$= 142\frac{1}{4}.$$

4. **Check.** We perform a partial check by estimating the total area as $11 \cdot 9 + 7 \cdot 7 = 99 + 49 = 148$ ft², Our answer, $142\frac{1}{4}$ ft², seems reasonable.

5. **State.** The total area of the carpet is $142\frac{1}{4}$ ft².

Do Exercise 13.

---

13. A room is $22\frac{1}{2}$ ft by $15\frac{1}{2}$ ft. A 9-ft by 12-ft Oriental rug is placed in the center of the room. How much area is not covered by the rug?

Answer on page A-10

4.7 Multiplication and Division Using Mixed Numerals; Applications

14. After two weeks, Kurt's tomato seedlings measure $9\frac{1}{2}$ in., $10\frac{3}{4}$ in., $10\frac{1}{4}$ in., and 9 in. tall. Find their average height.

Example 12 Melody has had three children. Their birth weights were $7\frac{1}{2}$ lb, $7\frac{3}{4}$ lb, and $6\frac{3}{4}$ lb. What was the average weight of her babies?

1. **Familiarize.** Recall that to compute an *average*, we add the values and then divide the sum by the number of values. We let $w =$ the average weight, in pounds.

2. **Translate.** We have
$$w = \frac{7\frac{1}{2} + 7\frac{3}{4} + 6\frac{3}{4}}{3}.$$

3. **Solve.** We first add:
$$7\frac{1}{2} + 7\frac{3}{4} + 6\frac{3}{4} = 7\frac{2}{4} + 7\frac{3}{4} + 6\frac{3}{4}$$
$$= 20\frac{8}{4} = 22. \quad 20\frac{8}{4} = 20 + \frac{8}{4} = 20 + 2$$

Then we divide:
$$w = \frac{7\frac{1}{2} + 7\frac{3}{4} + 6\frac{3}{4}}{3} = \frac{22}{3} = 7\frac{1}{3}. \quad \text{Dividing by 3.}$$

4. **Check.** As a partial check, we note that the average is smaller than the largest individual value and larger than the smallest individual value. We could also repeat our calculations.

5. **State.** The average weight of the three babies is $7\frac{1}{3}$ lb.

Answers on page A-10

Do Exercise 14.

Calculator Spotlight

Calculators equipped with a key, often labeled $\boxed{a^b/_c}$, allow for computations with fractional notation and mixed numerals. To calculate

$$\frac{2}{3} + \frac{4}{5}$$

with such a calculator, press:

$\boxed{2}\ \boxed{a^b/_c}\ \boxed{3}\ \boxed{+}\ \boxed{4}\ \boxed{a^b/_c}\ \boxed{5}\ \boxed{=}$.

The display that appears,

$\boxed{1\ \lrcorner\ 7\ \lrcorner\ 15}$,

represents the mixed numeral $1\frac{7}{15}$.

To express the answer in fractional notation, we press $\boxed{\text{Shift}}\ \boxed{d/c}$ and $22\ \lrcorner\ 15$ appears, representing $\frac{22}{15}$.

To enter a mixed numeral like $3\frac{2}{5}$ on a fraction calculator equipped with an $\boxed{a^b/_c}$ key, we press

$\boxed{3}\ \boxed{a^b/_c}\ \boxed{2}\ \boxed{a^b/_c}\ \boxed{5}$.

The calculator's display is in the form

$\boxed{3\ \lrcorner\ 2\ \lrcorner\ 5}$.

Some calculators are capable of displaying mixed numerals in the way in which we write them, as shown below.

Exercises

Calculate using a fraction calculator. Write the answers in fractional notation.

1. $\dfrac{3}{8} + \dfrac{1}{4}$ 2. $\dfrac{5}{12} + \dfrac{7}{10} - \dfrac{5}{12}$ 3. $\dfrac{15}{7} \cdot \dfrac{1}{3}$

4. $\dfrac{19}{20} \div \dfrac{17}{35}$ 5. $\dfrac{29}{30} - \dfrac{18}{25} \cdot \dfrac{2}{3}$ 6. $\dfrac{1}{2} + \dfrac{13}{29} \cdot \dfrac{3}{4}$

Calculate using a fraction calculator. Write the answers as mixed numerals.

7. $4\dfrac{1}{2} \cdot 5\dfrac{3}{7}$ 8. $7\dfrac{2}{3} \div 9\dfrac{4}{5}$

9. $8\dfrac{3}{7} + 5\dfrac{2}{9}$ 10. $13\dfrac{4}{9} - 7\dfrac{5}{8}$

11. $13\dfrac{1}{4} - 2\dfrac{1}{5} \cdot 4\dfrac{3}{8}$ 12. $2\dfrac{5}{6} + 5\dfrac{1}{6} \cdot 3\dfrac{1}{4}$

Exercise Set 4.7

a Multiply. Write a mixed numeral for each answer.

1. $10 \cdot 2\frac{5}{6}$
2. $5 \cdot 3\frac{3}{4}$
3. $6\frac{2}{3} \cdot 1\frac{1}{4}$
4. $9 \cdot 2\frac{3}{5}$

5. $-10 \cdot 7\frac{1}{3}$
6. $7\frac{3}{8} \cdot 4\frac{1}{3}$
7. $3\frac{1}{2} \cdot 4\frac{2}{3}$
8. $4\frac{1}{5} \cdot 5\frac{1}{4}$

9. $-2\frac{3}{10} \cdot 4\frac{2}{5}$
10. $4\frac{7}{10} \cdot 5\frac{3}{10}$
11. $6\frac{3}{10} \cdot 5\frac{7}{10}$
12. $-20\frac{1}{2} \cdot \left(-10\frac{1}{5}\right)$

b Divide. Write a mixed numeral for each answer whenever possible.

13. $20 \div 2\frac{3}{5}$
14. $18 \div 2\frac{1}{4}$
15. $8\frac{2}{5} \div 7$
16. $3\frac{3}{8} \div 3$

17. $4\frac{3}{4} \div 1\frac{1}{3}$
18. $5\frac{4}{5} \div 2\frac{1}{2}$
19. $-1\frac{7}{8} \div 1\frac{2}{3}$
20. $-4\frac{3}{8} \div 2\frac{5}{6}$

21. $5\frac{1}{10} \div 4\frac{3}{10}$
22. $4\frac{1}{10} \div 2\frac{1}{10}$
23. $20\frac{1}{4} \div (-90)$
24. $12\frac{1}{2} \div (-50)$

c Evaluate.

25. lw, for $l = 2\frac{3}{5}$ and $w = 9$

26. mv, for $m = 7$ and $v = 3\frac{2}{5}$

27. rs, for $r = 5$ and $s = 3\frac{1}{7}$

28. rt, for $r = 5\frac{2}{3}$ and $t = -2\frac{3}{8}$

29. mt, for $m = 6\frac{2}{9}$ and $t = -4\frac{3}{5}$

30. $M \div NP$, for $M = 2\frac{1}{4}$, $N = -5$, and $P = 2\frac{1}{3}$

31. $R \cdot S \div T$, for $R = 4\frac{2}{3}$, $S = 1\frac{3}{7}$, and $T = -5$

32. $a - bc$, for $a = 18$, $b = 2\frac{1}{5}$, and $c = 3\frac{3}{4}$

33. $r + ps$, for $r = 5\frac{1}{2}$, $p = 3$, and $s = 2\frac{1}{4}$

34. $s + rt$, for $s = 3\frac{1}{2}$, $r = 5\frac{1}{2}$, and $t = 7\frac{1}{2}$

35. $m + n \div p$, for $m = 7\frac{2}{5}$, $n = 4\frac{1}{2}$, and $p = 6$

36. $x - y \div z$, for $x = 9$, $y = 2\frac{1}{2}$, and $z = 3\frac{3}{4}$

d Solve.

37. *Home Furnishings.* Each shelf in June's entertainment center is 27 in. long. A videocassette is $1\frac{1}{8}$ in. thick. How many cassettes can she place on each shelf?

38. *Exercise.* At one point during a spinning class at Ray's health club, his bicycle wheel was completing $76\frac{2}{3}$ revolutions per minute. How many revolutions did the wheel complete in 6 min?

39. *Sodium Consumption.* The average American woman consumes $1\frac{1}{3}$ tsp of sodium each day (**Source**: *Nutrition Action Health Letter*, March 1994, p. 6. 1875 Connecticut Ave., N.W., Washington, DC 20009-5728). How much sodium do 10 average American women consume in one day?

40. *Aeronautics.* Most space shuttles orbit the earth once every $1\frac{1}{2}$ hr. How many orbits are made every 24 hr?

Chapter 4 Fractions: Addition & Subtraction

41. A serving of filleted fish is generally considered to be about $\frac{1}{3}$ lb. How many servings can be prepared from $5\frac{1}{2}$ lb of flounder fillet?

42. A serving of fish steak (cross section) is generally $\frac{1}{2}$ lb. How many servings can be prepared from a cleaned $18\frac{3}{4}$-lb tuna?

43. The weight of water is $62\frac{1}{2}$ lb per cubic foot. What is the weight of $5\frac{1}{2}$ cubic feet of water?

44. The weight of water is $62\frac{1}{2}$ lb per cubic foot. What is the weight of $2\frac{1}{4}$ cubic feet of water?

45. *Video Recording.* The tape in a VCR operating in the short-play mode travels at a rate of $1\frac{3}{8}$ in. per second. How many inches of tape are used to record for 60 sec in the short-play mode?

46. *Audio Recording.* The tape in an audio cassette is played at the rate of $1\frac{7}{8}$ in. per second. How many inches of tape are used when a cassette is played for $5\frac{1}{2}$ sec?

47. *Temperatures.* Fahrenheit temperature can be obtained from Celsius (centigrade) temperature by multiplying by $1\frac{4}{5}$ and adding 32°. What Fahrenheit temperature corresponds to a Celsius temperature of 20°?

48. *Temperature.* Fahrenheit temperature can be obtained from Celsius (centigrade) temperature by multiplying by $1\frac{4}{5}$ and adding 32°. What Fahrenheit temperature corresponds to the Celsius temperature of boiling water, which is 100°?

49. *Weightlifting.* In 1997, weightlifter Gao Shihong of China snatched $103\frac{1}{2}$ kg (**Source:** *The Guinness Book of Records*, 1998). This amount was about $1\frac{1}{2}$ times her body weight. How much did Shihong weigh?

50. *Weightlifting.* In 1983, weightlifter Stefan Topurov of Bulgaria hoisted $396\frac{3}{4}$ lb over his head (**Source:** *The Guinness Book of Records*, 1998). This amount was about three times his body weight. How much did Topurov weigh?

Exercise Set 4.7

51. *Birth Weights.* The Piper quadruplets of Great Britain weighed $2\frac{9}{16}$ lb, $2\frac{9}{32}$ lb, $2\frac{1}{8}$ lb, and $2\frac{5}{16}$ lb at birth. (**Source:** *The Guinness Book of Records,* 1998). Find their average birth weight.

52. *Vertical Leaps.* Eight-year-old Zachary registered vertical leaps of $12\frac{3}{4}$ in., $13\frac{3}{4}$ in., $13\frac{1}{2}$ in., and 14 in. Find his average vertical leap.

53. *Manufacturing.* A test of five light bulbs showed that they burned for the lengths of time given on the graph below. For how many days, on average, did the bulbs burn?

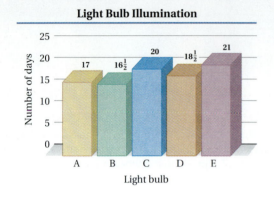

54. *Packaging.* A sample of four bags of beef jerky showed the weights given on the graph below. What was the average weight?

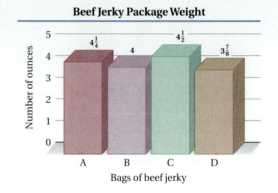

Find the area of each shaded region.

55.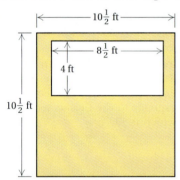

56.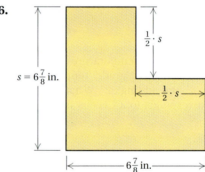

57. *Word Processing.* Kelly wants to create a table using Microsoft® Word software for word processing. She needs to have two columns, each $1\frac{1}{2}$ in. wide, and five columns, each $\frac{3}{4}$ in. wide. Will this table fit on a piece of standard paper that is $8\frac{1}{2}$ in. wide? If so, how wide will each margin be if her margins on each side are to be of equal width?

58. *Construction.* A rectangular lot has dimensions of $302\frac{1}{2}$ ft by $205\frac{1}{4}$ ft. A building with dimensions of 100 ft by $25\frac{1}{2}$ ft is built on the lot. How much area is left over?

Skill Maintenance

59. Multiply: $-8(x - 3)$ [2.7a]

60. On a winter night, the temperature dropped from 7°F to −12°F. How many degrees did it drop?   [2.3b]

61. Solve: $-9x = 189$.   [2.8b]

62. Solve: $-9 + x = 189$.   [2.8a]

63. Divide: $-198 \div (-6)$.   [2.5a]

64. Multiply: $(-7)(185)(0)$.   [2.4a]

Synthesis

65. ◈ If Kate and Jessie are both less than 5 ft, $6\frac{1}{2}$ in. tall, but Dot is over 5 ft, $6\frac{1}{2}$ in. tall, is it possible that the average height of the three exceeds 5 ft, $6\frac{1}{2}$ in.? Why or why not?

66. ◈ Write a problem for a classmate to solve. Design the problem so that its solution is found by performing the multiplication $4\frac{1}{2} \cdot 33\frac{1}{3}$.

67. ◈ Under what circumstances is a pair of mixed numerals more easily added than multiplied?

Simplify. Write each answer as a mixed numeral whenever possible.

68. 🖩 $15\frac{2}{11} \cdot 23\frac{31}{43}$

69. 🖩 $17\frac{23}{31} \cdot 19\frac{13}{15}$

70. $-8 \div \frac{1}{2} + \frac{3}{4} + \left(-5 - \frac{5}{8}\right)^2$

71. $\left(\frac{5}{9} - \frac{1}{4}\right)(-12) + \left(-4 - \frac{3}{4}\right)^2$

72. $\frac{1}{3} \div \left(\frac{1}{2} - \frac{1}{5}\right) \times \frac{1}{4} + \frac{1}{6}$

73. $\frac{7}{8} - 1\frac{1}{8} \times \frac{2}{3} + \frac{9}{10} \div \frac{3}{5}$

Exercise Set 4.7

Evaluate.

74. $ab + ac$ and $a(b + c)$, for $a = 3\frac{1}{4}$, $b = 5\frac{1}{3}$, and $c = 4\frac{5}{8}$

75. $a^3 + a^2$ and $a^2(a + 1)$, for $a = -3\frac{1}{2}$

76. Use a calculator to determine what whole number a must be in order for the following to be true:

$$\frac{a}{17} + \frac{10 + a}{23} = \frac{330}{391}.$$

77. *Heights.* Find the average height of the following NBA stars:

Shawn Kemp	6 ft, 10 in.
Grant Hill	6 ft, 7 in.
Damon Stoudamire	5 ft, 10 in.
Kobe Bryant	6 ft, 6 in.
Shaquille O'Neal	7 ft, 1 in.

Recipes. The following heart-healthy recipes serve four. How much of each ingredient would you use to serve ten?

78.

79.

Summary and Review: Chapter 4

Important Properties and Formulas

The Addition Principle: $a = b$ is equivalent to $a + c = b + c$.
The Multiplication Principle: For $c \neq 0$, $a = b$ is equivalent to $c \cdot a = c \cdot b$.

Review Exercises

Find the LCM. [4.1a]

1. 12 and 18
2. 18 and 45
3. 3, 6, and 30

Use $<$ or $>$ for ▢ to form a true sentence. [4.2c]

12. $\dfrac{4}{7}$ ▢ $\dfrac{5}{9}$

13. $-\dfrac{8}{9}$ ▢ $-\dfrac{11}{13}$

Perform the indicated operation and, if possible, simplify. [4.2a, b], [4.3a]

4. $\dfrac{2}{9} + \dfrac{5}{9}$

5. $\dfrac{3}{a} + \dfrac{4}{a}$

Solve. [4.3b], [4.4a]

14. $x + \dfrac{2}{5} = \dfrac{7}{8}$

15. $7a - 2 = 26$

6. $-\dfrac{6}{5} + \dfrac{11}{15}$

7. $\dfrac{5}{16} + \dfrac{3}{24}$

16. $5 + \dfrac{16}{3}x = \dfrac{5}{9}$

17. $\dfrac{22}{5} = \dfrac{16}{5} + \dfrac{5}{2}x$

8. $\dfrac{5}{9} - \dfrac{2}{9}$

9. $\dfrac{3}{4} - \dfrac{7}{8}$

Convert to fractional notation. [4.5a]

18. $7\dfrac{1}{2}$

19. $8\dfrac{3}{8}$

10. $\dfrac{11}{27} - \dfrac{2}{9}$

11. $\dfrac{5}{6} - \dfrac{2}{9}$

20. $-4\dfrac{1}{3}$

21. $10\dfrac{5}{7}$

Summary and Review: Chapter 4

Convert to a mixed numeral. [4.5b]

22. $\dfrac{7}{3}$

23. $\dfrac{-27}{4}$

24. $\dfrac{63}{5}$

25. $\dfrac{7}{2}$

26. Divide. Write a mixed numeral for the answer. [4.5c]

$7896 \div (-9)$

27. Gina's golf scores were 79, 81, and 84. What was her average score? [4.5c]

Perform the indicated operation. Write a mixed numeral for each answer. [4.6a, b, d]

28. $\;\;5\dfrac{3}{5}$
$\;\;\;\;+\,4\dfrac{4}{5}$

29. $\;\;8\dfrac{1}{3}$
$\;\;\;\;+\,3\dfrac{2}{5}$

30. $-5\dfrac{5}{6} + \left(-3\dfrac{1}{6}\right)$

31. $-2\dfrac{3}{4} + 4\dfrac{1}{2}$

32. $\;\;12$
$\;\;-\,4\dfrac{2}{9}$

33. $\;\;9\dfrac{3}{5}$
$\;\;-\,4\dfrac{13}{15}$

34. $4\dfrac{5}{8} - 9\dfrac{3}{4}$

35. $-7\dfrac{1}{2} - 6\dfrac{3}{4}$

Combine like terms. [4.2a], [4.6b]

36. $\dfrac{4}{9}x + \dfrac{1}{3}x$

37. $5\dfrac{3}{10}a - 2\dfrac{1}{8}a$

Perform the indicated operation. Write a mixed numeral or integer for each answer. [4.7a, b]

38. $6 \cdot 2\dfrac{2}{3}$

39. $-5\dfrac{1}{4} \cdot \dfrac{2}{3}$

40. $2\dfrac{1}{5} \cdot 1\dfrac{1}{10}$

41. $2\dfrac{2}{5} \cdot 2\dfrac{1}{2}$

Chapter 4 Fractions: Addition & Subtraction

42. $27 \div 2\frac{1}{4}$

43. $2\frac{2}{5} \div \left(-1\frac{7}{10}\right)$

44. $3\frac{1}{4} \div 26$

45. $4\frac{1}{5} \div 4\frac{2}{3}$

Evaluate. [4.7c]

46. $5x - y$, for $x = 3\frac{1}{5}$ and $y = 2\frac{2}{7}$

47. $2a \div b$, for $a = 5\frac{2}{11}$ and $b = 3\frac{4}{5}$

Solve.

48. A curtain requires $2\frac{3}{5}$ m of material. How many curtains can be made from 39 m of material? [4.7d]

49. On the first day of trading on the stock market, stock in Alcoa opened at $67\frac{3}{4}$ and rose by $2\frac{5}{8}$ at the close of trading. What was the stock's closing price? [4.6c]

50. Mica pedals up a $\frac{1}{10}$-mi hill and then coasts for $\frac{1}{2}$ mi down the other side. How far has she traveled? [4.2d]

51. A wedding-cake recipe requires 12 cups of shortening. Being calorie-conscious, the wedding couple decides to reduce the shortening by $3\frac{5}{8}$ cups and replace it with prune purée. How many cups of shortening are used in their new recipe? [4.6c]

52. What is the sum of the areas in the figure below? [4.6c], [4.7d]

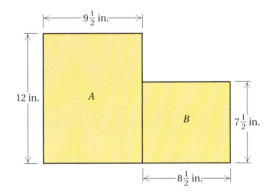

53. In the figure in Exercise 52, how much larger is the area of rectangle A than the area of rectangle B? [4.6c], [4.7d]

Summary and Review: Chapter 4

Skill Maintenance

54. Multiply and simplify: $\dfrac{9}{10} \cdot \left(-\dfrac{4}{3}\right)$. [3.6a]

55. Divide and simplify: $\dfrac{5}{4} \div \left(-\dfrac{5}{6}\right)$. [3.7b]

56. Bright Sunshine Landscaping made $230 Monday but lost $150 on Tuesday and $110 on Wednesday. Find the total profit or loss. [2.3b]

57. Multiply: $5(a - 9)$. [2.7a]

Synthesis

58. ◈ Rachel insists that $3\frac{2}{5} \cdot 1\frac{3}{7} = 3\frac{6}{35}$. What mistake is she probably making and how should she have proceeded instead? [4.7a]

59. ◈ Do least common multiples play any role in the addition or subtraction of mixed numerals? Why or why not? [4.6a, b]

60. ▦ Find the LCM of 141, 2419, and 1357. [4.1a]

61. Find r if
$$\dfrac{1}{r} = \dfrac{1}{100} + \dfrac{1}{150} + \dfrac{1}{200}.$$ [4.2b]

62. Find the smallest integer for which each fraction is greater than $\frac{1}{2}$. [4.2c]

a) $\dfrac{\blacksquare}{11}$ b) $\dfrac{\blacksquare}{8}$

c) $\dfrac{\blacksquare}{23}$ d) $\dfrac{\blacksquare}{35}$

e) $\dfrac{-51}{\blacksquare}$ f) $\dfrac{-78}{\blacksquare}$

g) $\dfrac{-2}{\blacksquare}$ h) $\dfrac{-1}{\blacksquare}$

63. Find the largest integer for which each fraction is greater than 1. [4.2c]

a) $\dfrac{7}{\blacksquare}$ b) $\dfrac{11}{\blacksquare}$

c) $\dfrac{47}{\blacksquare}$ d) $\dfrac{\frac{9}{8}}{\blacksquare}$

e) $\dfrac{\blacksquare}{-13}$ f) $\dfrac{\blacksquare}{-27}$

g) $\dfrac{\blacksquare}{-1}$ h) $\dfrac{\blacksquare}{-\frac{1}{2}}$

Test: Chapter 4

1. Find the LCM of 12 and 16.

Perform the indicated operation and, if possible, simplify.

2. $\dfrac{1}{2} + \dfrac{5}{2}$

3. $-\dfrac{7}{8} + \dfrac{2}{3}$

4. $\dfrac{5}{t} - \dfrac{3}{t}$

5. $\dfrac{5}{6} - \dfrac{3}{4}$

6. $\dfrac{5}{8} - \dfrac{17}{24}$

7. Use < or > for ▇ to form a true sentence.

    $\dfrac{6}{7}$ ▇ $\dfrac{21}{25}$

Solve.

8. $x + \dfrac{2}{3} = \dfrac{11}{12}$

9. $-5x - 2 = 10$

10. $32 = 2 + \dfrac{5}{3}x$

Convert to fractional notation.

11. $3\dfrac{1}{2}$

12. $-9\dfrac{7}{8}$

13. Convert to a mixed numeral:

    $-\dfrac{74}{9}$.

14. Divide. Write a mixed numeral for the answer.

    $11\overline{)1789}$

Perform the indicated operation. Write a mixed numeral for each answer.

15. $6\dfrac{2}{5}$
    $+ 7\dfrac{4}{5}$

16. $9\dfrac{1}{4}$
    $+ 5\dfrac{1}{6}$

17. $10\dfrac{1}{6}$
    $- 5\dfrac{7}{8}$

18. $14 + \left(-5\dfrac{3}{7}\right)$

19. $3\dfrac{4}{5} - 9\dfrac{1}{2}$

Combine like terms.

20. $\dfrac{3}{8}x - \dfrac{1}{2}x$

21. $5\dfrac{2}{11}a - 3\dfrac{1}{5}a$

Answers

1. _____
2. _____
3. _____
4. _____
5. _____
6. _____
7. _____
8. _____
9. _____
10. _____
11. _____
12. _____
13. _____
14. _____
15. _____
16. _____
17. _____
18. _____
19. _____
20. _____
21. _____

Perform the indicated operation. Write a mixed numeral for each answer.

22. $9 \cdot 4\frac{1}{3}$

23. $6\frac{3}{4} \cdot \left(-2\frac{2}{3}\right)$

24. $33 \div 5\frac{1}{2}$

25. $2\frac{1}{3} \div 1\frac{1}{6}$

Evaluate.

26. $\frac{2}{3}ab$, for $a = 7$ and $b = 4\frac{1}{5}$

27. $4 + mn$, for $m = 7\frac{2}{5}$ and $n = 3\frac{1}{4}$

Solve.

28. One batch of low-cholesterol turkey chili calls for $1\frac{1}{2}$ lb of roasted turkey breast. How much turkey is needed for 5 batches?

29. An order of books for a math course weighs 220 lb. Each book weighs $2\frac{3}{4}$ lb. How many books are in the order?

30. Marilyn weighs 123 lb. Her twin brother Mike weighs 174 lb. What is the average of their weights?

31. A standard piece of paper is $\frac{43}{200}$ m by $\frac{7}{25}$ m. By how much does the length exceed the width?

Skill Maintenance

32. Multiply: $9(x - 6)$.

33. Divide and simplify:

$$\left(-\frac{4}{3}\right) \div \left(-\frac{5}{6}\right).$$

34. Multiply and simplify: $\frac{4}{3} \cdot \frac{5}{6}$.

35. A rock climber descended from an altitude of 720 ft to a depth of 470 ft below sea level. How many feet did the climber descend?

Synthesis

36. Yuri and Olga are orangutans who perform in a circus by riding bicycles around a circular track. It takes Yuri $\frac{6}{25}$ min and Olga $\frac{8}{25}$ min to complete one lap. They start their act together at one point and complete their act when they are next together at that point. How long does the act last?

37. Dolores runs 17 laps at her health club. Terence runs 17 laps at his health club. If the track at Dolores's health club is $\frac{1}{7}$ mi long, and the track at Terence's is $\frac{1}{8}$ mi long, who runs farther? How much farther?

38. The students in a math class can be organized into study groups of 8 each such that no students are left out. The same class of students can also be organized into groups of 6 such that no students are left out.

a) Find some class sizes for which this will work.

b) Find the smallest such class size.

39. Simplify each of the following, using fractional notation. Try to answer part (e) by recognizing a pattern in parts (a) through (d).

a) $\dfrac{1}{1 \cdot 2}$

b) $\dfrac{1}{1 \cdot 2} + \dfrac{1}{2 \cdot 3}$

c) $\dfrac{1}{1 \cdot 2} + \dfrac{1}{2 \cdot 3} + \dfrac{1}{3 \cdot 4}$

d) $\dfrac{1}{1 \cdot 2} + \dfrac{1}{2 \cdot 3} + \dfrac{1}{3 \cdot 4} + \dfrac{1}{4 \cdot 5}$

e) $\dfrac{1}{1 \cdot 2} + \dfrac{1}{2 \cdot 3} + \dfrac{1}{3 \cdot 4} + \dfrac{1}{4 \cdot 5} + \dfrac{1}{5 \cdot 6}$
$+ \dfrac{1}{6 \cdot 7} + \dfrac{1}{7 \cdot 8} + \dfrac{1}{8 \cdot 9} + \dfrac{1}{9 \cdot 10}$

Cumulative Review: Chapters 1–4

1. In the number 2753, what digit names tens?

2. Write expanded notation for 6075.

3. Write a word name for the number in the following sentence. The diameter of Uranus is 29,500 miles.

Add and, if possible, simplify.

4. $\begin{array}{r} 375 \\ +248 \\ \hline \end{array}$

5. $29 + (-37)$

6. $\frac{3}{8} + \frac{1}{24}$

7. $\begin{array}{r} 2\frac{3}{4} \\ +5\frac{1}{2} \\ \hline \end{array}$

Subtract and, if possible, simplify.

8. $\begin{array}{r} 7469 \\ -2345 \\ \hline \end{array}$

9. $-9 - (-25)$

10. $\frac{1}{3} - \frac{3}{4}$

11. $\begin{array}{r} 2\frac{1}{3} \\ -1\frac{1}{6} \\ \hline \end{array}$

Multiply and, if possible, simplify.

12. $\begin{array}{r} 278 \\ \times\ 18 \\ \hline \end{array}$

13. $29(-5)$

14. $\frac{9}{10} \cdot \frac{5}{3}$

15. $18\left(-\frac{5}{6}\right)$

16. $2\frac{1}{3} \cdot 3\frac{1}{7}$

Divide. Write the answer with the remainder in the form 34 R 7.

17. $731 \div 15$

18. $45\overline{)2531}$

19. In Question 18, write a mixed numeral for the answer.

Divide and, if possible, simplify.

20. $\frac{2}{5} \div \left(-\frac{7}{10}\right)$

21. $2\frac{1}{5} \div \frac{3}{10}$

22. Round 38,478 to the nearest hundred.

23. Find the LCM of 24 and 36.

24. Without performing the division, determine whether 4296 is divisible by 6.

25. Find all factors of 16.

26. What part is shaded?

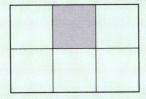

Use <, >, or = for ▪ to form a true sentence.

27. $\frac{4}{5}$ ▪ $\frac{4}{6}$

28. $-\frac{3}{7}$ ▪ $-\frac{5}{12}$

Simplify.

29. $\dfrac{36}{45}$

30. $-\dfrac{420}{30}$

31. Convert to fractional notation: $4\dfrac{5}{8}$.

32. Convert to a mixed numeral: $-\dfrac{17}{3}$.

Solve.

33. $x + 24 = 117$

34. $x + \dfrac{7}{9} = \dfrac{4}{3}$

35. $\dfrac{7}{9} \cdot t = -\dfrac{4}{3}$

36. $\dfrac{5}{7} = \dfrac{1}{3} + 4a$

37. Evaluate $\dfrac{t + p}{3}$ for $t = -4$ and $p = 16$.

38. Multiply: $7(b - 5)$.

39. Multiply: $-3(x - 2 + z)$.

40. Combine like terms: $x - 5 - 7x - 4$.

Solve.

41. A jacket costs $87 and a coat costs $148. How much does it cost to buy both?

42. The emergency soup kitchen fund contains $978. From this fund, $148 and $167 are withdrawn for expenses. How much is left in the fund?

43. A rectangular lot measures 27 ft by 11 ft. What is its area?

44. How many people can get equal $16 shares from a total of $496?

45. A recipe calls for $\dfrac{4}{5}$ tsp of salt. How much salt should be used in $\dfrac{1}{2}$ recipe?

46. A book weighs $2\dfrac{3}{5}$ lb. How much do 15 books weigh?

47. How many pieces, each $2\dfrac{3}{8}$ cm long, can be cut from a piece of wire 38 cm long?

48. How long is the shortest bolt that will pass through a $\dfrac{1}{16}$-in. thick washer, a $\dfrac{3}{4}$-in. thick backboard, and a $\dfrac{3}{8}$-in thick nut? Disregard the head of the bolt.

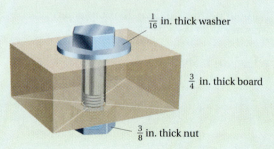

Synthesis

49. Solve: $7x - \dfrac{2}{3}(x - 6) = 6\dfrac{5}{7}$.

50. Each floor of a seven-story office building is 25 m by $22\dfrac{1}{2}$ m, with a 5-m by $4\dfrac{1}{2}$-m elevator/stairwell. How many square meters of office space are in the building?

5

Decimal Notation

Introduction

In this chapter, we first consider the meaning of decimal notation. This will enable us to study addition, subtraction, multiplication, and division using decimals. Also discussed are rounding, estimating, equation solving, and problem solving involving decimals.

5.1 Decimal Notation
5.2 Addition and Subtraction with Decimals
5.3 Multiplication with Decimals
5.4 Division with Decimals
5.5 More with Fractional Notation and Decimal Notation
5.6 Estimating
5.7 Solving Equations
5.8 Applications and Problem Solving

An Application

Upon graduation from college, Jannette will be faced with repaying a Stafford loan that totals $23,334. The loan is to be paid back over 10 yr in equal payments. Find the amount of each payment.

This problem appears as Example 3 in Section 5.8.

The Mathematics

This is a multistep problem. First, we determine the number of payments ($10 \cdot 12$, or 120), and then we divide:

Monthly payment size **is** Total owed **Divided by** Number of payments

$$m = 23{,}334 \div 120.$$

Division like this is most easily performed using *decimal notation*.

For more information, visit us at www.mathmax.com

Pretest: Chapter 5

1. Write a word name for 17.369.

2. Write $625.27 in words, as on a check.

Write fractional notation. Do not simplify.

3. 0.21

4. 5.408

Write decimal notation.

5. $\dfrac{379}{100}$

6. $-\dfrac{79}{10{,}000}$

7. Round 21.0448 to the nearest tenth.

Perform the indicated operation.

8. $\begin{array}{r} 6\ 0\ 1.3 \\ 5.8\ 1 \\ +\ \ \ \ 0.1\ 0\ 9 \\ \hline \end{array}$

9. $\begin{array}{r} 9\ 4.0\ 6\ 1 \\ -\ \ \ \ 2.3\ 2\ 9 \\ \hline \end{array}$

10. $\begin{array}{r} 7.3\ 2\ 5 \\ \times\ \ \ 0.6\ 4 \\ \hline \end{array}$

11. $91.6851 - 344.6788$

12. $-6.6\,\overline{)\,2\ 0\ 0.6\ 4\ }$

13. Combine like terms: $8.3a + 4.6a$.

14. Combine like terms: $-2.7x + 5.1 - 4.2x + 1.7$.

15. Simplify: $(2 - 1.7)^2 - 4.1 \times 3.1$.

16. Estimate the sum $3.649 + 4.038$ to the nearest tenth.

17. Multiply by 1 to find decimal notation for $\tfrac{7}{5}$.

18. Use division to find decimal notation for $\tfrac{29}{7}$.

19. Calculate: $\tfrac{3}{4} \times 2.378$.

20. Find the area of a triangular sail that is 2.8 m wide at the base and 3.1 m tall.

Solve.

21. $x + 3.91 = 7.26$

22. $-9.6y = 808.896$

23. $4.2x - 3.8 = 18.88$

24. $4.7a - 1.9 = 3.2a + 7.1$

25. $2.3(t + 4) - 0.5t = 5.8t - 9$

Solve.

26. A checking account contained $434.19. After a $148.24 check was drawn, how much was left in the account?

27. On a three-day trip, Doris drove the following distances: 432.6 mi, 179.2 mi, and 469.8 mi. What was the total number of miles driven?

28. What is the cost of 6 compact discs at $14.95 each?

29. Costas Construction paid $47,567.89 for 14 acres of land. How much did 1 acre of land cost? Round to the nearest cent.

30. Jorge filled the gas tank of his Ford Explorer and noted that the odometer read 52,091.7. At the next fillup, when the odometer read 52,214.9, it took 8 gal to fill the tank. How many miles per gallon did Jorge's Explorer get?

5.1 Decimal Notation

The set of **rational numbers** consists of the **integers**

$$\ldots, -3, -2, -1, 0, 1, 2, 3, \ldots,$$

and fractions like

$$\frac{1}{2}, \frac{2}{3}, \frac{-7}{8}, \frac{17}{-10}, \text{ and so on.}$$

We used fractional notation for rational numbers in Chapters 3 and 4. In Chapter 5, we will use *decimal notation*. We will still consider the same set of numbers, but now with a different notation. For example, instead of using fractional notation for $\frac{7}{8}$, we use decimal notation, 0.875.

a | Decimal Notation and Word Names

Decimal notation for the women's shotput record is 74.249 ft. To understand what 74.249 means, we use a **place-value chart**. The value of each place is $\frac{1}{10}$ as large as the one to its left. To the right of the decimal point, each place value ends with *ths*.

PLACE-VALUE CHART							
Hundreds	Tens	Ones	Ten*ths*	Hundred*ths*	Thousand*ths*	Ten-Thousand*ths*	Hundred-Thousand*ths*
100	10	1	$\frac{1}{10}$	$\frac{1}{100}$	$\frac{1}{1000}$	$\frac{1}{10,000}$	$\frac{1}{100,000}$
7	4 .	2	4	9			

The decimal notation 74.249 means

      7 tens + 4 ones + 2 tenths + 4 hundredths + 9 thousandths,

or $74 + \frac{2}{10} + \frac{4}{100} + \frac{9}{1000}$

or $74 + \frac{200}{1000} + \frac{40}{1000} + \frac{9}{1000}$, or $74\frac{249}{1000}$.

A mixed numeral for 74.249 is $74\frac{249}{1000}$. We read 74.249 as "seventy-four and two hundred forty-nine thousandths." When we come to the decimal point (that is, the dot in front of the 2), we read "and." We can also read 74.249 as "seventy-four *point* two four nine."

To write a word name from decimal notation,

a) write the name of the integer that appears to the left of the decimal point,

          397.685 → Three hundred ninety-seven

b) write the word "and" for the decimal point, and

          397.685   Three hundred ninety-seven **and**

c) write the name of the integer that appears to the right of the decimal point, followed by the place value of the last digit.

          397.**685**   Three hundred ninety-seven and **six hundred eighty-five *thousandths***

Objectives

a. Given decimal notation, write a word name, and write a word name for an amount of money.

b. Convert from decimal notation to fractional notation.

c. Convert from fractional notation and mixed numerals to decimal notation.

d. Given a pair of numbers in decimal notation, tell which is larger.

e. Round to the nearest thousandth, hundredth, tenth, one, ten, hundred, or thousand.

For Extra Help

TAPE 9   MAC WIN   CD-ROM

Write a word name for each number.

1. Each person in this country consumes an average of 20.5 gallons of coffee per year (**Source:** Department of Agriculture).

2. The racehorse *Swale* won the Belmont Stakes in a time of 2.4533 minutes.

3. −453.27

4. 51,739.082

Write in words, as on a check.

5. $4217.56

6. $13.98

Answers on pages A-11

Example 1 Write a word name for the number in this sentence: Each person consumes an average of 41.2 gallons of water per year.

Forty-one and two tenths

Example 2 Write a word name for −413.07.

Negative four hundred thirteen and seven hundredths

Example 3 Write a word name for the number in this sentence: The world record in the men's marathon is 2.1008 hours.

Two and one thousand eight ten-thousandths

Example 4 Write a word name for the number in this sentence: The fastest time in the women's marathon is 2.341 hours.

Two and three hundred forty-one thousandths

Do Exercises 1–4.

Decimal notation is also used with money. It is common on a check to write "and ninety-five cents" as "and $\frac{95}{100}$ dollars."

Example 5 Write $5876.95 in words, as on a check.

Five thousand, eight hundred seventy-six and $\frac{95}{100}$ dollars

Do Exercises 5 and 6.

b Converting from Decimal Notation to Fractional Notation

We can find fractional notation as follows:

$$9.875 = 9 + \frac{875}{1000}$$
$$= \frac{9000}{1000} + \frac{875}{1000} = \frac{9875}{1000}$$

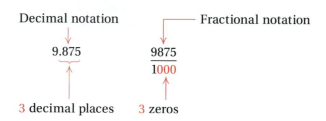

To convert from decimal to fractional notation,

a) count the number of decimal places,   4.98   (2 places)

b) move the decimal point that many places to the right, and   4.98.   Move 2 places,

c) write the answer over a denominator with a 1 followed by that number of zeros. $\dfrac{498}{100}$ 2 zeros

Example 6   Write fractional notation for 0.876. Do not simplify.

$$0.876 \qquad 0.876. \qquad 0.876 = \dfrac{876}{1000}$$

3 places   3 zeros

For a number like 0.876, we generally write a 0 before the decimal point to avoid forgetting or overlooking the decimal point.

Example 7   Write fractional notation for 56.23. Do not simplify.

$$56.23 \qquad 56.23. \qquad 56.23 = \dfrac{5623}{100}$$

2 places   2 zeros

Negative numbers written in decimal notation can also be converted.

Example 8   Write fractional notation for -1.5018. Do not simplify.

$$-1.5018 \qquad -1.5018. \qquad -1.5018 = -\dfrac{15{,}018}{10{,}000}, \text{ or } -\dfrac{15{,}018}{10{,}000}$$

4 places   4 zeros

Do Exercises 7–10.

c   Converting from Fractional Notation and Mixed Numerals to Decimal Notation

Suppose we wish to write $\dfrac{5328}{10}$ in decimal notation. In Section 4.5, we learned that division can be used to find an equivalent mixed numeral:

$$\dfrac{5328}{10} = 532\dfrac{8}{10}.$$

Note that

$$532\dfrac{8}{10} = 532 + \dfrac{8}{10}$$
$$= 532.8.$$

```
      5 3 2
 1 0 ) 5 3 2 8
       5 0
         3 2
         3 0
           2 8
           2 0
              8
```

This procedure can be generalized. It is the reverse of the procedure used in Examples 6–8.

Write fractional notation. Do not simplify.

7. 0.896

8. -39.08

9. 5.6789

10. -3.7

Answers on page A-11

5.1   Decimal Notation

277

Write decimal notation.

11. $\dfrac{743}{100}$

12. $\dfrac{48}{1000}$

13. $\dfrac{67{,}089}{10{,}000}$

14. $-\dfrac{9}{10}$

Write decimal notation.

15. $-7\dfrac{3}{100}$

16. $23\dfrac{47}{1000}$

Answers on page A-11

To convert from fractional notation to decimal notation when the denominator is 10, 100, 1000, and so on,

a) count the number of zeros, and

$\dfrac{8679}{1000}$ — 3 zeros

b) move the decimal point that number of places to the left. Leave off the denominator.

8.679. Move 3 places.

$\dfrac{8679}{1000} = 8.679$

Example 9 Write decimal notation for $\dfrac{47}{10}$.

$\dfrac{47}{10}$ — 1 zero   4.7. $\dfrac{47}{10} = 4.7$ The decimal point is moved 1 place.

Example 10 Write decimal notation for $\dfrac{123{,}067}{10{,}000}$.

$\dfrac{123{,}067}{10{,}000}$ — 4 zeros   12.3067. $\dfrac{123{,}067}{10{,}000} = 12.3067$ The decimal point is moved 4 places.

Example 11 Write decimal notation for $-\dfrac{9}{100}$.

$-\dfrac{9}{100}$ — 2 zeros $-0.09.$ $-\dfrac{9}{100} = -0.09$ The decimal point is moved 2 places.

Do Exercises 11–14.

For denominators other than 10, 100, and so on, we will usually perform long division. This is examined in Section 5.5.

If a mixed numeral has a fractional part with a denominator that is a power of ten, such as 10, 100, or 1000, and so on, we first write the mixed numeral as a sum of a whole number and a fraction. Then we convert to decimal notation.

Example 12 Write decimal notation for $23\dfrac{59}{100}$.

$23\dfrac{59}{100} = 23 + \dfrac{59}{100} = 23 \text{ and } \dfrac{59}{100} = 23.59$

Do Exercises 15 and 16.

Chapter 5 Decimal Notation

d Order

To compare numbers in decimal notation, consider 0.85 and 0.9. First note that $0.9 = 0.90$ because $\frac{9}{10} = \frac{90}{100}$. Since $0.85 = \frac{85}{100}$, it follows that $\frac{85}{100} < \frac{90}{100}$ and $0.85 < 0.9$. This leads us to a quick way to compare two numbers in decimal notation.

> To compare two positive numbers in decimal notation, start at the left and compare corresponding digits. When two digits differ, the number with the larger digit is the larger of the two numbers. To ease the comparison, extra zeros can be written to the right of the last decimal place.

Example 13 Which is larger: 2.109 or 2.1?

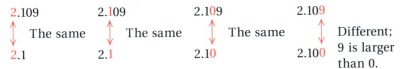

Thus, 2.109 is larger. In symbols, $2.109 > 2.1$.

Example 14 Which is larger: 0.09 or 0.108?

Thus, 0.108 is larger. In symbols, $0.108 > 0.09$.

As before, we can use a number line to visualize order. We illustrate Examples 13 and 14 below. Larger numbers are always to the right.

Note from the number line that $-2 < -1$. Similarly, $-1.57 < -1.52$.

> To compare two negative numbers in decimal notation, start at the left and compare corresponding digits. When two digits differ, the number with the smaller digit is the larger of the two numbers.

Example 15 Which is larger: -3.8 or -3.82?

Thus, -3.8 is larger. In symbols, $-3.8 > -3.82$. (See the graph above.)

Do Exercises 17–24.

Which number is larger?

17. 2.04, 2.039

18. 0.06, 0.008

19. 0.5, 0.58

20. 1, 0.9999

21. 0.8989, 0.09898

22. 21.006, 21.05

23. -34.01, -34.008

24. -9.12s, -8.98

Answers on page A-11

5.1 Decimal Notation

Round to the nearest tenth.
25. 2.76
26. 13.85

27. −234.448
28. 7.009

Round to the nearest hundredth.
29. 0.636
30. −7.834

31. 34.695
32. −0.025

Round to the nearest thousandth.
33. 0.9434
34. −8.0038

35. −43.1119
36. 37.4005

Round 7459.3598 to the nearest:
37. Thousandth.

38. Hundredth.

39. Tenth.

40. One.

41. Ten. (*Caution:* "Tens" are not "tenths.")

42. Hundred.

43. Thousand.

Answers on pages A-11 and A-12

e Rounding

Rounding is done as for whole numbers. To see how, we use a number line.

Example 16 Round 0.37 to the nearest tenth.

Here is part of a number line, magnified.

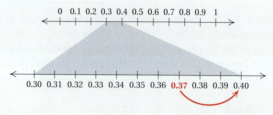

We see that 0.37 is closer to 0.40 than to 0.30. Thus, when 0.37 is rounded to the nearest tenth, we round *up* to 0.4.

To round to a certain place:
a) Locate the digit in that place.
b) Consider the next digit to the right.
c) If the digit to the right is 5 or greater, round up; if the digit to the right is 4 or less, round down.

Example 17 Round 72.3846 to the nearest hundredth.

a) Locate the digit in the hundredths place.

    7 2.3 **8** 4 6
           ↑

b) Consider the next digit to the right.

    7 2.3 8 **4** 6
         ↳↑

CAUTION! 72.39 is not a correct answer to Example 17. It is incorrect to round sequentially from right to left as follows:

72.3846, 72.385, 72.39.

c) Since that digit, 4, is less than 5, we round *down* from 72.3846 to 72.38.

Example 18 Round −0.06 to the nearest tenth.

a) Locate the digit in the tenths place.   −0.**0** 6
                                                      ↑

b) Consider the next digit to the right.   −0.0 **6**
                                                     ↳↑

c) Since that digit, 6, is greater than 5, round from −0.06 to −0.1.

The answer is −0.1. Since −0.1 < −0.06, we actually rounded *down*.

Do Exercises 25–43.

Exercise Set 5.1

a Write a word name for the number in each sentence.

1. The largest pumpkin ever grown weighed 481.27 kilograms (**Source:** *Guinness Book of Records,* 1998).

2. The average loss of daylight in October in Anchorage, Alaska, is 5.63 min per day.

3. Recently, one British pound was worth about $1.5599 in U.S. currency.

4. The cost of a fast modem for a computer was $289.95.

Write a word name.

5. 34.891

6. 27.1245

Write in words, as on a check.

7. $326.48

8. $125.99

9. $36.72

10. $0.67

b Write fractional notation. Do not simplify.

11. 8.3

12. 0.17

13. 203.6

14. -57.32

15. -2.703

16. 0.00013

17. 0.0109

18. 1.0008

19. -6.004

20. -9.012

c Write decimal notation.

21. $\dfrac{8}{10}$

22. $\dfrac{51}{10}$

23. $-\dfrac{59}{100}$

24. $-\dfrac{67}{100}$

25. $\dfrac{3798}{1000}$

26. $\dfrac{780}{1000}$

27. $\dfrac{78}{10{,}000}$

28. $\dfrac{56{,}788}{100{,}000}$

29. $\dfrac{-18}{100{,}000}$

30. $\dfrac{-2347}{100}$

31. $\dfrac{376{,}193}{1{,}000{,}000}$

32. $\dfrac{8{,}953{,}074}{1{,}000{,}000}$

33. $99\dfrac{44}{100}$

34. $4\dfrac{909}{1000}$

35. $-8\dfrac{431}{1000}$

36. $-49\dfrac{32}{1000}$

37. $2\dfrac{1739}{10{,}000}$

38. $9243\dfrac{1}{10}$

39. $8\dfrac{953{,}073}{1{,}000{,}000}$

40. $2256\dfrac{3059}{10{,}000}$

d Which number is larger?

41. 0.06, 0.58

42. 0.008, 0.8

43. 0.905, 0.91

44. 42.06, 42.1

45. -5.046, -5.043

46. -324.19, -325.19

47. 234.07, 235.07

48. 0.99999, 1

49. 0.004, $\dfrac{4}{100}$

50. $\dfrac{73}{10}$, 0.73

51. -0.872, -0.873

52. -0.8437, -0.84384

e Round to the nearest tenth.

53. 0.11

54. 0.85

55. -0.37

56. -0.26

57. 2.951

58. 4.98

59. -327.2347

60. -8.749

Round to the nearest hundredth.

61. 0.893

62. 0.675

63. -0.6666

64. -7.525

65. 0.995

66. 207.9976

67. -0.0348

68. -9.2748

Round to the nearest thousandth.

69. 0.3246 **70.** 0.6666 **71.** 17.0015 **72.** 123.4562

73. −20.20202 **74.** −0.10346 **75.** 9.9848 **76.** 67.100602

Round 809.4732 to the nearest:
77. Tenth. **78.** Thousandth. **79.** Hundredth. **80.** One.

Skill Maintenance

81. Simplify: $\dfrac{0}{-19}$. [3.3c]

82. Add: $\dfrac{2}{15} + \dfrac{5}{9}$. [4.2b]

83. Subtract: $\dfrac{4}{9} - \dfrac{2}{3}$. [4.3a]

84. Solve: $7x + 5 = 3$. [4.4a]

85. Solve: $3x - 8 = 21$. [4.4a]

86. Subtract: $\dfrac{3}{14} - \dfrac{2}{7}$. [4.3a]

Synthesis

87. ◆ Brian rounds 536.447 to the nearest one and, incorrectly, gets 537. How might he have made this mistake?

88. ◆ Explain why −73.69 is smaller than −73.67.

89. ◆ Describe in your own words a procedure for converting from decimal notation to fractional notation.

Global Warming. The graph below is based on the average global temperatures from January through May of 1880 through 1998. Each bar indicates, in Fahrenheit degrees, how much above or below average the temperature was for the year.

90. For what year(s) was the yearly temperature more than 0.4 degree above average?

91. What was the last year in which the yearly temperature was more than 0.6 degree below average?

92. What was the last year in which the yearly temperature was below average?

93. For what year(s) was the yearly temperature more than 1.0 degree above average?

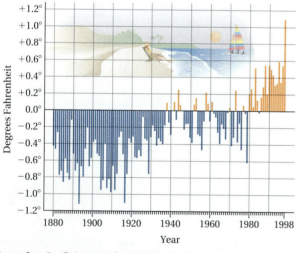

A Warming Trend

Degrees above or below the average global temperature between 1880 and 1998 for January through May. In degrees Fahrenheit.

Source: Council on Environmental Quality, The New York Times.

There are other methods of rounding decimal notation. A computer often uses a method called **truncating**. To round using truncating, simply drop all decimal places past the rounding place, which is the same as changing all digits to the right to zeros. For example, truncating 6.78163 to the third decimal place gives 6.781. Use truncating to round each of the following to the fifth decimal place.

94. 6.78346123

95. 0.07070707

5.2 Addition and Subtraction with Decimals

a | Addition

Adding with decimal notation is similar to adding whole numbers. First we line up the decimal points so that we can add corresponding place-value digits. Then we add digits from the right. For example, we add the thousandths, then the hundredths, and so on, carrying if necessary. If desired, we can write extra zeros to the right of the last decimal place so that the number of places is the same.

Example 1 Add: $56.314 + 17.78$.

```
   5 6 . 3 1 4      Lining up the decimal points in order to add
 + 1 7 . 7 8 0      Writing an extra zero to the right
                    of the last decimal place
```

```
   5 6 . 3 1 4      Adding thousandths
 + 1 7 . 7 8 0
             4
```

```
   5 6 . 3 1 4      Adding hundredths
 + 1 7 . 7 8 0
           9 4
```

```
       1
   5 6 . 3 1 4      Adding tenths
 + 1 7 . 7 8 0      Write a decimal point in the answer.
       . 0 9 4      We get 10 tenths = 1 one + 0 tenths,
                    so we carry the 1 to the ones column.
```

```
     1 1
   5 6 . 3 1 4      Adding ones
 + 1 7 . 7 8 0
     4 . 0 9 4      We get 14 ones = 1 ten + 4 ones,
                    so we carry the 1 to the tens column.
```

```
     1 1
   5 6 . 3 1 4      Adding tens
 + 1 7 . 7 8 0
   7 4 . 0 9 4
```

Do Exercises 1 and 2.

Remember, we can write extra zeros to the right of the last decimal place to get the same number of decimal places.

Example 2 Add: $3.42 + 0.237 + 14.1$.

```
     3.4 2 0      Lining up the decimal points
     0.2 3 7      and writing extra zeros
 + 1 4.1 0 0
   1 7.7 5 7      Adding
```

Do Exercises 3–5.

Objectives

a | Add using decimal notation.
b | Subtract using decimal notation.
c | Add and subtract negative decimals.
d | Combine like terms with decimal coefficients.

For Extra Help

TAPE 9   MAC   CD-ROM
           WIN

Add.

1.   0.8 4 7
   $+$ 1 0.0 7

2.   2.1
        0.7 3 9
   $+$ 3 1.3 6 8 9

Add.

3. $0.02 + 4.3 + 0.649$

4. $0.12 + 3.006 + 0.4357$

5. $0.4591 + 0.2374 + 8.70894$

Answers on page A-12

Add.

6. 789 + 123.67

7. 45.78 + 2467 + 1.993

Subtract.

8. 37.428 − 26.674

9. 0.347
      − 0.008

Answers on page A-12

Chapter 5   Decimal Notation

Consider the addition 3456 + 19.347. Keep in mind that an integer, such as 3456, has an "unwritten" decimal point at the right, with 0 fractional parts. When adding, we can always write in that decimal point and extra zeros if desired.

Example 3 Add: 3456 + 19.347.

$$
\begin{array}{r}
1\\
3456.000\\
+19.347\\
\hline
3475.347
\end{array}
$$

    Writing in the decimal point and extra zeros
    Lining up the decimal points
    Adding

Do Exercises 6 and 7.

b Subtraction

Subtracting with decimal notation is similar to subtracting integers. First we line up the decimal points so that we can subtract corresponding place-value digits. Then we subtract digits from the right. For example, we subtract the thousandths, then the hundredths, the tenths, and so on, borrowing if necessary.

Example 4 Subtract: 56.314 − 17.78.

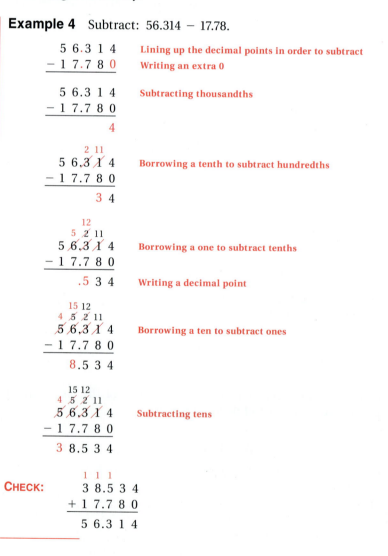

Do Exercises 8 and 9.

Example 5 Subtract: $23.08 - 5.0053$.

$$\begin{array}{r} \overset{1}{\overset{}{2}}\overset{13}{\overset{}{3}}.0\overset{7}{\overset{}{8}}\overset{9}{\overset{}{0}}\overset{10}{\overset{}{0}} \\ -5.053 \\ \hline 18.0747 \end{array}$$ Writing two extra zeros

Subtracting

Do Exercises 10–12.

As with addition, when subtraction involves an integer, there is an "unwritten" decimal point that can be written in. Extra zeros can then be written in to the right of the decimal point.

Example 6 Subtract: $456 - 2.467$.

$$\begin{array}{r} \overset{5}{\overset{}{}}\overset{9}{\overset{}{}}\overset{9}{\overset{}{}}\overset{10}{\overset{}{}} \\ 456.000 \\ -2.467 \\ \hline 453.533 \end{array}$$ Writing in the decimal point and extra zeros

Subtracting

Do Exercises 13 and 14.

c | Adding and Subtracting with Negatives

Negative numbers in decimal notation are added or subtracted just like integers.

> To add a negative number and a positive number:
> a) Determine which number has the greater absolute value.
> b) Subtract the smaller absolute value from the larger one.
> c) The answer is the difference from part (b) with the sign from part (a).

Example 7 Add: $-13.82 + 4.69$.

a) Note that $|-13.82| = 13.82$, and $|4.69| = 4.69$. Since $|-13.82| > |4.69|$, the answer is *negative*.

b) $$\begin{array}{r} \overset{7}{\overset{}{}}\overset{12}{\overset{}{}} \\ 13.\overset{}{\overset{}{8}}\overset{}{\overset{}{2}} \\ -4.69 \\ \hline 9.13 \end{array}$$ Finding the difference of the absolute values

Next, use the result of part (a).

c) $-13.82 + 4.69 = -9.13$

Do Exercises 15 and 16.

> To add two negative numbers:
> a) Add the absolute values.
> b) Make the answer negative.

Example 8 Add: $-2.306 + (-3.125)$.

a) $$\left.\begin{array}{r} 2.306 \\ +3.125 \\ \hline 5.431 \end{array}\right\}$$ Note that $|-2.306| = 2.306$ and $|-3.125| = 3.125$.

Adding the absolute values

b) $-2.306 + (-3.125) = -5.431$ The sum of two negatives is negative.

Do Exercise 17.

Subtract.
10. $1.7 - 0.23$

11. $0.43 - 0.18762$

12. $7.37 - 0.00008$

Subtract.
13. $1277 - 82.78$

14. $5 - 0.0089$

Add.
15. $7.42 + (-9.38)$

16. $-4.201 + 7.36$

17. Add: $-4.95 + (-3.6)$.

Answers on page A-12

Subtract.
18. $9.25 - 13.41$

19. $-4.26 - 3.18$

20. $9.8 - (-2.6)$

21. $-5.9 - (-3.2)$

Combine like terms.
22. $7.9x - 3.2x$

23. $-5.9a + 7.6a$

24. $-4.8y + 7.5 + 2.1y - 2.1$

Answers on page A-12

Chapter 5 Decimal Notation

To subtract, we add the opposite of the number being subtracted.

Example 9 Subtract: $-3.1 - 4.8$.
$$-3.1 - 4.8 = -3.1 + (-4.8) \quad \text{Adding the opposite of 4.8}$$
$$= -7.9 \quad \text{The sum of two negatives is negative.}$$

Example 10 Subtract: $-7.9 - (-8.5)$.
$$-7.9 - (-8.5) = -7.9 + 8.5 \quad \text{Adding the opposite of } -8.5$$
$$= 0.6 \quad \text{Subtracting absolute values. The answer is positive since 8.5 has the larger absolute value.}$$

Do Exercises 18–21.

d Combining Like Terms

Recall that like, or similar, terms have exactly the same variable factors. When we combine like terms, we add or subtract coefficients to form an equivalent expression.

Example 11 Combine like terms: $3.2x + 4.6x$.
$$3.2x + 4.6x = (3.2 + 4.6)x \quad \text{These are the coefficients.}$$
$$\text{Using the distributive law}$$
$$\text{Try to do this step mentally}$$
$$= 7.8x \quad \text{Adding}$$

A similar procedure is used when subtracting like terms.

Example 12 Combine like terms: $4.13a - 7.56a$.
$$4.13a - 7.56a = (4.13 - 7.56)a \quad \text{Using the distributive law}$$
$$= (4.13 + (-7.56))a \quad \text{Adding the opposite of 7.56}$$
$$= -3.43a \quad \text{Subtracting absolute values. The coefficient is negative since } |-7.56| > |4.13|.$$

When more than one pair of like terms is present, we can rearrange the terms and then simplify.

Example 13 Combine like terms: $5.7x - 3.9y - 2.4x + 4.5y$.
$$5.7x - 3.9y - 2.4x + 4.5y$$
$$= 5.7x + (-3.9y) + (-2.4x) + 4.5y \quad \text{Rewriting as addition}$$
$$= 5.7x + (-2.4x) + 4.5y + (-3.9y) \quad \text{Using the commutative law to rearrange}$$
$$= 3.3x + 0.6y \quad \text{Combining like terms}$$

With practice, you will be able to perform many of the above steps mentally.

Do Exercises 22–24.

Exercise Set 5.2

a Add.

1. 316.25
    $+18.12$

2. 41.823
    $+614.915$

3. 659.403
    $+916.812$

4. 3.25
    $+1123.39$

5. 9.104
    $+123.456$

6. 4.1523
    $+3.2778$

7. 61.006
    $+3.407$

8. $0.8096 + 0.7856$

9. $20.0124 + 30.0124$

10. $0.263 + 0.8$

11. $0.83 + 0.005$

12. $0.347 + 10.04$

13. $0.34 + 3.5 + 0.127 + 768$

14. $2.3 + 0.729 + 23$

15. $17 + 3.24 + 0.256 + 0.3689$

16. 47.8
    219.852
    43.59
    $+666.713$

17. 2.703
    78.33
    28.0009
    $+118.4341$

18. 13.72
    9.112
    6542.7908
    $+23.901$

b Subtract.

19. 5.2
    -3.1

20. 11.345
    -2.105

21. 51.31
    -2.29

22. 37.45
    -6.32

23. 2.5
    − 0.0 0 2 5

24. 2 8.0
    −   0.2 8

25. 9 2.3 4 1
    −   6.4 2

26. 0.3 4 6
    − 0.0 3 4 6

27. 3.0 0 7 4
    − 1.3 4 0 8

28. 3 2.7 9 7 8
    −   0.0 5 9 2

29.   6.0 7
    − 2.0 0 7 8

30. 1.0
    − 0.9 9 9 9

31. 28.2 − 19.35

32. 100.12 − 0.112

33. 34.07 − 30.7

34. 36.2 − 16.28

35. 8.45 − 7.405

36. 3.801 − 2.81

37. 6.003 − 2.3

38. 1 − 0.0098

39. 2 − 1.0908

40. 100 − 0.34

41. 624 − 18.79

42. 7.48 − 2.6

43. 3 − 2.006

44. 25.008 − 12.4

45. 263.7 − 102.08

46. 19 − 1.198

47. 45 − 0.999

48. 10.056 − 0.392

c Add or subtract, as indicated.

49. −8.02 + 9.73

50. −4.31 + 7.66

51. 12.9 − 15.4

52. 27.2 − 31.9

53. −2.9 + (−4.3)

54. −5.7 + (−1.9)

55. −4.301 + 7.68

56. −5.952 + 7.98

57. $-13.4 - 9.2$

58. $-8.7 - 12.4$

59. $-2.1 - (-4.6)$

60. $-4.3 - (-2.5)$

61. $14.301 + (-17.82)$

62. $13.45 + (-18.701)$

63. $7.201 - (-2.4)$

64. $2.901 - (-5.7)$

65. $23.9 + (-9.4)$

66. $43.2 + (-10.9)$

67. $-8.9 - (-12.7)$

68. $-4.5 - (-7.3)$

69. $-4.9 - 5.392$

70. $89.3 - 92.1$

71. $14.7 - 23.5$

72. $-7.201 - 1.9$

d Combine like terms.

73. $5.1x + 3.6x$

74. $7.9x + 1.8x$

75. $17.59a - 12.73a$

76. $23.28a - 15.79a$

77. $15.2t + 7.9 + 5.9t$

78. $29.5t - 4.8 + 7.6t$

79. $9.208t - 14.519t$

80. $6.317t - 9.429t$

81. $4.906y - 7.1 + 3.2y$

82. $9.108y + 4.2 + 3.7y$

83. $4.8x + 1.9y - 5.7x + 1.2y$

84. $3.2r - 4.1t + 5.6t + 1.9r$

85. $4.9 - 3.9t + 2.3 - 4.5t$

86. $5.8 + 9.7x - 7.2 - 12.8x$

Exercise Set 5.2

Skill Maintenance

87. Simplify: $\dfrac{0}{-92}$. [3.3c]

88. Add: $\dfrac{-2}{7} + \dfrac{5}{21}$. [4.2b]

89. Subtract: $\dfrac{3}{5} - \dfrac{7}{10}$. [4.3a]

90. Solve: $9x - 16 = 5$. [4.4a]

91. Solve: $7x + 19 = 40$. [4.4a]

92. Subtract: $\dfrac{2}{9} - \dfrac{2}{3}$. [4.3a]

Synthesis

93. ◆ Explain the error in the following:
Subtract.

$$\begin{array}{r} 73.089 \\ -5.0061 \\ \hline 2.3028 \end{array}$$

94. ◆ A student claims to be able to add negative numbers but not subtract them. What advice would you give this student?

95. ◆ Although the step in which it is used may not always be written out, the commutative law is often used when combining like terms. Under what circumstances would the commutative law *not* be needed for combining like terms?

Combine like terms.

96. ▦ $-3.928 - 4.39a + 7.4b - 8.073 + 2.0001a - 9.931b - 9.8799a + 12.897b$

97. ▦ $79.02x + 0.0093y - 53.14z - 0.02001y - 37.987z - 97.203x - 0.00987y$

98. ▦ $39.123a - 42.458b - 72.457a + 31.462b - 59.491 + 37.927a$

99. Fred presses the wrong key when using a calculator and adds 235.7 instead of subtracting it. The incorrect answer is 817.2. What is the correct answer?

100. Millie presses the wrong key when using a calculator and subtracts 349.2 instead of adding it. The incorrect answer is −836.9. What is the correct answer?

101. ▦ Find the errors, if any, in the balances in this checkbook.

DATE	CHECK NUMBER	TRANSACTION DESCRIPTION	✓T	AMOUNT OF PAYMENT OR DEBIT (−)	(+ OR −) OTHER	AMOUNT OF CREDIT (+)	BALANCE FORWARD
							8767 73
8/16	432	Burch Laundry		23 56			8744 16
8/19	433	Rogers TV		20 49			8764 65
8/20		Deposit				85 00	8848 65
8/21	434	Galaxy Records		48 60			8801 05
8/22	435	Electric Works		267 95			8533 09

Chapter 5 Decimal Notation

5.3 Multiplication with Decimals

a Multiplication

Objectives

a. Multiply using decimal notation.
b. Convert from dollars to cents and cents to dollars, and from notation like 45.7 million to standard notation.
c. Evaluate algebraic expressions using decimal notation.

For Extra Help

TAPE 9   MAC WIN   CD-ROM

To develop an understanding of how decimals are multiplied, consider

$$2.3 \times 1.12.$$

One way to find this product is to first convert each factor to fractional notation:

$$2.3 \times 1.12 = \frac{23}{10} \times \frac{112}{100}.$$

Next, we multiply the fractions and then return to decimal notation:

$$\frac{23}{10} \times \frac{112}{100} = \frac{2576}{1000} = 2.576.$$

Note that the number of decimal places in the product is the sum of the number of decimal places in the factors.

$$\begin{array}{rl} 1.12 & \text{(2 decimal places)} \\ \times\ 2.3 & \text{(1 decimal place)} \\ \hline 2.576 & \text{(3 decimal places)} \end{array}$$

Now consider 0.02×3.412:

$$0.02 \times 3.412 = \frac{2}{100} \times \frac{3412}{1000}$$

$$= \frac{6824}{100{,}000} = 0.06824.$$

Again, note the number of decimal places in the product is the sum of the number of decimal places in the factors.

$$\begin{array}{rl} 3.412 & \text{(3 decimal places)} \\ \times\ 0.02 & \text{(2 decimal places)} \\ \hline 0.06824 & \text{(5 decimal places)} \end{array}$$

It is important to write in this zero.

We have the following rule for multiplying decimals.

To multiply using decimals:

a) Ignore the decimal points, for the moment, and multiply as though both factors were integers.

b) Then place the decimal point in the result. The number of decimal places in the product is the sum of the number of places in the factors (count places from the right).

0.8×0.43

$$\begin{array}{r} 2 \\ 0.43 \\ \times\ 0.8 \\ \hline 344 \end{array}$$ Ignore the decimal points for now.

$$\begin{array}{rl} 0.43 & \text{(2 decimal places)} \\ \times\ 0.8 & \text{(1 decimal place)} \\ \hline 0.344 & \text{(3 decimal places)} \end{array}$$

1. Multiply.

$$\begin{array}{r} 7\,6.3 \\ \times\quad 8.2 \\ \hline \end{array}$$

Multiply.

2.
$$\begin{array}{r} 4\,2\,1\,3 \\ \times\,0.0\,0\,5\,1 \\ \hline \end{array}$$

3. 2.3×0.0041

4. $5.2014 \times (-2.41)$

Answers on page A-12

Chapter 5   Decimal Notation

Example 1   Multiply: 8.3×74.6.

a) Ignore the decimal points and multiply as if both factors were integers:

$$\begin{array}{r} \overset{3}{\underset{1}{}}\overset{4}{\underset{1}{}} \\ 7\,4.6 \\ \times\quad 8.3 \\ \hline 2\,2\,3\,8 \\ 5\,9\,6\,8\,0 \\ \hline 6\,1\,9\,1\,8 \quad \text{We are not yet finished.} \end{array}$$

b) Place the decimal point in the result. The number of decimal places in the product is the sum, $1 + 1$, of the number of decimal places in the factors.

$$\begin{array}{rl} 7\,4.6 & \text{(1 decimal place)} \\ \times\quad 8.3 & \text{(1 decimal place)} \\ \hline 2\,2\,3\,8 & \\ 5\,9\,6\,8\,0 & \\ \hline 6\,1\,9.1\,8 & \text{(2 decimal places)} \end{array}$$

Do Exercise 1.

As we catch on to the skill, we can combine the two steps.

Example 2   Multiply: 0.0032×2148.

$$\begin{array}{rl} 2\,1\,4\,8 & \text{(0 decimal places)} \\ \times\,0.0\,0\,3\,2 & \text{(4 decimal places)} \\ \hline 4\,2\,9\,6 & \\ 6\,4\,4\,4\,0 & \\ \hline 6.8\,7\,3\,6 & \text{(4 decimal places)} \end{array}$$

Example 3   Multiply: -0.14×0.867.

Multiplying the absolute values, we have

$$\begin{array}{rl} 0.8\,6\,7 & \text{(3 decimal places)} \\ \times\quad 0.1\,4 & \text{(2 decimal places)} \\ \hline 3\,4\,6\,8 & \\ 8\,6\,7\,0 & \\ \hline 0.1\,2\,1\,3\,8 & \text{(5 decimal places)} \end{array}$$

Since the product of a negative and a positive is negative, the answer is -0.12138.

Do Exercises 2–4.

Suppose that a product involves multiplication by a tenth, hundredth, thousandth, and so on. From the following products, a pattern emerges.

$$\begin{array}{cccc} 4\,5.6 & 4\,5.6 & 4\,5.6 & 4\,5.6 \\ \times\quad 0.1 & \times\,0.0\,1 & \times\,0.0\,0\,1 & \times\,0.0\,0\,0\,1 \\ \hline 4.5\,6 & 0.4\,5\,6 & 0.0\,4\,5\,6 & 0.0\,0\,4\,5\,6 \end{array}$$

Note the location of the decimal point in each product. In each case, the product is *smaller* than 45.6 and contains the digits 456.

To multiply any number by a tenth, hundredth, or thousandth, and so on,

a) count the number of decimal places in the tenth, hundredth, or thousandth, and

b) move the decimal point that many places to the left. Use zeros as placeholders if necessary.

0.001 × 34.45678
→ 3 places

0.001 × 34.45678 = 0.034.45678

Move 3 places to the left.

0.001 × 34.45678 = 0.03445678

Examples Multiply.

4. 0.1 × 45 = 4.5   Moving the decimal point one place to the left

5. 0.01 × 243.7 = 2.437   Moving the decimal point two places to the left

6. 0.001 × (−8.2) = −0.0082   Moving the decimal point three places to the left. This requires writing two extra zeros.

7. 0.0001 × 536.9 = 0.05369   Moving the decimal point four places to the left. This requires writing one extra zero.

Do Exercises 5–8.

Next we consider multiplication of a decimal by a power of ten such as 10, 100, 1000, and so on. From the following products, a pattern emerges.

```
   5.2 3 7            5.2 3 7             5.2 3 7
 ×     1 0          ×   1 0 0           × 1 0 0 0
   0 0 0 0            0 0 0 0             0 0 0 0
   5 2 3 7            0 0 0 0             0 0 0 0
   ───────            5 2 3 7             0 0 0 0
   5 2.3 7 0          ───────             5 2 3 7
                      5 2 3.7 0 0         ───────
                                          5 2 3 7.0 0 0
```

Note the location of the decimal point in each product. In each case, the product is *larger* than 5.237 and contains the digits 5237.

To multiply any number by a power of ten, such as 10, 100, 1000, and so on,

a) count the number of zeros, and

b) move the decimal point that many places to the right. Use zeros as placeholders if necessary.

1000 × 34.45678
→ 3 zeros

1000 × 34.45678 = 34.456.78

Move 3 places to the right.

1000 × 34.45678 = 34,456.78

Multiply.

5. 0.1 × 359

6. 0.001 × 732.4

7. (−0.01) × 5.8

8. 0.0001 × 723.6

Answers on page A-12

5.3 Multiplication with Decimals

293

Multiply.

9. 10 × 53.917

10. 100 × (−62.417)

11. 1000 × 64.7

12. 10,000 × 43.01

Answers on page A-12

Chapter 5 Decimal Notation

Examples Multiply.

8. 10 × 32.98 = 329.8 — Moving the decimal point one place to the right

9. 100 × 4.7 = 470 — Moving the decimal point two places to the right. The 0 in 470 is a placeholder.

10. 1000 × (−2.4167) = −2416.7 — Moving the decimal point three places to the right

11. 10,000 × 7.52 = 75,200 — Moving the decimal point four places to the right and using two zeros as placeholders

Do Exercises 9–12.

b | Applications Using Multiplication with Decimal Notation

Naming Large Numbers

We often see notation like the following in newspapers, magazines and on television.

O'Hare International Airport handles 67.3 million passengers per year.

In 1995, the Internal Revenue Service collected $1.39 trillion.

The population of the world is 6.6 billion.

To understand such notation, it helps to consider the following table.

1 hundred = 100 = 10^2 → 2 zeros

1 thousand = 1000 = 10^3 → 3 zeros

1 million = 1,000,000 = 10^6 → 6 zeros

1 billion = 1,000,000,000 = 10^9 → 9 zeros

1 trillion = 1,000,000,000,000 = 10^{12} → 12 zeros

To convert to standard notation, we proceed as follows.

Example 12 Convert the number in this sentence to standard notation: O'Hare International Airport handles 67.3 million passengers per year.

$$67.3 \text{ million} = 67.3 \times 1 \text{ million}$$
$$= 67.3 \times 1,000,000$$
$$= 67,300,000$$

Do Exercises 13 and 14.

Money Conversion

Converting from dollars to cents is like multiplying by 100. To see why, consider $19.43.

$19.43 = 19.43 \times \$1$ We think of $19.43 as 19.43 × 1 dollar, or 19.43 × $1.

$= 19.43 \times 100¢$ Substituting 100¢ for $1: $1 = 100¢

$= 1943¢$ Multiplying

> To convert from dollars to cents, move the decimal point two places to the right and change from the $ sign in front to the ¢ sign at the end.

Examples Convert from dollars to cents.

13. $189.64 = 18,964¢
14. $0.75 = 75¢

Do Exercises 15 and 16.

Converting from cents to dollars is like multiplying by 0.01. To see why, consider 65¢.

$65¢ = 65 \times 1¢$ We think of 65¢ as 65 × 1 cent, or 65 × 1¢.

$= 65 \times \$0.01$ Substituting $0.01 for 1¢: 1¢ = $0.01

$= \$0.65$ Multiplying

> To convert from cents to dollars, move the decimal point two places to the left and change from the ¢ sign at the end to the $ sign in front.

Examples Convert from cents to dollars.

15. 395¢ = $3.95
16. 8503¢ = $85.03

Do Exercises 17 and 18.

Convert the number in each sentence to standard notation.

13. In a recent year, there were more than 4.3 million skateboarders in the United States (**Source:** *Statistical Abstract of the United States, 1997*).

14. In a recent year, the U.S. trade deficit with Japan was $44.1 billion.

Convert from dollars to cents.
15. $15.69

16. $0.17

Convert from cents to dollars.
17. 35¢

18. 577¢

Answers on page A-12

19. Evaluate lwh for $l = 3.2$, $w = 2.6$, and $h = 0.8$.
(This is the formula for the volume of a rectangular box.)

20. Find the area of the stamp in Example 18.

21. Evaluate $6.28rh + 3.14r^2$ for $r = 1.5$ and $h = 5.1$.
(This is the formula for the area of an open can.)

Answers on page A-12

c Evaluating

Algebraic expressions are often evaluated using numbers written in decimal notation.

Example 17 Evaluate Prt for $P = 80$, $r = 0.12$, and $t = 0.5$.

We will see in Chapter 8 that this product could be used to determine the interest earned on $80, invested at 12 percent simple interest, for half a year. We substitute as follows:

$$Prt = 80 \cdot 0.12 \cdot 0.5 = 80 \cdot 0.06 = 4.8.$$

Do Exercise 19.

Example 18 Find the perimeter of a stamp that is 3.25 cm long and 2.5 cm wide.

Recall that the perimeter, P, of a rectangle of length l and width w is given by the formula

$$P = 2l + 2w.$$

Thus we evaluate $2l + 2w$ for $l = 3.25$ and $w = 2.5$:

$$2l + 2w = 2 \cdot 3.25 + 2 \cdot 2.5$$
$$= 6.5 + 5.0 \qquad \text{Remember the rules for order of operations.}$$
$$= 11.5.$$

The perimeter is 11.5 cm.

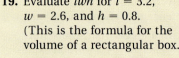

Example 19 Evaluate $4.9t^2$ for $t = 5.1$.

This formula is used in physics to find the distance, in meters, traveled by a falling body. We substitute as follows:

$$4.9t^2 = 4.9 \cdot 5.1^2$$
$$= 4.9 \cdot (5.1)(5.1) \qquad \text{Square first, then multiply.}$$
$$= 4.9 \cdot 26.01 = 127.449.$$

Do Exercises 20 and 21.

Calculator Spotlight

Most scientific and graphing calculators are equipped with a key labeled $\boxed{x^2}$. On a scientific calculator, pressing $\boxed{x^2}$ usually squares whatever number was last displayed. For instance, to compute 7^2, we press

$\boxed{7}\ \boxed{x^2}$.

To compute 7^2 on most graphing calculators, we press

$\boxed{7}\ \boxed{x^2}\ \boxed{\text{ENTER}}$.

Exercises

1. Use a scientific or graphing calculator to check Example 19.

2. Use a scientific or graphing calculator to evaluate Pm^2 for each of the following.
   a) $P = 7536$ and $m = 1.046$
   b) $P = 927.45$ and $m = 1.057$
   c) $P = 10{,}475$ and $m = 1.062$

Exercise Set 5.3

a Multiply.

1. 6.8
   $\underline{\times7}$

2. 5.7
   $\underline{\times0.9}$

3. 0.84
   $\underline{\times8}$

4. 7.3
   $\underline{\times0.6}$

5. 6.3
   $\underline{\times\,0.04}$

6. 7.8
   $\underline{\times\,0.09}$

7. 17.2
   $\underline{\times\,0.006}$

8. 8.7
   $\underline{\times\,0.06}$

9. 10×42.63

10. 100×2.8793

11. -1000×783.686852

12. -0.34×1000

13. -7.8×100

14. $0.00238 \times (-10)$

15. 0.1×79.18

16. 0.01×789.235

17. 0.001×97.68

18. 8976.23×0.001

19. $28.7 \times (-0.01)$

20. $0.0325 \times (-0.1)$

21. 2.73
    $\underline{\times16}$

22. 8.27
    $\underline{\times5.4}$

23. 0.984
    $\underline{\times3.3}$

24. 7.489
    $\underline{\times8.2}$

25. $(-37.4)(-2.4)$

26. $569(-1.05)$

27. $749(-0.43)$

28. $(-876)(-20.4)$

29. 0.87
    $\underline{\times64}$

30. 7.25
    $\underline{\times60}$

31. 46.50
    $\underline{\times75}$

32. 8.24
    $\underline{\times\,703}$

33. $(-0.231)(-0.5)$

34. $(-12.3)(-1.08)$

35. $9.42 \times (-1000)$

36. $-7.6 \times (-1000)$

37. $-95.3 \times (-0.0001)$

38. $-4.23 \times (-0.001)$

b Convert from dollars to cents.

39. $28.88

40. $67.43

41. $0.66

42. $1.78

Convert from cents to dollars.

43. 34¢

44. 95¢

45. 3445¢

46. 933¢

Convert the number in each sentence to standard notation.

47. AMTRAK operating revenues for 1995 were $32.279 billion.

48. AMTRAK operating expenses for 1995 were $27.897 billion.

49. In a recent year, the daily circulation of the *Los Angeles Times* was 1.03 million.

50. The total surface area of Earth is 196.8 million square miles.

c Evaluate.

51. $P + Prt$, for $P = 10,000$, $r = 0.04$, and $t = 2.5$
(*A formula for adding interest*)

52. $6.28r(h + r)$, for $r = 10$ and $h = 17.2$
(*Surface area of a cylinder*)

53. $vt + 0.5at^2$, for $v = 10$, $t = 1.5$, and $a = 9.8$
(*A physics formula*)

54. $4lh + 2h^2$, for $l = 3.5$ and $h = 1.2$
(*Surface area of a rectangular prism*)

Find **(a)** the perimeter and **(b)** the area of a rectangular room with the given dimensions.

55. 12.5 ft long, 9.5 ft wide

56. 10.25 ft long, 8 ft wide

Skill Maintenance

57. Simplify: $\dfrac{-109}{-109}$. [3.3c]

58. Add: $\dfrac{-2}{10} + \dfrac{4}{15}$. [4.2b]

59. Subtract: $\dfrac{2}{9} - \dfrac{5}{18}$. [4.3a]

60. Solve: $7x - 4 = -2$. [4.4a]

61. Add: $-\dfrac{3}{20} + \dfrac{3}{4}$. [4.2b]

62. Simplify: $\dfrac{0}{-19}$. [3.3c]

Synthesis

63. ◆ Is it easier to multiply numbers written in decimal notation or fractional notation? Why?

64. ◆ If two rectangles have the same perimeter, will they also have the same area? Why?

65. ◆ A student insists that 346.708×0.1 is 3467.08. How can you convince the student that a mistake has been made?

Evaluate using a calculator.

66. $d + vt + at^2$, for $d = 79.2$, $v = 3.029$, $t = 7.355$, and $a = 4.9$ (*A physics formula*)

67. $3.14r^2 + 6.28rh$, for $r = 5.756$ and $h = 9.047$
(*Surface area of a silo*)

Express as a power of 10.

68. (1 trillion) · (1 billion)

69. (1 million) · (1 billion)

Electric Bills. Recently, electric bills from the Central Vermont Public Service Corporation consisted of a "customer charge" of $0.35 per day plus an "energy charge" of $0.10470 per kilowatt-hour (kWh) for the first 250 kWh used and $0.09079 per kilowatt-hour for each kilowatt-hour in excess of 250 (*Source*: 1999 CVPS monthly statement).

70. From April 20 to May 20, the Coy-Bergers used 480 kWh of electricity. What was their bill for the period?

71. From June 20 to July 20, the D'Amicos used 430 kWh of electricity. What was their bill for the period?

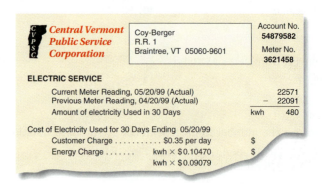

5.4 Division with Decimals

a Division

Objectives

a Divide using decimal notation.

b Simplify expressions using the rules for order of operations.

For Extra Help

TAPE 9   MAC WIN   CD-ROM

Whole-Number Divisors

Now that we have studied multiplication of decimals, we can develop a procedure for division. The following divisions are justified by the multiplication in each *check*:

This is the dividend. → $\dfrac{651}{7} = 93$ ← This is the quotient.   *Check*: $7 \cdot 93 = 651$.

This is the divisor. ↗

$\dfrac{65.1}{7} = 9.3$ *Check*: $7 \cdot 9.3 = 65.1$.

$\dfrac{6.51}{7} = 0.93$ *Check*: $7 \cdot 0.93 = 6.51$.

$\dfrac{0.651}{7} = 0.093$ *Check*: $7 \cdot 0.093 = 0.651$.

Note that the number of decimal places in each quotient is the same as the number of decimal places in the dividend.

To divide by a whole number,
a) place the decimal point directly above the decimal point in the dividend, and
b) divide as though dividing whole numbers.

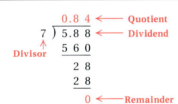

Divide.

1. $9 \overline{)\, 5.4}$

Example 1 Divide: $82.08 \div 24$.

We have

```
         3.4 2    ← Place the decimal point.
    ┌─────────
 24 ) 8 2.0 8
      7 2 0 0
      ───────
      1 0 0 8    ⎫
        9 6 0    ⎬ Divide as though dividing whole numbers.
        ─────    ⎭
          4 8
          4 8
          ───
            0
```

Estimation can be used as a partial check: $24 \approx 25$ and $82.08 \approx 75$; since $75 \div 25 = 3$ and $3 \approx 3.42$, we have at least a partial check.

2. $15 \overline{)\, 2\,5.5}$

3. $82 \overline{)\, 3\,8.5\,4}$

Do Exercises 1–3.

Sometimes it helps to write some extra zeros to the right of the dividend's decimal point. They don't change the number.

Answers on page A-12

5.4 Division with Decimals

299

Example 2 Divide: 30 ÷ 8.

$$\begin{array}{r} 3. \\ 8\,\overline{)3\,0.} \\ \underline{2\,4} \\ 6 \end{array}$$ Place the decimal point and divide to find how many ones.

$$\begin{array}{r} 3. \\ 8\,\overline{)3\,0.0} \\ \underline{2\,4}\downarrow \\ 6\,0 \end{array}$$ Write an extra zero. This does not change the number.

$$\begin{array}{r} 3.7 \\ 8\,\overline{)3\,0.0} \\ \underline{2\,4} \\ 6\,0 \\ \underline{5\,6} \\ 4 \end{array}$$ Divide to find how many tenths.

$$\begin{array}{r} 3.7 \\ 8\,\overline{)3\,0.0\,0} \\ \underline{2\,4} \\ 6\,0 \\ \underline{5\,6}\downarrow \\ 4\,0 \end{array}$$ Write an extra zero.

$$\begin{array}{r} 3.7\,5 \\ 8\,\overline{)3\,0.0\,0} \\ \underline{2\,4} \\ 6\,0 \\ \underline{5\,6}\downarrow \\ 4\,0 \\ \underline{4\,0} \\ 0 \end{array}$$ Repeat the procedure: Divide to find how many hundredths are in the quotient.

Since the remainder is 0, we are finished.

Calculator Spotlight

It is possible to use a calculator to find whole-number remainders when doing division. To see how one method works, consider the quotient 17 ÷ 8. We know that

$$17 \div 8 = 2.125.$$

To check, we can multiply:

$$8 \times 2.125 = 17,$$

or

$$8 \times (2 + 0.125) = 8 \times 2 + 8 \times 0.125$$
$$= 16 + 1 = 17.$$

Note that 17 ÷ 8 = 2 R 1. Thus we can find a whole-number remainder by multiplying the decimal portion of a quotient by the divisor.

To find the quotient and the whole-number remainder for 567 ÷ 13, we can use a calculator to find that

$$567 \div 13 \approx 43.61538462.$$ To isolate the decimal part, we can subtract 43.

When the decimal part of the quotient is multiplied by the divisor, we have

$$0.61538462 \times 13 = 8.00000006.$$

The rounding error in the result may vary, depending on the calculator used. We see that 567 ÷ 13 = 43 R 8.

Exercises Find the quotient and the whole-number remainder for each of the following.

1. 478 ÷ 17
2. 815 ÷ 7
3. 824 ÷ 11
4. 7888 ÷ 19

Example 3 Divide: $-4 \div 25$.

We first consider $4 \div 25$:

$$
\begin{array}{r}
0.1\,6 \\
25\overline{\smash{)}4.0\,0} \\
\underline{2\,5} \\
1\,5\,0 \\
\underline{1\,5\,0} \\
0
\end{array}
$$

← We can write as many extra zeros as needed.

← Since the remainder is 0, we are finished.

Since a negative number divided by a positive numer is negative, the answer is -0.16.

Do Exercises 4–6.

Divisors That Are Not Whole Numbers

Consider the division

$$0.24\overline{\smash{)}8.208}$$

We write the division as $\dfrac{8.208}{0.24}$. Multiplying by a form of 1, we can find an equivalent division with a whole-number divisor, as in Examples 1–3:

$$\dfrac{8.208}{0.24} = \dfrac{8.208}{0.24} \times \dfrac{100}{100} = \dfrac{820.8}{24}.$$

We chose to use 100 in order to move the decimal point in 0.24 two places.

Since the divisor is now a whole number, we have effectively traded a "new" problem for an equivalent problem that is more familiar:

$$0.24\overline{\smash{)}8.208}$$

is equivalent to

$$24\overline{\smash{)}820.8}$$

To divide when the divisor is not a whole number,

a) move the decimal point (multiply by 10, 100, and so on) to make the divisor a whole number;

$0.24\overline{\smash{)}8.208}$

Move 2 places to the right.

b) move the decimal point the same number of places (multiply the same way) in the dividend; and

$0.24\overline{\smash{)}8.208}$

Move 2 places to the right.

c) place the decimal point for the answer directly above the new decimal point in the dividend and divide as though dividing whole numbers.

$$
\begin{array}{r}
34.2 \\
0.24\overline{\smash{)}8.20{}_\wedge8} \\
\underline{7\,2\,0\,0} \\
1\,0\,0\,8 \\
\underline{9\,6\,0} \\
4\,8 \\
\underline{4\,8} \\
0
\end{array}
$$

(The new decimal point in the dividend is indicated by a caret.)

Divide.

4. $25\overline{\smash{)}8}$

5. $-23 \div 4$

6. $86\overline{\smash{)}21.5}$

Answers on page A-12

5.4 Division with Decimals

7. a) Complete.

$$\frac{3.75}{0.25} = \frac{3.75}{0.25} \times \frac{100}{100}$$

$$= \frac{()}{25}$$

b) Divide.

$$0.25 \overline{)3.75}$$

Divide.

8. $0.83 \overline{)4.067}$

9. $-44.8 \div (-3.5)$

10. Divide.

$$1.6 \overline{)25}$$

Answer on page A-12

Example 4 Divide: $5.848 \div 8.6$.

$$8.6 \overline{)5.848}$$

Multiply the divisor by 10 (move the decimal point 1 place). Multiply the same way in the dividend (move 1 place).

$$\begin{array}{r} 0.68 \\ 8.6 \overline{)5.8\,48} \\ \underline{5\,16\,0} \\ 6\,8\,8 \\ \underline{6\,8\,8} \\ 0 \end{array}$$

Then divide.

Note: $\dfrac{5.848}{8.6} = \dfrac{5.848}{8.6} \cdot \dfrac{10}{10} = \dfrac{58.48}{86}$.

Do Exercises 7–9.

Example 5 Divide: $12 \div 0.64$.

$$0.64 \overline{)12.}$$

Put a decimal point at the end of the whole number.

$$0.64 \overline{)12.00}$$

Multiply the divisor by 100 (move the decimal point 2 places). Multiply the same way in the dividend (move 2 places).

$$\begin{array}{r} 18.75 \\ 0.64 \overline{)12.00\,00} \\ \underline{6\,4\,0} \\ 5\,6\,0 \\ \underline{5\,1\,2} \\ 4\,8\,0 \\ \underline{4\,4\,8} \\ 3\,2\,0 \\ \underline{3\,2\,0} \\ 0 \end{array}$$

Then divide.

Since the remainder is 0, we are finished.

Do Exercise 10.

To divide quickly by a thousandth, hundredth, tenth, ten, hundred, and so on, consider

$$\frac{43.9}{100} \quad \text{and} \quad \frac{43.9}{0.001}.$$

$$\begin{array}{r} .439 \\ 100 \overline{)43.900} \\ \underline{400} \\ 390 \\ \underline{300} \\ 900 \\ \underline{900} \\ 0 \end{array} \qquad \begin{array}{r} 43900. \\ 0.001 \overline{)43.900} \end{array}$$

Division of 43.9 by a number greater than 1 results in a quotient *smaller* than 43.9, whereas division by a positive number less than 1 results in a quotient that is *larger* than 43.9.

Chapter 5 Decimal Notation

To divide by a power of ten, such as 10, 100, or 1000, and so on,

a) count the number of zeros in the divisor, and

$$\frac{713.495}{100}$$

↳ 2 zeros

b) move the decimal point that number of places to the left.

$$\frac{713.495}{100}, \quad 7.13.495 \quad \frac{713.495}{100} = \frac{7.13495}{1.00} = 7.13495$$

2 places to the left

To divide by a tenth, hundredth, or thousandth, and so on,

a) count the number of decimal places in the divisor, and

$$\frac{89.12}{0.001}$$

↳ 3 places

b) move the decimal point that number of places to the right.

$$\frac{89.12}{0.001}, \quad 89.120. \quad \frac{89.12}{0.001} = \frac{89120}{1.0} = 89,120$$

3 places to the right

Example 6 Divide: $\frac{0.0732}{10}$.

$$\frac{0.0732}{10}, \quad 0.0.0732, \quad \frac{0.0732}{10} = 0.00732$$

1 zero   1 place to the left to change 10 to 1

Example 7 Divide: $\frac{23.738}{0.001}$.

$$\frac{23.738}{0.001}, \quad 23.738. \quad \frac{23.738}{0.001} = 23,738$$

3 places   3 places to the right to change 0.001 to 1

Do Exercises 11–14.

b   Order of Operations: Decimal Notation

The same rules for order of operations used with integers apply when we are simplifying expressions involving decimal notation.

> **RULES FOR ORDER OF OPERATIONS**
> 1. Do all calculations within parentheses before operations outside.
> 2. Evaluate all exponential expressions.
> 3. Do all multiplications and divisions in order from left to right.
> 4. Do all additions and subtractions in order from left to right.

Divide.

11. $\frac{0.1278}{0.01}$

12. $\frac{0.1278}{100}$

13. $\frac{98.47}{1000}$

14. $\frac{6.7832}{-0.1}$

Answers on page A-12

5.4   Division with Decimals

Simplify.

15. $0.25 \cdot (1 + 0.08) - 0.0274$

16. $[(19.7 - 17.2)^2 + 3] \div (-1.25)$

17. *Movie Attendance.* The number of tickets sold at the movies, in billions, in each of the four years from 1993 to 1996 is shown in the bar graph below. Find the average number of tickets sold.

Source: Motion Picture Association of America

Answers on page A-12

Chapter 5 Decimal Notation

Example 8 Simplify: $(5 - 0.06) \div 2 + 3.42 \times 0.1$.

$(5 - 0.06) \div 2 + 3.42 \times 0.1 = 4.94 \div 2 + 3.42 \times 0.1$ Carrying out operations inside parentheses

$= 2.47 + 0.342$ Doing all multiplications and divisions in order from left to right

$= 2.812$

Example 9 Simplify: $13 - [5.4(1.3^2 + 0.21) \div 0.6]$.

$13 - [5.4(1.3^2 + 0.21) \div 0.6]$
$= 13 - [5.4(1.69 + 0.21) \div 0.6]$ Working in the innermost parentheses first
$= 13 - [5.4 \times 1.9 \div 0.6]$
$= 13 - [10.26 \div 0.6]$ Multiplying
$= 13 - 17.1$ Dividing
$= -4.1$

Do Exercises 15 and 16.

Example 10 *Movie Revenue.* The bar graph shows movie box-office revenue (money taken in), in billions of dollars, in each of the four years from 1993 to 1996. Find the average revenue.

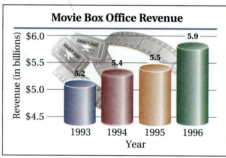

Source: Motion Picture Association of America

To find the average of a set of numbers, we add them. Then we divide by the number of addends. In this case, we are finding the average of 5.2, 5.4, 5.5, and 5.9. The average is given by

$(5.2 + 5.4 + 5.5 + 5.9) \div 4.$

Thus,

$(5.2 + 5.4 + 5.5 + 5.9) \div 4 = 22 \div 4 = 5.5.$

The average box-office revenue was $5.5 billion.

Do Exercise 17.

Exercise Set 5.4

a Divide.

1. $5\overline{)8\,2}$

2. $5\overline{)1\,8}$

3. $4\overline{)9\,5.1\,2}$

4. $8\overline{)2\,5.9\,2}$

5. $1\,2\overline{)8\,9.7\,6}$

6. $2\,3\overline{)2\,5.0\,7}$

7. $3\,3\overline{)2\,3\,7.6}$

8. $12.4 \div (-4)$

9. $9.144 \div (-8)$

10. $3.6 \div 4$

11. $-5.4 \div 6$

12. $0.0\,4\overline{)1.6\,8}$

13. $0.1\,2\overline{)8.4}$

14. $3.2\overline{)1\,2\,8}$

15. $2.6\overline{)1\,0\,4}$

16. $6 \div (-15)$

17. $1.8 \div (-12)$

18. $3\,6\overline{)1\,4.7\,6}$

19. $2.7\overline{)1\,2\,9.6}$

20. $6.2\overline{)4\,6.5}$

21. $8.5\overline{)2\,7.2}$

22. $39.06 \div (-4.2)$

23. $-5 \div (-8)$

24. $-7 \div (-8)$

25. $0.4\,7\overline{)0.1\,2\,2\,2}$

26. $0.5\,4\overline{)0.2\,7}$

27. $0.0\,3\,2\overline{)0.0\,7\,4\,8\,8}$

28. $0.0\,1\,7\overline{)1.5\,8\,1}$

29. $-24.969 \div 82$

30. $-25.221 \div 42$

31. $\dfrac{-213.4567}{100}$

32. $\dfrac{-213.4567}{10}$

33. $\dfrac{1.0237}{0.001}$

34. $\dfrac{1.0237}{-0.01}$

35. $\dfrac{56.78}{-0.001}$

36. $\dfrac{0.5678}{1000}$

37. $\dfrac{0.97}{0.1}$

38. $\dfrac{0.97}{0.001}$

39. $\dfrac{75.3}{-0.001}$

40. $\dfrac{-75.3}{1000}$

41. $\dfrac{23{,}001}{100}$

42. $\dfrac{23{,}001}{0.01}$

b Simplify.

43. $14 \times (82.6 + 67.9)$

44. $(26.2 - 14.8) \times 12$

45. $0.003 + 3.03 \div (-0.01)$

46. $42 \times (10.6 + 0.024)$

47. $(4.9 - 18.6) \times 13$

48. $4.2 \times 5.7 + 0.7 \div 3.5$

49. $123.3 - 4.24 \times 1.01$

50. $-9.0072 + 0.04 \div 0.1^2$

51. $12 \div (-0.03) - 12 \times 0.03^2$

52. $(5 - 0.04)^2 \div 4 + 8.7 \times 0.4$

53. $(4 - 2.5)^2 \div 100 + 0.1 \times 6.5$

54. $4 \div 0.4 - 0.1 \times 5 + 0.1^2$

Chapter 5 Decimal Notation

55. $6 \times 0.9 - 0.1 \div 4 + 0.2^3$

56. $5.5^2 \times [(6 - 7.8) \div 0.06 + 0.12]$

57. $12^2 \div (12 + 2.4) - [(2 - 2.4) \div 0.8]$

58. $0.01 \times \{[(4 - 0.25) \div 2.5] - (4.5 - 4.025)\}$

59. *World Population.* Using the information in the following bar graph, determine the average population of the world for the years 1950 through 2000.

60. *Manufacturing.* Using the information in the following bar graph, determine the average amount of aluminum in a particular automobile during four recent years.

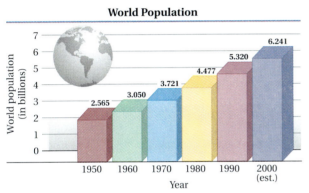

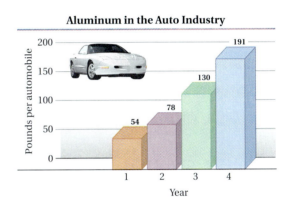

The following table lists the global average temperature for the years 1987 through 1998. Use the table for Exercises 61 and 62.

YEAR	1987	1988	1989	1990	1991	1992	1993	1994	1995	1996	1997	1998
Global temperature (in degrees Fahrenheit)	59.58°	59.63°	59.45°	59.85°	59.74°	59.23°	59.36°	59.56°	59.72°	59.58°	59.74°	60.26°

Sources: Lester R. Brown et al., *Vital Signs 1997* and the Council of Environmental Quality.

61. Find the average temperature for the years 1992 through 1996.

62. Find the average temperature for the years 1987 through 1991.

Skill Maintenance

63. Add: $-4\frac{1}{3} + 7\frac{5}{6}$. [4.6a]

64. Subtract: $7 - 8\frac{1}{3}$. [4.6b]

Solve. [4.4a]

65. $-3x + 7 = 31$.

66. $38 = 9 - 2x$

67. $-28 = 5 - 3x$

68. $9 = \frac{2}{3}x - 1$

Synthesis

69. ◆ Which is easier and why: Dividing a decimal by a decimal or dividing a fraction by a fraction?

70. ◆ Maurice insists that $0.247 \div 0.1$ is 0.0247. How could you convince him that a mistake has been made?

71. ◆ Ellie insists that $0.247 \div 10$ is 2.47. How could you convince her that a mistake has been made?

Simplify.

72. 🖩 $9.0534 - 2.041^2 \times 0.731 \div 1.043^2$

73. 🖩 $23.042(7 - 4.037 \times 1.46 - 0.932^2)$

In Exercises 74–76, find the missing value.

74. $439.57 \times 0.01 \div 1000 \times \square = 4.3957$

75. $5.2738 \div 0.01 \times 1000 \div \square = 52.738$

76. $0.0329 \div 0.001 \times 10^4 \div \square = 3290$

77. *Television Ratings.* A television rating point represents 980,000 households. The 1998 NBA finals was viewed in approximately 18.5 million households, which was a record for the NBA (**Source:** *Burlington Free Press,* 6/16/98). How many rating points did the finals receive? Round to the nearest tenth.

Electric Bills. Recently, electric bills from the Central Vermont Public Service Corporation consisted of a "customer charge" of $0.35 per day plus an "energy charge" of $0.10470 per kilowatt-hour (kWh) for the first 250 kWh used and $0.09079 per kWh for each kWh in excess of 250.

78. From August 20 to September 20, the Kaufmans' bill was $54.28. How many kilowatt-hours of electricity did they use?

79. From July 20 to August 20, the McGuires' bill was $67.89. How many kilowatt-hours of electricity did they use?

5.5 More with Fractional Notation and Decimal Notation

Now that we know how to divide using decimal notation, we can express *any* fraction as a decimal.

a Using Division to Find Decimal Notation

Recall that the expression $\frac{a}{b}$ means $a \div b$. This gives us one way of converting fractional notation to decimal notation.

Example 1 Find decimal notation for $\frac{3}{20}$.

We have

$$\frac{3}{20} = 3 \div 20 \qquad \begin{array}{r} 0.1\,5 \\ 20\overline{)3.0\,0} \\ \underline{2\,0} \\ 1\,0\,0 \\ \underline{1\,0\,0} \\ 0 \end{array} \qquad \frac{3}{20} = 0.15$$

We are finished when the remainder is 0.

Example 2 Find decimal notation for $\frac{7}{8}$.

We have

$$\frac{7}{8} = 7 \div 8 \qquad \begin{array}{r} 0.8\,7\,5 \\ 8\overline{)7.0\,0\,0} \\ \underline{6\,4} \\ 6\,0 \\ \underline{5\,6} \\ 4\,0 \\ \underline{4\,0} \\ 0 \end{array} \qquad \frac{7}{8} = 0.875$$

Do Exercises 1 and 2.

Note that the fractional notation in Examples 1 and 2 had already been simplified as much as possible. Furthermore, note that because each denominator contained only 2's and/or 5's as prime factors, each division eventually led to a remainder of 0. When the division ends, or *terminates*, with a remainder of 0, we have what is called a **terminating decimal.**

Often the denominator of a number written in simplified fractional form contains a prime factor other than 2 or 5. In such cases we can still divide to get decimal notation, but answers will be *repeating* decimals. For example, $\frac{5}{6}$ can be converted to decimal notation, but since 6 contains 3 as a prime factor, the answer will be a repeating decimal, as follows.

Objectives

a Use division to convert fractional notation to decimal notation.

b Round numbers named by repeating decimals.

c Convert certain fractions to decimal notation by using equivalent fractions.

d Simplify expressions that contain both fractional and decimal notation.

For Extra Help

TAPE 10   MAC WIN   CD-ROM

Find decimal notation.

1. $\frac{2}{5}$

2. $\frac{3}{8}$

Answers on page A-12

5.5 More with Fractional Notation and Decimal Notation

309

Find decimal notation.

3. $\dfrac{1}{6}$

4. $\dfrac{2}{3}$

Find decimal notation.

5. $\dfrac{5}{11}$

6. $-\dfrac{12}{11}$

Answers on page A-12

Example 3 Find decimal notation for $\dfrac{5}{6}$.

We have

$$\dfrac{5}{6} = 5 \div 6 \qquad 6\overline{)\begin{array}{r}0.8\,3\,3\\5.0\,0\,0\\\underline{4\,8}\\2\,0\\\underline{1\,8}\\2\,0\\\underline{1\,8}\\2\end{array}}$$

Since 2 keeps reappearing as a remainder, the digits repeat and will continue to do so; therefore,

$$\dfrac{5}{6} = 0.83333\ldots.$$

The dots indicate an endless sequence of digits in the quotient. When there is a repeating pattern, the dots are often replaced by a bar to indicate the repeating part—in this case, only the 3:

$$\dfrac{5}{6} = 0.8\overline{3}.$$

Do Exercises 3 and 4.

Example 4 Find decimal notation for $-\dfrac{4}{11}$.

First consider $\dfrac{4}{11}$. Because 11 is not a product of 2's and/or 5's, we expect a repeating decimal:

$$\dfrac{4}{11} = 4 \div 11 \qquad 11\overline{)\begin{array}{r}0.3\,6\,3\,6\\4.0\,0\,0\,0\\\underline{3\,3}\\7\,0\\\underline{6\,6}\\4\,0\\\underline{3\,3}\\7\,0\\\underline{6\,6}\\4\end{array}}$$

Since 7 and 4 keep reappearing as remainders, the sequence of digits "36" repeats in the quotient, and

$$\dfrac{4}{11} = 0.363636\ldots, \quad \text{or} \quad 0.\overline{36}.$$

Thus, $-\dfrac{4}{11} = -0.\overline{36}$.

Do Exercises 5 and 6.

Example 5 Find decimal notation for $\frac{3}{7}$.

Because 7 is not a product of 2's and/or 5's, we again expect a repeating decimal:

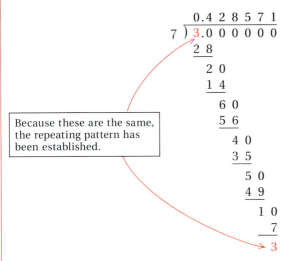

Because these are the same, the repeating pattern has been established.

Since we have already divided 7 into 3, the sequence of digits "428571" repeats in the quotient, and

$\frac{3}{7} = 0.428571428571\ldots$, or $0.\overline{428571}$.

It is possible for the repeating part of a repeating decimal to be so long that it will not fit on a calculator. For example, when $\frac{5}{97}$ is written in decimal form, its repeating part is 96 digits long! Most calculators round off repeating decimals to 9 or 10 decimal places.

Do Exercise 7.

b Rounding Repeating Decimals

In applied problems, repeating decimals are generally rounded to a predetermined degree of accuracy.

Example 6 Round $4.\overline{27}$ to the nearest thousandth.

We first rewrite the decimal without the bar. The repeating part is rewritten until we have passed the thousandths place:

$4.\overline{27} = 4.2727\ldots.$

Now we round as in Section 5.1.

a) Locate the digit in the thousandths place.   4.2 7 2 7…

b) Consider the next digit to the right.   4.2 7 2 7…

c) Since that digit, 7, is greater than or equal to 5, round up.

   4.273   **This is the answer.**

7. Find decimal notation for $\frac{5}{7}$.

Answer on page A-12

5.5 More with Fractional Notation and Decimal Notation

311

Round each to the nearest tenth, hundredth, and thousandth.

8. $0.\overline{6}$

9. $0.6\overline{08}$

10. $-7.3\overline{49}$

11. $2.6\overline{891}$

Find decimal notation. Use multiplying by 1.

12. $\dfrac{4}{5}$

13. $-\dfrac{9}{20}$

14. $\dfrac{7}{200}$

15. $\dfrac{33}{25}$

Answers on page A-12

Examples Round each to the nearest tenth, hundredth, and thousandth.

	Nearest tenth	Nearest hundredth	Nearest thousandth
7. $0.8\overline{3} = 0.83333\ldots$	0.8	0.83	0.833
8. $3.\overline{09} = 3.090909\ldots$	3.1	3.09	3.091
9. $-4.1\overline{763} = -4.1763763\ldots$	-4.2	-4.18	-4.176

Do Exercises 8–11.

c More with Conversions

Recall that fractional notation like $\dfrac{3}{10}$ or $-\dfrac{71}{1000}$ can be converted quickly to decimal notation, without performing long division. When a denominator is a factor of 10, 100, and so on, we can convert to decimal notation by finding (perhaps mentally) an equivalent fraction in which the denominator is a power of 10.

Example 10 Find decimal notation for $\dfrac{3}{20}$.

Note that 20 is a factor of 100 ($20 \cdot 5 = 100$). Thus, by using $\dfrac{5}{5}$ as an expression for 1, we can easily find an equivalent fraction with a denominator that is a power of 10:

$$\dfrac{3}{20} = \dfrac{3}{20} \cdot \dfrac{5}{5} = \dfrac{15}{100} = 0.15.$$

To perform this mentally, you might say to yourself "20 goes into 100 five times; 5 times 3 is 15; $\dfrac{15}{100}$ is 0.15."

Example 11 Find decimal notation for $-\dfrac{7}{500}$.

Since $500 \cdot 2 = 1000$, a power of 10, we use $\dfrac{2}{2}$ as an expression for 1:

$$-\dfrac{7}{500} = -\dfrac{7}{500} \cdot \dfrac{2}{2} = -\dfrac{14}{1000} = -0.014.$$

Example 12 Find decimal notation for $\dfrac{9}{25}$.

$$\dfrac{9}{25} = \dfrac{9}{25} \cdot \dfrac{4}{4} = \dfrac{36}{100} = 0.36 \quad \text{Using } \tfrac{4}{4} \text{ for 1 to get a denominator of 100}$$

As a check, we can divide:

```
       0.3 6
  2 5 ) 9.0 0
         7 5
         1 5 0
         1 5 0
             0
```

Note that multiplication by 1 is much faster.

Example 13 Find decimal notation for $\dfrac{7}{4}$.

$$\dfrac{7}{4} = \dfrac{7}{4} \cdot \dfrac{25}{25} = \dfrac{175}{100} = 1.75 \quad \text{Using } \tfrac{25}{25} \text{ for 1 to get a denominator of 100. You might also note that 7 quarters is \$1.75.}$$

Do Exercises 12–15.

d Calculations with Fractional and Decimal Notation Together

In certain kinds of calculations, fractional and decimal notation might occur together. In such cases, there are at least three ways in which we might proceed.

Example 14 Calculate: $\frac{2}{3} \times 0.576$.

METHOD 1. Perhaps the quickest method is to treat 0.576 as $\frac{0.576}{1}$. Then we multiply 0.576 by 2, and divide the result by 3.

$$\frac{2}{3} \times 0.576 = \frac{2}{3} \times \frac{0.576}{1}$$
$$= \frac{2 \times 0.576}{3} = \frac{1.152}{3}$$
$$= 0.384$$

```
    0.3 8 4
3 ) 1.1 5 2
    9
    2 5
    2 4
      1 2
      1 2
        0
```

METHOD 2. A second way to do this calculation is to convert the fractional notation to decimal notation so that both numbers are in decimal notation. Since $\frac{2}{3}$ converts to repeating decimal notation, it is first rounded to some chosen decimal place. We choose three decimal places. Then, using decimal notation, we multiply. Note that the answer is not as accurate as that found by Method 1, due to the rounding.

$$\frac{2}{3} \times 0.576 = 0.\overline{6} \times 0.576$$
$$\approx 0.667 \times 0.576 = 0.384192$$

METHOD 3. Another way to do this calculation is to convert the decimal notation to fractional notation so that both numbers are in fractional notation. The answer can be left in fractional notation and simplified, or we can convert back to decimal notation and, if appropriate, round.

$$\frac{2}{3} \times 0.576 = \frac{2}{3} \cdot \frac{576}{1000} = \frac{2 \cdot 576}{3 \cdot 1000}$$
$$= \frac{2 \cdot 2 \cdot 2 \cdot 2 \cdot 2 \cdot 2 \cdot 3 \cdot 3}{2 \cdot 2 \cdot 2 \cdot 3 \cdot 5 \cdot 5 \cdot 5} \quad \text{Factoring}$$
$$= \frac{2 \cdot 2 \cdot 2 \cdot 3}{2 \cdot 2 \cdot 2 \cdot 3} \cdot \frac{2 \cdot 2 \cdot 2 \cdot 3}{5 \cdot 5 \cdot 5} \quad \text{Removing a factor equal to 1: } \frac{2 \cdot 2 \cdot 2 \cdot 3}{2 \cdot 2 \cdot 2 \cdot 3} = 1$$
$$= \frac{2 \cdot 2 \cdot 2 \cdot 2 \cdot 3}{5 \cdot 5 \cdot 5} = \frac{48}{125}, \text{ or } 0.384$$

Do Exercises 16 and 17.

Calculate.

16. Calculate: $\frac{5}{6} \times 0.864$

17. $\frac{1}{3} \times 0.384 + \frac{5}{8} \times 0.6784$

Answers on page A-12

18. Find the area of a triangular window that is 3.25 ft wide and 2.6 ft tall.

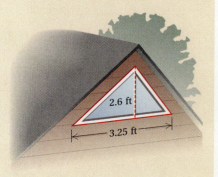

Example 15 *Boating.* A triangular sail from a single-sail day cruiser is 3.4 m wide and 4.2 m tall. Find the area of the sail.

1. **Familiarize.** We first make a drawing and recall that the formula for the area, A, of a triangle with base b and height h is $A = \frac{1}{2}bh$.

2. **Translate.** We substitute 3.4 for b and 4.2 for h:

$$A = \frac{1}{2}bh = \frac{1}{2}(3.4)(4.2). \quad \text{Evaluating}$$

3. **Solve.** We simplify as follows:

$$A = \frac{1}{2}(3.4)(4.2)$$

$$= \frac{3.4}{2}(4.2) \quad \text{Multiplying } \frac{1}{2} \text{ and } \frac{3.4}{1}$$

$$= 1.7(4.2) \quad \text{Dividing}$$

$$= 7.14. \quad \text{Multiplying}$$

4. **Check.** To check, we repeat the calculations, using the commutative law to multiply in a different order. We also rewrite $\frac{1}{2}$ as 0.5:

$$\frac{1}{2}(4.2)(3.4) = 0.5(4.2)(3.4)$$

$$= (2.1)(3.4) = 7.14.$$

Our answer checks.

5. **State.** The area of the sail is 7.14 m² (square meters).

Do Exercise 18.

Calculator Spotlight

 Many geometric applications of decimal notation involve the number π (see Exercises 86–89 of this section).

Most calculators are now equipped with a key that provides an approximation of π to at least six decimal places. Often a key labeled SHIFT or 2nd must be pressed first.

To calculate the value of an expression like $2\pi(7.5)$, on most calculators we simply press

2 × 2nd π × 7 . 5

and then = or ENTER .

Exercises

1. Calculate $4\pi(9.8)$.
2. Evaluate $2\pi r$ for $r = 8.37$.

Answer on page A-12

Exercise Set 5.5

a, c Find decimal notation.

1. $\dfrac{5}{16}$
2. $\dfrac{9}{20}$
3. $\dfrac{19}{40}$
4. $\dfrac{1}{16}$

5. $-\dfrac{1}{5}$
6. $-\dfrac{3}{20}$
7. $\dfrac{13}{20}$
8. $\dfrac{3}{40}$

9. $\dfrac{17}{40}$
10. $-\dfrac{39}{40}$
11. $\dfrac{49}{40}$
12. $\dfrac{13}{40}$

13. $-\dfrac{13}{25}$
14. $-\dfrac{21}{125}$
15. $\dfrac{2502}{125}$
16. $\dfrac{121}{200}$

17. $\dfrac{-1}{4}$
18. $\dfrac{-1}{2}$
19. $\dfrac{23}{40}$
20. $\dfrac{11}{20}$

21. $-\dfrac{5}{8}$
22. $-\dfrac{19}{16}$
23. $\dfrac{37}{25}$
24. $\dfrac{18}{25}$

25. $\dfrac{8}{15}$　　26. $\dfrac{7}{9}$　　27. $\dfrac{1}{3}$　　28. $\dfrac{1}{9}$

29. $\dfrac{-4}{3}$　　30. $\dfrac{-8}{9}$　　31. $\dfrac{7}{6}$　　32. $\dfrac{7}{11}$

33. $\dfrac{4}{7}$　　34. $\dfrac{14}{11}$　　35. $-\dfrac{11}{12}$　　36. $-\dfrac{5}{12}$

b

37.–47. Round each answer of the odd-numbered
(odd)　Exercises 25–35 to the nearest tenth, hundredth, and thousandth.

38.–48. Round each answer of the even-numbered
(even)　Exercises 26–36 to the nearest tenth, hundredth, and thousandth.

Round each of the following to the nearest tenth, hundredth, and thousandth.

49. $0.\overline{74}$　　50. $0.\overline{38}$　　51. $-7.9\overline{6}$　　52. $-3.09\overline{7}$

d Calculate.

53. $\dfrac{7}{8}(10.84)$

54. $\dfrac{4}{5}(264.8)$

55. $\dfrac{47}{9}(-79.95)$

56. $\dfrac{7}{11}(-2.7873)$

57. $\left(\dfrac{1}{6}\right)0.0765 + \left(\dfrac{3}{4}\right)0.1124$

58. $\left(\dfrac{2}{5}\right)6384.1 - \left(\dfrac{5}{8}\right)156.56$

59. $\dfrac{3}{4} \times 2.56 - \dfrac{7}{8} \times 3.94$

60. $\dfrac{2}{5} \times 3.91 - \dfrac{7}{10} \times 4.15$

61. $5.2 \times 1\dfrac{7}{8} \div 0.4$

62. $4\dfrac{3}{4} \times 0.5 \div 0.1$

Solve.

63. Find the area of a triangular shawl that is 1.8 m long and 1.2 m wide.

64. Find the area of a triangular sign that is 1.5 m wide and 1.5 m tall.

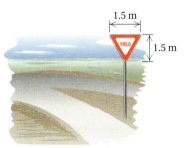

65. Find the area of a triangular stamp that is 3.4 cm wide and 3.4 cm tall.

66. Find the area of a triangular reflector that is 7.4 cm wide and 9.1 cm tall.

Skill Maintenance

67. Subtract: $20 - 16\frac{3}{5}$. [4.6b]

68. Add: $14\frac{3}{5} + 16\frac{1}{10}$. [4.6a]

69. Simplify: $\frac{95}{-1}$. [3.3c]

70. Solve: $5x - 9 = 7x + 11$. [4.4a]

71. Simplify: $9 - 4 + 2 \div (-1) \cdot 6$. [2.6b]

72. Simplify: $\frac{-9}{-9}$. [3.3c]

Synthesis

73. ◆ When is long division *not* the fastest way of converting a fraction to decimal notation?

74. ◆ Examine Example 14 of this section. How could the problem be changed so that method 2 would give a result that is completely accurate?

75. ◆ Are the numbers $6.2\overline{35}$ and $6.2\overline{3535}$ equal? Why or why not?

▦ Find decimal notation. Save the answers for Exercise 81.

76. $\frac{1}{7}$

77. $\frac{2}{7}$

78. $\frac{3}{7}$

79. $\frac{4}{7}$

80. $\frac{5}{7}$

81. ▦ From the pattern of Exercises 76–80, predict the decimal notation for $\frac{6}{7}$. Check your answer on a calculator.

Find decimal notation. Save the answers for Exercise 85.

82. $\frac{1}{9}$

83. $\frac{1}{99}$

84. $\frac{1}{999}$

85. ▦ From the pattern of Exercises 82–84, predict the decimal notation for $\frac{1}{9999}$. Check your answer on a calculator.

The formula $A = \pi r^2$ is used to find the area, A, of a circle with radius r. For Exercises 86 and 87, find the area of a circle with the given radius, using $\frac{22}{7}$ for π. For Exercises 88 and 89, use 3.14 for π or a calculator with a π key.

86. $r = 2.1$ cm

87. $r = 1.4$ cm

88. $r = \frac{3}{4}$ ft

89. $r = 4\frac{1}{2}$ yd

90. ◆ ▦ A scientific calculator indicates that

$$\frac{5}{6} = 0.833333333 \quad \text{and} \quad \frac{4{,}999{,}999{,}998}{6{,}000{,}000{,}000} = 0.833333333.$$

a) Is it true that $\frac{5}{6} = \frac{4{,}999{,}999{,}998}{6{,}000{,}000{,}000}$? Why or why not?

b) Should decimal notation for $\frac{4{,}999{,}999{,}998}{6{,}000{,}000{,}000}$ repeat? Why or why not?

5.6 Estimating

a Estimating Sums, Differences, Products and Quotients

Estimating has many uses. It can be done before a problem is even attempted in order to get an idea of the answer. It can be done afterward as a check, even when we are using a calculator. In many situations, an estimate is all we need. We usually estimate by rounding the numbers so that there are one or two nonzero digits. Consider the following advertisements for Examples 1–4.

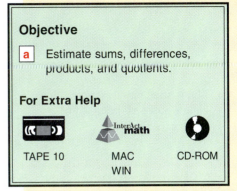

Objective

a Estimate sums, differences, products, and quotients.

For Extra Help

TAPE 10   MAC WIN   CD-ROM

Example 1 Estimate to the nearest ten the total cost of one fax machine and one TV/VCR.

We are estimating the sum

$219.99 + $349.95 = Total cost.

The estimate to the nearest ten is

$220 + $350 = $570.   (Estimated total cost)

We rounded $219.99 to the nearest ten and $349.95 to the nearest ten. The estimated sum is $570.

Do Exercise 1.

Example 2 About how much more does the TV/VCR cost than the fax machine? Estimate to the nearest ten.

We are estimating the difference

$349.95 − $219.99 = Price difference.

The estimate to the nearest ten is

$350 − $220 = $130.   (Estimated price difference)

Do Exercise 2.

1. Estimate to the nearest ten the total cost of one TV/VCR and one vacuum cleaner.

2. About how much more does the TV/VCR cost than the vacuum cleaner?

Answers on page A-13

5.6 Estimating

319

3. Estimate the total cost of 6 fax machines.

4. About how many vacuum cleaners can be bought for $830?

Estimate each product. Do not find the actual product.

5. 2.1×8.02

6. 36×0.54

7. 0.93×472

8. 0.72×0.1

9. 0.12×180.3

10. 24.359×5.2

Answers on page A-13

Chapter 5   Decimal Notation

Example 3   Estimate the total cost of 4 vacuum cleaners. (See p. 319.)

We are estimating the product

$4 \times \$189.95 =$ Total cost.

The estimate is found by rounding $189.95 to the nearest ten:

$4 \times \$190 = \$760$.

Do Exercise 3.

Example 4   About how many fax machines can be bought for $1480? (See p. 319.)

To estimate $1480 \div 219.99$, we mentally search for a number near 1480 that is a multiple of a number near 219.99. Rounding $219.99 to the nearest hundred, we get $200. Since $1480 is close to $1400, which is a multiple of 200, we have

$1480 \div 219.99 \approx 1400 \div 200 = 7$,

so the answer is 7.

Do Exercise 4.

Example 5   Estimate: 4.8×52. Do not find the actual product.

We round 4.8 to the nearest one and 52 to the nearest ten. This gives us two easy numbers with which to work, 5 and 50. Since

$4.8 \times 52 \approx 5 \times 50$

and

$5 \times 50 = 250$,

the estimated product is 250.

Compare these estimates for the product 4.94×38:

$5 \times 40 = 200$, $5 \times 38 = 190$, $4.9 \times 40 = 196$, $4.9 \times 38 = 186.2$.

The first estimate was the easiest. You could probably do it mentally. The others had more nonzero digits and were more accurate but required more work.

When making estimates, we usually look for numbers that are easy to work with mentally. For example, if multiplying, we might round 0.43 to 0.5 and 8.9 to 10, because 0.5 and 10 are such convenient numbers by which to multiply.

Do Exercises 5–10.

Example 6 Which of the following is the best estimate of 82.08 ÷ 24?

a) 400   b) 16   c) 40   d) 4

This is about 80 ÷ 20, so the answer is about 4. We could also estimate the division as 75 ÷ 25, or 3. In any case, of the choices listed, (d) is the most appropriate.

Example 7 Which of the following is the best estimate of 94.18 ÷ 3.2?

a) 30   b) 300   c) 3   d) 60

This is about 90 ÷ 3, so the answer is about 30. Thus the most appropriate choice is (a).

Example 8 Which of the following is the best estimate of 0.0156 ÷ 1.3?

a) 0.2   b) 0.002   c) 0.02   d) 20

This is about 0.02 ÷ 1, so the answer is about 0.02. Thus the most appropriate choice is (c).

Do Exercises 11–13.

In some cases, it is easier to estimate a quotient by checking products than by rounding the divisor and the dividend.

Example 9 Which of the following is the best estimate of 0.0074 ÷ 0.23?

a) 0.3   b) 0.03   c) 300   d) 3

Note that 0.23 is close to 0.25 and that 0.25 is easier to multiply by than divide by. Thus we use 0.25 to check some products.
We first try 3:

$0.23 \times 3 \approx 0.25 \times 3 = 0.75.$ This is too large.

We try a smaller estimate, 0.3:

$0.23 \times 0.3 \approx 0.25 \times 0.3 = 0.075.$ This is also too large.

We make the estimate smaller still, 0.03:

$0.23 \times 0.03 \approx 0.25 \times 0.03 = 0.0075.$

This is close to 0.0074, so the quotient is close to 0.03. Thus the most appropriate choice is (b).

Do Exercise 14.

Select the most appropriate estimate for each quotient.

11. 59.78 ÷ 29.1
   a) 200   b) 20
   c) 2   d) 0.2

12. 82.08 ÷ 2.4
   a) 40   b) 4.0
   c) 400   d) 0.4

13. 0.1768 ÷ 0.08
   a) 8   b) 10
   c) 2   d) 20

14. Which of the following is an appropriate estimate of 0.0069 ÷ 0.15?
   a) 0.5   b) 50
   c) 0.05   d) 23.4

Answers on page A-13

5.6 Estimating

Improving Your Math Study Skills

Other study tips appear on pages 50, 90, 100, 150, 368, and 596.

Study Tips for Trouble Spots

By now you have probably encountered certain topics that gave you more difficulty than others. It is important to know that this happens to every person who studies mathematics. Unfortunately, frustration is often part of the learning process and it is important not to give up when difficulty arises.

One source of frustration for many students is not being able to set aside sufficient time for studying. Family commitments, work schedules, and athletics are just a few of the time demands that many students face. Couple these demands with a math lesson that seems to require a greater than usual amount of study time, and it is no wonder that many students often feel frustrated. Below are some study tips that might be useful if and when troubles arise.

- **Realize that everyone—even your instructor—has been stymied at times when studying math.** You are not the first person, nor will you be the last, to encounter a "roadblock."
- **Whether working alone or with a classmate, try to allow enough study time so that you won't need to constantly glance at a clock.** Difficult material is best mastered when your mind is completely focused on the subject matter. Thus, if you are tired, it is usually best to study early the next morning or to take a ten-minute "power-nap" in order to make the most productive use of your time.
- **Talk about your trouble spot with a classmate.** It is possible that she or he is also having difficulty with the same material. If that is the case, perhaps the majority of your class is confused and your instructor's coverage of the topic is not yet finished. If your classmate *does* understand the topic that is troubling you, patiently allow him or her to explain it to you. By verbalizing the math in question, your classmate may help clarify the material for both of you. Perhaps you will be able to return the favor for your classmate when he or she is struggling with a topic that you understand.
- **Try to study in a "controlled" environment.** What we mean by this is that you can often put yourself in a setting that will enable you to maximize your powers of concentration. For example, whereas some students may succeed in studying at home or in a dorm room, for many these settings are filled with distractions. Consider a trip to a library, classroom building, or perhaps the attic or basement if such a setting is more conducive to studying. If you plan on working with a classmate, try to find a location in which conversation will not be bothersome to others.
- **When working on difficult material, it is often helpful to first "back up" and review the most recent material that *did* make sense.** This can build your confidence and create a momentum that can often carry you through the roadblock. Sometimes a small piece of information that appeared in a previous section is all that is needed for your problem spot to disappear. When the difficult material is finally mastered, try to make use of what is fresh in your mind by taking a "sneak preview" of what your next topic for study will be.
- Consider keeping a mathematical journal and/or toolbox. In a mathematical journal, you write study tips and observations that have been gleaned from the most recent lesson. In a mathematical toolbox, you use a designated notebook or portion of a notebook to list important formulas and concepts as they are encountered throughout the course.

Maintaining your journal or toolbox can often help you to clarify your understanding of course material.

Exercise Set 5.6

a Consider the following advertisements for Exercises 1–8. Estimate the sums, differences, products, or quotients involved in these problems.

1. Estimate the total cost of one entertainment center and one mini system.

2. Estimate the total cost of one entertainment center and one TV.

3. About how much more does the TV cost than the mini system?

4. About how much more does the TV cost than the entertainment center?

5. Estimate the total cost of 9 TVs.

6. Estimate the total cost of 16 mini systems.

7. About how many TVs can be bought for $1700?

8. About how many mini systems can be bought for $1300?

Estimate by rounding as directed.

9. 0.02 + 1.31 + 0.34; nearest tenth

10. 0.88 + 2.07 + 1.54; nearest one

11. 6.03 + 0.007 + 0.214; nearest one

12. 1.11 + 8.888 + 99.94; nearest one

13. 52.367 + 1.307 + 7.324; nearest one

14. 12.9882 + 1.0115; nearest tenth

15. 2.678 − 0.445; nearest tenth

16. 12.9882 − 1.0115; nearest one

17. 198.67432 − 24.5007; nearest ten

Estimate. Choose a rounding digit that gives one or two nonzero digits. Indicate which of the choices is an appropriate estimate.

18. 234.12321 − 200.3223
 a) 600 b) 60
 c) 300 d) 30

19. 49 × 7.89
 a) 400 b) 40
 c) 4   d) 0.4

20. 7.4 × 8.9
 a) 95   b) 63
 c) 124 d) 6

21. 98.4 × 0.083
 a) 80   b) 14
 c) 8   d) 0.8

22. 78 × 5.3
 a) 400 b) 800
 c) 40   d) 8

23. 3.6 ÷ 4
 a) 10   b) 1
 c) 0.1 d) 0.01

24. 0.0713 ÷ 1.94
 a) 4   b) 0.4
 c) 0.04 d) 40

25. 74.68 ÷ 24.7
 a) 9   b) 3
 c) 12   d) 120

26. 914 ÷ 0.921
 a) 9   b) 90
 c) 900 d) 0.9

27. *Movie Revenue.* Total summer box-office revenue (money taken in) for the movie *Eraser* was $53.6 million (**Source:** *Hollywood Reporter Magazine*). Each theater showing the movie averaged $6716 in revenue. Estimate how many screens were showing this movie.

28. *Nintendo and the Sears Tower.* The Nintendo Game Boy portable video game is 4.5 in. (0.375 ft) tall (**Source:** Nintendo of America). Estimate how many game units it would take to reach the top of the Sears Tower, which is 1454 ft tall. Round to the nearest one.

Skill Maintenance

Find the prime factorization. [3.2c]

29. 108 **30.** 400 **31.** 325 **32.** 666

Simplify. [2.5a]

33. $\dfrac{125}{400}$ **34.** $\dfrac{3225}{6275}$ **35.** $\dfrac{72}{81}$ **36.** $\dfrac{325}{625}$

Synthesis

37. ◆ Under what circumstance(s) would you round down from 5.8 to 5.0 when rounding to the nearest one?

38. ◆ A roll of fiberglass insulation costs $21.95. Describe two situations involving estimating and the cost of fiberglass insulation. Devise one situation so that $21.95 is rounded to $22. Devise the other situation so that $21.95 is rounded to $20.

39. ◆ Describe a situation in which an estimation is made by rounding to the nearest 10,000 and then multiplying.

The following were done on a calculator. Estimate to see if the decimal point was placed correctly.

40. 178.9462 × 61.78 = 11,055.29624

41. 14,973.35 ÷ 298.75 = 501.2

42. 19.7236 − 1.4738 × 4.1097 = 1.366672414

43. 28.46901 ÷ 4.9187 − 2.5081 = 3.279813473

5.7 Solving Equations

In Section 4.4, we saw how the addition and multiplication principles can be used to solve equations like $5x + 7 = -3$. We now use those same properties to solve similar equations involving decimals.

a   Equations with One Variable Term

Recall that to solve equations like $3x = 12$, we can use the multiplication principle to multiply by $\frac{1}{3}$ on both sides. This is the same as dividing by 3 on both sides. The same procedure works with decimals.

Example 1   Solve: $3.4x = 6.97$.

We have

$$3.4x = 6.97$$

$$\frac{3.4x}{3.4} = \frac{6.97}{3.4} \quad \text{Dividing by 3.4 (or multiplying by } \tfrac{1}{3.4}\text{) on both sides}$$

$$x = 2.05.$$

```
       2.0 5
3.4. ) 6.9.7 0
       6 8 0 0
         1 7 0
         1 7 0
             0
```

To check, we can approximate: Note that $3.4 \approx 3.5$ and $2.05 \approx 2$. Since $3.5 \cdot 2 = 7.0 \approx 6.97$, we have a partial check. The solution is 2.05.

Do Exercises 1 and 2.

To solve equations like $x + 7 = -3$, we use the addition principle to add -7 on both sides or, equivalently, to subtract 7 on both sides. The same approach is used with decimals.

Example 2   Solve: $x + 7.4 = -3.1$.

We have

$$x + 7.4 = -3.1$$
$$x + 7.4 + (-7.4) = -3.1 + (-7.4) \quad \text{Adding } -7.4 \text{ on both sides}$$
$$x = -10.5.$$

To check, note that $-10.5 + 7.4 = -3.1$. The solution is -10.5.

In Examples 1 and 2, we used the addition and multiplication principles to produce equivalent equations from which we could easily read the solution. A similar approach, involving more steps, will be used in the examples that follow.

Do Exercises 3 and 4.

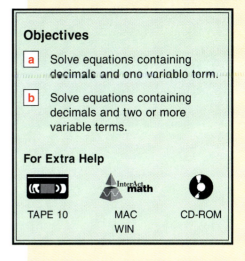

Objectives

a   Solve equations containing decimals and one variable term.

b   Solve equations containing decimals and two or more variable terms.

For Extra Help

TAPE 10   MAC WIN   CD-ROM

Solve.
1. $23x = 96.6$

2. $1.25t = 7.125$

Solve.
3. $x + 9.8 = 12.4$

4. $6.5 + t = -4.3$

Answers on page A-13

Solve.

5. $7.4t + 1.25 = 27.89$

6. $-2.7 + 4.8x = -11.82$

Answers on page A-13

As we saw in Section 4.4, when an equation requires both multiplication *and* addition, we generally "undo" the addition before we "undo" the multiplication. This reverses the order of operations in which we multiply first and then add.

Example 3 Solve: $4.2x + 3.7 = -26.12$.

$$4.2x + 3.7 = -26.12$$
$$4.2x + 3.7 - 3.7 = -26.12 - 3.7 \quad \text{Subtracting 3.7 or adding } -3.7 \text{ on both sides}$$
$$4.2x = -29.82 \quad \text{Simplifying}$$
$$\frac{4.2x}{4.2} = \frac{-29.82}{4.2} \quad \text{Multiplying by } \frac{1}{4.2} \text{ or dividing by 4.2 on both sides}$$
$$x = -7.1 \quad \text{Simplifying}$$

CHECK:
$$\begin{array}{c|c} 4.2x + 3.7 = -26.12 \\ \hline 4.2(-7.1) + 3.7 \;?\; -26.12 \\ -29.82 + 3.7 \\ -26.12 & -26.12 \quad \text{TRUE} \end{array}$$

The solution is -7.1.

Do Exercises 5 and 6.

Calculator Spotlight

 To check the solution of Example 3 with a scientific calculator, we press the following sequence of keys:

[4] [.] [2] [×] [7] [.] [1] [+/−] [+] [3] [.] [7] [=]

The result, -26.12, shows that -7.1 is a solution of $4.2x + 3.7 = -26.12$. Note that a negative number is entered as a positive number, followed by the [+/−] key. On most graphing calculators, the [(−)] and [ENTER] keys take the place of the [+/−] and [=] keys.

Exercises

1. Use a calculator to check the solution of Examples 1 and 2.
2. Use a calculator to show that -1.9 is a solution of $-2.7 + 4.8x = -11.82$ (Margin Exercise 6).
3. Use a calculator to show that -3.6 is *not* a solution of $7.4t + 1.25 = 27.89$ (Margin Exercise 5).

b Equations with Two or More Variable Terms

Some equations have variable terms on both sides. To solve such an equation, we use the addition principle to get all variable terms on one side of the equation and all constant terms on the other side.

Example 4 Solve: $10x - 7 = 2x + 13$.

We begin by subtracting $2x$ (or adding $-2x$) on both sides. This will group all variable terms on one side of the equation:

$10x - 7 - 2x = 2x + 13 - 2x$ Adding $-2x$ on both sides

$8x - 7 = 13.$ Combining like terms

This last equation is similar to Example 3. As in that example, we use the addition principle to isolate all constant terms on one side:

$8x - 7 = 13$

$8x - 7 + 7 = 13 + 7$ Adding 7 on both sides

$8x = 20$ Simplifying (combining like terms)

$\dfrac{8x}{8} = \dfrac{20}{8}$ Dividing on both sides, as in Example 1

$x = 2.5.$

CHECK:

$10x - 7 = 2x + 13$

$10(2.5) - 7 \;?\; 2(2.5) + 13$
$25 - 7 \;|\; 5 + 13$
$18 \;|\; 18$ TRUE

The solution is 2.5.

Sometimes it may be easier to combine all variable terms on the right side and all constant terms on the left side.

Example 5 Solve: $11 - 3t = 7t + 8$.

We can combine all variable terms on the right side by adding $3t$ on both sides:

$11 - 3t = 7t + 8$

$11 - 3t + 3t = 7t + 8 + 3t$ Using the addition principle

$11 = 10t + 8$ Combining like terms

$11 - 8 = 10t + 8 - 8$ Adding -8 on both sides

$3 = 10t$

$\dfrac{3}{10} = \dfrac{10t}{10}$ Dividing by 10 on both sides

$0.3 = t.$

CHECK:

$11 - 3t = 7t + 8$

$11 - 3(0.3) \;?\; 7(0.3) + 8$
$11 - 0.9 \;|\; 2.1 + 8$
$10.1 \;|\; 10.1$ TRUE

The solution is 0.3.

Solve.

7. $10t - 3 = 4t + 18$

8. $8 + 4x = 9x - 3$

9. $2.1x - 45.3 = 17.3x + 23.1$

10. Solve:

$3(x + 4) = 17 - x.$

Answers on page A-13

Because equations are reversible, it does not matter whether the variable is isolated on the right or left side. What is important is that you have a clear direction to your work as you proceed from step to step.

Do Exercises 7–9.

Note that after we had combined all variable terms on one side of the equation, Examples 4 and 5 were similar to Example 3.

> **PROBLEM-SOLVING TIP**
>
> When faced with a new type of problem, see if you can change the problem into an equivalent problem that you find easier to solve.

Example 6 Solve: $5(x + 1) = 3x + 12$.

By using the distributive law, we can find an equivalent equation:

$5(x + 1) = 3x + 12$

$5 \cdot x + 5 \cdot 1 = 3x + 12$ Using the distributive law to remove parentheses

$5x + 5 = 3x + 12.$ Simplifying

We now solve as we did in Examples 4 and 5:

$5x + 5 - 3x = 3x + 12 - 3x$ Subtracting $3x$ on both sides

$2x + 5 = 12$ Simplifying

$2x + 5 - 5 = 12 - 5$ Subtracting 5 on both sides

$2x = 7$

$\dfrac{2x}{2} = \dfrac{7}{2}$ Dividing by 2 on both sides

$x = 3.5.$

CHECK:

$\begin{array}{c|c} 5(x + 1) = 3x + 12 \\ \hline 5(3.5 + 1) \ ? \ 3(3.5) + 12 \\ 5(4.5) \ | \ 10.5 + 12 \\ 22.5 \ | \ 22.5 \end{array}$ TRUE

The solution is 3.5.

Do Exercise 10.

Chapter 5   Decimal Notation

Exercise Set 5.7

a Solve. Remember to check.

1. $5x = 27$

2. $36 \cdot y = 14.76$

3. $100t = 52.39$

4. $789.23 = 0.25 \cdot q$

5. $-23.4 = 5.2a$

6. $-40.74 = 4.2x$

7. $-9.2x = -94.76$

8. $-7.6a = -29.64$

9. $t - 19.27 = 24.51$

10. $t - 3.012 = 10.478$

11. $4.1 = -3.6 + n$

12. $2.7 = -5.31 + m$

13. $x + 13.9 = 4.2$

14. $x + 15.7 = 3.1$

15. $4x - 7 = 13$

16. $5x - 8 = 22$

17. $7.1x - 9.3 = 8.45$

18. $6.9x - 8.4 = 4.02$

19. $12.4 + 3.7t = 2.04$

20. $21.6 + 4.1t = 6.43$

21. $-26.05 = 7.5x + 9.2$

22. $-43.42 = 8.7x + 5.3$

23. $-4.2x + 3.04 = -4.1$

24. $-2.9x - 2.24 = -17.9$

b Solve. Remember to check.

25. $9x - 6 = 5x + 30$

26. $8x - 5 = 6x + 9$

27. $3x + 15 = 11x + 5$

28. $2x + 18 = 7x + 2$

29. $6y - 5 = 8 + 10y$

30. $5y - 3 = 4 + 9y$

31. $5.9x + 67 = 7.6x + 16$

32. $2.1x + 42 = 5.2x - 20$

33. $7.8a + 2 = 2.4a + 19.28$

34. $7.5a - 5.16 = 3.1a + 12$

35. $5(x + 2) = 3x + 18$

36. $6(x + 2) = 4x + 30$

37. $2(x + 3) = 4x - 11$

38. $5(x + 3) = 15x - 6$

39. $2a + 17 = 12(a - 1)$

40. $7a + 6 = 15(a - 2)$

41. $2(x + 7.3) = 6x - 0.83$

42. $2.9(x + 8.1) = 7.8x - 3.95$

43. $-7.37 - 3.2t = 4.9(t + 6.1)$

44. $-6.21 - 4.3t = 9.8(t + 2.1)$

45. $9(x - 4) + 8 = 4x + 7$

46. $4(x - 2) - 9 = 2x + 9$

47. $34(5 - 3.5x) = 12(3x - 8) + 653.5$

48. $43(7 - 2x) + 34 = 50(x - 4.1) + 744$

Skill Maintenance

49. Simplify: $\dfrac{-43}{-43}$. [3.3a]

50. Add: $\dfrac{4}{9} + \dfrac{5}{6}$. [4.2b]

51. Subtract: $\dfrac{3}{25} - \dfrac{7}{10}$. [4.3a]

52. Simplify: $\dfrac{0}{-18}$. [3.3a]

53. Add: $-17 + 24 + (-9)$. [2.2a]

54. Solve: $3x - 10 = 14$. [4.4a]

Synthesis

55. ◆ When solving linear equations, can the multiplication principle be used before the addition principle? Why or why not?

56. ◆ Is it possible for an equation like $x + 3 = x + 5$ to have a solution? Why or why not?

57. ◆ Describe a method for constructing an equation similar in form to Exercise 17, but with 8.3 as the solution.

Solve.

58. $7.035(4.91x - 8.21) + 17.401 = 23.902x - 7.372815$

59. $8.701(3.4 - 5.1x) - 89.321 = 5.401x + 74.65787$

60. $5(x - 4.2) + 3[2x - 5(x + 7)] = 39 + 2(7.5 - 6x) + 3x$

61. $14(2.5x - 3) + 9x + 5 = 4(3.25 - x) + 2[5x - 3(x + 1)]$

Create and solve equations as a group.

5.8 Applications and Problem Solving

a Solving applied problems with decimals is like solving applied problems with integers. We translate first to an equation that corresponds to the situation. Then we solve the equation.

> **Objective**
>
> **a** Solve applied problems involving decimals.
>
> **For Extra Help**
>
>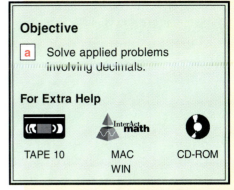
>
> TAPE 10   MAC WIN   CD-ROM

Example 1 *Dining Out.* More and more Americans are eating meals outside the home. The following graph compares the average check for meals of various types for the years 1997 and 1998. How much more is the average check for fast food in 1998 than in 1997?

Source: Sandelman and Associates, Brea, California

1. **Familiarize.** We use the bar graph to visualize the situation and to obtain the appropriate data. We let c = the additional amount spent on fast food in 1998.

2. **Translate.** This is a "how-much-more" situation. We translate as follows, using the data from the bar graph.

Average check in 1997	plus	Additional amount	is	Average check in 1998
↓	↓	↓	↓	↓
$10.56	+	c	=	$11.01

3. **Solve.** We solve the equation by subtracting 10.56 from both sides:

   $$10.56 + c - 10.56 = 11.01 - 10.56$$
   $$c = 0.45.$$

   $$\begin{array}{r} \overset{0\ 9\ 11}{1\,\cancel{1}.\cancel{0}\,\cancel{1}} \\ -\ 1\ 0.5\ 6 \\ \hline 0.4\ 5 \end{array}$$

4. **Check.** We can check by adding 0.45 to 10.56 to get 11.01.

5. **State.** The average check for fast food in 1998 was 45¢ more than in 1997.

Do Exercise 1.

1. *Body Temperature.* Normal body temperature is 98.6°F. When fevered, most people will die if their bodies reach 107°F. This is a rise of how many degrees?

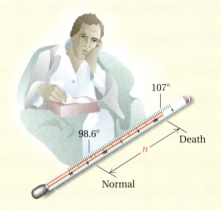

Answer on page A-13

2. At Copylot Printing, the cost of copying is 8 cents per page. How much, in dollars, would it cost to make 466 copies?

Example 2 *IRS Driving Allowance.* In 1997, the Internal Revenue Service (IRS) allowed a tax deduction of 31.5¢ per mile for mileage driven for business purposes. What deduction, in dollars, could Jill take for driving 640 work-related miles?

1. **Familiarize.** We first make a drawing or at least visualize the situation. Repeated addition fits this situation. We let d = the deduction, in dollars, allowed for driving 640 mi. Note that 31.5¢ is $0.315.

2. **Translate.** We translate as follows.

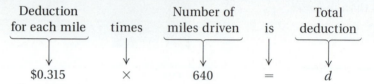

3. **Solve.** To solve the equation, we carry out the multiplication.

$$\begin{array}{r} 0.3\,1\,5 \\ \times\quad\;\,6\,4\,0 \\ \hline 0\,0\,0 \\ 1\,2\,6\,0 \\ 1\,8\,9\,0 \\ \hline 2\,0\,1.6\,0\,0 \end{array}$$

Thus, $d = 201.60$.

4. **Check.** We can obtain a partial check by rounding and estimating:

$$0.315 \times 640 \approx 0.3 \times 650$$
$$= 195 \approx 201.6.$$

5. **State.** Jill could take a tax deduction of $201.60.

Do Exercise 2.

Multistep Problems

Example 3 *Student Loans.* Upon graduation from college, Jannette will be faced with repaying a Stafford loan that totals $23,334. The loan is to be paid back over 10 yr in equal payments. Find the amount of each payment.

Answer on page A-13

1. **Familiarize.** We imagine the situation as one in which money that was borrowed is then repaid in monthly checks that are always the same amount. Since we are not told how many checks there will be, one part of solving this problem is to determine how many months are in 10 yr. We let $m =$ the size of each monthly payment.

2. **Translate.** To find the amount of the monthly payment, we note that the amount owed is split up, or *divided,* into payments of equal size. The size of each payment will depend on how many payments there are. To find the number of payments, we first determine that in 10 yr there are

    $10 \cdot 12 = 120$ months. **There are 12 months in a year.**

    We have

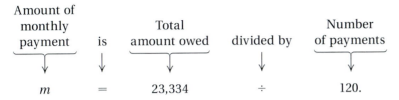

3. **Solve.** To solve, we carry out the division.

    ```
              1 9 4.4 5
      1 2 0 ) 2 3,3 3 4.0 0      m = 194.45
              1 2 0
              1 1 3 3
              1 0 8 0
                  5 3 4
                  4 8 0
                    5 4 0
                    4 8 0
                      6 0 0
                      6 0 0
                          0
    ```

4. **Check.** To check, we first verify that there are 120 months in 10 yr. We can do this with division:

    120 months ÷ 12 months per year = 10 years.

    To check that the amount of the monthly payment is correct, we can estimate the product:

    $194.45 \cdot 120 \approx 200 \cdot 120 = 24{,}000 \approx 23{,}334.$

5. **State.** Jannette's monthly payments will be $194.45.

Do Exercise 3.

3. *Car Payments.* Kevin's car loan totals $11,370 and is to be paid over 5 yr in monthly payments of equal size. Find the amount of each payment?

Answer on page A-13

5.8 Applications and Problem Solving

335

4. *Gas Mileage.* Ivan filled his Dodge Stratus and noted that the odometer read 38,320.8. After the next filling, the odometer read 38,735.5. It took 14.5 gal to fill the tank. How many miles per gallon (mpg) did Ivan's Dodge get?

Example 4 *Gas Mileage.* Emma filled her Ford Contour with gas and noted that the odometer read 67,507.8. After the next filling, the odometer read 67,890.3. It took 12.5 gal to fill the tank. How many miles per gallon did Emma's Ford get?

1. **Familiarize.** We make a drawing.

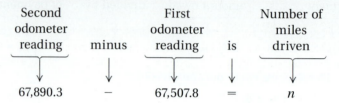

This is a two-step problem. First, we find the number of miles driven between fillups. We let $n = $ the number of miles driven.

2., 3. **Translate** and **Solve.** To find the number of miles driven, we translate and solve as follows.

Second odometer reading	minus	First odometer reading	is	Number of miles driven
↓	↓	↓	↓	↓
67,890.3	−	67,507.8	=	n

To solve the equation, we simplify on the left side:

$67,890.3 - 67,507.8 = n$

$382.5 = n.$

```
  6 7,8 9 0.3
− 6 7,5 0 7.8
  ─────────
      3 8 2.5
```

Next, we divide the number of miles driven by the number of gallons used. This gives us $m = $ the number of miles per gallon—that is, the mileage. The division that corresponds to the situation is

$382.5 \div 12.5 = m.$

To find the number m, we divide.

```
            3 0.6
   1 2.5. ) 3 8 2.5ˬ0
              3 7 5 0
              ─────
                7 5 0
                7 5 0
                ─────
                    0
```

Thus, $m = 30.6.$

4. **Check.** To check, we first multiply the number of miles per gallon times the number of gallons:

$12.5 \times 30.6 = 382.5.$ **12.5 gal would take Emma 382.5 mi.**

Then we add 382.5 to 67,507.8:

$67,507.8 + 382.5 = 67,890.3.$

The mileage 30.6 checks.

5. **State.** Emma's Ford Contour got 30.6 miles per gallon.

Do Exercise 4.

Answer on page A-13

Some problems may require us to recall important formulas. Example 5 involves a formula from geometry that is worth remembering.

> In any circle, a **diameter** is a segment that passes through the center of the circle with endpoints on the circle. A **radius** is a segment with one endpoint on the center and the other endpoint on the circle. The area, A, of a circle with radius of length r is given by
> $$A = \pi \cdot r^2,$$
> where $\pi \approx 3.14$.

Example 5 The Northfield Tap and Die Company stamps 6-cm–wide discs out of metal squares that are 6 cm by 6 cm. How much metal remains after the disc has been punched out?

1. **Familiarize.** We make, and label, a drawing. The question deals with discs, squares, and leftover material, so we list the relevant area formulas.

   For a square with sides of length s,

   Area $= s^2$.

   For a circle with radius of length r,

   Area $= \pi \cdot r^2$,

   where $\pi \approx 3.14$.

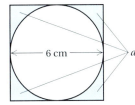

2. **Translate.** To find the amount left over, we subtract the area of the disc from the area of the square. Note that a circle's radius is half of its diameter, or width.

   $$\underbrace{6^2}_{\text{Area of square}} \underbrace{-}_{\text{minus}} \underbrace{3.14 \times \left(\frac{6}{2}\right)^2}_{\text{Area of disc}} \underbrace{=}_{\text{is}} \underbrace{a}_{\text{Area left over}}$$

3. **Solve.** We simplify as follows:

   $$6^2 - 3.14\left(\frac{6}{2}\right)^2 = a$$
   $$36 - 3.14(3)^2 = a$$
   $$36 - 3.14 \cdot 9 = a$$
   $$36 - 28.26 = a$$
   $$7.74 = a.$$

4. **Check.** We can repeat our calculation as a check. Note that 7.74 is less than the area of the disc, which in turn is less than the area of the square. This agrees with the impression given by our drawing.

5. **State.** The amount of material left over is 7.74 cm².

Do Exercise 5.

---

5. Suppose that an 8-in.–wide disc is punched out of an 8-in. by 8-in. sheet of metal. How much material is left over?

Answer on page A-13

6. Yardbird Landscaping charges customers $25 plus $20 per hour to rototill a garden. For how many hours can Emily hire Yardbird if she has budgeted $50 for rototilling?

Example 6 *Truck Rentals.* Yardbird Landscaping has rented a 22-ft truck at a daily rate of $49.95 plus 35 cents a mile. They have budgeted $125 for renting a truck to deliver trees to customers around the county. How many miles can a one-day rental truck be driven without exceeding the budget?

1. **Familiarize.** Suppose the landscapers drive 100 mi. Then the cost would be

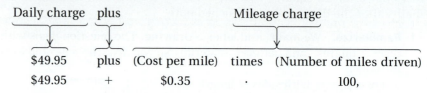

which is $49.95 + $35, or $84.95. This familiarizes us with the way in which a calculation is made. Note that we convert 35 cents to $0.35 so that only one unit, dollars, is used. Note also that the landscapers can exceed 100 mi and still be within budget. To see just how many miles the budget allows for, we could make and check more guesses, but this would be very time-consuming. Instead we let $m =$ the number of miles that can be driven within a $125 budget.

2. **Translate.** The problem can be rephrased and translated as follows.

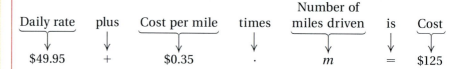

3. **Solve.** We solve the equation:

$$49.95 + 0.35m = 125$$
$$0.35m = 75.05 \quad \text{Subtracting 49.95 on both sides}$$
$$m = \frac{75.05}{0.35} \quad \text{Dividing by 0.35 on both sides}$$
$$m \approx 214.4. \quad \text{Rounding to the nearest tenth}$$

4. **Check.** We check in the original problem. We multiply 214.4 by $0.35, getting $75.04. Then we add $75.04 to $49.95 and get $124.99, which is just about the $125 allotted. Our answer is not exact because we rounded.

5. **State.** Yardbird Landscaping can drive the truck about 214.4 mi without exceeding the budget.

Do Exercise 6.

Answer on page A-13

Exercise Set 5.8

a Solve.

1. What is the cost of 7 shirts at $32.98 each?

2. What is the cost of 8 pairs of socks at $4.95 each?

3. What is the cost, in dollars, of 20.4 gal of gasoline at 129.9 cents per gallon? (129.9 cents = $1.299) Round the answer to the nearest cent.

4. What is the cost, in dollars, of 17.7 gal of gasoline at 119.9 cents per gallon? (119.9 cents = $1.199) Round the answer to the nearest cent.

5. Madeleine buys a book for $44.68 and pays with a $50 bill. How much change does she receive?

6. Roberto bought a CD for $16.99 and paid with a $20 bill. How much change was there?

7. *Nursing.* A nurse draws 17.85 mg of blood and uses 9.68 mg in a blood test. How much is left?

8. *Medicine.* Normal body temperature is 98.6°F. During an illness, a patient's temperature rose 4.2°. What was the new temperature?

9. *Finance.* A car loan totaling $4425 is to be paid off in 12 monthly payments of equal size. How much is each payment?

10. *Culinary Arts.* One pound of crabmeat makes three servings at the Key West Seafood Restaurant. If the crabmeat costs $16.95 per pound, what is the cost per serving?

11. *Medicine.* After being tested for allergies, Mike was given allergy shots of 0.25 mg, 0.4 mg, 0.5 mg, and 0.5 mg over a 2-month period. What was the total amount of the injections?

12. *Beverage Consumption.* Each year, the average American drinks about 49.0 gal of soft drinks, 41.2 gal of water, 25.3 gal of milk, 24.8 gal of coffee, and 7.8 gal of fruit juice. What is the total amount that the average American drinks?

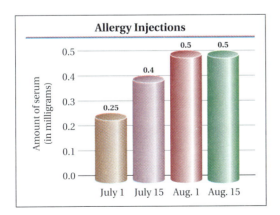

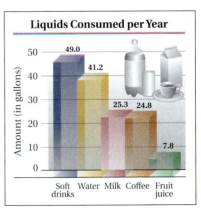

Source: U.S. Department of Agriculture

13. *Lotteries.* In Texas, one of the state lotteries is called "Cash 5." In a recent weekly game, the lottery prize of $127,315 was shared equally by 6 winners. How much was each winner's share? Round to the nearest cent.

14. A group of 4 students pays $40.76 for lunch. What is each person's share?

15. A rectangular poster measures 73.2 cm by 61.8 cm. What is its area?

16. A rectangular fenced yard measures 40.3 yd by 65.7 yd. What is its area?

17. The length of the Panama Canal is 81.6 kilometers (km). The length of the Suez Canal is 175.5 km. How much longer is the Suez Canal?

18. In 1970, the median age of a first-time bride was 20.8 yr. In 1992, the median age was 24.4 yr. How much greater was the median age of a bride in 1992 than in 1970?

19. *Taxes.* The Colavitos own a house with an assessed value of $124,500. For every $1000 of assessed value, they pay $7.68 in taxes. How much in taxes do they pay?

20. *Chemistry.* The water in a filled tank weighs 748.45 lb. One cubic foot of water weighs 62.5 lb. How many cubic feet of water does the tank hold?

Find the distance around (perimeter of) each figure.

21.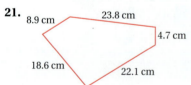

22.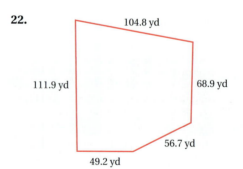

Chapter 5 Decimal Notation

340

23.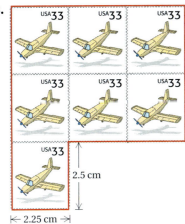

24.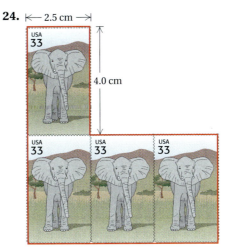

Find the length *d* in each figure.

25.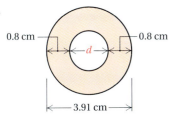

26.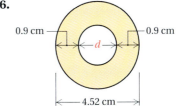

27. *Mileage.* Peggy filled her van's gas tank and noted that the odometer read 26,342.8. After the next filling, the odometer read 26,736.7. It took 19.5 gal to fill the tank. How many miles per gallon (mpg) did the van get?

28. *Mileage.* Peter filled his Honda's gas tank and noted that the odometer read 18,943.2. After the next filling, the odometer read 19,306.2. It took 13.2 gal to fill the tank. How many miles per gallon did the car get?

29. Roberto bought 3 CDs at $12.99 (tax included). He paid with a $50 bill. How much change did he receive?

30. Natalie Clad had $185.00 to spend for fall clothes: She spent $44.95 for shoes, $71.95 for a jacket, and $55.35 for pants. How much was left?

Exercise Set 5.8

31. *Carpentry.* A round, 6-ft–wide, hot tub is being built into a 12-ft by 30-ft rectangular deck. How much decking is needed for the surface of the deck?

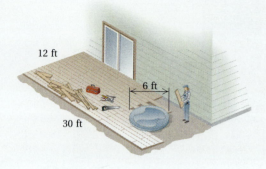

32. *Landscaping.* A rectangular yard is 20 ft by 15 ft. The yard is covered with grass except for a circular flower garden with an 8-ft diameter. How much grass is in the yard?

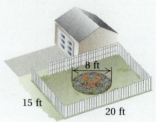

33. A 4-ft by 6-ft table top is cut from a round table top that is 9 ft wide. How much wood will be left over?

34. A 4-ft by 4-ft tablecloth is cut from a round tablecloth that is 6 ft wide. How much cloth will be left over?

35. Zachary worked 53 hr during a week one summer. He earned $8.50 per hour for the first 40 hr and $12.75 per hour for overtime. How much did Zachary earn during the week?

36. It costs $24.95 a day plus 27 cents per mile to rent a compact car at Shuttles Rent-A-Car. How much, in dollars, would it cost to drive the car 120 mi in 1 day?

37. *Car Rentals.* Badger Rent-A-Car rents a midsize car at a daily rate of $34.95 plus 10 cents per mile. A businessperson is allotted $80 for car rental. How many miles can the businessperson travel on the $80 budget?

38. *Car Rentals.* Badger also rents a full-size car at $43.95 plus 10 cents per mile. A businessperson has a car rental allotment of $90. How many miles can the businessperson travel on the $90 budget?

39. *Bike Rentals.* Mike's Bikes rents mountain bikes. The shop charges $5.50 insurance for each rental plus $2.40 per hour. For how many hours can a person rent a bike with $25.00?

40. *Phone Bills.* Auritech Communication charges 50¢ for the first minute and 32¢ for each additional minute for a certain phone call. For how long could two parties speak if the bill for the call cannot exceed $5.30?

41. *Service Calls.* JoJo's Service Center charges $30 for a house call plus $37.50 for each hour the job takes. For how long has a repairperson worked on a house call if the bill comes to $123.75?

42. *Electric Rates.* Southeast Electric charges 9¢ per kilowatt-hour for the first 200 kWh. The company charges 11¢ per kilowatt-hour for all electrical usage in excess of 200 kWh. How many kilowatt-hours were used if a monthly electric bill was $57.60?

43. *Field Dimensions.* The dimensions of a World Cup soccer field are 114.9 yd by 74.4 yd. The dimensions of a standard football field are 120 yd by 53.3 yd. How much greater is the area of a World Cup soccer field?

44. *Overtime Pay.* A construction worker earned $17 per hour for the first 40 hr of work and $25.50 per hour for work in excess of 40 hr. One week she earned $896.75. How much overtime did she work?

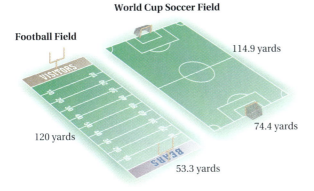

45. Frank has been sent to the store with $40 to purchase 6 lb of cheese at $4.79 a pound and as many bottles of seltzer, at $0.64 a bottle, as possible. How many bottles of seltzer should Frank buy?

46. Janice has been sent to the store with $30 to purchase 5 pt of salsa at $2.49 a pint and as many bags of chips, at $1.39 a bag, as possible. How many bags of chips should Janice buy?

Exercise Set 5.8

Skill Maintenance

47. Simplify: $\dfrac{0}{-13}$. [3.3a]

48. Add: $-\dfrac{4}{5} + \dfrac{7}{10}$. [4.2b]

49. Subtract: $\dfrac{8}{11} - \dfrac{4}{3}$. [4.3a]

50. Solve: $4x - 7 = 9x + 13$. [5.7b]

51. Add: $4\dfrac{1}{3} + 2\dfrac{1}{2}$. [4.6a]

52. Simplify: $\dfrac{-72}{-72}$. [3.3a]

Synthesis

53. ◆ Write a problem for a classmate to solve. Design the problem so that the solution is "The larger field is 200 m² bigger."

54. ◆ Write a problem for a classmate to solve. Design the problem so that the solution is "Mona's Buick got 23.5 mpg."

55. ◆ 🖩 Which is a better deal and why: a 14-in. pizza that costs $9.95 or a 16-in. pizza that costs $11.95?

56. You can drive from home to work using either of two routes:

Route A: Via interstate highway, 7.6 mi, with a speed limit of 65 mph.
Route B: Via a country road, 5.6 mi, with a speed limit of 50 mph.

Assuming you drive at the posted speed limit, how much time can you save by taking the faster route?

57. 🖩 A 25-ft by 30-ft yard contains an 8-ft–wide, round fountain. How many 1-lb bags of grass seed should be purchased to seed the lawn if 1 lb of seed covers 300 ft²?

58. If the daily rental for a car is $18.90 plus a certain price per mile and Lindsey must drive 190 mi and still stay within a $55.00 budget, what is the highest price per mile that Lindsey can afford?

59. Find the shaded area. What assumptions must you make?

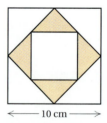

60. *Fast-Food Meals.* In 1995, the average fast-food meal cost $9.42; in 1996, it was $10.06; in 1997, it was $10.56; and in 1998, it was $11.01 (**Source**: Sandelman and Associates, Brea, California). Determine the average yearly increase in the cost of a fast-food meal.

61. *Family Dining.* In 1995, the average cost of a family's dinner outside the home was $25.39; in 1996, it was $27.40; in 1997, it was $27.00; and in 1998, it was $29.35 (**Source**: Sandelman and Associates, Brea, California). Determine the average yearly increase in what it costs a family to eat out.

Prepare a budget for redecorating the classroom.

Summary and Review Exercises: Chapter 5

1. Write a word name for 3.47. [5.1a]

2. Write $597.25 in words, as on a check. [5.1a]

Write fractional notation. [5.1b]

3. 0.09
4. −3.0227

Write decimal notation. [5.1c]

5. $-\dfrac{34}{1000}$
6. $\dfrac{2791}{100}$

Which number is larger? [5.1d]

7. 0.034, 0.0185
8. −0.67, −0.19

Round 39.4287 to the nearest: [5.1e]

9. Tenth.
10. Hundredth.

Perform the indicated operation.

11.
```
   2 3 6.2 3 1
   2 6 3.4
 +     0.1 9 8    [5.2a]
```

12.
```
   3 7.6 4 5
 −     8.4 9 7   [5.2b]
```

13. 219.3 + 2.8 + 7 [5.2a]

14. 745.0109 − 59.959 [5.2b]

15. −37.8 + (−19.5) [5.2c]

16. −7.52 − (−9.89) [5.2c]

17.
```
      4 8
  × 0.2 7    [5.3a]
```

18. −3.7(0.29) [5.3a]

19.
```
   2 4.6 8
 × 1 0 0 0   [5.3a]
```

20. $25\overline{)80}$ [5.4a]

21. 11.52 ÷ (−7.2) [5.4a]

22. $\dfrac{276.3}{1000}$ [5.4a]

Combine like terms. [5.2d]

23. $3.7x - 5.2y - 1.5x - 3.9y$

24. $7.94 - 3.89a + 4.63 + 1.05a$

25. Evaluate $P - Prt$ for $P = 1000$, $r = 0.05$, and $t = 1.5$. (*A formula for depreciation*) [5.3c]

26. Simplify: $9 - 3.2(-1.5) + 5.2^2$. [5.4b]

27. Estimate the sum 7.298 + 3.961 to the nearest tenth. [5.6a]

28. About how many videotapes, at $2.45 each, can be purchased with $49.95? [5.6a]

29. Which of the following is an appropriate estimate of 7.9 × 4.8? [5.6a]

   a) 240 b) 24
   c) 40   d) 4

30. Convert 1549 cents to dollars. [5.3b]

31. Round 248.$\overline{27}$ to the nearest hundredth. [5.5b]

Find decimal notation. Use multiplying by 1. [5.5c]

32. $\dfrac{13}{5}$

33. $\dfrac{32}{25}$

Find decimal notation. Use division. [5.5a]

34. $\dfrac{13}{4}$

35. $-\dfrac{7}{6}$

36. Calculate: $\dfrac{4}{15} \times 79.05$. [5.5d]

Solve. Remember to check.

37. $t - 4.3 = -7.5$ [5.7a]

38. $4.1x + 5.6 = -6.7$ [5.7a]

39. $6x - 11 = 8x + 4$ [5.7b]

40. $3(x + 2) = 5x - 7$ [5.7b]

Solve. [5.8a]

41. In the United States, there are 51.81 telephone poles for every 100 people. In Canada, there are 40.65. How many more telephone poles for every 100 people are there in the United States?

42. Zack's times in the quarter-mile run were 89.3 sec, 88.9 sec, and 90.0 sec. What was his average time?

43. The McCoys have 4 corn fields. One year the harvest in each field was 1419.3 bushels, 1761.8 bushels, 1095.2 bushels, and 2088.8 bushels. What was the year's total harvest?

44. A florist sold 13 potted palms for a total of $423.65. What was the cost for each palm? Round to the nearest cent.

45. Worldwide, the average person drinks 3.48 cups of tea per day. How many cups of tea does the average person drink in a week? in a month (30 days)?

46. A taxi driver charges $7.25 plus 95 cents a mile for out-of-town fares. How far can an out-of-towner travel on $15.23?

Skill Maintenance

47. Simplify: $\dfrac{-29}{-29}$. [3.3c]

48. Add and simplify: $-\dfrac{1}{9} + \dfrac{1}{6}$. [4.2b]

49. Subtract and simplify: $\dfrac{4}{5} - \dfrac{1}{2}$. [4.3a]

50. Simplify: $\dfrac{1}{2}x + \dfrac{3}{4}y - \dfrac{3}{4}x - y$. [4.3a]

Synthesis

51. ◆ Explain how fractional notation can be used when explaining why we add decimal places when multiplying with decimal notation. [5.3a]

52. ◆ Explain what mistake is made in the following calculation. [5.2a]

    $$\begin{array}{r} 1\,3.0\,7 \\ +\ 9.2\,0\,5 \\ \hline 1\,0.5\,1\,2 \end{array}$$

53. ▦ Arrange from smallest to largest: [5.1d], [5.5a]

    $-\dfrac{2}{3},\ -\dfrac{15}{19},\ -\dfrac{11}{13},\ \dfrac{-5}{7},\ \dfrac{-13}{15},\ \dfrac{-17}{20}.$

54. The Fit Fiddle health club generally charges a $79 membership fee and $42.50 a month. Alayn has a coupon that will allow her to join the club for $299 for six months. How much will Alayn save if she uses the coupon? [5.8a]

Test: Chapter 5

1. Write a word name for 6.0401.

2. Write $1234.78 in words, as on a check.

Write fractional notation. Do not simplify.

3. −0.2

4. 7.308

Write decimal notation.

5. $\dfrac{49}{10{,}000}$

6. $-\dfrac{528}{100}$

Which number is larger?

7. 0.07, 0.162

8. −0.173, −0.25

Round 9.4523 to the nearest:

9. Tenth.

10. Thousandth.

Perform the indicated operation.

11.
```
   4 0 2.3
       2.8 1
 + 0.1 0 9
```

12.
```
     0.1 2 5
 ×     0.2 4
```

13.
```
   2 1 3.4 5
 ×   0.0 0 1
```

14.
```
   5 2.0 9 1
 −      7.3 4 5
```

15. 342.9 + 8.1 + 5.37

16. −9.5 + 7.3

17. 2 − 0.0054

18. $2\,5\,\overline{)\,1\,1\,}$

19. $3.3\,\overline{)\,1\,0\,0.3\,2\,}$

20. $\dfrac{-346.82}{1000}$

21. Convert $179.82 to cents.

22. Combine like terms:
$4.1x + 5.2 - 3.9y + 5.7x - 9.8.$

23. Evaluate $2l + 4w + 2h$ for $l = 2.4$, $w = 1.3$, and $h = 0.8$.
   (*The total girth of a postal package*)

24. Simplify: $20 \div 5(-2)^2 - 8.4.$

25. About how many gallons of gasoline, at $1.269 per gallon, can be bought with $10? Round to the nearest gallon.

26. Round $48.\overline{74}$ to the nearest tenth.

Find decimal notation. Use multiplying by 1.

27. $\dfrac{8}{5}$

28. $\dfrac{21}{4}$

Answers

1.
2.
3.
4.
5.
6.
7.
8.
9.
10.
11.
12.
13.
14.
15.
16.
17.
18.
19.
20.
21.
22.
23.
24.
25.
26.
27.
28.

Test: Chapter 5

Find decimal notation. Use division.

29. $-\dfrac{7}{16}$

30. $\dfrac{11}{9}$

31. Calculate: $\dfrac{3}{8} \times 45.6 - \dfrac{1}{5} \times 36.9$.

Solve. Remember to check.

32. $17y - 3.12 = -58.2$

33. $9t - 4 = 6t + 26$

34. $4 + 2(x - 3) = 7x - 9$

Solve.

35. Raul wrote checks of $123.89, $56.78, and $3446.98. What was the average amount of the checks? Round to the nearest cent.

36. In 1896, Alfred Hajos set the world record in the 100-m freestyle swim with a time of 82.2 sec. A hundred years later, Aleksandr Popov set a new record of 48.74 sec. How much better was Popov's time?

37. Alexia bought 6 books at $19.95 each. How much was spent?

38. The three Szmansky sisters commute 3 mi, 5.2 mi, and 16.4 mi to their jobs each day. How far do they commute on average?

39. Air-Tight Heating Service charges $35 for a house call plus $32.50 an hour for labor. A family's bill is $83.75 for repairs to their furnace. How long did the repairs take?

Skill Maintenance

40. Simplify: $\dfrac{0}{57}$.

41. Add and simplify: $\dfrac{2}{7} + \dfrac{3}{21}$.

42. Subtract and simplify: $\dfrac{2}{3} - \dfrac{7}{10}$.

43. Simplify: $\dfrac{4}{5}x + \dfrac{2}{3}y - \dfrac{1}{10}x - \dfrac{3}{5}y$.

Synthesis

44. Use one of the words *sometimes*, *never*, or *always* to complete each of the following.

   a) The product of two numbers greater than 0 and less than 1 is _____ less than 1.

   b) The product of two numbers greater than 1 is _____ less than 1.

   c) The product of a number greater than 1 and a number less than 1 is _____ equal to 1.

   d) The product of a number greater than 1 and a number less than 1 is _____ equal to 0.

Cumulative Review: Chapters 1–5

1. Write expanded notation: 12,758.

2. Write $802.53 in words, as on a check.

3. Write fractional notation: 10.09.

4. Convert to fractional notation: $3\frac{3}{8}$.

5. Write decimal notation: $\frac{-35}{1000}$.

6. List all the factors of 66.

7. Find the prime factorization of 66.

8. Find the LCM of 28 and 35.

9. Round 6962.4721 to the nearest hundred.

10. Round 6962.4721 to the nearest hundredth.

Add and, if possible, simplify.

11. $3\frac{2}{3}$
    $+ 2\frac{5}{9}$

12. 110.863
    0.73
    121.9
    $+ 1.904$

13. 529
    215
    $+ 31$

14. $-\frac{4}{15} + \frac{7}{30}$

Subtract.

15. $29 - (-17)$

16. $9010 - 563.47$

17. $\frac{8}{9} - \frac{7}{8}$

18. $7\frac{1}{5} - 3\frac{4}{5}$

Multiply and simplify.

19. 23.9
    $\times\ 0.2$

20. $-\frac{3}{5} \times \frac{10}{21}$

21. $3\frac{2}{11} \cdot 4\frac{2}{7}$

22. $5 \cdot \frac{3}{10}$

Divide and simplify.

23. $2\frac{4}{5} \div 1\frac{13}{15}$

24. $\frac{6}{5} \div \frac{7}{8}$

25. $-43.795 \div 0.001$

26. $2.1 \overline{)43.26}$

Use <, >, or = for ▢ to write a true sentence.

27. $\frac{5}{7}$ ▢ $\frac{2}{3}$

28. -3 ▢ -5

29. Evaluate $\dfrac{x}{3} - y$ for $x = 15$ and $y = 7$.

30. Multiply: $4(x - y + 3)$.

Combine like terms.

31. $-4p + 28 + 11p - 33$

32. $x - 9 + 13x - 2$

Solve. Remember to check.

33. $8.32 + x = 9.1$

34. $-75 \cdot x = 2100$

35. $y \cdot 9.47 = 81.6314$

36. $1062 - y = -368{,}313$

37. $t + \dfrac{5}{6} = \dfrac{8}{9}$

38. $\dfrac{7}{8} \cdot t = \dfrac{7}{16}$

39. $2.4x - 7.1 = 2.05$

40. $9 = -\dfrac{2}{3}x - 1$

Solve.

41. In 1996, there were 2344 heart transplants, 12,080 kidney transplants, 4064 liver transplants, and 811 lung transplants.* How many transplants of these four organs were performed?

42. After making a $150 down payment on a motorcycle, $\dfrac{3}{10}$ of the total cost was paid. How much did the motorcycle cost?

43. There are 60 seconds in a minute and 60 minutes in an hour. How many seconds are in a day?

44. Claude's college tuition was $4200. A loan was obtained for $\dfrac{2}{3}$ of the tuition. For how much was the loan?

45. The balance in a checking account is $314.79. After a check is written for $56.02, what is the balance in the account?

46. A clerk in a deli sold $1\dfrac{1}{2}$ lb of ham, $2\dfrac{3}{4}$ lb of turkey, and $2\dfrac{1}{4}$ lb of roast beef. How many pounds of meat were sold altogether?

47. A triangular sail has a height of 16 ft and a base of 11 ft. Find its area.

48. A rectangular billboard measures 19.8 ft by 23.6 ft. Find its area.

Synthesis

49. A box of Jello®-mix packages weighs $15\dfrac{3}{4}$ lb. Each package weighs $1\dfrac{3}{4}$ oz. How many packages are in the box?

50. In the Newton Market, Brenda used a manufacturer's coupon to buy juice. With the coupon, if 5 cartons of juice were purchased, the sixth carton was free. The price of each carton was $1.89. What was the cost per carton with the coupon? Round to the nearest cent.

---

*United Network for Organ Sharing, cited in *The Statistical Abstract of the United States, 1998*.

6

Introduction to Graphing and Statistics

An Application

Meg earned grades of B, A, A, C, and D in courses that carried credit values of 3, 4, 3, 4, and 1, respectively. Find Meg's grade point average if an A corresponds to 4.0, a B corresponds to 3.0, a C corresponds to 2.0, and a D corresponds to 1.0.

The Mathematics

We first multiply each grade point value by the number of credit hours in the course. Then we add and divide by the total number of credits to find the grade point average (GPA).

This problem appears as Example 3 in Section 6.5.

Introduction

There are many ways in which data can be represented and analyzed. One way is to use graphs. In this chapter, we examine several kinds of graphs: pictographs, bar graphs, line graphs, and graphs drawn from equations in two variables. Another way to analyze data is to study certain numbers, or *statistics*, that are related to the data. We will consider three statistics in this chapter: the *mean*, the *median*, and the *mode*.

- 6.1 Tables and Pictographs
- 6.2 Bar Graphs and Line Graphs
- 6.3 Ordered Pairs and Equations in Two Variables
- 6.4 Graphing Linear Equations
- 6.5 Means, Medians, and Modes
- 6.6 Predictions and Probability

World Wide Web For more information, visit us at www.mathmax.com

Pretest: Chapter 6

1. The table at right shows the comparison of the cost of a $100,000 life insurance policy for female smokers and nonsmokers at certain ages.
   a) How much does it cost a female smoker, age 32, for insurance?
   b) How much does it cost a female nonsmoker, age 32, for insurance?
   c) How much more does it cost a female smoker, age 35, than a nonsmoker of the same age?

LIFE INSURANCE: FEMALE		
Age	Cost (Smoker)	Cost (Nonsmoker)
31	$294	$170
32	298	172
33	302	176
34	310	178
35	316	182

2. Using the data in Question 1, draw a vertical bar graph showing the cost of insurance for a female smoker, ages 31–35. Use age on the horizontal scale and cost on the vertical scale.

3. Using the data in Question 1, draw a line graph showing the cost of insurance for a female smoker, ages 31–35. Use age on the horizontal scale and cost on the vertical scale.

The line graph at right shows the relationship between blood cholesterol level and risk of coronary heart disease.

4. Which of the cholesterol levels listed has the highest risk?

5. About how much higher is the risk at 260 than at 200?

6. Plot these points:
   $(-2, 4), (5, -3), (0, 6), (3, 0), (-4, -1), (2, 4)$.

7. In which quadrant is the point $(-5, 7)$ located?

8. Determine whether the ordered pair $(-2, 4)$ is a solution of the equation $2x - y = 0$.

Graph on a plane.

9. $y = -x$

10. $x + 2y = 9$

11. $y = \dfrac{2}{3}x - 1$

In Questions 12–14, find (a) the mean, (b) the median, and (c) any modes that exist.

12. 46, 50, 53, 55

13. 5, 5, 3, 1, 1

14. 4, 17, 4, 18, 4, 17, 18, 20

15. Wynton drove 660 mi in 12 hr. What was his average rate of travel?

16. To get a B in chemistry, Delia must average 80 on four tests. Scores on the first three tests were 78, 81, and 75. What is the lowest score that she can receive on the last test and still get a B?

17. Use the graph from Question 4 to estimate the frequency of occurrence of heart disease among people whose blood cholesterol level is 207.

18. A die is about to be rolled. Find the probability that a 2 will be rolled.

Chapter 6 Introduction to Graphing and Statistics

6.1 Tables and Pictographs

a | Reading and Interpreting Tables

A **table** is often used to present data in rows and columns.

Example 1 *Cereal Data.* Let's assume that you generally have a 2-cup bowl of cereal each morning. The following table lists nutritional information for five name-brand cereals. (It does not consider the use of milk, sugar, or sweetener.) The data have been determined by doubling the information given for a 1-cup serving that is found in the Nutrition Facts panel on a box of cereal.

Cereal	Calories	Fat	Total Carbohydrate	Sodium
Ralston Rice Chex	240	0 g	54 g	460 mg
Kellogg's Complete Bran Flakes	240	1.3 g	64 g	613.3 mg
Kellogg's Special K	220	0 g	44 g	500 mg
Honey Nut Cheerios	240	3 g	48 g	540 mg
Wheaties	220	2 g	48 g	440 mg

a) Which cereal has the least amount of sodium per serving?
b) Which cereal has the greatest amount of fat?
c) Which cereal has the least amount of fat?
d) Find the average total carbohydrate in the cereals.

Careful examination of the table will give the answers.

a) To determine which cereal has the least amount of sodium, look down the column headed "Sodium" until you find the smallest number. That number is 440 mg. Then look across that row to find the brand of cereal, Wheaties.

b) To determine which cereal has the greatest amount of fat, look down the column headed "Fat" until you find the largest number. That number is 3 g. Then look across that row to find the cereal, Honey Nut Cheerios.

c) To determine which cereal has the least amount of fat, look down the column headed "Fat" until you find the smallest number. There are two listings of 0 g. Then look across those rows to find the cereals, Ralston Rice Chex and Kellogg's Special K.

d) Find the average of all the numbers in the column labeled "Total Carbohydrate":

$$\frac{54 + 64 + 44 + 48 + 48}{5} = 51.6.$$

The average total carbohydrate content is 51.6 g.

Do Exercises 1–5.

Objectives

a | Extract and interpret data from tables.
b | Extract and interpret data from pictographs.
c | Draw simple pictographs.

For Extra Help

TAPE 11   MAC WIN   CD-ROM

Use the table in Example 1 to answer each of the following.

1. Which cereal has the most total carbohydrate?

2. Which cereal has the least total carbohydrate?

3. Which cereal has the least number of calories?

4. Which cereal has the greatest number of calories?

5. Find the average amount of sodium in the cereals.

Answers on page A-14

Use the table in Example 2 to answer each of the following.

6. Which plan has the highest off-peak rate?

7. Which plan has a peak rate of 14¢/min.

8. Using an off-peak rate of 8¢/min for MCI, find the average off-peak rate for the seven companies.

9. What is the average of the three most expensive off-peak rates? Round to the nearest tenth of a cent.

Answers on page A-14

Example 2 *Phone Rates.* The following table shows the rates for several popular long-distance calling plans during a recent year.

Company	Plan	Peak Rate	Off-Peak Rate	Info Number
AT&T	One rate [1]	15¢/min	15¢/min	800 222-0300
GTE	One Price	14	14	800 483-3737
Sprint	Sprint Sense	25	10	800 366-1044
LCI	One rate [2]	15	15	800 860-2255
Matrix	SmartWorld	19.9	9.9	800 282-0242
MCI	MCI One	25	5–10	800 444-3333
WorldCom	Home Adv.	25	10	800 275-0100

[1] With Plus plan, U.S. calls cost 10¢/min anytime, with a $4.95 monthly fee.
[2] This plan costs an additional $1.00 per month.

a) Of the plans listed, which has the least expensive peak rate?

b) Of the plans listed, which has the least expensive off-peak rate?

c) Shannon has the Sprint Sense plan. She spoke for 50 min at the peak rate and for 75 min at the off-peak rate. What was the total cost for these calls?

d) For the plans listed, what is the average peak rate?

Careful examination of the table will give the answers.

a) To determine the least expensive peak rate, we look down the column labeled "Peak Rate" for the lowest rate listed. When we find that rate (14¢/min), we read across to the left and find that this rate is for the GTE One Price plan.

b) To determine which plan has the least expensive off-peak rate, we look down the column labeled "Off-Peak Rate" for the lowest rate listed. We note there that the lowest rate is part of a range, from 5¢ to 10¢ per minute. Next, we read across that line, to the left, and find that the MCI One plan has the lowest off-peak rate, 5¢/min.

c) Under the Sprint Sense plan, if Shannon spoke for 50 min at the peak rate, her bill for those minutes would be (50 min)(25¢/min), or $12.50. If she spoke for 75 min at the off-peak rate of 10¢/min, her bill for those minutes would be (75 min)(10¢/min), or $7.50. Altogether, 50 min at the peak rate and 75 min at the off-peak rate would cost $12.50 + $7.50, or $20.

d) To find the average peak rate of all the plans, we add the peak rates for the seven companies and then divide by 7:

$$\text{Average peak rate} = \frac{15¢ + 14¢ + 25¢ + 15¢ + 19.9¢ + 25¢ + 25¢}{7}$$

$$\approx 19.8¢/\text{min}. \quad \text{Rounding to the nearest tenth}$$

Do Exercises 6–9.

b Reading and Interpreting Pictographs

Pictographs (or *picture graphs*) are another way to show information. Instead of actually listing the amounts to be considered, a **pictograph** uses symbols to represent the amounts. In addition, a *key* is given telling what each symbol represents.

Example 3 *Coffee Consumption.* For selected countries, the following pictograph shows approximately how many cups of coffee each person (per capita) drinks annually. A key indicates that each symbol ☕ represents 100 cups.

Coffee Consumption (per Capita)

Germany	☕☕☕☕☕☕☕☕☕☕☕▏
United States	☕☕☕☕▎
Switzerland	☕☕☕☕☕☕☕☕☕☕☕☕▏
France	☕☕☕☕☕☕☕☕
Italy	☕☕☕☕☕☕☕

☕ = 100 cups

Source: Beverage Marketing Corporation and *The Statistical Abstract of the United States, 1997*

a) Determine the approximate annual coffee consumption per capita of Germany.

b) Which two countries have the greatest difference in coffee consumption? Estimate that difference.

We use the data from the pictograph as follows.

a) Germany's consumption is represented by 11 whole symbols (1100 cups) and, though it is visually debatable, about $\frac{1}{8}$ of another symbol (about 13 cups), for a total of 1113 cups.

b) Visually, we see that, of the countries listed, Switzerland has the most consumption and the United States the least. Switzerland's annual coffee consumption per capita is represented by 12 whole symbols (1200 cups) and about $\frac{1}{5}$ of another symbol (20 cups), for a total of 1220 cups. U.S. consumption is represented by 4 whole symbols (400 cups) and about $\frac{1}{3}$ of another symbol (33 cups), for a total of 433 cups. The difference between these amounts is 1220 − 433, or 787 cups.

One advantage of pictographs is that the appropriate choice of a symbol will tell you, at a glance, the kind of measurement being made. Another advantage is that the comparison of amounts represented in the graph can be expressed more easily by just counting symbols. For instance, in Example 3, we can tell at a glance that the annual coffee consumption per capita in Germany is more than twice that of the United States.

One disadvantage of pictographs is that, in order to make a pictograph easy to read, we must generally round amounts to the unit that a symbol represents. Another disadvantage is that it is difficult to determine how much a partial symbol represents. A third disadvantage is that we must usually multiply to compute the amount represented, since the total amounts are rarely listed.

Do Exercises 10–12.

Use the pictograph in Example 3 to answer each of the following.

10. Approximate the annual coffee consumption per capita in France.

11. Approximate the annual coffee consumption per capita in Italy.

12. The approximate annual coffee consumption in Finland is about the same as the combined coffee consumption in Switzerland and the United States. What is the approximate coffee consumption in Finland?

Answers on page A-14

13. *Concert Revenue.* The following is a list of three other groups active during the same time and their total gross revenues. Draw a pictograph to represent the data.

Boyz II Men $43.2 million   (1995)
REM $38.7 million   (1995)
Kiss $43.6 million   (1996)

c | Drawing Pictographs

Example 4   *North American Concert Revenue.* The following is a list of the gross revenue (money taken in) during one year by five of the top concert acts for the years 1992–1998 (**Source:** *The World Almanac*). Draw a pictograph to represent the data. Let the symbol represent $10,000,000.

The Rolling Stones	$121.2 million	(1994)
Pink Floyd	$103.5 million	(1994)
The Eagles	$79.4 million	(1994)
U2	$67.0 million	(1992)
The Grateful Dead	$52.4 million	(1994)

Some computation is necessary before we can draw the pictograph.

The Rolling Stones: Note that $121.2 \div 10 = 12.12$. Thus we need 12 whole symbols and 0.12 of another symbol. Now 0.12 is hard to draw, but we estimate it to be about $\frac{1}{10}$ of a symbol.

Pink Floyd: Note that $103.5 \div 10 = 10.35$. Thus we need 10 whole symbols and 0.35, or about $\frac{1}{3}$ of another symbol.

The Eagles: Note that $79.4 \div 10 = 7.94$. Thus we need 7 whole symbols and 0.94, or about $\frac{9}{10}$ of another symbol.

U2: Note that $67 \div 10 = 6.7$. Thus we need 6 whole symbols and $\frac{7}{10}$ of another symbol.

The Grateful Dead: Note that $52.4 \div 10 = 5.24$. Thus we need 5 whole symbols and 0.24, or about $\frac{1}{4}$ of another symbol.

The pictograph can now be drawn as follows. We list the concert act in one column, draw the monetary amounts using symbols, and title the overall graph "Total Gross Revenue."

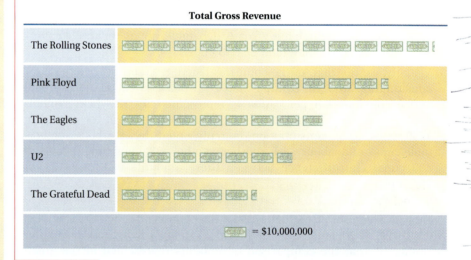

Do Exercise 13.

Answer on page A-14

Exercise Set 6.1

a *Astronomy.* Use the following table, which lists information about the planets, for Exercises 1–10.

Planet	Average Distance from Sun (in miles)	Diameter (in miles)	Length of Planet's Day in Earth Time (in days)	Time of Revolution in Earth Time (in years)
Mercury	35,983,000	3,031	58.82	0.24
Venus	67,237,700	7,520	224.59	0.62
Earth	92,955,900	7,926	1.00	1.00
Mars	141,634,800	4,221	1.03	1.88
Jupiter	483,612,200	88,846	0.41	11.86
Saturn	888,184,000	74,898	0.43	29.46
Uranus	1,782,000,000	31,763	0.45	84.01
Neptune	2,794,000,000	31,329	0.66	164.78
Pluto	3,666,000,000	1,423	6.41	248.53

Source: *Handy Science Answer Book,* Gale Research, Inc.

1. Find the average distance from the sun to Jupiter.

2. How long is a day on Venus?

3. Which planet has a time of revolution of 164.78 yr?

4. Which planet has a diameter of 4221 mi?

5. Which planets have an average distance from the sun that is greater than 1,000,000 mi?

6. Which planets have a diameter that is less than 100,000 mi?

7. About how many earth diameters would it take to equal the diameter of Jupiter?

8. How much longer is the longest time of revolution than the shortest?

9. What is the average diameter of a planet?

10. What is the average distance from the sun to a planet?

Global Warming. Ecologists are increasingly concerned about global warming, that is, the trend of average global temperatures to rise over recent years. One possible effect is the melting of the polar icecaps. Use the following table for Exercises 11–18.

Year	Average Global Temperature (°F)
1986	59.29°
1987	59.58°
1988	59.63°
1989	59.45°
1990	59.85°
1991	59.74°
1992	59.23°
1993	59.36°
1994	59.56°
1995	59.72°
1996	59.58°

Source: Vital Signs, 1997

11. In what year was the average global temperature the lowest?

12. In what year was the average global temperature the highest?

13. Find the average global temperatures for 1986 and 1987.

14. Find the average global temperatures for 1992 and 1993.

15. Find the average of the average global temperatures for the years 1986 through 1988. Find the eight-year average global temperature for the years 1989 through 1996. By how many degrees does the latter average exceed the former?

16. Find the average of the average global temperatures for the years 1994 to 1996. Find the ten-year average global temperature for the years 1987 to 1996. By how many degrees does the former average exceed the latter?

17. Between which two years did the average global temperature increase the most?

18. Between which two years did the average global temperature decrease the most?

Global Warming.

Chapter 6 Introduction to Graphing and Statistics

b *World Population Growth.* The following pictograph shows world population in various years. Use the pictograph for Exercises 19–26.

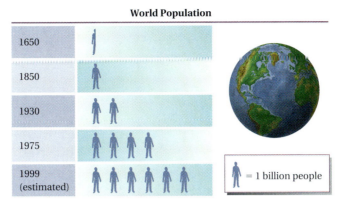

19. What was the world population in 1850?

20. What was the world population in 1975?

21. In which year is the population the greatest?

22. In which year was the population the least?

23. Between which two years was the amount of growth the least?

24. Between which two years was the amount of growth the greatest?

25. How much greater will the world population in 1999 be than in 1975?

26. How much greater will the world population be in 1999 than in 1930?

TV News Magazine Programs. The number of network news magazine programs has increased dramatically since the early 1980s when there were just two—ABC's "20/20" and CBS's "60 Minutes." In the pictograph below, each symbol represents a 1-hr prime-time news magazine in the network's weekly fall schedule. Use the pictograph for Exercises 27–34.

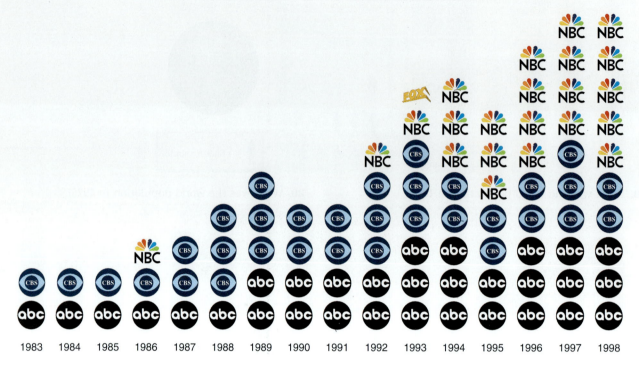

Sources: "Total Television," by Alex McNeil. Penguin Books; The Hollywood Reporter; Fox Broadcasting. *The New York Times*, June 8, 1998.

27. In which year was there exactly 7 hr of prime-time news magazine programming per week?

28. In which year was there exactly 5 hr of prime-time news magazine programming per week?

29. How many hours per week of prime-time news magazine programming were there in 1994?

30. How many hours per week of prime-time news magazine programming were there in 1997?

31. How much did prime-time news magazine programming increase from 1991 through 1993?

32. How much did prime-time news magazine programming increase from 1991 through 1997?

33. Between which two years did weekly prime-time news magazine programming increase the most?

34. Between which two years did weekly news magazine programming first decrease?

Chapter 6 Introduction to Graphing and Statistics

35. *Lettuce Sales.* The sales of lettuce have experienced a tremendous increase in recent years due to the convenience of prepackaged, prewashed, and prechopped lettuce. Sales for recent years are listed below (*Source*: International Fresh-Cut Produce Association). Draw a pictograph to represent lettuce sales for these years. Use the symbol to represent $100,000,000.

1992 $168,000,000
1993 $312,000,000
1994 $577,000,000
1995 $889,000,000
1996 $1,100,000,000

Skill Maintenance

Solve.

36. $9x - 5 = -23$ [4.4a]

37. $3x - 2 = 7x + 10$ [5.7b]

38. $-4x = 3x - 7$ [5.7b]

Convert to decimal notation.   [5.5a]

39. $\dfrac{3}{8}$

40. $\dfrac{3}{50}$

41. $\dfrac{29}{25}$

42. $\dfrac{5}{6}$

Synthesis

43. ◆ Loreena is drawing a pictograph in which dollar bills are used as symbols to represent the tuition at various private colleges. Should each dollar bill represent $10,000, $5000, or $500? Why?

44. ◆ Suppose you are drawing a pictograph in which stopwatches are used as symbols to represent the number of minutes each of three professors spends on the phone each month. Should each stopwatch represent 1 min, 10 min, or 100 min? Why?

45. ◆ What advantage(s) does a table have over a pictograph?

46. Redraw the pictograph appearing in Example 3 as one in which each symbol represents 150 cups of coffee.

47. Use the pictograph from Exercises 27–34 to determine the average yearly increase in prime-time news magazine programming for the years 1983 to 1998. (For example, the yearly increase between 1991 and 1992 was 2 hr.)

48. Refer to the table in Example 2. Bridget spoke the same number of minutes at the peak rate as she did at the off-peak rate. If her Sprint Sense bill was for $42, how many minutes did she speak for at each rate?

6.2 Bar Graphs and Line Graphs

Beginning in Chapter 1, we have used *bar graphs* to convey information (see pages 3, 61, 262, 282, 304, and 333). In this section, we make further use of bar graphs and also introduce *line graphs*.

a Reading and Interpreting Bar Graphs

Example 1 *Fat Content in Fast Foods.* Wendy's Hamburgers is a national chain of fast-food restaurants. The following bar graph shows the fat content of various sandwiches sold by Wendy's.

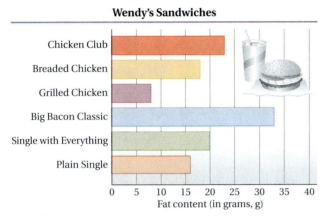

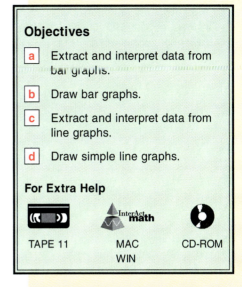

Objectives

a Extract and interpret data from bar graphs.

b Draw bar graphs.

c Extract and interpret data from line graphs.

d Draw simple line graphs.

For Extra Help

TAPE 11   MAC WIN   CD-ROM

Source: Wendy's International

a) About how much fat is in a Chicken Club sandwich?
b) Which sandwich contains the least amount of fat?
c) Which sandwich contains about 20 g of fat?

We look at the graph to answer the questions.

a) We move to the right along the bar representing Chicken Club sandwiches. We can read, fairly accurately, that there is approximately 23 g of fat in the Chicken Club sandwich.

b) The shortest bar is for the Grilled Chicken sandwich. Thus that sandwich contains the least amount of fat.

c) We locate the line representing 20 g and then go up until we reach a bar that ends at approximately 20 g. We then go across to the left and read the name of the sandwich, which is the Single with Everything.

Do Exercises 1–3.

Use the bar graph in Example 1 to answer each of the following.

1. About how much fat is in the plain single sandwich?

2. Which sandwich contains the greatest amount of fat?

3. Which sandwiches contain 20 g or more of fat?

Answers on page A-15

6.2 Bar Graphs and Line Graphs

363

Use the bar graph in Example 2 to answer each of the following.

4. Approximately how many women, per 100,000, develop breast cancer between the ages of 35 and 39?

5. In what age group is the mortality rate the highest?

6. In what age group do about 350 out of every 100,000 women develop breast cancer?

7. Does the breast-cancer mortality rate seem to increase from the youngest to the oldest age group?

Bar graphs are often drawn vertically and sometimes a double bar graph is used to make comparisons.

Example 2 *Breast Cancer.* The following graph indicates the incidence and mortality rates of breast cancer for women of various age groups.

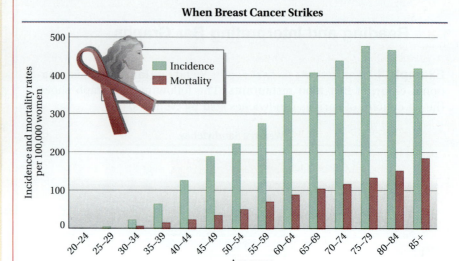

Source: National Cancer Institute

a) Approximately how many women, per 100,000, develop breast cancer between the ages of 40 and 44?

b) In what age range is the mortality rate for breast cancer approximately 100 for every 100,000 women?

c) In what age range is the incidence of breast cancer the highest?

d) Does the incidence of breast cancer always increase from the younger to older age groups?

We look at the graph to answer the questions.

a) We go to the right, across the bottom, to the green bar above the age group 40–44. Next, we go up to the top of that bar and, from there, back to the left to read approximately 130 on the vertical scale. About 130 out of every 100,000 women develop breast cancer between the ages of 40 and 44.

b) We read up the vertical scale to the number 100. From there we move to the right until we come to the top of a red bar. Moving down that bar, we find that in the 65–69 age group, about 100 out of every 100,000 women die of breast cancer.

c) We look for the tallest green bar and read the age range below it. The incidence of breast cancer is highest for women in the 75–79 age group.

d) Looking at the heights of the bars, we see that the incidence of breast cancer actually *decreases* after ages 75–79. Thus the incidence of breast cancer does not always increase from the younger to older age groups.

Do Exercises 4–7.

Answers on page A-15

b Drawing Bar Graphs

Example 3 *Centenarians.* The number of centenarians—that is, people 100 yr or older—is growing rapidly. Projections from the U.S. Bureau of the Census and the National Center for Health Statistics are shown below. Use the projections to form a bar graph.

Year	Projected Number of Centenarians
2000	72,000
2010	131,000
2020	214,000
2030	324,000
2040	447,000
2050	834,000

Source: The New York Times, 6/22/98, p. A14

First, we draw a horizontal scale with six equally spaced intervals and the different years listed. We title that scale "Year." (See the figure on the left below.) Next, we label the vertical scale "Projected Number (in thousands)." Note that the largest number (in thousands) is 834 and the smallest is 72. If we count by 100s, we can range from 0 to 900 with just nine marks. Finally, we draw vertical bars to represent the number of centenarians projected for each year and title the graph. (See the figure on the right below.)

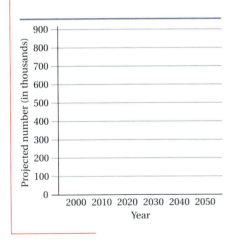
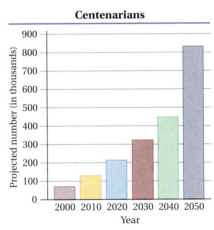

Do Exercise 8.

8. *Planetary Moons.* Make a horizontal bar graph to show the number of moons orbiting the various planets.

Planet	Number of Moons
Earth	1
Mars	2
Jupiter	16
Saturn	18
Uranus	15
Neptune	8
Pluto	1

Answer on page A-15

Use the line graph in Example 4 to answer each of the following.

9. For which month were new home sales lowest?

10. Between which months did new home sales decrease?

11. For which months were new home sales below 700 thousand?

Answers on page A-15

c Reading and Interpreting Line Graphs

Line graphs are often used to show a change over time as well as to indicate patterns or trends.

Example 4 *New Home Sales.* The following line graph shows the number of new home sales, in thousands, over a recent twelve-month period. The jagged line at the base of the vertical scale indicates an unnecessary portion of the scale. Note that the vertical scale differs from the horizontal scale so that the data can be easily shown.

Source: U.S. Department of Commerce

a) For which month were new home sales the greatest?
b) Between which months did new home sales increase?
c) For which months were new home sales about 700 thousand?

We look at the graph to answer the questions.

a) The greatest number of new home sales was about 825 thousand in month 1.

b) Reading the graph from left to right, we see that new home sales increased from month 2 to month 3, from month 3 to month 4, from month 5 to month 6, from month 7 to month 8, from month 8 to month 9, from month 9 to month 10, and from month 10 to month 11.

c) We look from left to right along the line at 700.

We see that points are closest to 700 thousand at months 3, 6, 10, 11, and 12.

Do Exercises 9–11.

d Drawing Line Graphs

Example 5 *Movie Releases.* Draw a line graph to show how the number of movies released each year has changed over a period of 6 yr. Use the following data (**Source:** Motion Picture Association of America).

 1991: 164 movies
 1992: 150 movies
 1993: 161 movies
 1994: 184 movies
 1995: 234 movies
 1996: 260 movies

First, we indicate on the horizontal scale the different years and title it "Year." (See the graph below.) Then we mark the vertical scale appropriately by 50s to show the number of movies released and title it "Number per Year." We also give the overall title "Movies Released" to the graph.

Next, we mark at the appropriate level above each year the points that indicate the number of movies released. Then we draw line segments connecting the points. The change over time can now be observed easily from the graph.

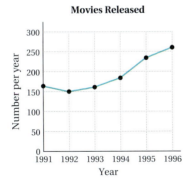

Do Exercise 12.

12. *SAT Scores.* Draw a line graph to show how the average combined verbal–math SAT score has changed over a period of 6 yr. Use the following data (**Source:** The College Board).

 1991: 999
 1992: 1001
 1993: 1003
 1994: 1003
 1995: 1010
 1996: 1013
 1997: 1016

Answer on page A-15

6.2 Bar Graphs and Line Graphs

Improving Your Math Study Skills

How Many Women Have Won the Ultimate Math Contest?

Although this Study Skill feature does not contain specific tips on studying mathematics, we hope that you will find this 1997 article both challenging and encouraging.

Every year on college campuses across the United States and Canada, the most brilliant math students face the ultimate challenge. For six hours, they struggle with problems from the merely intractable to the seemingly impossible.

Every spring, five are chosen winners of the William Lowell Putnam Mathematical Competition, the Olympics of college mathematics. Every year for 56 years, all have been men.

Until this year.

This spring, Ioana Dumitriu (pronounced yo-AHN-na doo-mee-TREE-oo), 20, a New York University sophomore from Romania, became the first woman to win the award.

Ms. Dumitriu, the daughter of two electrical engineering professors in Romania, who as a girl solved math puzzles for fun, was identified as a math talent early in her schooling in Bucharest. At 11, Ms. Dumitriu was steered into years of math training camps as preparation for the Romanian entry in the International Mathematics Olympiad.

It was this training, and a handsome young coach, that led her to New York City. He was several years older. They fell in love. He chose N.Y.U. for its graduate school in mathematics, and at 19 she joined him in New York.

The test Ms. Dumitriu won is dauntingly difficult, even for math majors. About half of the 2,407 test-takers scored 2 or less of a possible 120, and a third scored 0. Some students simply walk out after staring at the questions for a while.

Ms. Dumitriu said that in the six hours allotted, she had time to do 8 of the 12 problems, each worth a maximum of 10 points. The last one she did in 10 minutes. This year, Ms. Dumitriu and her five co-winners (there was a tie for fifth place) scored between 76 and 98. She does not know her exact score or rank because the organizers do not announce them.

"I didn't ever tell myself that I was unlikely to win, that no woman before had ever won and therefore I couldn't," she said. "It is not that I forget that I'm a woman. It's just that I don't see it as an obstacle or a ——."

Her English is near-perfect, but she paused because she could not find the right word. "The mathematics community is made up of persons, and that is what I am primarily."

Prof. Joel Spencer, who was a Putnam winner himself, said her work for his class in problem solving last year was remarkable. "What really got me was her fearlessness," he said. "To be good at math, you have to go right at it and start playing around with it, and she had that from the start."

In the graduate lounge in the Courant Institute of Mathematical Sciences at N.Y.U., Ms. Dumitriu, a tall, striking redhead, stands out. Instead of jeans and T-shirts, she wears gray pin-striped slacks and a rust-colored turtleneck and vest.

"There is a social perception of women and math, a stereotype," Ms. Dumitriu said during an interview. "What's happening right now is that the stereotype is defied. It starts breaking."

Still, even as women began to flock to sciences, math has remained largely a male bastion.

"Math remains the bottom line of sex differences for many," said Sheila Tobias, author of "Overcoming Math Anxiety" (W.W. Norton & Company, 1994). "It's one thing for women to write books, negotiate bills through Congress, litigate, fire missiles; quite another for them to do math."

Besides collecting the $1,000 awarded to each Putnam fellow, Ms. Dumitriu also won the $500 Elizabeth Lowell Putnam prize for the top woman finisher for the second year in a row, a prize created five years ago to encourage women to take the test. This year 414 did.

In her view, there are never too many problems, never too much practice.

Besides, each new problem holds its own allure: "When you have all the pieces and you put them together and you see the puzzle, that moment always amazes me."

Copyright © 1997 by The New York Times Co. Reprinted by permission. Article by Karen W. Arenson.

Exercise Set 6.2

a *Chocolate Sweets.* The following horizontal bar graph shows the average caloric content of various kinds of chocolate foods or beverages. Use the bar graph for Exercises 1–12.

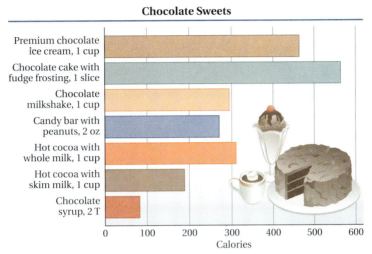

Source: *Better Homes and Gardens,* December 1996

1. Estimate how many calories there are in 1 cup of hot cocoa with skim milk.

2. Estimate how many calories there are in 1 cup of premium chocolate ice cream.

3. Which food or beverage has the highest caloric content?

4. Which food or beverage has the lowest caloric content?

5. Which food or beverage contains about 460 calories?

6. Which foods or beverages contain about 300 calories?

7. How many more calories are there in 1 cup of hot cocoa made with whole milk than in 1 cup of hot cocoa made with skim milk?

8. Fred generally drinks a 4-cup chocolate milkshake. How many calories does he consume?

9. Kristin likes to eat 2 cups of premium chocolate ice cream in the course of a weekend. How many calories does she consume?

10. Barney likes to eat a 6-oz chocolate bar with peanuts for lunch. How many calories does he consume?

11. Paul adds a 2-oz chocolate bar with peanuts to his diet each day for 1 yr (365 days) and makes no other changes in his eating or exercise habits. Consumption of 3500 extra calories will add about 1 lb to his body weight. How many pounds will he gain?

12. Tricia adds one slice of chocolate cake with fudge frosting to her diet each day for 1 yr (365 days) and makes no other changes in her eating or exercise habits. Consumption of 3500 extra calories will add about 1 lb to her body weight. How many pounds will she gain?

Deforestation. The world is gradually losing its tropical forests. The following triple bar graph shows the amount of forested land of three tropical regions in the years 1980 and 1990. Use the bar graph for Exercises 13–20.

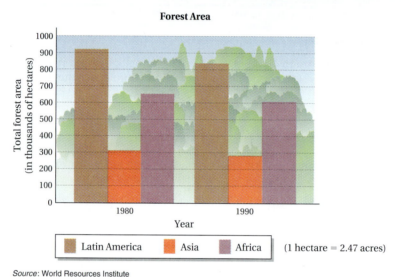

13. What was the forest area of Latin America in 1980?

14. What was the forest area of Africa in 1990?

15. Which region experienced the greatest loss of forest area from 1980 to 1990?

16. Which region experienced the smallest loss of forest area from 1980 to 1990?

17. Which region had a forest area of about 600 thousand hectares in 1990?

18. Which region had a forest area of about 300 thousand hectares in 1980?

19. What was the average forest area in Latin America for the two years?

20. What was the average forest area in Asia for the two years?

Chapter 6   Introduction to Graphing and Statistics

21. *Commuting Time.* The following table lists the average commuting time in several metropolitan areas with more than 1 million people. Make a vertical bar graph to illustrate the data.

City	Commuting Time (in minutes)
New York	30.6
Los Angeles	26.4
Phoenix	23.0
Dallas	24.1
Indianapolis	21.9
Orlando	22.9

Source: Census Bureau

Use the data and the bar graph in Exercise 21 for Exercises 22–25.

22. Which city has the greatest commuting time?

23. Which city has the least commuting time?

24. What is the average commuting time for all six cities? Round to the nearest tenth.

25. What is the average commuting time for New York and Los Angeles?

26. *Calorie Expenditure.* Use the following information to make a horizontal bar graph showing the number of calories burned during each activity by a person weighing 152 lb.

   Tennis: 420 calories per hour
   Jogging: 650 calories per hour
   Hiking: 590 calories per hour
   Office work: 180 calories per hour
   Sleeping: 70 calories per hour

Exercise Set 6.2

371

Use the data and the bar graph in Exercise 26 for Exercises 27–30.

27. What is the difference in the number of calories burned per hour between sleeping and jogging?

28. Suppose you were trying to lose weight by exercising and had to choose one of these exercises. If your doctor told you not to jog, what would be the most beneficial exercise?

29. Ryan works at the office for 8 hr and then sleeps for 7 hr. How many calories does Ryan burn doing this?

30. Nancy hiked for 6 hr and then slept for 8 hr. How many calories did she burn doing this?

c *Average Salary of Major-League Baseball Players.* The following graph shows the average salary of major-league baseball players over a recent 8-yr period. Use the graph for Exercises 31–36.

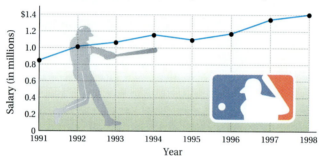

31. In which year was the average salary the highest?

32. In which year was the average salary the lowest?

33. What was the difference in salary between the highest and lowest salaries?

34. Between which two years was the increase in salary the greatest?

35. Between which two years did the salary decrease?

36. In what year was the average salary about $1.2 million?

Chapter 6 Introduction to Graphing and Statistics

37. *Ozone Layer.* Make a line graph of the data, listing years on the horizontal scale.

Year	Ozone Level (in parts per billion)
1991	2981
1992	3133
1993	3148
1994	3138
1995	3124

Source: National Oceanic and Atmospheric Administration

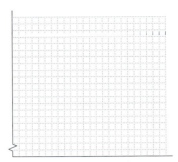

Use the data and the line graph in Exercise 37 for Exercises 38–41.

38. Between which two years was the increase in the ozone level the greatest?

39. Between which two years was the decrease in the ozone level the greatest?

40. What was the average ozone level over the 5-yr period?

41. What was the average ozone level from 1992 through 1995?

42. *Motion Picture Expense.* Make a line graph of the data, listing years on the horizontal scale.

Year	Average Expense per Picture (in millions)
1991	$38.2
1992	42.4
1993	44.0
1994	50.4
1995	54.1
1996	61.0

Source: Motion Picture Association of America

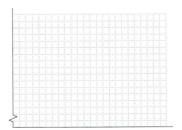

Exercise Set 6.2

373

Use the data and the line graph in Exercise 42 for Exercises 43–46.

43. Between which two years was the increase in motion-picture expense the greatest?

44. Between which two years was the increase in motion-picture expense the least?

45. What was the average motion-picture expense over the 6-yr period?

46. What was the average motion-picture expense from 1994 through 1996? Round to the nearest tenth of a million dollars.

Skill Maintenance

47. How many 12-oz bottles can be filled from a vat containing 408 oz of catsup?   [1.8a]

48. It is known to operators of pizza restaurants that if 50 pizzas are ordered in an evening, people will request extra cheese on 9 of them. What fraction of the pizzas sold are ordered with extra cheese?   [3.3b]

49. A can of Coca-Cola contains 12 fluid ounces. How many fluid ounces are in a six-pack?   [1.8a]

50. 24 is $\frac{3}{4}$ of what number?   [3.4c]

51. $\frac{2}{3}$ of 75 is what number?   [3.4c]

52. $\frac{3}{5}$ of 30 is what number?   [3.4c]

Synthesis

53. ◆ Can bar graphs always, sometimes, or never be converted to line graphs? Why?

54. ◆ Consider the graph in Example 4. Sam states that the initial drop shows that sales were nearly cut in half over the first month of the year. What mistake is Sam making?

55. ◆ Using the data in Exercise 42, how could someone make a reasonable estimate of the average expense per motion picture in 1998?

56. Bonnie eats a 700-Cal breakfast, jogs for 45 min, works in her office for $2\frac{1}{2}$ hr, and eats a 615-Cal lunch. She then works for another $5\frac{1}{2}$ hr before eating a 235-Cal snack and playing tennis for 40 min. If Bonnie weighs 152 lb (see Exercise 26), how many calories has she lost or gained in the course of the day?

57. Use the information in Example 2 to approximate the average rate of incidence of breast cancer for all women above the age of 24.

6.3 Ordered Pairs and Equations in Two Variables

Bar graphs and line graphs are used to illustrate relationships between the items or quantities listed along the bottom and the side of the graph. The horizontal and vertical sides of a bar graph or line graph are often called the **axes** (pronounced ăk´ sēz; singular: **axis**). By using two perpendicular number lines as axes, we can use points to represent solutions of certain equations. First, however, we must learn to graph points.

Objectives

a Plot a point, given its coordinates. Find coordinates, given a point.

b Determine the quadrant in which a point lies.

c Determine whether an ordered pair is a solution of an equation with two variables.

For Extra Help

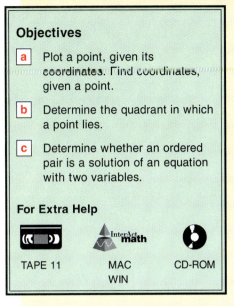

TAPE 11   MAC WIN   CD-ROM

a Points and Ordered Pairs

When two number lines are used as axes, a grid can be formed. The grid provides a helpful way of locating any point on the plane. Just as a location in a city might be given as the intersection of an avenue and a side street, a point on a plane can be regarded as the intersection of a vertical line and a horizontal line. In the figure below, these lines pass through 3 on the horizontal axis and 4 on the vertical axis. Thus the **first coordinate** of this point is 3 and the **second coordinate** is 4. **Ordered pair** notation, (3, 4), provides a quick way of listing these coordinates.

CAUTION! When writing an ordered pair, you should always list the coordinate from the horizontal axis first.

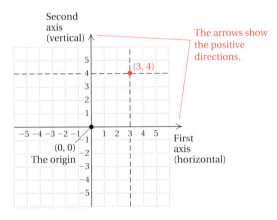

Plot these points on the graph below.

1.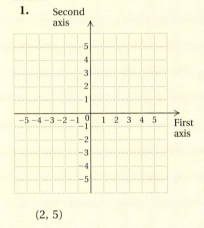

(2, 5)

The point (0, 0), where the axes cross each other, is called the **origin**. To graph, or *plot,* the point (3, 4), we can begin at the origin and move horizontally (along the first axis) to the number 3. From there, we move up 4 units vertically and make a "dot."

It is important to always make sure that the first coordinate matches the number that would be below (or above) the point on the horizontal axis. Similarly, the second coordinate should always match the number that would be to the left (or right) of the point on the vertical axis.

Do Exercises 1 and 2.

2. (4, 1)

Answers on page A-15

Plot these points on the graph below.

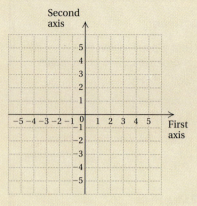

3. $(-2, 5)$

4. $(-3, -4)$

5. $(5, -3)$

6. $(-2, -1)$

7. $(0, -3)$

8. $(2, 0)$

9. Find the coordinates of points A, B, C, D, E, F, and G on the graph below.

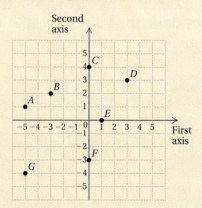

Answers on page A-16

Example 1 Plot the points $(-5, 2)$ and $(2, -5)$.

To plot $(-5, 2)$, we locate -5 on the first, or horizontal, axis. Then we go up 2 units and make a dot.

To plot $(2, -5)$, we locate 2 on the first, or horizontal, axis. Then we go down 5 units and make a dot. Note that the order of the numbers within a pair is important: $(2, -5) \neq (-5, 2)$.

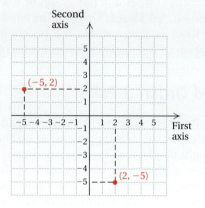

Do Exercises 3–8.

To find the coordinates of a given point, we first look above (or below) the point and list the point's horizontal coordinate. Then we look to the left (or right) of the point and list the vertical coordinate.

Example 2 Find the coordinates of points A, B, C, D, E, F, and G.

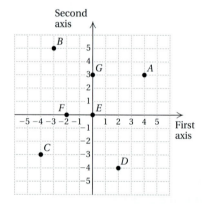

We look below point A to see that its first coordinate is 4. Looking to the left of point A, we find that its second coordinate is 3. Thus the coordinates of point A are (4, 3). The coordinates of the other points are given below.

B: $(-3, 5)$;   C: $(-4, -3)$;   D: $(2, -4)$;

E: $(0, 0)$;   F: $(-2, 0)$;   G: $(0, 3)$.

Do Exercise 9.

b Quadrants

The axes divide the plane into four regions, or **quadrants**. In region I (the *first quadrant*), both coordinates of any point are positive. In region II (the *second quadrant*), the first coordinate is negative and the second coordinate is positive. In region III (the *third quadrant*), both coordinates are negative. In region IV (the *fourth quadrant*), the first coordinate is positive and the second coordinate is negative.

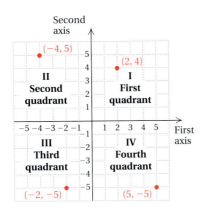

As the figure above illustrates, the point (2, 4) is in the first quadrant, (−4, 5) is in the second quadrant, (−2, −5) is in the third quadrant, and (5, −5) is in the fourth quadrant.

Do Exercises 10–15.

c Solutions of Equations

The coordinate system we have just introduced is called the **Cartesian** coordinate system, in honor of the great mathematician and philosopher René Descartes (1596–1650). Descartes devised this coordinate system in part as a method of presenting solutions of equations containing two variables. Equations like $3x + 2y = 8$ have ordered pairs as solutions. In Section 6.4, we will find solutions and graph them. Here we simply practice checking to see if an ordered pair is a solution.

To determine whether an ordered pair is a solution of an equation, we substitute the first coordinate for the letter that comes first alphabetically and the second coordinate for the letter that is last alphabetically. The letters x and y are used most often.

Example 3 Determine whether the ordered pair (2, 1) is a solution of the equation $3x + 2y = 8$.

We substitute:

$$\begin{array}{c|c} 3x + 2y = 8 & \\ \hline 3 \cdot 2 + 2 \cdot 1 \; ? \; 8 & \text{Substituting 2 for } x \text{ and 1 for } y \\ 6 + 2 & \text{(alphabetical order of variables)} \\ 8 \; | \; 8 & \text{TRUE} \end{array}$$

Since the equation becomes true, (2, 1) is a solution.

In a similar manner, we can show that (0, 4) and (4, −2) are also solutions of $3x + 2y = 8$. In fact, there is an infinite number of solutions of $3x + 2y = 8$.

---

10. What can you say about the coordinates of a point in the third quadrant?

11. What can you say about the coordinates of a point in the fourth quadrant?

In which quadrant is the point located?

12. (5, 3)

13. (−6, −4)

14. (10, −14)

15. (−13, 9)

Answers on page A-16

16. Determine whether (5, 1) is a solution of $y = 2x + 3$.

Example 4 Determine whether the ordered pair $(-2, 3)$ is a solution of the equation $2t = 4s - 8$.

We substitute:

$$\begin{array}{c|c} 2t = 4s - 8 \\ \hline 2 \cdot 3 \; ? \; 4(-2) - 8 & \text{Substituting } -2 \text{ for } s \text{ and } 3 \text{ for } t \\ 6 & -8 - 8 \\ 6 & -16 \quad \text{FALSE} \end{array}$$

Since the equation becomes false, $(-2, 3)$ is not a solution.

Do Exercises 16 and 17.

Calculator Spotlight

Solutions of equations in two variables can be easily checked on a calculator. For instance, to show that $(5.1, -3.65)$ is a solution of $3x + 2y = 8$, on many calculators we press

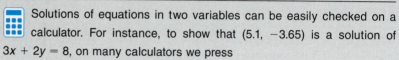

The result, 8, shows that $(5.1, -3.65)$ is a solution.

Most calculators now have memory keys. These keys enable us to store and recall a number as needed. Any number being displayed can be stored by pressing a particular key. On many calculators, this key is labeled [STO], [M], or [Min]. Once a number has been stored, we can retrieve the number by pressing a key labeled [RCL] or [MR].

To show that (7.35, 10.7) is a solution of $2t = 4s - 8$, we can first evaluate the right side of the equation:

The result, 21.4, has been stored in the calculator's memory, so we need not worry about writing it down. To complete the check, we clear the calculator and evaluate the left side of the equation:

[2] [×] [1] [0] [.] [7] [=] .

To show that this result matches the earlier number that was stored, we do not clear the display, but instead press

[−] [RCL] [=] .

The result, 0, indicates that 2×10.7 and $4 \times 7.35 - 8$ are equal. A result other than 0 would indicate that the ordered pair in question does not check.

Exercises

Determine whether each point is a solution of the given equation.

1. (7.9, 3.2); $5x + 4y = 52.3$ **2.** (1.9, 2.3); $7x - 8y = 5.1$
3. (4.3, 4.75); $5y = 6x - 7$ **4.** (3.8, -4.3); $9a = 17 - 4b$
5. (9.4, -3.9); $3a - 15 = 29 + 4b$ **6.** (5.6, 8.8); $4y + 23 = 7x + 19$
7. (-2.4, 8.5625); $3.5x + 17.4 = 3.2y - 18.4$ **8.** (1.8, 2.6); $9.2x - 15.3 = 4.8y - 13.7$

17. Determine whether $(4, -1)$ is a solution of $3x + 2y = 10$.

Answers on page A-16

Exercise Set 6.3

a Plot each group of points on the given graph below.

1. (2, 5)   (−1, 3)   (3, −2)   (−2, −4)
   (0, 4)   (0, 5)   (5, 0)   (−5, 0)

2. (4, 4)   (−2, 4)   (5, −3)   (−5, −5)
   (0, 4)   (0, −4)   (3, 0)   (−4, 0)

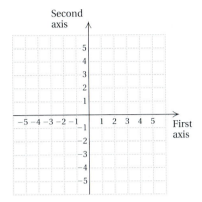

3. (−3, −1)   (5, 1)   (−1, −5)   (0, 0)
   (0, 1)   (−4, 0) $(2, 3\frac{1}{2})$ $(4\frac{1}{2}, -2)$

4. (−2, −4)   (5, −4) $(0, 3\frac{1}{2})$ $(4, 3\frac{1}{2})$
   (−1, −3)   (−1, 5)   (4, −1)   (−2, 0)

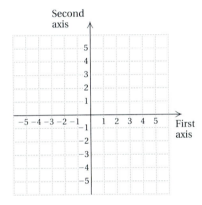
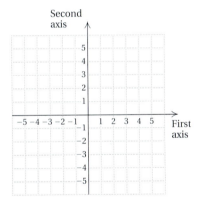

Find the coordinates of points A, B, C, D, E, and F.

5.
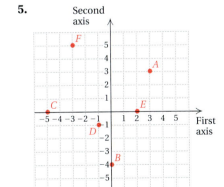

6.

7.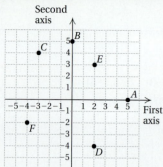

8.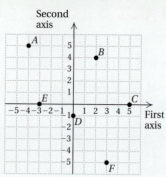

b In which quadrant is each point located?

9. (−5, 3)

10. (−12, 1)

11. (100, −1)

12. $\left(35\frac{1}{2}, -2\frac{1}{2}\right)$

13. (−6.5, −1.9)

14. (−3.4, −5.9)

15. $\left(3\frac{7}{10}, 9\frac{1}{11}\right)$

16. (1895, 1492)

Complete each sentence using the words *positive* or *negative* or the numerals I, II, III, or IV.

17. In quadrant IV, first coordinates are always _____ and second coordinates are always _____.

18. In quadrant III, first coordinates are always _____ and second coordinates are always _____.

19. In quadrant _____, both coordinates are always negative.

20. In quadrant _____, both coordinates are always positive.

21. In quadrants I and _____, the first coordinate is always _____.

22. In quadrants II and _____, the second coordinate is always _____.

c Determine whether each ordered pair is a solution of the given equation.

23. (2, 7); $y = 3x + 1$

24. (1, 7); $y = 2x + 5$

25. (2, −3); $3x − y = 4$

26. (−1, 4); $2x + y = 6$

27. (−2, −1); $2c + 3d = −7$

28. (0, −4); $4p + 2q = −9$

29. (5, −4); $3x + y = 19$

30. (−1, 7); $x − y = −8$

31. $\left(2\frac{1}{3}, 5\right)$; $2q − 3p = 3$

32. $\left(3, 1\frac{1}{4}\right)$; $2p − 4q = 1$

33. (2.4, 0.7); $y = 5x − 6.3$

34. (1.8, 7.4); $y = 3x + 2$

Exercise Set 6.3

Skill Maintenance

Solve.

35. $3x - 4 = 17$ [4.4a]

36. $7 + 2x = 25$ [4.4a]

37. $5(x - 2) = 3x - 4$ [5.7b]

38. Simplify: $\dfrac{90}{51}$. [3.5b]

39. Combine like terms: [4.6b]
$$7\tfrac{2}{11}a - 5\tfrac{1}{3}a.$$

40. Simplify: [2.7b]
$$3(x - 5) + 4x - 9.$$

Synthesis

41. ◆ Under what conditions will the points (a, b) and (b, a) be in the same quadrant?

42. ◆ Describe in your own words how to plot the point (a, b).

43. ◆ In which quadrant, if any, is the point $(5, 0)$? Why?

Determine whether each ordered pair is a solution of the given equation.

44. ▦ $(-2.37, 1.23)$; $5.2x + 6.1y = -4.821$

45. ▦ $(4.16, -9.35)$; $6.5x - 7.2y = -94.36$

In Exercises 46–49, tell in which quadrant(s) the point could be located.

46. The first coordinate is positive.

47. The second coordinate is negative.

48. The first and second coordinates are equal.

49. The first coordinate is the opposite of the second coordinate.

50. The points $(-1, 1)$, $(4, 1)$, and $(4, -5)$ are three vertices of a rectangle. Find the coordinates of the fourth vertex.

51. A parallelogram is a four-sided polygon with two pairs of parallel sides. Three parallelograms share the vertices $(-2, -3)$, $(-1, 2)$, and $(4, -3)$. Find the fourth vertex of each parallelogram.

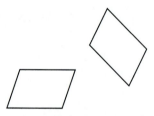

52. Graph eight points such that the sum of the coordinates in each pair is 6.

53. Graph eight points such that for each point the first coordinate minus the second coordinate is 1.

54. Find the perimeter of a rectangle with vertices at $(5, 3)$, $(5, -2)$, $(-3, -2)$, and $(-3, 3)$.

55. Find the area of a rectangle with vertices at $(0, 9)$, $(0, -4)$, $(5, -4)$, and $(5, 9)$.

Practice finding and plotting ordered pairs by playing a variation of the game Battleship.

6.4 Graphing Linear Equations

In Section 6.3, we saw how to determine whether an ordered pair is a solution of an equation in two variables. We now develop a way of finding such solutions on our own. Once we are able to find a few ordered pairs that solve an equation, we will be able to graph the equation.

a Finding Solutions

To solve an equation with one variable, like $3x + 2 = 8$, we isolate the variable, x, on one side of the equation. To solve an equation with two variables, we will first replace one variable with some number choice and then solve the resulting equation.

Example 1 Find a solution of $x + y = 7$. Let $x = 5$.

If x is 5, then $x + y = 7$ can be rewritten as

$$5 + y = 7.$$

We solve as follows:

$$5 + y = 7$$
$$5 + y - 5 = 7 - 5 \quad \text{Subtracting 5 or adding } -5 \text{ on both sides}$$
$$y = 2.$$

The ordered pair (5, 2) is a solution of $x + y = 7$.

Do Exercise 1.

Example 2 Complete these solutions of $2x + 3y = 8$: (, 2); $(-2,\ )$.

To complete the pair (, 2), we replace y with 2 and solve for x:

$$2x + 3y = 8$$
$$2x + 3 \cdot 2 = 8 \quad \text{Substituting 2 for } y$$
$$2x + 6 = 8$$
$$2x + 6 - 6 = 8 - 6 \quad \text{Subtracting 6 on both sides}$$
$$2x = 2$$
$$\tfrac{1}{2} \cdot 2x = \tfrac{1}{2} \cdot 2 \quad \text{Multiplying by } \tfrac{1}{2} \text{ on both sides}$$
$$x = 1.$$

Thus, (1, 2) is a solution of $2x + 3y = 8$.

To complete the pair $(-2,\ )$, we replace x with -2 and solve for y:

$$2x + 3y = 8$$
$$2(-2) + 3y = 8 \quad \text{Substituting } -2 \text{ for } x$$
$$-4 + 3y = 8$$
$$3y = 12 \quad \text{Adding 4 on both sides}$$
$$y = 4. \quad \text{Dividing by 3 on both sides}$$

Thus, $(-2, 4)$ is also a solution of $2x + 3y = 8$.

Do Exercise 2.

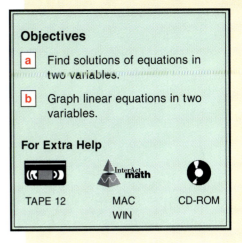

Objectives

a Find solutions of equations in two variables.

b Graph linear equations in two variables.

For Extra Help

TAPE 12   MAC WIN   CD-ROM

1. Find a solution of $x - y = 3$. Let $y = 5$.

2. Complete these solutions of $5x + y = 10$: $(1,\ )$; $(\ , -5)$.

Answers on page A-16

3. Find three solutions of
$x + 2y = 7$. Answers may vary.

Example 3 Find three solutions of $2x - y = 5$.

We are free to choose *any* number as a replacement for x or y. To find one solution, we choose to replace x with 1. We then solve for y:

$2x - y = 5$
$2 \cdot 1 - y = 5$ Substituting 1 for *x*. Other choices are possible.
$2 - y = 5$
$-y = 3$ Subtracting 2 on both sides
$-1y = 3$ Recall that $-a = -1 \cdot a$.
$y = -3$.   Dividing by -1 on both sides

Thus, $(1, -3)$ is one solution of $2x - y = 5$.

To find a second solution, we choose to replace y with 3 and solve for x:

$2x - y = 5$
$2x - 3 = 5$ Substituting 3 for *y*. Other choices are possible.
$2x = 8$ Adding 3 on both sides
$x = 4$.   Dividing by 2 on both sides

Thus, $(4, 3)$ is a second solution of $2x - y = 5$.

To find a third solution, we can replace x with 0 and solve for y:

$2x - y = 5$
$2 \cdot 0 - y = 5$ Substituting 0 for *x*. Other choices are possible.
$0 - y = 5$
$-y = 5$
$-1y = 5$ Try to do this step mentally.
$y = -5$.   Dividing by -1 on both sides

The pair $(0, -5)$ is a third solution of $2x - y = 5$.

Note that three different choices for x or y would have given three different solutions. There is an infinite number of ordered pairs that are solutions, so it is unlikely for two students to have solutions that match entirely.

4. Find three solutions of
$y = -2x + 7$. Answers may vary.

Do Exercises 3 and 4.

b Graphing Equations

Equations like those considered in Examples 1–3 are in the form $Ax + By = C$. All equations that can be written this way are said to be **linear** because the solutions of each equation, when graphed, form a straight line. When the appropriate line is drawn, we say that we have *graphed* the equation.

Example 4 Graph: $2x - y = 5$.

In Example 3, we found that $(1, -3)$, $(4, 3)$, and $(0, -5)$ are solutions of $2x - y = 5$. Had we not known that, before graphing we would need to calculate two or three solutions, much as we did in Example 3.

Answer on page A-16

Next, we plot the points and look for a pattern. As expected, the points describe a straight line. We draw the line, as shown on the right below.

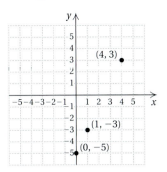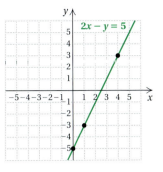

Note that two points are enough to determine a line, but if either point is calculated incorrectly, the wrong line will be drawn. For this reason, we generally calculate at least three ordered pairs before drawing a line. If when we plot these pairs we note that they do not all line up, we know that a mistake has been made.

Do Exercise 5.

Equations like $y = 2x$ or $y = x + 2$ are also linear. To find pairs that solve such equations, it is usually easiest to substitute for x and then find y.

Example 5 Graph: $y = 2x$.

First, we find some ordered pairs that are solutions. Suppose we choose 3 for x. Then

$$y = 2x = 2 \cdot 3 = 6,$$

so $(3, 6)$ is one solution.

To find a second solution, we can replace x with -2:

$$y = 2x = 2(-2) = -4.$$

Thus, $(-2, -4)$ is a solution.

To find a third solution, we can replace x with 0:

$$y = 2x = 2 \cdot 0 = 0.$$

Thus, $(0, 0)$ is a solution.

We can compute additional pairs if we wish and form a table.

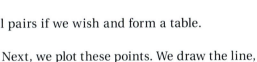

Next, we plot these points. We draw the line, or graph, with a ruler and label it $y = 2x$.

x	$y = 2x$	(x, y)
3	6	(3, 6)
−2	−4	(−2, −4)
0	0	(0, 0)
1	2	(1, 2)

(1) Choose x.
(2) Compute y.
(3) Form the pair (x, y).
(4) Plot the points.
(5) Draw and label the graph.

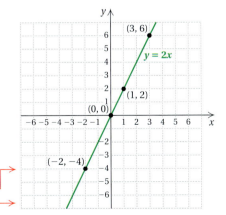

5. Graph $x + 2y = 7$. Use the results from Margin Exercise 3.

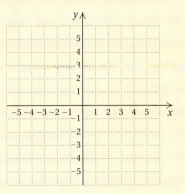

Graph.

6. $y = 3x$

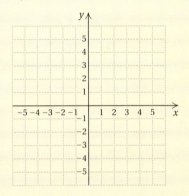

7. $y = \dfrac{1}{2}x$

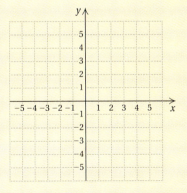

Answers on page A-16

6.4 Graphing Linear Equations

385

Graph.

8. $y = -x$ (or $y = -1 \cdot x$)

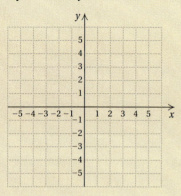

9. $y = -2x$

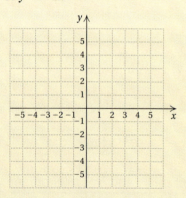

Do Exercises 6 and 7 on the preceding page.

Example 6 Graph: $y = -3x$.

We make a table of solutions. Then we plot the points, draw the line with a ruler, and label the line $y = -3x$.

If x is 0, then $y = -3 \cdot 0 = 0$.
If x is 1, then $y = -3 \cdot 1 = -3$.
If x is -2, then $y = -3(-2) = 6$.
If x is 2, then $y = -3 \cdot 2 = -6$.

x	y $y = -3x$	(x, y)
0	0	(0, 0)
1	-3	(1, -3)
-2	6	(-2, 6)
2	-6	(2, -6)

Don't forget the label!

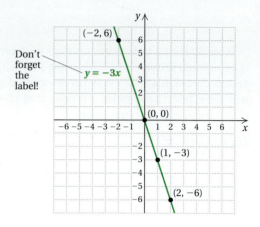

Do Exercises 8 and 9.

Example 7 Graph: $y = x + 2$.

We make a table of solutions. Then we plot the points, draw the line with a ruler, and label it.

If x is 0, then $y = 0 + 2 = 2$.
If x is 1, then $y = 1 + 2 = 3$.
If x is -1, then $y = -1 + 2 = 1$.
If x is 3, then $y = 3 + 2 = 5$.

x	y $y = x + 2$	(x, y)
0	2	(0, 2)
1	3	(1, 3)
-1	1	(-1, 1)
3	5	(3, 5)

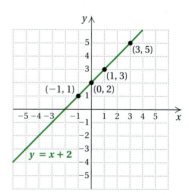

Answers on page A-16

Example 8 Graph: $y = \frac{2}{3}x$.

We make a table of solutions, plot the points, and draw and label the line. It is important to note that by selecting multiples of 3 as x-values, we avoid fractional values for y.

If x is 6, then $y = \frac{2}{3} \cdot 6 = 4$.
If x is 3, then $y = \frac{2}{3} \cdot 3 = 2$.
If x is 0, then $y = \frac{2}{3} \cdot 0 = 0$.
If x is -3, then $y = \frac{2}{3}(-3) = -2$.

x	$y = \frac{2}{3}x$	(x, y)
6	4	(6, 4)
3	2	(3, 2)
0	0	(0, 0)
-3	-2	$(-3, -2)$

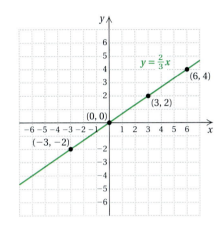

Do Exercises 10–12.

As is the case with many skills, your ability to graph linear equations will grow with practice. By carefully plotting at least three points for each equation, you can check your work for errors before any line is drawn. It is extremely unlikely for three points to line up if one of them was calculated incorrectly.

Graph.

10. $y = x + 1$

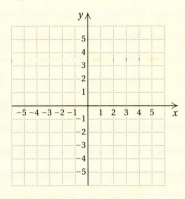

11. $y = -2x + 1$

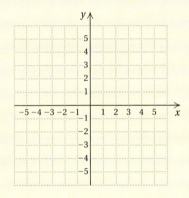

12. $y = \frac{3}{5}x$

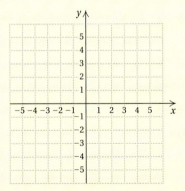

Answers on pages A-16 and A-17

Calculator Spotlight

Calculators or computers with graphing capability have become increasingly common. This technology is generally used for graphing equations that are more complicated than $y = x + 2$ and $y = \frac{2}{3}x$ (Examples 7 and 8) and *in no way decreases the importance of understanding how such equations are graphed by hand.* The purpose of the following discussion is to show how graphing technology can be used to check some of your work and how it might enable you to handle more challenging problems.

All graphing technology utilizes a *window,* the rectangular portion of the screen in which a graph appears. For our purposes, a window extending from -10 to 10 on the *x*- and *y*-axes will suffice. Such settings may be standard or can be easily adjusted with keystrokes that vary depending on the technology in use (consult a user's manual or an instructor for the exact procedure).

To graph the equation $y = x + 2$, we press a key (often labeled $\boxed{Y=}$) and then

$\boxed{X, T, \theta}$ $\boxed{+}$ $\boxed{2}$ $\boxed{\text{GRAPH}}$

(keystrokes will vary with the technology used). A graph similar to that shown on the left below should appear. To view some of the ordered pairs that are solutions, a TRACE key can be used to move a cursor along the line. Near the bottom of the window the cursor's coordinates appear (see the graph on the right below).

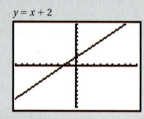
$y = x + 2$

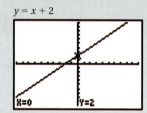
$y = x + 2$

Exercises

Use graphing technology to graph each of the following.

1. $y = \frac{2}{3}x$
   (Example 8)
2. $y = x + 1$
   (Margin Exercise 10)
3. $y = -2x + 1$
   (Margin Exercise 11)
4. $y = \frac{3}{5}x$
   (Margin Exercise 12)

Exercise Set 6.4

a For each equation, use the indicated value to find an ordered pair that is a solution. See Example 1.

1. $x - y = 3$; let $x = 8$
2. $x + y = 5$; let $x = 4$
3. $2x + y = 7$; let $x = 3$

4. $x + 2y = 9$; let $y = 4$
5. $y = 3x - 1$; let $x = 5$
6. $y = 2x + 7$; let $x = 3$

7. $x + 3y = 1$; let $x = 7$
8. $5x + y = 7$; let $y = 2$
9. $2x + 5y = 17$; let $x = 1$

10. $5x + 2y = 19$; let $x = 1$
11. $3x - 2y = 8$; let $y = -1$
12. $2x - 5y = 12$; let $y = -2$

For each equation, complete the given ordered pairs. See Example 2.

13. $x + y = 7$; (, 8); (4,)
14. $x - y = 6$; (, 2); (9,)
15. $x - y = 4$; (, 3); (10,)

16. $x + y = 10$; (, 8); (3,)
17. $2x + 3y = 15$; (3,); (, 1)
18. $3x + 2y = 16$; (4,); (, −1)

19. $5x + 2y = 11$; $(3, \square)$; $(\square, 3)$
20. $4x + 3y = 11$; $(5, \square)$; $(\square, 2)$
21. $y = 4x$; $(\square, 4)$; $(-2, \square)$

22. $y = 6x$; $(\square, 6)$; $(-2, \square)$
23. $2x + 5y = 3$; $(0, \square)$; $(\square, 0)$
24. $5x + 7y = 9$; $(0, \square)$; $(\square, 0)$

For each equation, find three solutions. Answers may vary.

25. $x + y = 12$
26. $x + y = 19$
27. $y = 4x$
28. $y = 5x$

29. $3x + y = 13$
30. $x + 5y = 12$
31. $y = 3x - 1$
32. $y = 2x + 5$

33. $y = -5x$
34. $y = -3x$
35. $4 + y = x$
36. $3 + y = x$

37. $3x + 2y = 12$
38. $2x + 3y = 18$
39. $y = \frac{1}{3}x + 2$
40. $y = \frac{1}{2}x + 5$

Chapter 6 Introduction to Graphing and Statistics

b Use your own graph paper. Draw and label x- and y-axes. Then graph each equation.

41. $x + y = 4$ **42.** $x + y = 6$ **43.** $x - 1 = y$ **44.** $x - 2 = y$

45. $y = x - 3$ **46.** $y = x - 5$ **47.** $y = \dfrac{1}{3}x$ **48.** $y = -\dfrac{1}{3}x$

49. $y = x$ **50.** $y = x - 7$ **51.** $y = 2x - 1$ **52.** $y = 2x - 3$

53. $y = 2x - 7$ **54.** $y = 3x - 7$ **55.** $y = \dfrac{2}{5}x$ **56.** $y = \dfrac{3}{4}x$

57. $y = -x + 4$ **58.** $y = -x + 5$

Exercise Set 6.4

Skill Maintenance

59. A recipe for a batch of chili calls for $\frac{3}{4}$ cup of red wine vinegar. How much vinegar is needed to make $2\frac{1}{2}$ batches of chili? [4.7d]

Simplify.

60. $-\dfrac{49}{77}$ [3.5b]

61. $-8 - 5^2 \cdot 2(3 - 4)$ [2.5b]

62. $\dfrac{3}{10}\left(-\dfrac{25}{12}\right)$ [3.4b]

Solve.

63. $4.8 - 1.5x = 0.9$ [5.7a]

64. $3x - 8 = 5x - 12$ [5.7b]

Synthesis

65. ◈ What is the greatest number of quadrants that a line can pass through? Why?

66. ◈ In Example 8, we found that by choosing multiples of 3 for x, we could avoid fractions. What is the advantage of avoiding fractions? Would it have been incorrect to substitute values for x that are *not* multiples of 3? Why or why not?

67. ◈ To graph a linear equation, a student plots three points and discovers that the points do not line up with each other. What should the student do next?

Find three solutions of each equation. Then graph the equation.

68. 🖩 $25x + 80y = 100$

69. 🖩 $50x + 75y = 180$

70. Use the graph in Example 4 to find three solutions of $2x - y = 5$. Do not use the ordered pairs already listed.

71. Use the graph in Example 7 to find three solutions of $y = x + 2$. Do not use the ordered pairs already listed.

72. Find all the whole-number solutions of $x + y = 6$.

73. Graph three solutions of $y = |x|$ in the second quadrant and another three solutions in the first quadrant.

6.5 Means, Medians, and Modes

We have seen that pictographs, bar graphs, and line graphs provide three ways of representing a collection of data *visually*. Sometimes it is useful to describe a set of data *numerically*, using *statistics*. A **statistic** is simply a number that is derived from a set of data. There are three statistics that are considered *center point*, that is, numbers that serve to represent the entire data set. Let's examine all three.

a | Means

The most commonly used center point is the *average* of the set of numbers. We have already computed an average several times in this book (see pages 76, 240, 258, and 304). Although the word "average" is often used in everyday speech, in math we more often use the word *mean* instead.

> To find the **mean** of a set of numbers, add the numbers and then divide by the number of items of data.

Objectives

a | Find the mean of a set of numbers and solve applied problems involving means.

b | Find the median of a set of numbers and solve applied problems involving medians.

c | Find the mode of a set of numbers and solve applied problems involving modes.

For Extra Help

TAPE 12   MAC WIN   CD-ROM

Example 1 *Golfing.* In 1997, Tiger Woods set the record for the lowest total score in a golf tournament consisting of four rounds. His scores were 70, 66, 65, and 69. Find his mean score per round.

To find the mean, we add the scores together and then divide by the number of scores, 4:

$$\frac{70 + 66 + 65 + 69}{4} = \frac{270}{4} = 67.5.$$

Tiger Woods' mean score was 67.5.

Do Exercises 1–4.

Example 2 *Food Waste.* Courtney is a typical American consumer. In the course of 1 yr, she discards 100 lb of food waste. What is the average number of pounds of food waste discarded each week? Round to the nearest tenth.

We already know the total amount of food waste for the year. Since there are 52 weeks in a year, we divide by 52 and round:

$$\frac{100}{52} \approx 1.9.$$

On average, Courtney discards 1.9 lb of food waste per week.

Do Exercise 5.

Find the mean.
1. 14, 175, 36

2. 75, 36.8, 95.7, 12.1

3. Wendy scored the following on five tests: 96, 85, 82, 74, 68. What was her mean score?

4. In the first five games, a basketball player scored points as follows: 26, 21, 13, 14, 23. Find the mean number of points scored per game.

5. *Food Waste.* Courtney also composts (converts to dirt) 5 lb of food waste each year. How much, on average, does Courtney compost per month? Round to the nearest tenth.

Answers on page A-18

6.5 Means, Medians, and Modes

393

6. *GPA.* Alex earned the following grades one semester.

Grade	Number of Credit Hours in Course
B	3
C	4
C	4
A	2

What was Alex's grade point average? Assume that the grade point values are 4.0 for an A, 3.0 for a B, and so on. Round to the nearest tenth.

Example 3 *GPA.* In most colleges, students are assigned grade point values for grades obtained. The **grade point average**, or **GPA**, is the average of the grade point values for each credit hour taken. At most colleges, grade point values are assigned as follows:

A: 4.0
B: 3.0
C: 2.0
D: 1.0
F: 0.0

Meg earned the following grades for one semester. What was her grade point average?

Course	Grade	Number of Credit Hours in Course
Colonial History	B	3
Basic Mathematics	A	4
English Literature	A	3
French	C	4
Physical Education	D	1

To find the GPA, we first multiply the grade point value (in color below) by the number of credit hours in the course and then add, as follows:

$$\begin{aligned}
\text{Colonial History} \quad & 3.0 \cdot 3 = 9 \\
\text{Basic Mathematics} \quad & 4.0 \cdot 4 = 16 \\
\text{English Literature} \quad & 4.0 \cdot 3 = 12 \\
\text{French} \quad & 2.0 \cdot 4 = 8 \\
\text{Physical Education} \quad & 1.0 \cdot 1 = \underline{1} \\
& \qquad\qquad\; 46 \;\; \text{(Total)}
\end{aligned}$$

The total number of credit hours taken is $3 + 4 + 3 + 4 + 1$, or 15. We divide 46 by 15 and round to the nearest tenth:

$$\text{GPA} = \frac{46}{15} \approx 3.1.$$

Meg's grade point average was 3.1.

Do Exercise 6.

Answers on page A-18

Example 4 To get a B in math, Geraldo must have a mean test score of at least 80. On the first four tests, his scores were 79, 88, 64, and 78. What is the lowest score that Geraldo can get on the last test and still get a B?

We can find the total of the five scores needed as follows:

$$80 + 80 + 80 + 80 + 80 = 5 \cdot 80, \text{ or } 400.$$

The total of the scores on the first four tests is

$$79 + 88 + 64 + 78 = 309.$$

Thus Geraldo needs to get at least

$$400 - 309, \text{ or } 91$$

in order to get a B. We can check this as follows:

$$\frac{79 + 88 + 64 + 78 + 91}{5} = \frac{400}{5}, \text{ or } 80.$$

Do Exercise 7.

b Medians

Another type of center-point statistic is the *median*. Medians are useful when we wish to de-emphasize unusually extreme scores. For example, suppose a small class scored as follows on an exam.

Phil: 78   Pat: 56
Jill: 81   Olga: 84
Matt: 82

Let's first list the scores in order from smallest to largest:

56, 78, 81, 82, 84.
↑
Middle score

The middle score—in this case, 81—is called the **median**. Note that because of the extremely low score of 56, the average of the scores is 76.2. In this example, the median may be more indicative of how the class as a whole performed.

Example 5 What is the median of this set of numbers?

99, 870, 91, 98, 106, 90, 98

We first rearrange the numbers in order from smallest to largest. Then we locate the middle number, 98.

90, 91, 98, 98, 99, 106, 870
↑
Middle number

The median is 98.

Do Exercises 8–10.

> Once a set of data is listed in order, from smallest to largest, the **median** is the middle number if there is an odd number of data items. If there is an even number of items, the median is the number that is the average of the two middle numbers.

---

7. To get an A in math, Rosa must have a mean test grade of at least 90. On the first three tests, her scores were 80, 100, and 86. What is the lowest score that Rosa can get on the last test and still get an A?

Calculator Spotlight

 Means can be easily computed on a calculator if we remember the order in which operations are performed. For example, to calculate

$$\frac{85 + 92 + 79}{3}$$

on most calculators, we press

[8] [5] [+] [9] [2] [+] [7] [9] [=]
[÷] [3] [=] ,

or

[(] [8] [5] [+] [9] [2] [+] [7] [9]
[)] [÷] [3] [=] .

Exercises

1. What would the result have been if we had not used parentheses in the latter sequence of keystrokes?
2. Use a calculator to solve Examples 1–5.

Find the median.

8. 17, 13, 18, 14, 19

9. 20, 14, 13, 19, 16, 18, 17

10. 78, 81, 83, 91, 103, 102, 122, 119, 88

Answers on page A-18

Find the median.

11. $1300, $2000, $3900, $1600, $1800, $1400

12. 68, 34, 67, 69, 58, 70

Find the modes.

13. 23, 45, 45, 45, 78

14. 34, 34, 67, 67, 68, 70

15. 13, 24, 27, 28, 67, 89

16. In a lab, Gina determined the mass, in grams, of each of five eggs:

15 g, 19 g, 19 g, 14 g, 18 g.

a) What is the mean?
b) What is the median?
c) What is the mode?

Answers on page A-18

Chapter 6 Introduction to Graphing and Statistics

Example 6 What is the median of this set of yearly salaries?

$35,000, $500,000, $28,000, $34,000, $27,000, $42,000

We rearrange the numbers in order from smallest to largest. The two middle numbers are $34,000 and $35,000. Thus the median is halfway between $34,000 and $35,000 (the average of $34,000 and $35,000):

$27000, $28,000, $34,000, $35,000, $42,000, $500,000

$$\text{Median} = \frac{\$34,000 + \$35,000}{2} = \frac{\$69,000}{2} = \$34,500.$$

Do Exercises 11 and 12.

c Modes

The final type of center-point statistic is the **mode**.

> The **mode** of a set of data is the number or numbers that occur most often. If each number occurs the same number of times, there is *no* mode.

Example 7 Find the mode of these data.

13, 14, 17, 17, 18, 19

The number that occurs most often is 17. Thus the mode is 17.

A set of data has just one mean and just one median, but it can have more than one mode. It is also possible for a set of data to have no mode—when all numbers are equally represented. For example, the set of data 5, 7, 11, 13, 19 has no mode.

Example 8 Find the modes of these data.

33, 34, 34, 34, 35, 36, 37, 37, 37, 38, 39, 40

There are two numbers that occur most often, 34 and 37. Thus the modes are 34 and 37.

Do Exercises 13–16.

Which statistic is best for a particular situation? If someone is bowling, the *average* from several games is a good indicator of that person's ability. If someone is applying for a job, the *median* salary at that business is often most indicative of what people are earning there. Finally, if someone is reordering for a clothing store, the *mode* of the waist sizes sold is probably the most important statistic.

Exercise Set 6.5

a, b, c For each set of numbers, find the mean, the median, and any modes that exist.

1. 16, 18, 29, 14, 29, 19, 15
2. 72, 83, 85, 88, 92
3. 5, 30, 20, 20, 35, 5, 25

4. 13, 32, 25, 27, 13
5. 1.2, 4.3, 5.7, 7.4, 7.4
6. 12.6, 13.4, 12.6, 43.7

7. 134, 128, 128, 129, 128, 178
8. $29.95, $28.79, $30.95, $29.95

9. *Basketball.* Lisa Leslie of the Los Angeles Spark once scored 23, 21, 19, 23, and 20 points in consecutive games. What was the mean for the five games? the median? the mode?

10. The following temperatures were recorded for seven days in Hartford:

    43°, 40°, 23°, 38°, 54°, 35°, 47°.

    What was the mean temperature? the median? the mode?

11. *Gas Mileage.* According to recent EPA estimates, an Achieva can be expected to travel 297 mi (highway) on 9 gal of gasoline (**Source:** *Motor Trend Magazine*). What is the mean number of miles expected per gallon?

12. *Gas Mileage.* According to recent EPA estimates an Aurora can be expected to travel 192 mi (highway) on 8 gal of gasoline (**Source:** *Motor Trend Magazine*). What is the mean number of miles expected per gallon?

GPA. In Exercises 13 and 14 are the grades of a student for one semester. In each case, find the grade point average. Assume that the grade point values are 4.0 for an A, 3.0 for a B, and so on. Round to the nearest tenth.

13.

Grades	Number of Credit Hours in Course
B	4
A	3
B	3
C	4

14.

Grades	Number of Credit Hours in Course
A	4
B	4
B	3
C	3

15. *Fish Prices.* The following prices per pound of Atlantic salmon were found at five fish markets:

    $7.99, $9.49, $9.99, $7.99, $10.49.

    What was the mean price per pound? the median price? the mode?

16. *Cheese Prices.* The following prices per pound of Vermont cheddar cheese were found at five supermarkets:

    $4.99, $5.79, $4.99, $5.99, $5.79.

    What was the mean price per pound? the median price? the mode?

17. *Grading.* To get a B in math, Rich must average at least 80 on five tests. Scores on the first four tests were 80, 74, 81, and 75. What is the lowest score that Rich can get on the last test and still receive a B?

18. *Grading.* To get an A in math, Cybil must average at least 90 on five tests. Scores on the first four tests were 90, 91, 81, and 92. What is the lowest score that Cybil can get on the last test and still receive an A?

19. *Length of Pregnancy.* Marta was pregnant 270 days, 259 days, and 272 days for her first three pregnancies. In order for Marta's mean pregnancy to equal the worldwide average of 266 days, how long must her fourth pregnancy last? (**Source**: David Crystal (ed.), *The Cambridge Factfinder.* Cambridge CB2 1RP: Cambridge University Press, 1997, p. 84.)

20. *Male Height.* Jason's brothers are 174 cm, 180 cm, 179 cm, and 172 cm tall. The average male is 176.5 cm tall. How tall is Jason if he and his brothers have a mean height of 176.5 cm?

Skill Maintenance

Multiply.

21. $14 \cdot 14$ [1.5b]
22. $-144 \div (-9)$ [2.5a]
23. 1.4×1.4 [5.3a]
24. $-\dfrac{4}{9} \cdot \dfrac{15}{22}$ [3.4b]

Solve.   [5.8a]

25. A disc jockey charges a $40 setup fee and $50 an hour. How long can the disc jockey work for $165?

26. To rent a floor sander costs $15 an hour plus a $10 supply fee. For how long can the machine be rented if $100 has been budgeted for the sander?

Synthesis

27. ◆ Why might a firm's median salary be more indicative of what employees earn than the average, or mean, salary would be? (*Hint*: See Example 6.)

28. ◆ The following is a list of the number of children in each family in a certain Glen View neighborhood: 0, 2, 3, 0, 5, 2, 2, 0, 0, 2, 0, 0. Explain why the mode might be the most indicative statistic for the number of children in a family.

29. ◆ Is it possible for a driver to average 20 mph on a 30-mi trip and still receive a ticket for driving 75 mph? Why or why not?

Bowling Averages. Bowling averages are always computed by rounding down to the nearest integer. For example, suppose a bowler gets a total of 599 for 3 games. To find the average, we divide 599 by 3 and drop the amount to the right of the decimal point:

$\dfrac{599}{3} \approx 199.67.$ The bowler's average is 199.

30. If Frances bowls 4176 in 23 games, what is her average?

31. If Eric bowls 4621 in 27 games, what is his average?

32. The ordered set of data 18, 21, 24, a, 36, 37, b has a median of 30 and an mean of 32. Find a and b.

33. *Hank Aaron.* Hank Aaron averaged $34\tfrac{7}{22}$ home runs per year over a 22-yr career. After 21 yr, Aaron had averaged $35\tfrac{10}{21}$ home runs per year. How many home runs did Aaron hit in his final year?

34. Because of a poor grade on the fifth and final test, Chris's mean test grade fell from 90.5 to 84.0. What did Chris score on the fifth test? Assume that all tests are equally important.

35. *Price Negotiations.* Amy offers $3200 for a used Ford Taurus advertised at $4000. The first offer from Jim, the car's owner, is to "split the diffence" and sell the car for $(3200 + 4000) \div 2$, or $3600. Amy's second offer is to split the difference between Jim's offer and her first offer. Jim's second offer is to split the difference between Amy's second offer and his first offer. If this pattern continues and Amy accepts Jim's third (and final) offer, how much will she pay for the car?

Use averages to negotiate the price of a car.

6.6 Predictions and Probability

a | Making Predictions

Sometimes we use data to make predictions or estimates of missing data points. One process for doing so is called **interpolation**. Interpolation enables us to estimate missing "in-between values" on the basis of known information.

Objectives

a | Make predictions from a set of data using interpolation or extrapolation.

b | Calculate the probability of an event occurring.

For Extra Help

TAPE 12 MAC WIN CD-ROM

Example 1 *Monthly Mortgage Payments.* When money is borrowed and then repaid in monthly installments, the size of the payments increases as the length of the loan, in years, decreases. The table below lists the size of a monthly payment when $110,000 is borrowed (at 9% interest) for various lengths of time. Use interpolation to estimate the monthly payment on a 35-yr loan.

Year	Monthly Payment
5	$2283.42
10	1393.43
15	1115.69
20	989.70
25	923.12
30	885.08
35	?
40	848.50

To use interpolation, we first plot the points and look for a trend. It seems reasonable to draw a line between the points corresponding to 30 and 40. We can "zoom-in" to better visualize the situation. To estimate the second coordinate that is paired with 35, we trace a vertical line up from 35 to the graph and then left to the vertical axis. Thus we estimate the value to be 867. We can also estimate this value by averaging $885.08 and $848.50:

$$\frac{\$885.08 + \$848.50}{2} = \$866.79.$$

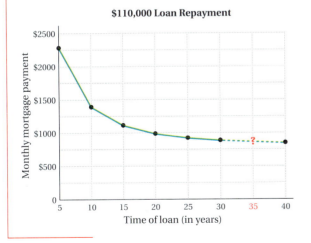
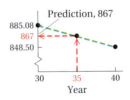

6.6 Predictions and Probability

1. *World Bicycle Production.* Use interpolation to estimate world bicycle production in 1994 from the information in the following table.

Year	World Bicycle Production (in millions)
1989	95
1990	90
1991	96
1992	103
1993	108
1994	?
1995	114

   Source: United Nations Interbike Directory

2. *Study Time and Test Scores.* A professor gathered the following data comparing study time and test scores. Use extrapolation to estimate the test score received when studying for 23 hr.

Study Time (in hours)	Test Grade (in percent)
19	83
20	85
21	88
22	91
23	?

Answer on page A-18

When we estimate in this way to find an in-between value, we are *interpolating*. Real-world information about the data might tell us that an estimate found in this way is unreliable. For example, data from the stock market might be too erratic for interpolation.

Do Exercise 1.

We often analyze data with the view of going "beyond" the data. One process for doing so is called **extrapolation**.

Example 2 *Movies Released.* The data in the following table and graphs show the number of movie releases over a period of years. Use extrapolation to estimate the number of movies released in 1997.

Year	Movies Released
1991	164
1992	150
1993	161
1994	184
1995	234
1996	260
1997	?

Source: Motion Picture Association of America

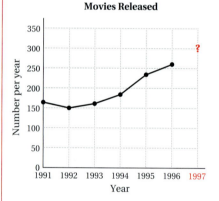
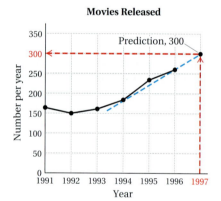

First, we analyze the data and note that they tend to follow a straight line past 1994. Keeping this trend in mind, we draw a "representative" line through the data and beyond. To estimate a value for 1997, we draw a vertical line up from 1997 until it hits the representative line. We go to the left and read off a value—about 300. When we estimate in this way to find a "go-beyond value," we are *extrapolating*. Answers found with this method can vary greatly depending on the points chosen to determine the "representative" line.

Do Exercise 2.

b Probability

The predictions made in Examples 1 and 2 have a good chance of being reasonably accurate. A branch of mathematics known as *probability* is used to attach a numerical value to the likelihood that a specific event will occur.

Suppose we were to flip a coin. Because the coin is just as likely to land heads as it is to land tails, we say that the *probability* of it landing heads is $\frac{1}{2}$. Similarly, if we roll a die (plural: dice), we are as likely to roll a ⚄ as we are to roll a ⚀, ⚁, ⚂, ⚃, or ⚅. Because of this, we say that the probability of rolling a ⚄ is $\frac{1}{6}$.

Example 3 A die is about to be rolled. Find the probability that a number greater than 4 will be rolled.

Since rolling a ⚀, ⚁, ⚂, ⚃, ⚄, or ⚅ are all equally likely to occur, and since two of these possibilities involve numbers greater than 4, we have

The probability of rolling a number greater than 4 $= \dfrac{2}{6}$ ← Number of ways to roll a 5 or 6
← Number of (equally likely) possible outcomes

$= \dfrac{1}{3}.$

The reasoning shown in Example 3 can be used in a variety of applications.

Example 4 A cloth bag contains 20 equally sized marbles: 5 are red, 7 are blue, and 8 are yellow. A marble is randomly selected. Find the probability that **(a)** a red marble is selected; **(b)** a blue marble is selected; **(c)** a yellow marble is selected.

a) Since all 20 marbles are equally likely to be selected we have

The probability of selecting a red marble $= \dfrac{\text{Number of ways to select a red marble}}{\text{Number of ways to select any marble}}$

$= \dfrac{5}{20} = \dfrac{1}{4}$, or 0.25.

b) The probability of selecting a blue marble $= \dfrac{\text{Number of ways to select a blue marble}}{\text{Number of ways to select any marble}}$

$= \dfrac{7}{20}$, or 0.35.

c) The probability of selecting a yellow marble $= \dfrac{\text{Number of ways to select a yellow marble}}{\text{Number of ways to select any marble}}$

$= \dfrac{8}{20} = \dfrac{2}{5}$, or 0.4.

Do Exercise 3.

---

3. A presentation of *The Lion King* is attended by 250 people: 40 children, 60 seniors, and 150 (nonsenior) adults. After everyone has been seated, one audience member is selected at random. Find the probability of each of the following.

a) A child is selected.

b) A senior is selected.

c) A (nonsenior) adult is selected.

Answers on page A-18

4. A card is randomly selected from a well-shuffled deck of cards. Find the probability of each of the following.

a) The card is a diamond.

b) The card is a king or queen.

Many probability problems involve a standard deck of 52 playing cards. Such a deck is made up as shown below.

A deck of 52 cards

Example 5 A card is randomly selected from a well-shuffled (mixed) deck of cards. Find the probability that **(a)** the card is a jack; **(b)** the card is a club.

a) The probability of selecting a jack $= \dfrac{\text{Number of ways to select a jack}}{\text{Number of ways to select any card}}$

$= \dfrac{4}{52} = \dfrac{1}{13}$

b) The probability of selecting a club $= \dfrac{\text{Number of ways to select a club}}{\text{Number of ways to select any card}}$

$= \dfrac{13}{52} = \dfrac{1}{4}$

Do Exercise 4.

In Examples 3–5, several "events" were discussed: rolling a number greater than 4, selecting a marble of a certain color, and selecting a certain type of playing card. The likelihood of each event occurring was determined by considering the total number of possible outcomes, using the principle formally stated below.

> **THE PRIMARY PRINCIPLE OF PROBABILITY**
>
> If an event E can occur m ways out of n possible equally likely outcomes, then
>
> The probability of E occurring $= \dfrac{m}{n}$.

Answers on page A-18

Exercise Set 6.6

a Use interpolation or extrapolation to find the missing data values.

1. *Study Time and Grades.* A math instructor asked her students to keep track of how much time each spent studying the chapter on decimal notation in her basic mathematics course. They collected the information together with test scores from that chapter's test. The data are given in the following table. Estimate the missing data value.

Study Time (in hours)	Test Grade
9	75
11	93
13	80
15	85
16	85
17	80
19	?
21	86
23	91

2. *Maximum Heart Rate.* A person's maximum heart rate depends on his or her gender, age, and resting heart rate. The following table relates resting heart rate and maximum heart rate for a 20-yr-old man. Estimate the missing data value.

Resting Heart Rate (in beats per minute)	Maximum Heart Rate (in beats per minute)
50	166
60	168
65	?
70	170
80	172

Source: American Heart Association

Estimate the missing data value in each of the following tables.

3. *Ozone Layer.*

Year	Ozone Level (in parts per billion)
1991	2981
1992	3133
1993	3148
1994	3138
1995	3124
1996	?

Source: National Oceanic and Atmospheric Administration

4. *Motion Picture Expense.*

Year	Average Expense per Picture (in millions)
1991	$38.2
1992	42.4
1993	44.0
1994	50.4
1995	54.1
1996	61.0
1997	?

Source: Motion Picture Association of America

5. *Credit-Card Spending.*

Year	Credit-Card Spending from Thanksgiving to Christmas (in billions)
1991	$ 59.8
1992	66.8
1993	79.1
1994	96.9
1995	116.3
1996	131.4
1997	?

Source: RAM Research Group, National Credit Counseling Services

6. *U.S. Book-Buying Growth.*

Year	Book Sales (in billions)
1992	$21
1993	23
1994	24
1995	25
1996	26
1997	?

Source: Book Industry Trends 1995

7. *FedEx Priority Rates.*

FedEx Letter up to 8 oz.	FedEx Priority Overnight®	Delivery by 10:00 a.m. next business day
	$ 13.25	
1 lb.	$ 18.30	
2 lbs.	19.20	
3	21.00	
4	22.80	
5	24.90	
6	?	
7	29.70	
8	31.80	
9	34.20	
10	36.80	
11	37.80	

All other packaging/Weight in lbs.

Source: Federal Express Corporation

8. *FedEx Standard Rates.*

FedEx Letter up to 8 oz.	FedEx Standard Overnight®	Delivery by 3:00 p.m. next business day
	$ 11.50	
1 lb.	$ 16.00	
2 lbs.	17.00	
3	18.00	
4	19.00	
5	20.00	
6	21.75	
7	?	
8	25.25	
9	27.00	
10	28.75	
11	30.75	

All other packaging/Weight in lbs.

Source: Federal Express Corporation

b Find each of the following probabilities.

Rolling a die. In Exercises 9–12, assume that a die is about to be rolled.

9. Find the probability that a ⚁ is rolled.

10. Find the probability that a ⚄ is rolled.

11. Find the probability that an odd number is rolled.

12. Find the probability that a number greater than 2 is rolled.

Playing Cards. In Exercises 13–18, assume that one card is randomly selected from a well-shuffled deck.

13. Find the probability that the card is the ace of spades.

14. Find the probability that the card is a picture card (jack, queen, or king).

15. Find the probability that an 8 or a 6 is selected.

16. Find the probability that a red 2 is selected.

17. Find the probability that a black picture card (jack, queen, or king) is selected.

18. Find the probability that a 10 is selected.

Candy Colors. Made by the Tootsie Industries of Chicago, Illinois, Mason Dots® is a gumdrop candy. A box was opened by the authors and found to contain the following number of gumdrops:

Strawberry	7
Lemon	8
Orange	9
Cherry	4
Lime	5
Grape	6

In Exercises 19–22, assume that one gumdrop is randomly choosen from the box.

19. Find the probability that a cherry gumdrop is selected.

20. Find the probability that an orange gumdrop is selected.

21. Find the probability that a gumdrop is *not* lime.

22. Find the probability that the gumdrop is *not* lemon.

Exercise Set 6.6

Skill Maintenance

Solve. [5.7b]

23. $-3x + 8 = 2x - 7$

24. $3(x - 4) = 7x - 2$

25. $-7 + 3x - 5 = 8x - 1$

Convert to decimal notation. [5.5a]

26. $\dfrac{4}{9}$

27. $\dfrac{17}{15}$

28. $-\dfrac{5}{8}$

Synthesis

29. ◆ Would a company considering expansion be more interested in interpolation or extrapolation? Why?

30. ◆ Would a bookkeeper who is lacking records from a firm's third year of operation be more interested in interpolation or extrapolation? Why?

31. ◆ Is it possible for the probability of an event occurring to exceed 1? Why or why not?

32. A coin is flipped twice. What is the probability that two heads will occur?

33. A coin is flipped twice. What is the probability that one head and one tail will occur?

34. A die is rolled twice. What is the probability that a ⚁ is rolled twice?

35. A day is chosen randomly during a leap year. What is the probability that the day is in July?

Summary and Review Exercises: Chapter 6

For Exercises 1–5, use the following tables, which list the ideal body weights for men and women over age 25.

DESIRABLE WEIGHT OF MEN			
Height	Small Frame (in pounds)	Medium Frame (in pounds)	Large Frame (in pounds)
5 ft, 7 in.	138	152	166
5 ft, 9 in.	146	160	174
5 ft, 11 in.	154	169	184
6 ft, 1 in.	163	179	194
6 ft, 3 in.	172	188	204

DESIRABLE WEIGHT OF WOMEN			
Height	Small Frame (in pounds)	Medium Frame (in pounds)	Large Frame (in pounds)
5 ft, 1 in.	105	113	122
5 ft, 3 in.	111	120	130
5 ft, 5 in.	118	128	139
5 ft, 7 in.	126	137	147
5 ft, 9 in.	134	144	155

1. What is the ideal weight for a 6 ft, 1 in. man with a medium frame? [6.1a]

2. What is the ideal weight for a 5 ft, 5 in. woman with a small frame? [6.1a]

3. What size woman has an ideal weight of 120 lb? [6.1a]

4. What size man has an ideal weight of 169 lb? [6.1a]

5. Use the information provided to draw a line graph showing the ideal body weight for a large-framed woman. Use height on the horizontal axis and weight, in pounds, on the vertical axis. [6.2d]

This pictograph shows the number of officers in the largest U.S. police forces. Use the pictorgraph for Exercises 6–9.

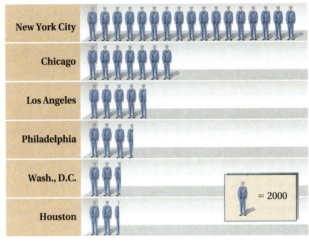

Source: International Association of Chiefs of Police

6. About how many officers are in the Chicago police force? [6.1b]

7. Which city has about 9000 officers on its force? [6.1b]

8. Of the cities listed, which has the smallest police force? [6.1b]

9. Estimate the average size of these six police forces. [6.1b], [6.5a]

The following bar graph shows the number of U.S. households that owned different types of pets in a recent year. Use the graph for Exercises 10–15. [6.2a]

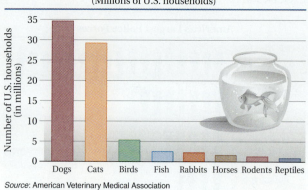

Source: American Veterinary Medical Association

10. About how many U.S. households have pet fish?

11. What type of pet is owned by about 5 million U.S. households?

12. Of the pets listed, which type is owned by the smallest number of U.S. households?

13. About how many more dog owners are there than cat owners?

14. True or false? There are more cat owners than bird, fish, rabbit, horse, rodent, and reptile owners combined.

15. True of false? There are more dog owners than all other pet owners combined.

The following line graph shows the number of accidents per 100 drivers, by age. Use the graph for Exercises 16–21. [6.2c]

16. Which age group has the most accidents per 100 drivers?

17. What is the fewest number of accidents per 100 in any age group?

18. How many more accidents do people over 75 yr of age have than those in the age range of 65–74?

19. Between what ages does the number of accidents stay basically the same?

20. How many fewer accidents do people 25–34 yr of age have than those 20–24 yr of age?

21. Which age group has accidents more than three times as often as people 55–64 yr of age?

Find the coordinates for each point. [6.3a]

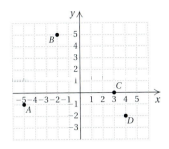

22. A **23.** B

24. C **25.** D

Plot these points using graph paper. [6.3a]

26. (2, 5) **27.** (0, −3) **28.** (−4, −2)

In which quadrant is the point located? [6.3b]

29. (3, −8) **30.** (−20, −14) **31.** $\left(4\frac{9}{10}, 1\frac{3}{10}\right)$

Determine whether the given point is a solution of the equation $2y - x = 10$. [6.3c]

32. (2, −6) **33.** (0, 5)

Graph on a plane. [6.4b]

34. $y = 2x - 5$ **35.** $y = -\frac{3}{4}x$

36. $x + y = 4$ **37.** $y = 3 - 4x$

In Exercises 38–43, find **(a)** the mean, **(b)** the median, and **(c)** any modes that exist. [6.5a, b, c]

38. 26, 51, 34, 26, 43 **39.** 11, 14, 17, 17, 21, 7, 11

40. 500, 25, 470, 190, 470, 280

41. 700, 700, 1900, 2700, 3000

42. $2, $14, $17, $17, $26, $29

43. $30,000, $75,000, $20,000, $25,000

Summary and Review Exercises: Chapter 6

409

44. One summer, a student earned the following amounts over a four-week period: $302, $312, $330, and $298. What was the mean of his weekly earnings? the median? [6.5a, b]

45. The following temperatures were recorded in St. Louis every 4 hr on a certain day in June: 63°, 58°, 66°, 72°, 71°, 67°. What was the mean temperature for that day? [6.5a]

46. To get an A in math, Sasha must average at least 90 on four tests. Scores on her first three tests were 94, 78, and 92. What is the lowest score that she can receive on the last test and still get an A? [6.5a]

47. Use interpolation and the graph in Exercises 16–21 to estimate the number of accidents per 100 drivers that are 22.5 to 27.5 yr old. [6.6a]

A deck of 52 playing cards is thoroughly shuffled and a car is randomly selected. [6.6b]

48. Find the probability that the five of clubs was selected.

49. Find the probability that a red card was selected.

Skill Maintenance

50. Divide: $-405 \div 3$. [2.5a]

51. A cutting board measures $\frac{2}{3}$ m by $\frac{1}{2}$ m. What is its area? [3.4c]

52. Simplify: $-\frac{32}{4}$. [3.5b]

53. Solve: $3.1 + 4x = -2.7$. [5.7a]

Synthesis

54. ◆ Is it possible for the graph of a linear equation to pass through all four quadrants? Why or why not? [6.4b]

55. ◆ Write a statistics problem for which the solution is
$$\frac{21{,}500 + 23{,}800 + 24{,}400 + 27{,}000}{4} = 24{,}175.$$
[6.5a]

56. ▦ Find three solutions and then graph $34x + 47y = 100$. [6.4a, b]

57. ▦ A typing pool consists of four senior typists who earn $12.35 per hour and nine other typists who earn $11.15 per hour. Find the mean hourly wage. [6.5a]

For Exercises 58–61, refer to the table and graph in Exercises 1–5 and 10–15.

58. If the average dog-owning household has 1.2 dogs, how many pet dogs are there in the United States? [6.2a]

59. If the average fish-owning household has 14.3 fish, how many pet fish are there in the United States? [6.2a]

60. Is it possible for a woman's ideal body weight to exceed the ideal body weight for a man of the same height? [6.1a]

61. Is it possible for a woman's ideal body weight to exceed the ideal body weight of a taller man? [6.1a]

Graph on a plane. [6.4b]

62. $1\frac{2}{3}x + \frac{3}{4}y = 2$ **63.** $\frac{3}{4}x - 2\frac{1}{2}y = 3$

Test: Chapter 6

This table lists the number of calories burned during various walking activities. Use it for Questions 1 and 2.

	Calories Burned in 30 Min		
Walking Activity	110 lb	132 lb	154 lb
Walking			
Fitness (5 mph)	183	213	246
Mildly energetic (3.5 mph)	111	132	159
Strolling (2 mph)	69	84	99
Hiking			
3 mph with 20-lb load	210	249	285
3 mph with 10-lb load	195	228	264
3 mph with no load	183	213	246

1. Which activity provides the greatest benefit in burned calories for a person who weighs 132 lb?

2. What is the least strenuous activity you must perform if you weigh 154 lb and you want to burn at least 250 calories every 30 min?

3. Draw a vertical bar graph using an appropriate scale, showing the recent starting salary of college graduates in various fields. Be sure to label the axes properly.

   Accounting: $28,500
   Marketing: $26,500
   Humanities: $24,000
   Computer science: $34,000
   Mathematics: $31,000

Answers

1. _____

2. _____

3. _____

The following pictograph shows the number of hits in 1998 for several major-league baseball players. Use the graph for Questions 4–7.

Number of Hits in 1998 for 5 Professional Players

Tony Gwynn	🍥🍥🍥🍥🍥🍥
Bernie Williams	🍥🍥🍥🍥🍥🍥🍥
Aaron Boone	🍥🍥🍥🍥🍥🍥
Alex Rodriguez	🍥🍥🍥🍥🍥🍥🍥🍥
Tony Fernandez	🍥🍥🍥🍥🍥

🍥 = 25 hits

4. How many hits did Bernie Williams have?

5. Who had the most hits?

6. Who had the fewest hits?

7. Who had 150 hits?

The following line graph shows the revenues of Nike, Inc. Use the graph for Questions 8–13.

Source: Nike, Inc., annual report and The World Almanac 1999.

8. How much revenue was earned in 1996?

9. What was the mean revenue for the six years?

10. How much more revenue was earned in 1997 than in 1992?

11. In which year did revenue increase the most?

12. What was the median revenue for the six years?

13. Use extrapolation to estimate the revenue in 1998.

In which quadrant is each point located?

14. $\left(-\frac{1}{2}, 7\right)$

15. $(-5, -6)$

Find the coordinates of each point.

16. A

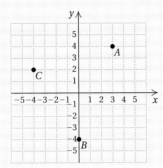

17. B

18. C

19. Determine whether $(2, -4)$ is a solution of the equation $y - 3x = -10$.

Graph.

20. $y = 2x - 1$

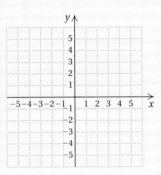

21. $y = -\dfrac{3}{2}x$

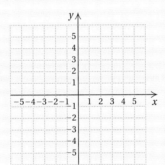

22. $x + 2y = 8$

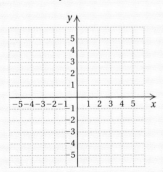

Answers

14. _____

15. _____

16. _____

17. _____

18. _____

19. _____

20. _____

21. _____

22. _____

Test: Chapter 6

413

Find the mean.

23. 45, 49, 52, 54

24. 1, 2, 3, 4, 5

25. 3, 17, 17, 18, 18, 20

Find the median and any modes that exist.

26. 45, 47, 54, 54

27. 1, 2, 3, 4, 5

28. 20, 17, 17, 18, 3, 18

29. Bill drove 754 km in 13 hr. What was the average number of kilometers per hour?

30. To get a C in chemistry, Ernie must score an average of 70 on four tests. Scores on his first three tests were 68, 71, and 65. What is the lowest score that Ernie can receive on the last test and still get a C?

31. A month of the year is randomly selected for a company's party. What is the probability that a month whose name begins with J is chosen?

Skill Maintenance

32. Divide: $\frac{-700}{35}$.

33. Simplify: $\frac{75}{45}$.

34. A recipe for nachos calls for $\frac{3}{4}$ lb of shredded cheese. How much cheese should be used to make $\frac{1}{3}$ of a recipe?

35. Solve: $-9.8 = 5x - 1.7$.

Synthesis

Graph.

36. $\frac{1}{4}x + 3\frac{1}{2}y = 1$

37. $\frac{5}{6}x - 2\frac{1}{3}y = 1$

38. Find the area of a rectangle whose vertices are $(-3, 1)$, $(5, 1)$, $(5, 8)$, and $(-3, 8)$.

Cumulative Review: Chapters 1–6

1. Write expanded notation for 3671.

2. Jonathan pedals 23 mi on each of 5 days. How many miles are bicycled in all?

3. Write standard notation for the number in this sentence: Experts predict the global population to surpass 8 billion by the year 2030.

4. Find the perimeter and the area of the rectangle.

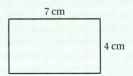

5. Kati poured 129 g of carbon and then 87 g of sodium chloride into a beaker. How many grams of chemicals were poured into the beaker altogether?

6. Write exponential notation: $7 \cdot 7 \cdot 7 \cdot 7$.

7. Tell which integers correspond to this situation: Monique lost 9 lb and Jacques gained 4 lb.

Use either < or > for ▨ to form a true statement.

8. $1 \; \square \; -7$

9. $\dfrac{4}{9} \; \square \; \dfrac{3}{7}$

10. $-4.8 \; \square \; -4.09$

11. Find $-x$ when $x = -5$.

12. Find $-(-x)$ when $x = 17$.

13. Evaluate $2x - y$ for $x = 3$ and $y = 8$.

14. Combine like terms: $6x + 4y - 8x - 3y$.

15. Find all the factors of 36.

16. Determine whether 732 is divisible by 6.

17. Write two different expressions for $\dfrac{-7}{x}$ with negative signs in different places.

18. Multiply: $5(2a - 3b + 1)$.

19. What part is shaded?

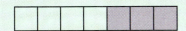

20. Find another name for the given number but with the denominator indicated.

$\dfrac{2}{7} = \dfrac{?}{35}$

Simplify, if possible. Assume that all variables are nonzero.

21. $427 - 398$

22. $17 \cdot 28$

23. $63 \div (-7)$

24. $-32 + (-83)$

25. $\dfrac{3}{7} + \dfrac{2}{7}$

26. $\dfrac{3}{7} \div \dfrac{9}{5}$

27. $\dfrac{5}{6} - \dfrac{1}{9}$

28. $\dfrac{-2}{15} + \dfrac{3}{10}$

29. $\dfrac{8}{11} \cdot \dfrac{11}{8}$

30. $3\dfrac{1}{4} + 5\dfrac{7}{8}$

31. $7\dfrac{2}{3}x - 5\dfrac{1}{4}x$

32. $4\dfrac{1}{5} \cdot 3\dfrac{1}{7}$

33. $39.72 + 43.56$

34. $1334.183 \div 21.4$

35. $17.4(-2.43)$

36. $\dfrac{9a}{9a}$

37. $\dfrac{4x}{1}$

38. $\dfrac{0}{7x}$

Solve.

39. $x + \dfrac{2}{3} = -\dfrac{2}{5}$

40. $\dfrac{3}{8}x + 2 = 14$

41. $3(x - 5) = 7x + 2$

42. In which quadrant is the point $(-4, 9)$ located?

43. Graph: $y = \dfrac{1}{2}x - 4$.

44. Find the mean:
19, 39, 34, 52.

45. Find the median:
7, 9, 12, 35.

46. Find the mode:
43, 56, 56, 43, 49, 49, 49.

47. A marathoner ran 59 km in 3 hr. What was the average number of kilometers per hour?

Synthesis

48. Simplify:
$\left(\dfrac{3}{4}\right)^2 - \dfrac{1}{8} \cdot \left(3 - 1\dfrac{1}{2}\right)^2$.

49. Add and write the answer as a mixed numeral:
$-5\dfrac{42}{100} + \dfrac{355}{100} + \dfrac{89}{10} + \dfrac{17}{1000}$.

50. A square with sides parallel to the axes has the point (2, 3) at its center. Find the coordinates of the square's vertices if each side is 8 units long.

7

Ratio and Proportion

Introduction

The equation shown below is an example of a *proportion*. Fractional expressions appearing in a proportion are examples of *ratios*. In this chapter, we will study ratios and proportions and then use them for problem solving. A topic of interest to consumers, unit pricing, is also examined.

- **7.1** Introduction to Ratios
- **7.2** Rates and Unit Prices
- **7.3** Proportions
- **7.4** Applications of Proportions
- **7.5** Geometric Applications

An Application

A scale model of an addition to an athletic facility is 12 cm wide at the base and rises to a height of 15 cm. If the actual base is to be 116 ft, what will be the actual height of the addition?

This problem appears as Example 5 in Section 7.5.

The Mathematics

We let $h =$ the height of the addition, in feet. Then we translate to a proportion and solve.

Each of these is a ratio.

Width $\longrightarrow \dfrac{12}{15} = \dfrac{116}{h} \longleftarrow$ Width
Height $\longrightarrow$ $\phantom{\dfrac{12}{15}}$ $\phantom{\dfrac{116}{h}}$ $\longleftarrow$ Height

This is a proportion.

World Wide Web   For more information, visit us at www.mathmax.com

Pretest: Chapter 7

1. Write fractional notation for the ratio 31 to 54.

2. In this rectangle, find the ratio of width to length and simplify.

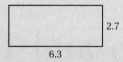

3. Craig's Ford Windstar can travel 336 mi on 14 gal of gasoline. What is the rate in miles per gallon?

4. A 16-oz bottle of Snapple iced tea costs $1.05. Find the unit cost in dollars per ounce. Round to the nearest hundredth of a cent.

5. Which has the lower unit price?

   MINERAL WATER
   Sparkle: 93¢ for 12 oz
   Cleary's: $1.07 for 16 oz

6. Determine whether the pairs 3, 5 and 21, 35 are proportional.

Solve.

7. $\dfrac{6}{5} = \dfrac{27}{x}$

8. $\dfrac{y}{0.25} = \dfrac{0.3}{0.1}$

9. The clock on Sharon's stove loses 5 min in 10 hr. At this rate, how many minutes will it lose in 24 hr?

10. On a California state map, 4 in. represents 225 mi. If two cities are 7 in. apart on the map, how far apart are they in reality?

11. The figures below represent two similar polygons. Find the missing lengths.

    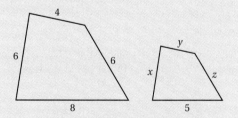

12. How high is a steeple that casts a 97.5-ft shadow at the same time that an 8-ft road sign casts a 13-ft shadow?

Chapter 7   Ratio and Proportion

7.1 Introduction to Ratios

a Ratio

> A ratio is the quotient of two quantities.

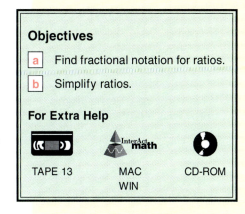

Objectives

a Find fractional notation for ratios.
b Simplify ratios.

For Extra Help

TAPE 13   MAC WIN   CD-ROM

For every 26 lb of waste produced in the United States, about 7 lb are recycled. The *ratio* of the amount of waste recycled to the amount of waste produced is shown by the fractional notation

$\frac{7}{26}$,   or by the notation $7:26$.

We can read such notation as "the ratio of 7 to 26," listing the numerator first and the denominator second. Note that a ratio indicates how important, or sizable, one quantity is in comparison to a second quantity.

1. Write fractional notation for the ratio of 5 to 11.

> The ratio of a to b is written $\frac{a}{b}$, or $a:b$.

2. Write fractional notation for the ratio of 57.3 to 86.1.

For most of our work, we will use fractional notation for ratios.

Example 1 Write fractional notation for the ratio of 7 to 8.

The ratio is $\frac{7}{8}$.

3. Write fractional notation for the ratio of $6\frac{3}{4}$ to $7\frac{2}{5}$.

Example 2 Write fractional notation for the ratio of 31.4 to 100.

The ratio is $\frac{31.4}{100}$.

Example 3 Write fractional notation for the ratio of $4\frac{2}{3}$ to $5\frac{7}{8}$.

The ratio is $\frac{4\frac{2}{3}}{5\frac{7}{8}}$.

4. *Household Economics.* A family earning $28,500 per year will spend abut $7410 for food. What is the ratio of food expenses to yearly income?

Do Exercises 1–3.

Example 4 *Drive-Through Fast Food.* For every 6 Americans who use a fast-food restaurant's drive-through window, 4 others order indoors. What is the ratio of drive-through orders to indoor orders?

The ratio is $\frac{6}{4}$.

5. *Beverage Consumption.* The average American drinks 182.5 gal of liquid each year. Of this, 21.1 gal is milk. Find the ratio of milk consumed to total amount of liquid consumed.

Example 5 *Car Expenses.* A family earning $42,800 per year allots about $6420 for car expenses. What is the ratio of car expenses to yearly income?

The ratio is $\frac{6420}{42,800}$.

Answers on page A-20

7.1 Introduction to Ratios

419

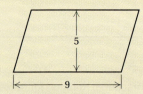

In the parallelogram above:

6. What is the ratio of the height to the length of the base?

7. What is the ratio of the length of the base to the height?

8. *Commuting by Car.* According to an Eno Transportation Foundation study, 73 of every 100 American workers drive to work alone. Find the ratio of single drivers to those who do not drive to work alone.

9. In Example 8, what is the ratio of the length of the shortest side of the television screen to the length of the longest side?

Write the ratio of the two given numbers. Then simplify each to find two other numbers in the same ratio.

10. 18 to 27

11. 3.6 to 12

12. 1.2 to 1.5

Answers on page A-20

Chapter 7 Ratio and Proportion

420

Example 6 In the triangle at right:

a) What is the ratio of the length of the longest side to the length of the shortest side?

The ratio is $\dfrac{5}{3}$.

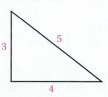

b) What is the ratio of the length of the shortest side to the length of the longest side?

The ratio is $\dfrac{3}{5}$.

Do Exercises 4–7. (Exercises 4 and 5 are on the preceding page.)

Example 7 *Packaging.* For every dollar spent on food, about 13 cents goes to pay for the package. What is the ratio of the cost of the package to the cost of the package's contents?

If 13 cents of each dollar pays for packaging, then $100 - 13 = 87$ cents of each dollar pays for the package's contents. Thus the ratio of the cost of the package to the cost of the contents is

$$\dfrac{13}{87}.$$

Do Exercise 8.

b Simplifying Notation for Ratios

Sometimes a ratio can be simplified. This provides a means of finding other numbers with the same ratio.

Example 8 Most television screens have the same ratio of length to width. Emilio's television has a screen that is 20 in. long and 15 in. wide. What is the ratio of length to width?

We write the ratio in fractional notation and then simplify:

$$\dfrac{20}{15} = \dfrac{5 \cdot 4}{5 \cdot 3} = \dfrac{5}{5} \cdot \dfrac{4}{3} = \dfrac{4}{3}.$$

Thus we can say that the ratio of length to width is 4 to 3.

Example 9 Write the ratio of 2.4 to 9.2. Then simplify and find two other numbers in the same ratio.

We first write the ratio. Next, we multiply by $\dfrac{10}{10}$, or 1, to clear the decimals from the numerator and the denominator. Then we simplify:

$$\dfrac{2.4}{9.2} = \dfrac{2.4}{9.2} \cdot \dfrac{10}{10} = \dfrac{24}{92} = \dfrac{4 \cdot 6}{4 \cdot 23} = \dfrac{4}{4} \cdot \dfrac{6}{23} = \dfrac{6}{23}.$$

We can say that 2.4 is to 9.2 as 6 is to 23.

Do Exercises 9–12.

Exercise Set 7.1

a Write fractional notation for each ratio. You need not simplify.

1. 4 to 5

2. 3 to 2

3. 178 to 572

4. 329 to 967

5. 0.4 to 12

6. 2.3 to 22

7. 3.8 to 7.4

8. 0.6 to 0.7

9. 56.78 to 98.35

10. 456.2 to 333.1

11. $8\frac{3}{4}$ to $9\frac{5}{6}$

12. $10\frac{1}{2}$ to $43\frac{1}{4}$

13. One person in four plays a musical instrument. In a typical group of people, what is the ratio of those who play an instrument to the total number of people? What is the ratio of those who do not play an instrument to those who do?

14. Of the 365 days in each year, it takes 107 days of work for the average person to pay his or her taxes. What is the ratio of days worked for taxes to total number of days in a year?

15. *Corvette Accidents.* Of every 5 fatal accidents involving a Corvette, 4 do not involve another vehicle (**Source:** *Harper's Magazine*). Find the ratio of fatal accidents involving just a Corvette to those involving a Corvette and at least one other vehicle.

16. *New York Commuters.* Of every 5 people who commute to work in New York City, 2 spend more than 90 min a day commuting (**Source:** *The Amicus Journal*). Find the ratio of people whose daily commute to New York exceeds 90 min a day to those whose commute is 90 min or less.

17. In this rectangle, find the ratios of length to width and of width to length.

18. In this right triangle, find the ratios of shortest length to longest length and of longest length to shortest length.

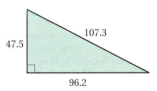

b Simplify each ratio.

19. 4 to 6

20. 6 to 10

21. 18 to 24

22. 28 to 36

23. 4.8 to 10

24. 5.6 to 10

25. 2.8 to 3.6

26. 4.8 to 6.4

27. 20 to 30

28. 40 to 60

29. 56 to 100

30. 42 to 100

31. 128 to 256   **32.** 232 to 116   **33.** 0.48 to 0.64   **34.** 0.32 to 0.96

35. The ratio of females to males worldwide is 51 to 49. Write this ratio in simplified fractional form.

36. The ratio of Americans aged 18–24 living with their parents to all Americans aged 18–24 is 54 to 100. Write this ratio in simplified fractional form.

37. In this right triangle, find the ratio of shortest length to longest length and simplify.

38. In this rectangle, find the ratio of width to length and simplify.

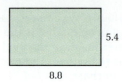

Skill Maintenance

Use < or > for ▩ to write a true sentence. [4.2c]

39. $-\dfrac{5}{6}$ ▩ $-\dfrac{3}{4}$ **40.** $\dfrac{12}{8}$ ▩ $\dfrac{7}{4}$ **41.** $\dfrac{5}{9}$ ▩ $\dfrac{6}{11}$ **42.** $-\dfrac{3}{4}$ ▩ $-\dfrac{2}{3}$

Solve. [4.6c]

43. Rocky is $187\tfrac{1}{10}$ cm tall and his daughter is $180\tfrac{3}{4}$ cm tall. How much taller is Rocky?

44. Aunt Louise is $168\tfrac{1}{4}$ cm tall and her son is $150\tfrac{7}{10}$ cm tall. How much taller is Aunt Louise?

Synthesis

45. ◆ Is it possible for the ratio of the lengths of the two legs of a right triangle to be 1 : 5? Why or why not?

46. ◆ Can every ratio be written as the ratio of some number to 1? Why or why not?

47. ◆ What can be concluded about the width of a rectangle if the ratio of length to perimeter is 1 to 3? Make some sketches and explain your reasoning.

48. ▦ In 1996, the total payroll of major league baseball teams was $937,905,284. The New York Yankees won the World Series that year. Their payroll was the highest at $61,511,870. Find the ratio in decimal notation of the Yankees payroll to the overall payroll.

49. ▦ See Exercise 48. In 1995, the total payroll of major league baseball teams was $927,334,416. Find the ratio of the payroll in 1996 to the payroll in 1995.

50. Write simplified fractional form for the ratio of $3\tfrac{3}{4}$ to $5\tfrac{7}{8}$.

Exercises 51 and 52 refer to a common fertilizer known as "5, 10, 15." This mixture contains 5 parts of potassium for every 10 parts of phosphorus and 15 parts of nitrogen (this is often denoted 5 : 10 : 15).

51. Find the ratio of potassium to nitrogen and of nitrogen to phosphorus.

52. Simplify the ratio 5 : 10 : 15.

Analyze the ratios of different colored M&M candies.

7.2 Rates and Unit Prices

a Rates

When a ratio is used to compare two different kinds of measure, we call it a **rate**. Suppose that a car is driven 200 km in 4 hr. The ratio

$$\frac{200 \text{ km}}{4 \text{ hr}}, \quad \text{or } 50\frac{\text{km}}{\text{hr}}, \quad \text{or } 50 \text{ kilometers per hour}, \quad \text{or } 50 \text{ km/h}$$

> Recall that "per" means "division," or "for each."

is the rate of travel in kilometers per hour, which is the division of the number of kilometers by the number of hours. A ratio of distance traveled to time is also called **speed**.

Example 1 Pierre's moped travels 145 km on 2.5 L of gas. What is the rate in kilometers per liter?

The rate is $\frac{145 \text{ km}}{2.5 \text{ L}}$, or $58\frac{\text{km}}{\text{L}}$.

Example 2 It takes 60 oz of grass seed to seed 3000 sq ft of lawn. What is the rate in ounces per square foot?

The rate is $\frac{60 \text{ oz}}{3000 \text{ sq ft}} = \frac{1}{50}\frac{\text{oz}}{\text{sq ft}}$, or $0.02\frac{\text{oz}}{\text{sq ft}}$.

Example 3 A cook buys 10 lb of potatoes for $3.69. What is the rate in cents per pound?

The rate is $\frac{\$3.69}{10 \text{ lb}} = \frac{369 \text{ cents}}{10 \text{ lb}}$, or $36.9\frac{\text{cents}}{\text{lb}}$.

Example 4 A student nurse working in a health center earned $3690 for working 3 months one summer. What was the rate of pay per month?

The rate of pay is the ratio of money earned per length of time worked, or

$$\frac{\$3690}{3 \text{ mo}} = 1230\frac{\text{dollars}}{\text{month}}, \quad \text{or } \$1230 \text{ per month}.$$

Example 5 *At-Bats to Home-Run Ratio.* At one point during the 1998 baseball season, slugger Mark McGwire had hit 50 home runs in 390 at-bats (times at bat). Find his at-bats per home-run rate.*

His rate was $\frac{390 \text{ at-bats}}{50 \text{ home runs}} = 7.8\frac{\text{at-bats}}{\text{home runs}}$.

Do Exercises 1–8.

---

*See also Exercise 53 on p. 428.

Objectives

a Give the ratio of two different kinds of measure as a rate.

b Find unit prices and use them to determine which of two possible purchases has the lower unit price.

For Extra Help

TAPE 13   MAC WIN   CD-ROM

What is the rate, or speed, in miles per hour?

1. 45 mi, 9 hr

2. 120 mi, 10 hr

3. 3 mi, 10 hr

What is the rate, or speed, in feet per second?

4. 2200 ft, 2 sec

5. 52 ft, 13 sec

6. 232 ft, 16 sec

7. A well-hit golf ball can travel 500 ft in 2 sec. What is the rate, or speed, of the golf ball in feet per second?

8. A leaky faucet can lose 14 gal of water in a week. What is the rate in gallons per day?

Answers on page A-20

9. Kate bought a 14-oz package of cereal for $2.89. What is the unit price in cents per ounce? Round to the nearest hundredth of a cent.

10. Which has the lower unit price? [*Note*: 1 qt = 32 fl oz (fluid ounces).]

Answers on page A-20

Chapter 7   Ratio and Proportion

424

b Unit Pricing

> A **unit price** or **unit rate** is the ratio of price to the number of units.

By carrying out the division indicated by the ratio, we can find the price per unit.

Example 6 Ruby bought a 40-lb bag of Nutro™ dog food for $34. What is the unit price in dollars per pound?

The unit price is the price in dollars for each pound.

$$\text{Unit price} = \frac{\text{Price}}{\text{Number of units}}$$

$$= \frac{\$34}{40 \text{ lb}} = \frac{34}{40} \cdot \frac{\$}{\text{lb}}$$

$$= 0.85 \text{ dollar per pound}$$

Do Exercise 9.

For comparison shopping, it helps to find unit prices. Make sure that the same units are being used in all items being compared.

Example 7 Which has the lower unit price?

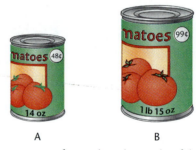

To find out, we compare the unit prices—in this case, the price per ounce.

For can A: $\frac{48 \text{ cents}}{14 \text{ oz}} \approx 3.429 \frac{\text{cents}}{\text{oz}}$.

For can B: We need to find the total number of ounces:

1 lb, 15 oz = 16 oz + 15 oz = 31 oz.

Then

$\frac{99 \text{ cents}}{31 \text{ oz}} \approx 3.194 \frac{\text{cents}}{\text{oz}}$.

Thus can B has the lower unit price.

In supermarkets, unit prices are usually listed below the items on the shelves.

Do Exercise 10.

Exercise Set 7.2

a In Exercises 1–6, find the rate as a ratio of distance to time. Use the units given.

1. 120 km, 3 hr

2. 18 mi, 9 hr

3. 440 m, 40 sec

4. 200 mi, 25 sec

5. 342 yd, 2.25 days

6. 492 m, 60 sec

7. A delivery van is driven 500 mi in 20 hr. What is the rate in miles per hour? in hours per mile?

8. Franny eats 3 hamburgers in 15 min. What is the rate in hamburgers per minute? in minutes per hamburger?

9. A long-distance telephone call between two cities costs $5.75 for 10 min. What is the rate in cents per minute?

10. An 8-lb boneless ham contains 36 servings of meat. What is the ratio in servings per pound?

11. A thoroughly watered lawn requires 623 gal of water for every 1000 ft^2. What is the rate in gallons per square foot?

12. A limousine is driven 200 km on 40 L of gasoline. What is the rate in kilometers per liter?

13. Sound travels 66,000 ft in 1 min. What is its rate, or speed, in feet per second?

14. Light travels 11,160,000 mi in 1 min. What is its rate, or speed, in miles per second?

15. Impulses in nerve fibers can travel 310 km in 2.5 hr. What is the rate, or speed, in kilometers per hour?

16. A black racer snake can travel 4.6 km in 2 hr. What is its rate, or speed, in kilometers per hour?

17. A jet flew 2660 mi in 4.75 hr. What was its speed?

18. A turtle traveled 0.42 mi in 2.5 hr. What was its speed?

19. The fabric for a wedding gown costs $165.75 for 8.5 yd. Find the unit price.

20. An 8-oz tube of toothpaste costs $2.59. Find the unit price.

21. A 2-lb can of coffee costs $6.59. What is the unit price in cents per ounce? Round to the nearest hundredth of a cent.

22. A 24-can package of 12-oz cans of soda is on sale for $6.99. What is the unit price in cents per ounce? Round to the nearest hundredth of a cent.

23. A $\frac{2}{3}$-lb package of Monterey Jack cheese costs $2.89. Find the unit price in dollars per pound. Round to the nearest hundredth of a dollar.

24. A $1\frac{1}{4}$-lb container of cottage cheese costs $1.62. Find the unit price in dollars per pound. Round to the nearest hundredth of a dollar.

Which has the lower unit price?

25.

SALSA	
Tico's:	18 oz for $3.79
Sure Fire:	16 oz for $3.49

26.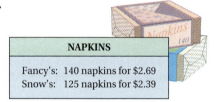

NAPKINS	
Fancy's:	140 napkins for $2.69
Snow's:	125 napkins for $2.39

27.

GRAPEFRUIT JUICE	
Sunbeam:	$2.79 for 2 qt
Dell's:	$2.09 for 48 oz

28.

EVAPORATED MILK	
Meyer's:	79 cents for 12 oz
Barton's:	$2.69 for 1 qt, 8 oz

Chapter 7 Ratio and Proportion

29.

SOAP

Shine: $2.19 for 3 bars
Pristine: $1.58 for 2 bars

30.

BROCCOLI SOUP

Big House: 9.25 oz for 96 cents
Chet's: 10.75 oz for $1.11

31.

FANCY TUNA

Tina's: $1.19 for $6\frac{1}{8}$ oz
Big Net: $1.11 for 6 oz

32.

FLOUR

Nift Sift: $1.25 for 3 lb, 2 oz
Marlo's: $2.05 for 5 lb

33.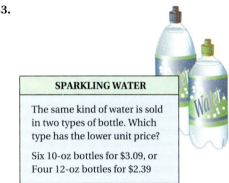

SPARKLING WATER

The same kind of water is sold in two types of bottle. Which type has the lower unit price?

Six 10-oz bottles for $3.09, or
Four 12-oz bottles for $2.39

34.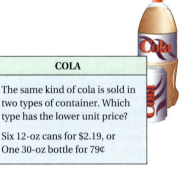

COLA

The same kind of cola is sold in two types of container. Which type has the lower unit price?

Six 12-oz cans for $2.19, or
One 30-oz bottle for 79¢

35.

INSTANT PUDDING (SAME BRAND)

Package A: 5.9 oz for 99¢
Package B: 3.9 oz for 63¢

36.

TACO SHELLS (SAME BRAND)

Family Pack: 18 in a box for $2.49
Regular: 12 in a box for $1.99

37.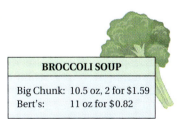

BROCCOLI SOUP

Big Chunk: 10.5 oz, 2 for $1.59
Bert's: 11 oz for $0.82

38.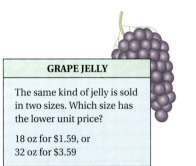

GRAPE JELLY

The same kind of jelly is sold in two sizes. Which size has the lower unit price?

18 oz for $1.59, or
32 oz for $3.59

Exercise Set 7.2

Skill Maintenance

Solve.

39. There are 20.6 million people in this country who play the piano and 18.9 million who play the guitar. How many more play the piano than the guitar? [5.8a]

40. A serving of fish steak (cross section) is generally $\frac{1}{2}$ lb. How many servings can be prepared from a cleaned $12\frac{3}{4}$-lb salmon? [4.7d]

Multiply. [5.3a]

41. 4 5.6 7
   ×   2.4

42. 6 7 8.1 9
   ×   1 0 0

In which quadrant is each point located? [6.3b]

43. (3, −8)

44. (−5, 4.1)

Synthesis

45. ◆ The unit price of an item generally drops when larger packages of that item are purchased. Why?

46. ◆ Suppose that the same type of juice is available in two sizes and that the larger bottle has the lower unit price. If the larger bottle costs $3.79 and contains twice as much juice, what can you conclude about the price of the smaller bottle? Why?

47. ◆ Manufacturers of laundry detergent sometimes charge a higher unit price for their larger packages. Why do you think they do so?

48. Recently, certain manufacturers have been changing the size of their containers in such a way that the consumer thinks the price of a product has been lowered when, in reality, a higher unit price is being charged.

a) Some aluminum juice cans are now concave (curved in) on the bottom. Suppose the volume of the can in the figure has been reduced from a fluid capacity of 6 oz to 5.5 oz, and the price of each can has been reduced from 65¢ to 60¢. Find the unit price of each container in cents per ounce.

b) Suppose that at one time the cost of a certain kind of paper towel was $0.89 for a roll containing 78 ft² of absorbent surface. Later the surface area was changed to 65 ft² and the price was decreased to $0.79. Find the unit price of each product in cents per square foot.

49. In 1994, Coca-Cola introduced a 20-oz soda bottle. At first it was sold for 64¢ a bottle, the same price as their 16-oz bottle. After about a month, the price of a 20-oz bottle rose to 80¢. How did the unit price change for a consumer who made the switch from the 16-oz to the 20-oz bottle?

50. Suppose that a pasta manufacturer shrinks the size of a box from 1 lb to 14 oz, but keeps the price at 85 cents a box. How does the unit price change?

51. ▦ Use the formula for the area of a circle, $A = \pi r^2$, to determine which is a better deal: a 14-in. pizza for $10.50 or a 16-in. pizza for $11.95. Use 3.14 for π.

52. ▦ Suppose that, 25 mi from where you're standing, a bolt of lightning splits a tree. How long will it take for you to hear the accompanying crack of thunder? How long will it take for you to see the flash of light? (*Hint*: Use the information in Exercises 13 and 14.)

53. ▦ In 1998, Mark McGwire set a major-league record, hitting 70 home runs in 509 at-bats. Find his at-bats per home run rate.

7.3 Proportions

a Proportion

When two pairs of numbers (such as 3, 2 and 6, 4) have the same ratio, we say that they are **proportional**. The equation

$$\frac{3}{2} = \frac{6}{4}$$

states that the pairs 3, 2 and 6, 4 are proportional. Such an equation is called a **proportion**. We sometimes read $\frac{3}{2} = \frac{6}{4}$ as "3 is to 2 as 6 is to 4." Because proportions arise frequently and in many fields of study, being able to solve them is an extremely useful skill. Several important applications of proportions are considered in Section 7.4.

Checking to see whether two pairs of numbers are proportional is the same as checking to see whether two fractions are equal. To develop a quick way of doing this, consider a/b and c/d and assume that neither b nor d is 0.

Note that if

$$\frac{a}{b} = \frac{c}{d}$$

is true, then (using the multiplication principle)

$$bd \cdot \frac{a}{b} = bd \cdot \frac{c}{d} \qquad \text{Multiplying by } bd \text{ on both sides}$$

is also true. Simplifying, we have

$$da = bc.$$

If neither b nor d is 0, the above equations are equivalent. Thus any time that $da = bc$ (with $d \neq 0$ and $b \neq 0$), it follows that

$$\frac{a}{b} = \frac{c}{d}.$$

> **A TEST FOR EQUALITY**
>
> We multiply these two numbers: $3 \cdot 4$.   We multiply these two numbers: $2 \cdot 6$.
>
> $$\frac{3}{2} \stackrel{?}{=} \frac{6}{4}$$
>
> Since $3 \cdot 4 = 2 \cdot 6$, we know that
>
> $$\frac{3}{2} = \frac{6}{4}.$$
>
> We call $3 \cdot 4$ and $2 \cdot 6$ *cross products*.

Example 1 Determine whether 1, 2 and 3, 6 are proportional.

We can use cross products to check an equivalent equation:

$$1 \cdot 6 = 6 \qquad \frac{1}{2} \stackrel{?}{=} \frac{3}{6} \qquad 2 \cdot 3 = 6$$

$$1 \cdot 6 \stackrel{?}{=} 2 \cdot 3$$

$$6 = 6.$$

Since this last equation is true, we know that the first equation is also true and the numbers 1, 2 and 3, 6 are proportional.

Objectives

a Determine whether two pairs of numbers are proportional.

b Solve proportions.

For Extra Help

TAPE 13   MAC WIN   CD-ROM

Determine whether the two pairs of numbers are proportional.

1. 3, 4 and 6, 8

2. 1, 4 and 10, 39

3. 1, 2 and 20, 39

Example 2 Determine whether 2, 5 and 4, 7 are proportional.

We can check an equivalent equation using cross products:

$2 \cdot 7 = 14 \qquad \dfrac{2}{5} \stackrel{?}{=} \dfrac{4}{7} \qquad 5 \cdot 4 = 20$

$2 \cdot 7 \stackrel{?}{=} 5 \cdot 4$

$14 \neq 20.$

Since $14 \neq 20$, we know that $\dfrac{2}{5} \neq \dfrac{4}{7}$, so 2, 5 and 4, 7 are not proportional.

Do Exercises 1–3.

Example 3 Determine whether $1\dfrac{1}{4}, \dfrac{1}{2}$ and $1, \dfrac{2}{5}$ are proportional.

We can use cross products:

$1\dfrac{1}{4} \cdot \dfrac{2}{5} = \dfrac{5}{4} \cdot \dfrac{2}{5} \qquad \dfrac{1\frac{1}{4}}{\frac{1}{2}} \stackrel{?}{=} \dfrac{1}{\frac{2}{5}} \qquad \dfrac{1}{2} \cdot 1 = \dfrac{1}{2}$

$\dfrac{5}{4} \cdot \dfrac{2}{5} \stackrel{?}{=} \dfrac{1}{2}$

$\dfrac{10}{20} = \dfrac{1}{2}.$

Since $\dfrac{10}{20} = \dfrac{1}{2}$, we know that

$\dfrac{1\frac{1}{4}}{\frac{1}{2}} = \dfrac{1}{\frac{2}{5}}.$

The numbers $1\dfrac{1}{4}, \dfrac{1}{2}$ and $1, \dfrac{2}{5}$ are proportional.

Do Exercises 4 and 5.

Determine whether the two pairs of numbers are proportional.

4. $4\dfrac{2}{3}, 5\dfrac{1}{2}$ and $14, 16\dfrac{1}{2}$

5. 7.4, 6.8 and 4.2, 3.6

b   Solving Proportions

Often one of the four numbers in a proportion is unknown. Cross products can be used to find the missing number and "solve" the proportion.

Example 4 Solve: $\dfrac{x}{8} = \dfrac{3}{5}$.

We form an equivalent equation by equating cross products. Then we solve for x.

$5 \cdot x = 8 \cdot 3$ Equating cross products

$\dfrac{5x}{5} = \dfrac{24}{5}$ Dividing by 5 on both sides

$x = \dfrac{24}{5}$ Simplifying

$= 4.8.$ Dividing

To check that 4.8 is the solution, we replace x with 4.8 and use cross products:

$4.8 \cdot 5 = 24 \qquad \dfrac{4.8}{8} \stackrel{?}{=} \dfrac{3}{5} \qquad 8 \cdot 3 = 24.$ The cross products are the same.

Answers on page A-20

Since the cross products are the same, it follows that $\frac{4.8}{8} = \frac{3}{5}$. Thus, 4.8, 8 and 3, 5 are proportional, and 4.8 is the solution of the equation.

> To solve $\frac{a}{b} = \frac{c}{d}$ for a specific variable, equate cross products and then divide on both sides to get that variable alone. (Assume $b, d \neq 0$.)

Do Exercise 6.

6. Solve: $\frac{x}{63} = \frac{2}{9}$.

Example 5 Solve: $\frac{x}{7} = \frac{5}{3}$. Write a mixed numeral for the answer.

We have

$$\frac{x}{7} = \frac{5}{3}$$

$3 \cdot x = 7 \cdot 5$ Equating cross products

$\frac{3x}{3} = \frac{35}{3}$ Dividing by 3 on both sides

$x = \frac{35}{3}$, or $11\frac{2}{3}$.

The solution is $11\frac{2}{3}$.

Do Exercise 7.

7. Solve: $\frac{x}{9} = \frac{5}{4}$.

Example 6 Solve: $\frac{7.7}{15.4} = \frac{y}{2.2}$. Write decimal notation for the answer.

We have

$$\frac{7.7}{15.4} = \frac{y}{2.2}$$

$(7.7)(2.2) = 15.4y$ Equating cross products

$\frac{(7.7)(2.2)}{15.4} = \frac{15.4y}{15.4}$ Dividing by 15.4 on both sides

$\frac{16.94}{15.4} = y$ Simplifying

$1.1 = y$.   Dividing:
$$\begin{array}{r} 1.1 \\ 15.4\overline{)16.9\,4} \\ \underline{1\,5\,4\,0} \\ 1\,5\,4 \\ \underline{1\,5\,4} \\ 0 \end{array}$$

The solution is 1.1.

Do Exercise 8.

8. Solve: $\frac{21}{5} = \frac{n}{2.5}$.

Answers on page A-20

7.3 Proportions

431

9. Solve: $\dfrac{2}{3} = \dfrac{6}{x}$.

Example 7 Solve: $\dfrac{3}{x} = \dfrac{6}{4}$.

We have

$$\dfrac{3}{x} = \dfrac{6}{4}$$

$3 \cdot 4 = x \cdot 6$ **Equating cross products**

$\dfrac{12}{6} = \dfrac{6x}{6}$ **Dividing by 6 on both sides**

$2 = x.$ **Simplifying**

The solution is 2.

Do Exercise 9.

Example 8 Solve: $\dfrac{3.4}{4.93} = \dfrac{10}{n}$.

We have

$$\dfrac{3.4}{4.93} = \dfrac{10}{n}$$

$(n)(3.4) = (4.93)(10)$ **Equating cross products**

$\dfrac{3.4n}{3.4} = \dfrac{(4.93)(10)}{3.4}$ **Dividing by 3.4 on both sides**

$n = \dfrac{49.3}{3.4}$ **Multiplying**

$= 14.5.$ **Dividing:**

```
        1 4.5
 3.4 ) 4 9.3₀0
        3 4 0 0
        1 5 3 0
        1 3 6 0
          1 7 0
          1 7 0
              0
```

10. Solve: $\dfrac{0.4}{0.9} = \dfrac{4.8}{t}$.

The solution is 14.5.

Do Exercise 10.

Answer on page A-20

Exercise Set 7.3

a Determine whether the two pairs of numbers are proportional.

1. 5, 6 and 7, 9
2. 7, 5 and 6, 4
3. 1, 2 and 10, 20
4. 7, 3 and 21, 9

5. 2.4, 3.6 and 1.8, 2.7
6. 4.5, 3.8 and 6.7, 5.2

7. $5\frac{1}{3}, 8\frac{1}{4}$ and $2\frac{1}{5}, 9\frac{1}{2}$
8. $2\frac{1}{3}, 3\frac{1}{2}$ and 14, 21

b Solve.

9. $\dfrac{18}{4} = \dfrac{x}{10}$
10. $\dfrac{x}{45} = \dfrac{20}{25}$
11. $\dfrac{x}{8} = \dfrac{9}{6}$
12. $\dfrac{8}{10} = \dfrac{n}{5}$

13. $\dfrac{t}{12} = \dfrac{5}{6}$
14. $\dfrac{12}{4} = \dfrac{x}{3}$
15. $\dfrac{2}{5} = \dfrac{8}{n}$
16. $\dfrac{10}{6} = \dfrac{5}{x}$

17. $\dfrac{n}{15} = \dfrac{10}{30}$
18. $\dfrac{2}{24} = \dfrac{x}{36}$
19. $\dfrac{16}{12} = \dfrac{24}{x}$
20. $\dfrac{7}{11} = \dfrac{2}{x}$

21. $\dfrac{6}{11} = \dfrac{12}{x}$
22. $\dfrac{8}{9} = \dfrac{32}{n}$
23. $\dfrac{20}{7} = \dfrac{80}{x}$
24. $\dfrac{5}{x} = \dfrac{4}{10}$

25. $\dfrac{12}{9} = \dfrac{x}{7}$
26. $\dfrac{x}{20} = \dfrac{16}{15}$
27. $\dfrac{x}{13} = \dfrac{2}{9}$
28. $\dfrac{1.2}{4} = \dfrac{x}{9}$

29. $\dfrac{t}{0.16} = \dfrac{0.15}{0.40}$
30. $\dfrac{x}{11} = \dfrac{7.1}{2}$
31. $\dfrac{100}{25} = \dfrac{20}{n}$
32. $\dfrac{35}{125} = \dfrac{7}{m}$

33. $\dfrac{7}{\frac{1}{4}} = \dfrac{28}{x}$ 　　34. $\dfrac{x}{6} = \dfrac{1}{6}$ 　　35. $\dfrac{\frac{1}{4}}{\frac{1}{2}} = \dfrac{\frac{1}{2}}{x}$ 　　36. $\dfrac{1}{7} = \dfrac{x}{4\frac{1}{2}}$

37. $\dfrac{x}{\frac{4}{5}} = \dfrac{0}{\frac{9}{11}}$ 　　38. $\dfrac{\frac{2}{7}}{\frac{3}{4}} = \dfrac{\frac{5}{6}}{y}$ 　　39. $\dfrac{2\frac{1}{2}}{3\frac{1}{3}} = \dfrac{x}{4\frac{1}{4}}$ 　　40. $\dfrac{5\frac{1}{5}}{6\frac{1}{6}} = \dfrac{y}{3\frac{1}{2}}$

41. $\dfrac{1.28}{3.76} = \dfrac{4.28}{y}$ 　　42. $\dfrac{10.4}{12.4} = \dfrac{6.76}{t}$ 　　43. $\dfrac{10\frac{3}{8}}{12\frac{2}{3}} = \dfrac{5\frac{3}{4}}{y}$ 　　44. $\dfrac{12\frac{7}{8}}{20\frac{3}{4}} = \dfrac{5\frac{2}{3}}{y}$

Skill Maintenance

In which quadrant is each point located? [6.3b]

45. $(-3.2, -5.7)$ 　　46. $\left(-7\frac{1}{2}, 13\right)$ 　　47. $(9, -0.1)$ 　　48. $(19, 57)$

Divide. Write decimal notation for the answer. [5.4a]

49. $260 \div (-5)$ 　　50. $395 \div (-20)$ 　　51. $4648 \div 16$ 　　52. $3427 \div 2.25$

Synthesis

53. ◆ Instead of equating cross products, a student solves $\frac{x}{7} = \frac{5}{3}$ (see Example 5) by multiplying on both sides by the least common denominator, 21. Is the student's approach a good one? Why or why not?

54. ◆ An instructor predicts that a student's test grade will be proportional to the amount of time the student spends studying. What is meant by this? Write an example of a proportion that involves the grades of two students and their study times.

55. ◆ Joaquin argues that $\frac{0}{0}$ is equal to $\frac{3}{4}$ because $0 \cdot 3 = 4 \cdot 0$. Is he correct? Why or why not?

Solve.

56. 🖩 $\dfrac{1728}{5643} = \dfrac{836.4}{x}$ 　　57. 🖩 $\dfrac{328.56}{627.48} = \dfrac{y}{127.66}$

58. $\dfrac{x}{4} = \dfrac{x-1}{6}$ 　　59. $\dfrac{x+3}{5} = \dfrac{x}{7}$

60. Show using a sequence of steps—each of which can be justified—that for $a, b, c, d, \neq 0$,

$\dfrac{a}{b} = \dfrac{c}{d}$ is equivalent to $\dfrac{d}{b} = \dfrac{c}{a}$.

Chapter 7　Ratio and Proportion

7.4 Applications of Proportions

a Problem Solving

Proportions have applications in business, the natural and social sciences, home economics, and many areas of daily life.

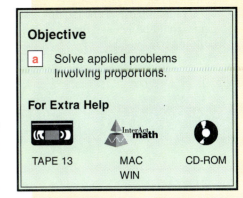

Objective

a Solve applied problems involving proportions.

For Extra Help

TAPE 13   MAC WIN   CD-ROM

Example 1 *Calories Burned.* Heather's stairmaster tells her that if she exercises for 24 min, she will burn 356 calories. How many calories will she burn if she exercises for 30 min?

We let n = the number of calories that Heather will burn in 30 min. Then we translate to a proportion in which each side is the ratio of the number of calories burned to the number of minutes spent exercising.

Calories burned in 30 min → $\dfrac{n}{30} = \dfrac{356}{24}$ ← Calories burned in 24 min
Time →   ← Time

Each side of the equation represents the same ratio. It may help to verbalize the proportion as "the unknown number of calories is to 30 min, as 356 calories is to 24 min."

Solve: $24 \cdot n = 30 \cdot 356$ Equating cross products

A calculator can shorten these steps
$$\dfrac{24n}{24} = \dfrac{30 \cdot 356}{24}$$ Dividing by 24 on both sides

$$n = \dfrac{6 \cdot 5 \cdot 4 \cdot 89}{6 \cdot 4}$$ Factoring

$$= 5 \cdot 89$$ Removing a factor equal to 1: $\dfrac{6 \cdot 4}{6 \cdot 4} = 1$

$$= 445$$

Heather will burn 445 calories in 30 min.

Do Exercise 1.

Proportion problems can be solved in more than one way. For Example 1, any one of the following equations could have been used:

$$\dfrac{356}{24} = \dfrac{n}{30}, \quad \dfrac{30}{n} = \dfrac{24}{356}, \quad \dfrac{30}{24} = \dfrac{n}{356}, \quad \dfrac{356}{n} = \dfrac{24}{30}.$$

Example 2 To control a fever, a doctor suggests that a child weighing 28 kg be given 420 mg of Tylenol™. Under similar circumstances, how much Tylenol could be recommended for a child weighing 35 kg?

We let t = the number of milligrams of Tylenol and form a proportion.

Tylenol suggested → $\dfrac{420}{28} = \dfrac{t}{35}$ ← Tylenol suggested
Child's weight →   ← Child's weight

1. Brian drives his delivery van 800 mi in 5 days. At this rate, how far will he travel in 7 days?

Answer on page A-20

2. Campus Painting, Inc., can paint 1700 ft² of clapboard with 4 gal of paint. How much paint would be needed for a building with 6800 ft² of clapboard?

Solve: $420 \cdot 35 = 28 \cdot t$ **Equating cross products**

$$\frac{420 \cdot 35}{28} = \frac{28t}{28}$$ **Dividing by 28 on both sides**

$$\frac{4 \cdot 105 \cdot 5 \cdot 7}{4 \cdot 7} = t$$ **Removing a factor equal to 1: $\frac{4 \cdot 7}{4 \cdot 7} = 1$**

$$525 = t.$$

Thus, 525 mg could be recommended for a 35-kg child.

Do Exercise 2.

Example 3 *Ticket Purchases.* Rosa bought 8 tickets to an international food festival for $52. How many tickets could she purchase with $90?

We let n = the number of tickets that can be purchased with $90. Then we translate to a proportion.

$$\text{Cost} \rightarrow \frac{52}{8} = \frac{90}{n} \leftarrow \text{Cost}$$
$$\text{Tickets} \rightarrow \phantom{\frac{52}{8}} \phantom{\frac{90}{n}} \leftarrow \text{Tickets}$$

3. *Purchasing Shirts.* If 2 shirts can be bought for $47, how many shirts can be bought with $200?

Solve: $52n = 8 \cdot 90$ **Equating cross products**

$$n = \frac{8 \cdot 90}{52}$$ **Dividing by 52 on both sides**

$$\approx 13.8.$$ **Simplifying and rounding**

Because it is impossible to buy a fractional part of a ticket, we must round our answer *down* to 13. As a check, we use a different approach: We find the cost per ticket and then divide $90 by that price. Since $52 \div 8 = 6.50$ and $90 \div 6.50 \approx 13.8$, we have a check. Rosa could purchase 13 tickets with $90.

Do Exercise 3.

Example 4 *Waist-to-Hip Ratio.* For improved health, it is recommended that a woman's waist-to-hip ratio be 0.85 (or lower) (**Source**: David Schmidt, "Lifting Weight Myths," *Nutrition Action Newsletter* 20, no. 4, October 1993). Marta's hip measurement is 40 in. To meet the recommendation, what should Marta's waist measurement be?

4. *Waist-to-Hip Ratio.* It is also recommended that a man's waist-to-hip ratio be 0.95 (or lower). Uri's hip measurement is 45 in. To meet the recommendation, what should Uri's waist measurement be?

Hip measurement is the largest measurement around the widest part of the buttocks.

Waist measurement is the smallest measurement below the ribs but above the navel.

Note that $0.85 = \frac{85}{100}$. We let w = Marta's waist measurement and translate to a proportion.

$$\text{Waist measurement} \rightarrow \frac{w}{40} = \frac{85}{100} \leftarrow \text{Recommended}$$
$$\text{Hip measurement} \rightarrow \phantom{\frac{w}{40}} \phantom{\frac{85}{100}} \leftarrow \text{waist-to-hip ratio}$$

Solve: $100w = 40 \cdot 85$ **Equating cross products**

$$w = \frac{40 \cdot 85}{100}$$ **Dividing by 100 on both sides**

$$= 34$$

Marta's recommended waist measurement is 34 in. (or less).

Answers on page A-20

Do Exercise 4.

Exercise Set 7.4

a Solve.

1. *Gasoline Mileage.* Nancy's van traveled 84 mi on 6.5 gal of gasoline. At this rate, how many gallons would be needed to travel 126 mi?

2. *Bicycling.* Roy bicycled 234 mi in 14 days. At this rate, how far would Roy travel in 42 days?

3. *Quality Control.* A quality-control inspector examined 100 lightbulbs and found 7 of them to be defective. At this rate, how many defective bulbs will there be in a lot of 2500?

4. *Grading.* A professor must grade 32 essays in a literature class. She can grade 5 essays in 40 min. At this rate, how long will it take her to grade all 32 essays?

5. *Painting.* Fred uses 3 gal of paint to cover 1275 ft^2 of siding. How much siding can Fred paint with 7 gal of paint?

6. *Waterproofing.* Bonnie can waterproof 450 ft^2 of decking with 2 gal of sealant. How many gallons should Bonnie buy for a 1200-ft^2 deck?

7. *Publishing.* Every 6 pages of an author's manuscript corresponds to 5 published pages. How many published pages will a 540-page manuscript become?

8. *Turkey Servings.* An 8-lb turkey breast contains 36 servings of meat. How many pounds of turkey breast would be needed for 54 servings?

9. *Lefties.* In a class of 40 students, on average, 6 will be left-handed. If a class includes 9 "lefties," how many students would you estimate are in the class?

10. *Sugaring.* When 38 gal of maple sap are boiled down, the result is 2 gal of maple syrup. How much sap is needed to produce 9 gal of syrup?

11. *Mileage.* Jean bought a new car. In the first 8 months, it was driven 9000 mi. At this rate, how many miles will the car be driven in 1 yr?

12. *Coffee Production.* Coffee beans from 14 trees are required to produce the 17 lb of coffee that the average person in the United States drinks each year. How many trees are required to produce 375 lb of coffee?

13. *Metallurgy.* In a metal alloy, the ratio of zinc to copper is 3 to 13. If there are 520 lb of copper, how many pounds of zinc are there?

14. *Class Size.* A college advertises that its student-to-faculty ratio is 14 to 1. If 56 students register for Introductory Spanish, how many sections of the course would you expect to see offered?

15. *Painting.* Helen can paint 950 ft^2 with 2 gal of paint. How many 1-gal cans does she need in order to paint a 30,000-ft^2 wall?

16. *Snow to Water.* Under typical conditions, $1\frac{1}{2}$ ft of snow will melt to 2 in. of water. To how many inches of water will $5\frac{1}{2}$ ft of snow melt?

17. *Map Scaling.* On a map, $\frac{1}{4}$ in. represents 50 mi. If two cities are $3\frac{1}{4}$ in. apart on the map, how far apart are they in reality?

18. *Map Scaling.* On a road atlas map, 1 in. represents 16.6 mi. If two cities are 3.5 in. apart on the map, how far apart are they in reality?

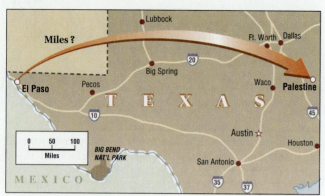

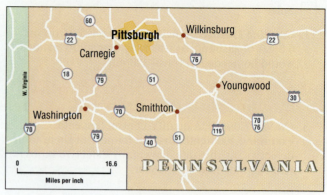

19. At the Bertocinis' church, two pews can seat 14 people. How many pews will be needed for a wedding party of 44 people?

20. Halo grass seed is sold in 1-lb bags. Each bag covers 800 ft² of lawn. How many bags must be purchased in order to cover a 4300 ft² lawn?

Skill Maintenance

21. Plot the following points. [6.3a]
    $(-3, 2)$, $(4, 5)$, $(-4, -1)$, $(0, 3)$

22. Multiply: $-19.3(4.1)$. [5.3a]

23. Divide: $-13.11 \div 5.7$. [5.4a]

24. Divide: $169.36 \div (-23.2)$. [5.4a]

25. Add: $-19.7 + 12.5$. [5.2c]

26. Subtract: $-3.7 - (-1.9)$. [5.2c]

Synthesis

27. ◆ Polly solved Example 1 by forming the proportion
    $$\frac{24}{30} = \frac{356}{n},$$
    whereas Rudy wrote
    $$\frac{24}{n} = \frac{356}{30}.$$
    Are both approaches valid? Why or why not?

28. ◆ Rob's waist and hips measure 35 in. and 33 in., respectively (see Margin Exercise 4). Suppose that Rob can either gain or lose 1 in. from one of his measurements. Where should the inch come from or go to? Why?

29. ◆ Can unit prices be used to solve proportions that involve money? Why or why not?

30. Carlson College is expanding from 850 to 1050 students. To avoid any rise in the student-to-faculty ratio, the faculty of 69 professors must also increase. How many new faculty positions should be created?

31. In recognition of Sheri's outstanding work, her salary has been increased from $26,000 to $29,380. Tim is earning $23,000 and is requesting a proportional raise. How much more should he ask for?

32. *Baseball Statistics.* Cy Young, one of the greatest baseball pitchers of all time, gave up an average of 2.63 earned runs every 9 innings. Young pitched 7356 innings, more than anyone in the history of baseball. How many earned runs did he give up?

33. *Real-Estate Values.* According to Coldwell Banker Real Estate Corporation, a home selling for $89,000 in Austin, Texas, would sell for $286,000 in San Francisco. How much would a $450,000 home in San Francisco sell for in Austin? Round to the nearest $1000.

34. The ratio 1:3:2 is used to estimate the relative costs of a CD player, receiver, and speakers when shopping for a stereo. That is, the receiver should cost three times the amount spent on the CD player and the speakers should cost twice as much as the amount spent on the CD player. If you had $900 to spend, how would you allocate the money, using this ratio?

Use proportions to make predictions of your college's student population.

7.5 Geometric Applications

a | Proportions and Similar Triangles

Look at the pair of triangles below. Note that they appear to have the same shape, but their sizes are different. These are examples of **similar triangles**. By using a magnifying glass, you could imagine enlarging the smaller triangle to get the larger. This process works because the corresponding sides of each triangle have the same ratio. That is, the following proportion is true.

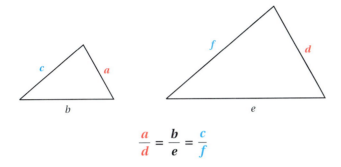

$$\frac{a}{d} = \frac{b}{e} = \frac{c}{f}$$

> **Similar triangles** have the same shape. The lengths of their corresponding sides have the same ratio—that is, they are proportional.

Example 1 The triangles at right are similar triangles. Find the missing length x.

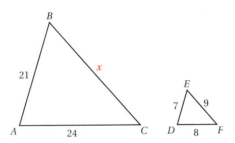

The ratio of x to 9 is the same as the ratio of 24 to 8 or 21 to 7. We get the proportions

$$\frac{x}{9} = \frac{24}{8} \quad \text{and} \quad \frac{x}{9} = \frac{21}{7}.$$

We can solve either one of these proportions as follows:

$\frac{x}{9} = 3$ Simplifying

$x = 3 \cdot 9$ Multiplying by 9 on both sides

$ = 27.$ Simplifying

The missing length x is 27. Other proportions could also be used.

Do Exercise 1.

Similar triangles and proportions can often be used to find lengths that would ordinarily be difficult to measure. For example, we could find the height of a flagpole without climbing it or the distance across a river without crossing it.

Objectives

a | Find lengths of sides of similar triangles using proportions.

b | Use proportions to find lengths in pairs of figures that differ only in size.

For Extra Help

TAPE 13   MAC WIN   CD-ROM

1. This pair of triangles is similar. Find the missing length x.

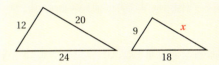

Answer on page A-20

7.5 Geometric Applications

439

2. How high is a flagpole that casts a 45-ft shadow at the same time that a 5.5-ft woman casts a 10-ft shadow?

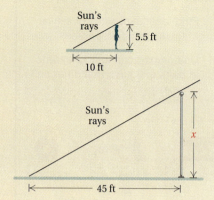

Example 2 How high is a flagpole that casts a 56-ft shadow at the same time that a 6-ft man casts a 5-ft shadow?

If we use the sun's rays to represent the third side of the triangle in our drawing of the situation, we see that we have similar triangles. Let p = the height of the flagpole. The ratio of 6 to p is the same as the ratio of 5 to 56. Thus we have the proportion

Height of man → $\dfrac{6}{p} = \dfrac{5}{56}$ ← Length of shadow of man
Height of pole →   ← Length of shadow of pole

Solve: $6 \cdot 56 = 5 \cdot p$ **Equating cross products**

$\dfrac{6 \cdot 56}{5} = p$ **Dividing by 5 on both sides**

$67.2 = p$ **Simplifying**

The height of the flagpole is 67.2 ft.

Do Exercise 2.

3. *F-106 Blueprint.* Referring to Example 3, find the length x of the wing.

Example 3 *F-106 Blueprint.* A blueprint for an F-106 Delta Dart military plane is a scale drawing. Each wing of the plane has a triangular shape. The blueprint shows similar triangles. Find the length of side a of the wing.

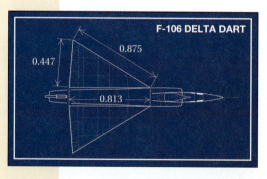
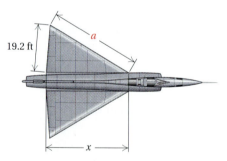

We let a = the length of the wing. Thus we have the proportion

Length on the blueprint → $\dfrac{0.447}{19.2} = \dfrac{0.875}{a}$ ← Length on the blueprint
Length of the wing →   ← Length of the wing

Solve: $0.447 \cdot a = 19.2 \cdot 0.875$ **Equating cross products**

$a = \dfrac{19.2 \cdot 0.875}{0.447}$ **Dividing by 0.447 on both sides**

≈ 37.6 ft

The length of side a of the wing is about 37.6 ft.

Do Exercise 3.

Answers on page A-20

b Proportions and Other Geometric Shapes

When one geometric figure is a magnification of another, the figures are similar. Thus the corresponding lengths are proportional.

Example 1 The sides in the negative and photograph below are proportional. Find the width of the photograph.

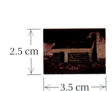

We let $x =$ the width of the photograph. Then we translate to a proportion.

Photo width → $\dfrac{x}{2.5} = \dfrac{10.5}{3.5}$ ← Photo length
Negative width →   ← Negative length

Solve: $\dfrac{x}{2.5} = 3$ **Simplifying**

$x = 3(2.5)$ **Multiplying by 2.5 on both sides**

$ = 7.5$ **Simplifying**

Thus the width of the photograph is 7.5 cm.

Do Exercise 4.

Example 5 A scale model of an addition to an athletic facility is 12 cm wide at the base and rises to a height of 15 cm. If the actual base is to be 116 ft, what will be the actual height of the addition?

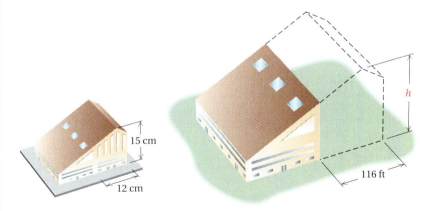

We let $h =$ the height of the addition. Then we translate to a proportion.

Width in model → $\dfrac{12}{116} = \dfrac{15}{h}$ ← Height in model
Actual width →   ← Actual height

4. The sides in the photographs below are proportional. Find the width of the larger photograph.

Answer on page A-20

7.5 Geometric Applications

441

5. Refer to the figures in Example 5. If a model skylight is 3 cm wide, how wide will the actual skylight be?

Solve: $12h = 116 \cdot 15$

$h = \dfrac{116 \cdot 15}{12}$ Dividing by 12 on both sides

$= \dfrac{\cancel{4} \cdot 29 \cdot \cancel{3} \cdot 5}{\cancel{4} \cdot \cancel{3}}$ Removing a factor equal to 1: $\dfrac{4 \cdot 3}{4 \cdot 3} = 1$

$= 145.$ Multiplying

Equating cross products

Thus the height of the addition will be 145 ft. Note that the proportion on page 417 will yield the same result.

Do Exercise 5.

Calculator Spotlight

 Proportions can be solved easily with the aid of a calculator. For example, earlier in this section we solved

$$\dfrac{0.447}{19.2} = \dfrac{0.875}{a}$$

by equating cross products and then dividing. The solution,

$$\dfrac{19.2 \cdot 0.875}{0.447},$$

uses the three numbers that were given: The numerator is a cross product that uses two of those numbers, and the denominator is the remaining number. This pattern can be used to solve *any* proportion. For instance, to solve

$$\dfrac{5}{8} = \dfrac{x}{20},$$

we can press [5] [×] [2] [0] [÷] [8] [=]. When a calculator is not available, it is usually easiest to simplify first by removing a factor equal to 1, as in Example 5.

Exercises

Solve each proportion.

1. $\dfrac{15.75}{20} = \dfrac{a}{35}$

2. $\dfrac{32}{x} = \dfrac{25}{20}$

3. $\dfrac{t}{57} = \dfrac{17}{64}$

4. $\dfrac{16}{29} = \dfrac{23}{a}$

5. $\dfrac{71.2}{a} = \dfrac{42.5}{23.9}$

6. $\dfrac{29.6}{3.15} = \dfrac{x}{4.23}$

7. $\dfrac{0.023}{0.15} = \dfrac{0.401}{t}$

8. $\dfrac{a}{3.01} = \dfrac{1.7}{0.043}$

Answer on page A-20

Exercise Set 7.5

a The triangles in each exercise are similar. Find the missing lengths.

1.

2.

3.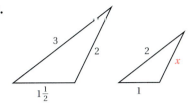

4.

5.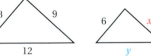

6.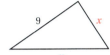

7.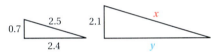

8.

9. When a tree 8 m high casts a shadow 5 m long, how long a shadow is cast by a person 2 m tall?

10. How high is a flagpole that casts a 42-ft shadow at the same time that a $5\frac{1}{2}$-ft woman casts a 7-ft shadow?

11. How high is a tree that casts a 27-ft shadow at the same time that a 4-ft fence post casts a 3-ft shadow?

12. How high is a tree that casts a 32-ft shadow at the same time that an 8-ft light pole casts a 9-ft shadow?

13. Find the height h of the wall.

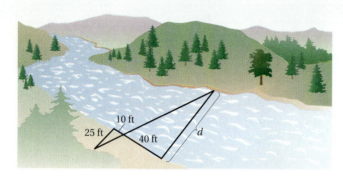

14. Find the length L of the lake.

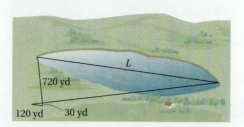

15. Find the distance across the river. Assume that the ratio of d to 25 ft is the same as the ratio of 40 ft to 10 ft.

16. To measure the height of a hill, a string is drawn tight from level ground to the top of the hill. A 3-ft stick is placed under the string, touching it at point P, a distance of 5 ft from point G, where the string touches the ground. The string is then detached and found to be 120 ft long. How high is the hill?

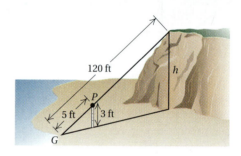

b The sides in each pair of figures are proportional. Find the missing lengths.

17.

18.

19.

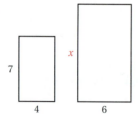

20.

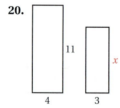

Chapter 7 Ratio and Proportion

444

21.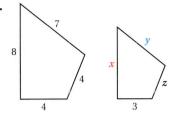

22.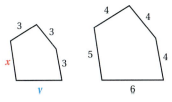

23.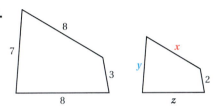

24.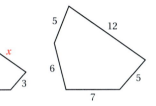

25.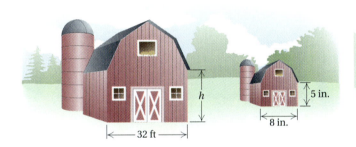

26.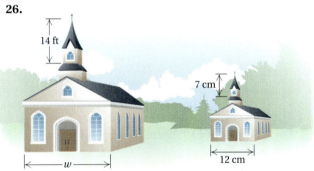

Skill Maintenance

27. Kathy has $34.97 to spend for a book at $49.95, a CD at $14.88, and a sweatshirt at $29.95. How much more money does she need to make these purchases? [5.8a]

28. Divide: $80.892 \div 8.4$. [5.4a]

Multiply. [5.3a]

29. -8.4×80.892

30. 0.01×274.568

31. 100×274.568

32. $-0.002(-274.568)$

Exercise Set 7.5

445

Synthesis

33. ◆ Is it possible for two triangles to have two pairs of sides that are proportional without the triangles being similar? Why or why not?

34. ◆ Suppose that all the sides in one triangle are half the size of the corresponding sides in a similar triangle. Does it follow that the area of the smaller triangle is half the area of the larger triangle? Why or why not? (*Hint*: $A = \frac{1}{2}bh$ for any triangle.)

35. ◆ Design for a classmate a problem involving similar triangles for which
$$\frac{18}{128.95} = \frac{x}{789.89}.$$

Hockey Goals. An official hockey goal is 6 ft wide. To make scoring more difficult, goalies often locate themselves far in front of the goal to "cut down the angle." In Exercises 36 and 37, suppose that a slapshot from point A is attempted and that the goalie is 2.7 ft wide. Determine how far from the goal the goalie should be located if point A is the given distance from the goal. (*Hint*: First find how far the goalie should be from point A.)

36. 25 ft

37. 35 ft

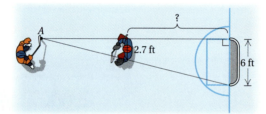

The triangles in each exercise are similar triangles. Find the lengths not given.

38.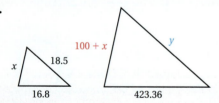

39.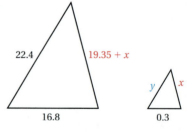

40. A miniature basketball hoop is built for the model referred to in Example 5. An actual hoop is 10 ft high. How high should the model hoop be? Round to the nearest thousandth of a centimeter.

41. A miniature baseball diamond is drawn to accompany the model used in Example 5. If the actual baseball diamond is a 90-ft by 90-ft square, what will the area of the model diamond be?

Summary and Review Exercises: Chapter 7

1. Write fractional notation for the ratio 37 to 85. [7.1a]

2. In this rectangle, find the ratio of width to length and simplify. [7.1b]

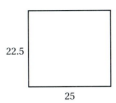

3. In 1996, approximately 66 million citizens were registered to vote. Of those registered, about 54 million voted. Write simplified fractional notation for the ratio of those who voted to those were registered. [7.1a, b]

4. An Olympic marathoner ran 20 km in 57 min. What is the rate in kilometers per minute? in minutes per kilometer? [7.2a]

5. A bicycle racer biked 25 mi in 1.25 hr. What was the racer's rate in minutes per mile? [7.2a]

6. A 12-oz container of Jones' Hot Pepper Jelly costs $3.24. Find the unit price in dollars per ounce. [7.2b]

7. Which has the lower unit cost? [7.2b]

CANNED PINEAPPLE JUICE
Delacorte: 12 oz for 99 cents
BiteFine: 18 oz for $1.26

8. Determine whether the pairs 7, 5 and 13, 11 are proportional. [7.3a]

Solve. [7.3b]

9. $\dfrac{8}{9} = \dfrac{x}{36}$

10. $\dfrac{120}{\frac{3}{7}} = \dfrac{7}{x}$

11. $\dfrac{6}{x} = \dfrac{48}{56}$

12. $\dfrac{4.5}{120} = \dfrac{0.9}{x}$

Solve. [7.4a]

13. If 3 dozen eggs cost $3.99, how much will 5 dozen eggs cost?

14. In a factory, it was discovered that 39 circuits in a lot of 65 were defective. At this rate, how many defective circuits can be expected in a lot of 585 circuits?

15. The ratio of children to adults at the Ever Ready Daycare Center cannot exceed 13 to 2. If there are 35 children at the center, how many adults must be present?

16. A train travels 448 km in 7 hr. At this rate, how far would it travel in 13 hr?

17. It has been estimated that, on average, 5 people produce 13 kg of garbage each day. San Antonio, Texas, has a population of 936,000 people. Estimate the quantity of garbage produced in San Antonio each day.

18. In Michigan, there are 2.3 lawyers for every 1000 people. The population of the Detroit metropolitan area is 4,307,000. Estimate the number of lawyers in Detroit.

Each pair of triangles in Exercises 19 and 20 is similar. Find the missing length(s). [7.5a]

19.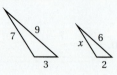

20.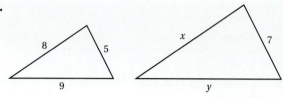

21. How high is a billboard that casts a 25-ft shadow at the same time that an 8-ft sapling casts a 5-ft shadow? [7.5a]

22. The lengths in the figures below are proportional. Find the missing lengths. [7.5b]

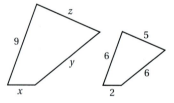

Skill Maintenance

23. A family has $2347.89 in its checking account. It writes checks for $678.95 and $38.54. How much is left in the checking account? [5.8a]

24. In which quadrant is each point located? [6.3b]
$(-4, -3)$, $(-2, 7)$, $(1, 4)$, and $(3, -2)$

25. Multiply: $456.1(-23.4)$. [5.3a]

26. Divide. Write decimal notation for the answer. [5.4a]

$5.6 \overline{)254.8}$

Synthesis

27. ◆ Write a proportion problem for a classmate to solve. Design the problem so that the solution is "Leslie would need 16 gal of gasoline in order to travel 368 mi." [7.4a]

28. ◆ If you were a college president, which would you prefer: a low or high faculty-to-student ratio? Why? What about the student-to-faculty ratio? [7.1a]

State Lottery Prizes. The chart below shows the prizes awarded in certain state lotteries in a recent year. (**Source:** *Statistical Abstract of the United States, 1998*). Use the information to do Exercises 29–32.

State	Prizes (in millions)
California	$1132
Colorado	192
*Connecticut	402
Florida	1024
Illinois	839
*Maine	84
Maryland	609
*Massachusetts	2140
Michigan	736
*New Hampshire	98
New Jersey	796
New York	1827
Ohio	1363
Pennsylvania	891
*Rhode Island	303
*Vermont	44
Washington	224

* = a New England State

29. 🧮 What is the ratio of the lottery prizes in New York to the total prizes awarded by all lotteries? [7.1a]

30. 🧮 What is the ratio of the lottery prizes in New England to the total prizes awarded by all lotteries? [7.1a]

31. 🧮 The population of California is 31,858,000. If all prize money were equally shared, how much would the average Californian receive? [7.2a]

32. 🧮 The population of New York is 18,134,000. If all prize money were equally shared, how much would the average New Yorker receive? [7.2a]

33. Shine-and-Glo Painters uses 2 gal of finishing paint for every 3 gal of primer. Each gallon of finishing paint covers 450 ft². If a surface of 4950 ft² needs both primer and finishing paint, how many gallons should be purchased altogether? [7.4a]

34. It takes Yancy Martinez 10 min to type two-thirds of a page of his term paper. At this rate, how long will it take him to type a 7-page term paper? [7.4a]

Test: Chapter 7

Write fractional notation for each ratio. Do not simplify.

1. 83 to 94
2. 0.34 to 124

3. Simplify the ratio of length to width in this rectangle.

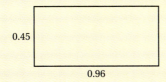

4. A diver descends 10 ft in 16 sec. What is the rate of descent in feet per second?

5. In 1 hr 20 min, a bicycle racer biked 48 km. What was the racer's rate in minutes per kilometer?

6. An 8-oz package of extra-sharp cheddar cheese costs $2.79. Find the unit price in dollars per ounce.

7. Determine whether the pairs 8, 6 and 15, 10 are proportional.

Solve.

8. $\dfrac{27}{x} = \dfrac{9}{4}$

9. $\dfrac{150}{2.5} = \dfrac{x}{6}$

10. An ocean liner traveled 432 km in 12 hr. At this rate, how far would the boat travel in 42 hr?

11. A watch loses 2 min in 10 hr. At this rate, how much will it lose in 24 hr.?

12. On a map, 3 in. represents 225 mi. If two cities are 7 in. apart on the map, how far are they apart in reality?

13. A birdhouse built on a pole that is 3 m high casts a shadow 5 m long. At the same time, the shadow of a tower is 110 m long. How high is the tower?

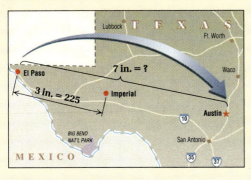

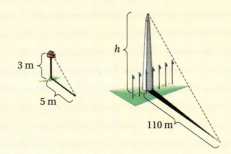

Answers

1. _____
2. _____
3. _____
4. _____
5. _____
6. _____
7. _____
8. _____
9. _____
10. _____
11. _____
12. _____
13. _____

The lengths in each pair of figures are proportional. Find the missing lengths.

14.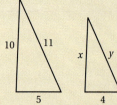

15.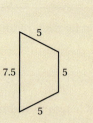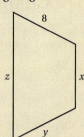

Skill Maintenance

16. In a recent year, Kellogg sold 146.2 million lb of Corn Flakes and 120.4 million lb of Frosted Flakes. How many more pounds of Corn Flakes did they sell than Frosted Flakes?

17. In which quadrant is each point located?

    $(-2, 4)$, $(5, -2)$, $(7, 1)$, and $(-1, -6)$

18. Multiply: $-24.13(-7.2)$.

19. Divide: $\dfrac{-99.44}{100}$.

Synthesis

Solve.

20. $\dfrac{5}{9} = \dfrac{2x - 1}{x + 3}$

21. $\dfrac{3x + 1}{7} = \dfrac{4x - 5}{6}$

22. Nancy Morano-Smith wants to guess the number of marbles in an 8-gal jar. Knowing that there are 128 oz in a gallon, Nancy goes home and fills an 8-oz jar with 46 marbles. How many marbles should she guess are in the 8-gal jar?

23. The Johnson triplets and the Solomini twins went out to dinner and decided to split the bill of $79.85 proportionately. How much will the Johnsons pay?

24. A soccer goalie wishing to block an opponent's shot moves toward the shooter to reduce the shooter's view of the goal. If the goalie can only defend a region 10 ft wide, how far in front of the goal should the goalie be (see the following figure)?

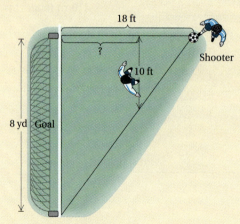

Cumulative Review: Chapters 1–7

Add and simplify.

1. $\quad 137.186$
   $\quad23.019$
   $\underline{+483.297}$

2. $\quad 2\frac{1}{3}$
   $\underline{+\, 4\frac{5}{12}}$

3. $\dfrac{6}{35} + \dfrac{5}{28}$

Perform the indicated operation and simplify.

4. $\quad 60.21$
   $\underline{-9.709}$

5. $-32 - (-15)$

6. $\dfrac{4}{15} - \dfrac{3}{20}$

7. $\quad 37.64$
   $\underline{\times 5.9}$

8. $-43(15)$

9. $2\dfrac{1}{3} \cdot 1\dfrac{2}{7}$

10. $2.3 \overline{)98.9}$

11. $-306 \div 6$

12. $\dfrac{7}{11} \div \dfrac{14}{33}$

13. Write expanded notation for 30,074.

14. Write a word name for 120.07.

Which number is larger?

15. 0.7, 0.698

16. -0.799, -0.8

17. Find the prime factorization of 144.

18. Find the LCM of 44 and 55.

19. What part is shaded?

20. Simplify: $\dfrac{90}{144}$.

Calculate.

21. $\dfrac{3}{5} \times 9.53$

22. $7.2 \div 0.4(-1.5) + (1.2)^2$

23. Find the mean: 23, 49, 52, 71.

24. Determine whether the pairs 3, 9 and 25, 75 are proportional.

25. Graph on a plane: $y = -x - 4$.

26. Evaluate $\dfrac{t-7}{w}$ for $t = -3$ and $w = -2$.

Solve.

27. $\dfrac{14}{25} = \dfrac{x}{54}$

28. $-423 = 16 \cdot t$

29. $9x - 7 = -43$

30. $2(x - 3) + 9 = 5x - 6$

31. $34.56 + n = 67.9$

32. $\dfrac{2}{3}x = \dfrac{16}{27}$

Solve.

33. A car travels 337.62 mi in 8 hr. How far does it travel in 1 hr?

34. A machine can stamp out 925 washers in 5 min. How much time would be needed to stamp out 1295 washers?

35. Elise drove 347.6 mi, 249.8 mi, and 379.5 mi on three separate trips. What was the total mileage?

36. In a recent year, 1,635,000 people camped at Yosemite National Park. In the same year, 1,221,000 people camped at Yellowstone National Park. How many more people camped at Yosemite than at Yellowstone?

37. A triangular window is 4 ft high and has a base that measures 5 ft. Find its area.

38. It takes Jose $\tfrac{2}{3}$ hr to hang a door. How many doors can he hang in 8 hr?

39. A 46-oz juice can contains $5\tfrac{3}{4}$ cups of juice. A recipe calls for $3\tfrac{1}{2}$ cups of juice. How many cups are left?

40. A space shuttle recently made 16 orbits a day during an 8.25-day mission. How many orbits were made during the entire mission?

Synthesis

41. A car travels 88 ft in 1 sec. What is the rate in miles per hour?

42. A 12-oz bag of shredded mozzarella cheese is on sale for $2.07. Blocks of mozzarella cheese are on sale for $2.79 per pound. Which is the better buy?

43. Hans attends a university where the academic year consists of two 16-week semesters. He budgets $1200 for incidental expenses for the academic year. After 3 weeks, Hans has spent $150 for incidental expenses. Assuming he continues to spend at the same rate, will the budget for incidental expenses be adequate? If not, when will the money be exhausted and how much more will be needed to complete the year?

44. A basic sound system consists of a CD player, a receiver, and two speakers. A standard rule of thumb on the relative investment in these components is 1:3:2. That is, the receiver should cost three times as much as the CD player and the speakers should cost twice as much as the CD player.
    a) You have $1200 to spend. How should you allocate the funds if you use this rule of thumb?
    b) How should you spend $3000?

Cumulative Review: Chapters 1–7

8

Percent Notation

Introduction

This chapter introduces a new kind of notation for numbers: percent notation. We will see that $\frac{1}{5}$, 0.2, and 20% are all names for the same number. Then we will use percent notation and equations to solve applied problems.

- 8.1 Percent Notation
- 8.2 Solving Percent Problems Using Proportions
- 8.3 Solving Percent Problems Using Equations
- 8.4 Applications of Percent
- 8.5 Consumer Applications: Sales Tax, Commission, and Discount
- 8.6 Consumer Applications: Interest

An Application

Recently, the United States was generating 81.5 million tons of paper waste per year, of which 32.6 million tons were recycled (**Source**: Franklin Associates, Ltd., Prairie Village, KS). What percent of paper waste was recycled?

This problem appears as Example 1 in Section 8.4.

The Mathematics

We let n = the percent of paper waste that was recycled. The problem can be translated as follows:

$$\underbrace{32.6 \text{ million}}_{32.6} \underbrace{\text{is}}_{=} \underbrace{\text{what percent}}_{n} \underbrace{\text{of}}_{\cdot} \underbrace{81.5?}_{81.5}$$

Once n has been found, we will change it from decimal notation to percent notation.

For more information, visit us at www.mathmax.com

Pretest: Chapter 8

1. Write decimal notation for 87%.

2. Write percent notation for 0.537.

3. Write percent notation for $\frac{3}{4}$.

4. Write fractional notation for 37%.

5. Translate to an equation. Then solve.
   What is 60% of 75?

6. Translate to a proportion. Then solve.
   What percent of 50 is 35?

Solve.

7. The weight of muscles in an adult male is 40% of total body weight. A man weighs 225 lb. What do the muscles weigh?

8. The population of Winchester increased from 3000 to 3600. Write the percent of increase in population.

9. The sales tax rate in California is 6%. How much tax is charged on a purchase of $286? What is the total price?

10. Anwar's commission rate is 28%. What is his commission from the sale of $18,400 worth of merchandise?

11. The marked price of a Panasonic camcorder is $450. This camera is on sale at Lowland Appliances for 25% off. What is the discount and what is the sale price?

12. What is the simple interest on $1200 principal at the interest rate of 8.3% for 1 year?

13. What is the simple interest on $500 at 8% for $\frac{1}{2}$ year?

14. Interest is compounded quarterly. Write the amount in an account if $6000 is invested at 8% for 6 months.

8.1 Percent Notation

a | Understanding Percent Notation

Of all wood harvested, 35% is used for paper. What does this mean? It means that, on average, of every 100 tons of wood harvested, 35 tons is made into paper. Thus, 35% is a ratio of 35 to 100, or $\frac{35}{100}$. Percent means parts of 100.

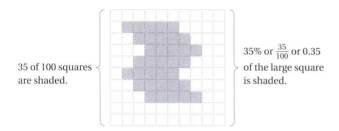

35 of 100 squares are shaded.

35% or $\frac{35}{100}$ or 0.35 of the large square is shaded.

Objectives

a | Write three kinds of notation for a percent.

b | Convert between percent notation and decimal notation.

c | Convert between fractional notation and percent notation.

For Extra Help

TAPE 14   MAC WIN   CD-ROM

Percent notation is used extensively in our lives. Here are some examples:

Astronauts lose 1% of their bone mass for each month of weightlessness.

95% of hair spray is alcohol.

55% of all baseball merchandise sold is purchased by women.

62.4% of all aluminum cans were recycled in a recent year.

56% of all fruit juice purchased is orange juice.

45.8% of us sleep between 7 and 8 hours per night.

Percent notation is often represented by pie charts to show how the parts of a quantity are related. For example, the chart below relates the length of time couples are engaged before marrying.

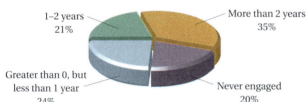

Length of Engagement Before Marriage

1–2 years 21%
More than 2 years 35%
Greater than 0, but less than 1 year 24%
Never engaged 20%

To draw a pie chart like the one above, think of a pie cut into 100 equally sized pieces. Then shade in a wedge equal in size to 20 of these pieces to represent 20%. Shade in a wedge equal in size to 35 of the 100 pieces to represent 35% and so on.

Do Exercise 1.

The notation $n\%$ arose historically meaning "n per hundred." This rate can be expressed as $n/100$, $n \times \frac{1}{100}$, or $n \times 0.01$.

Percent notation, $n\%$, is defined in three ways:

ratio ➜ $n\%$ = the ratio of n to 100 = $\frac{n}{100}$;

fractional notation ➜ $n\% = n \times \frac{1}{100}$;

decimal notation ➜ $n\% = n \times 0.01$.

---

1. The circle below is divided into 100 sections. Use the following information on the amounts of fruit juices sold to draw and label a circle graph.

Apple:	14%
Orange:	56%
Grapefruit:	4%
Blends:	6%
Grape:	5%
Other:	15%

Source: Beverage Marketing Corporation

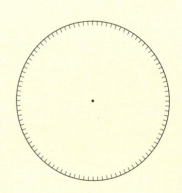

Answer on page A-21

Write three kinds of notation as in Examples 1 and 2.

2. 70%

3. 23.4%

4. 100%

It is thought that the Roman Emperor Augustus began percent notation by taxing goods sold at a rate of $\frac{1}{100}$. In time, the symbol "%" evolved by interchanging the parts of the symbol "100" to "0/0" and then to "%".

Answers on page A-21

Example 1 Write three kinds of notation for 35%.

Using ratio: $35\% = \dfrac{35}{100}$ A ratio of 35 to 100

Using fractional notation: $35\% = 35 \times \dfrac{1}{100}$ Replacing % with $\times \dfrac{1}{100}$

Using decimal notation: $35\% = 35 \times 0.01$ Replacing % with $\times 0.01$

Example 2 Write three kinds of notation for 67.8%.

Using ratio: $67.8\% = \dfrac{67.8}{100}$ A ratio of 67.8 to 100

Using fractional notation: $67.8\% = 67.8 \times \dfrac{1}{100}$ Replacing % with $\times \dfrac{1}{100}$

Using decimal notation: $67.8\% = 67.8 \times 0.01$ Replacing % with $\times 0.01$

Do Exercises 2–4.

b Converting Between Percent Notation and Decimal Notation

To write decimal notation for a number like 78%, we can replace "%" with "$\times 0.01$" and multiply:

$78\% = 78 \times 0.01$ Replacing % with $\times 0.01$

$= 0.78.$ Multiplying

Similarly,

$4.9\% = 4.9 \times 0.01$ Replacing % with $\times 0.01$

$= 0.049$ Multiplying

and

$265\% = 265 \times 0.01$ Replacing % with $\times 0.01$

$= 2.65.$ Multiplying

When we multiply by 0.01, the decimal point is moved two places to the left. Thus, to quickly convert from percent notation to decimal notation, we can drop the percent symbol and move the decimal point two places to the left.

To convert from percent notation to decimal notation,	29.7%
a) replace the percent symbol % with $\times 0.01$, and	29.7×0.01
b) multiply by 0.01, which means move the decimal point two places to the left.	0.29.7 Move 2 places to the left. $29.7\% = 0.297$

Chapter 8 Percent Notation

Example 3 Write decimal notation for 92.43%.

a) Replace the percent symbol with × 0.01.   92.43 × 0.01

b) Multiply to move the decimal point two places to the left.   0.92.43

Thus, 92.43% = 0.9243. With practice, you will be able to make this conversion mentally.

Sometimes it may be necessary to write zeros as placeholders.

Example 4 In 1997, the population of North America was 7.9% of the world population. Write decimal notation for 7.9%.

a) Replace the percent symbol with × 0.01.   7.9 × 0.01

b) Multiply to move the decimal point two places to the left.   0.07.9   — This zero serves as a placeholder.

Thus, 7.9% = 0.079.

Do Exercises 5–8.

The procedure used in Examples 3 and 4 can be reversed to write a decimal, like 0.38, in percent notation. To see why, consider the following:

0.38	= 0.38 × 100%	Multiplying by 100% or 1
	= 0.38 × 100 × 0.01	Replacing 100% with 100 × 0.01
	= (0.38 × 100) × 0.01	Using an associative law
	= 38 × 0.01	
	= 38%.	Replacing × 0.01 with %

To summarize, 0.38 = 0.38 × 100% = (0.38 × 100)% = 38%.

To convert from decimal notation to percent notation, multiply by 100%. That is,	0.675 = 0.675 × 100%
a) move the decimal point two places to the right, and	0.67.5 Move 2 places to the right.
b) write a % symbol.	67.5%
	0.675 = 67.5%

Example 5 Babies are born before their "due" date 0.3 of the time. Write percent notation for 0.3.

a) Multiply by 100 to move the decimal point two places to the right. Recall that 0.3 = 0.30 since $\frac{3}{10} = \frac{30}{100}$.   0.30.   — This zero serves as a placeholder.

b) Write a % symbol.   30%

Thus, 0.3 = 30%.

Write decimal notation.

5. 34%

6. 78.9%

7. Electricity prices in 1996 were 104.5% of the prices in 1990. (**Source:** *Statistical Abstract of the United States, 1997*). Write decimal notation for this number.

8. Soft drink sales in the United States have grown 4.2% annually over the past decade. Write decimal notation for this number.

Answers on page A-21

Write percent notation.

9. 0.24

10. 3.47

11. 1

12. Muscles make up 0.4 of an adult male's body. Find percent notation for 0.4.

13. The average television set is on 0.25 of the time. Find percent notation for 0.25.

Write percent notation.

14. $\dfrac{1}{4}$

15. $\dfrac{7}{8}$

Answers on page A-21

Chapter 8 Percent Notation

458

Example 6 Property values in Dorchester were multiplied by 1.23 when reassessment was performed. Find percent notation for 1.23.

a) Multiply by 100 to move the decimal point two places to the right.   1.23.

b) Write a % symbol.   123%

Thus, 1.23 = **123%**.

Do Exercises 9–13.

c Converting Between Fractional Notation and Percent Notation

To convert from fractional notation to percent notation,	$\dfrac{3}{5}$ Fractional notation
a) find decimal notation by division, and	$\begin{array}{r} 0.6 \\ 5\,\overline{)\,3.0} \\ \underline{3\,0} \\ 0 \end{array}$
b) convert the decimal notation to percent notation.	$0.6 = 0.60 = 60\%$ Percent notation $\dfrac{3}{5} = 60\%$

Example 7 Write percent notation for $\dfrac{3}{8}$.

a) Find decimal notation by division.

$$\begin{array}{r} 0.3\,7\,5 \\ 8\,\overline{)\,3.0\,0\,0} \\ \underline{2\,4} \\ 6\,0 \\ \underline{5\,6} \\ 4\,0 \\ \underline{4\,0} \\ 0 \end{array} \qquad \dfrac{3}{8} = 0.375$$

b) Convert the decimal notation to percent notation. To do so, multiply by 100 to move the decimal point two places to the right, and write a % symbol.

$$\dfrac{3}{8} = 0.375 = 37.5\%, \text{ or } 37\dfrac{1}{2}\%$$

Don't forget the % symbol.

Do Exercises 14 and 15.

Example 8 Of all meals, $\frac{1}{3}$ are eaten outside the home. Write percent notation for $\frac{1}{3}$.

a) Find decimal notation by division.

$$\begin{array}{r} 0.333 \\ 3\overline{)1.000} \\ \underline{9} \\ 10 \\ \underline{9} \\ 10 \\ \underline{9} \\ 1 \end{array}$$

Remember that to find percent notation, we will need to move the decimal point two places to the right.

We get a repeating decimal: $0.33\overline{3}$.

b) Convert the answer to percent notation.

$$0.33.\overline{3}$$

$$\frac{1}{3} = 33.\overline{3}\%, \text{ or } 33\frac{1}{3}\%$$

Do Exercises 16 and 17.

In some cases, division is not the fastest way to convert to percent notation. The following are some optional ways in which conversion might be done.

Example 9 Write percent notation for $\frac{69}{100}$.

We use the definition of percent as a ratio.

$$\frac{69}{100} = 69\%$$

Example 10 Write percent notation for $\frac{17}{20}$.

We multiply $\frac{17}{20}$ by 1 to get 100 in the denominator. We think of what we have to multiply 20 by in order to get 100. That number is 5, so we multiply by 1 using $\frac{5}{5}$.

$$\frac{17}{20} \cdot \frac{5}{5} = \frac{85}{100} = 85\% \qquad \text{Check:} \quad \begin{array}{r} .85 \\ 20\overline{)17.00} \\ \underline{160} \\ 100 \\ \underline{100} \\ 0 \end{array}$$

Note that this shortcut works only when the denominator is a factor of 100.

Do Exercises 18 and 19.

---

16. The human body is $\frac{2}{3}$ water. Write percent notation for $\frac{2}{3}$.

17. Write percent notation: $\frac{15}{8}$.

Write percent notation.

18. $\frac{57}{100}$

19. $\frac{19}{25}$

Calculator Spotlight

Conversion. Calculators are often used when we are converting fractional notation to percent notation. We simply perform the division on the calculator and then convert the decimal notation to percent notation. For example, percent notation for $\frac{17}{40}$ can be found by pressing

[1][7][÷][4][0][=]

and then converting the result, 0.425, to percent notation, 42.5%.

Exercises

Find percent notation. Round to the nearest hundredth of a percent.

1. $\frac{13}{25}$ 2. $\frac{5}{13}$

3. $\frac{42}{39}$ 4. $\frac{12}{7}$

5. $\frac{217}{364}$ 6. $\frac{2378}{8401}$

Answers on page A-21

Write fractional notation.

20. 180%

21. 3.25%

22. $66\frac{2}{3}\%$

23. Complete this table.

Fractional Notation	$\frac{1}{5}$		
Decimal Notation		$0.83\overline{3}$	
Percent Notation			$37\frac{1}{2}\%$

Answers on pages A-21 and A-22

The method used in Examples 9 and 10 is reversed when we convert from percent notation to fractional notation.

To convert from percent notation to fractional notation,	30% Percent notation
a) use the definition of percent as a ratio, and	$\frac{30}{100}$
b) simplify, if possible.	$\frac{3}{10}$ Fractional notation

Example 11 Write fractional notation for 75%.

$$75\% = \frac{75}{100} \quad \text{Using the definition of percent}$$
$$= \frac{3 \cdot 25}{4 \cdot 25} = \frac{3}{4} \cdot \frac{25}{25} \quad \Bigg\} \text{Simplifying}$$
$$= \frac{3}{4}$$

Example 12 Write fractional notation for 112.5%.

$$112.5\% = \frac{112.5}{100} \quad \text{Using the definition of percent}$$
$$= \frac{112.5}{100} \times \frac{10}{10} \quad \text{Multiplying by 1 to eliminate the decimal point in the numerator}$$
$$= \frac{1125}{1000}$$
$$= \frac{5 \cdot 225}{5 \cdot 200} = \frac{5}{5} \cdot \frac{225}{200} \quad \Bigg\} \text{Simplifying}$$
$$= \frac{225}{200} = \frac{25 \cdot 9}{25 \cdot 8} = \frac{9}{8}$$

Example 13 Write fractional notation for $16\frac{2}{3}\%$.

$$16\frac{2}{3}\% = \frac{50}{3}\% \quad \text{Converting from the mixed numeral to fractional notation}$$
$$= \frac{50}{3} \times \frac{1}{100} \quad \text{Using the definition of percent}$$
$$= \frac{50 \cdot 1}{3 \cdot 50 \cdot 2} = \frac{1}{6} \cdot \frac{50}{50} = \frac{1}{6}$$

Do Exercises 20–22.

Had we noticed that $16\frac{2}{3}\%$ is half of $33\frac{1}{3}\%$, and if we remembered that $33\frac{1}{3}\% = \frac{1}{3}$, Example 13 could have been solved as follows:

$$16\frac{2}{3}\% = \frac{1}{2} \times 33\frac{1}{3}\% = \frac{1}{2} \times \frac{1}{3} = \frac{1}{6}.$$

By memorizing the fractional, decimal, and percent equivalents that are listed on the inside back cover, you can greatly simplify some of your work.

Do Exercise 23.

Exercise Set 8.1

a Use the given information to complete a circle graph. Note that each circle is divided into 100 sections.

1. *Reasons for Drinking Coffee.*

To get going in the morning:	32%
Like the taste:	33%
Not sure:	2%
To relax:	4%
As a pick-me-up:	10%
It's a habit:	19%

Source: LMK Associates survey for Au Bon Pain Co., Inc.

2. *How Vacation Money is Spent.*

Transportation:	15%
Meals:	20%
Lodging:	32%
Recreation:	18%
Other:	15%

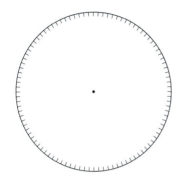
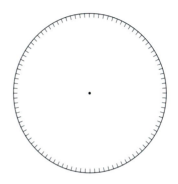

Write three kinds of notation as in Examples 1 and 2 on p. 456.

3. 90% **4.** 43.8% **5.** 12.5% **6.** 120%

b Write decimal notation.

7. 12% **8.** 42% **9.** 34.7% **10.** 69.7% **11.** 59.01%

12. 20.08% **13.** 10% **14.** 20% **15.** 1% **16.** 100%

17. 300% **18.** 700% **19.** 0.6% **20.** 0.2% **21.** 0.23%

22. 0.19% **23.** 105.24% **24.** 103.76%

Write decimal notation for the percent notation in each sentence.

25. On average, about 40% of the body weight of an adult male is muscle.

26. On average, about 23% of the body weight of an adult female is muscle.

27. A person's brain is 2.5% of his or her body weight.

28. It is known that 16% of all dessert orders in restaurants is for pie.

29. It is known that 62.2% of us think Monday is the worst day of the week.

30. Of all 18-year-olds, 68.4% have a driver's license.

Write percent notation.

31. 0.47   **32.** 0.87   **33.** 0.03   **34.** 0.01   **35.** 8.7

36. 4   **37.** 0.334   **38.** 0.889   **39.** 0.75   **40.** 0.99

41. 0.4   **42.** 0.5   **43.** 0.8925   **44.** 0.0258

Write percent notation for the decimal notation in each sentence.

45. Around the fourth of July, about 0.000104 of all children aged 15 to 19 suffer injuries from fireworks.

46. With a relative humidity of 0.80, a temperature of 75°F feels like 88°F.

47. It is known that 0.24 of us go to the movies once a month.

48. It is known that 0.458 of us sleep between 7 and 8 hours.

49. Of all money spent on sound recordings, 0.326 is spent on rock music.

50. About 0.026 of all college football players go on to play professional football.

Chapter 8   Percent Notation

c Write percent notation.

51. $\frac{41}{100}$ **52.** $\frac{36}{100}$ **53.** $\frac{5}{100}$ **54.** $\frac{1}{100}$ **55.** $\frac{2}{10}$ **56.** $\frac{7}{10}$

57. $\frac{7}{25}$ **58.** $\frac{1}{20}$ **59.** $\frac{3}{20}$ **60.** $\frac{3}{4}$ **61.** $\frac{5}{8}$ **62.** $\frac{13}{50}$

63. $\frac{4}{5}$ **64.** $\frac{2}{5}$ **65.** $\frac{2}{3}$ **66.** $\frac{1}{3}$ **67.** $\frac{29}{50}$ **68.** $\frac{13}{25}$

Write percent notation for the fractional notation in each sentence.

69. Bread is $\frac{9}{25}$ water.

70. Milk is $\frac{7}{8}$ water.

Write fractional notation. Try to remember the information in the table that appears on the inside back cover.

71. 85%   **72.** 55%   **73.** 62.5%   **74.** 12.5%

75. $33\frac{1}{3}\%$ **76.** $83\frac{1}{3}\%$ **77.** $16.\overline{6}\%$ **78.** $66.\overline{6}\%$

79. 7.25%   **80.** 4.85%   **81.** 0.8%   **82.** 0.2%

Exercise Set 8.1

Write percent notation for the fractions in this pie chart.

Engagement Times of Married Couples

83. $\dfrac{21}{100}$

84. $\dfrac{1}{5}$

85. $\dfrac{6}{25}$

86. $\dfrac{7}{20}$

Use this bar graph to write fractional notation for the percentage of people who "greatly enjoy" each type of food.

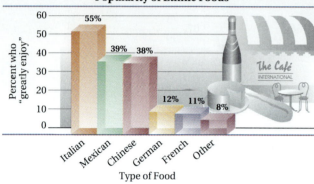

Popularity of Ethnic Foods

Note that there can be multiple responses.

87. Italian

88. Mexican

89. Chinese

90. German

91. French

92. Other

93. *Energy Consumption.* The United States uses 26.4% of the world's energy (**Source**: U.S. Energy Information Administration, *International Energy Annual*). Write fractional notation for 26.4%.

94. *World Population.* The United States has 4.5% of the world's population. Write fractional notation for 4.5%.

95. *Election Turnouts.* In the 1996 presidential election, 45.7% of the voting-age population voted. Write fractional notation for 45.7%.

96. *Recordings Purchased.* Of the money spent on all sound recordings, 84.5% comes from people less than 45 yr old (**Source**: Recording Industry Association of America, Inc., *1997 Consumer Profile*). Write fractional notation for 84.5%.

Complete the table.

97.

Fractional Notation	Decimal Notation	Percent Notation
$\frac{1}{8}$		$12\frac{1}{2}\%$, or 12.5%
$\frac{1}{6}$		
		20%
	0.25	
		$33\frac{1}{3}\%$, or $33.\overline{3}\%$
		$37\frac{1}{2}\%$, or 37.5%
		40%
$\frac{1}{2}$	0.5	50%

98.

Fractional Notation	Decimal Notation	Percent Notation
$\frac{3}{5}$		
	0.625	
$\frac{2}{3}$		
	0.75	75%
$\frac{4}{5}$		
$\frac{5}{6}$		$83\frac{1}{3}\%$, or $83.\overline{3}\%$
$\frac{7}{8}$		$87\frac{1}{2}\%$, or 87.5%
		100%

99.

Fractional Notation	Decimal Notation	Percent Notation
	0.5	
$\frac{1}{3}$		
		25%
		$16\frac{2}{3}\%$, or $16.\overline{6}\%$
	0.125	
$\frac{3}{4}$		
	$0.8\overline{3}$	
$\frac{3}{8}$		

100.

Fractional Notation	Decimal Notation	Percent Notation
		40%
		$62\frac{1}{2}\%$, or 62.5%
	0.875	
$\frac{1}{1}$		
	0.6	
	$0.\overline{6}$	
$\frac{1}{5}$		
	0.8	

Exercise Set 8.1

Skill Maintenance

Convert to a mixed numeral. [4.5b]

101. $\dfrac{100}{3}$

102. $-\dfrac{75}{2}$

Solve.

103. $0.05 \times b = 20$ [5.7a]

104. $3 = 0.16 \times b$ [5.7a]

105. $\dfrac{24}{37} = \dfrac{15}{x}$ [7.3b]

106. $\dfrac{17}{18} = \dfrac{x}{27}$ [7.3b]

Synthesis

107. ◆ Tammy remembers that $\frac{1}{4} = 25\%$. Explain how she can use this to (a) write $\frac{1}{8}$ in percent notation and (b) write $\frac{5}{8}$ in percent notation.

108. ◆ Is it always best to convert from fractional notation to percent notation by first finding decimal notation? Why or why not?

109. ◆ Athletes sometimes speak of "giving 110%" effort. Does this make sense? Why or why not?

Write percent notation.

110. ▦ $\dfrac{41}{369}$

111. ▦ $\dfrac{54}{999}$

112. $2.5\overline{74631}$

113. $3.2\overline{93847}$

Write decimal notation.

114. $\dfrac{14}{9}\%$

115. $\dfrac{19}{12}\%$

116. $\dfrac{729}{7}\%$

117. $\dfrac{637}{6}\%$

8.2 Solving Percent Problems Using Proportions*

a Translating to Proportions

A percent is a ratio of some number to 100. For example, 75% is the ratio $\frac{75}{100}$. The numbers 3 and 4 have the same ratio as 75 and 100. Thus,

$$75\% = \frac{75}{100} = \frac{3}{4}.$$

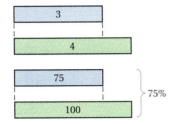

Objectives

a Translate percent problems to proportions.

b Solve basic percent problems.

For Extra Help

TAPE 14   MAC WIN   CD-ROM

To solve a percent problem using a proportion, we translate as follows:

$$\begin{array}{r}\text{Number} \rightarrow \\ 100 \longrightarrow\end{array} \frac{N}{100} = \frac{a}{b} \begin{array}{l}\leftarrow \text{Amount} \\ \leftarrow \text{Base}\end{array}$$

You might find it helpful to read this as "part is to whole as part is to whole."

For example,

60% of 25 is 15

translates to

$$\frac{60}{100} = \frac{15}{25}. \begin{array}{l}\leftarrow \text{Amount} \\ \leftarrow \text{Base}\end{array}$$

A clue in translating is that the base, b, corresponds to 100 and usually follows the wording "percent of." Also, $N\%$ always translates to $N/100$. Another aid in translating is to make a comparison drawing. To do this, we start with the percent side and list 0% at the top and 100% near the bottom. Then we estimate where the specified percent—in this case, 60%—is located. The corresponding quantities are then filled in. The base—in this case, 25—always corresponds to 100% and the amount—in this case, 15—corresponds to the specified percent.

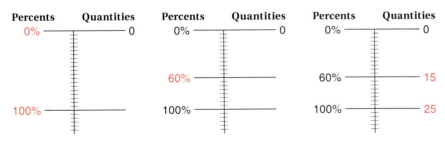

The proportion can then be read easily from the drawing.

*Note: Sections 8.2 and 8.3 present two methods for solving percent problems. You may prefer one method over the other, or your instructor may specify the method to be used. Section 8.2 is used as a check in Section 8.4, but otherwise it is not used in future sections of this book.

Translate to a proportion. Do not solve.

1. 12% of 50 is what?

2. What is 40% of 60?

3. 130% of 72 is what?

Translate to a proportion. Do not solve.

4. 45 is 20% of what?

5. 120% of what is 60?

Example 1 Translate to a proportion.

23% of 5 is what?

↓ ↓ ↘
number of   base   amount,
hundredths a

$$\frac{23}{100} = \frac{a}{5}$$

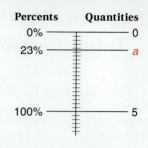

Example 2 Translate to a proportion.

What is 124% of 49?

↙ ↓ ↓
amount,   number of   base
a hundredths

$$\frac{124}{100} = \frac{a}{49}$$

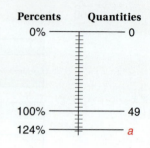

Do Exercises 1–3.

Example 3 Translate to a proportion.

3 is 10% of what?

↓ ↓ ↘
amount   number of   base,
          hundredths b

$$\frac{10}{100} = \frac{3}{b}$$

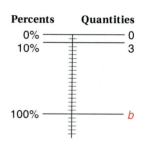

Example 4 Translate to a proportion.

45% of what is 23?

↓ ↓ ↓
number of   base,   amount
hundredths b

$$\frac{45}{100} = \frac{23}{b}$$

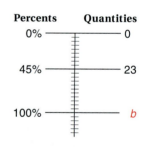

Do Exercises 4 and 5.

Example 5 Translate to a proportion.

10 is what percent of 20?

↓ ↙ ↘
amount   number of   base
         hundredths, N

$$\frac{N}{100} = \frac{10}{20}$$

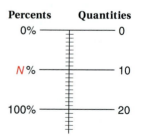

Answers on page A-22

Example 6 Translate to a proportion.

What percent of 50 is 7?

↓ ↓ ↓
number of   base   amount
hundredths, N

$$\frac{N}{100} = \frac{7}{50}$$

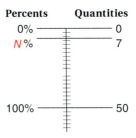

Do Exercises 6 and 7.

b Solving Percent Problems

After a percent problem has been translated to a proportion, we solve as in Section 7.3.

Example 7 5% of what is $20?

↓ ↓ ↓
number of   base,   amount
hundredths b

Translate: $\dfrac{5}{100} = \dfrac{20}{b}$

Solve: $5 \cdot b = 100 \cdot 20$ **Equating cross products**

$\dfrac{5b}{5} = \dfrac{2000}{5}$ **Dividing by 5**

$b = 400$ **Simplifying**

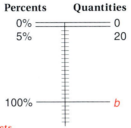

Thus, 5% of $400 is $20. The answer is $400.

Do Exercise 8.

Example 8 120% of 42 is what?

↓ ↓ ↓
number of   base   amount,
hundredths a

Translate: $\dfrac{120}{100} = \dfrac{a}{42}$

Solve: $120 \cdot 42 = 100 \cdot a$ **Equating cross products**

$\dfrac{5040}{100} = \dfrac{100a}{100}$ **Dividing by 100**

$50.4 = a$ **Simplifying**

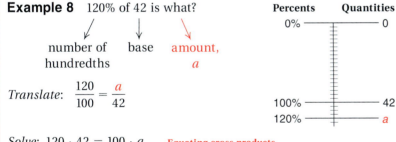

Thus, 120% of 42 is 50.4. The answer is 50.4.

Do Exercises 9 and 10.

Translate to a proportion. Do not solve.

6. 16 is what percent of 40?

7. What percent of 84 is 10.5?

8. Solve:

    20% of what is $45?

Solve.

9. 64% of 55 is what?

10. What is 12% of 50?

Answers on page A-22

11. Solve:

60 is 120% of what?

12. Solve:

$12 is what percent of $40?

13. Solve:

What percent of 84 is 10.5?

Answers on page A-22

Example 9 3 is 16% of what?

  ↓ ↓ ↓
amount number of base, b
   hundredths

Translate: $\dfrac{3}{b} = \dfrac{16}{100}$

Solve: $3 \cdot 100 = b \cdot 16$ Equating cross products

$\dfrac{300}{16} = \dfrac{16b}{16}$ Dividing by 16

$18.75 = b$ Simplifying

Thus, 3 is 16% of 18.75. The answer is 18.75.

Do Exercise 11.

Example 10 $10 is what percent of $20?

  ↓ ↓ ↓
amount number of base
   hundredths, N

Translate: $\dfrac{10}{20} = \dfrac{N}{100}$

Solve: $10 \cdot 100 = 20 \cdot N$ Equating cross products

$\dfrac{1000}{20} = \dfrac{20N}{20}$ Dividing by 20

$50 = N$ Simplifying

Always "look before you leap." Many students can solve this problem mentally: $10 is half, or 50%, or $20.

Thus, $10 is 50% of $20. The answer is 50%.

Do Exercise 12.

Example 11 What percent of 50 is 16?

  ↓ ↓ ↓
number of base amount
hundredths, N

Translate: $\dfrac{N}{100} = \dfrac{16}{50}$

Solve: $50 \cdot N = 100 \cdot 16$ Equating cross products

$\dfrac{50 \cdot N}{50} = \dfrac{100 \cdot 16}{50}$ Dividing by 50

$N = \dfrac{2 \cdot 50 \cdot 16}{50}$ Removing a factor equal to 1: $\dfrac{50}{50} = 1$

$N = 32$ Multiplying

Thus, 32% of 50 is 16. The answer is 32%. A quick alternative approach is to note that, since $50 \cdot 2 = 100$, we have $\dfrac{16}{50} = \dfrac{16 \cdot 2}{50 \cdot 2} = \dfrac{32}{100}$, or 32%.

Do Exercise 13.

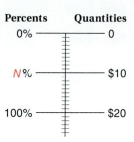

Exercise Set 8.2

a Translate to a proportion. Do not solve.

1. What is 37% of 74?

2. 66% of 74 is what?

3. 4.3 is what percent of 5.9?

4. What percent of 6.8 is 5.3?

5. 14 is 25% of what?

6. 133% of what is 40?

7. 9% of what is 37?

8. What is 132% of 75?

9. 70% of 660 is what?

10. 17 is 23% of what?

b Solve.

11. What is 4% of 1000?

12. What is 6% of 2000?

13. 4.8% of 60 is what?

14. 63.1% of 80 is what?

15. $24 is what percent of $96?

16. $14 is what percent of $70?

17. 102 is what percent of 100?

18. 103 is what percent of 100?

19. What percent of $480 is $120?

20. What percent of $80 is $60?

21. What percent of 160 is 150?

22. What percent of 33 is 11?

23. $18 is 25% of what?

24. $75 is 20% of what?

25. 60% of what is 54?

26. 80% of what is 96?

27. 65.12 is 74% of what?

28. 63.7 is 65% of what?

29. 80% of what is 16?

30. 80% of what is 10?

31. What is $62\frac{1}{2}$% of 40?

32. What is $43\frac{1}{4}$% of 2600?

33. What is 9.4% of $8300?

34. What is 8.7% of $76,000?

35. 9.48 is 120% of what?

36. 8.45 is 130% of what?

Skill Maintenance

Graph. [6.4b]

37. $y = -\frac{1}{2}x$

38. $y = 3x$

39. $y = 2x - 4$

40. $y = \frac{1}{2}x - 3$

Solve.

41. A recipe for pancakes calls for $\frac{1}{2}$ qt of buttermilk, $\frac{1}{3}$ qt of skim milk, and $\frac{1}{16}$ qt of oil. How many quarts of liquid ingredients does the recipe call for? [4.2d]

42. Guilford Gardeners purchased $\frac{3}{4}$ ton (T) of top soil. If the soil is to be shared equally among 6 gardeners, how much will each gardener receive? [3.7c]

Synthesis

43. ◆ In your own words, list steps that a classmate could use to solve any percent problem in this section.

44. ◆ In solving Example 10, a student simplifies $\frac{10}{20}$ before solving. Is this a good idea? Why or why not?

45. ◆ Can "comparison drawings," like those used in this section, be used to solve *any* proportion? Why or why not?

Solve.

46. ▦ What is 8.85% of $12,640?
Estimate _____
Calculate _____

47. ▦ 78.8% of what is 9809.024?
Estimate _____
Calculate _____

Chapter 8 Percent Notation

8.3 Solving Percent Problems Using Equations

a Translating to Equations

A second method for solving percent problems is to translate a problem's wording directly to an equation. For example, "23% of 5 is what?" translates as follows.

$$23\% \cdot 5 = a$$

Note how the key words are translated.

> "**Of**" translates to "$\cdot$", or "$\times$". "**Is**" translates to "$=$".
> "**What**" translates to a variable. **%** translates to "$\times \frac{1}{100}$" or "$\times 0.01$".

Example 1 Translate:

What is 11% of 49?
$$a = 11\% \cdot 49$$

Any letter can be used as a variable.

Do Exercises 1 and 2.

Example 2 Translate:

3 is 10% of what?
$$3 = 10\% \cdot b$$

Example 3 Translate:

45% of what is 23?
$$45\% \cdot b = 23$$

Do Exercises 3 and 4.

Example 4 Translate:

10 is what percent of 20?
$$10 = n \cdot 20$$

Example 5 Translate:

What percent of 50 is 7?
$$n \cdot 50 = 7$$

Do Exercises 5 and 6.

Objectives

a Translate percent problems to equations.

b Solve basic percent problems.

For Extra Help

TAPE 14   MAC WIN   CD-ROM

Translate to an equation. Do not solve.

1. 12% of 50 is what?

2. What is 40% of 60?

Translate to an equation. Do not solve.

3. 45 is 20% of what?

4. 120% of what is 60?

Translate to an equation. Do not solve.

5. 16 is what percent of 40?

6. What percent of 84 is 10.5?

Answers on page A-23

7. Solve:

What is 12% of 50?

b Solving Percent Problems

In solving percent problems, we use the *Translate* and *Solve* steps in the problem-solving strategy used throughout this text.

Percent problems are actually of three different types. Although the method we present does *not* require that you be able to identify which type we are studying, it is helpful to know them.

We know that

15 is 25% of 60,   or

$15 = 25\% \cdot 60$.

We can think of this as:

> Amount = Percent number · Base.

Each of the three types of percent problems depends on which of the three pieces of information is missing.

1. **Finding the *amount* (the result of taking the percent)**

   Example: What is 25% of 60?

   Translation: y =   25%   ·   60

2. **Finding the *base* (the number you are taking the percent of)**

   Example: 15 is 25% of what number?

   Translation: 15   =   25%   · y

3. **Finding the *percent number* (the percent itself)**

   Example: 15 is what percent of 60?

   Translation: 15   = y ·   60

Finding the Amount

Example 6   What is 11% of 49?

Translate: $a = 11\% \cdot 49$.

Solve: The variable is by itself. To solve the equation, we just convert 11% to decimal notation and multiply.

$a = 0.11(49)$

```
    4 9
  × 0.1 1
  ─────
    4 9
   4 9 0
  ─────
   5.3 9
```

$a = 5.39$

> A way of checking answers is by estimating as follows:
>
> $11\% \times 49 \approx 10\% \cdot 50$
> $= 0.10(50) = 5.$
>
> Since 5 is close to 5.39, our answer is reasonable.

Thus, 5.39 is 11% of 49. The answer is 5.39.

Do Exercise 7.

Answer on page A-23

Example 7 120% of $42 is what?

Translate: $120\% \cdot 42 = a$.

Solve: The variable is by itself. To solve the equation, we convert 120% to 1.2 and carry out the calculation.

$1.2(42) = a$

42
$\times 1.2$
84
420
$50.4 = a \qquad 50.4$

Thus, 120% of $42 is $50.40. The answer is $50.40.

Do Exercise 8.

Finding the Base

Example 8 5% of what is 20?

Translate: $5\% \cdot b = 20$.

Solve: This time the variable is *not* by itself. To solve the equation, we convert 5% to 0.05 and divide by 0.05 on both sides:

$$\frac{0.05 \cdot b}{0.05} = \frac{20}{0.05}$$ Dividing by 0.05 on both sides

$$b = \frac{20}{0.05}$$

$$= 400.$$

$0.05 \overline{)20.00}$ gives 400.
2000
0

Thus, 5% of 400 is 20. The answer is 400.

Example 9 $3 is 16% of what?

Translate: 3 is 16% of what?
$\downarrow\downarrow\downarrow\downarrow\downarrow$
$3 = 16\% \cdot b$.

Solve: Again, the variable is not by itself. To solve the equation, we convert 16% to 0.16 and divide by 0.16 on both sides:

$$\frac{3}{0.16} = \frac{0.16 \cdot b}{0.16}$$ Dividing by 0.16 on both sides

$$\frac{3}{0.16} = b$$

$18.75 = b$.

$0.16 \overline{)3.0000}$ gives 18.75
16
140
128
120
112
80
80
0

Thus, $3 is 16% of $18.75. The answer is $18.75.

Do Exercises 9 and 10.

8. Solve:

64% of $55 is what?

Solve.

9. 20% of what is 45?

10. $60 is 120% of what?

Answers on page A-23

11. Solve:

16 is what percent of 40?

Finding the Percent

In solving these problems, you *must* remember to convert to percent notation after you have solved the equation.

Example 10 17 is what percent of 20?

Translate:
$$17 = n \cdot 20$$

Solve: To solve the equation, we divide by 20 on both sides and convert the result to percent notation:

$$17 = n \cdot 20$$

$$\frac{17}{20} = \frac{n \cdot 20}{20}$$ Dividing by 20 on both sides

$$0.85 = n, \text{ or } n = 85\%.$$

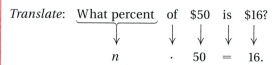

Thus, 17 is 85% of 20. The answer is 85%.

Do Exercise 11.

Example 11 What percent of $50 is $16?

Translate: What percent of $50 is $16?

$$n \cdot 50 = 16.$$

Solve: To solve the equation, we divide by 50 on both sides and convert the answer to percent notation:

$$\frac{n \cdot 50}{50} = \frac{16}{50}$$ Dividing by 50 on both sides

$$n = \frac{16}{50}$$

$$= \frac{16}{50} \cdot \frac{2}{2}$$ Multiplying by $\frac{2}{2}$, or 1, to get a denominator of 100

$$= \frac{32}{100}$$

$$= 32\%.$$ Converting to percent notation

Thus, 32% of $50 is $16. The answer is 32%.

Do Exercise 12.

CAUTION! When a question asks "what percent?", be sure to give the answer in percent notation.

---

12. Solve:

What percent of $84 is $10.50?

Answers on page A-23

Exercise Set 8.3

a Translate to an equation. Do not solve.

1. What is 32% of 78?
2. 98% of 57 is what?
3. 89 is what percent of 99?

4. What percent of 25 is 8?
5. 13 is 25% of what?
6. 21.4% of what is 20?

b Solve.

7. What is 85% of 276?
8. What is 74% of 53?

9. 150% of 30 is what?
10. 100% of 13 is what?

11. What is 6% of $300?
12. What is 4% of $45?

13. 3.8% of 50 is what?
14. $33\frac{1}{3}$% of 480 is what? (*Hint*: $33\frac{1}{3}\% = \frac{1}{3}$.)

15. $39 is what percent of $50?
16. $16 is what percent of $90?

17. 20 is what percent of 10?
18. 60 is what percent of 20?

19. What percent of $300 is $150?
20. What percent of $50 is $40?

21. What percent of 80 is 100?
22. What percent of 60 is 15?

23. 20 is 50% of what?

24. 57 is 20% of what?

25. 40% of what is $16?

26. 100% of what is $74?

27. 56.32 is 64% of what?

28. 71.04 is 96% of what?

29. 70% of what is 14?

30. 70% of what is 35?

31. What is $62\frac{1}{2}$% of 10?

32. What is $35\frac{1}{4}$% of 1200?

33. What is 8.3% of $10,200?

34. What is 9.2% of $5600?

35. $66\frac{2}{3}$% of what is 27.4?
(Hint: $66\frac{2}{3}\% = \frac{2}{3}$.)

36. $33\frac{1}{3}$% of what is 17.2?

Skill Maintenance

Write fractional notation. [5.1b]

37. 0.623

38. 1.9

39. 2.37

Write decimal notation. [5.1c]

40. $\dfrac{9}{1000}$

41. $\dfrac{39}{100}$

42. $\dfrac{57}{10}$

Synthesis

43. ◆ Write a percent problem that could be translated to the equation
$30 = n \cdot 80$.

44. ◆ Write a percent problem that could be translated to the equation
$25 = 4\% \cdot b$.

45. ◆ To calculate a 15% tip on a $24 bill, a customer adds $2.40 and half of $2.40, or $1.20, to get $3.60. Is this procedure valid? Why or why not?

Solve.

46. What is 7.75% of $10,880?
Estimate _____
Calculate _____

47. 50,951.775 is what percent of 78,995?
Estimate _____
Calculate _____

48. *Recyclables.* It is estimated that 40% to 50% of all trash is recyclable. If a community produces 270 tons of trash, how much of their trash is recyclable?

49. 40% of $18\frac{3}{4}$% of $25,000 is what?

8.4 Applications of Percent

a | Applied Problems Involving Percent

Applied problems involving percent are not always stated in a manner easily translated to an equation. In such cases, it is helpful to rephrase the problem before translating. Sometimes it also helps to make a drawing.

Objectives

a | Solve applied problems involving percent.

b | Solve applied problems involving percent of increase or decrease.

For Extra Help

TAPE 15   MAC WIN   CD-ROM

Example 1 *Paper Recycling.* In a recent year, the United States generated 81.5 million tons of paper waste, of which about 32.6 million tons were recycled (**Source:** Environmental Protection Agency). What percent of paper waste was recycled?

1. **Familiarize.** The question asks for a percent. We know that 10% of 81.5 is approximately 8. Since $8 \cdot 4 = 32$, which is close to 32.6, we expect the answer to be close to 40%. We let n = the percent of paper waste that was recycled.

2. **Translate.** We can rephrase the question and translate as follows:

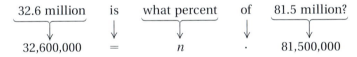

32.6 million   is   what percent   of   81.5 million?
32,600,000   = n ·   81,500,000

3. **Solve.** We solve as we did in Section 8.3:

$32{,}600{,}000 = n \cdot 81{,}500{,}000$

$\dfrac{32{,}600{,}000}{81{,}500{,}000} = \dfrac{n \cdot 81{,}500{,}000}{81{,}500{,}000}$ Dividing by 81,500,000 on both sides

$0.4 = n$

$40\% = n.$ Remember to write percent notation.

4. **Check.** To check, we note that 40% is just what we predicted in the *Familiarize* step.

5. **State.** About 40% of the paper waste was recycled.

Do Exercise 1.

Example 2 *"Junk" Mail.* The U.S. Postal Service estimates that we read 78% of the junk mail we receive. Suppose that a business sends out 9500 advertising brochures. How many brochures can the business expect to be opened and read?

1. **Familiarize.** Since 78% ≈ 75% and 9500 ≈ 10,000, we can estimate that about 75% of 10,000, or 7500 brochures will be opened and read. To find a more exact answer, we let a = the number of brochures that will be opened and read.

1. *Desserts.* If a restaurant sells 250 desserts in an evening, it is typical that 40 of them will be pie. What percent of the desserts sold will be pie?

Answer on page A-23

8.4 Applications of Percent

479

2. *Human Anatomy.* The weight of a human brain is 2.5% of total body weight. A person weighs 200 lb. What does the brain weigh?

Brain Weight vs. Body Weight

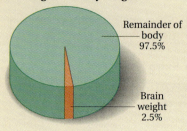

2. **Translate.** The question can be rephrased and translated as follows.*

$$\text{What number is } 78\% \text{ of } 9500?$$
$$a \quad = \quad 78\% \quad \cdot \quad 9500$$

3. **Solve.** We convert 78% to decimal notation and multiply:

$$a = 78\% \cdot 9500 = 0.78(9500) = 7410.$$

4. **Check.** To check, we note that our answer, 7410, is not far from 7500, our estimate in the *Familiarize* step. The student can also check that $7410 \div 9500$ is 0.78, or 78%.

5. **State.** The business can expect 7410 of its brochures to be opened and read.

Do Exercise 2.

b | Percent of Increase or Decrease

What do we mean when we say that the price of Swiss cheese has decreased 8%? If the price was $5.00 per pound and it went down to $4.60 per pound, then the decrease is $0.40, which is 8% of the original price. We can see this in the following figure.

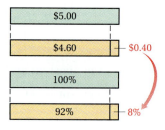

Example 3 *Energy Consumption.* With proper furnace maintenance, a family that pays a monthly fuel bill of $78.00 can reduce their bill to $70.20. What is the percent of decrease?

1. **Familiarize.** We find the amount of decrease and then make a drawing.

$$\begin{array}{r} 78.00 \\ -70.20 \\ \hline 7.80 \end{array} \quad \begin{array}{l} \text{Original bill} \\ \text{New bill} \\ \text{Decrease} \end{array}$$

We let n = the percent of decrease.

2. **Translate.** We rephrase and translate as follows.*

$$7.80 \text{ is what percent of } 78.00?$$
$$7.80 \quad = \quad n \quad \cdot \quad 78.00$$

Answer on page A-23

---

*We can also use the proportion method of Section 8.2 to solve.

3. **Solve.** To solve the equation, we divide by 78 on both sides:

$$\frac{7.80}{78.00} = \frac{n \cdot 78.00}{78.00}$$ Dividing by 78 on both sides

$$0.1 = n$$ You may have noticed earlier that 7.8 is 10% of 78.

$$10\% = n.$$ Changing from decimal to percent notation

4. **Check.** To check, we note that, with a 10% decrease, the reduced bill should be 90% of the original bill. Since 90% of 78 = 0.9(78) = 70.20, our answer checks.

5. **State.** The percent of decrease of the fuel bill is 10%.

Do Exercise 3.

Example 4 *Wages.* A sixth-grade teacher earns $27,000 one year and receives a 6% raise the next. What is the new salary?

1. **Familiarize.** We note that the amount of the raise can be found and then added to the old salary. A drawing can help us visualize this. We let x = the new salary.

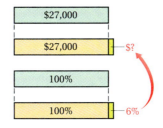

2. **Translate.** We rephrase the question and translate as follows.

What	is	the old salary	plus	6%	of	the old salary?
↓	↓	↓	↓	↓	↓	↓
x	=	27,000	+	6%	·	27,000

3. **Solve.** To solve, we convert 6% to a decimal and simplify:

$$x = 27,000 + 0.06(27,000)$$
$$= 27,000 + 1620 \quad \text{The value of the raise is \$1620.}$$
$$= 28,620.$$

4. **Check.** To check, we note that the new salary is 100% of the old salary plus 6% of the old salary. Thus the new salary is 106% of the old salary. Since 1.06(27,000) = 28,620, our answer checks.

5. **State.** The new salary is $28,620.

Do Exercise 4.

Percents provide a way to measure increases or decreases relative to the original amount. For example, the average salary of an NBA basketball player increased from $1.558 million in 1994 to $1.867 million in 1995. To find the *percent of increase* in salary, we first subtract to find out how much more the salary was in 1995: $1.867 million − $1.558 million = $0.309 million. Then we determine what percent of the original amount the increase was.

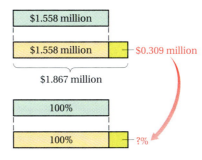

3. *Energy Consumption.* By using only cold water in the washing machine, a household with a monthly fuel bill of $78.00 can reduce their bill to $74.88. What is the percent of decrease?

4. *Wages.* A daycare worker earns $17,500 one year and receives a 9% raise the next. What is the new salary?

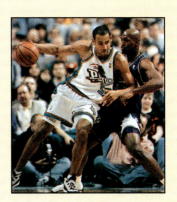

Answers on page A-23

8.4 Applications of Percent

481

5. *Automobile Price.* The price of an automobile increased from $15,800 to $17,222. What was the percent of increase?

Since $0.309 million = $309,000, we are asking

$309,000 is what percent of $1,558,000?

This translates to the following:

$309,000 = x \cdot 1,558,000.$

This is an equation of the type studied in Sections 8.2 and 8.3. Solving the equation, we can confirm that $0.309 million, or $309,000, is about 19.8% of $1.558 million. Thus the percent of increase in salary was 19.8%.

> To find a percent of increase or decrease:
> a) Find the amount of increase or decrease.
> b) Then determine what percent this is of the original amount.

Example 5 *Digital-Camera Screen Size.* The diagonal of the display screen of a digital camera was recently increased from 1.8 in. to 2.5 in. What was the percent of increase in the diagonal?

1. **Familiarize.** We note that the increase in the diagonal was $2.5 - 1.8$, or 0.7 in. A drawing can help us to visualize the situation. We let n = the percent of increase.

2. **Translate.** We rephrase the question and translate.

   0.7 in. is what percent of 1.8 in.?

   $0.7 = n \cdot 1.8$

3. **Solve.** To solve the equation, we divide by 1.8 on both sides:*

   $$\frac{0.7}{1.8} = \frac{n \cdot 1.8}{1.8}$$

   $0.389 \approx n$ **Rounded to the nearest thousandth**

   $38.9\% \approx n.$ **Remember to write percent notation.**

4. **Check.** To check, we take 38.9% of 1.8:

   $38.9\% \cdot 1.8 = 0.389(1.8) = 0.7002.$

   Since we rounded the percent, this approximation is close enough to 0.7 to be a good check.

5. **State.** The percent of increase of the screen diagonal is 38.9%.

Do Exercise 5.

---

*We can also use the proportion method of Section 8.2 and solve:

$$\frac{0.7}{1.8} = \frac{N}{100}.$$

Answer on page A-23

Exercise Set 8.4

a Solve.

1. *Left-handed Professional Bowlers.* It has been determined by sociologists that 17% of the population is left-handed. Each tournament conducted by the Professional Bowlers Association has 120 entrants. How many would you expect to be left-handed? not left-handed? Round to the nearest one.

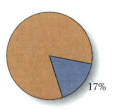

Total: 120

2. *Advertising Budget.* A common guideline for businesses is to use 5% of their operating budget for advertising. Ariel Electronics has an operating budget of $8000 per week. How much should it spend each week for advertising? for other expenses?

Total: $8000

3. Of all moviegoers, 67% are in the 12–29 age group. A five-screen cinema complex is filled with 600 moviegoers. How many would you expect to find in the 12–29 age group?

4. Deming, New Mexico, claims to have the purest drinking water in the world. It is 99.9% pure. If you had 240 L of water from Deming, how much of it, in liters, would be pure? impure?

5. A baseball player gets 13 hits in 40 at-bats. What percent are hits? not hits?

6. On a test of 80 items, Erika had 76 correct. What percent were correct? incorrect?

7. A lab technician has 680 mL of a solution of water and acid; 3% is acid. How many milliliters are acid? water?

8. A lab technician has 540 mL of a solution of alcohol and water; 8% is alcohol. How many milliliters are alcohol? water?

9. *TV Usage.* Of the 8760 hr in a year, most television sets are on for 2190 hr. What percent is this?

10. *Colds from Kissing.* In a medical study, it was determined that if 800 people kiss someone who has a cold, only 56 will actually catch a cold. What percent is this?

11. *Maximum Heart Rate.* Treadmill tests are often administered to diagnose heart ailments. A guideline in such a test is to try to get you to reach your *maximum heart rate,* in beats per minute. The maximum heart rate is found by subtracting your age from 220 and then multiplying by 85%. What is the maximum heart rate of someone whose age is 25? 36? 48? 55? 76? Round to the nearest one.

12. *Tampoline Injuries.* As the number of home trampolines has increased, so too has the number of trampoline-related injuries. In 1995, there were 58,400 injuries (**Source:** *The New York Times,* March 3, 1998).

   a) Research shows that 93% of all trampoline-related injuries occur at home. How many occurred at home in 1995?
   b) From 1990 to 1995, a total of 249,400 trampoline-related injuries occurred. What percentage of these occurred in 1995?

b Solve.

13. The amount in a savings account increased from $150 to $162. What was the percent of increase?

14. The population of South Creek increased from 840 to 882. What was the percent of increase?

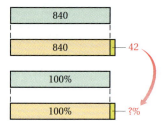

15. During a sale, the price of a dress decreased from $90 to $72. What was the percent of decrease?

16. A person on a diet goes from a weight of 125 lb to a weight of 110 lb. What is the percent of decrease?

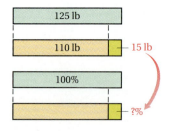

17. Rachel earns $28,600 one year and receives a 5% raise in salary. What is the new salary?

18. Derek earns $20,400 one year and receives an 8% raise in salary. What is the new salary?

19. The value of a car typically decreases by 30% in the first year. A car is bought for $18,000. What is its value one year later?

20. One year the pilots of an airline shocked the business world by taking an 11% pay cut. The former salary was $55,000. What was the reduced salary?

21. *World Population.* World population is increasing by 1.6% each year. In 1999, it was 6.0 billion. How much will it be in 2000? 2001? 2002?

22. *Cooling Costs.* By increasing the thermostat from 72° to 78°, a family can reduce its cooling bill by 50%. If the cooling bill was $106.00, what would the new bill be? By what percent has the temperature been increased?

23. *Car Depreciation.* A car generally depreciates 30% of its original value in the first year. A car is worth $25,480 after the first year. What was its original cost?

24. *Car Depreciation.* Given normal use, an American-made car will depreciate 30% of its original cost the first year and 14% of its remaining value in the second year. What is the value of a car at the end of the second year if its original cost was $36,400? $28,400? $26,800?

25. *Tipping.* Diners frequently add a 15% tip when charging a meal to a credit card. What is the total amount charged if the cost of the meal, without tip, is $15? $34? $49?

26. *Carpentry.* A cross section of a standard "two-by-four" board actually measures $1\frac{1}{2}$ in. by $3\frac{1}{2}$ in. The rough board is 2 in. by 4 in. but is planed and dried to the finished size. What percent of the wood is removed in planing and drying?

27. *Drunk Driving.* Despite efforts by groups such as MADD (Mothers Against Drunk Driving), the number of alcohol-related deaths rose in 1995 after many years of decline. The data in the table show the number of deaths from 1986 to 1995.
   a) What is the percent of increase in the number of alcohol-related deaths from 1994 to 1995?
   b) What is the percent of decrease in the number of alcohol-related deaths from 1986 to 1994?

   Alcohol-Related Traffic Deaths Back on the Increase!!

Year	Deaths
1986	24,045
1987	23,641
1988	23,626
1989	22,436
1990	22,084
1991	19,887
1992	17,859
1993	17,473
1994	16,589
1995	17,274

   Source: National Highway Traffic Safety Administration

28. *Fetal Acoustic Stimulation.* Each year there are about 4 million births in the United States. Of these, about 120,000 births occur in breech position (delivery of a fetus with the buttocks or feet appearing first). A new technique, called *fetal acoustic stimulation (FAS)*, uses sound directed through a mother's abdomen in order to stimulate movement of the fetus to a safer position. In a recent study of this low-risk and low-cost procedure, FAS enabled doctors to turn the baby in 34 of 38 cases (**Source:** Johnson and Elliott, "Fetal Acoustic Stimulation, an Adjunct to External Cephalic Versions: A Blinded, Randomized Crossover Study," American Journal of Obstetrics & Gynecology **173**, no. 5 (1995): 1369–1372).
   a) What percent of U.S. births are breech?
   b) What percent (rounded to the nearest tenth) of cases showed success with FAS?
   c) About how many breech babies yearly might be turned if FAS could be implemented in all breech births in the United States?
   d) Breech position is one reason for performing Caesarean section (or C-section) birth surgery. Researchers expect that FAS alone can eliminate the need for about 2000 C-sections yearly in the United States. Given this information, how many C-sections per year are due to breech position alone?

Exercise Set 8.4

29. *Strike Zone.* In baseball, the *strike zone* is normally a 17-in. by 40-in. rectangle. Some batters give the pitcher an advantage by swinging at pitches thrown out of the strike zone. By what percent is the area of the strike zone increased if a 2-in. border is added to the outside?

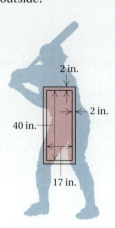

30. *Gardening.* Tony is planting grass on a 24-ft by 36-ft area in his back yard. He installs a 6-ft by 8-ft garden. By what percent has he reduced the area he has to mow?

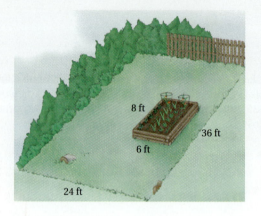

Skill Maintenance

Convert to decimal notation. [5.1c]

31. $\dfrac{25}{11}$

32. $\dfrac{11}{25}$

33. $\dfrac{27}{8}$

34. $\dfrac{43}{9}$

35. $\dfrac{23}{25}$

36. $\dfrac{20}{24}$

37. $\dfrac{14}{32}$

38. $\dfrac{2317}{1000}$

39. $\dfrac{34{,}809}{10{,}000}$

40. $\dfrac{27}{40}$

Synthesis

41. ◆ Which is better for a wage earner, and why: a 10% raise followed by a 5% raise a year later, or a 5% raise followed by a 10% raise a year later?

42. ◆ Write a problem for a classmate to solve. Design the problem so that the solution is "Jackie's raise was $7\tfrac{1}{2}\%$."

43. ◆ The former baseball player and manager Yogi Berra once said that "ninety percent of the game is half mental." If this is true, what percent of the game is mental? Explain your reasoning.

44. ◆ A workers' union is offered either a 5% "across-the-board" raise in which all salaries would increase 5%, or a flat $1650 raise for each worker. If the total payroll for the 123 workers is $4,213,365, which offer should the union select? Why?

45. *Adult Height.* It has been determined that at the age of 10, a girl has reached 84.4% of her final adult growth. Cynthia is 4 ft, 8 in. at the age of 10. What will be her final adult height?

46. *Adult Height.* It has been determined that at the age of 15, a boy has reached 96.1% of his final adult height. Claude is 6 ft, 4 in. at the age of 15. What will be his final adult height?

47. If p is 120% of q, then q is what percent of p?

48. A coupon allows a couple to have dinner and then have $10 subtracted from the bill. Before subtracting $10, however, the restaurant adds a tip of 15%. If the couple is presented with a bill for $44.05, how much would the dinner (without tip) have cost without the coupon?

49. A worker receives raises of 3%, 6%, and then 9%. By what percent has the original salary increased?

8.5 Consumer Applications: Sales Tax, Commission, and Discount

Objectives

a Solve applied problems involving sales tax and percent.

b Solve applied problems involving commission and percent.

c Solve applied problems involving discount and percent.

For Extra Help

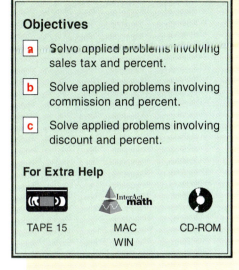

TAPE 15   MAC WIN   CD-ROM

a Sales Tax

Sales tax computations represent a special type of percent of increase problem. The sales tax rate in Colorado is 3%. This means that the tax is 3% of the purchase price. Suppose the purchase price on a coat is $124.95. The sales tax is then

$$3\% \text{ of } \$124.95, \quad \text{or} \quad 0.03 \cdot 124.95,$$

or

$$3.7485, \quad \text{or about} \quad \$3.75.$$

$124.95 + 3% sales tax

The total that you pay is the price plus the sales tax:

$$\$124.95 + \$3.75, \quad \text{or} \quad \$128.70.$$

Bill:		
Purchase price	=	$124.95
Sales tax (3% of $124.95)	=	+ 3.75
Total price		$128.70

> **Sales tax** = Sales tax rate · Purchase price
>
> **Total price** = Purchase price + Sales tax

Example 1 *New Jersey Sales Tax.* In 1998, the sales tax rate in New Jersey was 6%. How much tax was charged on the purchase of 3 CDs at $13.95 each? What was the total price?

a) We first find the cost of the CDs. It is

$$3 \cdot \$13.95 = \$41.85.$$

b) The sales tax on items costing $41.85 is

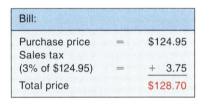

Sales tax rate · Purchase price

6% · $41.85,

or 0.06 · 41.85, or 2.511. Thus the tax is $2.51.

c) The total price is given by the purchase price plus the sales tax:

$$\$41.85 + \$2.51, \quad \text{or} \quad \$44.36.$$

To check, note that the total price is the purchase price plus 6% of the purchase price. Thus the total price is 106% of the purchase price. Since 1.06 · 41.85 = 44.361 and 44.361 rounds to 44.36, we have a check. The total price was $44.36.

Do Exercises 1 and 2.

1. *Connecticut Sales Tax.* In 1998, the sales tax rate in Connecticut was 6%. How much tax was charged on the purchase of a refrigerator that sold for $668.95? What was the total price?

2. *Rhode Island Sales Tax.* Morris buys 5 blank audiocassettes in Rhode Island, where the sales tax rate is 7%. If each tape costs $2.95, how much tax will be charged? What is the total price?

Answers on page A-23

3. The sales tax is $33 on the purchase of a $550 washer. What is the sales tax rate?

Example 2 The sales tax is $32 on the purchase of an $800 sofa. What is the sales tax rate?

Rephrase: Sales tax is what percent of purchase price?

Translate: $32 = r \cdot 800$

To solve the equation, we divide by 800 on both sides:

$$\frac{32}{800} = \frac{r \cdot 800}{800}$$

$$0.04 = r$$

$$4\% = r.$$

The sales tax rate is 4%.

Do Exercise 3.

Example 3 The sales tax on a laser printer is $31.74 and the sales tax rate is 5%. Find the purchase price (the price before taxes are added).

4. The sales tax on a 27-inch television is $25.20 and the sales tax rate is 6%. Find the purchase price (the price before taxes are added).

Rephrase: Sales tax is 5% of what?

Translate: $31.74 = 5\% \cdot b$, or $31.74 = 0.05 \cdot b$.

To solve, we divide by 0.05 on both sides:

$$\frac{31.74}{0.05} = \frac{0.05 \cdot b}{0.05}$$

$$634.8 = b.$$

```
           6 3 4.8
  0.0 5 ) 3 1.7 4ˏ0
           3 0 0 0
             1 7 4
             1 5 0
               2 4
               2 0
                 4 0
                 4 0
                   0
```

The purchase price is $634.80.

Answers on page A-23

Do Exercise 4.

b Commission

When you work for a **salary**, you receive the same amount of money each week or month. When you work for a **commission**, you are paid a percentage of the total sales for which you are responsible.

> **Commission** = Commission rate · Sales

Example 4 *Stereo Equipment Sales.* A salesperson's commission rate is 20%. What is the commission from the sale of $25,560 worth of stereo equipment?

$$\begin{array}{ccccc} \text{Commission} & = & \text{Commission rate} & \cdot & \text{Sales} \\ C & = & 20\% & \cdot & 25{,}560 \end{array}$$

This tells us what to do. We multiply.

$$\begin{array}{r} 25{,}560 \\ \times 0.2 \\ \hline 5112.0 \end{array}$$ 20% = 0.20 = 0.2

The commission is $5112.

Do Exercise 5.

Example 5 *Farm Machinery Sales.* Dawn earns a commission of $3000 selling $60,000 worth of farm machinery. What is the commission rate?

$$\begin{array}{ccccc} \text{Commission} & = & \text{Commission rate} & \cdot & \text{Sales} \\ 3000 & = & r & \cdot & 60{,}000 \end{array}$$

To solve this equation, we divide by 60,000 on both sides:

$$\frac{3000}{60{,}000} = \frac{r \cdot 60{,}000}{60{,}000} = r.$$

---

5. Raul's commission rate is 30%. What is the commission from the sale of $18,760 worth of air conditioners?

Answer on page A-23

8.5 Consumer Applications: Sales Tax, Commission, and Discount

489

6. Liz earns a commission of $6000 selling $24,000 worth of refrigerators. What is the commission rate?

We can divide, but this time we simplify by removing a factor equal to 1:

$$r = \frac{3000}{60,000} = \frac{1}{20} \cdot \frac{3000}{3000} = \frac{1}{20} = 0.05 = 5\%.$$

The commission rate is 5%.

Do Exercise 6.

Example 6 *Motorcycle Sales.* Joyce's commission rate is 25%. She receives a commission of $425 on the sale of a motorcycle. How much did the motorcycle cost?

Commission = Commission rate · Sales
425 = 25% · S

To solve this equation, we divide by 0.25 on both sides:

$$\frac{425}{0.25} = \frac{0.25 \cdot S}{0.25}$$

$$1700 = S.$$

```
         1 7 0 0.
0.2 5 ) 4 2 5.0 0
         2 5 0
         1 7 5
         1 7 5
             0
```

The motorcycle cost $1700.

7. Ben's commission rate is 16%. He receives a commission of $268 from sales of clothing. How many dollars worth of clothing were sold?

Do Exercise 7.

c Discount

Suppose that the regular price of a rug is $60, and the rug is on sale at 25% off. Since 25% of $60 is $15, the sale price is $60 − $15, or $45. We call $60 the **original**, or **marked price**, 25% the **rate of discount**, $15 the **discount**, and $45 the **sale price**. Note that discount problems are a type of percent of decrease problem.

> **Discount** = Rate of discount · Original price
> **Sale price** = Original price − Discount

Answers on page A-23

Example 7 *Rug Prices.* A rug marked $240 is on sale at 25% off. What is the discount? the sale price?

a) Discount = Rate of discount · Original price
   D = 25% · 240

This tells us what to do. We convert 25% to decimal notation and multiply.

```
    2 4 0
  × 0.2 5      25% = 0.25
  -------
    1 2 0 0
    4 8 0 0
  -------
    6 0.0 0
```

The discount is $60.

b) Sale price = Marked price − Discount
   S = 240 − 60

This tells us what to do. We subtract.

```
    2 4 0
  −   6 0
  -------
    1 8 0
```

To check, note that the sale price is 75% of the marked price: 0.75 × 240 = 180.

The sale price is $180.

Do Exercise 8.

Example 8 *Antique Pricing.* An antique table is marked down from $620 to $527. What is the rate of discount?

We first find the discount by subtracting the sale price from the original price:

```
    6 2 0
  − 5 2 7
  -------
      9 3.
```

The discount is $93.

Next, we use the equation for discount:

   Discount = Rate of discount · Original price
   93 = r · 620.

---

8. A suit marked $140 is on sale at 24% off. What is the discount? the sale price?

Answer on page A-23

8.5 Consumer Applications: Sales Tax, Commission, and Discount

491

9. A pair of running shoes is reduced from $75 to $60. Find the rate of discount.

To solve, we divide by 620 on both sides:

$$\frac{93}{620} = \frac{r \cdot 620}{620}$$

$$0.15 = r$$

$$15\% = r.$$

```
        0.1 5
6 2 0 ) 9 3.0 0
        6 2 0
        3 1 0 0
        3 1 0 0
              0
```

The discount rate is 15%.

> To check, note that a 15% discount rate means that 85% of the original price is paid:
> $0.85 \cdot 620 = 527.$

Do Exercise 9.

Calculator Spotlight

% Key. Many calculators have a percent key. This key can be useful in calculations like 20% · $25,560, as in Example 4, but you may need to change the order to 25,560 · 20%. To do the calculation, press

| 2 | 5 | 5 | 6 | 0 | × | 2 | 0 | SHIFT | % |.

The displayed result is

| 5112 |.

Check your manual for other procedures for determining percents.

Exercises

Calculate.

1. 250 · 20%
2. 37% · 18,924
3. 67.2% · 124,898
4. 56,788.22 · 64.2%

On many calculators, there is a fast way to increase or decrease a number by any given percentage. In Example 1, the result of taking 6% of $41.85 and adding it to $41.85 can often be found by pressing

| 4 | 1 | . | 8 | 5 | × | 6 | SHIFT | % | + |.

The displayed result would be

| 44.361 |.

If the price had been *reduced* by 6%, the computation would be

| 4 | 1 | . | 8 | 5 | × | 6 | SHIFT | % | − |.

The displayed result would be

| 39.339 |.

Check your manual for other procedures for determining percents.

Exercise

Use a calculator with a | % | key to confirm your answers to Margin Exercises 5 and 8.

Answer on page A-23

Exercise Set 8.5

a Solve.

1. *Indiana Sales Tax.* The sales tax rate in Indiana is 5%. How much tax is charged on a generator costing $586? What is the total price?

2. *New York City Sales Tax.* The sales tax rate in New York City is 8.25%. How much tax is charged on photo equipment costing $248? What is the total price?

3. *Illinois Sales Tax.* The sales tax rate in Illinois is 6.25%. How much tax is charged on a purchase of 5 telephones at $53 apiece? What is the total price?

4. *California Sales Tax.* The sales tax rate in California is 6%. How much tax is charged on a purchase of 5 teapots at $37.99 apiece? What is the total price?

5. *Mountain Bike Sales.* The sales tax is $15.96 on the purchase of a mountain bike that sells for $399. What is the sales tax rate?

6. *Jewelry Sales.* The sales tax is $15 on the purchase of a diamond ring that sells for $500. What is the sales tax rate?

7. The sales tax is $44.75 on the purchase of a fiberglass canoe that sells for $895. What is the sales tax rate?

8. The sales tax is $9.12 on the purchase of a patio set that sells for $456. What is the sales tax rate?

9. *Truck Sales.* The sales tax on a used pickup truck is $250 and the sales tax rate is 5%. Find the purchase price (the price before taxes are added).

10. *Motorboat Sales.* The sales tax on the purchase of a motorboat is $112 and the sales tax rate is 2%. Find the purchase price.

11. The sales tax on a dining room set is $28 and the sales tax rate is 3.5%. Find the purchase price.

12. The sales tax on a home-theater speaker system is $66 and the sales tax rate is 5.5%. Find the purchase price.

13. The sales tax rate in Dallas is 2% for the city and 6.25% for the state. Find the total amount paid for 2 shower units at $332.50 apiece.

14. The sales tax rate in Omaha is 1% for the city and 5% for the state. Find the total amount paid for 3 air conditioners at $260 apiece.

15. The sales tax is $1030.40 on the purchase of a used Chevrolet Camaro for $18,400. What is the sales tax rate?

16. The sales tax is $979.60 on the purchase of a used Dodge Caravan for $15,800. What is the sales tax rate?

b Solve.

17. Kelly is about to receive her commission of 35% for selling $2580 of Amstar products. How much commission will Kelly receive?

18. Jose's commission rate is 32%. What is the commission from the sale of $12,500 worth of sailboards?

19. Bernie receives $87 as commission for selling $174 worth of cosmetics. What is the commission rate?

20. Donna earns $408 selling $3400 worth of running shoes. What is the commission rate?

21. An art gallery's commission rate is 40%. They receive a commission of $392. How many dollars worth of artwork were sold?

22. A real estate agent's commission rate is 7%. She receives a commission of $5600 on the sale of a home. How much did the home sell for?

23. A real estate commission is 6%. What is the commission on the sale of a $98,000 home?

24. A real estate commission is 8%. What is the commission on the sale of a piece of land for $68,000?

25. Bonnie earns $280.80 selling $2340 worth of tee shirts. What is the commission rate?

26. Chuck earns $1147.50 selling $7650 worth of ski passes. What is the commission rate?

27. Miguel's commission is increased according to how much he sells. He receives a commission of 5% for the first $2000 and 8% on the amount over $2000. What is the total commission on sales of $6000?

28. Lucinda earns a salary of $500 a month, plus a 2% commission on software sales. One month, she sold $8700 worth of software. What were her wages that month?

Chapter 8 Percent Notation

c Find what is missing.

29.
Marked Price	Rate of Discount	Discount	Sale Price
$300	10%		

30.
Marked Price	Rate of Discount	Discount	Sale Price
$2000	40%		

31.
$17.00	15%		

32.
$20.00	25%		

33.
	10%	$12.50	

34.
	15%	$65.70	

35.
$600		$240	

36.
$12,800		$1920	

37. Find the discount and the rate of discount for the ring in this ad.

1/2 CARAT T.W.
DIAMOND, 14K GOLD
LADY'S BRIDAL SET
was $1275.00
$888

38. Find the discount and the rate of discount for the calculator in this ad.

Calc-U-Sure C96
Graphing Calculator
- 8 line × 16 character display
- Pull-down menus
- Uses 3 "AAA" batteries
- Sliding plastic cover
- Model C96
- Mfr. List $115.00

69⁹⁸

39. Find the marked price and the rate of discount for the camcorder in this ad.

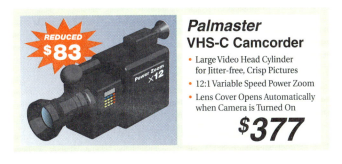

REDUCED $83

Palmaster
VHS-C Camcorder
- Large Video Head Cylinder for Jitter-free, Crisp Pictures
- 12:1 Variable Speed Power Zoom
- Lens Cover Opens Automatically when Camera is Turned On

$377

40. Find the marked price and the rate of discount for the cedar chest in this ad.

Lane Cedar Chest with Decorative Decal

249⁹⁹
Save $50

ALL CEDAR CHESTS ON SALE!
Largest selection of Lane cedar chests in stock!

Exercise Set 8.5

Skill Maintenance

Solve. [7.3b]

41. $\dfrac{x}{12} = \dfrac{24}{16}$

42. $\dfrac{7}{2} = \dfrac{11}{x}$

Graph. [6.4b]

43. $y = \dfrac{4}{3}x$

44. $y = -\dfrac{4}{3}x + 1$

Write decimal notation. [5.1c]

45. $\dfrac{5}{9}$

46. $\dfrac{23}{11}$

47. $-\dfrac{11}{12}$

48. $-\dfrac{13}{7}$

Synthesis

49. ◆ Carl's Car Care mistakenly charged Dawn 5% tax for a cleaning job that was all labor. To correct the mistake, they subtracted 5% from what Dawn paid. Was this correct? Why or why not?

50. ◆ Is the following ad mathematically correct? Why or why not?

51. ◆ An item that is no longer on sale at "25% off" receives a price tag that is $33\frac{1}{3}$% more than the sale price. Has the item price been restored to its original price? Why or why not?

52. ◆ Which is better, a discount of 40% on a book's list price or a discount of 20% on list price followed by another discount of 20% on the reduced price? Explain.

53. A real estate commission rate is 7.5%. A house sells for $98,500. How much does the seller get for the house after paying the commission?

54. *People Magazine.* In a recent subscription drive, *People* offered a subscription of 52 weekly issues for a price of $1.89 per issue. They advertised that this was a savings of 29.7% off the newsstand price. What was the newsstand price?

55. Gordon receives a 10% commission on the first $5000 in sales and 15% on all sales beyond $5000. If Gordon receives a commission of $2405, how much did he sell? Use a calculator and trial and error if you wish.

56. Tee shirts are being sold at the mall for $5 each, or 3 for $10. If you buy three tee shirts, what is the rate of discount?

57. Herb collects baseball memorabilia. He bought two autographed plaques, but became short of funds and had to sell them quickly for $200 each. On one, he made a 20% profit and on the other, he lost 20%. Did he make or lose money on the sale?

Calculate the costs associated with the purchase of a car or truck.

8.6 Consumer Applications: Interest

a Simple Interest

Suppose you put $100 into an investment for 1 year. The $100 is called the **principal**. If the **interest rate** is 8%, in addition to the principal, you will get back 8% of the principal, which is

   8% of $100,   or   0.08 · 100,   or $8.00.

The $8.00 is called the **interest**, or more precisely, the **simple interest**. It is, in effect, the price that a financial institution pays for the use of the money over time.

> The **simple interest** I on principal P, invested for t years at interest rate r, is given by
> $$I = P \cdot r \cdot t.$$

Example 1 What is the interest on $2500 invested at an interest rate of 6% for 1 year?

We use the formula $I = P \cdot r \cdot t$:

$I = P \cdot r \cdot t = \$2500 \cdot 6\% \cdot 1$
$ = \$2500 \cdot 0.06$
$ = \$150.$

```
    2 5 0 0
  ×   0.0 6
  ─────────
  1 5 0.0 0
```

The interest for 1 year is $150.

Do Exercise 1.

Example 2 What is the interest on a principal of $2500 invested at an interest rate of 6% for $\frac{1}{4}$ year?

We use the formula $I = P \cdot r \cdot t$:

$I = P \cdot r \cdot t = \$2500 \cdot 6\% \cdot \dfrac{1}{4}$

$ = \dfrac{\$2500 \cdot 0.06}{4}$

$ = \$37.50.$

```
         3 7.5
     4 ) 1 5 0.0
         1 2 0
         ─────
           3 0
           2 8
           ───
             2 0
             2 0
             ───
               0
```

> We could have instead found $\frac{1}{4}$ of 6% and then multiplied by 2500.

The interest for $\frac{1}{4}$ year is $37.50.

Do Exercise 2.

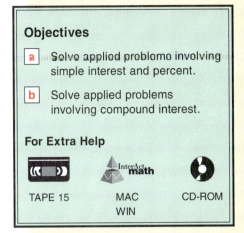

Objectives

a Solve applied problems involving simple interest and percent.

b Solve applied problems involving compound interest.

For Extra Help

TAPE 15   MAC WIN   CD-ROM

1. What is the interest on $4300 invested at an interest rate of 14% for 1 year?

2. What is the interest on a principal of $4300 invested at an interest rate of 14% for $\frac{3}{4}$ year?

Answers on page A-23

3. The Glass Nook borrows $4800 at 7% for 30 days. Find (a) the amount of simple interest due and (b) the total amount that must be paid after 30 days.

Unless specified otherwise, it is understood that interest rates refer to the *yearly* rate at which interest is paid. This means that when a time period is given in days, we must divide it by 365 to express the time as a fractional part of a year.

Example 3 To pay for a shipment of tee shirts, New Wave Designs borrows $8000 at 9% for 60 days. Find (a) the amount of simple interest that is due and (b) the total amount that must be paid after 60 days.

a) We express 60 days as a fractional part of a year:

$$I = P \cdot r \cdot t = \$8000 \cdot 9\% \cdot \frac{60}{365}$$ Note that 60 days = $\frac{60}{365}$ year.

$$= \$8000 \cdot 0.09 \cdot \frac{60}{365}$$

 $\$118.36.$ Using a calculator

The interest due for 60 days is $118.36.

b) The total amount to be paid after 60 days is the principal plus the interest:

$$\$8000 + \$118.36 = \$8118.36.$$

The total amount due is $8118.36.

Do Exercise 3.

b   Compound Interest

Simple interest is interest paid on principal only. When interest is paid on the accumulated interest as well as on the principal, we call it **compound interest.** This is the type of interest usually paid on investments. Suppose you have $5000 in a savings account at 6%. In 1 year, the account will contain the original $5000 plus 6% of $5000. Thus the total in the account after 1 year will be

106% of $5000,   or   1.06 · $5000,   or $5300.

Now suppose that the total of $5300 remains in the account for another year. At the end of this second year, the account will contain the $5300 plus 6% of $5300. The total in the account would thus be

106% of $5300,   or   1.06 · $5300,   or $5618.

Note that in the second year, interest is earned on the first year's interest. When this happens, we say that interest is **compounded annually.**

Answer on page A-23

Example 4 Find the amount in an account if $2000 is invested at 8%, compounded annually, for 2 years.

a) After 1 year, the account will contain 108% of $2000:

$1.08 \cdot \$2000 = \$2160.$

```
    2 0 0 0
  ×   1.0 8
  ─────────
    1 6 0 0 0
    0 0 0 0
  2 0 0 0
  ─────────
  2 1 6 0.0 0
```

b) At the end of the second year, the account will contain 108% of $2160:

$1.08 \cdot \$2160 = \$2332.80.$

```
      2 1 6 0
  ×     1.0 8
  ───────────
    1 7 2 8 0
    0 0 0 0
  2 1 6 0
  ───────────
  2 3 3 2.8 0
```

The amount in the account after 2 years is $2332.80.

Do Exercise 4.

Suppose that the interest in Example 4 were **compounded semi-annually**—that is, every half year. Interest would then be calculated twice a year at a rate of 8% ÷ 2, or 4%, each time. The approach used in Example 4 can then be adapted, as follows.

After the first $\frac{1}{2}$ year, the account will contain 104% of $2000:

$1.04(\$2000) = \$2080.$ *These calculations can be confirmed with a calculator.*

After a second $\frac{1}{2}$ year (1 full year), the account will contain 104% of $2080:

$1.04(\$2080) = \$2163.20.$

After a third $\frac{1}{2}$ year $\left(1\frac{1}{2} \text{ full years}\right)$, the account will contain 104% of $2163.20:

$1.04(\$2163.20) = \$2249.728$

$\approx \$2249.73.$ *Rounding to the nearest cent*

Finally, after a fourth $\frac{1}{2}$ year (2 full years), the account will contain 104% of $2249.73:

$1.04(\$2249.73) = \$2339.7192$

$\approx \$2339.72.$ *Rounding to the nearest cent*

Note that each multiplication was by 1.04 and that

$\$2000 \cdot 1.04^4 = \$2339.72.$ *Using a calculator and rounding to the nearest cent*

---

4. Find the amount in an account if $2000 is invested at 11%, compounded annually, for 2 years.

Answer on page A-23

8.6 Consumer Applications: Interest

5. A couple invests $7000 in an account paying 10%, compounded semiannually. Find the amount in the account after $1\frac{1}{2}$ years.

We have illustrated the following result.

> If a principal P has been invested at interest rate r, compounded n times a year, in t years it will grow to an amount A given by
> $$A = P \cdot \left(1 + \frac{r}{n}\right)^{n \cdot t},$$ where r is written in decimal notation.
>
> (Here $n \cdot t$ is the number of compounding periods and $\frac{r}{n}$ is the interest rate for each period.)

Example 5 The Ibsens invest $7000 in an account paying 8%, compounded quarterly. Find the amount in the account after $2\frac{1}{2}$ years.

We substitute $7000 for P, 0.08 for r, 4 for n, and $2\frac{1}{2}$ for t and solve for A:

$$A = P \cdot \left(1 + \frac{r}{n}\right)^{n \cdot t}$$
$$= 7000 \cdot \left(1 + \frac{0.08}{4}\right)^{4 \cdot (5/2)} \quad \text{Writing } 2\tfrac{1}{2} \text{ as } \tfrac{5}{2}$$
$$= 7000 \cdot (1 + 0.02)^{10} \quad 4 \cdot \tfrac{5}{2} = \tfrac{20}{2} = 10$$
$$= 7000 \cdot (1.02)^{10}$$
$$\approx 8532.96. \quad \text{Using a calculator}$$

The amount in the account after $2\frac{1}{2}$ years is $8532.96.

Do Exercise 5.

Calculator Spotlight

When using a calculator for interest computations, it is important to remember the order in which operations are performed and to minimize "round-off error." For example, to find the amount due on a $20,000 loan made for 25 days at 11%, compounded daily, we press the following sequence of keys:

[2][0][0][0][0][×][( ][1][+][0][.][1][1][÷][3][6][5][ )][x^y][2][5][=].

This key may appear differently. See p. 75.

Without parentheses keys, we would press

[1][+][0][.][1][1][÷][3][6][5][=][x^y][2][5][=][×][2][0][0][0][0][=].

Note that in both sequences of keystrokes we raise $1 + \frac{0.11}{365}$, not just $\frac{0.11}{365}$, to the power. After 25 days, $20,151.23 is due.

Exercises

1. Find the amount due on a $16,000 loan made for 62 days at 13%, compounded daily.

2. An investment of $12,500 is made for 90 days at 8.5%, compounded daily. How much is the investment worth after 90 days?

Answer on page A-23

Exercise Set 8.6

a Find the *simple* interest.

	Principal	Rate of Interest	Time
1.	$200	13%	1 year
2.	$450	18%	1 year
3.	$2000	12.4%	$\frac{1}{2}$ year
4.	$200	7.7%	$\frac{1}{2}$ year
5.	$4300	14%	$\frac{1}{4}$ year
6.	$2000	15%	$\frac{1}{4}$ year

Solve. Assume that simple interest is being calculated in each case.

7. CopiPix, Inc., borrows $10,000 at 9% for 60 days. Find (a) the amount of interest due and (b) the total amount that must be paid after 60 days.

8. Sal's Laundry borrows $8000 at 10% for 90 days. Find (a) the amount of interest due and (b) the total amount that must be paid after 90 days.

9. Animal Instinct, a pet supply shop, borrows $6500 at 8% for 90 days. Find (a) the amount of interest due and (b) the total amount that must be paid after 90 days.

10. Andante's Cafe borrows $4500 at 9% for 60 days. Find (a) the amount of interest due and (b) the total amount that must be paid after 60 days.

11. Jean's Garage borrows $5600 at 10% for 30 days. Find (a) the amount of interest due and (b) the total amount that must be paid after 30 days.

12. Shear Delights, a hair salon, borrows $3600 at 8% for 30 days. Find (a) the amount of interest due and (b) the total amount that must be paid after 30 days.

b Interest is compounded annually. Find the amount in the account after the given length of time. Round to the nearest cent.

Principal	Rate of interest	Time
13. $400	10%	2 years
14. $400	7.7%	2 years
15. $200	8.8%	2 years
16. $1000	15%	2 years

Interest is compounded semiannually. Find the amount in the account after the given length of time. Round to the nearest cent.

	Principal	Rate of interest	Time
17.	$4000	7%	1 year
18.	$1000	5%	1 year
19.	$2000	9%	3 years
20.	$5000	8%	30 months

Solve.

21. A family invests $4000 in an account paying 6%, compounded monthly. How much is in the account after 5 months?

22. A couple invests $2500 in an account paying 9%, compounded monthly. How much is in the account after 6 months?

23. A couple invests $1200 in an account paying 10%, compounded quarterly. How much is in the account after 1 year?

24. The O'Hares invest $6000 in an account paying 8%, compounded quarterly. How much is in the account after 18 months?

Skill Maintenance

Solve. [7.3b]

25. $\dfrac{9}{10} = \dfrac{x}{5}$

26. $\dfrac{7}{x} = \dfrac{4}{5}$

27. $\dfrac{3}{4} = \dfrac{6}{x}$

28. $\dfrac{7}{8} = \dfrac{x}{100}$

Convert to a mixed numeral. [4.5b]

29. $-\dfrac{64}{17}$

30. $\dfrac{38}{11}$

Convert from a mixed numeral to fractional notation. [4.5a]

31. $1\dfrac{1}{17}$

32. $20\dfrac{9}{10}$

Synthesis

33. ◆ Which is a better investment and why: $1000 invested at $14\dfrac{3}{4}$% simple interest for 1 year, or $1000 invested at 14% compounded monthly for 1 year?

34. ◆ A firm must choose between borrowing $5000 at 10% for 30 days and borrowing $10,000 at 8% for 60 days. Give arguments in favor of and against each option.

35. ◆ Without performing the multiplications, determine which gives the most interest: $1000 × 8% × $\dfrac{1}{12}$, or $1000 × 8% × $\dfrac{30}{365}$? How did you decide?

36. ▦ What is the simple interest on $24,680 at 7.75% for $\dfrac{3}{4}$ year?

37. ▦ Interest is compounded semiannually. Find the value of the investment if $24,800 is invested at 6.4% for 5 years.

38. ▦ Interest is compounded quarterly. Find the value of the investment if $125,000 is invested at 9.2% for $2\dfrac{1}{2}$ years.

Effective Yield. The *effective yield* is the yearly rate of simple interest that corresponds to an interest rate that is compounded two or more times a year. For example, if P is invested at 12%, compounded quarterly, we would multiply P by $(1 + 0.12/4)^4$, or 1.03^4. Since $1.03^4 \approx 1.126$ or 112.6%, the 12% compounded quarterly corresponds to an effective yield of approximately 12.6%. In Exercises 39 and 40, find the effective yield for the indicated account.

39. ▦ The account pays 9% compounded monthly.

40. ▦ The account pays 10% compounded daily.

41. ▦ Rather than spend $20,000 on a new car that will lose 30% of its value in 1 year, the Coniglios invest the money at 9%, compounded daily. After 1 year, how much have the Coniglios saved by not buying the car?

Prepare an amortization table for a car loan.

Summary and Review Exercises: Chapter 8

Important Properties and Formulas

Commission = Commission rate × Sales
Sale price = Original price − Discount
Compound Interest: $A = P \cdot \left(1 + \frac{r}{n}\right)^{n \cdot t}$

Discount = Rate of discount × Original price
Simple Interest: $I = P \cdot r \cdot t$

Write percent notation. [8.1b]
1. 0.483
2. 0.36

Write percent notation. [8.1c]
3. $\frac{3}{8}$
4. $\frac{1}{3}$

Write decimal notation. [8.1b]
5. 73.5%
6. $6\frac{1}{2}$%

Write fractional notation. [8.1c]
7. 24%
8. 6.3%

Translate to an equation. Then solve. [8.3a, b]
9. 30.6 is what percent of 90?

10. 63 is 84 percent of what?

11. What is $38\frac{1}{2}$% of 168?

Translate to a proportion. Then solve. [8.2a, b]
12. 24 percent of what is 16.8?

13. 42 is what percent of 30?

14. What is 10.5% of 84?

Solve. [8.4a, b]
15. Food expenses account for 26% of the average family's budget. A family makes $2300 one month. How much do they spend for food?

16. The price of a television set was reduced from $350 to $308. Find the percent of decrease in price.

17. Jerome County has a population that is increasing 3% each year. This year the population is 80,000. What will it be next year?

18. The price of a box of cookies increased from $1.70 to $2.04. What was the percent of increase in the price?

19. Carney College has a student body of 960 students. Of these, 17.5% are seniors. How many students are seniors?

Solve. [8.5a, b, c]
20. A city charges a meals tax of $4\frac{1}{2}$%. What is the meals tax charged on a dinner party costing $320?

21. In Massachusetts, a sales tax of $378 is collected on the purchase of a used car for $7560. What is the sales tax rate?

22. Kim earns $753.50 selling $6850 worth of televisions. What is the commission rate?

23. An air conditioner has a marked price of $350. It is placed on sale at 12% off. What are the discount and the sale price?

24. A fax machine priced at $305 is discounted at the rate of 14%. What are the discount and the sale price?

25. An insurance salesperson receives a 7% commission. If $42,000 worth of life insurance is sold, what is the commission?

Solve. [8.6a, b]

26. What is the simple interest on $1800 at 6% for $\frac{1}{3}$ year?

27. The Dress Shack borrows $24,000 at 10% simple interest for 60 days. Find (a) the amount of interest due and (b) the total amount that must be paid after 60 days.

28. What is the simple interest on $2200 principal at the interest rate of 5.5% for 1 year?

29. The Kleins invest $7500 in an investment account paying 12%, compounded monthly. How much is in the account after 3 months?

30. Find the amount in an investment account if $8000 is invested at 9%, compounded annually, for 2 years.

31. Find the rate of discount. [8.5c]

Skill Maintenance

Solve. [7.3b]

32. $\frac{3}{8} = \frac{7}{x}$

33. $\frac{1}{6} = \frac{7}{x}$

Graph. [6.4b]

34. $y = 2x - 4$

35. $y = -\frac{1}{3}x - 4$

Convert to decimal notation. [5.5a]

36. $\frac{11}{3}$

37. $\frac{11}{7}$

Convert to a mixed numeral. [4.5b]

38. $\frac{11}{3}$

39. $\frac{121}{7}$

Synthesis

40. ◆ Ollie buys a microwave oven during a 10%-off sale. The sale price that Ollie paid was $162. To find the original price, Ollie calculates 10% of $162 and adds that to $162. Is this correct? Why or why not? [8.5c]

41. ◆ Which is a better deal for a consumer and why: a discount of 40% or a discount of 20% followed by another of 22%? [8.5c]

42. ▦ *Land Area of the United States.* When Hawaii and Alaska became states, the total land area of the United States increased from 2,963,681 mi² to 3,540,939 mi². What was the percent of increase? [8.4b]

43. Rhonda's Dress Shop reduces the price of a dress by 40% during a sale. By what percent must the store increase the sale price, after the sale, to get back to the original price? [8.5c]

44. A $200 coat is marked up 20%. After 30 days, it is marked down 30% and sold. What was the final selling price of the coat? [8.5c]

45. How many successive 10% discounts are necessary to lower the price of an item to below 50% of its original price? [8.5c]

Test: Chapter 8

1. Write decimal notation for 89%.

2. Write percent notation for 0.674.

3. Write percent notation for $\frac{11}{8}$.

4. Write fractional notation for 65%.

5. Translate to an equation. Then solve.
   What is 40% of 55?

6. Translate to a proportion. Then solve.
   What percent of 80 is 65?

Solve.

7. *Weight of Muscles.* The weight of muscles in an adult woman is about 23% of total body weight. A woman weighs 125 lb. What do the muscles weigh?

8. *Population Growth.* The population of Rippington increased from 1500 to 3600. Write the percent of increase in population.

9. *Arizona Tax Rate.* The sales tax rate in Arizona is 5%. How much tax is charged on a purchase of $324? What is the total price?

10. *Sales Commissions.* Gwen's commission rate is 15%. What is the commission from the sale of $4200 worth of merchandise?

11. The marked price of a CD player is $200 and the item is on sale at 20% off. What are the discount and the sale price?

12. What is the simple interest on a principal of $120 at the interest rate of 7.1% for 1 year?

13. The Burnham Parents–Teachers Association invests $5200 at 6% simple interest. How much is in the account after $\frac{1}{2}$ year?

14. Write the amount in an account if $1000 is invested at 5%, compounded annually, for 2 years.

Answers

1. _____
2. _____
3. _____
4. _____
5. _____
6. _____
7. _____
8. _____
9. _____
10. _____
11. _____
12. _____
13. _____
14. _____

15. The Suarez family invests $10,000 at 9%, compounded monthly. How much is in the account after 3 months?

16. Find the discount and the discount rate of the bed in this ad.

WHITE IRON DAYBED
WITH BRASS ACCENTS
100 TO SELL
FANTASTIC VALUE!
MARKET VALUE $249.95
Choice of finish!
$118 Springs Included!

Skill Maintenance

17. Graph: $y = -x + 2$.

18. Solve: $\dfrac{5}{8} = \dfrac{10}{x}$.

19. Convert to decimal notation: $\dfrac{17}{12}$.

20. Convert to a mixed numeral: $\dfrac{153}{44}$.

Synthesis

21. By selling a home without using a realtor, Juan and Marie can avoid paying a 7.5% commission. They receive an offer of $109,000 from a potential buyer. In order to give a comparable offer, for what price would a realtor need to sell the house? Round to the nearest hundred.

22. Karen's commission rate is 16%. She invests her commission from the sale of $15,000 worth of merchandise at the interest rate of 12%, compounded quarterly. How much is Karen's investment worth after 6 months?

23. A housing development is constructed on a dead-end road along a river and ends in a cul-de-sac, as shown in the figure.

The property owners agree to share the cost of maintaining the road in the following manner. The first fifth of the road in front of lot 1 is to be shared equally among all five lot owners. The cost of the second fifth in front of lot 2 is to be shared equally among the owners of lots 2–5, and so on. Assume that all five sections of the road cost the same to maintain.

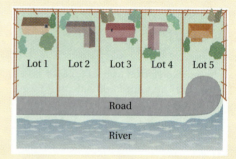

a) What fractional part of the cost is paid by each owner?
b) What percent of the cost is paid by each owner?
c) If lots 3, 4, and 5 were all owned by the same person, what percent of the cost of maintenance would this person pay?

Chapter 8 Percent Notation

Cumulative Review: Chapters 1–8

1. Write fractional notation for 0.091.

2. Write decimal notation for $\frac{13}{6}$.

3. Write decimal notation for 3%.

4. Write percent notation for $\frac{9}{8}$.

5. Write fractional notation for the ratio 5 to 0.5.

6. Write the rate in kilometers per hour.

   350 km, 15 hr

Use <, >, or = for ▮ to write a true sentence.

7. $\frac{5}{7}$ ▮ $\frac{6}{8}$

8. -3.78 ▮ -37.8

Estimate the sum or difference by rounding to the nearest hundred.

9. $263{,}961 + 32{,}090 + 127.89$

10. $73{,}510 - 23{,}450$

11. Calculate: $46 - [4(6 + 4 \div 2) + 2 \times 3 - 5]$

12. Combine like terms: $5x - 9 - 7x - 5$.

Peform the indicated operation and simplify.

13. $\frac{6}{5} + 1\frac{5}{6}$

14. $-46.9 + 32.7$

15. $487{,}094$
    $6{,}936$
    $+\ \ 21{,}120$

16. $35 - 34.98$

17. $3\frac{1}{3} - 2\frac{2}{3}$

18. $-\frac{8}{9} - \frac{6}{7}$

19. $\frac{7}{9} \cdot \frac{3}{14}$

20. $(-32)(-4)(-3)$

21. 46.012
    $\times\ \ \ \ \ 0.03$

22. $6\frac{3}{5} \div 4\frac{2}{5}$

23. $431.2 \div 35.2$

24. $15 \overline{)1850}$

Solve.

25. $36 \cdot x = 3420$

26. $y + 142.87 = 151$

27. $\frac{2}{15} \cdot t = -\frac{6}{5}$

28. $\frac{3}{4} + x = \frac{5}{6}$

29. $3(x - 7) + 2 = 12x - 3$

30. $\frac{16}{n} = \frac{21}{11}$

31. In what quadrant does the point $(-3, -5)$ lie?

32. Graph on a plane: $y = -\frac{3}{5}x$.

33. Find the mean: 19, 29, 34, 39, 45.

34. Find the median: 7, 7, 12, 15, 19.

35. Find the perimeter of a 15-in. by 15-in. chessboard.

36. Find the area of a 40-yd by 80-yd soccer field.

Solve.

37. A 12-oz box of cereal costs $1.80. Find the unit price in cents per ounce.

38. A bus travels 456 km in 6 hr. At this rate, how far would the bus travel in 8 hr?

39. In a recent year, Americans recycled 37 million lb of paper. It is projected that this will increase to 53 million lb in 2005. Find the percent of increase.

40. The state of Utah has an area of 1,722,850 mi^2. Of this area, 60% is owned by the government. How many square miles are owned by the government?

41. How many pieces of ribbon $1\frac{4}{5}$ yd long can be cut from a length of ribbon 9 yd long?

42. Bobbie walked $\frac{7}{10}$ km to school and then $\frac{8}{10}$ km to the library. How far did she walk?

Synthesis

On a trip through the mountains, a Dodge hatchback traveled 240 mi on $7\frac{1}{2}$ gal of gasoline. Going across the plains, the same car averaged 36 miles per gallon.

43. What was the percent of increase or decrease in miles per gallon when the car left the mountains for the plains?

44. How many miles per gallon did the Dodge average over the entire trip if it used 5 gal of gas to cross the plains?

The bar graph below shows how the winning times in the Olympic marathon and the qualifying times for the Boston Marathon have changed over time.

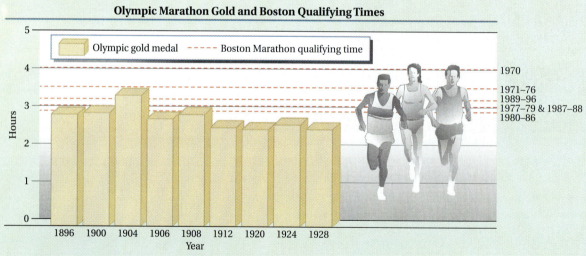

45. During which of the years listed would the winner of the Olympic marathon have qualified for the 1985 Boston Marathon?

46. What was the first year in which the Boston Marathon's qualifying time was lower than a time that had been good enough to win an Olympic gold medal?

Cumulative Review: Chapters 1–8

9
Geometry and Measures

Introduction

This chapter introduces American and metric systems used to measure length, area, volume, weight, mass, and temperature. Parallelograms, trapezoids, circles, and angles are also studied.

- **9.1** Systems of Linear Measurement
- **9.2** More with Perimeter and Area
- **9.3** Converting Units of Area
- **9.4** Angles
- **9.5** Square Roots and the Pythagorean Theorem
- **9.6** Volume and Capacity
- **9.7** Weight, Mass, and Temperature

An Application

The medicine Albuterol is used for the treatment of asthma. It typically comes in an inhaler that contains 18 g. If one actuation, or spray, delivers 90 mg, how many actuations are in one inhaler?

This problem appears as Exercise 81 in Section 9.7.

The Mathematics

Before dividing by 90, we convert 18 g to milligrams:

$$18 \text{ g} = 18 \text{ g} \cdot \frac{1000 \text{ mg}}{1 \text{ g}}.$$

These are units of measure.

For more information, visit us at www.mathmax.com

Pretest: Chapter 9

Complete.

1. 9 ft = _____ in.
2. 8 in. = _____ ft
3. 7.32 km = _____ m
4. 7.5 mm = _____ cm
5. 3 ft^2 = _____ in^2
6. 108 ft^2 = _____ yd^2
7. 4.2 cm^2 = _____ mm^2
8. 7 gal = _____ oz
9. 7 T = _____ lb
10. 48 oz = _____ lb
11. 423 g = _____ kg
12. 0.4 kg = _____ mg

Find each area. Use 3.14 for π.

13.
14.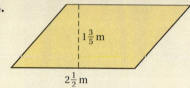
15.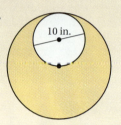

16. Find the length of a diameter of a circle with a radius of 9.4 m.

17. Find the circumference of a 70-cm–diameter bicycle wheel. Use $\frac{22}{7}$ for π.

18. Find the area of a 20-cm–wide circular mat. Use 3.14 for π.

19. A nurse administers a 45-mL allergy shot. How many liters are injected?

In a right triangle, find the length of the side not given. Assume that c represents the length of the hypotenuse and a and b are lengths of legs. Find an exact answer and an approximation to three decimal places.

20. $a = 12$, $b = 16$
21. $a = 2$, $c = 7$

Find each volume. Use $\frac{22}{7}$ for π.

22.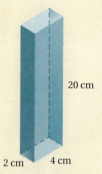
23.
24.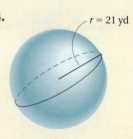

25. Find the measure of a complement of a 42° angle.

26. Find the measure of a supplement of a 25° angle.

27. Convert 77°F to Celsius.

28. Convert 37°C to Fahrenheit.

Chapter 9 Geometry and Measures

9.1 Systems of Linear Measurement

Length, or distance, is one kind of measure. To find lengths, we start with some **unit segment** and assign to it a measure of 1. Suppose $\overline{AB}$ below is a unit segment.

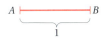

Let's measure segment $\overline{CD}$ below, using $\overline{AB}$ as our unit segment.

Since 4 unit segments fit end to end along $\overline{CD}$, the measure of $\overline{CD}$ is 4.

Sometimes we have to use parts of units. For example, the measure of the segment $\overline{MN}$ below is $1\frac{1}{2}$.

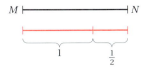

Do Exercises 1–4.

Objectives

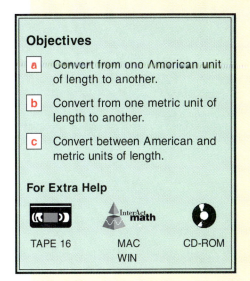

a. Convert from one American unit of length to another.

b. Convert from one metric unit of length to another.

c. Convert between American and metric units of length.

For Extra Help

TAPE 16   MAC WIN   CD-ROM

Use the unit below to measure the length of each segment or object.

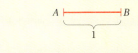

1. ⊢─────────⊣

a American Measures

American units of length are related as follows.

(Actual size, in inches)

> **AMERICAN UNITS OF LENGTH**
>
> 12 inches (in.) = 1 foot (ft)   3 feet = 1 yard (yd)
> 36 inches = 1 yard   5280 feet = 1 mile (mi)

The symbolism 13 in. = 13" and 27 ft = 27' is also used for inches and feet. American units have also been called "English," or "British–American," because at one time they were used by both countries. Today, both Canada and England have officially converted to the metric system. However, if you travel in England, you will still see units such as "miles" on road signs.

Example 1   Complete: 5 yd = _____ in.

$$5 \text{ yd} = 5 \cdot 1 \text{ yd}$$
$$= 5 \cdot 36 \text{ in.} \quad \text{Substituting 36 in. for 1 yd}$$
$$= 180 \text{ in.} \quad \text{Multiplying}$$

2.

3.

4.

Answers on page A-24

9.1 Systems of Linear Measurement

Complete.

5. 8 yd = _____ in.

6. 14.5 yd = _____ ft

7. 3.8 mi = _____ in.

Complete.

8. 72 in. = _____ ft

9. 24 ft = _____ yd

Complete.

10. 18 yd = _____ ft

11. 35 ft = _____ yd

Answers on page A-24

Chapter 9 Geometry and Measures

Example 2 Complete: 2 mi = _____ in.

$$\begin{aligned}
2 \text{ mi} &= 2 \cdot 1 \text{ mi} \\
&= 2 \cdot 5280 \text{ ft} \quad \text{Substituting 5280 ft for 1 mi} \\
&= 10{,}560 \cdot 1 \text{ ft} \\
&= 10{,}560 \cdot 12 \text{ in.} \quad \text{Substituting 12 in. for 1 ft} \\
&= 126{,}720 \text{ in.} \quad \text{The student should check the multiplication.}
\end{aligned}$$

Do Exercises 5–7.

Sometimes it helps to use multiplying by 1 when making conversions. For example, 12 in. = 1 ft, so we might choose to write 1 as

$$\frac{12 \text{ in.}}{1 \text{ ft}} \quad \text{or} \quad \frac{1 \text{ ft}}{12 \text{ in.}}.$$

Example 3 Complete: 48 in. = _____ ft.

To convert from "in." to "ft," we multiply by 1 using a symbol for 1 with "in." on the bottom and "ft" on the top. This process introduces feet and at the same time eliminates inches.

$$\begin{aligned}
48 \text{ in.} &= \frac{48 \text{ in.}}{1} \cdot \frac{1 \text{ ft}}{12 \text{ in.}} \quad \text{Multiplying by 1 using } \frac{1 \text{ ft}}{12 \text{ in.}} \text{ to eliminate in.} \\
&= \frac{48 \text{ in.}}{12 \text{ in.}} \cdot 1 \text{ ft} \quad \text{Pay careful attention to the units.} \\
&= \frac{48}{12} \cdot \frac{\text{in.}}{\text{in.}} \cdot 1 \text{ ft} \quad \text{The } \frac{\text{in.}}{\text{in.}} \text{ acts like 1, so we can omit it.} \\
&= 4 \cdot 1 \text{ ft} \quad \text{Dividing by 12} \\
&= 4 \text{ ft.}
\end{aligned}$$

The conversion can also be regarded as "canceling" units:

$$48 \text{ in.} = \frac{48 \text{ \cancel{in.}}}{1} \cdot \frac{1 \text{ ft}}{12 \text{ \cancel{in.}}} = \frac{48}{12} \cdot 1 \text{ ft} = 4 \text{ ft.}$$

Do Exercises 8 and 9.

In Examples 4 and 5, we will use only the "canceling" method.

Example 4 Complete: 75 yd = _____ ft.

Since we are converting from "yd" to "ft," we choose a symbol for 1 with "ft" on the top and "yd" on the bottom:

$$\begin{aligned}
75 \text{ yd} &= 75 \text{ \cancel{yd}} \cdot \frac{3 \text{ ft}}{1 \text{ \cancel{yd}}} \\
&= 75 \cdot 3 \text{ ft} \\
&= 225 \text{ ft.} \quad \text{Multiplying by 3}
\end{aligned}$$

Do Exercises 10 and 11.

Example 5 Complete: 23,760 ft = _____ mi.

We choose a symbol for 1 with "mi" on the top and "ft" on the bottom:

$$23{,}760 \text{ ft} = 23{,}760 \text{ ft} \cdot \frac{1 \text{ mi}}{5280 \text{ ft}} \qquad 5280 \text{ ft} = 1 \text{ mi, so } \frac{1 \text{ mi}}{5280 \text{ ft}} = 1.$$

$$= \frac{23{,}760}{5280} \cdot 1 \text{ mi}$$

$$= 4.5 \cdot 1 \text{ mi} \qquad \text{Dividing by 5280}$$

$$= 4.5 \text{ mi}.$$

Do Exercises 12 and 13.

b │ The Metric System

The **metric system** is used in most countries of the world, and the United States is now making greater use of it as well. The metric system does not use inches, feet, pounds, and so on, although units for time and electricity are the same as those you use now.

An advantage of the metric system is that it is easier to convert from one unit to another. That is because the metric system is based on the number 10.

The basic unit of length is the **meter**. It is just over a yard. In fact, 1 meter ≈ 1.1 yd.

(Comparative sizes are shown.)

1 Meter

1 Yard

The other units of length are multiples of the length of a meter:

10 times a meter, 100 times a meter, 1000 times a meter,

or fractions of a meter:

$\frac{1}{10}$ of a meter, $\frac{1}{100}$ of a meter, $\frac{1}{1000}$ of a meter.

> **METRIC UNITS OF LENGTH**
> 1 *kilo*meter (km) = 1000 meters (m)
> 1 *hecto*meter (hm) = 100 meters (m)
> 1 *deka*meter (dam) = 10 meters (m)
> 1 meter (m)
> 1 *deci*meter (dm) = $\frac{1}{10}$ meter (m)
> 1 *centi*meter (cm) = $\frac{1}{100}$ meter (m)
> 1 *milli*meter (mm) = $\frac{1}{1000}$ meter (m)
>
> *hm, dam* and *dm* are not often used.

It is important to remember these names and abbreviations. The prefixes *kilo-* for 1000, *deci-* for $\frac{1}{10}$, *centi-* for $\frac{1}{100}$, and *milli-* for $\frac{1}{1000}$ are used the most. These prefixes are also used when measuring capacity (volume) and mass (weight).

Complete.

12. 26,400 ft = _____ mi

13. 6 mi = _____ ft

Answers on page A-24

9.1 Systems of Linear Measurement

Complete.

14. 23 km = _____ m

To familiarize yourself with metric units, consider the following.

1 kilometer (1000 meters)	is slightly more than $\frac{1}{2}$ mile ($\approx$0.6 mi).
1 meter	is just over a yard ($\approx$1.1 yd).
1 centimeter (0.01 meter)	is a little more than the width of a jumbo paperclip ($\approx$0.3937 inch).
1 millimeter	is about the diameter of a paperclip wire.

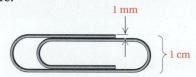

1 inch is about 2.54 centimeters.

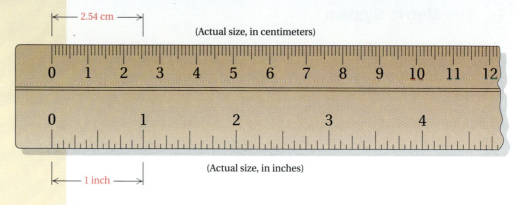

Example 6 Complete: 4 km = _____ m.

$$4 \text{ km} = 4 \cdot 1 \text{ km}$$
$$= 4 \cdot 1000 \text{ m} \quad \text{Substituting 1000 m for 1 km}$$
$$= 4000 \text{ m} \quad \text{Multiplying by 1000}$$

15. 4 hm = _____ m

Do Exercises 14 and 15.

Since

$$\frac{1}{10} \text{ m} = 1 \text{ dm}, \quad \frac{1}{100} \text{ m} = 1 \text{ cm}, \quad \text{and} \quad \frac{1}{1000} \text{ m} = 1 \text{ mm},$$

it follows that

> 1 m = 10 dm,   1 m = 100 cm,   and   1 m = 1000 mm.

Memorizing these equations will help you to write forms of 1 when canceling to make conversions. The procedure is the same as that used in Examples 4 and 5.

Example 7 Complete: 93.4 m = _____ cm.

To convert from "m" to "cm," we multiply by 1 using a symbol for 1 with "m" on the bottom and "cm" on the top. This process introduces centimeters and at the same time eliminates meters.

$$93.4 \text{ m} = 93.4 \text{ m} \cdot \frac{100 \text{ cm}}{1 \text{ m}} \quad \text{Multiplying by 1 using } \frac{100 \text{ cm}}{1 \text{ m}}$$
$$= 93.4 \text{ m} \cdot \frac{100 \text{ cm}}{1 \text{ m}} = 93.4 \cdot 100 \text{ cm} = 9340 \text{ cm}$$

Answers on page A-24

Example 8 Complete: 0.248 m = _____ mm.

We are converting from "m" to "mm," so we choose a symbol for 1 with "mm" on the top and "m" on the bottom:

$$0.248 \text{ m} = 0.248 \text{ m} \cdot \frac{1000 \text{ mm}}{1 \text{ m}} = 0.248 \cdot 1000 \text{ mm} = 248 \text{ mm}.$$

Do Exercises 16 and 17.

Example 9 Complete: 2347 m = _____ km.

We multiply by 1 using $\frac{1 \text{ km}}{1000 \text{ m}}$:

$$2347 \text{ m} = 2347 \text{ m} \cdot \frac{1 \text{ km}}{1000 \text{ m}} = \frac{2347}{1000} \cdot 1 \text{ km} = 2.347 \text{ km}.$$

Do Exercises 18 and 19.

It is helpful to remember that 1000 mm = 100 cm and, more simply, 10 mm = 1 cm.

Example 10 Complete: 8.42 mm = _____ cm.

We can multiply by 1 using either $\frac{1 \text{ cm}}{10 \text{ mm}}$ or $\frac{100 \text{ cm}}{1000 \text{ mm}}$. Both expressions for 1 will eliminate mm and leave cm:

$$8.42 \text{ mm} = 8.42 \text{ mm} \cdot \frac{1 \text{ cm}}{10 \text{ mm}} = \frac{8.42}{10} \cdot 1 \text{ cm} = 0.842 \text{ cm}.$$

Do Exercises 20 and 21.

Mental Conversion

Note in Examples 6–10 that changing from one unit to another in the metric system involves moving a decimal point. This occurs because the metric system is based on 10. To find a faster way to convert, consider these equivalent ways of expressing the width of a standard sheet of paper.

> Width of a standard sheet of paper = 216 mm = 21.6 cm = 2.16 dm = 0.216 m = 0.0216 dam = 0.00216 hm = 0.000216 km

Each unit in the box above has a value that is ten times as large as the next smaller unit. Thus converting to the next larger unit means moving the decimal point one place to the left.

Example 11 Complete: 35.7 mm = _____ cm.

Think: Centimeters is the next larger unit after millimeters. Thus we move the decimal point one place to the left.

35.7   3.5.7   35.7 mm = 3.57 cm

Converting to the next *smaller* unit means moving the decimal point one place to the right.

Complete.

16. 1.78 m = _____ cm

17. 9.04 m = _____ mm

Complete.

18. 7814 m = _____ km

19. 7814 m = _____ dam

Complete.

20. 87.2 mm = _____ cm

21. 89 km = _____ cm

Answers on page A-24

Complete. Try to do this mentally using the table on page 515.

22. 6780 m = _____ km

23. 9.74 cm = _____ mm

24. 1 mm = _____ cm

25. 845.1 mm = _____ dm

Complete.

26. 100 yd = _____ m
(The length of a football field)

27. 500 mi = _____ km
(The Indianapolis 500-mile race)

28. 2383 km = _____ mi
(The distance from St. Louis to Phoenix)

Answers on page A-24

Chapter 9 Geometry and Measures

Example 12 Complete: 3 m = _____ cm.

Think: A meter is 100 times as large as a centimeter (100 cm = 1 m). Thus we move the decimal point two places to the right. To do so, we write two additional zeros.

3   3.00.   3 m = 300 cm

Example 13 Complete: 4.37 km = _____ cm.

Think: Kilometers are 100,000 times as large as centimeters (100,000 cm = 1 km). Thus we move the decimal point five places to the right. This requires writting three additional zeros.

4.37   4.37000.   4.37 km = 437,000 cm

> The most commonly used metric units of length are km, m, cm, and mm. We have purposely used these more often than the others in the exercises and examples.

Do Exercises 22–25.

c Converting Between American and Metric Units

We can make conversions between American and metric units by using the following table. Again, we either make a substitution or multiply by 1 appropriately.

Metric	American
1 m	39.37 in.
1 m	3.3 ft
0.303 m	1 ft
2.54 cm	1 in.
1 km	0.621 mi
1.609 km	1 mi

Example 14 Complete: 26.2 mi = _____ km. (This is the approximate length of the Olympic marathon.)

26.2 mi = 26.2 · 1 mi
≈ 26.2 · 1.609 km
≈ 42.1558 km

Example 15 Complete: 100 m = _____ yd. (This is the length of a dash in track.)

100 m = 100 · 1 m ≈ 100 · 3.3 ft ≈ 330 ft   **Converting to feet**

≈ 330 ft · $\frac{1 \text{ yd}}{3 \text{ ft}}$ ≈ $\frac{330}{3}$ yd ≈ 110 yd   **Converting feet to yards**

Do Exercises 26–28.

Exercise Set 9.1

a Complete.

1. 1 ft = _____ in.

2. 1 yd = _____ ft

3. 1 in. = _____ ft

4. 1 mi = _____ yd

5. 1 mi = _____ ft

6. 1 ft = _____ yd

7. 4 yd = _____ in.

8. 3 yd = _____ ft

9. 84 in. = _____ ft

10. 48 ft = _____ yd

11. 18 in. = _____ ft

12. 29 ft = _____ yd

13. 5 mi = _____ ft

14. 5 mi = _____ yd

15. 48 in. = _____ ft

16. 11,616 ft = _____ mi

17. 19 ft = _____ yd

18. 5.2 yd = _____ ft

19. 10 mi = _____ ft

20. 15,840 ft = _____ mi

21. $7\frac{1}{2}$ ft = _____ yd

22. 36 in. = _____ ft

23. 360 in. = _____ yd

24. 7.2 ft = _____ in.

25. 330 ft = _____ yd

26. 1760 yd = _____ mi

27. 3520 yd = _____ mi

28. 25 mi = _____ ft

29. 100 yd = _____ ft

30. 240 in. = _____ ft

31. 63,360 in. = _____ mi

32. 2 mi = _____ in.

b Complete. Do as much as possible mentally.

33. a) 1 km = _____ m

b) 1 m = _____ km

34. a) 1 hm = _____ m

b) 1 m = _____ hm

35. a) 1 dam = _____ m

b) 1 m = _____ dam

36. a) 1 dm = _____ m

b) 1 m = _____ dm

37. a) 1 cm = _____ m

b) 1 m = _____ cm

38. a) 1 mm = _____ m

b) 1 m = _____ mm

39. 6.7 km = _____ m

40. 27 km = _____ m

41. 98 cm = _____ m

42. 53 cm = _____ m

43. 8921 m = _____ km

44. 8664 m = _____ km

45. 56.66 m = _____ km

46. 4.733 m = _____ km

47. 5666 m = _____ cm

48. 869 m = _____ cm

49. 477 cm = _____ m

50. 6.27 mm = _____ m

51. 6.88 m = _____ cm

52. 6.88 m = _____ dm

53. 1 mm = _____ cm

54. 1 cm = _____ km

55. 1 km = _____ cm

56. 2 km = _____ cm

Chapter 9 Geometry and Measures

57. 14.2 cm = _____ mm **58.** 25.3 cm = _____ mm **59.** 8.2 mm = _____ cm

60. 9.7 mm = _____ cm **61.** 4500 mm = _____ cm **62.** 8,000,000 m = _____ km

63. 0.024 mm = _____ m **64.** 60,000 mm = _____ dam **65.** 6.88 m = _____ dam

66. 7.44 m = _____ hm **67.** 2.3 dam = _____ dm **68.** 9 km = _____ hm

c Complete. Answers mary vary slightly, depending on the conversion used.

69. 10 km = _____ mi
(A common running distance)

70. 5 mi = _____ km
(A common running distance)

71. 14 in. = _____ cm
(A common paper length)

72. 400 m = _____ yd
(A common race distance)

73. 65 mph = _____ km/h
(A common speed limit in the United States)

74. 100 km/h = _____ mph
(A common speed limit in Canada)

75. 330 ft = _____ m
(The length of most baseball foul lines)

76. 165 cm = _____ in.
(A common height for a woman)

77. 180 cm = _____ in.
(A common snowboard length)

78. 450 ft = _____ m
(The length of a long home run in baseball)

79. 36 yd = _____ m
(A common length for a roll of tape)

80. 70 in. = _____ cm
(A common height for a man)

Exercise Set 9.1

Skill Maintenance

Solve.

81. $-7x - 9x = 24$ [5.7b]

82. $-2a + 9 = 5a + 23$ [5.7b]

83. If 3 calculators cost $43.50, how much would 7 calculators cost?   [7.4a]

84. A principal of $500 is invested at a rate of 8.9% for 1 year. Find the simple interest.   [8.6a]

Convert to percent notation.

85. 0.47   [8.1b]

86. $\dfrac{7}{20}$ [8.1c]

Synthesis

87. ◆ A student writes the following conversion:
$$23 \text{ in.} = 23 \cdot (12 \text{ ft}) = 276 \text{ ft.}$$
What mistake has been made?

88. ◆ Explain in your own words why metric units are easier to work with than American units.

89. ◆ Would you expect the world record for the 100-m dash to be longer or shorter than the record for the 100-yd dash? Why?

Complete. Answers may vary, depending on the conversion used.

90. 2 mi = _____ cm

91. 10 km = _____ in.

92. Audio cassettes are generally played at a rate of $1\frac{7}{8}$ in. per second. How many meters of tape are used for a 60-min cassette? (*Note*: A 60-min cassette has 30 min of playing time on each side.)

93. In a recent year, the world record for the 100-m dash was 9.86 sec. How fast is this in miles per hour? Round to the nearest tenth of a mile per hour.

94. *National Debt.* Recently the national debt was $5.103 trillion. To get an idea of this amount, picture that if that many $1 bills were stacked on top of each other, they would reach 1.382 times the distance to the moon. The distance to the moon is 238,866 mi. How thick, in inches, is a $1 bill?

Use < or > to complete the following. Perform only approximate, mental calculations.

95. 59 in. ▪ 59 cm

96. 35 yd ▪ 35 m

97. 7 km ▪ 6 mi

98. 9 mi ▪ 18 km

99. 24 ft ▪ 6 m

100. 30 in. ▪ 90 cm

Practice conversions with old British monetary units.

9.2 More with Perimeter and Area

We have already studied how to find the perimeter of polygons and the area of squares, rectangles, and triangles. In this section, we learn how to find the area of *parallelograms, trapezoids,* and *circles*. We also learn how to calculate the perimeter, or *circumference,* of a circle.

Objectives

a Find the area of a parallelogram or trapezoid.

b Find the circumference, area, radius, or diameter of a circle, given the length of a radius or diameter.

For Extra Help

TAPE 16   MAC WIN   CD-ROM

a Parallelograms and Trapezoids

A **parallelogram** is a four-sided figure with two pairs of parallel sides, as shown below.

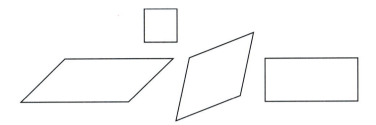

To find the area of a parallelogram, consider the one below.

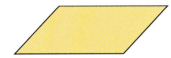

If we cut off a piece and move it to the other end, we get a rectangle.

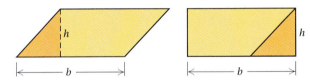

We can find the area by multiplying the length b, called a **base**, by h, called the **height**.

> The **area of a parallelogram** is the product of the length of a base b and the height h:
> $A = b \cdot h.$

Example 1 Find the area of this parallelogram.

$A = b \cdot h$
$ = 7 \text{ km} \cdot 5 \text{ km}$
$ = 35 \text{ km}^2$

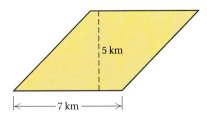

9.2 More with Perimeter and Area

Find the area.

1.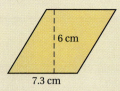

2.

Example 2 Find the area of this parallelogram.

$A = b \cdot h$
$= (1.2 \text{ m}) \cdot (6 \text{ m})$
$= 7.2 \text{ m}^2$

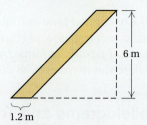

Do Exercises 1 and 2.

Trapezoids

A **trapezoid** is a polygon with four sides, two of which, the **bases**, are parallel to each other.*

To find the area of a trapezoid, think of cutting out another just like it.

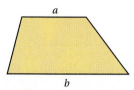

Then place the second one like this.

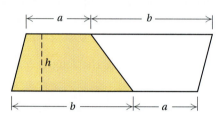

The resulting figure is a parallelogram with an area of

$h \cdot (a + b)$.   **The base is $a + b$.**

The trapezoid we started with has half the area of the parallelogram, or

$\frac{1}{2} \cdot h \cdot (a + b)$.

> The **area of a trapezoid** is half the product of the height and the sum of the lengths of the parallel sides, or the product of the height and the average length of the bases:
>
> $A = \frac{1}{2} \cdot h \cdot (a + b) = h \cdot \frac{a + b}{2}.$

---

*Some definitions of trapezoid specify *exactly* two parallel sides. We refrain from doing so. Thus we consider a parallelogram a special type of trapezoid.

Example 3 Find the area of this trapezoid.

$$A = \frac{1}{2} \cdot h \cdot (a + b)$$

$$= \frac{1}{2} \cdot 7 \text{ cm} \cdot (12 + 18) \text{ cm}$$

$$= \frac{7 \cdot 30}{2} \cdot \text{cm}^2 = \frac{7 \cdot 15 \cdot 2}{1 \cdot 2} \text{ cm}^2$$

$$= 105 \text{ cm}^2 \quad \text{Removing a factor equal to 1: } \tfrac{2}{2} = 1$$

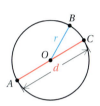

Do Exercises 3 and 4.

b | Circles

Radius and Diameter

At right is a circle with center O. Segment $\overline{AC}$ is a *diameter*. A **diameter** is a segment that passes through the center of the circle and has endpoints on the circle. Segment $\overline{OB}$ is called a *radius*. A **radius** is a segment with one endpoint on the center and the other endpoint on the circle. The words *radius* and *diameter* are also used to represent the lengths of a circle's radius and diameter, respectively.

> Suppose that d is the diameter of a circle and r is the radius. Then
> $$d = 2 \cdot r \quad \text{or} \quad r = \frac{d}{2}.$$

Example 4 Find the length of a radius of this circle.

$$r = \frac{d}{2}$$

$$= \frac{12 \text{ m}}{2}$$

$$= 6 \text{ m}$$

The radius is 6 m.

Example 5 Find the length of a diameter of this circle.

$$d = 2 \cdot r$$

$$= 2 \cdot \frac{1}{4} \text{ ft}$$

$$= \frac{1}{2} \text{ ft}$$

The diameter is $\frac{1}{2}$ ft.

Do Exercises 5 and 6.

Find the area.

3.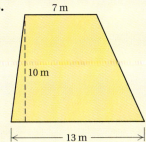

4.

5. Find the length of a radius.

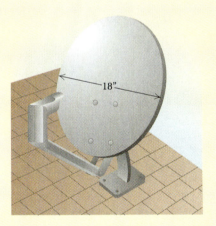

6. Find the length of a diameter.

Answers on page A-25

9.2 More with Perimeter and Area

525

7. Find the circumference of this circle. Use 3.14 for π.

Circumference

The perimeter of a circle is called its **circumference**. Take a 12-oz soda can and measure its circumference C and diameter d. Next, consider the ratio C/d:

$$\frac{C}{d} = \frac{7.8 \text{ in.}}{2.5 \text{ in.}} \approx 3.1.$$

Suppose we found this ratio for cans and circles of several sizes. We would always get a number close to 3.1. Any time we divide the circumference C by the diameter d, we get the same number. We call this number π (pi).

▶ $\dfrac{C}{d} = \pi$ or $C = \pi \cdot d$. The number π is about 3.14, or about $\dfrac{22}{7}$.

Example 6 Find the circumference of this circle. Use 3.14 for π.

$C = \pi \cdot d$
$\approx 3.14 \cdot 6 \text{ cm}$
$\approx 18.84 \text{ cm}$

The circumference is about 18.84 cm.

Do Exercise 7.

Since $d = 2 \cdot r$, where r is the length of a radius, it follows that

$C = \pi \cdot d = \pi \cdot (2 \cdot r).$

▶ $C = 2 \cdot \pi \cdot r$

Example 7 Find the circumference of this circle. Use $\frac{22}{7}$ for π.

$C = 2 \cdot \pi \cdot r$
$\approx 2 \cdot \dfrac{22}{7} \cdot 70 \text{ in.}$
$\approx 2 \cdot 22 \cdot \dfrac{70}{7} \text{ in.}$
$\approx 44 \cdot 10 \text{ in.}$
$\approx 440 \text{ in.}$

The circumference is about 440 in.

Answer on page A-25

Example 8 Find the perimeter of this figure. Use 3.14 for π.

We let $P =$ the perimeter. We see that we have half a circle attached to three sides of a square. Thus we add half the circumference to the lengths of the three line segments.

$$P = 3 \cdot 9.4 \text{ km} + \frac{1}{2} \cdot 2 \cdot \pi \cdot 4.7 \text{ km}$$
$$\approx 28.2 \text{ km} + 3.14 \cdot 4.7 \text{ km}$$
$$\approx 28.2 \text{ km} + 14.758 \text{ km}$$
$$\approx 42.958 \text{ km}$$

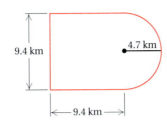

The perimeter is about 42.958 km.

Do Exercises 8–10.

Area

To find the area of a circle, consider cutting half a circular region into small slices and arranging them as shown below.

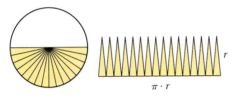

Then imagine slicing the other half of the circular region and arranging the pieces in between the others as shown below.

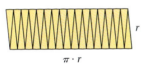

This is almost a parallelogram. The base has length $\frac{1}{2} \cdot 2 \cdot \pi \cdot r$, or $\pi \cdot r$ (half the circumference) and the height is r. Thus the area is

$(\pi \cdot r) \cdot r$.

This is the area of a circle.

> The **area of a circle** with radius of length r is given by
> $A = \pi \cdot r \cdot r,$ or $A = \pi \cdot r^2.$
>

Example 9 Find the area of this circle. Use $\frac{22}{7}$ for π.

$$A = \pi \cdot r^2 = \pi \cdot r \cdot r$$
$$\approx \frac{22}{7} \cdot 14 \text{ cm} \cdot 14 \text{ cm}$$
$$\approx \frac{22}{\cancel{7}} \cdot \frac{\cancel{7} \cdot 2}{1} \text{ cm} \cdot 14 \text{ cm}$$
$$\approx 616 \text{ cm}^2 \quad \text{Note: } r^2 \neq 2r.$$

The area is about 616 cm².

Do Exercise 11.

---

8. Find the circumference of this circle. Use 3.14 for π.

9. Find the circumference of this bicycle wheel. Use $\frac{22}{7}$ for π.

10. Find the perimeter of this figure. Use 3.14 for π.

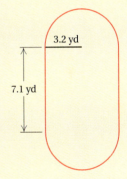

11. Find the area of this circle. Use $\frac{22}{7}$ for π.

Answer on page A-25

9.2 More with Perimeter and Area

12. Find the area of this circle. Use 3.14 for π.

10.4 cm

13. Which is larger and by how much: a 10-ft square flower bed or a 12-ft diameter flower bed?

Calculator Spotlight

On certain calculators, there is a pi key, $\boxed{\pi}$. You can use a $\boxed{\pi}$ key for most computations instead of stopping to round the value of π. Rounding, if necessary, is done at the end.

Exercises

1. If you have a $\boxed{\pi}$ key on your calculator, to how many decimal places does this key give the value of π?
2. Find the circumference and the area of a circle with a radius of 225.68 in.
3. Find the area of a circle with a diameter of $46\frac{12}{13}$ in.
4. Find the area of a large irrigated farming circle with a diameter of 400 ft.

Answers on page A-25

Example 10 Find the area of this circle. Use 3.14 for π. Round to the nearest hundredth.

$A = \pi \cdot r \cdot r$

$\approx 3.14 \cdot 2.1 \text{ m} \cdot 2.1 \text{ m}$

$\approx 3.14 \cdot 4.41 \text{ m}^2$

$\approx 13.8474 \text{ m}^2 \approx 13.85 \text{ m}^2$

2.1 m

The area is about 13.85 m².

Do Exercise 12.

Example 11 *Area of a Pizza Pan.* Which makes a larger pizza and by how much: a 16-in. square pizza pan or a 16-in. diameter circular pizza pan?

1. **Familiarize.** From examining a picture of each, we see that the square pan has the larger area. We let D = the difference in area.

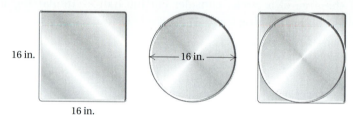

16 in.   16 in.

16 in.

2. **Translate.** The problem can be rephrased as follows.

Area of square pan	minus	Area of circular pan	is	Difference in area
$s \cdot s$	$-$	$\pi \cdot r \cdot r$	$=$	D

3. **Solve.** We use 3.14 for π and substitute:

16 in. · 16 in. $-$ 3.14 · 8 in. · 8 in. $\approx D$ Substituting

256 in² $-$ 200.96 in² $\approx D$

55.04 in² $\approx D$.

4. **Check.** We can check by repeating our calculations. Note also that the area of the square pan is larger, as we expected.

5. **State.** The square pan is larger by about 55.04 in².

Do Exercise 13.

Area of a parallelogram: $A = b \cdot h$

Area of a trapezoid: $A = \frac{1}{2} \cdot h \cdot (a + b)$

or $A = h \cdot \frac{a + b}{2}$

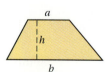

Circumference of a circle: $C = \pi \cdot d$

or $C = 2 \cdot \pi \cdot r$

Area of a circle: $A = \pi \cdot r^2$

Note: $d = 2r$ and $\pi \approx 3.14 \approx \frac{22}{7}$.

Chapter 9 Geometry and Measures

Exercise Set 9.2

a Find the area of each parallelogram or trapezoid.

1.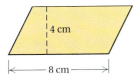

2.

3.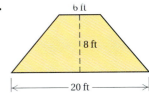

4.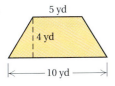

5.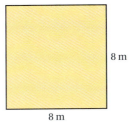

6.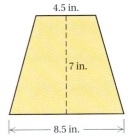

7.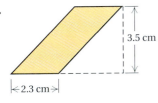

8.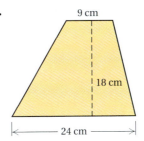

9.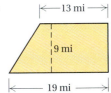

10.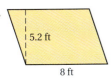

11.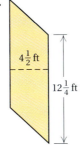

12.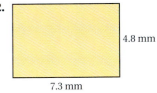

13.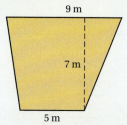

14.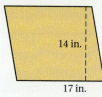

15.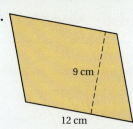

16.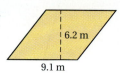

17.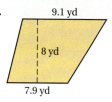

18.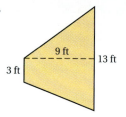

b Find the length of a diameter of each circle.

19.

20.

21.

22.

Find the length of a radius of each circle.

23. 24. 25. 26.

Find the circumference of each circle in Exercises 19–22. Use $\frac{22}{7}$ for π.

27. Exercise 19 **28.** Exercise 20 **29.** Exercise 21 **30.** Exercise 22

Find the circumference of each circle in Exercises 23–26. Use 3.14 for π.

31. Exercise 23 **32.** Exercise 24 **33.** Exercise 25 **34.** Exercise 26

Find the area of each circle in Exercises 19–22. Use $\frac{22}{7}$ for π.

35. Exercise 19 **36.** Exercise 20 **37.** Exercise 21 **38.** Exercise 22

Find the area of each circle in Exercises 23–26. Use 3.14 for π.

39. Exercise 23 **40.** Exercise 24 **41.** Exercise 25 **42.** Exercise 26

Exercise Set 9.2

Solve. Use 3.14 for π.

43. A penny has a 1-cm radius. What is its diameter? circumference? area?

44. The top of a soda can has a 6-cm diameter. What is its radius? circumference? area?

45. A radio station is allowed by the FCC to broadcast over an area with a radius of 220 mi. How much area is this?

46. Which is larger and by how much: a 12-in. circular pizza or a 12-in. square pizza?

47. The diameter of a quarter is 2.5 cm. What is the circumference? area?

48. The diameter of a dime is 1.8 cm. What is the circumference? area?

49. *Botany.* To protect an elm tree, a 47.1-in. gypsy moth tape is wrapped once around the trunk. What is the diameter of the tree?

50. *Farming.* The circumference of a silo is 62.8 ft. What is the diameter of the silo?

51. *Track and Field.* Track meets take place on a track similar to the one shown below. Find the shortest distance around the track.

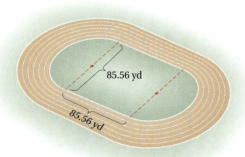

52. *Masonry.* Iris plans to install a 1-yd–wide walk around a circular swimming pool. The diameter of the pool is 8 yd. What will the area of the walk be?

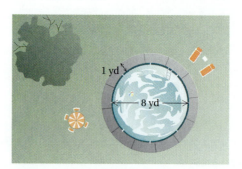

Chapter 9 Geometry and Measures

Find the perimeter of each figure. Use 3.14 for π.

53.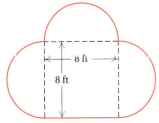

54.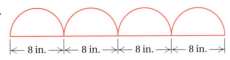

55.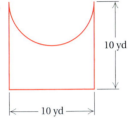

56.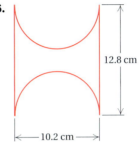

Find the area of the shaded region in each figure. Use 3.14 for π.

57.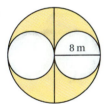

58.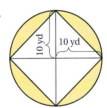

59.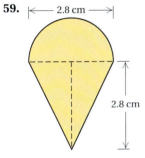

60.

Exercise Set 9.2

533

Skill Maintenance

Convert to fractional notation. [8.1c]

61. 9.25%

62. $87\frac{1}{2}\%$

Convert to percent notation. [8.1c]

63. $\frac{11}{8}$

64. $\frac{2}{3}$

65. $\frac{5}{4}$

66. $\frac{8}{5}$

Synthesis

67. ◆ Explain why a 16-in.–diameter pizza that costs $16.25 is a better buy than a 10-in.–diameter pizza that costs $7.85.

68. ◆ The radius of one circle is twice the length of another circle's radius. Is the area of the first circle twice the area of the other circle? Why or why not?

69. ◆ The radius of one circle is twice the size of another circle's radius. Is the circumference of the first circle twice the circumference of the other circle? Why or why not?

70. ▦ Calculate the surface area of an unopened steel can that has a height of 3.5 in. and a diameter of 2.5 in. (*Hint*: Make a sketch and "unroll" the sides of the can.) Use 3.14 for π.

71. ▦ The sides of a cake box are trapezoidal, as shown in the figure. Determine the surface area of the box.

72. ▦ $\pi \approx \frac{3927}{1250}$ is another approximation for π. Find decimal notation using a calculator. Round to the nearest thousandth.

73. ▦ The distance from Kansas City to Indianapolis is 500 mi. A car was driven this distance using tires with a radius of 14 in. How many revolutions of each tire occurred on the trip? Use $\frac{22}{7}$ for π.

74. ◆ ▦ *Urban Planning.* Years ago, when a 12-in.–diameter tree was cut down in New York City, new trees with a combined diameter of 12 in. had to be planted. Now, instead of being able to use four 3-in.–diameter trees as replacement, a total of *sixteen* 3-in.–diameter trees must be planted. (**Source**: *The New York Times* 7/24/88, p. 6; article by David W. Dunlap). Consider area and explain why the new replacement calculation is more correct mathematically.

75. *Sports Marketing.* Tennis balls are generally packed vertically, three in a can, one on top of another. Without using a calculator, determine the larger measurement: the can's circumference or the can's height.

76. ▦ *Landscaping.* Seed is needed for the field surrounded by the track in Exercise 51. If seed comes in 3-lb boxes, and each pound covers 120 ft^2, how many boxes must be purchased?

Verify the formulas for the area of a parallelogram, triangle, and trapezoid. Estimate the value of π.

9.3 Converting Units of Area

a | American Units

It is often necessary to convert units of area. First we will convert from one American unit of area to another.

Example 1 Complete: $1 \text{ yd}^2 = \underline{} \text{ ft}^2$.

We recall that 1 yd = 3 ft and make a sketch. Note that $1 \text{ yd}^2 = 9 \text{ ft}^2$. The same result can be found as follows:

$1 \text{ yd}^2 = 1 \cdot (3 \text{ ft})^2$ Substituting 3 ft for 1 yd
$= 3 \text{ ft} \cdot 3 \text{ ft}$
$= 9 \text{ ft}^2$.   Note that $\text{ft} \cdot \text{ft} = \text{ft}^2$.

Example 2 Complete: $2 \text{ ft}^2 = \underline{} \text{ in.}^2$.

$2 \text{ ft}^2 = 2 \cdot (12 \text{ in.})^2$ Substituting 12 in. for 1 ft
$= 2 \cdot 12 \text{ in.} \cdot 12 \text{ in.}$
$= 288 \text{ in.}^2$ Note that $\text{in.} \cdot \text{in.} = \text{in.}^2$.

Do Exercises 1–3.

American units of area are related as follows.

> 1 square yard (yd^2) = 9 square feet (ft^2)
> 1 square foot (ft^2) = 144 square inches (in^2)
> 1 square mile (mi^2) = 640 acres
> 1 acre = 43,560 ft^2

Example 3 Complete: $36 \text{ ft}^2 = \underline{} \text{ yd}^2$.

We are converting from "ft^2" to "yd^2". Thus we choose a symbol for 1 with yd^2 on top and ft^2 on the bottom.

$36 \text{ ft}^2 = 36 \text{ ft}^2 \cdot \dfrac{1 \text{ yd}^2}{9 \text{ ft}^2}$ Multiplying by 1 using $\dfrac{1 \text{ yd}^2}{9 \text{ ft}^2}$
$= \dfrac{36}{9} \cdot \text{yd}^2 = 4 \text{ yd}^2$

Example 4 Complete: $7 \text{ mi}^2 = \underline{}$ acres.

$7 \text{ mi}^2 = 7 \cdot 640$ acres   Substituting 640 acres for 1 mi^2
$= 4480$ acres

Had we used canceling, we could have multiplied 7 mi^2 by $\dfrac{640 \text{ acres}}{1 \text{ mi}^2}$:

$7 \text{ mi}^2 = 7 \text{ mi}^2 \cdot \dfrac{640 \text{ acres}}{1 \text{ mi}^2} = 4480$ acres.

Do Exercises 4 and 5.

Objectives

a | Convert from one American unit of area to another.

b | Convert from one metric unit of area to another.

For Extra Help

TAPE 16   MAC WIN   CD-ROM

Complete.
1. $1 \text{ ft}^2 = \underline{} \text{ in.}^2$

2. $10 \text{ ft}^2 = \underline{} \text{ in.}^2$

3. $7 \text{ yd}^2 = \underline{} \text{ ft}^2$

Complete.
4. $360 \text{ in.}^2 = \underline{} \text{ ft}^2$

5. $5 \text{ mi}^2 = \underline{}$ acres

Answers on page A-25

9.3 Converting Units of Area

Complete.

6. $1\ m^2 =$ _____ mm^2

7. $1\ cm^2 =$ _____ mm^2

Complete.

8. $2.88\ m^2 =$ _____ cm^2

9. $4.3\ mm^2 =$ _____ cm^2

10. $678{,}000\ m^2 =$ _____ km^2

Answers on page A-25

b Metric Units

We next convert from one metric unit of area to another.

Example 5 Complete: $1\ km^2 =$ _____ m^2.

$$1\ km^2 = 1 \cdot (1000\ m)^2 \quad \text{Substituting 1000 m for 1 km}$$
$$= 1000\ m \cdot 1000\ m$$
$$= 1{,}000{,}000\ m^2 \quad \text{Note that } m \cdot m = m^2.$$

Example 6 Complete: $1\ m^2 =$ _____ cm^2.

$$1\ m^2 = 1 \cdot (100\ cm)^2 \quad \text{Substituting 100 cm for 1 m}$$
$$= 100\ cm \cdot 100\ cm$$
$$= 10{,}000\ cm^2 \quad \text{Note that } cm \cdot cm = cm^2.$$

Do Exercises 6 and 7.

Mental Conversion

Note in Example 5 that whereas it takes 1000 m to make 1 km, it takes 1,000,000 m^2 to make 1 km^2. Similarly, in Example 6, we saw that although it takes 100 cm to make 1 m, it takes 10,000 cm^2 to make 1 m^2. In general, if a *length* conversion requires moving the decimal point n places, the corresponding *area* conversion requires moving the decimal point $2n$ places. For example, below we list four equivalent ways of expressing the area of a standard sheet of paper.

Area of a standard sheet of paper = 60,264 mm^2
= 602.64 cm^2
= 0.060264 m^2
≈ 0.00000006 km^2

Example 7 Complete: $3.48\ km^2 =$ _____ m^2.

Think: A kilometer is 1000 times as big as a meter, so 1 km^2 is 1,000,000 times as big as 1 m^2. We shift the decimal point *six* places to the right.

3.48   3.480000. $3.48\ km^2 = 3{,}480{,}000\ m^2$

Example 8 Complete: $586.78\ cm^2 =$ _____ m^2.

Think: To convert from cm to m, we shift the decimal point two places to the left. To convert from cm^2 to m^2, we shift the decimal point *four* places to the left.

586.78   0.0586.78 $586.78\ cm^2 = 0.058678\ m^2$

Do Exercises 8–10.

Exercise Set 9.3

a Complete.

1. $4 \text{ yd}^2 = \underline{} \text{ ft}^2$

2. $5 \text{ ft}^2 = \underline{} \text{ in}^2$

3. $7 \text{ ft}^2 = \underline{} \text{ in}^2$

4. $2 \text{ acres} = \underline{} \text{ ft}^2$

5. $432 \text{ in}^2 = \underline{} \text{ ft}^2$

6. $54 \text{ ft}^2 = \underline{} \text{ yd}^2$

7. $22 \text{ yd}^2 = \underline{} \text{ ft}^2$

8. $40 \text{ ft}^2 = \underline{} \text{ in}^2$

9. $44 \text{ yd}^2 = \underline{} \text{ ft}^2$

10. $144 \text{ ft}^2 = \underline{} \text{ yd}^2$

11. $20 \text{ mi}^2 = \underline{} \text{ acres}$

12. $576 \text{ in}^2 = \underline{} \text{ ft}^2$

13. $69 \text{ ft}^2 = \underline{} \text{ yd}^2$

14. $1 \text{ mi}^2 = \underline{} \text{ yd}^2$

15. 720 in.² = _____ ft²

16. 27 ft² = _____ yd²

17. 1 in.² = _____ ft²

18. 72 in.² = _____ ft²

19. 1 acre = _____ mi²

20. 4 acres = _____ ft²

b Complete.

21. 17 km² = _____ m²

22. 65 km² = _____ m²

23. 6.31 m² = _____ cm²

24. 2.7 m² = _____ mm²

25. 2345.6 mm² = _____ cm²

26. 8.38 cm² = _____ mm²

27. 349 cm² = _____ m²

28. 125 mm² = _____ m²

Chapter 9 Geometry and Measures

29. 250,000 mm² = _____ cm²

30. 2400 mm² = _____ cm²

31. 472,800 m² = _____ km²

32. 1.37 cm² = _____ mm²

Find the area of the shaded region of each figure. Give the answer in square feet. (Figures are not drawn to scale.)

33.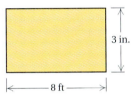

34.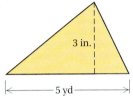

35.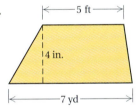

36.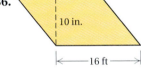

Find the area of the shaded region of each figure.

37.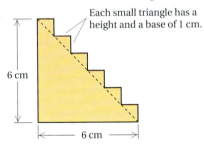
Each small triangle has a height and a base of 1 cm.

38.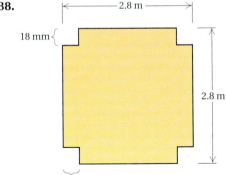

Exercise Set 9.3

539

Skill Maintenance

Find the simple interest. [8.6a]

	Principal	Rate of interest	Time
39.	$700	5%	$\frac{1}{2}$ year
40.	$450	6%	$\frac{1}{4}$ year
41.	$1200	8.9%	30 days
42.	$1800	12%	60 days

Synthesis

43. ◆ Which is larger and why: one square meter or nine square feet?

44. ◆ What advantage do metric units offer over American units when we are converting area measurements?

45. ◆ Why might a scientist choose to give area measurements in mm^2 rather than cm^2?

46. A 30-ft by 60-ft ballroom is to be turned into a nightclub by placing an 18-ft by 42-ft dance floor in the middle and carpeting the rest of the room. The new dance floor is laid in tiles that are 8 in. by 8 in. squares. How many such tiles are needed? What percent of the area is the dance floor?

Complete. Answers may vary slightly, depending on the conversion used.

47. $1\ m^2 =$ _____ ft^2

48. $1\ in^2 =$ _____ cm^2

49. $2\ yd^2 =$ _____ m^2

50. $1\ acre =$ _____ m^2

51. The president's family has about 20,175 ft^2 of living area in the White House. Estimate the living area in square meters.

52. A handwoven scarf is 2 m long and 10 in. wide. Find its area in square centimeters.

53. In order to remodel an office, a carpenter needs to purchase carpeting, at $8.45 a square yard, and molding for the base of the walls, at $0.87 a foot. If the room is 9 ft by 12 ft, with a 3-ft doorway, what will the materials cost?

Verify the conversions between American units of area.

9.4 Angles

a Measuring Angles

An **angle** is a set of points consisting of two **rays**, or half-lines, with a common endpoint. The endpoint is called the **vertex**.

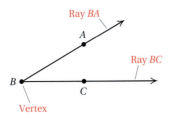

The rays are called the *sides*. The angle above can be named

angle *ABC*,   angle *CBA*,   angle *B*,   ∠*ABC*,   ∠*CBA*,   or   ∠*B*.

Note that the name of the vertex is either in the middle or, if no confusion results, listed by itself.

Do Exercises 1 and 2.

To measure angles, we start with some unit angle and assign to it a measure of 1. Suppose that ∠*U*, below, is a unit angle. To measure ∠*DEF*, we find that 3 copies of ∠*U* will "fill up" ∠*DEF*. Thus the measure of ∠*DEF* would be 3.

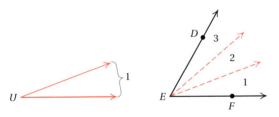

The unit most commonly used for angle measure is the degree. Below is such a unit. Its measure is 1 degree, or 1°.

A device called a **protractor** is used to measure angles. Protractors have two scales. To measure an angle like ∠*Q* below, we place the protractor's ▲ at the vertex and line up one of the angle's sides at 0°. Then we check where the angle's other side crosses the scale. In the figure below, 0° is on the inside scale, so we check where the angle's other side crosses the inside scale. We see that $m\angle Q = 145°$. The notation $m\angle Q$ is read "the measure of angle *Q*."

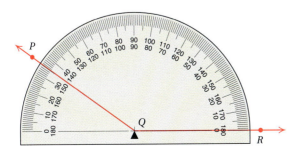

Do Exercise 3.

Objectives

a Name a given angle in four different ways and given an angle, measure it with a protractor.

b Classify an angle as right, straight, acute, or obtuse.

c Identify complementary and supplementary angles and find the measure of a complement or a supplement of a given angle.

For Extra Help

TAPE 16   MAC WIN   CD-ROM

Name each angle in six different ways.

1.
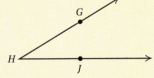

2.
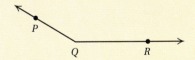

3. Use a protractor to measure this angle.

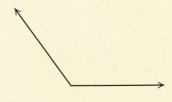

Answers on page A-25

9.4 Angles

541

4. Use a protractor to measure this angle.

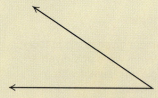

5. *Lengths of Engagement of Married Couples.* The data below relate the percent of married couples who were engaged for a certain time period before marriage (**Source**: Bruskin Goldring Research). Use this information to draw a circle graph.

Less than 1 yr:	24%
1–2 yr:	21%
More than 2 yr:	35%
Never engaged:	20%

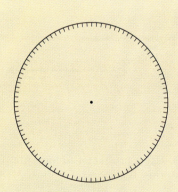

Let's find the measure of $\angle ABC$. This time we will use the 0° on the outside scale. We see that $m\angle ABC = 42°$.

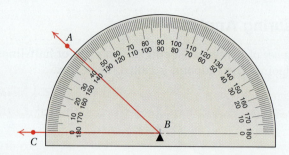

Do Exercise 4.

Protractors are needed when drawing circle graphs by hand.

Example 1 *Water Supplies.* Predictions indicate that by the year 2050, water supplies will be scarce in 18% of the world, stressed in 24% of the world, and sufficient in just 58% of the world (**Source**: Simon, Paul, *Tapped Out*, New York, 1998, Welcome Rain Publishers). Draw a circle graph to represent these figures.

Every circle graph contains a total of 360°. Thus,

24% of the circle is a 0.24(360°), or 86.4° angle;

18% of the circle is a 0.18(360°), or 64.8° angle; and

58% of the circle is a 0.58(360°), or 208.8° angle.

To draw an 86.4° angle, we first draw a horizontal segment and use a progractor to mark off an 86.4° angle. From that mark, we draw a segment to complete the angle. From that segment, we repeat the procedure to draw a 64.8° angle.

To confirm that the remainder of the circle is indeed 208.8°, we measure 180°, make a mark, and from there measure another 28.8°.

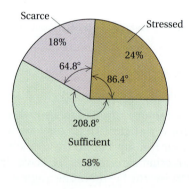

Do Exercise 5.

Answers on page A-25

Chapter 9 Geometry and Measures

b Classifying Angles

The following are ways in which we classify angles.

> **Right angle:** An angle that measures 90°.
> **Straight angle:** An angle that measures 180°.
> **Acute angle:** An angle that measures more than 0° and less than 90°.
> **Obtuse angle:** An angle that measures more than 90° and less than 180°.

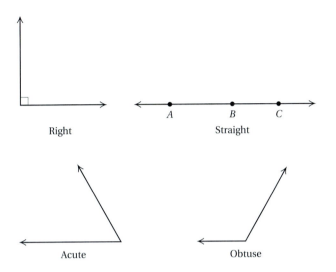

Do Exercises 6–9.

c Complementary and Supplementary Angles

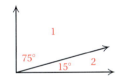

∠1 and ∠2 above are **complementary** angles.

$$m\angle 1 + m\angle 2 = 90°$$
$$75° + 15° = 90°$$

> Two angles are **complementary** if the sum of their measures is 90°. Each angle is called a **complement** of the other.

If two angles are complementary, each is an acute angle. When complementary angles are adjacent to each other, they form a right angle.

Classify each angle as right, straight, acute, or obtuse. Use a protractor if necessary.

6.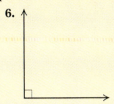

7.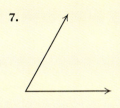

8.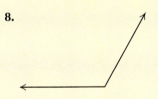

9.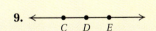

Answers on page A-25

9.4 Angles

543

10. Identify each pair of complementary angles.

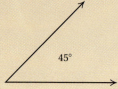

Find the measure of a complement of the angle.

11.

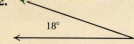

12.

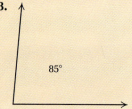

13.

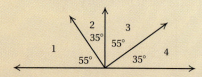

14. Identify each pair of supplementary angles.

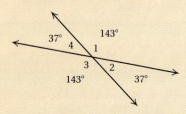

Find the measure of a supplement of an angle with the given measure.

15. 38°

16. 157°

17. 90°

Answers on page A-25

Chapter 9 Geometry and Measures

544

Example 2 Identify each pair of complementary angles.

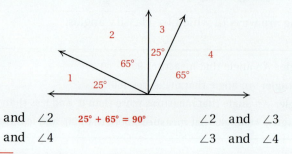

∠1 and ∠2   25° + 65° = 90°   ∠2 and ∠3
∠1 and ∠4   ∠3 and ∠4

Example 3 Find the measure of a complement of an angle of 39°.

    90° − 39° = 51°

The measure of a complement is 51°.

Do Exercises 10–13.

Next, consider ∠1 and ∠2 as shown below. Because the sum of their measures is 180°, ∠1 and ∠2 are said to be **supplementary**. Note that when supplementary angles are adjacent, they form a straight angle.

$m\angle 1 + m\angle 2 = 180°$
$30° + 150° = 180°$

> Two angles are **supplementary** if the sum of their measures is 180°. Each angle is called a **supplement** of the other.

Example 4 Identify each pair of supplementary angles.

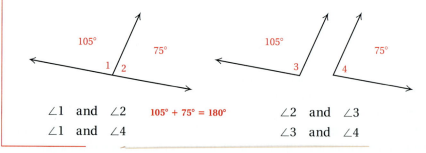

∠1 and ∠2   105° + 75° = 180°   ∠2 and ∠3
∠1 and ∠4   ∠3 and ∠4

Example 5 Find the measure of a supplement of an angle of 112°.

   180° − 112° = 68°

The measure of a supplement is 68°.

Do Exercises 14–17.

Exercise Set 9.4

a Name each angle in six different ways.

1.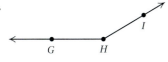

2.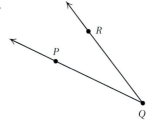

Use a protractor to measure each angle.

3.

4.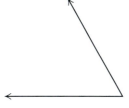

5.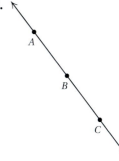

6.

7.

8.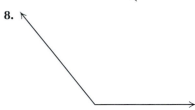

Use the given information and a protractor to draw a circle graph.

9. *Sporting Goods Purchases.* Below is a list of educational backgrounds of people who purchase sporting goods.

Less than high school:	10%
High school:	26%
Some college:	35%
College graduate:	29%

   Source: *The Sporting Goods Market in 1997.* Mt. Prospect, IL: National Sporting Goods Association (copyright).

10. *Lottery Sales.* Below is a list of how money was spent on lottery games in the United States in a recent year.

Lotto:	28%	Instant:	40%
4-digit:	6%	Other:	10%
3-digit:	16%		

    Source: 1998 *World Lottery Almanac* annual. Boyds, MD: TLF Publications, Inc.; *LaFleur's Fiscal 1997 Lottery Special Report*; and *LaFleur's Lottery World Government Profits Report* (copyright).

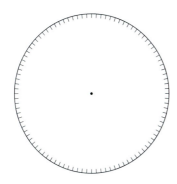

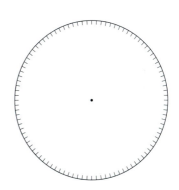

b

11.–18. Classify each of the angles in Exercises 1–8 as right, straight, acute, or obtuse.

19.–22. Classify each of the angles in Margin Exercises 1–4 as right, straight, acute, or obtuse.

c Find the measure of a complement of an angle with the given measure.

23. 11°   24. 83°   25. 67°   26. 5°

27. 58°   28. 32°   29. 29°   30. 54°

Find the measure of a supplement of an angle with the given measure.

31. 3°   32. 54°   33. 139°   34. 13°

35. 85°   36. 129°   37. 102°   38. 45°

Skill Maintenance

39. Convert to decimal notation: 56.1%. [8.1b]

40. Convert to percent notation: 0.6734. [8.1b]

41. Solve: $3.1x + 4.3 = x + 9.55$. [5.7b]

42. Convert to percent notation: $\dfrac{9}{8}$. [8.1c]

43. Add: $-9.7 + 3.8$. [5.2c]

44. Subtract: $-4.3 - (-9.8)$. [5.2c]

Synthesis

45. ◆ Do parallelograms always contain two acute and two obtuse angles? Why or why not?

46. ◆ Explain a procedure that could be used to determine the measure of an angle's supplement from the measure of the angle's complement.

47. ◆ Is it possible that both an angle and its supplement can be obtuse? Why or why not?

48. ▦ In the figure, $m\angle 1 = 79.8°$ and $m\angle 6 = 33.07°$. Find $m\angle 2$, $m\angle 3$, $m\angle 4$, and $m\angle 5$.

49. ▦ In the figure, $m\angle 2 = 42.17°$ and $m\angle 3 = 81.9°$. Find $m\angle 1$, $m\angle 4$, $m\angle 5$, and $m\angle 6$.

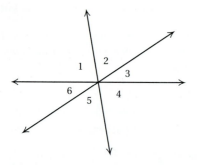

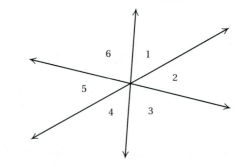

50. For any triangle, the sum of the measures of the angles is 180°. Use this fact to help find $m\angle ACB$, $m\angle CAB$, $m\angle EBC$, $m\angle EBA$, $m\angle AEB$, and $m\angle ADB$ in the rectangle shown at right.

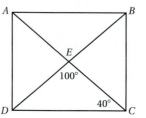

9.5 Square Roots and the Pythagorean Theorem

a | Square Roots

Objectives

a | Simplify square roots of squares such as $\sqrt{25}$.

b | Approximate square roots.

c | Given the lengths of any two sides of a right triangle, find the length of the third side.

For Extra Help

TAPE 17   MAC WIN   CD-ROM

> If a number is a product of a factor times itself, then that factor is a **square root** of the number. (If $c^2 = a$, then c is a square root of a.)

For example, 36 has two square roots, 6 and -6. To see this, note that $6 \cdot 6 = 36$ and $(-6) \cdot (-6) = 36$.

Example 1 Find the square roots of 25.

The square roots of 25 are 5 and -5, because $5^2 = 25$ and $(-5)^2 = 25$.

CAUTION! To find the *square* of a number, multiply the number by itself. To find a *square root* of a number, find a number that, when squared, gives the original number.

Find each square.

1. 9^2
2. $(-10)^2$
3. 11^2
4. 12^2

Do Exercises 1–12.

Since every positive number has two square roots, the symbol $\sqrt{}$ (called a *radical* sign) is used to denote the positive square root of the number underneath. Thus, $\sqrt{9}$ means 3, not -3. When we refer to *the* square root of some number n, we mean the positive square root, $\sqrt{n}$.

> It would be helpful to memorize the squares of numbers from 1 to 25.

5. 13^2
6. 14^2
7. 15^2
8. 16^2

Examples Simplify.

2. $\sqrt{36} = 6$ The square root of 36 is 6 because $6^2 = 36$ and 6 is positive.
3. $\sqrt{25} = 5$ Note that $5^2 = 25$.
4. $\sqrt{144} = 12$ Note that $12^2 = 144$.
5. $\sqrt{256} = 16$ Note that $16^2 = 256$.

Find all square roots. Use the results of Exercises 1–8 above, if necessary.

9. 100
10. 81
11. 49
12. 196

Do Exercises 13–22.

b | Approximating Square Roots

Many square roots can't be written as whole numbers or fractions. For example,

$$\sqrt{2}, \quad \sqrt{3}, \quad \sqrt{39}, \quad \text{and} \quad \sqrt{70}$$

cannot be precisely represented in decimal notation. To see this, consider the following decimal approximations for $\sqrt{2}$. Each gives a closer approximation, but none is exactly $\sqrt{2}$:

$\sqrt{2} \approx 1.4$ because $(1.4)^2 = 1.96$;
$\sqrt{2} \approx 1.41$ because $(1.41)^2 = 1.9881$;
$\sqrt{2} \approx 1.414$ because $(1.414)^2 = 1.999396$;
$\sqrt{2} \approx 1.4142$ because $(1.4142)^2 = 1.99996164$.

Decimal approximations like these are commonly found by using a calculator.

Simplify. Use the results of Exercises 1–8 above, if necessary.

13. $\sqrt{49}$
14. $\sqrt{16}$
15. $\sqrt{121}$
16. $\sqrt{100}$
17. $\sqrt{81}$
18. $\sqrt{64}$
19. $\sqrt{225}$
20. $\sqrt{169}$
21. $\sqrt{1}$
22. $\sqrt{0}$

Answers on page A-25

Approximate to three decimal places.

23. $\sqrt{5}$

24. $\sqrt{78}$

25. $\sqrt{168}$

Example 6 Approximate $\sqrt{3}$, $\sqrt{27}$, and $\sqrt{180}$ to three decimal places. Use a calculator.

We use a calculator to find each square root. Since more than three decimal places are given, we round back to three places.

$$\sqrt{3} \approx 1.732,$$
$$\sqrt{27} \approx 5.196,$$
$$\sqrt{180} \approx 13.416$$

As a check, note that because $1 \cdot 1 = 1$ and $2 \cdot 2 = 4$, we expect $\sqrt{3}$ to be between 1 and 2. Similarly, we expect $\sqrt{27}$ to be between 5 and 6 and $\sqrt{180}$ to be between 13 and 14.

Do Exercises 23–25.

c | The Pythagorean Theorem

A **right triangle** is a triangle with a 90° angle, as shown here.

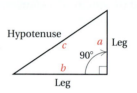

In a right triangle, the longest side is called the **hypotenuse**. It is also the side opposite the right angle. The other two sides are called **legs**. We generally use the letters a and b for the lengths of the legs and c for the length of the hypotenuse. They are related as follows.

> **THE PYTHAGOREAN THEOREM**
>
> In any right triangle, if a and b are the lengths of the legs and c is the length of the hypotenuse, then
> $$a^2 + b^2 = c^2, \text{ or}$$
> $$(\text{Leg})^2 + (\text{Other leg})^2 = (\text{Hypotenuse})^2.$$
>
>
>
> The equation $a^2 + b^2 = c^2$ is called the **Pythagorean equation.***

It is important to remember this theorem because it is extremely useful. By using the Pythagorean theorem, we can find the length of any side in a right triangle if the lengths of the other sides are known.

---

*The *converse* of the Pythagorean theorem is also true. That is, if $a^2 + b^2 = c^2$, then the triangle is a right triangle.

Calculator Spotlight

Most calculators have a square root key, $\boxed{\sqrt{\ }}$. On some calculators, square roots are found by pressing the $\boxed{x^2}$ key after first pressing a key labeled $\boxed{2\text{nd}}$ or $\boxed{\text{SHIFT}}$. (On these calculators, finding square roots is a secondary function.)

To find an approximation for $\sqrt{30}$, we simply press

$\boxed{3}\ \boxed{0}\ \boxed{\sqrt{\ }}$.

The value 5.477225575 appears.

It is always best to wait until calculations are complete before rounding off. For example, to round $9 \cdot \sqrt{5}$ to the nearest tenth, we do *not* first determine that $\sqrt{5} \approx 2.2$. Rather, we press

$\boxed{9}\ \boxed{\times}\ \boxed{5}\ \boxed{\sqrt{\ }}\ \boxed{=}$.

The result, 20.1246118, is then rounded to 20.1.

Exercises

Round to the nearest tenth.

1. $\sqrt{43}$ 2. $\sqrt{94}$
3. $7 \cdot \sqrt{8}$ 4. $5 \cdot \sqrt{12}$
5. $\sqrt{35} + 19$ 6. $17 + \sqrt{57}$
7. $13 \cdot \sqrt{68} + 14$
8. $24 \cdot \sqrt{31} - 18$
9. $5 \cdot \sqrt{30} - 3 \cdot \sqrt{14}$
10. $7 \cdot \sqrt{90} + 3 \cdot \sqrt{40}$

Answers on pages A-25 and A-26

Example 7 Find the length of the hypotenuse of this right triangle. Give an exact answer and an approximation to three decimal places.

We substitute in the Pythagorean equation:

$$a^2 + b^2 = c^2$$
$$4^2 + 7^2 = c^2 \quad \text{Substituting}$$
$$16 + 49 = c^2$$
$$65 = c^2.$$

The solutions of this equation are the square roots of 65. Thus, we would ordinarily say that the numbers $\sqrt{65}$ and $-\sqrt{65}$ are solutions. In this case, however, since we are solving for a length, only positive answers are acceptable.

Exact answer: $c = \sqrt{65}$

Approximate answer: $c \approx 8.062$ Using a calculator

Do Exercise 26.

Example 8 Find the length b for the right triangle shown. Give an exact answer and an approximation to three decimal places.

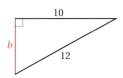

We substitute in the Pythagorean equation. Then we solve for b^2 and b, as follows:

$$a^2 + b^2 = c^2$$
$$10^2 + b^2 = 12^2 \quad \text{Substituting}$$
$$100 + b^2 = 144$$
$$100 + b^2 - 100 = 144 - 100 \quad \text{Subtracting 100 on both sides}$$
$$b^2 = 144 - 100$$
$$b^2 = 44$$

Exact answer: $b = \sqrt{44}$

Approximation: $b \approx 6.633$.   Using a calculator

Do Exercises 27–29.

26. Find the length of the hypotenuse of this right triangle. Give an exact answer and an approximation to three decimal places.

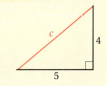

Find the length of the unknown leg of each right triangle. Give an exact answer and an approximation to three decimal places.

27.

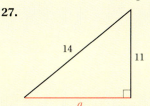

28.

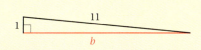

29.

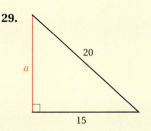

Answers on page A-26

30. How long is a guy wire reaching from the top of an 18-ft pole to a point on the ground 10 ft from the pole? Give an exact answer and an approximation to the nearest tenth of a foot.

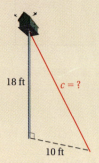

Example 9 *Height of Ladder.* A 12-ft ladder leans against a building. The bottom of the ladder is 7 ft from the building. How high is the top of the ladder? Give an exact answer and an approximation to the nearest tenth of a foot.

1. **Familiarize.** We first make a drawing. In it we see a right triangle. We let $h =$ the unknown height.

2. **Translate.** We substitute 7 for a, h for b, and 12 for c in the Pythagorean equation:

$$a^2 + b^2 = c^2 \quad \text{Pythagorean equation}$$
$$7^2 + h^2 = 12^2.$$

3. **Solve.** We solve for h^2 and then h:

$$49 + h^2 = 144$$
$$49 + h^2 - 49 = 144 - 49$$
$$h^2 = 95$$

Exact answer: $h = \sqrt{95}$

Approximation: $h \approx 9.7$ ft.

4. **Check.** $7^2 + (\sqrt{95})^2 = 49 + 95 = 144 = 12^2$.

5. **State.** The top of the ladder is $\sqrt{95}$, or about 9.7 ft from the ground.

Do Exercise 30.

The Pythagorean theorem is named for the Greek mathematician Pythagoras (569?–500? B.C.). In the diagram below, we show one way in which the theorem can be visualized.

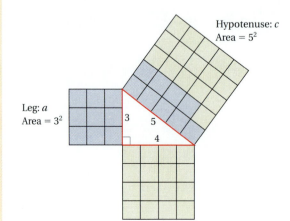

$a^2 + b^2 = c^2$
$3^2 + 4^2 = 5^2$
$9 + 16 = 25$

Answer on page A-26

Exercise Set 9.5

a Find all square roots.

1. 16
2. 9
3. 121
4. 49

5. 169
6. 144
7. 6400
8. 3600

Simplify.

9. $\sqrt{49}$
10. $\sqrt{4}$
11. $\sqrt{81}$
12. $\sqrt{64}$

13. $\sqrt{225}$
14. $\sqrt{121}$
15. $\sqrt{625}$
16. $\sqrt{900}$

17. $\sqrt{400}$
18. $\sqrt{169}$
19. $\sqrt{10,000}$
20. $\sqrt{1,000,000}$

b Approximate each number to three decimal places.

21. $\sqrt{48}$
22. $\sqrt{17}$
23. $\sqrt{8}$
24. $\sqrt{7}$

25. $\sqrt{3}$
26. $\sqrt{6}$
27. $\sqrt{12}$
28. $\sqrt{18}$

29. $\sqrt{19}$
30. $\sqrt{75}$
31. $\sqrt{110}$
32. $\sqrt{10}$

c Find the length of each third side of each right triangle. Give an exact answer and, when appropriate, an approximation to three decimal places.

33.

34.

35.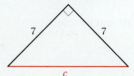

36.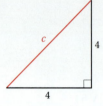

37.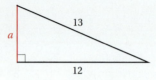

38.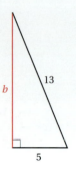

39.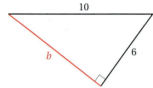

40.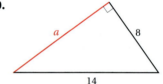

For each right triangle, find the length of the side not given. Assume that c represents the length of the hypotenuse. Give an exact answer and, when appropriate, an approximation to three decimal places.

41. $a = 10$, $b = 24$

42. $a = 5$, $b = 12$

43. $a = 9$, $c = 15$

44. $a = 18$, $c = 30$

45. $a = 1$, $c = 32$

46. $b = 1$, $c = 20$

47. $a = 4$, $b = 3$

48. $a = 1$, $c = 15$

Chapter 9 Geometry and Measures

In Exercises 49–56, give an exact answer and an approximation to the nearest tenth.

49. How long is a string of lights reaching from the top of a 12-ft pole to a point 8 ft from the base of the pole?

50. How long must a wire be in order to reach from the top of a 13-m telephone pole to a point on the ground 9 m from the base of the pole?

51. *Baseball Diamond.* A baseball diamond is actually a square 90 ft on a side. How far is it from home plate to second base?

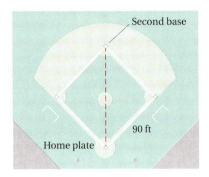

52. *Softball Diamond.* A slow-pitch softball diamond is actually a square 65 ft on a side. How far is it from home plate to second base?

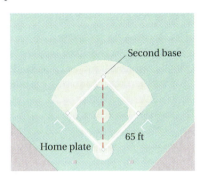

53. How tall is this tree?

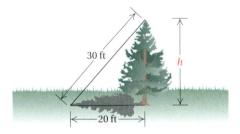

54. How far is the base of the fence post from point A?

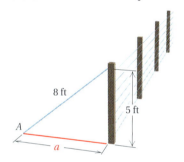

55. An airplane is flying at an altitude of 4100 ft. The slanted distance directly to the airport is 15,100 ft. How far is the airplane horizontally from the airport?

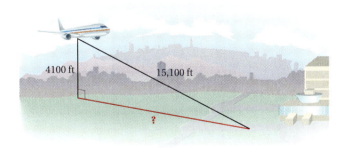

56. A surveyor had poles located at points P, Q, and R around a lake. The distances that the surveyor was able to measure are marked on the drawing. What is the distance from P to R across the lake?

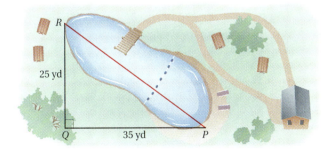

Exercise Set 9.5

Skill Maintenance

Solve.

57. Food expenses account for 26% of the average family's budget. A family makes $1800 one month. How much do they spend for food? [8.4a]

58. The price of a cellular phone was reduced from $350 to $308. Find the percent of decrease in price. [8.4b]

59. A county has a population that is increasing by 4% each year. This year the population is 180,000. What will it be next year? [8.4b]

60. The price of a box of cookies increased from $2.85 to $3.99. What was the percent of increase in the price? [8.4b]

61. A college has a student body of 1850 students. Of these, 17.5% are seniors. How many students are seniors? [8.4a]

62. A state charges a meals tax of $4\frac{1}{2}$%. What is the meals tax charged on a dinner party costing $540? [8.5a]

Synthesis

63. ◆ Explain how the Pythagorean theorem can be used to prove that a triangle is a *right* triangle.

64. ◆ Write a problem similar to Exercises 49–52 for a classmate to solve. Design the problem so that its solution involves the length $\sqrt{58}$ m.

65. ◆ Give an argument that could be used to convince a classmate that $\sqrt{2501}$ is not a whole number. Do not use a calculator.

66. ▦ Find the area of the trapezoid shown. Round to the nearest hundredth.

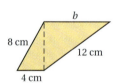

67. Which of the triangles below has the larger area?

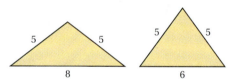

68. A 19-in. television set has a rectangular screen that measures 19 in. diagonally. The ratio of length to width in a conventional television set is 4 to 3. Find the length and the width of the screen.

69. A Philips 42-in. plasma television has a rectangular screen that measures 42 in. diagonally. The ratio of length to width is 16 to 9. Find the length and the width of the screen.

9.6 Volume and Capacity

a | Volume

The **volume** of a **rectangular solid** is the number of unit cubes needed to fill it.

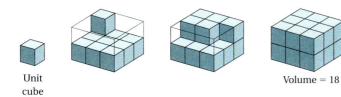

Unit cube

Volume = 18

Two other units are shown below.

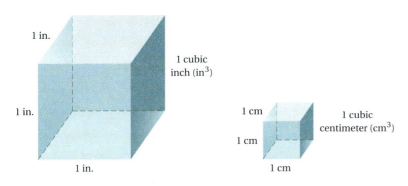

1 cubic inch (in³)

1 cubic centimeter (cm³)

Example 1 Find the volume.

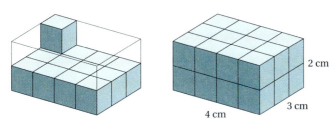

The figure is made up of 2 layers of 12 cubes each, so its volume is 24 cubic centimeters (cm³).

Do Exercise 1.

The volume of a rectangular solid can be thought of as the product of the area of the base times the height:

$$V = l \cdot w \cdot h$$
$$= (l \cdot w) \cdot h$$
$$= (\text{Area of the base}) \cdot h$$
$$= B \cdot h,$$

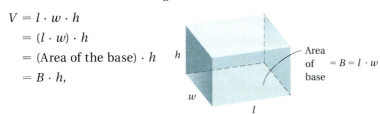

Area of base $= B = l \cdot w$

where B is the area of the base.

Objectives

- **a** Find the volume of a rectangular solid, a cylinder, and a sphere.
- **b** Convert from one unit of capacity to another.
- **c** Solve applied problems involving capacity.

For Extra Help

TAPE 17   MAC WIN   CD-ROM

1. Find the volume.

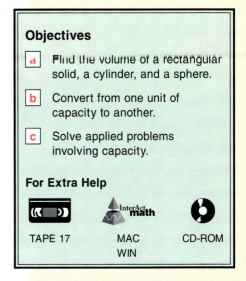

2 cm
2 cm
3 cm

Answer on page A-26

2. In a recent year, people in the United States bought enough unpopped popcorn to provide every person in the country with a bag of popped corn measuring 2 ft by 2 ft by 5 ft. Find the volume of such a bag.

3. *Cord of Wood.* A cord of firewood measures 4 ft by 4 ft by 8 ft. What is the volume of a cord of firewood?

Answers on page A-26

The **volume of a rectangular solid** is found by multiplying length by width by height:

$$V = \underbrace{l \cdot w}_{\text{Area of base}} \cdot \underbrace{h}_{\text{Height}}.$$

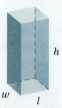

Example 2 The largest piece of luggage that you can carry on an airplane measures 23 in. by 10 in. by 13 in. Find the volume of this solid.

$V = l \cdot w \cdot h$
$ = 23 \text{ in.} \cdot 10 \text{ in.} \cdot 13 \text{ in.}$
$ = 230 \cdot 13 \text{ in}^3$
$ = 2990 \text{ in}^3$

Do Exercises 2 and 3.

Cylinders

Like rectangular solids, **circular cylinders** have tops and bottoms of equal area that lie in parallel planes. The bases of circular cylinders are circular regions.

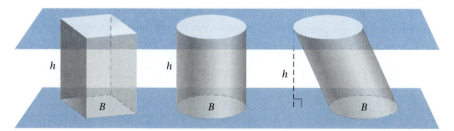

The volume of a circular cylinder is also the product of the area of the base times the height. The height is always measured perpendicular to—that is, at a 90° angle from—the base.

The volume of a circular cylinder of radius r is the product of the area of the base B and the height h:
$$V = B \cdot h, \quad \text{or} \quad V = \pi \cdot r^2 \cdot h.$$

Example 3 Find the volume of this circular cylinder. Use 3.14 to approximate π.

$$V = Bh = \pi \cdot r^2 \cdot h$$
$$\approx 3.14 \cdot 4 \text{ cm} \cdot 4 \text{ cm} \cdot 12 \text{ cm}$$
$$\approx 602.88 \text{ cm}^3$$

Do Exercises 4 and 5.

Spheres

A **sphere** is the three-dimensional counterpart of a circle. It is the set of all points in space that are a given distance (the radius) from a given point (the center). The volume of a sphere depends on its radius.

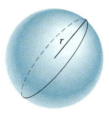

> The volume of a sphere of radius r is given by
> $$V = \frac{4}{3} \cdot \pi \cdot r^3.$$

Example 4 *Bowling.* The radius of a standard-sized bowling ball is about 11 cm. Find the volume of a standard-sized bowling ball. Round to the nearest cubic centimeter. Use 3.14 for π.

$$V = \frac{4}{3} \cdot \pi \cdot r^3 \approx \frac{4}{3} \cdot 3.14 \cdot (11 \text{ cm})^3$$
$$\approx \frac{4 \cdot 3.14 \cdot 1331 \text{ cm}^3}{3} \approx 5572 \text{ cm}^3$$

Do Exercises 6 and 7.

b | Capacity

Since many substances come in containers that have irregular shape, to answer a question like "How much soda is in the bottle?" we need measures of **capacity**. American units of capacity are ounces, cups, pints, quarts, and gallons. These units are related as follows.

> **AMERICAN UNITS OF CAPACITY**
> 1 gallon (gal) = 4 quarts (qt)   1 pt = 2 cups = 16 ounces (oz)
> 1 qt = 2 pints (pt)   1 cup = 8 oz

---

4. Find the volume of the cylinder. Use 3.14 to approximate π.

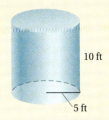

5. Find the volume of the cylinder. Use $\frac{22}{7}$ to approximate π.

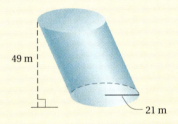

6. Find the volume of the sphere. Use 3.14 for π.

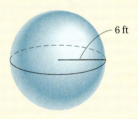

7. The radius of a standard-sized golf ball is about $\frac{4}{5}$ in. Find its volume. Use $\frac{22}{7}$ for π.

Answers on page A-26

Complete

8. 80 qt = _____ gal.

Example 5 Complete: 24 qt = _____ gal.

In this case, we multiply by 1 using 1 gal in the numerator, since we are converting to gallons, and 4 qt in the denominator, since we are converting from quarts.

$$24 \text{ qt} = 24 \text{ qt} \cdot \frac{1 \text{ gal}}{4 \text{ qt}} = \frac{24}{4} \cdot 1 \text{ gal} = 6 \text{ gal}$$

To check that our answer is reasonable, note that since we are converting from smaller to larger units, our answer is a smaller number than the one with which we started.

Do Exercise 8.

Example 6 Complete: 9 gal = _____ oz.

First, we multiply by 1 using 4 qt on the top and 1 gal on the bottom:

$$9 \text{ gal} = 9 \text{ gal} \cdot \frac{4 \text{ qt}}{1 \text{ gal}} \quad \text{This converts from gallons to quarts.}$$

$$= 9 \cdot 4 \text{ qt} = 36 \text{ qt}.$$

Next, we convert 36 qt to ounces by multiplying by 32 oz/1 qt:

$$9 \text{ gal} = 36 \text{ qt} = 36 \text{ qt} \cdot \frac{32 \text{ oz}}{1 \text{ qt}}$$

$$= 36 \cdot 32 \text{ oz} = 1152 \text{ oz}.$$

This conversion could have been done in one step had we known that 1 gal = 128 oz. We then would have multiplied 9 gal by 128 oz/1 gal.

Do Exercise 9.

Complete.

9. 4 gal = _____ pt.

The basic unit of capacity for the metric system is the **liter**. A liter is just a bit more than a quart (1 liter = 1.06 quarts). It is defined as follows.

1 liter   1 quart

> **METRIC UNITS OF CAPACITY**
>
> 1 liter (L) = 1000 cubic centimeters (1000 cm³)
>
> The script letter ℓ is also used for "liter."

Answers on page A-26

Chapter 9   Geometry and Measures

Metric prefixes are also used with liters. The most common is **milli-**. The milliliter (mL) is, then, $\frac{1}{1000}$ liter. Thus,

> 1 L = 1000 mL = 1000 cm³;
> 0.001 L = 1 mL = 1 cm³.

A preferred unit for drug dosage is the milliliter (mL) or the cubic centimeter (cm³). The notation "cc" is also used for cubic centimeter, especially in medicine. A milliliter and a cubic centimeter are the same size. Each is about the size of a sugar cube.

> 1 mL = 1 cm³ = 1 cc

Volumes for which quarts and gallons are used are expressed in liters. Large volumes may be expressed using cubic meters (m³).

Do Exercises 10–13.

Example 7 Complete: 4.5 L = _____ mL.

$$4.5 \text{ L} = 4.5 \text{ L} \cdot \frac{1000 \text{ mL}}{1 \text{ L}}$$
$$= 4.5 \cdot 1000 \text{ mL}$$
$$= 4500 \text{ mL}$$

Example 8 Complete: 280 mL = _____ L.

$$280 \text{ mL} = 280 \text{ mL} \cdot \frac{1 \text{ L}}{1000 \text{ mL}}$$
$$= \frac{280}{1000} \text{ L}$$
$$= 0.28 \text{ L}$$

Do Exercises 14 and 15.

c Solving Problems

Example 9 At a self-service gasoline station, 89-octane gasoline sells for 28.3¢ a liter. Estimate the price of 1 gal in dollars.

Since 1 liter is about 1 quart and there are 4 quarts in a gallon, the price of a gallon is about 4 times the price of a liter:

$$4 \cdot 28.3¢ = 113.2¢ = \$1.132.$$

Thus 89-octane gasoline sells for about $1.13 a gallon.

Do Exercise 16.

Complete with mL or L.

10. To prevent infection, a patient received an injection of 2 _____ of penicillin.

11. There are 250 _____ in a coffee cup.

12. The gas tank holds 80 _____ .

13. Bring home 8 _____ of milk.

Complete.
14. 0.97 L = _____ mL

15. 8990 mL = _____ L

16. At the same station, the price of 87-octane gasoline is 26.9 cents a liter. Estimate the price of 1 gallon in dollars.

Answers on page A-26

17. **Medicine Capsule.** A cold capsule is 8 mm long and 4 mm in diameter. Find the volume of the capsule. Use 3.14 for π. (*Hint*: First find the length of the cylindrical section.)

Example 10 *Propane Gas Tank.* A propane gas tank is shaped like a circular cylinder with half of a sphere at each end. Find the volume of the tank if the cylindrical section is 5 ft long with a 4-ft diameter. Use 3.14 for π.

1. **Familiarize.** We first make a drawing.

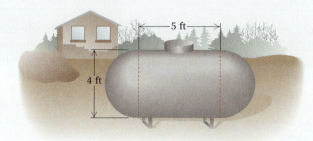

2. **Translate.** This is a two-step problem. We first find the volume of the cylindrical portion. Then we find the volume of the two ends and add. Note that the radius is 2 ft and that together the two ends make a sphere. We let $V =$ the total volume.

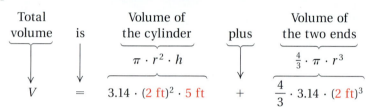

3. **Solve.** The volume of the cylinder is approximately

$$3.14 \cdot (2 \text{ ft})^2 \cdot 5 \text{ ft} \approx 3.14 \cdot 2 \text{ ft} \cdot 2 \text{ ft} \cdot 5 \text{ ft}$$
$$\approx 62.8 \text{ ft}^3.$$

The volume of the two ends is approximately

$$\frac{4}{3} \cdot 3.14 \cdot (2 \text{ ft})^3 \approx 1.33 \cdot 3.14 \cdot 2 \text{ ft} \cdot 2 \text{ ft} \cdot 2 \text{ ft}$$
$$\approx 33.4 \text{ ft}^3.$$

The total volume is approximately

$$62.8 \text{ ft}^3 + 33.4 \text{ ft}^3 = 96.2 \text{ ft}^3.$$

4. **Check.** The check is left to the student.
5. **State.** The volume of the tank is about 96.2 ft³.

Do Exercise 17.

Calculator Spotlight

 Exercises

1. *Measuring the volume of a cloud.* Using a calculator with a $\boxed{\pi}$ key, find the volume of a spherical cloud with a 1000-m diameter.

2. The box shown is just big enough to hold 3 golf balls. If the radius of a golf ball is 2.1 cm, how much air surrounds the three balls in the closed box?

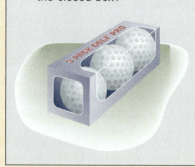

Answer on page A-26

Exercise Set 9.6

a Find each volume. Use 3.14 for π in Exercises 9–12 and 15–18. Use $\frac{22}{7}$ for π in Exercises 13, 14, 19, and 20.

1.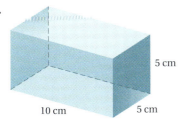

2.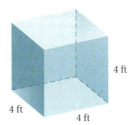

3.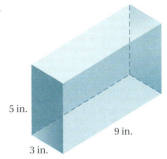

4.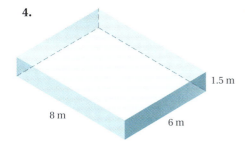

5.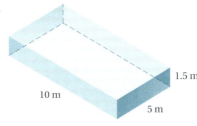

6.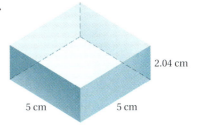

7.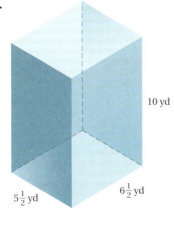

8.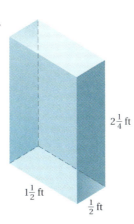

9.

10.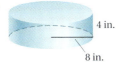

11.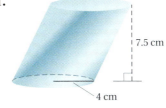

12.

13. 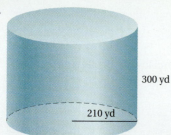 **14.** 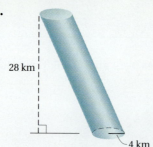 **15.**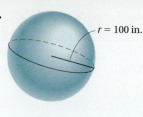

16. **17.** 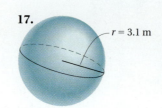 **18.**

19. 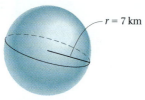 **20.**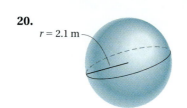

b Complete.

21. 1 L = _____ mL = _____ cm³

22. _____ L = 1 mL = _____ cm³

23. 87 L = _____ mL

24. 806 L = _____ mL

25. 49 mL = _____ L

26. 19 mL = _____ L

27. 27.3 L = _____ cm³

28. 49.2 L = _____ cm³

29. 5 gal = _____ pt

30. 48 oz = _____ pt

31. 10 qt = _____ oz

32. 2 gal = _____ cups

33. 24 oz = _____ cups

34. 20 cups = _____ pt

35. 8 gal = _____ qt

36. 5 gal = _____ cups

37. 3 gal = _____ cups

38. 72 oz = _____ cups

39. 15 pt = _____ gal

40. 9 qt = _____ gal

c Solve.

41. *Medicine.* Dr. Carey ordered 0.5 L of normal saline solution. How many milliliters were ordered?

42. *Medicine.* Ingrid receives 84 mL per hour of normal saline solution. How many liters did Ingrid receive in a 24-hr period?

43. *Medicine.* Dr. Norris tells a patient to purchase 0.5 L of hydrogen peroxide. Commercially, hydrogen peroxide is found on the shelf in bottles that hold 4 oz, 8 oz, and 16 oz. Which bottle comes closest to filling the prescription?

44. *Medicine.* Dr. Gomez wants a patient to receive 3 L of a normal glucose solution in a 24-hr period. How many milliliters per hour should the patient receive?

45. A rung of a ladder is 2 in. in diameter and 18 in. long. Find the volume. Use 3.14 for π.

46. The diameter of a cylindrical trashcan is 0.7 yd. The height is 1.1 yd. Find the volume. Use $\frac{22}{7}$ for π.

47. An oak log has a diameter of 12 cm and a length of 42 cm. Find the volume. Use 3.14 for π.

48. *Farming.* A barn silo, excluding the top, is a circular cylinder. The silo is 6 m in diameter and the height is 13 m. Find the volume. Use 3.14 for π.

49. *Merchandising.* Tennis balls are generally packaged in circular cylinders that hold 3 balls each. The diameter of a tennis ball is 6.5 cm. Find the volume of a can of tennis balls. Use 3.14 for π.

50. *Oceanography.* A research submarine is capsule-shaped. Find the volume of the submarine if it has a length of 10 m and a diameter of 8 m. Use 3.14 for π. (*Hint*: First find the length of the cylindrical section.)

51. *Metallurgy.* If all the gold in the world could be gathered together, it would form a cube 18 yd on a side. Find the volume of the world's gold.

52. *Astronomy.* The radius of Pluto's moon is about 500 km. Find the volume of this satellite. Use $\frac{22}{7}$ for π.

53. *Farming.* A water storage tank is a right circular cylinder with a radius of 14 cm and a height of 100 cm. What is the tank's volume? How many liters can it hold? Use $\frac{22}{7}$ for π.

54. *Conservation.* Many people leave the water running while brushing their teeth. Suppose that one person wastes 32 oz of water in such a manner each day. How much water, in gallons, would that person waste in a week? in 30 days? in a year? If each of the 261 million people in this country wastes water this way, estimate how much water is wasted in a year.

Skill Maintenance

55. Find the simple interest on $600 at 8% for $\frac{1}{2}$ yr. [8.6a]

56. Find the simple interest on $5000 at 7% for $\frac{1}{2}$ yr. [8.6a]

57. If 9 pens cost $8.01, how much would 12 pens cost? [7.4a]

58. Solve: $9(x - 1) = 3x + 5$. [5.7b]

59. Solve: $-5y + 3 = -12y - 4$. [5.7b]

60. A barge travels 320 km in 15 days. At this rate, how far will it travel in 21 days? [7.4a]

Synthesis

61. ◆ Which occupies more volume: two spheres, each with radius r, or one sphere with radius $2r$? Explain why.

62. ◆ What advantages do metric units of capacity have over American units?

63. ◆ How could you use the volume formulas in this section to help estimate the volume of an egg?

64. ▦ Audio-cassette cases are typically 7 cm by 10.75 cm by 1.5 cm and contain 90 min of music. Compact-disc cases are typically 12.4 cm by 14.1 cm by 1 cm and contain 50 min of music. Which container holds the most music per cubic centimeter?

65. ▦ A 2-cm–wide stream of water passes through a 30-m long garden hose. At the instant that the water is turned off, how many liters of water are in the hose? Use 3.141593 for π.

66. ▦ The volume of a basketball is 2304π cm³. Find the volume of a cube-shaped box that is just large enough to hold the ball.

67. ▦ The width of a dollar bill is 2.3125 in., the length is 6.0625 in., and the thickness is 0.0041 in. Find the volume occupied by one million one-dollar bills.

68. ▦ A sphere with diameter 1 m is circumscribed by a cube. How much more volume is in the cube?

69. ▦ A cube is circumscribed by a sphere with a 1-m diameter. How much more volume is in the sphere?

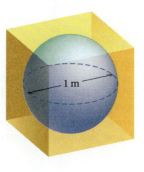

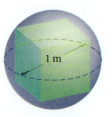

9.7 Weight, Mass, and Temperature

a | Weight: The American System

The American units of weight are as follows.

> **AMERICAN UNITS OF WEIGHT**
> 1 ton (T) = 2000 pounds (lb)
> 1 lb = 16 ounces (oz)

The term "ounce" used here for weight is different from the "ounce" we used for capacity in Section 9.6.

Example 1 A well-known hamburger is called a "quarter-pounder." Find its name in ounces: a "_____ ouncer."

$$\frac{1}{4} \text{ lb} = \frac{1}{4} \cdot 1 \text{ lb}$$
$$= \frac{1}{4} \cdot 16 \text{ oz} \quad \text{Substituting 16 oz for 1 lb}$$
$$= 4 \text{ oz}$$

A "quarter-pounder" can also be called a "four-ouncer."

Example 2 Complete: 15,360 lb = _____ T.

$$15{,}360 \text{ lb} = 15{,}360 \text{ lb} \cdot \frac{1 \text{ T}}{2000 \text{ lb}} \quad \text{Multiplying by 1}$$
$$= \frac{15{,}360}{2000} \text{ T} \quad \text{Dividing by 2000}$$
$$= 7.68 \text{ T}$$

Do Exercises 1–3.

b | Mass: The Metric System

There is a difference between **mass** and **weight**, but the terms are often used interchangeably. People sometimes use the word "weight" instead of "mass." Weight is related to the force of gravity. The farther you are from the center of the earth, the less you weigh. Your mass stays the same no matter where you are.

The basic unit of mass is the **gram** (g), which is the mass of 1 cubic centimeter (1 cm³ or 1 mL) of water. Since a cubic centimeter is small, a gram is a small unit of mass.

1 g = 1 gram = the mass of 1 cm³ (1 mL) of water

Objectives

a | Convert from one American unit of weight to another.

b | Convert from one metric unit of mass to another.

c | Convert temperatures from Celsius to Fahrenheit and from Fahrenheit to Celsius.

For Extra Help

TAPE 17   MAC WIN   CD-ROM

Complete.
1. 5 lb = _____ oz

2. 8640 lb = _____ T

3. 1 T = _____ oz

Answers on page A-26

Complete with mg, g, kg, or t.

4. A laptop computer has a mass of 2 _____.

5. Rosita has a body mass of 56 _____.

6. The athlete took a 200-_____ pain reliever.

7. A pen has a mass of 12 _____.

8. A pickup truck has a mass of 1.5 _____.

Answers on page A-26

Chapter 9   Geometry and Measures

566

The metric prefixes for mass are the same as those used for length.

> **METRIC UNITS OF MASS**
>
> 1 metric ton (t) = 1000 kilograms (kg)
> 1 *kilo*gram (kg) = 1000 grams (g)
> 1 *hecto*gram (hg) = 100 grams (g)
> 1 *deka*gram (dag) = 10 grams (g)
> 1 gram (g)
> 1 *deci*gram (dg) = $\frac{1}{10}$ gram (g)
> 1 *centi*gram (cg) = $\frac{1}{100}$ gram (g)
> 1 *milli*gram (mg) = $\frac{1}{1000}$ gram (g)

Thinking Metric

The mass of 1 raisin or 1 paperclip is approximately 1 gram (g). Since 1 metric ton is 1000 kg and 1 kg is about 2.2 lb, it follows that 1 metric ton (t) is about 2200 lb, or about 10% more than 1 American ton (T). The metric ton is used for very large masses, such as vehicles; the kilogram is used for masses of people or larger food packages; the gram is used for smaller food packages or objects like a coin or a ring; the milligram is used for even smaller masses like a dosage of medicine.

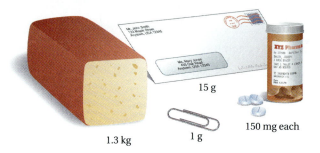

1.3 kg   15 g   1 g   150 mg each

Do Exercises 4–8.

Changing Units Mentally

As before, changing from one metric unit to another amounts to only the movement of a decimal point. Consider these equivalent masses.

> **MASS OF A STANDARD SHEET OF PAPER**
>
> 4260 mg = 426 cg = 4.26 g = 0.00426 kg

Example 3   Complete: 8 kg = _____ g.

Think: A kilogram is 1000 times the mass of a gram. Thus we move the decimal point three places to the right.

8.0   8.000.   8 kg = **8000 g**

Example 4 Complete: 4235 g = _____ kg.

Think: There are 1000 grams in 1 kilogram. Thus we move the decimal point three places to the left.

   4235.0   4.235.0   4235 g = 4.235 kg

Example 5 Complete: 6.98 cg = _____ mg.

Think: One centigram has the mass of 10 milligrams. Thus we move the decimal point one place to the right.

   6.98   6.9.8   6.98 cg = 69.8 mg

> The most commonly used metric units of mass are kg, g, cg, and mg. We have purposely used those more often than the others in the exercises.

Do Exercises 9–12.

c Temperature

Estimated Conversions

Below are two temperature scales: **Fahrenheit** for American measure and **Celsius** for metric measure.

By laying a straightedge horizontally between the scales, we can approximate conversions between Celsius and Fahrenheit.

Complete.

9. 6.2 kg = _____ g

10. 79.3 g = _____ kg

11. 7.7 cg = _____ mg

12. 2344 mg = _____ cg

Answers on page A-26

9.7 Weight, Mass, and Temperature

Use a straightedge to convert the following temperatures to either Celsius or Fahrenheit. Answers may be approximate.

13. 25°F (Cold day)

14. 30°C (Warm beach day)

15. −10°F (Extremely cold day)

16. 10°C (A cold bath)

Convert to Fahrenheit or Celsius.

17. 80°C

18. 35°C

19. 95°F

20. 113°F

Answers on page A-26

Examples Use a straightedge to approximate each of the following conversions. Use the scales on p. 567.

6. 110°F (Hot bath) ≈ 43°C

7. 50°C (Warm food) = 122°F

8. 160°F (Temperature of a sauna) ≈ 72°C

9. 0°C (Freezing point of water) = 32°F   This is exact.

Do Exercises 13–16.

Exact Conversions

A formula allows us to make exact conversions from Celsius to Fahrenheit.

$$F = \frac{9}{5} \cdot C + 32, \quad \text{or} \quad F = 1.8 \cdot C + 32$$

(Multiply the Celsius temperature by $\frac{9}{5}$, or 1.8, and add 32.)

Examples Convert to Fahrenheit.

10. 0°C $F = \frac{9}{5} \cdot 0 + 32 = 0 + 32 = 32°$ Substituting 0 for C

Thus, 0°C = 32°F.

11. 37°C $F = 1.8 \cdot 37 + 32 = 66.6 + 32 = 98.6°$ Substituting 37 for C

Thus, 37°C = 98.6°F. This is normal body temperature.

A second formula gives exact conversions from Fahrenheit to Celsius.

$$C = \frac{5}{9} \cdot (F - 32)$$

(Subtract 32 from the Fahrenheit temperature and multiply by $\frac{5}{9}$.)

Examples Convert to Celsius.

12. 212°F $C = \frac{5}{9} \cdot (F - 32)$

$= \frac{5}{9} \cdot (212 - 32) = \frac{5}{9} \cdot 180 = 100°$

Thus, 212°F = 100°C.

13. 77°F $C = \frac{5}{9} \cdot (F - 32)$

$= \frac{5}{9} \cdot (77 - 32) = \frac{5}{9} \cdot 45 = 25°$

Thus, 77°F = 25°C.

Do Exercises 17–20.

Exercise Set 9.7

a Complete.

1. 1 lb = _____ oz
2. 1 T = _____ lb
3. 6000 lb = _____ T

4. 5 T = _____ lb
5. 3 lb = _____ oz
6. 10 lb = _____ oz

7. 3.5 T = _____ lb
8. 2.5 T = _____ lb
9. 4800 lb = _____ T

10. 7500 lb = _____ T
11. 72 oz = _____ lb
12. 960 oz = _____ lb

b Complete.

13. 1 kg = _____ g
14. 6 kg = _____ g
15. 1 g = _____ kg

16. 1 dg = _____ g
17. 1 cg = _____ g
18. 1 mg = _____ g

19. 1 g = _____ mg
20. 1 g = _____ cg
21. 1 g = _____ dg

22. 45 kg = _____ g
23. 725 kg = _____ g
24. 678 g = _____ kg

25. 6345 g = _____ kg
26. 42.75 kg = _____ g
27. 897 mg = _____ kg

28. 45 cg = _____ g
29. 7.32 kg = _____ g
30. 0.439 cg = _____ mg

31. 6780 g = _____ kg
32. 5677 g = _____ kg
33. 69 mg = _____ cg

34. 76.1 mg = _____ cg
35. 8 kg = _____ cg
36. 0.02 kg = _____ mg

37. 1 t = _____ kg
38. 2 t = _____ kg

39. 3.4 cg = _____ dag
40. 9.34 g = _____ mg

c Convert to Celsius. Round the answer to the nearest ten degrees. Use the scales on p. 567.

41. 178°F
42. 195°F
43. 140°F
44. 107°F

45. 68°F
46. 45°F
47. 10°F
48. 120°F

Convert to Fahrenheit. Round the answer to the nearest ten degrees. Use the scales on p. 567.

49. 86°C
50. 93°C
51. 58°C
52. 33°C

53. −10°C
54. −5°C
55. 5°C
56. 15°C

Convert to Fahrenheit. Use the formula $F = \frac{9}{5} \cdot C + 32$.

57. 25°C   **58.** 85°C   **59.** 40°C   **60.** 90°C

61. 3000°C (melting point of iron)   **62.** 1000°C (melting point of gold)

Convert to Celsius. Use the formula $C = \frac{5}{9} \cdot (F - 32)$.

63. 86°F   **64.** 59°F   **65.** 131°F   **66.** 140°F

67. 98.6°F (normal body temperature)   **68.** 104°F (high-fevered body temperature)

Skill Maintenance

Convert to percent notation. [8.1b]

69. 0.0043   **70.** 2.31

71. If 2 cans of tomato paste cost $1.49, how many cans of tomato paste can you buy for $7.45? [7.4a]

72. *Sound Levels.* Make a horizontal bar graph to show the loudness of various sounds, as listed below. A decibel is a measure of the loudness of sounds. [6.2b]

Sound	Loudness (in decibels)
Whisper	15
Tick of watch	30
Speaking aloud	60
Noisy factory	90
Moving car	80
Car horn	98
Subway	104

73. Solve: $9(x - 3) = 4x - 5$. [5.7b]

74. Solve: $34.1 - 17.4x = 2.1x - 14.65$. [5.7b]

Exercise Set 9.7

Synthesis

75. ◆ Near the Canadian border, a radio forecast calls for an overnight low of 60. Was the temperature given in Celsius or Fahrenheit? Explain how you can tell.

76. ◆ Give at least two reasons why someone might prefer the use of grams to the use of ounces.

77. ◆ Describe a situation in which one object weighs 70 kg, another object weighs 3 g, and a third object weighs 125 mg.

Complete. Use 1 kg = 2.205 lb and 453.5 g = 1 lb. Round to four decimal places.

78. 🖩 1 lb = _____ kg

79. 🖩 1 g = _____ lb

80. Use the formula $F = \frac{9}{5} \cdot C + 32$ to find the temperature that is the same for both the Fahrenheit and Celsius scales.

81. *Medicine.* The medicine Albuterol is used for the treatment of asthma. It typically comes in an inhaler that contains 18 g. One actuation, or spray, is 90 mg.

    a) How many actuations are in one inhaler?
    b) Myra is going away for 4 months of college and wants to take enough Albuterol to last for that time. Assuming that Myra will need 4 actuations per day, estimate the number of inhalers she will need for the 4-month period.

82. *Chemistry.* Another temperature scale often used is the **Kelvin** scale. Conversions from Celsius to Kelvin can be carried out using the formula

    $K = C + 273.$

    A chemistry textbook describes an experiment in which a reaction takes place at a temperature of 400° Kelvin. A student wishes to perform the experiment, but has only a Fahrenheit thermometer. At what Fahrenheit temperature will the reaction take place?

83. 🖩 A large egg is about $5\frac{1}{2}$ cm tall with a diameter of 4 cm. Estimate the mass of such an egg by averaging the volumes of two spheres. (*Hint:* 1 cc of water has a mass of 1 g.)

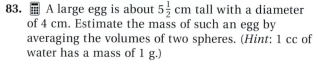

84. *Medicine.* Quinidine is a medicine that comes mixed with water. There are 80 mg of Quinidine in every milliliter of liquid. A standard dosage requires 200 mg of Quinidine. How much of the liquid mixture would be required in order to achieve this dosage?

85. *Medicine.* A medicine called cephalexin is available in a liquid mixture, part medicine and part water. There are 250 mg of cephalexin in 5 mL of liquid. A standard dosage is 400 mg. How much of the liquid would be required in order to achieve this dosage?

86. 🖩 *Track and Field.* A man's shot put weighs 16 lb and has a 5-in. diameter. Find its mass per cubic centimeter.

87. 🖩 *Track and Field.* A woman's shot put weighs 8.8 lb and has a 4.5-in. diameter. Find its mass per cubic centimeter.

Summary and Review: Chapter 9

Important Properties and Formulas

Area of a Parallelogram: $A = b \cdot h$

Area of a Trapezoid: $A = \frac{1}{2} \cdot h \cdot (a + b)$

Radius and Diameter of a Circle: $d = 2 \cdot r$, or $r = \frac{d}{2}$

Circumference of a Circle: $C = \pi \cdot d$, or $C = 2 \cdot \pi \cdot r$

Area of a Circle: $A = \pi \cdot r \cdot r$, or $A = \pi \cdot r^2$

Pythagorean Equation: $a^2 + b^2 = c^2$

Volume of a Rectangular Solid: $V = l \cdot w \cdot h$

Volume of a Circular Cylinder: $V = \pi \cdot r^2 \cdot h$

Volume of a Sphere: $V = \frac{4}{3} \cdot \pi \cdot r^3$

Temperature Conversion: $F = \frac{9}{5} \cdot C + 32$; $C = \frac{5}{9} \cdot (F - 32)$

See tables inside chapter for units of length, weight, mass, and capacity.

Review Exercises

Complete.

1. 10 ft = _____ yd [9.1a]
2. $\frac{5}{6}$ yd = _____ in. [9.1a]
3. 0.7 mm = _____ cm [9.1b]
4. 9 m = _____ km [9.1b]
5. 4 km = _____ cm [9.1b]
6. 14 in. = _____ ft [9.1a]
7. 9 lb = _____ oz [9.7a]
8. 4 g = _____ kg [9.7b]
9. 54 qt = _____ gal [9.6b]
10. 32 gal = _____ pt [9.6b]
11. 60 mL = _____ L [9.6b]
12. 0.4 L = _____ mL [9.6b]
13. 0.8 T = _____ lb [9.7a]
14. 0.2 g = _____ mg [9.7b]
15. 4.7 kg = _____ g [9.7b]
16. 4 cg = _____ g [9.7b]
17. 5 yd² = _____ ft² [9.3a]
18. 0.7 km² = _____ m² [9.3b]
19. 1008 in² = _____ ft² [9.3a]
20. 570 cm² = _____ m² [9.3b]

21. Find the circumference of a circle of radius 5 m. Use 3.14 for π. [9.2b]

Summary and Review: Chapter 9

22. Find the length of a radius of the circle. [9.2b]

23. Find the length of a diameter of the circle. [9.2b]

Find the area of each figure in Exercises 24–29. [9.2a, b]

24.

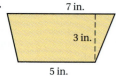

25.

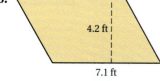

26.

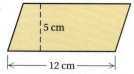

27.

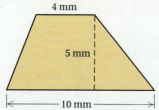

28. Use $\frac{22}{7}$ for π.

29. Use 3.14 for π.

30. A "Norman" window is designed with dimensions as shown. Find its area. Use 3.14 for π. [9.2b]

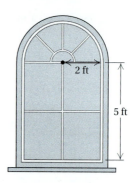

31. Find the measure of a complement of $\angle BAC$. [9.4c]

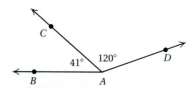

32. Find the measure of a supplement of a 44° angle. [9.4c]

For each right triangle, find the length of the side not given. Find an exact answer and an approximation to three decimal places. Assume that c represents the length of the hypotenuse.   [9.5c]

33. $a = 15$, $b = 25$ **34.** $a = 4$, $c = 10$

35.
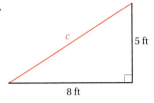

36.
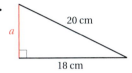

Find the volume of each figure. Use 3.14 for π.   [9.6a]

37.
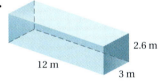

38.
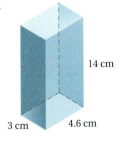

39.

40.
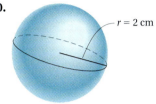

41.
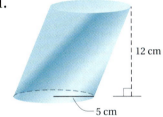

42. Convert 35°C to Fahrenheit.   [9.7c]

43. Convert 68°F to Celsius.   [9.7c]

44. A physician prescribed 650 mL per hour of a saline solution for a patient. How many liters of fluid did this patient receive in one day?   [9.6c]

Skill Maintenance

45. Find the simple interest on $5000 at 9.5% for 30 days. [8.6a]

46. Convert to percent notation: 0.47. [8.1b]

47. Convert to decimal notation: 56.7%. [8.1b]

48. Solve: $9x + 4 = 3x - 23$. [5.7b]

49. A hot air balloon travels 5 mi in $2\frac{1}{2}$ hr. At this rate, how far will the balloon travel in 12 hr? [7.4a]

50. Solve: $5(x - 1.5) = x + 2.5$. [5.7b]

Synthesis

51. ◆ Is a square a special type of parallelogram? Why or why not? [9.2a]

52. ◆ Which is a larger measure of volume: 1 m³ or 27 ft³? Explain how you can tell without using a calculator. [9.1c], [9.6a]

53. ◆ What weighs more: 32 oz or 1 kg? Explain how you can tell without using a calculator. [9.7a, b]

54. ▦ Find the area of the largest round pizza that can be baked in a 35-cm by 50-cm pan. [9.2b]

55. ▦ One lap around a standard running track is 440 yd. A marathon is 26 mi, 385 yd long. How many laps around a track does a marathon require? [9.1a]

56. Lumber that starts out at a certain measure must be trimmed to take out warps and get boards that are straight. Because of trimming, a "two-by-four" is trimmed to an actual size of $1\frac{1}{2}$ in. by $3\frac{1}{2}$ in. What percent of the wood in a 10-ft board is lost by trimming? [9.6c]

57. A community center has a rectangular swimming pool that is 50 ft wide, 100 ft long, and 10 ft deep. The center decides to fill the pool with water to a line that is 1 ft from the top. Water costs $2.25 per 1000 ft³. How much does it cost to fill the pool? [9.6c]

Test: Chapter 9

Complete.

1. 9 ft = _____ in.
2. 280 cm = _____ m
3. 2 yd² = _____ ft²
4. 5 km = _____ m
5. 8.7 mm = _____ cm
6. 4520 m² = _____ km²
7. 3080 mL = _____ L
8. 3.8 kg = _____ g
9. 10 gal = _____ oz
10. 0.24 L = _____ mL
11. 4 lb = _____ oz
12. 4.11 T = _____ lb

13. Find the length of a radius of this circle.

14. Find the area of a circle of radius 4 m. Use 3.14 for π.

15. Find the circumference of a circle of radius 14 ft. Use $\frac{22}{7}$ for π.

Find the area.

16.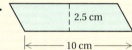

17.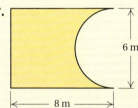

18.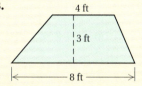

19. A 5-in. by 7-in. photo is mounted on matting board that is 6 in. by 8 in. What is the area of the border?

20. Find the measure of a supplement of $\angle CAD$.

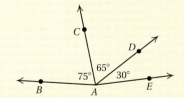

Answers

1. _____
2. _____
3. _____
4. _____
5. _____
6. _____
7. _____
8. _____
9. _____
10. _____
11. _____
12. _____
13. _____
14. _____
15. _____
16. _____
17. _____
18. _____
19. _____
20. _____

For each right triangle, find the length of the side not given. Find an exact answer and an approximation to three decimal places.

21.

22.

23. Find $\sqrt{121}$.

24. A ring box measures 6 cm by 8 cm by 4 cm. What is its volume?

In Exercises 25–27, find the volume in each figure. Use 3.14 for π.

25.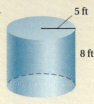

26.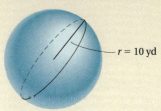

27.

28. Dr. Pietrofiro wants a patient to receive 0.5 L of a dextrose solution every 8 hr. How many milliliters will the patient have received after one 48-hr period?

29. Convert 86°F to Celsius.

30. Convert 45°C to Fahrenheit.

Skill Maintenance

31. Find the simple interest on $5000 at 8.8% for 1 year.

32. Convert to percent notation: 0.93.

33. Convert to decimal notation: 93.2%.

34. Solve: $5 - 2x = 7(3 - x) + 4$.

35. If 5 gal of gas costs $7.45, how much will 9 gal cost?

36. Solve: $7x - 3.6 = 3x + 6.4$.

Synthesis

37. The measure of $\angle SMC$ is three times that of its complement. Find the measure of $\angle SMC$.

38. A *board foot* is the amount of wood in a piece 12 in. by 12 in. by 1 in. A carpenter places the following order for a certain kind of lumber:

    25 pieces: 2 in. by 4 in. by 8 ft;
    32 pieces: 2 in. by 6 in. by 10 ft;
    24 pieces: 2 in. by 8 in. by 12 ft.

    The price of this type of lumber is $225 per thousand board feet. What is the total cost of the carpenter's order?

Cumulative Review: Chapters 1–9

Perform the indicated operations and simplify.

1. $4\frac{2}{3} + 5\frac{1}{2}$

2. $\left(\frac{1}{4}\right)^2 \div \left(\frac{1}{2}\right)^3 \times 2^4 + (10.3)(4)$

3. $120.5 - 32.98$

4. $-27{,}148 \div 22$

5. $14 \div [33 \div 11 + 8 \times 2 - (15 - 3)]$

6. $8^3 + 45 \cdot 24 - 9^2 \div 3$

Find fractional notation.

7. -6.23

8. 210%

Use $<$, $>$, or $=$ for ▮ to write a true sentence.

9. $\frac{5}{6}$ ▮ $\frac{7}{8}$

10. $\frac{5}{12}$ ▮ $\frac{3}{10}$

Complete.

11. 6 oz = _____ lb

12. 15°C = _____ °F

13. 0.087 L = _____ mL

14. 2.5 yd = _____ in.

15. 3 yd² = _____ ft²

16. 17 cm = _____ m

17. Find the perimeter and the area.

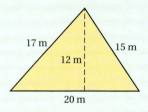

18. Combine like terms: $12a - 7 - 3a - 9$.

19. Graph: $y = -\frac{1}{3}x + 2$.

The line graph at right shows the average number of pounds of bananas eaten per person in the United States for the years 1991–1995 (*Source*: U.S. Department of Agriculture).

20. What was the average number of pounds of bananas that each person ate in 1995?

21. In what year did banana consumption peak?

Solve.

22. $\dfrac{12}{15} = \dfrac{x}{18}$

23. $1 - 7x = 4 - (x + 9)$

24. $-15x = 265$

25. $x + \dfrac{3}{4} = \dfrac{7}{8}$

26. A case of returnable bottles contains 24 bottles. Several students find that together they have 168 bottles. How many cases can they fill?

27. Americans own 52 million dogs, 56 million cats, 45 million birds, 250 million fish, and 125 million other creatures as house pets. How many pets do Americans own?

28. Find the mean: 49, 53, 60, 62, 69.

29. What is the simple interest on $800 at 12% for $\dfrac{1}{4}$ year?

30. How long must a rope be in order to reach from the top of an 8-m tree to a point on the ground 15 m from the bottom of the tree?

31. The sales tax on a purchase of $5.50 is $0.33. What is the sales tax rate?

32. A bolt of fabric in a fabric store has $10\tfrac{3}{4}$ yd on it. A customer purchases $8\tfrac{5}{8}$ yd. How many yards remain on the bolt?

33. What is the cost, in dollars, of 15.6 gal of gasoline at 108.9¢ per gallon? Round to the nearest cent.

34. A box of powdered milk that makes 20 qt costs $4.99. A box that makes 8 qt costs $1.99. Which size has the lower unit price?

35. It is $\dfrac{7}{10}$ km from Ida's dormitory to the library. She starts to walk there, changes her mind after going $\dfrac{1}{4}$ of the distance, and returns home. How far did Ida walk?

Synthesis

36. A house sits on a lot measuring 75 ft by 200 ft. The lot is a corner lot, so there are sidewalks on two sides of the lot. If the sidewalks are 3 ft wide and 4 in. of snow falls, what volume of snow must be shoveled?

37. The U.S. Postal Service will not ship a box if the sum of the box's lengthwise perimeter and widthwise perimeter exceeds 108 in. Will a 1-ft by 2-ft by 3-ft box be accepted for shipping? Support your answer mathematically.

10 Polynomials

Introduction

Algebraic expressions like $5x^7 + 3z - 2$ are called *polynomials*. One of the most important parts of algebra is the study of polynomials. In this chapter, we learn to add, subtract, and multiply polynomials. We also learn about integer exponents.

10.1 Addition and Subtraction of Polynomials

10.2 Introduction to Multiplying and Factoring Polynomials

10.3 More Multiplication of Polynomials

10.4 Integers as Exponents

An Application

The *polynomial*

$0.04x^3 - 0.23x^2 + 0.94x - 0.05$

can be used to estimate the number of cellular phones in use, in millions, *x* years after 1985.

This problem appears as Exercise 79 in Section 10.1.

The Mathematics

We can evaluate this polynomial to estimate how many millions of cellular phones will be in use in 2004.

For more information, visit us at www.mathmax.com

Pretest: Chapter 10

1. Add: $(5x^2 - 6x + 3) + (7x^2 + 4x - 13)$.

2. Find two equivalent expressions for the opposite of $6a^3b^2 - 7a^2b - 5$.

3. Subtract: $(3x^2 - 6x + 8) - (7x^2 + 11x - 13)$.

Evaluate.

4. 57^1

5. $(-23)^0$

6. $2x^3 - 5x^2$, for $x = 4$

Multiply.

7. $9a^3b(-2a^2b^2)$

8. $5x^2(7x^3 - 4x + 1)$

9. $(a + 4)(a - 7)$

10. $(a^2 - a)(2a^3 - a + 3)$

Factor.

11. $21a^3 + 30a^2 - 12a$

12. $15x^2y - 20xy^3$

Write an equivalent expression with positive exponents. Then simplify, if possible.

13. 7^{-2}

14. $5x^{-3}$

15. $\dfrac{a^{-8}}{b^{-5}c}$

16. $\left(\dfrac{5}{2}\right)^{-2}$

17. Write an expression equivalent to $\dfrac{1}{x^9}$ using a negative exponent.

Simplify. Use positive exponents in the answer.

18. $a^{-4} \cdot a^{-7}$

19. $(4x^7y^{-3})(-6x^2y^{-2})$

10.1 Addition and Subtraction of Polynomials

In Section 2.7, we defined a *term* as a number, a variable, a product of numbers and/or variables, or a quotient of numbers and/or variables. Thus expressions like

$$5x^2, \quad -34, \quad \frac{3}{4}ab^2, \quad xy^3z^5 \quad \text{and} \quad \frac{7n}{m}$$

are terms. A term is sometimes called a **monomial** if there is no division by a variable. Thus all of the expressions above, with the exception of $7n/m$, are monomials. Monomials are used to form **polynomials** like the following:

$$a^2b + c^3, \quad 5y + 3, \quad 3x^2 + 2x - 5, \quad -7a^3 + \tfrac{1}{2}a, \quad 37p^4, \quad x, \quad 0.$$

> A **polynomial** is a monomial or a combination of sums and/or differences of monomials.

Objectives

a. Add polynomials.
b. Find the opposite of a polynomial.
c. Subtract polynomials.
d. Evaluate a polynomial.

For Extra Help

TAPE 18   MAC WIN   CD-ROM

The following algebraic expressions are *not* polynomials:

(1) $\dfrac{x+3}{x-4}$, (2) $5x^3 - 2x^2 + \dfrac{1}{x}$, (3) $\dfrac{1}{x^3 - 2}$.

Expressions (1) and (3) are not polynomials because they represent quotients, not sums or differences. Expression (2) is not a polynomial because the term

$$\frac{1}{x}$$

is not a monomial.

a | Adding Polynomials

Recall that the commutative and associative laws are often used to make addition easier to perform. For example,

$$(9 + 17) + (1 + 13)$$

can be rewritten as the equivalent expression

$$(9 + 1) + (17 + 13), \quad \text{or} \quad 10 + 30.$$

A similar approach can be used for adding polynomials.

Example 1   Add: $(5x^3 + 4x^2 + 3x) + (2x^3 + 5x^2 - x)$.

$(5x^3 + 4x^2 + 3x) + (2x^3 + 5x^2 - x)$
$= (5x^3 + 2x^3) + (4x^2 + 5x^2) + (3x - x)$ Using the commutative and associative laws to pair up like terms

$= 7x^3 + 9x^2 + 2x$ Combining like terms. Remember that x means $1x$.

Add.

1. $(7a^2 + 2a + 8) + (2a^2 + a - 9)$

2. $(5x^2y + 3x^2 + 4) + (2x^2y + 4x)$

3. $(2a^3 + 17) + (2a^2 - 9a)$

Polynomials can be added even if their terms do not all form pairs of similar (like) terms.

Example 2 Add: $(3a^2 + 7a^2b) + (5a^2 - 6ab^2)$.

$(3a^2 + 7a^2b) + (5a^2 - 6ab^2)$
$= (3a^2 + 5a^2) + 7a^2b - 6ab^2$ Note that $7a^2b$ and $-6ab^2$ are *not* similar terms.
$= 8a^2 + 7a^2b - 6ab^2$ Combining like terms

Example 3 Add: $(7x^2 + 5) + (5x^3 + 4x)$.

$(7x^2 + 5) + (5x^3 + 4x) = 7x^2 + 5 + 5x^3 + 4x$ There are no similar terms here.
$= 5x^3 + 7x^2 + 4x + 5$ Rearranging the order

Note that in Example 3 we wrote the answer so that the powers of x decrease as we read from left to right. This **descending order** is the traditional way of expressing an answer, especially when the polynomials in the statement of the problem are given in descending order.

Do Exercises 1–3.

b | Opposites of Polynomials

To subtract a number, we can add its opposite. We can similarly subtract a polynomial by adding its opposite. To determine when two polynomials are opposites, recall that 5 and -5 are opposites, because $5 + (-5) = 0$.

> Two polynomials are **opposites**, or **additive inverses**, of each other if their sum is zero.

To develop a method for finding the opposite of a polynomial, consider that

$(5t^3 - 2) + (-5t^3 + 2) = 0$ and $(-9x^2 + x - 7) + (9x^2 - x + 7) = 0$.

Since $(5t^3 - 2) + (-5t^3 + 2) = 0$, we see that the opposite of $(5t^3 - 2)$ is $(-5t^3 + 2)$. This can be said with purely algebraic symbolism:

The opposite of $(5t^3 - 2)$ is $-5t^3 + 2$.

$-(5t^3 - 2) = -5t^3 + 2$.

Similarly,

The opposite of $(-9x^2 + x - 7)$ is $9x^2 - x + 7$.

$-(-9x^2 + x - 7) = 9x^2 - x + 7$.

> We can find an equivalent polynomial for the opposite, or additive inverse, of a polynomial by replacing each term with its opposite—that is, *changing the sign of every term.*

Answers on page A-27

Chapter 10 Polynomials

Example 4 Find two equivalent expressions for the opposite of
$$4x^5 - 7x^3 - 8x + \tfrac{5}{6}.$$

a) $-\left(4x^5 - 7x^3 - 8x + \tfrac{5}{6}\right)$ This is one expression for the opposite of $4x^5 - 7x^3 - 8x + \tfrac{5}{6}$.

b) $-4x^5 + 7x^3 + 8x - \tfrac{5}{6}$ Changing the sign of every term

Thus, $-\left(4x^5 - 7x^3 - 8x + \tfrac{5}{6}\right)$ is equivalent to $-4x^5 + 7x^3 + 8x - \tfrac{5}{6}$, and each is the opposite of the original polynomial $4x^5 - 7x^3 - 8x + \tfrac{5}{6}$.

Do Exercises 4–7.

Example 5 Simplify: $-\left(-7x^4 - \tfrac{5}{9}x^3 + 8x^2 - x + 67\right)$.

$$-\left(-7x^4 - \tfrac{5}{9}x^3 + 8x^2 - x + 67\right) = 7x^4 + \tfrac{5}{9}x^3 - 8x^2 + x - 67$$

Do Exercises 8–10.

c Subtracting Polynomials

We can now subtract a polynomial by adding the opposite of that polynomial. That is, for any polynomials p and q, $p - q = p + (-q)$.

Example 6 Subtract:

$$(9x^5 + x^3 - 2x^2 + 4) - (2x^5 + x^4 - 4x^3 - 3x^2).$$

We have

$(9x^5 + x^3 - 2x^2 + 4) - (2x^5 + x^4 - 4x^3 - 3x^2)$
$= (9x^5 + x^3 - 2x^2 + 4) + [-(2x^5 + x^4 - 4x^3 - 3x^2)]$ Adding the opposite
$= (9x^5 + x^3 - 2x^2 + 4) + [-2x^5 - x^4 + 4x^3 + 3x^2]$ Finding the opposite by changing the sign of *each* term
$= 9x^5 + x^3 - 2x^2 + 4 - 2x^5 - x^4 + 4x^3 + 3x^2$
$= 7x^5 - x^4 + 5x^3 + x^2 + 4.$ Combining like terms

Do Exercises 11 and 12.

To shorten our work, we often begin by changing the sign of each term in the polynomial being subtracted.

Example 7 Subtract:

$$(5a^4 - 7a^3 + 5a^2b) - (-3a^4 + 4a^2b + 6).$$

We have

$(5a^4 - 7a^3 + 5a^2b) - (-3a^4 + 4a^2b + 6)$
$= 5a^4 - 7a^3 + 5a^2b + 3a^4 - 4a^2b - 6$
$= 8a^4 - 7a^3 + a^2b - 6.$ Combining like terms

Do Exercise 13.

Find two equivalent expressions for the opposite of each polynomial.

4. $12x^4 - 3x^2 + 4x$

5. $-4x^4 + 3x^2 - 4x$

6. $-13x^6 + 2x^4 - 3x^2 + x - \tfrac{5}{13}$

7. $-8a^3b + 5ab^2 - 2ab$

Simplify.

8. $-(4x^3 - 6x + 3)$

9. $-(5x^3y + 3x^2y^2 - 7xy^3)$

10. $-\left(14x^{10} - \tfrac{1}{2}x^5 + 5x^3 - x^2 + 3x\right)$

Subtract.

11. $(7x^3 + 2x + 4) - (5x^3 - 4)$

12. $(-3x^2 + 5x - 4) - (-4x^2 + 11x - 2)$

13. Subtract

$(7x^3 + 3x^2 - xy) - (5x^3 + 3xy + 2)$.

Answers on page A-27

14. Evaluate each expression for $a = 2$. (See Margin Exercise 1.)
   a) $(7a^2 + 2a + 8) + (2a^2 + a - 9)$

   b) $9a^2 + 3a - 1$

15. In the situation of Example 9, what is the total number of games to be played in a league of 12 teams?

The perimeter of a square of side x is given by the polynomial $4x$.

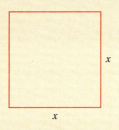

16. A baseball diamond is a square 90 ft on a side. Find the perimeter of a baseball diamond.

17. Find the perimeter of a softball diamond that is 65 ft on a side.

Answers on page A-27

d Evaluating Polynomials and Applications

It is important to keep in mind that when we are finding the sum or difference of two polynomials, we are *not* solving an equation. Rather, we are finding an equivalent expression that is usually more concise. One reason we do this is to make it easier to evaluate the original expression.

Example 8 Evaluate both $(5x^3 + 4x^2 + 3x) + (2x^3 + 5x^2 - x)$ and $7x^3 + 9x^2 + 2x$ for $x = 2$ (see Example 1).

a) When x is replaced by 2 in $(5x^3 + 4x^2 + 3x) + (2x^3 + 5x^2 - x)$, we have
$$5 \cdot 2^3 + 4 \cdot 2^2 + 3 \cdot 2 + 2 \cdot 2^3 + 5 \cdot 2^2 - 2,$$
or $5 \cdot 8 + 4 \cdot 4 + 6 + 2 \cdot 8 + 5 \cdot 4 - 2,$
or $40 + 16 + 6 + 16 + 20 - 2,$ which is 96.

b) Similarly, when x is replaced by 2 in $7x^3 + 9x^2 + 2x$, we have
$$7 \cdot 2^3 + 9 \cdot 2^2 + 2 \cdot 2,$$
or $7 \cdot 8 + 9 \cdot 4 + 4,$
or $56 + 36 + 4.$ As expected, this is also 96.

Do Exercise 14.

Polynomials are frequently evaluated in real-world situations.

Example 9 *Athletics.* In a sports league of n teams in which all teams play each other twice, the total number of games to be played is given by the polynomial
$$n^2 - n.$$
A women's softball league has 10 teams. If each team plays every other team twice, what is the total number of games to be played?

We evaluate the polynomial for $n = 10$:
$$n^2 - n = 10^2 - 10 = 100 - 10 = 90.$$

The league plays 90 games.

Do Exercises 15–17.

Chapter 10 Polynomials

Exercise Set 10.1

a Add.

1. $(2x + 7) + (-4x + 3)$

2. $(6x + 1) + (-7x + 2)$

3. $(-9x + 5) + (x^2 + x - 3)$

4. $(x^2 - 5x + 4) + (8x - 9)$

5. $(x^2 - 7) + (x^2 + 7)$

6. $(x^3 + x^2) + (2x^3 - 5x^2)$

7. $(6t^4 + 4t^3 - 1) + (5t^2 - t + 1)$

8. $(5t^2 - 3t + 12) + (2t^2 + 8t - 30)$

9. $(3 + 4x + 6x^2 + 7x^3) + (6 - 4x + 6x^2 - 7x^3)$

10. $(3x^4 - 6x - 5x^2 + 5) + (6x^2 - 4x^3 - 1 + 7x)$

11. $(9x^8 - 7x^4 + 2x^2 + 5) + (8x^7 + 4x^4 - 2x)$

12. $(4x^5 - 6x^3 - 9x + 1) + (6x^3 + 9x^2 + 9x)$

13. $(9t^4 + 6t^3 - t^2 + 3t) + (5t^4 - 2t^3 + t - 7)$

14. $(7t^5 - 3t^4 - 2t^2 + 5) + (3t^5 - 2t^4 + 4t^3 - t^2)$

15. $(-5x^4y^3 + 7x^3y^2 - 4xy^2) + (2x^3y^3 - 3x^3y^2 - 5xy)$

16. $(-9a^5b^4 + 7a^3b^3 + 2a^2b^2) + (2a^4b^4 - 5a^3b^3 - a^2b^2)$

17. $(8a^3b^2 + 5a^2b^2 + 6ab^2) + (5a^3b^2 - a^2b^2 - 4a^2b)$

18. $(6x^3y^3 - 4x^2y^2 + 3xy^2) + (x^3y^3 + 7x^3y^2 - 2xy^2)$

19. $(17.5abc^3 + 4.3a^2bc) + (-4.9a^2bc - 5.2abc)$

20. $(23.9x^3yz - 19.7x^2y^2z) + (-14.6x^3yz - 8x^2yz)$

b Find two equivalent expressions for the opposite of each polynomial.

21. $-5x$

22. $x^2 - 3x$

23. $-x^2 + 10x - 2$

24. $-4x^3 - x^2 - x$

25. $12x^4 - 3x^3 + 3$

26. $4x^3 - 6x^2 - 8x + 1$

Simplify.

27. $-(3x - 7)$

28. $-(-2x + 4)$

29. $-(4x^2 - 3x + 2)$

30. $-(-6a^3 + 2a^2 - 9a + 1)$

31. $-\left(-4x^4 + 6x^2 + \frac{3}{4}x - 8\right)$

32. $-(-5x^4 + 4x^3 - x^2 + 0.9)$

c Subtract.

33. $(3x + 2) - (-4x + 3)$

34. $(6x + 1) - (-7x + 2)$

35. $(9t^2 + 7t + 5) - (5t^2 + t - 1)$

36. $(8t^2 - 5t + 7) - (3t^2 - 2t + 1)$

37. $(-6x + 2) - (x^2 + x - 3)$

38. $(x^2 - 5x + 4) - (8x - 9)$

39. $(7a^2 + 5a - 9) - (2a^2 + 7)$

40. $(8a^2 - 6a + 5) - (2a^2 - 19a)$

41. $(6x^4 + 3x^3 - 1) - (4x^2 - 3x + 3)$

42. $(-4x^2 + 2x) - (3x^3 - 5x^2 + 3)$

43. $(1.2x^3 + 4.5x^2 - 3.8x) - (-3.4x^3 - 4.7x^2 + 23)$

44. $(0.5x^4 - 0.6x^2 + 0.7) - (2.3x^4 + 1.8x - 3.9)$

45. $\left(\frac{5}{8}x^3 - \frac{1}{4}x - \frac{1}{3}\right) - \left(-\frac{1}{8}x^3 + \frac{1}{4}x - \frac{1}{3}\right)$

46. $\left(\frac{1}{5}x^3 + 2x^2 - 0.1\right) - \left(-\frac{2}{5}x^3 + 2x^2 + 0.01\right)$

47. $(5x^3y^3 + 8x^2y^2 + 7xy) - (3x^3y^3 - 2x^2y + 3xy)$

48. $(3x^4y + 2x^3y - 7x^2y) - (5x^4y + 2x^2y^2 - 2x^2y)$

d Evaluate each polynomial for $x = 4$.

49. $-5x + 2$

50. $-3x + 1$

51. $2x^2 - 5x + 7$

52. $3x^2 + x + 7$

53. $x^3 - 5x^2 + x$

54. $7 - x + 3x^2$

Evaluate each polynomial for $x = -1$.

55. $3x + 5$

56. $6 - 2x$

57. $x^2 - 2x + 1$

58. $5x - 6 + x^2$

59. $-3x^3 + 7x^2 - 3x - 2$

60. $-2x^3 - 5x^2 + 4x + 3$

Daily Accidents. The daily number of accidents N (average number of accidents per day) involving drivers of age a is approximated by the polynomial

$$N = 0.4a^2 - 40a + 1039.$$

61. Evaluate the polynomial for $a = 18$ to find the daily number of accidents involving 18-year-old drivers.

62. Evaluate the polynomial for $a = 20$ to find the daily number of accidents involving 20-year-old drivers.

Falling Distance. The distance s, in feet, traveled by a body falling freely from rest in t seconds is approximated by the polynomial

$$s = 16t^2.$$

63. A stone is dropped from a cliff and takes 8 sec to hit the ground. How high is the cliff?

64. A brick is dropped from a building and takes 3 sec to hit the ground. How high is the building?

Total Revenue. Cutting Edge Electronics is marketing a new kind of stereo. *Total revenue* is the total amount of money taken in. The firm determines that when it sells x stereos, it will take in

$$280x - 0.4x^2 \text{ dollars}.$$

65. What is the total revenue from the sale of 75 stereos?

66. What is the total revenue from the sale of 100 stereos?

Total Cost. Cutting Edge Electronics determines that the total cost of producing x stereos is given by

$$5000 + 0.6x^2 \text{ dollars}.$$

67. What is the total cost of producing 500 stereos?

68. What is the total cost of producing 650 stereos?

Exercise Set 10.1

Skill Maintenance

69. A 10-lb fish serves 7 people. What is the ratio of servings to pounds? [7.1a]

70. A bicycle salesperson's commission rate is 22%. A commission of $783.20 is received. How many dollars worth of bicycles were sold? [8.5b]

71. In 1998, the sales tax rate in Vermont was 5%. How much tax would be paid in Vermont for a computer that sold for $1350? [8.5a]

72. Find the area of a rectangle that is 6.5 m by 4 m. [5.8a]

73. Find the area of a circle with radius 20 cm. Use 3.14 for π. [9.2b]

74. Melba earned $4740 for working 12 weeks. What was the rate of pay? [7.2a]

Synthesis

75. ◆ Is every term a monomial? Why or why not?

76. ◆ Suppose that two polynomials, each containing 3 terms, are added. Is it possible for the sum to contain more than 3 terms? fewer than 3 terms? exactly 3 terms? Explain.

77. ◆ Explain how the associative and commutative laws can be used when adding polynomials.

78. *Medicine.* When a person swallows 400 mg of ibuprofen, the number of milligrams in the bloodstream t hours later can be approximated by the polynomial
$$0.5t^4 + 3.45t^3 - 96.65t^2 + 347.7t$$
(where $0 \leq t \leq 6$).
Determine the amount of ibuprofen in the bloodstream **(a)** 1 hr after swallowing 400 mg; **(b)** 2 hr after swallowing 400 mg; **(c)** 6 hr after swallowing 400 mg.

79. *Cellular Phone Sales.* The polynomial
$$0.04x^3 - 0.23x^2 + 0.94x - 0.05$$
can be used to estimate the number of cellular phones in use, in millions, x years after 1985. Predict the number of cellular phones in use in 2004.

Perform the indicated operations and simplify.

80. $(7y^2 - 5y + 6) - (3y^2 + 8y - 12) + (8y^2 - 10y + 3)$

81. $(3x^2 - 4x + 6) - (-2x^2 + 4) + (-5x - 3)$

82. $(-y^4 - 7y^3 + y^2) + (-2y^4 + 5y - 2) - (-6y^3 + y^2)$

83. $(-4 + x^2 + 2x^3) - (-6 - x + 3x^3) - (-x^2 - 5x^3)$

84. Complete: $9x^4 + \_\_\_ + 5x^2 - 7x^3 + \_\_\_ - 9 + \_\_\_ = 12x^4 - 5x^3 + 5x^2 - 16$.

85. Complete: $8t^4 + \_\_\_ - 2t^3 + \_\_\_ - 2t^2 + t - \_\_\_ - 3 + \_\_\_ = 8t^4 + 7t^3 - 3t + 4$.

Determine the number of handshakes possible in a group.

10.2 Introduction to Multiplying and Factoring Polynomials

We now study how to multiply and factor certain polynomials.

a Multiplying Monomials

Recall that the area of a square with sides of length x is x^2.

If a rectangle is 3 times as long as it is wide, we can represent its width by x and its length by $3x$.

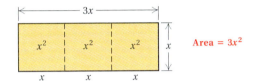

The area, $3x^2$, is the product of $3x$ and x. This product can be found using an associative law:

$(3x)x = 3(xx) = 3x^2$.

To find other products of monomials, we may need to use a commutative law as well.

Example 1 Multiply: $(4x)(5x)$.

$(4x)(5x) = 4 \cdot x \cdot 5 \cdot x$ **Using an associative law**
$= 4 \cdot 5 \cdot x \cdot x$ **Using a commutative law**
$= (4 \cdot 5)(xx)$ **Using an associative law**
$= 20x^2$

Example 1 can be regarded as finding the area of a rectangle of width $4x$ and length $5x$. Note that the area consists of 20 squares, each of which has area x^2.

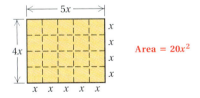

Do Exercises 1 and 2.

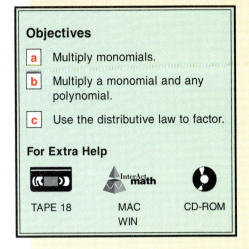

Objectives

a Multiply monomials.

b Multiply a monomial and any polynomial.

c Use the distributive law to factor.

For Extra Help

TAPE 18   MAC WIN   CD-ROM

Multiply.

1. $(6a)(3a)$

2. $(-7x)(2x)$

Answers on page A-27

Multiply.

3. $(4a)(12a)$

4. $(-m)(5m)$

5. $(-6a)(-7b)$

Multiply.

6. $a^5 \cdot a^4$

7. $(2x^8)(4x^5)$

8. $(-7m^4)(-5m^7)$

9. $(3a^5b^4)(5a^2b^8)$

Answers on page A-27

Usually the steps in Example 1 are combined: We multiply coefficients and we multiply variables.

Examples Multiply.

2. $(5x)(6x) = (5 \cdot 6)(x \cdot x)$ Multiplying the coefficients.
$= 30x^2$ Simplifying

3. $(3x)(-x) = (3x)(-1x)$ Rewriting $-x$ as $-1x$
$= (3)(-1)(x \cdot x)$
$= -3x^2$

4. $(7x)(4y) = (7 \cdot 4)(x \cdot y)$
$= 28xy$

Do Exercises 3–5.

Multiplying Powers with Like Bases

In later courses, you will likely learn several rules for manipulating exponents. The one rule that we develop now is useful when multiplying powers with like bases. Consider the following:

$$a^3 \cdot a^2 = \underbrace{(a \cdot a \cdot a)}_{3 \text{ factors}} \underbrace{(a \cdot a)}_{2 \text{ factors}} = \underbrace{a \cdot a \cdot a \cdot a \cdot a}_{5 \text{ factors}} = a^5.$$

Note that the exponent in a^5 is the sum of those in $a^3 \cdot a^2$. That is, $3 + 2 = 5$. Likewise,

$$b^4 \cdot b^3 = (b \cdot b \cdot b \cdot b)(b \cdot b \cdot b) = b^7, \text{ where } 4 + 3 = 7.$$

Adding the exponents gives the correct result.

> **THE PRODUCT RULE**
>
> For any number a and any positive integers m and n,
> $$a^m \cdot a^n = a^{m+n}.$$
>
> (When multiplying with exponential notation, if the bases are the same, keep the base and add the exponents.)

Examples Multiply and simplify.

5. $x^2 \cdot x^5 = x^{2+5}$ Adding exponents
$= x^7$

6. $(3a^4)(5a^2) = (3 \cdot 5)(a^4 \cdot a^2)$ Multiplying coefficients; adding exponents
$= 15a^6$

7. $(-4x^2y^3)(3x^6y^7) = (-4 \cdot 3)(x^2 \cdot x^6)(y^3 \cdot y^7)$
$= -12x^8y^{10}$

Do Exercises 6–9.

Chapter 10 Polynomials

We have not yet determined what the number 1 will mean when used as an exponent. Consider the following:

$$m \cdot m^2 = m \cdot m \cdot m = m^3;$$
$$x \cdot x^3 = x \cdot x \cdot x \cdot x = x^4;$$

and $\quad a \cdot a^4 = a \cdot a \cdot a \cdot a \cdot a = a^5.$

Note that if $m = m^1$, $x = x^1$, and $a = a^1$, the same results can be found using the product rule:

$$m \cdot m^2 = m^1 \cdot m^2 = m^3;$$
$$x \cdot x^3 = x^1 \cdot x^3 = x^4;$$

and $\quad a \cdot a^4 = a^1 \cdot a^4 = a^5.$

This suggests the following definition.

> $b^1 = b$ for any number b.

Example 8 Evaluate 5^1 and -23^1.

$5^1 = 5;$

$-23^1 = -23.$ We read -23^1 as "the opposite of 23^1."
To write -23 to the first, we would write $(-23)^1$.

Do Exercise 10.

b Multiplying a Monomial and Any Polynomial

When a polynomial contains two terms, it is called a **binomial**. The product of the monomial x and the binomial $x + 2$ can be visualized as the area of a rectangle with width x and length $x + 2$.

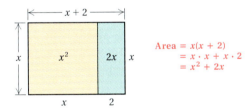

Although many products of monomials and polynomials are difficult to visualize geometrically, the distributive law can always be used to find these products algebraically (pronounced al-je-bray´-ik-ally).

Example 9 Multiply: $2x$ and $5x + 3$.

$2x(5x + 3) = 2x \cdot 5x + 2x \cdot 3$ Using the distributive law
$ = 10x^2 + 6x$ Multiplying the monomials

10. Evaluate 7^1 and -19^1.

Answer on page A-27

Multiply.

11. $4x$ and $3x + 5$

12. $3a(2a^2 - 5a + 7)$

13. $4a^3b^2(2a^2 + 5b^4)$

Example 10 Multiply: $5x(2x^2 - 3x + 4)$.

$$5x(2x^2 - 3x + 4) = 5x \cdot 2x^2 - 5x \cdot 3x + 5x \cdot 4$$
$$= 10x^3 - 15x^2 + 20x \quad \text{Note that } x \cdot x^2 = x^1 \cdot x^2 = x^3.$$

Example 11 Multiply: $-3r^2s(2r^3s^2 - 5rs)$.

$$-3r^2s(2r^3s^2 - 5rs) = -3r^2s \cdot 2r^3s^2 - (-3r^2s)5rs$$
$$= -6r^5s^3 + 15r^3s^2$$

Do Exercises 11–13.

c | Factoring

Factoring is the reverse of multiplying. To factor, we use the distributive law, beginning with a sum or a difference of two terms that contain a common factor:

$$ab + ac = a(b + c) \quad \text{and} \quad rs - rt = r(s - t).$$

> To **factor** an expression is to find an equivalent expression that is a product.

We have already factored polynomials in this text, although we did not call the process factoring at the time. For example, in Section 2.7, we factored when we first learned how to combine like terms (see p. 123).

$$5x + 3x = (5 + 3)x \quad \text{Here we factored } 5x + 3x.$$
$$= 8x. \quad \text{See also Sections 4.3 and 5.2.}$$

To *factor* an expression like $10y + 15$, we find an equivalent expression that is a product. To do this, we look to see if both of the terms have a factor in common. If there *is* a common factor, we can "factor it out" using the distributive law. Note the following:

$10y$ has the factors $10, 5, 2, 1, y, 2y, 5y,$ and $10y$;

15 has the factors $15, 5, 3, 1$.

We generally factor out the largest common factor. In this case, that factor is 5 (which is the *only* common factor here). Thus,

$$10y + 15 = 5 \cdot 2y + 5 \cdot 3 \quad \text{Try to do this step mentally.}$$
$$= 5(2y + 3). \quad \text{Using the distributive law}$$

Answers on page A-27

Examples Factor.

12. $5x - 10 = 5 \cdot x - 5 \cdot 2$ Try to do this step mentally.
$= 5(x - 2)$ You can check by multiplying.

13. $9x + 27y - 9 = 9 \cdot x + 9 \cdot 3y - 9 \cdot 1$
$= 9(x + 3y - 1)$

CAUTION! Note that although $3(3x + 9y - 3)$ is also equivalent to $9x + 27y - 9$, it is *not* factored "completely." However, we can complete the process by factoring out another factor of 3:

$9x + 27y - 9 = 3(3x + 9y - 3) = 3 \cdot 3(x + 3y - 1) = 9(x + 3y - 1).$

Remember to factor out the *largest common factor*.

Examples Factor. Try to write just the answer.

14. $-3x + 6y - 9z = -3(x - 2y + 3z)$

We generally factor out a negative when the first coefficient is negative. The way we factor can depend on the situation in which we are working. We might also factor as follows:

$-3x + 6y - 9z = 3(-x + 2y - 3z).$

15. $18z - 12x - 24 = 6(3z - 2x - 4)$ To check, multiply:
$6(3z - 2x - 4) = 6 \cdot 3z - 6 \cdot 2x - 6 \cdot 4$
$= 18z - 12x - 24.$

Remember: An expression is factored when it is written as a product.

Do Exercises 14–17.

When exponents appear, it is important to keep in mind the product rule.

Examples Factor.

16. $7x^2 + 7x = 7x(x + 1)$ The largest common factor is $7x$.

17. $10a^5 + 5a^3 + 15a^2 = 5a^2(2a^3 + a + 3)$ The largest common factor of 10, 5, and 15 is 5. The largest common factor of a^5, a^3, and a^2 is a^2.

18. $8xy^3 - 6x^2y = 2xy(4y^2 - 3x)$ To check, multiply:
$2xy(4y^2 - 3x) = 2xy \cdot 4y^2 - 2xy \cdot 3x$
$= 8xy^3 - 6x^2y.$

Note in Example 17 that the *largest* common factor of a is the *smallest* of the three powers that appear in the original polynomial.

Do Exercises 18–20.

Factor.

14. $6z - 12$

15. $3x - 6y + 9$

16. $16a - 36b + 42$

17. $-12x + 32y - 16z$

Factor.

18. $5a^3 + 5a$

19. $14x^3 - 7x^2 + 21x$

20. $9a^2b - 6ab^2$

Answers on page A-27

10.2 Introduction to Multiplying and Factoring Polynomials

595

Improving Your Math Study Skills

Preparing for a Final Exam

Best Scenario: Two Weeks of Study Time

The best scenario for preparing for a final exam is to do so over a period of at least two weeks. Work in a diligent, disciplined manner, doing some final-exam preparation *each* day. Here is a detailed plan that many find useful.

1. **Begin by browsing through each chapter, reviewing the highlighted or boxed information regarding important formulas in both the text and the Summary and Review.** There may be some formulas that you will need to memorize.
2. **Retake all chapter tests that you took, assuming your instructor has returned them. Otherwise, use the chapter tests in the book.** Restudy the objectives in the text that correspond to each question you missed.
3. **Then work the Cumulative Review that covers all chapters up to that point.** Be careful to avoid any questions corresponding to objectives not covered. Again, restudy the objectives in the text that correspond to each question you missed.
4. **If you are still having difficulty, use the supplements for extra review.** For example, you might check out the videotapes, the *Student's Solutions Manual,* or the InterAct Math Tutorial Software.
5. **For remaining difficulties, see your instructor, go to a tutoring session, or participate in a study group.**
6. **Check for former final exams that may be on file in the math department or a study center, or with students who have already taken the course.** Use them for practice, being alert to trouble spots.
7. **Take the Final Examination in the text during the last couple of days before the final.** Set aside the same amount of time that you will have for the final. See how much of the final exam you can complete under test-like conditions.

Moderate Scenario: Three Days to Two Weeks of Study Time

1. **Begin by browsing through each chapter, reviewing the highlighted or boxed information regarding important formulas in both the text and the Summary and Review.** There may be some formulas that you will need to memorize.
2. **Retake all chapter tests that you took, assuming your instructor has returned them. Otherwise, use the chapter tests in the book.** Restudy the objectives in the text that correspond to each question you missed.
3. **Then work the last Cumulative Review in the portion of the text that you covered.** Avoid any questions corresponding to objectives not covered. Again, restudy the objectives in the text that correspond to each question you missed.
4. **For remaining difficulties, see your instructor, go to a tutoring session, or participate in a study group.**
5. **Take the Final Examination in the text during the last couple of days before the final.** Set aside the same amount of time that you will have for the final. See how much of the final exam you can complete under test-like conditions.

Worst Scenario: One or Two Days of Study Time

1. **Begin by browsing through each chapter, reviewing the highlighted or boxed information regarding important formulas in both the text and the Summary and Review.** There may be some formulas that you will need to memorize.
2. **Then work the last Cumulative Review in the portion of the text that you covered.** Avoid any questions corresponding to objectives not covered. Restudy the objectives in the text that correspond to each question you missed.
3. **Attend a final-exam review session if one is available.**
4. **Take the Final Examination in the text as preparation for the final.** Set aside the same amount of time that you will have for the final. See how much of the final exam you can complete under test-like conditions.

Promise yourself that next semester you will allow a more appropriate amount of time for final exam preparation.

Exercise Set 10.2

a Multiply.

1. $(5a)(9a)$

2. $(7x)(6x)$

3. $(-4x)(15x)$

4. $(-9a)(10a)$

5. $(7x^5)(4x^3)$

6. $(10a^2)(3a^2)$

7. $(-0.1x^6)(0.2x^4)$

8. $(0.3x^3)(-0.4x^6)$

9. $(5x^2y^3)(7x^4y^9)$

10. $(6a^3b^4)(2a^4b^7)$

11. $(4a^3b^4c^2)(3a^5b^4)$

12. $(7x^3y^5z^2)(8x^3z^4)$

13. $(3x^2)(-4x^3)(2x^6)$

14. $(-2y^5)(10y^4)(-3y^3)$

b Multiply.

15. $3x(-x + 5)$

16. $2x(4x - 6)$

17. $-3x(x - 1)$

18. $-5x(-x - 1)$

19. $x^2(x^3 + 1)$

20. $-2x^3(x^2 - 1)$

21. $3x(2x^2 - 6x + 1)$

22. $-4x(2x^3 - 6x^2 - 5x + 1)$

23. $4xy(3x^2 + 2y)$

24. $5xy(3x^2 - 6y^2)$

25. $3a^2b(4a^5b^2 - 3a^2b^2)$

26. $4a^2b^2(2a^3b - 5ab^2)$

c Factor. Check by multiplying.

27. $2x + 6$

28. $3x + 12$

29. $7a - 21$

30. $9a - 18$

31. $14x + 21y$ **32.** $8x - 10y$ **33.** $9a - 27b + 81$ **34.** $5x + 10 + 15y$

35. $24 - 6m$ **36.** $32 - 4y$ **37.** $-16 - 8x + 40y$ **38.** $-35 + 14x - 21y$

39. $3x^5 + 3x$ **40.** $5x^6 + 5x$ **41.** $a^3 - 8a^2$ **42.** $a^5 - 9a^2$

43. $8x^3 - 6x^2 + 2x$ **44.** $9x^4 - 12x^3 + 3x$ **45.** $12a^4b^3 + 18a^5b^2$ **46.** $15a^5b^2 + 20a^2b^3$

Skill Maintenance

47. When used for a singles match, a regulation tennis court is 27 ft by 78 ft. Find its perimeter. [2.7c]

48. The Floral Doctor's delivery van traveled 147 mi on 10.5 gal of gas. How many miles per gallon did the van get? [7.2a]

49. Ramon's new truck gets 21 miles per gallon. This is 20% more than the mileage his old truck got. What mileage did the old truck get? [8.4b]

50. A 5% sales tax is added to the price of a two-speed washing machine. If the machine is priced at $399, find the total amount paid. [8.5a]

51. Of the 8 fish Mac caught, 3 were trout. What percentage were not trout? [8.4a]

52. The diameter of a compact disc is 12 cm. What is its circumference? (Use 3.14 for π.) [9.2b]

Synthesis

53. ◆ Describe a method for creating a binomial that has $5x^2$ as its largest common factor.

54. ◆ If all of a polynomial's coefficients are prime, is it still possible to factor the polynomial? Why or why not?

55. ◆ Explain in your own words why the product rule "works."

Factor.

56. $391x^{391} + 299x^{299}$

57. $703a^{437} + 437a^{703}$

58. $84a^7b^9c^{11} - 42a^8b^6c^{10} + 49a^9b^7c^8$

59. Draw a figure similar to those preceding Examples 1 and 9 to show that $2x \cdot 3x = 6x^2$.

10.3 More Multiplication of Polynomials

a Multiplying Two Binomials

To find an equivalent expression for the product of two binomials, we use the distributive law more than once. In the example that follows, the distributive law is used three times.

Example 1 Multiply: $x + 5$ and $x + 4$.

$(x + 5)(x + 4) = (x + 5)x + (x + 5)4$ Using the distributive law

$\qquad = x \cdot x + 5 \cdot x + x \cdot 4 + 5 \cdot 4$ Using the distributive law on each part

$\qquad = x^2 + 5x + 4x + 20$ Multiplying the monomials

$\qquad = x^2 + 9x + 20$ Combining like terms

Objectives

a Multiply two binomials.

b Multiply any two polynomials.

For Extra Help

TAPE 18   MAC WIN   CD-ROM

Multiply.
1. $x + 8$ and $x + 5$

Do Exercises 1 and 2.

We can visualize the product $(x + 5)(x + 4)$ as the area of a rectangle with width $x + 4$ and length $x + 5$. Note that the total area is the sum of the four smaller areas.

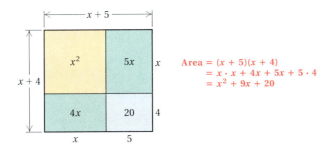

Area $= (x + 5)(x + 4)$
$= x \cdot x + 4x + 5x + 5 \cdot 4$
$= x^2 + 9x + 20$

2. $(x + 5)(x - 4)$

More complicated products of binomials are not as easily visualized, but can be simplified using the steps of Example 1.

Example 2 Multiply: $4x + 3$ and $x - 2$.

$(4x + 3)(x - 2) = (4x + 3)(x + (-2))$ Rewriting $x - 2$ as $x + (-2)$

$\qquad = (4x + 3)x + (4x + 3)(-2)$ Using the distributive law

$\qquad = 4x \cdot x + 3 \cdot x + 4x(-2) + 3(-2)$ Using the distributive law on each part

$\qquad = 4x^2 + 3x + (-8x) + (-6)$ Multiplying monomials

$\qquad = 4x^2 - 5x - 6$ Combining like terms

Multiply.
3. $5x + 3$ and $x - 4$

Do Exercises 3 and 4.

Note in Example 1 that four products were found: $x \cdot x$, $x \cdot 4$, $5 \cdot x$, and $5 \cdot 4$. Similarly, in Example 2 we found $4x \cdot x$, $4x(-2)$, $3 \cdot x$, and $3(-2)$. These products are found by multiplying the First terms, the Outer terms, the Inner terms, and the Last terms in the binomials. We use the word FOIL to help remember these products.

4. $(2x - 3)(3x - 5)$

Answers on page A-28

10.3 More Multiplication of Polynomials

599

Use FOIL to multiply.

5. $(x + 3)(x + 5)$

6. $(x - 3)(x - 8)$

Multiply.

7. $(x^2 + 3x - 4)(x^2 + 5)$

8. $(3y^2 - 7)(2y^3 - 2y + 5)$

Multiply.

9. $3x^2 - 2x + 4$
$ x + 5$

Answers on page A-28

Chapter 10 Polynomials

600

Example 3 Use FOIL to multiply $(x + 3)(x + 7)$.

We have

$\qquad$ First   Last
$\qquad (x + 3)(x + 7)$
$\qquad\quad$ Inner
$\qquad\quad$ Outer

$\qquad\qquad\qquad\qquad\qquad$ F $\quad\;\;$ O $\quad\;\;$ I $\quad\;\;$ L
$(x + 3)(x + 7) = x \cdot x + 7 \cdot x + 3 \cdot x + 3 \cdot 7$
$\qquad\qquad\quad\;\; = x^2 + 7x + 3x + 21$
$\qquad\qquad\quad\;\; = x^2 + 10x + 21.$

Do Exercises 5 and 6.

b Multiplying Any Polynomials

A polynomial containing three terms is called a **trinomial**. To find the product of a binomial and a trinomial, we again use the distributive law.

Example 4 Multiply: $(x^2 + 2x - 3)(x^2 + 4)$.

$(x^2 + 2x - 3)(x^2 + 4) = (x^2 + 2x - 3)x^2 + (x^2 + 2x - 3)4$
$\qquad\qquad\qquad\qquad\;\; = x^2 x^2 + 2xx^2 - 3x^2 + x^2 \cdot 4 + 2x \cdot 4 - 3 \cdot 4$
$\qquad\qquad\qquad\qquad\;\; = x^4 + 2x^3 - 3x^2 + 4x^2 + 8x - 12$
$\qquad\qquad\qquad\qquad\;\; = x^4 + 2x^3 + x^2 + 8x - 12 \qquad$ Combining like terms

Do Exercises 7 and 8.

> To multiply two polynomials P and Q, select one of the polynomials—say, P. Then multiply each term of P by every term of Q and combine like terms.

Columns can be used for long multiplication. To do so, we multiply each term at the top by every term below. We write like terms in columns and add the results. Such multiplication is like multiplying numbers:

$\quad\;\; 2\;3\;1 \qquad\qquad\qquad\quad 2\;3\;1 \qquad\qquad = 200 + 30 + 1$
$\times\quad 3\;2 \qquad\qquad \times \qquad\qquad 3\;2 \qquad\quad = 30 + 2$
$\quad\;\; 4\;6\;2 \qquad\qquad\qquad 400 + 60 + 2 \qquad = 2(231) = 2(200 + 30 + 1)$
$\;\; 6\;9\;3\;0 \qquad\qquad 6000 + 900 + 30 \qquad\quad = 30(231) = 30(200 + 30 + 1)$
$\;\; 7\;3\;9\;2 \qquad\qquad 6000 + 1300 + 90 + 2 \quad = 7392$

Example 5 Multiply: $(4x^2 - 2x + 3)(x + 2)$.

$\qquad\qquad 4x^2 - 2x + 3 \qquad$ It helps that both polynomials are in descending order.
$\qquad\qquad\qquad\;\;\; x + 2$
$\qquad\qquad 8x^2 - 4x + 6 \qquad$ Multiplying the top row by 2
$\qquad 4x^3 - 2x^2 + 3x \qquad$ Multiplying the top row by x
$\qquad 4x^3 + 6x^2 - x + 6 \qquad$ Combining like terms
$\qquad\qquad\qquad\qquad\qquad$ Line up like terms in columns.

Do Exercise 9.

Exercise Set 10.3

a Multiply.

1. $(x + 7)(x + 2)$
2. $(x + 5)(x + 2)$
3. $(x + 5)(x - 2)$
4. $(x + 1)(x - 3)$

5. $(x + 6)(x - 2)$
6. $(x - 4)(x - 3)$
7. $(x - 7)(x - 3)$
8. $(x + 3)(x - 3)$

9. $(x + 6)(x - 6)$
10. $(5 - x)(5 - 2x)$
11. $(3 + x)(6 + 2x)$
12. $(2x + 5)(2x + 5)$

13. $(3x - 4)(3x - 4)$
14. $(5x - 1)(5x + 2)$
15. $\left(x - \frac{5}{2}\right)\left(x + \frac{2}{5}\right)$
16. $\left(x + \frac{4}{3}\right)\left(x + \frac{3}{2}\right)$

b Multiply.

17. $(x^2 + x + 1)(x - 1)$
18. $(x^2 - x + 2)(x + 2)$
19. $(2x + 1)(2x^2 + 6x + 1)$

20. $(3x - 1)(4x^2 - 2x - 1)$
21. $(y^2 - 3)(3y^2 - 6y + 2)$
22. $(3y^2 - 3)(y^2 + 6y + 1)$

23. $(x^3 + x^2)(x^3 + x^2 - x)$
24. $(x^3 - x^2)(x^3 - x^2 + x)$
25. $(2t^2 - t - 4)(3t^2 + 2t - 1)$

26. $(3a^2 - 5a + 2)(2a^2 - 3a + 4)$
27. $(x - x^3 + x^5)(x^2 - 1 + x^4)$
28. $(x - x^3 + x^5)(3x^2 + 3x^6 + 3x^4)$

Skill Maintenance

29. A sidewalk of uniform width is built around three sides of a store, as shown in the figure. What is the area of the sidewalk? [1.8a]

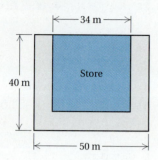

30. A real estate agent's commission rate is 6%. A commission of $7380 is received on the sale of a home. For how much did the home sell? [8.5b]

31. What percent of 24 is 32? [8.3b]

32. 39 is 150% of what number? [8.3b]

33. In 1998, the New York Yankees won 111 of their 162 games. What percentage of their games did the Yankees win? [7.4a]

34. The Sanchez's flower garden covers a 14-ft-wide circular region of their yard. Find the garden's area. Use $\frac{22}{7}$ for π. [9.2b]

Synthesis

35. ◆ A student insists that since $x \cdot x$ is x^2 and $5 \cdot 4 = 20$, it follows that $(x + 5)(x + 4) = x^2 + 20$. How could you convince the student that this is not correct?

36. ◆ Is the product of two binomials always a trinomial? Why or why not?

37. ◆ Ron says that since $(xy)^2 = (xy) \cdot (xy) = x^2y^2$, it follows that $(x + y)^2 = x^2 + y^2$. Is he correct? Why or why not?

38. ▣ (See Example 4.) Check that the expressions $(x^2 + 2x - 3)(x^2 + 4)$ and $x^4 + 2x^3 + x^2 + 8x - 12$ are equivalent by evaluating both expressions for $x = 5$, $x = 3.5$, and $x = -1.2$.

39. Simplify: $(x + 2)(x + 3) + (x - 4)^2$.

For each figure below, find a simplified expression for **(a)** the perimeter and **(b)** the area.

40.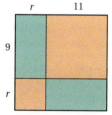

41.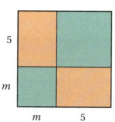

42. Find a polynomial for the shaded area.

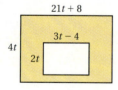

43. A box with a square bottom is to be made from a 12-in.-square piece of cardboard. Squares with side x are cut out of the corners and the sides are folded up. Find polynomials for the volume and the outside surface area of the box.

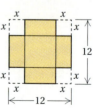

Multiply.

44. $(3x - 5)^2$

45. $(9x + 4)^2$

Visualize polynomials for factoring.

Chapter 10 Polynomials

10.4 Integers as Exponents

We have already used the numbers 1, 2, 3,..., as exponents. Here we consider 0, as well as negative integers, as exponents.

a   Zero as an Exponent

Look for a pattern in the following:

$8 \cdot 8 \cdot 8 \cdot 8 = 8^4$ We divide by 8 each time.
$8 \cdot 8 \cdot 8 = 8^3$
$8 \cdot 8 = 8^2$
$8 = 8^1$
$1 = 8^?.$

The exponents decrease by 1 each time. To continue the pattern, we would say that

$1 = 8^0.$

We make the following definition.

> $b^0 = 1$, for any nonzero number b.

We leave 0^0 undefined.

Example 1   Evaluate 3^0 and $(-7.3)^0$.

$3^0 = 1;$
$(-7.3)^0 = 1$ Note that $-7.3^0 = -1 \cdot 7.3^0 = -1 \cdot 1 = -1 \neq (-7.3)^0.$

Do Exercises 1–3.

Example 2   Evaluate $m^0 + 5$ for $m = 9$.

$m^0 + 5 = 9^0 + 5 = 1 + 5 = 6$

Example 3   Evaluate $(3x + 2)^0$ for $x = -5$.

We substitute -5 for x and follow the rules for order of operations:

$(3x + 2)^0 = (3(-5) + 2)^0$ Substituting
$= (-15 + 2)^0$ Multiplying
$= (-13)^0$
$= 1.$

CAUTION!   Students often confuse the powers 1 and 0. Be careful: $8^0 = 1$, whereas $8^1 = 8$.

Do Exercises 4 and 5.

Objectives

a   Evaluate algebraic expressions containing whole-number exponents.

b   Express exponential expressions involving negative exponents as equivalent expressions containing positive exponents.

For Extra Help

TAPE 18   MAC WIN   CD-ROM

Evaluate.
1. 7^0

2. 6^1

3. $(-47)^0$

4. Evaluate $t^0 - 4$ for $t = 7$.

5. Evaluate $(2x - 9)^0$ for $x = 3$.

Answers on page A-28

10.4   Integers as Exponents

603

Write an equivalent expression with positive exponents. Then simplify.

6. 4^{-3}

7. 5^{-2}

8. 2^{-4}

9. $(-2)^{-3}$

10. $\left(\dfrac{5}{3}\right)^{-2}$

Answers on page A-28

Chapter 10 Polynomials

b Negative Integers as Exponents

The pattern used to help define the exponent 0 can also be used to define negative-integer exponents:

$$8 \cdot 8 \cdot 8 = 8^3$$
$$8 \cdot 8 = 8^2$$
$$8 = 8^1$$
$$1 = 8^0$$
$$\dfrac{1}{8} = 8^?$$
$$\dfrac{1}{8 \cdot 8} = 8^?$$

We divide by 8 each time.

The exponents decrease by 1 each time. To continue the pattern, we would say that

$$\dfrac{1}{8} = 8^{-1}$$

and $\dfrac{1}{8 \cdot 8} = 8^{-2}$.

Thus, if we are to preserve the above pattern, we must have

$$\dfrac{1}{8^1} = 8^{-1} \quad \text{and} \quad \dfrac{1}{8^2} = 8^{-2}.$$

This leads to our definition of negative exponents:

> For any nonzero numbers a and b, and any integer n,
> $$a^{-n} = \dfrac{1}{a^n}, \quad \text{and} \quad \left(\dfrac{a}{b}\right)^{-n} = \left(\dfrac{b}{a}\right)^n.$$
> (A base raised to a negative exponent is equal to the reciprocal of the base raised to a positive exponent.)

Examples Write an equivalent expression using positive exponents. Then simplify.

4. $4^{-2} = \dfrac{1}{4^2} = \dfrac{1}{16}$ Note that 4^{-2} represents a *positive* number.

5. $(-3)^{-2} = \dfrac{1}{(-3)^2} = \dfrac{1}{(-3)(-3)} = \dfrac{1}{9}$

6. $m^{-3} = \dfrac{1}{m^3}$

7. $ab^{-1} = a\left(\dfrac{1}{b^1}\right) = a\left(\dfrac{1}{b}\right) = \dfrac{a}{b}$ Think of a as $\dfrac{a}{1}$ if you wish.

8. $\left(\dfrac{5}{6}\right)^{-2} = \left(\dfrac{6}{5}\right)^2$ $\left(\dfrac{a}{b}\right)^{-n} = \left(\dfrac{b}{a}\right)^n$
$= \dfrac{6}{5} \cdot \dfrac{6}{5} = \dfrac{36}{25}$

CAUTION! Note in Example 4 that

$$4^{-2} \neq 4(-2) \quad \text{and} \quad \frac{1}{4^2} \neq 4(-2).$$

Similarly, in Example 5,

$$(-3)^{-2} \neq (-3)(-2) \quad \text{and} \quad \frac{1}{(-3)^2} \neq (-3)(-2).$$

In general, $a^{-n} \neq a(-n)$. The negative exponent also does not mean to take the opposite of the denominator. That is,

$$4^{-2} = \frac{1}{16}, \quad \text{not} \quad \frac{1}{-16}.$$

Do Exercises 6–10 on the preceding page.

Examples Write an equivalent expression using negative exponents.

9. $\dfrac{1}{7^2} = 7^{-2}$ Reading $a^{-n} = \dfrac{1}{a^n}$ from right to left: $\dfrac{1}{a^n} = a^{-n}$

10. $\dfrac{5}{x^8} = 5 \cdot \dfrac{1}{x^8} = 5x^{-8}$

Do Exercises 11 and 12.

Consider an expression like

$$\frac{a^2}{b^{-3}},$$

in which the denominator is a negative power. We can simplify as follows:

$$\frac{a^2}{b^{-3}} = \frac{a^2}{\dfrac{1}{b^3}} \quad \text{Rewriting } b^{-3} \text{ as } \dfrac{1}{b^3}$$

$$= a^2 \cdot \frac{b^3}{1} \quad \text{To divide by a fractional expression, we multiply by its reciprocal.}$$

$$= a^2 b^3.$$

Do Exercises 13 and 14.

Our work above indicates that to divide by a base raised to a negative power, we can instead multiply by the opposite power of the same base. This will shorten our work.

Examples Write an equivalent expression using positive exponents.

11. $\dfrac{x^3}{y^{-2}} = x^3 y^2$ Instead of dividing by y^{-2}, multiply by y^2.

12. $\dfrac{a^2 b^5}{c^{-6}} = a^2 b^5 c^6$ Instead of dividing by c^{-6}, multiply by c^6.

13. $\dfrac{x^{-2} y}{z^{-3}} = x^{-2} y z^3 = \dfrac{y z^3}{x^2}$

Do Exercises 15–17.

Write an equivalent expression with negative exponents.

11. $\dfrac{1}{9^2}$

12. $\dfrac{7}{x^4}$

Write an equivalent expression with positive exponents.

13. $\dfrac{m^3}{n^{-5}}$

14. $\dfrac{ab}{c^{-1}}$

Write an equivalent expression with positive exponents.

15. $\dfrac{a^4}{b^{-6}}$

16. $\dfrac{x^7 y}{z^{-4}}$

17. $\dfrac{a^4 b^{-7}}{c^{-3}}$

Answers on page A-28

Simplify. Use positive powers in the answer.

18. $5^{-2} \cdot 5^4$

The product rule, developed in Section 10.2, still holds when exponents are zero or negative.

Examples Simplify. Use positive powers in the answer.

14. $7^{-3} \cdot 7^6 = 7^{-3+6}$ Adding exponents
$= 7^3$

15. $x^4 \cdot x^{-3} = x^{4+(-3)} = x^1 = x$

16. $(2a^3b^{-4})(3a^2b^7) = 2 \cdot 3 \cdot a^3 \cdot a^2 \cdot b^{-4} \cdot b^7$ Using the commutative and associative laws
$= 6a^{3+2}b^{-4+7}$ Using the product rule
$= 6a^5b^3$

17. $(x^{-4}y^5)(x^7y^{-11}) = x^{-4+7}y^{5+(-11)}$
$= x^3y^{-6}$
$= \dfrac{x^3}{y^6}$

Do Exercises 18–21.

19. $x^{-3} \cdot x^{-4}$

20. $(5x^{-3}y)(4x^{12}y^5)$

21. $(a^{-9}b^{-4})(a^2b^7)$

Calculator Spotlight

  Most calculators can evaluate expressions like

$$\left(-\dfrac{2}{5}\right)^{-6}$$

with just a few keystrokes. Although the keys may be labeled differently, depending on the calculator, the following keystrokes will generally work.

[2] [÷] [5] [=] [+/−] [x^y] [6] [+/−] [=]

The same result, 244.140625, is found when evaluating $\left(-\dfrac{5}{2}\right)^6$.

Exercises

Calculate the value of each pair of expressions.

1. $\left(\dfrac{3}{4}\right)^{-5}$; $\left(\dfrac{4}{3}\right)^5$

2. $\left(\dfrac{4}{5}\right)^{-7}$; $\left(\dfrac{5}{4}\right)^7$

3. $\left(-\dfrac{5}{4}\right)^{-7}$; $\left(-\dfrac{4}{5}\right)^7$

4. $\left(-\dfrac{7}{5}\right)^{-9}$; $\left(-\dfrac{5}{7}\right)^9$

5. $\left(-\dfrac{5}{6}\right)^{-4}$; $\left(-\dfrac{6}{5}\right)^4$

6. $\left(-\dfrac{9}{8}\right)^{-6}$; $\left(-\dfrac{8}{9}\right)^6$

Answers on page A-28

Exercise Set 10.4

a Evaluate.

1. 9^0
2. 17^0
3. 3.14^0
4. 2.67^1

5. $(-19.57)^1$
6. $(-34.6)^0$
7. $(-5.43)^0$
8. $(-98.6)^1$

9. x^0, $x \neq 0$
10. a^0, $a \neq 0$
11. $(3x - 17)^0$, for $x = 10$
12. $(7x - 45)^0$, for $x = 8$

13. $(5x - 3)^1$, for $x = 4$
14. $(35 - 4x)^1$, for $x = 8$
15. $(4m - 19)^0$, for $m = 3$
16. $(9 - 2x)^0$, for $x = 5$

17. $3x^0 + 4$, for $x = -2$
18. $7x^0 + 6$, for $x = -3$
19. $(3x)^0 + 4$, for $x = -2$
20. $(7x)^0 + 6$, for $x = -3$

21. $(5 - 3x^0)^1$, for $x = 19$
22. $(5x^1 - 29)^0$, for $x = 4$

b Write an equivalent expression with positive exponents. Then simplify, if possible.

23. 3^{-2}
24. 2^{-3}
25. 10^{-4}
26. 5^{-6}

27. a^{-3}
28. x^{-2}
29. $(-5)^{-2}$
30. $(-4)^{-3}$

31. $3x^{-7}$
32. $-6y^{-2}$
33. $\dfrac{x}{y^{-4}}$
34. $\dfrac{r}{t^{-7}}$

35. $\dfrac{a^3}{b^{-4}}$
36. $\dfrac{x^7}{y^{-5}}$
37. $-7a^{-9}$
38. $9p^{-4}$

39. $\left(\dfrac{2}{5}\right)^{-2}$
40. $\left(\dfrac{3}{7}\right)^{-2}$
41. $\left(\dfrac{5}{a}\right)^{-3}$
42. $\left(\dfrac{x}{3}\right)^{-4}$

Write an equivalent expression using negative exponents.

43. $\dfrac{1}{4^3}$ **44.** $\dfrac{1}{5^2}$ **45.** $\dfrac{9}{x^3}$ **46.** $\dfrac{4}{y^2}$

Simplify. Do not use negative exponents in the answer.

47. $x^{-2} \cdot x$ **48.** $x \cdot x^{-1}$ **49.** $x^4 \cdot x^{-4}$

50. $x^9 \cdot x^{-9}$ **51.** $x^{-7} \cdot x^{-6}$ **52.** $y^{-5} \cdot y^{-8}$

53. $(3a^2b^{-7})(2ab^9)$ **54.** $(5xy^8)(3x^4y^{-5})$ **55.** $(-2x^{-3}y^8)(3xy^{-2})$

56. $(5a^{-1}b^{-7})(-2a^4b^2)$ **57.** $(3a^{-4}bc^2)(2a^{-2}b^{-5}c)$ **58.** $(5x^2y^{-7}z)(-4xy^{-3}z^{-4})$

Skill Maintenance

59. George's Geo is driven 450 km in 9 hr. What is the rate in kilometers per hour? [7.2a]

60. A field hockey team won 18 of its 30 games. What percentage of its games did it win? [8.4a]

61. The Jets once won 14 of their 16 games. What percentage of their games did they win? [8.4a]

62. The sales tax is $27.60 on a purchase of $460. What is the sales tax rate? [8.5a]

63. A circle has radius 5 cm. Find its circumference. Use 3.14 for π. [9.2b]

64. Becky drove 326 mi on 14.5 gal of gas. What was her mileage? [7.2a]

Synthesis

65. ◆ Is there any choice of y for which $(5y)^0$ and $5y^0$ give the same result? Why or why not?

66. ◆ Consider the expression x^{-3}. When evaluated, will the expression always be negative? Will it *ever* be negative? Explain.

67. ◆ What number is larger and why: 5^{-8} or 6^{-8}? Do not use a calculator.

68. ▦ Evaluate $\dfrac{3^x}{3^{x-1}}$ for $x = -4$ and then for $x = -40$.

69. ▦ Evaluate $\dfrac{5^x}{5^{x+1}}$ for $x = -3$ and then for $x = -30$.

70. ◆ How can negative exponents and the product rule be used to answer Exercises 68 and 69 without using a calculator?

Simplify.

71. $(y^{2x})(y^{3x})$ **72.** $a^{5k} \div a^{3k}$ **73.** $\dfrac{a^{6t}(a^{7t})}{a^{9t}}$

Discover the pattern for negative integer exponents.

Summary and Review: Chapter 10

Important Properties and Formulas

The Product Rule: $a^m \cdot a^n = a^{m+n}$

Negative exponents: $a^{-n} = \dfrac{1}{a^n}$ and $\left(\dfrac{a}{b}\right)^{-n} = \left(\dfrac{b}{a}\right)^n$

Review Exercises

Perform the indicated operation. [10.1a, c]

1. $(-5x + 9) + (7x - 13)$

2. $(5x^4 - 7x^3 + 3x - 5) + (3x^3 - 4x + 3)$

3. $(9a^5 + 8a^3 + 4a + 7) - (a^5 - 4a^3 + a^2 - 2)$

4. $(7a^3b^3 + 9a^2b^3) - (2a^3b^3 - 3a^2b^3 + 7)$

5. Find two equivalent expressions for the opposite of $12x^3 - 4x^2 + 9x - 3$. [10.1b]

Evaluate.

6. $(-72)^1$ [10.2a]

7. $(4x - 17)^0$, for $x = 5$ [10.4a]

8. $5t^3 + t$, for $t = -2$ [10.1d]

Multiply.

9. $(5x^3)(6x^4)$ [10.2a]

10. $3x(6x^3 - 4x - 1)$ [10.2b]

11. $2a^4b(7a^3b^3 + 5a^2b^3)$ [10.2b]

12. $(x - 7)(x + 9)$ [10.3a]

13. $(2x - 1)(5x - 3)$ [10.3a]

14. $(a^2 - 1)(a^2 + 2a - 1)$ [10.3b]

Factor. [10.2c]

15. $45x^3 - 10x$

16. $7a - 35b - 49ac$

17. $6x^3y - 9x^2y^5$

Write an equivalent expression using positive exponents. Then simplify, if possible. [10.4b]

18. 12^{-2}

19. $8a^{-7}$

20. $\dfrac{x^{-3}}{y^5 z^{-6}}$

21. $\left(\dfrac{4}{5}\right)^{-2}$

22. Write an expression equivalent to $\dfrac{1}{x^7}$ using a negative exponent. [10.4b]

Simplify. Use positive exponents in the answer. [10.4b]

23. $x^{-5} \cdot x^{-12}$

24. $(-5x^4 y^{-7})(-3x^5 y^{-2})$

Skill Maintenance

25. Willie eats 4 slices of pizza in 30 min. What is his rate in slices per minute? in minutes per slice? [7.2a]

26. In a sample of 40 tapes, 12 were defective. What percent were defective? What percent were not defective? [8.4a]

27. Phi purchases 4 tires at $45 apiece. If the sales tax rate is 5%, what will the total price be? [8.5a]

28. A round serving platter has a diameter of 21 in. Find its area. Use $\frac{22}{7}$ for π. [9.2b]

Synthesis

29. ◆ Can x^{-2} represent a negative number? Why or why not? [10.4b]

30. ◆ A student claims that
$(3x^{-5})(-4x^{-2}) = -x^{10}$.
What mistake(s) is the student probably making? [10.4b]

Simplify.

31. ▦ $(2349x^7 - 357x^2)(493x^{10} + 597x^5)$ [10.3a]

32. $-3x^5 \cdot 3x^3 - x^6(2x)^2 + (3x^4)^2 + (2x^4)^2 - 40x^2(x^3)^2$ [10.2a]

Factor. [10.2c]

33. $39a^3b^7c^6 - 130a^2b^5c^8 + 52a^4b^6c^5$

34. $w^5x^6y^4z^5 - w^7x^3y^7z^3 + w^6x^2y^5z^6 - w^6x^7y^3z^4$

Test: Chapter 10

1. Add: $(10a^3 - 9a^2 + 7) + (7a^3 + 4a^2 - a)$.

2. Find two equivalent expressions for the opposite of $-9a^4 + 7b^2 - ab + 3$.

3. Subtract: $(13x^4 + 7x^2 - 8) - (9x^4 + 8x^2 + 5)$.

Evaluate.

4. 193^1

5. $(3x - 7)^0$, for $x = 2$

6. The height h, in meters, of a ball t seconds after it has been thrown is approximated by the polynomial $h = -4.9t^2 + 15t + 2$. How high is the ball 2 sec after it has been thrown?

Multiply.

7. $(-5x^4y^3)(2x^2y^5)$

8. $5a(7a^2 - 4a + 3)$

9. $(x - 6)(x + 7)$

10. $(2a + 1)(a^2 - 3a + 2)$

Factor.

11. $35x^6 - 25x^3 + 15x^2$

12. $6ab - 9bc + 12ac$

Answers

1. _____
2. _____
3. _____
4. _____
5. _____
6. _____
7. _____
8. _____
9. _____
10. _____
11. _____
12. _____

Write an equivalent expression with positive exponents. Then simplify, if possible.

13. 5^{-3}

14. $\dfrac{5a^{-3}}{b^{-2}}$

15. $\left(\dfrac{3}{5}\right)^{-3}$

Simplify. Use positive exponents in the answer.

16. $x^{-7} \cdot x^{-9}$

17. $(3a^{-7}b^9)(-2a^{10}b^{-12})$

Skill Maintenance

18. Charlene hammers 25 nails in 15 min. What is her rate in nails per minute? in minutes per nail?

19. Of 55 people surveyed, 22 felt satisfied with their diets. What percent felt satisfied? What percent did not feel satisfied?

20. Gina pays $340 in tax on the purchase of a used car. If the sales tax rate is 4%, what was the price of the car before tax?

21. A juggler's hoop has a diameter of 30 cm. Find its circumference. Use 3.14 for π.

Synthesis

22. The polynomial
$$0.041h - 0.018A - 2.69$$
can be used to estimate the lung capacity, in liters, of a female of height h, in centimeters, and age A, in years. Find the lung capacity of a 30-yr-old woman who is 150 cm tall.

23. Write an equivalent expression with positive exponents and then simplify: $12a^6(2a^3 - 6a)^{-2}$.

Final Examination

This exam reviews the entire textbook. A question may arise as to what notation to use for a particular problem or exercise. Although there is no hard-and-fast rule, especially as you use mathematics outside the classroom, here is the guideline that we follow: Use the notation given in the problem. That is, if the problem is given using mixed numerals, give the answer in mixed numerals. If the problem is given in decimal notation, give the answer in decimal notation.

1. In 46,301, what digit tells the number of thousands?

2. Write expanded notation for 8409.

Add and, if possible, simplify.

3. $7\,4\,3$
    $+\,2\,7\,5$

4. $4\,9\,0\,3$
    $5\,2\,7\,8$
    $6\,3\,9\,1$
    $+\,4\,5\,1\,3$

5. $\dfrac{4}{13} + \dfrac{1}{26}$

6. $5\dfrac{4}{9}$
   $+\,3\dfrac{1}{3}$

7. $-29 + 53$

8. $-543 + (-219)$

9. $-34.56 + 2.783 + 0.433 + (-13.02)$

10. $(4x^5 + 7x^4 - 3x^2 + 9) + (6x^5 - 8x^4 + 2x^3 - 7)$

Subtract and, if possible, simplify.

11. $6\,7\,4$
    $-\,4\,3\,1$

12. $-7x - 12x$

13. $\dfrac{2}{5} - \dfrac{7}{8}$

14. $4\dfrac{1}{3}$
    $-\,1\dfrac{5}{8}$

15. $2\,0.0$
    $-\,0.0\,0\,2\,7$

16. $(7x^3 + 2x^2 - x) - (5x^3 - 3x^2 - 8x)$

17. $(9a^2b + 3ab) - (13a^2b - 4ab)$

Multiply and, if possible, simplify.

18. $2\,9\,7$
    $\times\,1\,6$

19. $349 \cdot (-213)$

20. $2\dfrac{3}{4} \cdot 1\dfrac{2}{3}$

Answers

1. _____
2. _____
3. _____
4. _____
5. _____
6. _____
7. _____
8. _____
9. _____
10. _____
11. _____
12. _____
13. _____
14. _____
15. _____
16. _____
17. _____
18. _____
19. _____
20. _____

Final Examination

Answers

21. _____
22. _____
23. _____
24. _____
25. _____
26. _____
27. _____
28. _____
29. _____
30. _____
31. _____
32. _____
33. _____
34. _____
35. _____
36. _____
37. _____
38. _____
39. _____
40. _____
41. _____
42. _____
43. _____
44. _____

21. $-\dfrac{9}{7} \cdot \dfrac{14}{15}$

22. $12 \cdot \dfrac{5}{6}$

23. $\begin{array}{r} 34.09 \\ \times7.6 \\ \hline \end{array}$

24. $3(8x - 5)$

25. $(9a^3b^2)(3a^5b)$

26. $7x^2(3x^3 - 2x + 8)$

27. $(x + 2)(x - 7)$

28. $(a + 3)(a^2 - 5a + 4)$

Divide and simplify. State the answer using a remainder when appropriate.

29. $6\overline{)3438}$

30. $34\overline{)1914}$

Divide and simplify.

31. $\dfrac{4}{5} \div \left(-\dfrac{8}{15}\right)$

32. $-2\dfrac{1}{3} \div (-30)$

33. $2.7\overline{)105.3}$

34. Write a mixed numeral for the quotient in Question 30.

Simplify.

35. $10 \div 2 \times 20 - 5^2$

36. $\dfrac{|3^2 - 5^2|}{2 - 2 \cdot 5}$

37. Write exponential notation: $17 \cdot 17 \cdot 17 \cdot 17$.

38. Round 68,489 to the nearest thousand.

39. Round $21.\overline{83}$ to the nearest hundredth.

40. Determine whether 1368 is divisible by 3.

41. Find all the factors of 15.

42. Find the LCM of 15 and 35.

Simplify.

43. $\dfrac{21}{30}$

44. $\dfrac{-290}{15}$

Final Examination

614

45. Convert to a mixed numeral: $-\dfrac{18}{5}$.

46. Use < or > for ▓ to write a true sentence:

-17 ▓ -29.

47. Use < or > for ▓ to write a true sentence:

$\dfrac{4}{7}$ ▓ $\dfrac{3}{5}$.

48. Which number is greater, 1.001 or 0.9976?

49. Evaluate $\dfrac{a^2 - b}{3}$ for $a = -9$ and $b = -6$.

Factor.

50. $40 - 5t$

51. $18a^3 - 15a^2 + 6a$

52. What part is shaded?

Convert to decimal notation.

53. $\dfrac{37}{1000}$

54. $-\dfrac{13}{25}$

55. $\dfrac{8}{9}$

56. 7%

Convert to fractional notation.

57. 6.71

58. $-7\dfrac{1}{4}$

59. 40%

Convert to percent notation.

60. $\dfrac{17}{20}$

61. 1.5

62. Estimate the sum $9.389 + 4.2105$ to the nearest tenth.

Solve.

63. $234 + y = 789$

64. $3.9a = 249.6$

65. $\dfrac{2}{3} \cdot t = \dfrac{5}{6}$

66. $\dfrac{8}{17} = \dfrac{36}{x}$

67. $7x - 9 = 26$

68. $-2(x - 5) = 3x + 12$

Answers

45. _____
46. _____
47. _____
48. _____
49. _____
50. _____
51. _____
52. _____
53. _____
54. _____
55. _____
56. _____
57. _____
58. _____
59. _____
60. _____
61. _____
62. _____
63. _____
64. _____
65. _____
66. _____
67. _____
68. _____

Final Examination

Answers

69. _____

70. _____

71. _____

72. _____

73. _____

74. _____

75. _____

76. _____

77. _____

78. _____

79. _____

80. _____

81. _____

82. _____

83. _____

84. _____

85. _____

86. _____

87. _____

Final Examination

Solve.

69. Margie donated $20 to the Humane Society, $30 to the Red Cross, $25 to the Salvation Army, and $20 to Amnesty International. What was the average size of the donations?

70. A machine wraps 134 candy bars per minute. How long does it take this machine to wrap 8710 bars?

71. A share of IDX stock bought for 29\frac{5}{8}$ dropped 3\frac{7}{8}$ before it was resold. What was the price when it was resold?

72. At the start of a trip, the odometer on the Montgomery's Toyota read 27,428.6 mi and at the end of the trip the reading was 27,914.5 mi. How long was the trip?

73. From an income of $32,000, amounts of $6400 and $1600 are paid for federal and state taxes. How much remains after these taxes have been paid?

74. Shannon is paid $85 a day for 7 days work as a lifeguard. How much will she be paid?

75. A toddler walks $\frac{3}{5}$ km per hour. At this rate, how far would the child walk in $\frac{1}{2}$ hr?

76. Eight identical dresses cost a total of $679.68. What is the cost of each dress?

77. Eight gallons of paint covers 2000 ft^2. How much paint is needed to cover 3250 ft^2?

78. Eighteen ounces of a fruit "smoothie" costs $3.06. Find the unit price in cents per ounce.

79. What is the simple interest on $4000 principal at 8% for $\frac{3}{4}$ year?

80. Baldacci Real Estate received $5880 commission on the sale of an $84,000 home. What was the rate of commission?

81. The population of Bridgeton is 29,000 this year and is increasing at 4% per year. What will the population be next year?

82. Ace Car Rentals charges $35 a day plus 15 cents a mile for a van rental. If a couple's one-day van rental cost $68, how many miles did they drive?

Evaluate.

83. 18^2

84. 37^0

85. $\sqrt{121}$

Express with positive exponents. Then simplify, if possible.

86. 4^{-3}

87. $\left(\dfrac{5}{4}\right)^{-2}$

The following line graph show the driver fatality rate during a recent year. Use the line graph for Questions 88 and 89.

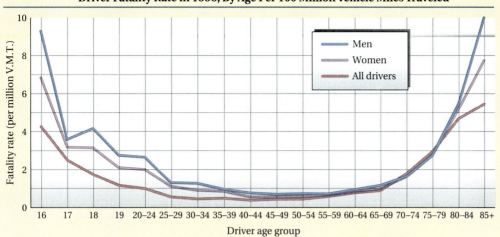

88. When all drivers are considered, what age group has the highest fatality rate? What is the rate?

89. Approximate the fatality rate for drivers aged 20–24.

90. The ages of students in a community college math lab are as follows:

18, 20, 27, 35, 20, 52, 26.

Find the mean, the median, and the mode of the ages.

91. In Sam's writing lab, 3 of the 20 students are left-handed. If a student is randomly selected, what is the probability that he or she is left-handed?

92. Plot the following points:

$(-5, 2)$, $(4, 0)$, $(3, -4)$, $(0, 2)$.

93. Graph: $y = -\dfrac{1}{3}x$.

94. These triangles are similar. Find the missing lengths.

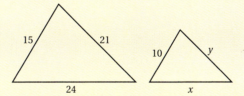

Complete.

95. $\dfrac{1}{3}$ yd = _____ in.

96. 4280 mm = _____ cm

97. 3.7 km = _____ m

98. 20,000 g = _____ kg

99. 10 lb = _____ oz

100. 0.008 cg = _____ mg

101. 8190 mL = _____ L

102. 20 qt = _____ gal

Answers

88. _____
89. _____
90. _____
91. _____
92. _____
93. _____
94. _____
95. _____
96. _____
97. _____
98. _____
99. _____
100. _____
101. _____
102. _____

Final Examination

Answers

103. _____

104. _____

105. _____

106. _____

107. _____

108. _____

109. _____

110. _____

111. _____

112. _____

103. A rectangular picture frame measures 20 in. by 24 in. Find its perimeter.

Find the area of each figure.

104.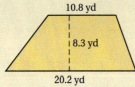

105.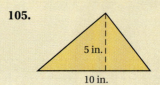

106.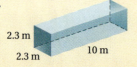

107.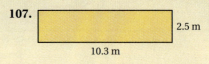

108. Find the diameter, the circumference, and the area of this circle. Use 3.14 for π.

Find the volume of each shape. Use 3.14 for π.

109.   **110.**   **111.**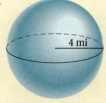

112. Find the length of the third side of this right triangle. Give an exact answer and an approximation to three decimal places.

Final Examination

Developmental Units

Introduction

These developmental units are meant to provide extra instruction for students who have difficulty with any of Sections 1.2, 1.3, 1.5, or 1.6. After reading one of these developmental units and doing the exercises in its exercise set, the student should restudy the appropriate section in Chapter 1.

A Addition
S Subtraction
M Multiplication
D Division

For more information, visit us at www.mathmax.com

Objectives

a Add any two of the numbers 0, 1, 2, 3, 4, 5, 6, 7, 8, 9.

b Find certain sums of three numbers such as 1 + 7 + 9.

c Add two whole numbers when carrying is not necessary.

d Add two whole numbers when carrying is necessary.

Add; think of joining sets of objects.

1. 4 + 5
2. 3 + 4

3. 9
   + 5

4. 8
   + 8

5. 9
   + 7

6. 7
   + 9

The first printed use of the + symbol was in a book by a German, Johann Widmann, in 1498.

Answers on page A-29

Developmental Units

620

A Addition

a Basic Addition

Basic addition can be explained by counting. The sum

$$3 + 4$$

can be found by counting out a set of 3 objects and a separate set of 4 objects, putting them together, and counting all the objects.

A set of 3   +   A set of 4   =   A set of 7

The numbers to be added are called **addends**. The result is the **sum**.

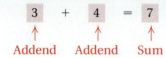

3   +   4   =   7
↑   ↑   ↑
Addend   Addend   Sum

Examples   Add. Think of putting sets of objects together.

1. 5 + 6 = 11

      5
   + 6
   ――
   11

2. 8 + 5 = 13

      8
   + 5
   ――
   13

We can also do these problems by counting up from one of the numbers. For example, in Example 1, we start at 5 and count up 6 times: 6, 7, 8, 9, 10, 11.

Do Exercises 1–6.

What happens when we add 0? Think of a set of 5 objects. If we add 0 objects to it, we still have 5 objects. Similarly, if we have a set with 0 objects in it and add 5 objects to it, we have a set with 5 objects. Thus,

$$5 + 0 = 5 \quad \text{and} \quad 0 + 5 = 5.$$

> Adding 0 to a number does not change the number:
> $$a + 0 = 0 + a = a.$$
> We say that 0 is the **additive identity**.

Examples Add.

3. $0 + 9 = 9$

$$\begin{array}{r} 0 \\ +\,9 \\ \hline 9 \end{array}$$

4. $0 + 0 = 0$

$$\begin{array}{r} 0 \\ +\,0 \\ \hline 0 \end{array}$$

5. $97 + 0 = 97$

$$\begin{array}{r} 97 \\ +\,0 \\ \hline 97 \end{array}$$

Do Exercises 7–12.

Your objective for this part of the section is to be able to add any of the numbers 0, 1, 2, 3, 4, 5, 6, 7, 8, 9. Adding 0 is easy. The rest of the sums are listed in this table. Memorize the table by saying it to yourself over and over or by using flash cards.

+	1	2	3	4	5	6	7	8	9
1	2	3	4	5	6	7	8	9	10
2	3	4	5	6	7	8	9	10	11
3	4	5	6	7	8	9	10	11	12
4	5	6	7	8	9	10	11	12	13
5	6	7	8	9	10	11	12	13	14
6	7	8	9	10	11	12	13	14	15
7	8	9	10	11	12	13	14	15	16
8	9	10	11	12	13	14	15	16	17
9	10	11	12	13	14	15	16	17	18

$6 + 7 = 13$
Find 6 at the left, and 7 at the top.

$7 + 6 = 13$
Find 7 at the left, and 6 at the top.

It is very important that you *memorize* the basic addition facts! If you do not, you will always have trouble with addition.

Note the following.

$3 + 4 = 7$ $\quad$ $7 + 6 = 13$ $\quad$ $7 + 2 = 9$
$4 + 3 = 7$ $\quad$ $6 + 7 = 13$ $\quad$ $2 + 7 = 9$

We can add whole numbers in any order. This is the *commutative law of addition*. Because of this law, you need to learn only about half the table above, as shown by the shading.

Do Exercises 13 and 14.

b Certain Sums of Three Numbers

To add $3 + 5 + 4$, we can add 3 and 5, then 4:

$3 + 5 + 4$
$8 + 4$
$12.$

We can also add 5 and 4, then 3:

$3 + 5 + 4$
$3 + 9$
$12.$

Either way we get 12.

Add.

7. $8 + 0$ $\qquad$ **8.** $0 + 8$

9. $\begin{array}{r} 7 \\ +\,0 \end{array}$ $\qquad$ **10.** $\begin{array}{r} 46 \\ +\,0 \end{array}$

11. $0 + 13$ $\qquad$ **12.** $58 + 0$

Complete the table.

13.

+	1	2	3	4	5
1			4		
2					
3				7	
4					
5					

14.

+	6	5	7	4	9
7		14			
9					
5			9		
8					
4					

Answers on page A-29

A Addition

Add from the top mentally.

15. 1
     6
   + 9

16. 2
     3
   + 4

17. 6
     1
   + 4

18. 5
     2
   + 8

Add.

19.   2 4
    + 3 5

20.   3 4 6
    + 2 0 3

21.   8 3 2 7
    + 1 6 5 2

22.   3 4 6 1
    + 2 0 3 5

Answers on page A-29

Developmental Units

622

Example 6 Add from the top mentally.

  1   We first add 1 and 7,   1
  7   getting 8. Then we add   7 → 8
+ 9   8 and 9, getting 17.   + 9   9 → 17

Example 7 Add from the top mentally.

  2
  4 → 6
+ 8   8 → 14

Do Exercises 15–18.

c Addition (No Carrying)

We now move to a more gradual, conceptual development of the addition procedure you considered in Section 1.2. It is intended to provide you with a greater understanding so that your skill level will increase.

To add larger numbers, we can add the ones first, then the tens, then the hundreds, and so on.

Example 8 Add: 5722 + 3234.

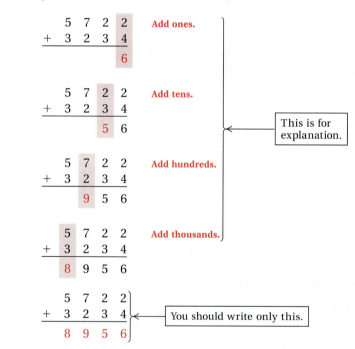

Do Exercises 19–22.

d Addition (With Carrying)

Carrying Tens

Example 9 Add: 18 + 27.

```
   1 8      Add ones.        Think:      8
 + 2 7                               +   7     15 ones = 10 ones + 5 ones
   ───                                ─────            = 1 ten + 5 ones
     ?                                  1 5
```

```
   1
   1 8      Write 5 in the ones column.
 + 2 7      Write 1 for a reminder above the tens.
   ───      This is called carrying.
       5
```

```
   1
   1 8      Add tens.
 + 2 7
   ───
   4 5
```

We can use money to help explain Example 9.

```
   1 8¢  ⟶ 1 dime and 8 pennies
 + 2 7¢  ⟶ 2 dimes and 7 pennies
   ─────
     15¢      We first add the pennies.
```

```
   1 dime
   1 8
 + 2 7          We exchange ten pennies for a dime.
   ───
   5 pennies
```

```
   1
   1 8          We now add the dimes. The result is
 + 2 7          4 dimes and 5 pennies.
   ───
   4 5
```

Do Exercises 23 and 24.

Carrying Hundreds

Example 10 Add: 256 + 391.

```
   2 5 6        Add ones.
 + 3 9 1
   ─────
         7
```

```
     1
   2 5 6        Add tens. We get 14 tens. Now 14 tens = 10 tens + 4 tens =
 + 3 9 1        1 hundred + 4 tens. Write 4 in the tens column and a 1 above
   ─────        the hundreds.
     4 7
```

> The carrying here is like exchanging 14 dimes for a 1 dollar bill and 4 dimes.

```
   1
   2 5 6        Add hundreds.
 + 3 9 1
   ─────
   6 4 7
```

Add.

23.   1 9
    + 3 7

24.   4 6
    + 3 9

Add.

25.   3 4 1
    + 4 8 8

26.   7 3 0
    + 2 9 6

Answers on page A-29

A Addition

623

27. Add.

$$\begin{array}{r} 7850 \\ +4848 \\ \hline \end{array}$$

Add.

28.
$$\begin{array}{r} 7989 \\ +5672 \\ \hline \end{array}$$

29.
$$\begin{array}{r} 56{,}789 \\ +14{,}539 \\ \hline \end{array}$$

To the student: *If you had trouble with Section 1.2 and have studied Developmental Unit A, you should go back and work through Section 1.2 after completing Exercise Set A.*

Answers on page A-29

Developmental Units

Do Exercises 25 and 26 on the preceding page.

Carrying Thousands

Example 11 Add: 4803 + 3792.

$$\begin{array}{r} 4\,8\,0\,\boxed{3} \\ +\;3\,7\,9\,\boxed{2} \\ \hline \boxed{5} \end{array}$$ Add ones.

$$\begin{array}{r} 4\,8\,\boxed{0}\,3 \\ +\;3\,7\,\boxed{9}\,2 \\ \hline \boxed{9}\,5 \end{array}$$ Add tens.

$$\begin{array}{r} 1 \\ 4\,\boxed{8}\,0\,3 \\ +\;3\,\boxed{7}\,9\,2 \\ \hline \boxed{5}\,9\,5 \end{array}$$ Add hundreds. We get 15 hundreds. Now 15 hundreds = 10 hundreds + 5 hundreds = 1 thousand + 5 hundreds. Write 5 in the hundreds column and 1 above the thousands.

$$\begin{array}{r} 1 \\ \boxed{4}\,8\,0\,3 \\ +\;\boxed{3}\,7\,9\,2 \\ \hline \boxed{8}\,5\,9\,5 \end{array}$$ Add thousands.

Do Exercise 27.

Carrying More Than Once

Sometimes we must carry more than once.

Example 12 Add: 5767 + 4993.

$$\begin{array}{r} 1 \\ 5\,7\,6\,\boxed{7} \\ +\;4\,9\,9\,\boxed{3} \\ \hline \boxed{0} \end{array}$$ Add ones. We get 10 ones. Now 10 ones = 1 ten + 0 ones. Write 0 in the ones column and 1 above the tens.

$$\begin{array}{r} 1\,1 \\ 5\,7\,\boxed{6}\,7 \\ +\;4\,9\,\boxed{9}\,3 \\ \hline \boxed{6}\,0 \end{array}$$ Add tens. We get 16 tens. Now 16 tens = 1 hundred + 6 tens. Write 6 in the tens column and 1 above the hundreds.

$$\begin{array}{r} 1\,1\,1 \\ 5\,\boxed{7}\,6\,7 \\ +\;4\,\boxed{9}\,9\,3 \\ \hline \boxed{7}\,6\,0 \end{array}$$ Add hundreds. We get 17 hundreds. Now 17 hundreds = 1 thousand + 7 hundreds. Write 7 in the hundreds column and 1 above the thousands.

$$\begin{array}{r} 1\,1\,1 \\ \boxed{5}\,7\,6\,7 \\ +\;\boxed{4}\,9\,9\,3 \\ \hline \boxed{1\,0}\,7\,6\,0 \end{array}$$ Add thousands. We get 10 thousands.

Do Exercises 28 and 29.

Exercise Set A

a Add. Try to do these mentally. If you have trouble, think of putting objects together.

1. 8 + 9	2. 8 + 7	3. 6 + 7	4. 9 + 5	5. 5 + 7	6. 5 + 6
7. 9 + 8	8. 9 + 7	9. 8 + 4	10. 9 + 1	11. 8 + 2	12. 3 + 8
13. 0 + 7	14. 4 + 3	15. 2 + 9	16. 0 + 0	17. 3 + 0	18. 9 + 9
19. 8 + 6	20. 3 + 7	21. 2 + 2	22. 7 + 7	23. 6 + 5	24. 7 + 8
25. 8 + 8	26. 8 + 1	27. 5 + 8	28. 5 + 9	29. 4 + 7	30. 6 + 1

31. 6 + 7
32. 7 + 7
33. 3 + 9
34. 6 + 0
35. 6 + 4

36. 9 + 3
37. 5 + 5
38. 5 + 3
39. 1 + 1
40. 4 + 5

41. 9 + 4
42. 0 + 8
43. 4 + 6
44. 2 + 7
45. 3 + 7

46. 3 + 3
47. 5 + 8
48. 3 + 6
49. 4 + 4
50. 4 + 7

b Add from the top mentally.

51. 1, 8, + 3	52. 1, 7, + 5	53. 3, 2, + 5	54. 4, 3, + 5	55. 1, 7, + 9
56. 5, 2, + 6	57. 4, 5, + 1	58. 1, 9, + 6	59. 1, 8, + 7	60. 1, 6, + 8

c Add.

61. 23
    +16

62. 54
    +35

63. 67
    +20

64. 496
    +503

65. 700
    +200

66. 801
    + 67

67. 666
    +333

68. 523
    +325

69. 747
    +130

70. 8250
    +9430

71. 6552
    +4321

72. 3406
    +1293

73. 7340
    +3527

74. 4825
    +5070

75. 2073
    +1925

76. 9111
    +9111

77. 7889
    +9000

78. 52,433
    +12,056

79. 43,723
    +56,276

80. 51,670
    +26,107

d Add.

81. 38
    + 8

82. 17
    + 9

83. 17
    +38

84. 95
    + 6

85. 862
    +781

86. 613
    +799

87. 355
    +491

88. 280
    +348

89. 814
    +390

90. 274
    +333

91. 9990
    + 10

92. 999
    + 11

93. 999
    +111

94. 839
    +388

95. 909
    +202

96. 808
    +909

97. 8718
    +1420

98. 3854
    +2700

99. 4828
    +1283

100. 6995
    +1432

101. 9889
    +   1

102. 6889
    +4723

103. 9128
    +1997

104. 8898
    +6645

105. 9989
    +6785

106. 46,889
    +21,786

107. 23,448
    +10,989

108. 67,658
    +98,786

109. 77,548
    +23,767

110. 44,684
    + 4,765

Developmental Units

S Subtraction

a   Basic Subtraction

Subtraction can be explained by taking away part of a set.

Example 1   Subtract: $7 - 3$.

We can do this by counting out 7 objects and then taking away 3 of them. Then we count the number that remain: $7 - 3 = 4$.

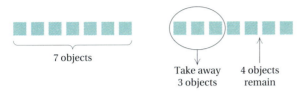

We could also do this mentally by starting at 7 and counting down 3 times: 6, 5, 4.

Objectives

a   Find basic differences such as $5 - 3$, $13 - 8$, and so on.

b   Subtract one whole number from another when borrowing is not necessary.

c   Subtract one whole number from another when borrowing is necessary.

Subtract.

1. $10 - 6$

Examples   Subtract. Think of "take away."

2. $11 - 6 = 5$ *Take away: "11 take away 6 is 5."*

$$\begin{array}{r} 11 \\ -6 \\ \hline 5 \end{array}$$

2. $11 - 4$

3. $17 - 9 = 8$

$$\begin{array}{r} 17 \\ -9 \\ \hline 8 \end{array}$$

Do Exercises 1–4.

In Developmental Unit A, you memorized an addition table. That table will enable you to subtract also. First, let's recall how addition and subtraction are related.

3. $\begin{array}{r} 16 \\ -8 \\ \hline \end{array}$

An addition:

Two related subtractions.

A.

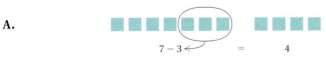

4. $\begin{array}{r} 10 \\ -7 \\ \hline \end{array}$

B.

Answers on page A-30

For each addition fact, write two subtraction facts.

5. 8 + 4 = 12

6. 6 + 7 = 13

Subtract. Try to do these mentally.

7. 14 − 6

8. 12 − 5

9. 1 3
   −   4
   ———

10. 1 1
    −   7
    ———

Answers on page A-30

Developmental Units

Since we know that

4 + 3 = 7,   **A basic addition fact**

we also know the two subtraction facts

7 − 3 = 4   and   7 − 4 = 3.

Example 4 From 8 + 9 = 17, write two subtraction facts.

a) The addend 8 is subtracted from the sum 17.

8 + 9 = 17   The related sentence is   17 − 8 = 9.

b) The addend 9 is subtracted from the sum 17.

8 + 9 = 17   The related sentence is   17 − 9 = 8.

Do Exercises 5 and 6.

We can use the idea that subtraction is defined in terms of addition to think of subtraction as "how much more."

Example 5 Find: 13 − 6.

To find 13 − 6, we ask, "6 plus what number is 13?"

6 + ▨ = 13

+	1	2	3	4	5	6	7	8	9
1	2	3	4	5	6	7	8	9	10
2	3	4	5	6	7	8	9	10	11
3	4	5	6	7	8	9	10	11	12
4	5	6	7	8	9	10	11	12	13
5	6	7	8	9	10	11	12	13	14
6	7	8	9	10	11	12	13	14	15
7	8	9	10	11	12	13	14	15	16
8	9	10	11	12	13	14	15	16	17
9	10	11	12	13	14	15	16	17	18

13 − 6 = 7

Using the addition table above, we find 13 inside the table and 6 at the left. Then we read the answer 7 from the top. Thus we have 13 − 6 = 7. Strive to do this kind of thinking mentally as fast as you can, without having to use the table.

Do Exercises 7–10.

b Subtraction (No Borrowing)

We now move to a more gradual, conceptual development of the subtraction procedure you considered in Section 1.3. It is intended to provide you with a greater understanding so that your skill level will increase.

To subtract larger numbers, we can subtract the ones first, then the tens, then the hundreds, and so on.

Example 6 Subtract: 5787 − 3214.

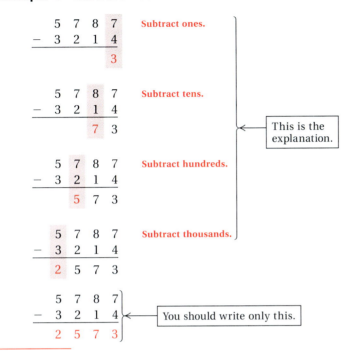

Do Exercises 11–14.

c Subtraction (with Borrowing)

We now consider subtraction when borrowing is necessary.

Borrowing from the Tens Place

Example 7 Subtract: 37 − 18.

```
   3 7     Try to subtract ones: 7 − 8 is not a whole number.
 − 1 8
     ?

   2 17
   3̸ 7̸    Borrow a ten. That is, 1 ten = 10 ones, and 10 ones + 7 ones = 17 ones.
 − 1 8    Write 2 above the tens column and 17 above the ones.

   2 17
   3̸ 7̸    Subtract ones.
 − 1 8
     9
```

The borrowing here is like exchanging 3 dimes and 7 pennies for 2 dimes and 17 pennies.

Subtract.

11.   7 8
    − 6 4

12.   2 9
    −   9

13.   5 4 2
    − 3 0 1

14.   6 8 9 6
    − 4 8 7 1

Answers on page A-30

Subtract.

15. 4 6
     − 2 9

16. 7 4
     − 3 8

Subtract.

17. 6 4 6
     − 1 9 2

18. 7 3 3
     − 4 8 3

Answers on page A-30

Developmental Units

$\overset{2}{}\overset{17}{}$
$\overset{\diagup}{3}\overset{\diagup}{7}$ Subtract tens.
− 1 8
1 9

$\overset{2}{}\overset{17}{}$
$\overset{\diagup}{3}\overset{\diagup}{7}$ You should write only this.
− 1 8
1 9

Do Exercises 15 and 16.

Borrowing Hundreds

Example 8 Subtract: 538 − 275.

  5 3 8   Subtract ones.
− 2 7 5
         3

  5 3 8   Try to subtract tens: 3 tens − 7 tens is not a whole number.
− 2 7 5
     ? 3

$\overset{4}{}\overset{13}{}$
$\overset{\diagup}{5}\overset{\diagup}{3}$ 8   Borrow a hundred. That is, 1 hundred = 10 tens, and
− 2 7 5   10 tens + 3 tens = 13 tens. Write 4 above the hundreds
         3   column and 13 above the tens.

| The borrowing is like exchanging 5 dollars, 3 dimes, and 8 pennies for 4 dollars, 13 dimes, and 8 pennies. |

$\overset{4}{}\overset{13}{}$
$\overset{\diagup}{5}\overset{\diagup}{3}$ 8   Subtract tens.
− 2 7 5
     6   3

$\overset{4}{}\overset{13}{}$
$\overset{\diagup}{5}\overset{\diagup}{3}$ 8   Subtract hundreds.
− 2 7 5
  2 6   3

$\overset{4}{}\overset{13}{}$
$\overset{\diagup}{5}\overset{\diagup}{3}$ 8   You should write only this.
− 2 7 5
  2 6   3

Do Exercises 17 and 18.

Borrowing More Than Once

Sometimes we must borrow more than once.

Example 9 Subtract: 672 − 394.

$$\begin{array}{r} 612 \\ 6\,\cancel{7}\,\cancel{2} \\ -3\,9\,4 \\ \hline 8 \end{array}$$ Borrowing a ten to subtract ones

$$\begin{array}{r} 16 \\ 5\cancel{6}12 \\ \cancel{6}\,\cancel{7}\,\cancel{2} \\ -3\,9\,4 \\ \hline 2\,7\,8 \end{array}$$ Borrowing a hundred to subtract tens

Do Exercises 19 and 20.

Example 10 Subtract: 6357 − 1769.

$$\begin{array}{r} 4\,17 \\ 6\,3\,\cancel{5}\,\cancel{7} \\ -1\,7\,6\,9 \\ \hline 8 \end{array}$$ We cannot subtract 9 from 7.
We borrow a ten.

$$\begin{array}{r} 14 \\ 2\,\cancel{4}\,17 \\ 6\,\cancel{3}\,\cancel{5}\,\cancel{7} \\ -1\,7\,6\,9 \\ \hline 8\,8 \end{array}$$ We cannot subtract 6 tens from 4 tens.
We borrow a hundred.

$$\begin{array}{r} 12\,14 \\ 5\,\cancel{2}\,\cancel{4}\,17 \\ \cancel{6}\,\cancel{3}\,\cancel{5}\,\cancel{7} \\ -1\,7\,6\,9 \\ \hline 4\,5\,8\,8 \end{array}$$ We cannot subtract 7 hundreds from 2 hundreds.
We borrow a thousand.

We can always check by adding the answer to the number being subtracted.

Example 11 Subtract: 8341 − 2673. Check by adding.

We check by adding 5668 and 2673.

$$\begin{array}{r} 12\,13 \\ 7\,\cancel{2}\,\cancel{3}\,11 \\ \cancel{8}\,\cancel{3}\,\cancel{4}\,\cancel{1} \\ -2\,6\,7\,3 \\ \hline 5\,6\,6\,8 \end{array}$$

Check:
$$\begin{array}{r} 1\,1\,1 \\ 5\,6\,6\,8 \\ +2\,6\,7\,3 \\ \hline 8\,3\,4\,1 \end{array}$$

Do Exercises 21 and 22.

Zeros in Subtraction

Before subtracting, note the following:

50 is 5 tens;

70 is 7 tens.

Then

100 is 10 tens;

200 is 20 tens.

Do Exercises 23–26.

Subtract.

19. $\begin{array}{r} 5\,6\,3 \\ -1\,8\,7 \\ \hline \end{array}$

20. $\begin{array}{r} 7\,3\,3 \\ -4\,8\,8 \\ \hline \end{array}$

Subtract. Check by adding.

21. $\begin{array}{r} 4\,2\,3\,6 \\ -1\,6\,7\,9 \\ \hline \end{array}$

22. $\begin{array}{r} 7\,5\,4\,1 \\ -3\,8\,6\,7 \\ \hline \end{array}$

Complete.

23. 80 = _____ tens

24. 60 = _____ tens

25. 300 = _____ tens

26. 900 = _____ tens

Answers on page A-30

S Subtraction

Complete.

27. 5000 = _____ tens

28. 9000 = _____ tens

29. 5380 = _____ tens

30. 6770 = _____ tens

Subtract.

31. 60
   −18

32. 480
   −256

Subtract.

33. 602
   −464

34. 408
   −364

Subtract.

35. 4006
   −1238

36. 9001
   −7804

Subtract.

37. 3000
   −1754

38. 8017
   −3289

To the student: *If you had trouble with Section 1.3 and have studied Developmental Unit S, you should go back and work through Section 1.3 after completing Exercise Set S.*

Answers on page A-30

Developmental Units

Also,

    230 is 2 hundreds + 3 tens

    or 20 tens + 3 tens

    or 23 tens.

Similarly,

    1000 is 100 tens;

    2000 is 200 tens;

    4670 is 467 tens.

Do Exercises 27–30.

Example 12   Subtract: 50 − 37.

```
  4 10
  5 0    We have 5 tens.
− 3 7    We keep 4 of them in the tens column.
  1 3    We put 1 ten, or 10 ones, with the ones.
```

Do Exercises 31 and 32.

Example 13   Subtract: 803 − 547.

```
  7 9 13
  8 0 3    We have 8 hundreds, or 80 tens.
− 5 4 7    We keep 79 tens.
  2 5 6    We put 1 ten, or 10 ones, with the ones.
```

Do Exercises 33 and 34.

Example 14   Subtract: 9003 − 2789.

```
  8 9 9 13
  9 0 0 3    We have 9 thousands, or 900 tens.
− 2 7 8 9    We keep 899 tens.
  6 2 1 4    We put 1 ten, or 10 ones, with the ones.
```

Do Exercises 35 and 36.

Examples   Subtract.

15.
```
  4 9 9 10
  5 0 0 0
− 2 8 6 1
  2 1 3 9
```

16.
```
         10
  4 9 0 13
  5 0 1 3    We have 5 thousands,
− 1 8 5 7    or 49 hundreds and
  3 1 5 6    10 tens.
```

Do Exercises 37 and 38.

Exercise Set S

a Subtract. Try to do these mentally.

1. 7 − 0	**2.** 8 − 8	**3.** 7 − 7	**4.** 8 − 3	**5.** 5 − 2
6. 16 − 8	**7.** 17 − 9	**8.** 12 − 6	**9.** 11 − 4	**10.** 12 − 9
11. 14 − 7	**12.** 18 − 9	**13.** 13 − 7	**14.** 15 − 9	**15.** 9 − 7

16. 7 − 3   **17.** 4 − 1   **18.** 2 − 0   **19.** 3 − 3   **20.** 6 − 3

21. 7 − 6   **22.** 9 − 8   **23.** 10 − 3   **24.** 6 − 6   **25.** 11 − 7

26. 12 − 8   **27.** 5 − 0   **28.** 4 − 0   **29.** 13 − 9   **30.** 14 − 9

31. 11 − 2   **32.** 12 − 3   **33.** 16 − 9   **34.** 18 − 9   **35.** 11 − 5

36. 10 − 4   **37.** 10 − 8   **38.** 14 − 8   **39.** 15 − 8   **40.** 10 − 2

b Subtract.

41. 64 − 31	**42.** 55 − 34	**43.** 548 − 301	**44.** 596 − 403	**45.** 700 − 200

Exercise Set S

46. 765 − 111
47. 525 − 323
48. 747 − 130
49. 988 − 700
50. 9450 − 8230

51. 6552 − 4321
52. 7547 − 3421
53. 5875 − 2111
54. 38,695 − 37,004
55. 67,899 − 66,673

56. 99,999 − 1
57. 56,780 − 56,770
58. 42,111 − 32,010
59. 77,654 − 66,611
60. 23,456 − 12,345

c Subtract.

61. 93 − 28
62. 42 − 13
63. 86 − 78
64. 98 − 89
65. 625 − 317

66. 735 − 609
67. 853 − 236
68. 961 − 747
69. 787 − 698
70. 6769 − 2367

71. 6431 − 2876
72. 7654 − 1765
73. 5246 − 2859
74. 6328 − 2679
75. 7641 − 3809

76. 8743 − 599
77. 12,647 − 4,897
78. 16,222 − 5,777
79. 46,781 − 12,988
80. 470 − 189

81. 690 − 235
82. 703 − 132
83. 6406 − 258
84. 2309 − 109
85. 3406 − 1293

86. 6807 − 3059
87. 8000 − 2794
88. 8002 − 6543
89. 38,000 − 37,695
90. 16,043 − 11,588

Developmental Units

M Multiplication

a Basic Multiplication

To multiply, we begin with two numbers, called **factors**, and get a third number, called a **product**. Multiplication can be explained by counting. The product 3×5 can be found by counting out 3 sets of 5 objects each, joining them (in a rectangular array if desired), and counting all the objects.

We can also think of multiplication as repeated addition.

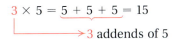

Objectives

a. Multiply any two of the numbers 0, 1, 2, 3, 4, 5, 6, 7, 8, 9.

b. Multiply by multiples of 10, 100, and 1000.

c. Multiply larger numbers by 0, 1, 2, 3, 4, 5, 6, 7, 8, 9.

d. Multiply by multiples of 10, 100, and 1000.

Examples Multiply. If you have trouble, think either of putting sets of objects together in a rectangular array or of repeated addition.

1. $5 \times 6 = 30$

    6
    $\underline{\times\ 5}$
    30

2. $8 \times 4 = 32$

    4
    $\underline{\times\ 8}$
    32

Do Exercises 1–4.

Multiplying by 0

How do we multiply by 0? Consider $4 \cdot 0$. Using repeated addition, we see that

$$4 \cdot 0 = \underbrace{0 + 0 + 0 + 0}_{\text{4 addends of 0}} = 0.$$

We can also think of this using sets. That is, $4 \cdot 0$ is 4 sets with 0 objects in each set, so the total is 0.

Consider $0 \cdot 4$. Using repeated addition, we say that this is 0 addends of 4, which is 0. Using sets, we say that this is 0 sets with 4 objects in each set, which is 0. Thus we have the following.

▶ Multiplying by 0 gives 0.

Examples Multiply.

3. $13 \times 0 = 0$

    0
    $\underline{\times 13}$
    0

4. $0 \cdot 11 = 0$

    11
    $\underline{\times\ 0}$
    0

5. $0 \cdot 0 = 0$

    0
    $\underline{\times 0}$
    0

Do Exercises 5 and 6.

Multiply. Think of joining sets in a rectangular array or of repeated addition.

1. $7 \cdot 8$ (The dot "·" means the same as "×".)

2. 9
   $\underline{\times\ 4}$

3. $4 \cdot 7$

4. 7
   $\underline{\times\ 6}$

Multiply.

5. $8 \cdot 0$

6. 17
   $\underline{\times\ \ 0}$

Answers on page A-30

Multiply.

7. 8 · 1

8. 2 3
   ×   1

9. Complete the table.

×	2	3	4	5
2				
3			12	
4				
5		15		
6				

10.

×	6	7	8	9
5				
6			48	
7				
8		56		
9				

Answers on page A-30

Developmental Units

Multiplying by 1

How do we multiply by 1? Consider 5 · 1. Using repeated addition, we see that

$$5 \cdot 1 = \underbrace{1 + 1 + 1 + 1 + 1}_{5 \text{ addends of } 1} = 5.$$

We can also think of this using sets. That is, 5 · 1 is 5 sets with 1 object in each set, so the total is 5.

Consider 1 · 5. Using repeated addition, we say that this is 1 addend of 5, which is 5. Using sets, we say that this is 1 set of 5 objects, which is 5. Thus we have the following.

> Multiplying a number by 1 does not change the number:
> $a \cdot 1 = 1 \cdot a = a$.
> We say that 1 is the **multiplicative identity.**

This is a very important property.

Examples Multiply.

6. 13 · 1 = 13
     1
  × 13
    13

7. 1 · 7 = 7
    7
  × 1
    7

8. 1 · 1 = 1
    1
  × 1
    1

Do Exercises 7 and 8.

You should be able to multiply any of the numbers 0, 1, 2, 3, 4, 5, 6, 7, 8, 9. Multiplying by 0 and 1 is easy. The rest of the products are listed in the following table.

×	2	3	4	5	6	7	8	9
2	4	6	8	10	12	14	16	18
3	6	9	12	15	18	21	24	27
4	8	12	16	20	24	28	32	36
5	10	15	20	25	30	35	40	45
6	12	18	24	30	36	42	48	54
7	14	21	28	35	42	49	56	63
8	16	24	32	40	48	56	64	72
9	18	27	36	45	54	63	72	81

5 × 7 = 35
Find 5 at the left, and 7 at the top.

8 · 4 = 32
Find 8 at the left, and 4 at the top.

It is *very* important that you have the basic multiplication facts *memorized*. If you do not, you will always have trouble with multiplication.

The *commutative law of multiplication* says that we can multiply numbers in any order. Thus you need to learn only about half the table, as shown by the shading.

Do Exercises 9 and 10.

b Multiplying Multiples of 10, 100, and 1000

We now move to a more gradual, conceptual development of the multiplication procedure you considered in Section 1.5. It is intended to provide you with a greater understanding so that your skill level will increase.

We begin by considering multiplication by multiples of 10, 100, and 1000. These are numbers such as 10, 20, 30, 100, 400, 1000, and 7000.

Multiplying by a Multiple of 10

We know that

$50 = 5$ tens $340 = 34$ tens   and $2340 = 234$ tens
$= 5 \cdot 10,$ $= 34 \cdot 10,$ $= 234 \cdot 10.$

Turning this around, we see that to multiply any number by 10, all we need do is write a 0 on the end of the number.

▶ To multiply a number by 10, write 0 on the end of the number.

Examples Multiply.

9. $10 \cdot 6 = 60$
10. $10 \cdot 47 = 470$
11. $10 \cdot 583 = 5830$

Do Exercises 11–15.

Let's find $4 \cdot 90$. This is $4 \cdot (9$ tens$)$, or 36 tens. The procedure is the same as multiplying 4 and 9 and writing a 0 on the end. Thus, $4 \cdot 90 = 360$.

Examples Multiply.

12. $5 \cdot 70 = 350$ $5 \cdot 7$, then write a 0
13. $8 \cdot 80 = 640$
14. $5 \cdot 60 = 300$

Do Exercises 16 and 17.

Multiplying by a Multiple of 100

Note the following:

$300 = 3$ hundreds $4700 = 47$ hundreds   and $56,800 = 568$ hundreds
$= 3 \cdot 100,$ $= 47 \cdot 100,$ $= 568 \cdot 100.$

Turning this around, we see that to multiply any number by 100, all we need do is write two 0's on the end of the number.

▶ To multiply a number by 100, write two 0's on the end of the number.

Multiply.
11. $10 \cdot 7$

12. $10 \cdot 45$

13. $10 \cdot 273$

14. $10 \cdot 10$

15. $10 \cdot 100$

Multiply.
16.   70
  × 8

17.   60
  × 6

Answers on page A-30

M Multiplication

Multiply.

18. 100 · 7   **19.** 100 · 23

20. 100 · 723   **21.** 100 · 100

22. 100 · 1000

Multiply.

23.   7 0 0
     ×   8

24.   4 0 0
     ×   4

Multiply.

25. 1000 · 9   **26.** 1000 · 852

27. 1000 · 10   **28.** 3 · 4000

29. 9 · 8000

Answers on page A-30

Developmental Units

Examples Multiply.

15. 100 · 6 = 600

16. 100 · 39 = 3900

17. 100 · 448 = 44,800

Do Exercises 18–22.

Let's find 4 · 900. This is 4 · (9 hundreds), or 36 hundreds. The procedure is the same as multiplying 4 and 9 and writing two 0's on the end. Thus, 4 · 900 = 3600.

Examples Multiply.

18. 6 · 800 = 4800   — 6 · 8, then write 00

19. 9 · 700 = 6300

20. 5 · 500 = 2500

Do Exercises 23 and 24.

Multiplying by a Multiple of 1000

Note the following:

   6000 = 6 thousands   and   19,000 = 19 thousands
        = 6 · 1000   = 19 · 1000.

Turning this around, we see that to multiply any number by 1000, all we need do is write three 0's on the end of the number.

> To multiply a number by 1000, write three 0's on the end of the number.

Examples Multiply.

21. 1000 · 8 = 8000

22. 2000 · 13 = 26,000

23. 1000 · 567 = 567,000

Do Exercises 25–29.

Multiplying Multiples by Multiples

Let's multiply 50 and 30. This is 50 · (3 tens), or 150 tens, or 1500. The procedure is the same as multiplying 5 and 3 and writing two 0's on the end.

> To multiply multiples of tens, hundreds, thousands, and so on:
> a) Multiply the one-digit numbers.
> b) Count the number of zeros.
> c) Write that many 0's on the end.

Examples Multiply.

24.
80 — 1 zero at end
$\times\ 60$ — 1 zero at end
$\overline{48\text{00}}$
↑ — 6 · 8, then write 00

25.
800 — 2 zeros at end
$\times\ 60$ — 1 zero at end
$\overline{48,\text{000}}$
↑ — 6 · 8, then write 000

26.
800 — 2 zeros at end
$\times\ 600$ — 2 zeros at end
$\overline{480,\text{000}}$
↑ — 6 · 8, then write 0,000

27.
800 — 2 zeros at end
$\times\ 50$ — 1 zero at end
$\overline{40,\text{000}}$
↑ — 5 · 8, then write 000

Do Exercises 30–33.

c Multiplying Larger Numbers

The product 3×24 can be represented as

$$3 \times (2 \text{ tens} + 4) = (2 \text{ tens} + 4) + (2 \text{ tens} + 4) + (2 \text{ tens} + 4)$$
$$= 6 \text{ tens} + 12$$
$$= 6 \text{ tens} + 1 \text{ ten} + 2$$
$$= 7 \text{ tens} + 2$$
$$= 72.$$

We multiply the 4 ones by 3, getting 12
We multiply the 2 tens by 3, getting $+60$
Then we add: 72

Example 28 Multiply: 3×24.

$$
\begin{array}{r}
2\ 4 \\
\times 3 \\
\hline
1\ 2 \\
6\ 0 \\
\hline
7\ 2
\end{array}
$$
← Multiply the 4 ones by 3.
← Multiply the 2 tens by 3.
← Add.

Do Exercises 34–36.

Example 29 Multiply: 5×734.

$$
\begin{array}{r}
7\ 3\ 4 \\
\times 5 \\
\hline
2\ 0 \\
1\ 5\ 0 \\
3\ 5\ 0\ 0 \\
\hline
3\ 6\ 7\ 0
\end{array}
$$
← Multiply the 4 ones by 5.
← Multiply the 3 tens by 5.
← Multiply the 7 hundreds by 5.
← Add.

Do Exercises 37 and 38.

Multiply.

30. $9\ 0\ 0\ 0$
$\times6$

31. $8\ 0$
$\times\ 7\ 0$

32. $8\ 0\ 0$
$\times7\ 0$

33. $6\ 0\ 0$
$\times3\ 0$

Multiply.

34. $1\ 4$
$\times2$

35. $5\ 8$
$\times2$

36. $3\ 7$
$\times4$

Multiply.

37. $8\ 2\ 3$
$\times6$

38. $1\ 3\ 4\ 8$
$\times5$

Answers on page A-30

M Multiplication

Multiply using the short form.

39. 58
   × 2

40. 37
   × 4

41. 823
   ×   6

42. 1348
   ×   5

Multiply.

43. 746
   × 8

44. 746
   × 80

45. 746
   ×800

To the student: *If you had trouble with Section 1.5 and have studied Developmental Unit M, you should go back and work through Section 1.5 after completing Exercise Set M.*

Answers on page A-30

Developmental Units

Let's look at Example 29 again. Instead of writing each product on a separate line, we can use a shorter form.

Example 30 Multiply: 5×734.

$$\begin{array}{r} \overset{2}{}73\boxed{4} \\ \times 5 \\ \hline 0 \end{array}$$

Multiply the ones by 5: $5 \cdot (4 \text{ ones}) = 20 \text{ ones} = 2 \text{ tens} + 0 \text{ ones}$. Write 0 in the ones column and 2 above the tens.

$$\begin{array}{r} \overset{1}{}\overset{2}{}7\boxed{3}4 \\ \times 5 \\ \hline 70 \end{array}$$

Multiply the 3 tens by 5 and add 2 tens: $5 \cdot (3 \text{ tens}) = 15 \text{ tens}$, $15 \text{ tens} + 2 \text{ tens} = 17 \text{ tens} = 1 \text{ hundred} + 7 \text{ tens}$. Write 7 in the tens column and 1 above the hundreds.

$$\begin{array}{r} \overset{1}{}\overset{2}{}\boxed{7}34 \\ \times 5 \\ \hline 3670 \end{array}$$

Multiply the 7 hundreds by 5 and add 1 hundred: $5 \cdot (7 \text{ hundreds}) = 35 \text{ hundreds}$, $35 \text{ hundreds} + 1 \text{ hundred} = 36 \text{ hundreds}$.

$$\left.\begin{array}{r} \overset{1}{}\overset{2}{}734 \\ \times 5 \\ \hline 3670 \end{array}\right\} \text{You should write only this.}$$

Try to avoid writing the reminders unless necessary.

Do Exercises 39–42.

d Multiplying by Multiples of 10, 100, and 1000

To multiply 327 by 50, we multiply by 10 (write a 0), and then multiply 327 by 5.

$$\begin{array}{r} 327 \\ \times 5\boxed{0} \\ \hline 16{,}35\boxed{0} \end{array}$$

← Write a 0.
↑ Multiply $5 \cdot 327$.

Example 31 Multiply: 400×289.

$$\begin{array}{r} 289 \\ \times 4\boxed{00} \\ \hline 00 \end{array}$$

← Write two 0's.

$$\begin{array}{r} 289 \\ \times 400 \\ \hline 115{,}600 \end{array}$$

Multiply 4 and 289:

$$\begin{array}{r} \overset{3}{}\overset{3}{}289 \\ \times 4 \\ \hline 1156 \end{array}$$

$$\left.\begin{array}{r} \overset{3}{}\overset{3}{}289 \\ \times 400 \\ \hline 115{,}600 \end{array}\right\} \text{You should write only this.}$$

Do Exercises 43–45.

Exercise Set M

a Multiply. Try to do these mentally.

1. 3 × 4	2. 6 × 0	3. 7 × 1	4. 0 × 2	5. 10 × 1	6. 6 × 5
7. 5 × 2	8. 9 × 7	9. 9 × 6	10. 2 × 6	11. 7 × 0	12. 8 × 9
13. 1 × 8	14. 8 × 0	15. 4 × 7	16. 3 × 8	17. 5 × 9	18. 2 × 9
19. 0 × 7	20. 5 × 7	21. 9 × 5	22. 5 × 8	23. 0 × 0	24. 2 × 8

25. 5 · 5   26. 9 · 9   27. 1 · 1   28. 0 · 0   29. 2 · 2

30. 6 · 6   31. 1 · 8   32. 0 · 1   33. 3 · 9   34. 2 · 9

35. 6 · 0   36. 10 · 1   37. 6 · 8   38. 9 · 6   39. 8 · 0

40. 9 · 8   41. 3 · 5   42. 1 · 8   43. 1 · 9   44. 2 · 1

45. 8 · 4   46. 3 · 2   47. 5 · 3   48. 1 · 6   49. 4 · 2

50. 4 · 5   51. 5 · 4   52. 4 · 4   53. 5 · 2   54. 8 · 0

Exercise Set M

b Multiply.

55. 10 × 8	56. 7 ×10	57. 20 × 8	58. 30 × 7	59. 45 ×10
60. 78 ×10	61. 80 × 7	62. 90 × 4	63. 100 × 8	64. 100 × 3
65. 100 × 9	66. 100 × 10	67. 3457 × 100	68. 400 × 3	69. 700 × 7
70. 500 × 8	71. 100 ×100	72. 1000 × 7	73. 1000 × 9	74. 1000 × 2
75. 457 ×1000	76. 6769 ×1000	77. 2000 × 9	78. 5000 × 4	79. 6000 × 8
80. 8000 × 2	81. 3000 × 2	82. 1000 ×1000	83. 40 ×30	84. 20 ×10
85. 80 ×50	86. 50 ×50	87. 400 × 30	88. 200 × 30	89. 700 × 90
90. 400 ×300	91. 4000 × 200	92. 6000 × 20	93. 4000 ×4000	94. 8000 × 10

c Multiply.

95. 49 × 3	96. 74 × 6	97. 593 × 5	98. 609 × 8	99. 899 × 7
100. 865 × 4	101. 8118 × 2	102. 6754 × 2	103. 43,777 × 2	104. 32,564 × 6

d Multiply.

105. 58 ×60	106. 93 ×30	107. 42 ×80	108. 78 ×90	109. 346 × 60
110. 267 × 40	111. 897 ×400	112. 366 ×300	113. 834 ×700	114. 333 ×900
115. 5673 ×2000	116. 4678 ×5000	117. 6788 ×9000	118. 9129 ×8000	

Developmental Units

D Division

a Basic Division

Division can be explained by arranging a set of objects in a rectangular array. This can be done in two ways.

Example 1 Divide: $18 \div 6$.

METHOD 1 We can do this division by taking 18 objects and determining into how many rows, each with 6 objects, we can arrange the objects.

3 rows of 6 objects

Since there are 3 rows of 6 objects, we have

$18 \div 6 = 3$.

METHOD 2 We can also arrange the objects into 6 rows and determine how many objects are in each row.

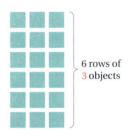
6 rows of 3 objects

Since there are 3 objects in each of the 6 rows, we have

$18 \div 6 = 3$.

We can also use fractional notation for division. That is,

$18 \div 6 = 18/6 = \dfrac{18}{6}$.

Examples Divide.

2. $9 \overline{)36}$ with quotient 4

   Think: 36 objects: How many rows, each with 9 objects? or 36 objects: How many objects in each of 9 rows?

3. $42 \div 7 = 6$

4. $\dfrac{24}{3} = 8$

Do Exercises 1–4.

Objectives

a Find basic quotients such as $20 \div 5$, $56 \div 7$, and so on.

b Divide using the "guess, multiply, and subtract" method.

c Divide by estimating multiples of thousands, hundreds, tens, and ones.

Divide.

1. $24 \div 6$

2. $64 \div 8$

3. $\dfrac{63}{7}$

4. $\dfrac{27}{9}$

Answers on page A-30

D Division

643

For each multiplication fact, write two division facts.

5. $6 \cdot 2 = 12$

6. $7 \times 6 = 42$

Answers on page A-30

Developmental Units

In Developmental Unit M, you memorized a multiplication table. That table will enable you to divide as well. First, let's recall how multiplication and division are related.

A multiplication: $5 \cdot 4 = 20$.

Two related divisions:

A.  $20 \div 5 = 4$.

4 rows of 5 objects

B. 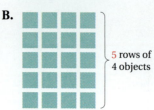 $20 \div 4 = 5$.

5 rows of 4 objects

Since we know that

$5 \cdot 4 = 20$,   **A basic multiplication fact**

we also know the two division facts

$20 \div 5 = 4$ and $20 \div 4 = 5$.

Example 5 From $7 \cdot 8 = 56$, write two division facts.

a) We have

$7 \cdot 8 = 56$ **Division sentence**

$7 = 56 \div 8$.   **Related multiplication sentence**

b) We also have

$7 \cdot 8 = 56$ **Division sentence**

$8 = 56 \div 7$.   **Related multiplication sentence**

Do Exercises 5 and 6.

We can use the idea that division is defined in terms of multiplication to do basic divisions.

Example 6 Find: $35 \div 5$.

To find $35 \div 5$, we ask, "5 times what number is 35?"

$5 \cdot \square = 35$

×	2	3	4	5	6	7	8	9
2	4	6	8	10	12	14	16	18
3	6	9	12	15	18	21	24	27
4	8	12	16	20	24	28	32	36
5	10	15	20	25	30	35	40	45
6	12	18	24	30	36	42	48	54
7	14	21	28	35	42	49	56	63
8	16	24	32	40	48	56	64	72
9	18	27	36	45	54	63	72	81

$35 \div 5 = 7$

Using the multiplication table above, we find 35 inside the table and 5 at the left. Then we read the answer 7 from the top. Thus we have $35 \div 5 = 7$. Strive to do this kind of thinking mentally as fast as you can, without having to use the table.

Do Exercises 7–10.

Division by 1

Note that

$3 \div 1 = 3$ because $3 = 3 \cdot 1$; $\quad \dfrac{14}{1} = 14$ because $14 = 14 \cdot 1$.

▶ Any number divided by 1 is that same number:

$a \div 1 = \dfrac{a}{1} = a.$

Examples Divide.

7. $\dfrac{8}{1} = 8$

8. $6 \div 1 = 6$

9. $34 \div 1 = 34$

Do Exercises 11–13.

Division by 0

Why can't we divide by 0? Suppose the number 4 could be divided by 0. Then if $\square$ were the answer,

$4 \div 0 = \square$

and since 0 times any number is 0, we would have

$4 = \square \cdot 0 = 0.$ False!

Divide.

7. $28 \div 4$

8. $81 \div 9$

9. $\dfrac{16}{2}$

10. $\dfrac{54}{6}$

Divide.

11. $6 \div 1$

12. $\dfrac{13}{1}$

13. $1 \div 1$

Answers on page A-30

D Division

Divide, if possible. If not possible, write "not defined."

14. $\dfrac{8}{4}$

15. $\dfrac{5}{0}$

16. $\dfrac{0}{5}$

17. $\dfrac{0}{0}$

18. $12 \div 0$

19. $100 \div 10$

20. $\dfrac{5}{3-3}$

21. $\dfrac{8-8}{4}$

Answers on page A-30

Developmental Units

646

Suppose 12 could be divided by 0. If ☐ were the answer,

$$12 \div 0 = \square$$

and since 0 times any number is 0, we would have

$$12 = \square \cdot 0 = 0. \quad \text{False!}$$

Thus, $a \div 0$ would be some number ☐ such that $a = \square \cdot 0 = 0$. So the only possible number that could be divided by 0 would be 0 itself.

But such a division would give us any number we wish, for

$$\left. \begin{array}{l} 0 \div 0 = 8 \quad \text{because} \quad 0 = 8 \cdot 0; \\ 0 \div 0 = 3 \quad \text{because} \quad 0 = 3 \cdot 0; \\ 0 \div 0 = 7 \quad \text{because} \quad 0 = 7 \cdot 0. \end{array} \right\} \text{All true!}$$

We avoid the preceding difficulties by agreeing to exclude division by 0.

> Division by 0 is not defined. (We agree not to divide by 0.)

Dividing 0 by Other Numbers

Note that

$$0 \div 3 = 0 \quad \text{because} \quad 0 = 0 \cdot 3; \qquad \frac{0}{12} = 0 \quad \text{because} \quad 0 = 0 \cdot 12.$$

> Zero divided by any number greater than 0 is 0:
> $$\frac{0}{a} = 0, \quad a > 0.$$

Examples Divide.

10. $0 \div 8 = 0$

11. $0 \div 22 = 0$

12. $\dfrac{0}{9} = 0$

Do Exercises 14–21.

Division of a Number by Itself

Note that

$$3 \div 3 = 1 \quad \text{because} \quad 3 = 1 \cdot 3; \qquad \frac{34}{34} = 1 \quad \text{because} \quad 34 = 1 \cdot 34.$$

> Any number greater than 0 divided by itself is 1:
> $$\frac{a}{a} = 1, \quad a > 0.$$

Examples Divide.

13. $8 \div 8 = 1$

14. $27 \div 27 = 1$

15. $\frac{32}{32} = 1$

Do Exercises 22–27.

b Dividing by "Guess, Multiply, and Subtract"

To understand the process of division, we use a method known as "guess, multiply, and subtract." We do this to develop a shorter way that is understandable.

Example 16 Divide $275 \div 4$. Use "guess, multiply, and subtract."

We *guess* a partial quotient of 35. We could guess *any* number—say, 4, 16, or 30. We *multiply* and *subtract* as follows:

```
        3 5  ← Partial quotient
    4 ) 2 7 5
        1 4 0  ← 35 · 4
        1 3 5  ← Remainder
```

Next, we look at 135 and *guess* another partial quotient—say, 20. Then we *multiply* and *subtract*:

```
        2 0    ← Second partial quotient
        3 5
    4 ) 2 7 5
        1 4 0
        1 3 5
          8 0  ← 20 · 4
          5 5  ← Remainder
```

Next, we look at 55 and *guess* another partial quotient—say, 13. Then we *multiply* and *subtract*:

```
        1 3    ← Third partial quotient
        2 0
        3 5
    4 ) 2 7 5
        1 4 0
        1 3 5
          8 0
          5 5
          5 2  ← 13 · 4
            3  ← Remainder is less than 4
```

Divide.

22. $23 \div 23$ **23.** $\frac{67}{67}$

24. $\frac{41}{41}$ **25.** $17 \div 17$

26. $17 \div 1$ **27.** $\frac{54}{54}$

Divide using the "guess, multiply, and subtract" method.

28. $6 \overline{) 4\ 5\ 4}$

29. $3\ 2 \overline{) 7\ 4\ 7}$

Answers on page A-30

D Division

Divide using the "guess, multiply, and subtract" method.

30. 7) 6 7 8 9

31. 6 4) 3 0 1 2

Since we cannot subtract any more 4's, the division is finished. We add our partial quotients.

```
      6 8  ← Quotient (sum of guesses)
      1 3
      2 0
      3 5
 4 ) 2 7 5
    1 4 0
    1 3 5
      8 0
      5 5
      5 2
        3
```

CHECK: $275 = (4 \times 68) + 3$
$275 \stackrel{?}{=} 272 + 3$
275

The answer is 68 R 3. This tells us that with 275 objects, we could make 68 rows of 4 and have 3 left over.

The partial quotients (guesses) can be made in any manner so long as subtraction is possible.

Do Exercises 28 and 29 on the preceding page.

Example 17 Divide: $1506 \div 32$.

```
           4 7  ← Quotient (sum of guesses)
          2 0 ⎫
            2 ⎪
          2 0 ⎬ ← Guesses
            5 ⎭
   3 2 ) 1 5 0 6
         1 6 0  ← 5 · 32
         1 3 4 6
           6 4 0  ← 20 · 32
             7 0 6
               6 4  ← 2 · 32
             6 4 2
             6 4 0  ← 20 · 32
                 2  ← Remainder: smaller than the divisor, 32
```

The answer is 47 R 2.

Remember, you can *guess any partial quotient* so long as subtraction is possible.

Do Exercises 30 and 31.

c | Dividing by Estimating Multiples

Let's refine the guessing process. We guess multiples of 10, 100, and 1000, and so on.

Answers on page A-30

Developmental Units

Example 18 Divide: 7643 ÷ 3.

a) Are there any thousands in the quotient? Yes, 3 · 1000 = 3000, which is less than 7643. To find how many thousands, we find products of 3 and multiples of 1000.

$$3 \cdot 1000 = 3000$$
$$3 \cdot 2000 = 6000$$
$$3 \cdot 3000 = 9000$$

← 7643 is here, so there are 2000 threes in the quotient.

```
      2 0 0 0
3 ) 7 6 4 3
    6 0 0 0
    -------
    1 6 4 3
```

b) Now go to the hundreds place. Are there any hundreds in the quotient?

$$3 \cdot 100 = 300$$
$$3 \cdot 200 = 600$$
$$3 \cdot 300 = 900$$
$$3 \cdot 400 = 1200$$
$$3 \cdot 500 = 1500$$
$$3 \cdot 600 = 1800$$

← 1643

```
          5 0 0
        2 0 0 0
    3 ) 7 6 4 3
        6 0 0 0
        -------
        1 6 4 3
        1 5 0 0
        -------
            1 4 3
```

c) Now go to the tens place. Are there any tens in the quotient?

$$3 \cdot 10 = 30$$
$$3 \cdot 20 = 60$$
$$3 \cdot 30 = 90$$
$$3 \cdot 40 = 120$$
$$3 \cdot 50 = 150$$

← 143

```
            4 0
          5 0 0
        2 0 0 0
    3 ) 7 6 4 3
        6 0 0 0
        -------
        1 6 4 3
        1 5 0 0
        -------
          1 4 3
          1 2 0
          -----
              2 3
```

d) Now go to the ones place. Are there any ones in the quotient?

$$3 \cdot 1 = 3$$
$$3 \cdot 2 = 6$$
$$3 \cdot 3 = 9$$
$$3 \cdot 4 = 12$$
$$3 \cdot 5 = 15$$
$$3 \cdot 6 = 18$$
$$3 \cdot 7 = 21$$
$$3 \cdot 8 = 24$$

← 23

```
        2 5 4 7
              7
            4 0
          5 0 0
        2 0 0 0
    3 ) 7 6 4 3
        6 0 0 0
        -------
        1 6 4 3
        1 5 0 0
        -------
          1 4 3
          1 2 0
          -----
            2 3
            2 1
            ---
              2
```

The answer is 2547 R 2.

Do Exercises 32 and 33.

Divide.

32. 4) 3 8 5

33. 7) 8 8 4 6

Answers on page A-30

D Division

Divide using the short form.

34. $2\overline{)648}$

35. $9\overline{)3758}$

Divide.

36. $11\overline{)415}$

37. $46\overline{)1075}$

To the student: *If you had trouble with Section 1.6 and have studied Developmental Unit D, you should go back and work through Section 1.6 after completing Exercise Set D.*

Answers on page A-30

A Short Form

Here is a shorter way to write Example 18.

Instead of this,

```
          2 5 4 7
              7
            4 0
          5 0 0
        2 0 0 0
    3 ) 7 6 4 3
        6 0 0 0
        1 6 4 3
        1 5 0 0
          1 4 3
          1 2 0
            2 3
            2 1
             2
```

Short form

we write this.
```
          2 5 4 7
    3 ) 7 6 4 3
        6 0 0 0
        1 6 4 3
        1 5 0 0
          1 4 3
          1 2 0
            2 3
            2 1
             2
```

We write a 2 above the thousands digit in the dividend to record 2000.
We write a 5 to record 500.
We write a 4 to record 40.
We write a 7 to record 7.

Do Exercises 34 and 35.

Example 19 Divide $2637 \div 41$. Use the short form.

```
            6
    4 1 ) 2 6 3 7
          2 4 6 0
            1 7 7


            6 4
    4 1 ) 2 6 3 7
          2 4 6 0
            1 7 7
            1 6 4
               1 3
```

The answer is 64 R 13.

Do Exercises 36 and 37.

In Section 1.6, the process of long division was refined with an estimation method. After doing Exercise Set D, you should restudy that procedure.

Exercise Set D

a Divide, if possible.

1. $24 \div 8$
2. $72 \div 9$
3. $28 \div 7$
4. $22 \div 22$
5. $32 \div 1$

6. $45 \div 5$
7. $14 \div 2$
8. $40 \div 8$
9. $37 \div 1$
10. $10 \div 2$

11. $36 \div 4$
12. $12 \div 3$
13. $54 \div 9$
14. $18 \div 2$
15. $20 \div 4$

16. $16 \div 2$
17. $72 \div 8$
18. $42 \div 7$
19. $12 \div 4$
20. $8 \div 4$

21. $54 \div 6$
22. $18 \div 9$
23. $9 \div 3$
24. $28 \div 4$
25. $56 \div 7$

26. $24 \div 6$
27. $14 \div 2$
28. $14 \div 7$
29. $21 \div 7$
30. $36 \div 6$

31. $8 \div 8$
32. $32 \div 8$
33. $30 \div 5$
34. $18 \div 6$
35. $49 \div 7$

36. $81 \div 9$
37. $0 \div 7$
38. $9 \div 0$
39. $16 \div 0$
40. $42 \div 6$

41. $\dfrac{48}{6}$
42. $\dfrac{35}{5}$
43. $\dfrac{9}{9}$
44. $\dfrac{45}{9}$
45. $\dfrac{0}{5}$
46. $\dfrac{0}{8}$

47. $\dfrac{6}{2}$
48. $\dfrac{3}{3}$
49. $\dfrac{8}{2}$
50. $\dfrac{7}{1}$
51. $\dfrac{5}{5}$
52. $\dfrac{6}{1}$

53. $\dfrac{2}{2}$
54. $\dfrac{25}{5}$
55. $\dfrac{4}{2}$
56. $\dfrac{24}{3}$
57. $\dfrac{0}{9}$
58. $\dfrac{0}{4}$

59. $\dfrac{40}{5}$
60. $\dfrac{3}{1}$
61. $\dfrac{16}{4}$
62. $\dfrac{9}{0}$
63. $\dfrac{32}{8}$
64. $\dfrac{9}{9}$

b Divide using the "guess, multiply, and subtract" method.

65. 4)277 **66.** 2)399 **67.** 8)737 **68.** 6)831

69. 5)8619 **70.** 3)8775 **71.** 9)7777 **72.** 8)4179

73. 7)3691 **74.** 2)5794 **75.** 20)875 **76.** 30)987

77. 21)999 **78.** 23)975 **79.** 85)7757 **80.** 54)2821

81. 111)3219 **82.** 102)5612 **83.** 346)78,910 **84.** 781)15,999

c Divide.

85. 5)105 **86.** 6)708 **87.** 9)820 **88.** 3)965

89. 5)4823 **90.** 8)5437 **91.** 7)9298 **92.** 41)1115

93. 46)1058 **94.** 24)7722 **95.** 38)8522 **96.** 81)2247

97. 94)2153 **98.** 82)4064 **99.** 117)44,902 **100.** 740)55,200

Developmental Units

Answers

Diagnostic Pretest, p. xxiii

1. [1.2b] 4313   2. [1.7b] 28   3. [1.5b] 16,761
4. [1.8a] 23; 2   5. [2.6a] 150   6. [2.1d] 17
7. [2.2a] -13 8. [2.2a] -3 9. [2.3a] -17
10. [2.3a] 5   11. [2.4a] 54   12. [2.5b] 7
13. [2.7c] 70 ft   14. [2.7b] $7x + 13$ 15. [2.8a] -17
16. [2.8b] 4   17. [3.6a] $-\frac{3}{2}$ 18. [3.8a] -40
19. [3.5a] $\frac{35}{40}$ 20. [3.7c] $\frac{3}{20}$ 21. [4.3b] $\frac{1}{10}$
22. [4.7b] $4\frac{2}{3}$ 23. [4.6a], [4.7b] $3\frac{1}{4}$ 24. [4.4a] $\frac{12}{25}$
25. [5.2a] 8.976   26. [5.4a] -101.8
27. [5.2d] $2.5x - 7.3y$ 28. [5.7a] 2.4
29. [6.4b]

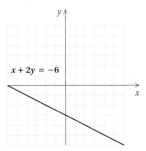

30. [6.5a, b, c] 14; 13; 12   31. [7.4a] 4   32. [7.5a] 7.5 ft
33. [8.1c] 62.5%   34. [8.1b] 0.024   35. [8.4b] 20%
36. [8.6a] $2090   37. [9.1a] 72   38. [9.1b] 0.00004
39. [9.7a] 4   40. [9.7b] 0.005   41. [9.3a] 18
42. [9.6b] 2.5   43. [9.2b] 36π cm²; 12π cm
44. [9.6a] 125 m³   45. [10.1a] $8x^3 - 3x^2 - 10$
46. [10.1c] $4x^4 - 8x^3 - x^2 - 2$
47. [10.3a] $2x^2 + 5x - 3$ 48. [10.2c] $2x^3(5x^2 + 4x - 6)$
49. [10.4a] 1   50. [10.4b] $10a^4b^5$

Chapter 1

Pretest: Chapter 1, p. 2

1. [1.1c] Three million, seventy-eight thousand, fifty-nine   2. [1.1a] 6 thousands + 9 hundreds + 8 tens + 7 ones   3. [1.1d] 2,047,398,589
4. [1.1e] 6 ten thousands   5. [1.4a] 956,000
6. [1.5c] 60,000   7. [1.2b] 10,216   8. [1.3d] 4108
9. [1.5b] 22,976   10. [1.6c] 503 R 11   11. [1.4c] <
12. [1.4c] >   13. [1.7b] 5542   14. [1.7b] 22
15. [1.7b] 34   16. [1.7b] 25   17. [1.8a] 12 lb
18. [1.8a] 128   19. [1.8a] 22,981,000
20. [1.8a] 2292 sq ft   21. [1.9b] 25   22. [1.9b] 64
23. [1.9c] 0   24. [1.9d] 0

Margin Exercises, Section 1.1, pp. 3–6

1. 1 thousand + 8 hundreds + 0 tens + 5 ones, or 1 thousand + 8 hundreds + 5 ones   2. 3 ten thousands + 6 thousands + 2 hundreds + 2 tens + 3 ones   3. 3 thousands + 2 hundreds + 1 ten
4. 2 thousands + 9 ones   5. 5 thousands + 7 hundreds   6. 5689   7. 87,128   8. 9003
9. Fifty-seven   10. Twenty-nine   11. Eighty-eight
12. Two hundred four   13. Seventy-nine thousand, two hundred four   14. One million, eight hundred seventy-nine thousand, two hundred four
15. Twenty-two billion, three hundred one million, eight hundred seventy-nine thousand, two hundred four   16. 213,105,329   17. 2 ten thousands
18. 2 hundred thousands   19. 2 millions
20. 2 ten millions   21. 6   22. 8   23. 4   24. 5

Exercise Set 1.1, p. 7

1. 5 thousands + 7 hundreds + 4 tens + 2 ones
3. 2 ten thousands + 7 thousands + 3 hundreds + 4 tens + 2 ones   5. 5 thousands + 6 hundreds + 9 ones   7. 2 thousands + 3 hundreds   9. 2475
11. 68,939   13. 7304   15. 1009   17. Eighty-five
19. Eighty-eight thousand   21. One hundred twenty-three thousand, seven hundred sixty-five
23. Seven billion, seven hundred fifty-four million, two hundred eleven thousand, five hundred seventy-seven
25. One million, eight hundred sixty-seven thousand
27. One billion, five hundred eighty-three million, one hundred forty-one thousand   29. 2,233,812
31. 8,000,000,000   33. 9,460,000,000,000
35. 2,974,600   37. 5 thousands   39. 5 hundreds
41. 3   43. 0   45. ◆   47. All 9's as digits. Answers may vary. For an 8-digit readout, it would be 99,999,999. This number has three periods.

Margin Exercises, Section 1.2, pp. 9–13

1. $8 + 2 = 10$ 2. $45 + $33 = 78 3. 100 mi + 93 mi = 193 mi   4. 5 ft + 7 ft = 12 ft   5. 4 in. + 5 in. + 9 in. + 6 in. + 5 in. = 29 in.   6. 5 ft + 6 ft + 5 ft + 6 ft = 22 ft   7. 30,000 sq ft + 40,000 sq ft = 70,000 sq ft   8. 8 sq yd + 9 sq yd = 17 sq yd
9. 6 cu yd + 8 cu yd = 14 cu yd   10. 80 gal + 56 gal = 136 gal   11. 9745   12. 13,465   13. 16,182
14. 27   15. 34   16. 27   17. 38   18. 47   19. 61
20. 27,474

Exercise Set 1.2, p. 15

1. 7 + 8 = 15 **3.** 500 acres + 300 acres = 800 acres
5. 114 mi **7.** 52 in. **9.** 1300 ft **11.** 387
13. 5188 **15.** 164 **17.** 100 **19.** 900 **21.** 1010
23. 8503 **25.** 5266 **27.** 4466 **29.** 8310 **31.** 6608
33. 16,784 **35.** 34,432 **37.** 101,310 **39.** 100,101
41. 28 **43.** 26 **45.** 67 **47.** 230 **49.** 130
51. 1349 **53.** 36,926 **55.** 18,424 **57.** 2320
59. 31,685 **61.** 11,679 **63.** 22,654 **65.** 12,765,097
67. 7992 **68.** Nine hundred twenty-four million, six hundred thousand **69.** 8 ten thousands
70. 23,000,000 **71.** ◆ **73.** 56,055,667
75. 1 + 99 = 100, 2 + 98 = 100, ..., 49 + 51 = 100. Then 49 · 100 = 4900 and 4900 + 50 + 100 = 5050.

Margin Exercises, Section 1.3, pp. 19–22

1. 67 cu yd − 5 cu yd = 62 cu yd **2.** 20,000 sq ft − 12,000 sq ft = 8,000 sq ft **3.** 7 = 2 + 5, or 7 = 5 + 2
4. 17 = 9 + 8, or 17 = 8 + 9 **5.** 5 = 13 − 8; 8 = 13 − 5 **6.** 11 = 14 − 3; 3 = 14 − 11
7. 67 + ▢ = 348; ▢ = 348 − 67
8. 800 + ▢ = 1200; ▢ = 1200 − 800
9. 3801 **10.** 6328 **11.** 4747 **12.** 56 **13.** 205
14. 658 **15.** 2851 **16.** 1546

Exercise Set 1.3, p. 23

1. $1260 − $450 = ▢ **3.** 16 oz − 5 oz = ▢
5. 7 = 3 + 4, or 7 = 4 + 3 **7.** 13 = 5 + 8, or 13 = 8 + 5 **9.** 23 = 14 + 9, or 23 = 9 + 14
11. 43 = 27 + 16, or 43 = 16 + 27 **13.** 6 = 15 − 9; 9 = 15 − 6 **15.** 8 = 15 − 7; 7 = 15 − 8
17. 17 = 23 − 6; 6 = 23 − 17 **19.** 23 = 32 − 9; 9 = 32 − 23 **21.** 17 + ▢ = 32; ▢ = 32 − 17
23. 10 + ▢ = 23; ▢ = 23 − 10 **25.** 12 **27.** 44
29. 533 **31.** 1126 **33.** 39 **35.** 298 **37.** 226
39. 234 **41.** 5382 **43.** 1493 **45.** 2187 **47.** 3831
49. 7748 **51.** 33,794 **53.** 2168 **55.** 43,028
57. 56 **59.** 36 **61.** 84 **63.** 454 **65.** 771
67. 2191 **69.** 3749 **71.** 7019 **73.** 5745 **75.** 95,974
77. 9989 **79.** 83,818 **81.** 4206 **83.** 10,305
85. 7 ten thousands **86.** Six million, three hundred seventy-five thousand, six hundred two **87.** 29,708
88. 22,692 **89.** ◆ **91.** 2,829,177 **93.** 3; 4

Margin Exercises, Section 1.4, pp. 27–30

1. 40 **2.** 50 **3.** 70 **4.** 100 **5.** 40 **6.** 80 **7.** 90
8. 140 **9.** 470 **10.** 240 **11.** 300 **12.** 600
13. 800 **14.** 800 **15.** 9300 **16.** 8000 **17.** 8000
18. 20,000 **19.** 69,000 **20.** 200 **21.** 1800
22. 2600 **23.** 11,000 **24.** < **25.** > **26.** >
27. < **28.** < **29.** >

Exercise Set 1.4, p. 31

1. 50 **3.** 70 **5.** 730 **7.** 900 **9.** 100 **11.** 1000
13. 9100 **15.** 32,900 **17.** 6000 **19.** 8000
21. 45,000 **23.** 373,000 **25.** 180 **27.** 5720
29. 220; incorrect **31.** 890; incorrect **33.** 16,500
35. 5200 **37.** 1600 **39.** 1500 **41.** 31,000
43. 69,000 **45.** < **47.** > **49.** < **51.** > **53.** >
55. > **57.** 86,754 **58.** 13,589 **59.** 48,824
60. 4415 **61.** ◆ **63.** 30,411 **65.** 69,594

Margin Exercises, Section 1.5, pp. 34–38

1. 8 · 7 = 56 **2.** 10 · 75 = 750 mL
3. 8 · 8 = 64 **4.** 4 · 6 = 24 sq ft **5.** 1035
6. 3024 **7.** 46,252 **8.** 205,065 **9.** 144,432
10. 287,232 **11.** 14,075,720 **12.** 391,760
13. 17,345,600 **14.** 56,200 **15.** 562,000
16. (a) 1081; (b) 1081; (c) same **17.** 40 **18.** 15
19. 210,000; 160,000

Exercise Set 1.5, p. 39

1. 21 · 21 = 441 **3.** 8 · 12 oz = 96 oz **5.** 18 sq ft
7. 121 sq yd **9.** 144 sq mm **11.** 870 **13.** 2,340,000
15. 520 **17.** 564 **19.** 65,200 **21.** 4,371,000
23. 1527 **25.** 64,603 **27.** 4770 **29.** 3995
31. 46,080 **33.** 14,652 **35.** 207,672 **37.** 798,408
39. 166,260 **41.** 11,794,332 **43.** 20,723,872
45. 362,128 **47.** 20,064,048 **49.** 25,236,000
51. 302,220 **53.** 49,101,136 **55.** 30,525
57. 298,738 **59.** 50 · 70 = 3500 **61.** 30 · 30 = 900
63. 900 · 300 = 270,000 **65.** 400 · 200 = 80,000
67. 6000 · 5000 = 30,000,000
69. 8000 · 6000 = 48,000,000 **71.** 4370 **72.** 3109
73. 2350; 2300; 2000 **75.** ◆

Margin Exercises, Section 1.6, pp. 44–49

1. 112 ÷ 14 = ▢ **2.** 112 ÷ 8 = ▢ **3.** 15 = 5 · 3, or 15 = 3 · 5 **4.** 72 = 9 · 8, or 72 = 8 · 9
5. 6 = 12 ÷ 2; 2 = 12 ÷ 6 **6.** 6 = 42 ÷ 7; 7 = 42 ÷ 6 **7.** 6; 6 · 9 = 54 **8.** 6 R 7; 6 · 9 = 54, 54 + 7 = 61 **9.** 4 R 5; 4 · 12 = 48, 48 + 5 = 53
10. 6 R 13; 6 · 24 = 144, 144 + 13 = 157 **11.** 59 R 3
12. 1475 R 5 **13.** 1015 **14.** 134 **15.** 63 R 12
16. 807 R 4 **17.** 1088 **18.** 360 R 4 **19.** 800 R 47

Calculator Spotlight, p. 48

1. 1475 R 5 **2.** 360 R 4 **3.** 800 R 47 **4.** 134

Exercise Set 1.6, p. 51

1. 760 ÷ 4 = ▢ **3.** 455 ÷ 5 = ▢ **5.** 18 = 3 · 6, or 18 = 6 · 3 **7.** 22 = 22 · 1, or 22 = 1 · 22
9. 54 = 6 · 9, or 54 = 9 · 6 **11.** 37 = 1 · 37, or 37 = 37 · 1 **13.** 9 = 45 ÷ 5; 5 = 45 ÷ 9

15. $37 = 37 \div 1$; $1 = 37 \div 37$ **17.** $8 = 64 \div 8$
19. $11 = 66 \div 6$; $6 = 66 \div 11$ **21.** 55 R 2   **23.** 108
25. 307   **27.** 753 R 3   **29.** 74 R 1   **31.** 92 R 2
33. 1703   **35.** 987 R 5   **37.** 12,700   **39.** 127
41. 52 R 52   **43.** 29 R 5   **45.** 40 R 12
47. 90 R 22   **49.** 29   **51.** 105 R 3   **53.** 1609 R 2
55. 1007 R 1   **57.** 23   **59.** 107 R 1   **61.** 370
63. 609 R 15   **65.** 304   **67.** 3508 R 219   **69.** 8070
71. 7 thousands + 8 hundreds + 8 tens + 2 ones
72. >   **73.** $21 = 16 + 5$, or $21 = 5 + 16$
74. $56 = 14 + 42$, or $56 = 42 + 14$ **75.** $47 = 56 - 9$;
$9 = 56 - 47$ **76.** $350 = 414 - 64$; $64 = 414 - 350$
77. ◆   **79.** 30

Margin Exercises, Section 1.7, pp. 55–58

1. 7   **2.** 5   **3.** No   **4.** Yes   **5.** 5   **6.** 10   **7.** 5
8. 22   **9.** 22,490   **10.** 9022   **11.** 570   **12.** 3661
13. 8   **14.** 45   **15.** 77   **16.** 3311   **17.** 6114   **18.** 8
19. 16   **20.** 644   **21.** 96   **22.** 94

Exercise Set 1.7, p. 59

1. 14   **3.** 0   **5.** 29   **7.** 0   **9.** 8   **11.** 14   **13.** 1035
15. 25   **17.** 450   **19.** 90,900   **21.** 32   **23.** 143
25. 79   **27.** 45   **29.** 324   **31.** 743   **33.** 37   **35.** 66
37. 15   **39.** 48   **41.** 175   **43.** 335   **45.** 104
47. 45   **49.** 4056   **51.** 17,603   **53.** 18,252   **55.** 205
57. $7 = 15 - 8$; $8 = 15 - 7$ **58.** $6 = 48 \div 8$;
$8 = 48 \div 6$ **59.** <   **60.** >   **61.** 142 R 5
62. 334 R 11   **63.** ◆   **65.** 347

Margin Exercises, Section 1.8, pp. 62–68

1. 1,424,000   **2.** $369   **3.** $18   **4.** $38,988
5. 9180 sq in.   **6.** 378 packages; 1 can left over
7. 37 gal   **8.** 70 min, or 1 hr, 10 min   **9.** 106

Exercise Set 1.8, p. 69

1. $12,276   **3.** $7,004,000,000   **5.** $64,000,000
7. 4007 mi   **9.** 384 in.   **11.** 4500   **13.** 7280
15. $247   **17.** 54 weeks; 1 episode left over
19. 168 hr   **21.** $400   **23.** (a) 4700 sq ft; (b) 288 ft
25. 44   **27.** 56 cartons; 11 books left over
29. 1600 mi; 27 in.   **31.** 18   **33.** 22   **35.** $704
37. 525 min, or 8 hr, 45 min   **39.** 3000 sq in.
41. 234,600   **42.** 235,000   **43.** 22,000   **44.** 16,000
45. 320,000   **46.** 720,000   **47.** ◆   **49.** 792,000 mi;
1,386,000 mi

Margin Exercises, Section 1.9, pp. 73–76

1. 5^4 **2.** 5^5 **3.** 10^2 **4.** 10^4 **5.** 10,000   **6.** 100
7. 512   **8.** 32   **9.** 51   **10.** 30   **11.** 584   **12.** 84
13. 4; 1   **14.** 52; 52   **15.** 29   **16.** 1880   **17.** 253
18. 93   **19.** 1880   **20.** 305   **21.** 93   **22.** 87 in.
23. 46   **24.** 4

Calculator Spotlight, p. 75

1. 1024   **2.** 40,353,607   **3.** 1,048,576   **4.** 49   **5.** 85
6. 135   **7.** 176

Exercise Set 1.9, p. 77

1. 3^4 **3.** 5^2 **5.** 7^5 **7.** 10^3 **9.** 49   **11.** 729
13. 20,736   **15.** 121   **17.** 22   **19.** 20   **21.** 100
23. 1   **25.** 49   **27.** 5   **29.** 434   **31.** 41   **33.** 88
35. 4   **37.** 303   **39.** 20   **41.** 70   **43.** 295   **45.** 32
47. 906   **49.** 62   **51.** 102   **53.** $94   **55.** 110   **57.** 7
59. 544   **61.** 708   **63.** 452   **64.** 13
65. 102,600 mi^2 **66.** 98 gal   **67.** ◆   **69.** 675
71. 24; $1 + 5 \cdot (4 + 3) = 36$
73. 7; $12 \div (4 + 2) \cdot 3 - 2 = 4$

Summary and Review: Chapter 1, p. 79

1. 2 thousands + 7 hundreds + 9 tens + 3 ones
2. 5 ten thousands + 6 thousands + 7 tens + 8 ones
3. 8669   **4.** 90,844   **5.** Sixty-seven thousand, eight
hundred nineteen   **6.** Two million, seven hundred
eighty-one thousand, four hundred twenty-seven
7. 476,588   **8.** 36,260,064   **9.** 8 thousands
10. 3   **11.** $406 + $78 = $484, or $78 + $406 = $484
12. 986 yd   **13.** 14,272   **14.** 66,024   **15.** 22,098
16. 98,921   **17.** 151 lb − 12 lb = 139 lb
18. $340 − $196 = $144   **19.** $10 = 6 + 4$, or
$10 = 4 + 6$ **20.** $8 = 11 - 3$; $3 = 11 - 8$ **21.** 5148
22. 1153   **23.** 2274   **24.** 17,757   **25.** 345,800
26. 345,760   **27.** 346,000   **28.** $41,300 + 19,700 = 61,000$ **29.** $38,700 - 24,500 = 14,200$
30. $400 \cdot 700 = 280,000$ **31.** >   **32.** <
33. $32 \cdot 15 = 480$ **34.** $125 \cdot 368 = 46,000$ yd^2
35. 420,000   **36.** 6,276,800   **37.** 506,748   **38.** 27,589
39. 5,331,810   **40.** $176 \div 4 =$ ▯   **41.** $222 \div 6 =$ ▯
42. $56 = 8 \cdot 7$, or $56 = 7 \cdot 8$ **43.** $4 = 52 \div 13$;
$13 = 52 \div 4$ **44.** 12 R 3   **45.** 5   **46.** 913 R 3
47. 384 R 1   **48.** 4 R 46   **49.** 54   **50.** 452
51. 5008   **52.** 4389   **53.** 8   **54.** 45   **55.** 546
56. $2413   **57.** 1982   **58.** $19,748   **59.** 137 beakers
filled; 13 mL of alcohol left over   **60.** 4^3 **61.** 10,000
62. 36   **63.** 65   **64.** 233   **65.** 56   **66.** 32
67. 260   **68.** 165
69. ◆ A vat contains 1152 oz of hot sauce. If
144 bottles are to be filled equally, how much will each
bottle contain? Answers may vary.
70. ◆ No; if subtraction were associative, then
$a - (b - c) = (a - b) - c$ for any a, b, and c. But, for
example,
$$12 - (8 - 4) = 12 - 4 = 8,$$
whereas
$$(12 - 8) - 4 = 4 - 4 = 0.$$
Since $8 \neq 0$, this example shows that subtraction is not
associative.
71. $d = 8$ **72.** $a = 8$, $b = 4$ **73.** 7 days

Test: Chapter 1, p. 81

1. [1.1a] 8 thousands + 8 hundreds + 4 tens + 3 ones
2. [1.1c] Thirty-eight million, four hundred three thousand, two hundred seventy-seven   3. [1.1e] 5
4. [1.2b] 9989   5. [1.2b] 63,791   6. [1.2b] 34
7. [1.2b] 10,515   8. [1.3d] 3630   9. [1.3d] 1039
10. [1.3d] 6848   11. [1.3d] 5175   12. [1.5b] 41,112
13. [1.5b] 5,325,600   14. [1.5b] 2405
15. [1.5b] 534,264   16. [1.6c] 3 R 3   17. [1.6c] 70
18. [1.6c] 97   19. [1.6c] 805 R 8   20. [1.8a] 1955
21. [1.8a] 92 packages, 3 cans left over
22. [1.8a] 62,811 mi^2 23. [1.8a] 120,000 m^2; 1600 m
24. [1.8a] 1808 lb   25. [1.8a] 20   26. [1.7b] 46
27. [1.7b] 13   28. [1.7b] 14   29. [1.4a] 35,000
30. [1.4a] 34,580   31. [1.4a] 34,600
32. [1.4b] 23,600 + 54,700 = 78,300
33. [1.4b] 54,800 − 23,600 = 31,200
34. [1.5c] 800 · 500 = 400,000   35. [1.4c] >
36. [1.4c] <   37. [1.9a] 12^4 38. [1.9b] 343
39. [1.9b] 8   40. [1.9c] 64   41. [1.9c] 96
42. [1.9c] 2   43. [1.9d] 216   44. [1.9c] 18
45. [1.9c] 92   46. [1.5a], [1.8a] 336 in^2 47. [1.8a] 80
48. [1.9c] 83   49. [1.9c] 9

Chapter 2

Pretest: Chapter 2, p. 84

1. [2.1a] −35; 67   2. [2.1b] <   3. [2.1b] >
4. [2.1b] <   5. [2.1c] 73   6. [2.1c] 57   7. [2.1d] 32
8. [2.1d] 17   9. [2.2a] −18   10. [2.2a] 6
11. [2.2a] −9   12. [2.2a] 0   13. [2.3a] −16
14. [2.3a] 2   15. [2.3a] 18   16. [2.3a] −6
17. [2.4a] 0   18. [2.4a] 48   19. [2.4b] −64
20. [2.5a] −5   21. [2.5a] 11   22. [2.5a] −80
23. [2.5a] 0   24. [2.5b] 0   25. [2.6a] −13
26. [2.7a] $3x + 15$ 27. [2.7a] $14x − 21y − 7$
28. [2.7b] $15a$ 29. [2.7b] $−10x + 15$ 30. [2.7c] 64 ft
31. [2.8b] −18   32. [2.8a] 22   33. [2.8b] 6

Margin Exercises, Section 2.1, pp. 86–89

1. 8; −5   2. 134; −80   3. −10; 148   4. −137; 289
5. >   6. >   7. <   8. >   9. 18   10. 9   11. 29
12. 52   13. −1   14. 2   15. 0   16. 4   17. 13
18. −28   19. 0   20. 7   21. 1   22. −6   23. −2

Calculator Spotlight, p. 86

1.–3. Keystrokes will vary by calculator.

Exercise Set 2.1, p. 91

1. −2   3. −1286; 29,028   5. 850; −432   7. >
9. <   11. <   13. <   15. >   17. <   19. 23
21. 0   23. 24   25. 53   27. 8   29. 8   31. 7
33. 0   35. 19   37. −42   39. 8   41. −7
43. 29   45. 22   47. −1   49. 3   51. −8
53. 2   55. 0   57. −34   59. 825   60. 125
61. 7106   62. 4   63. 81   64. 1550   65. ◆
67. ◆   69. [3][2][7][×][8][3][=][+/−].
Answers may vary.   71. −8   73. −7   75. −1, 0, 1
77. −100, −5, 0, |3|, 4, |−6|, 7^2, 10^2, 2^7, 2^{10}

Margin Exercises, Section 2.2, pp. 93–94

1. −1   2. −8   3. 4   4. 0   5. 4 + (−5) = −1
6. −2 + (−4) = −6   7. −3 + 8 = 5   8. −11
9. −12   10. −34   11. −22   12. 2   13. −4
14. −2   15. 3   16. 0   17. 0   18. 0   19. 0
20. −58   21. −56   22. −12

Exercise Set 2.2, p. 95

1. −5   3. −4   5. 6   7. 0   9. −4   11. −15
13. −11   15. 0   17. 0   19. 8   21. −8
23. −25   25. −27   27. 0   29. 0   31. 3
33. −9   35. −5   37. 9   39. −3   41. 0
43. −10   45. −24   47. −5   49. −21   51. 2
53. −26   55. −21   57. 25   59. −17   61. 6
63. 8   65. −160   67. −62   69. 3 ten thousands + 9 thousands + 4 hundreds + 1 ten + 7 ones   70. 700
71. 33,000   72. 2352   73. 32   74. 3500   75. ◆
77. ◆   79. 17   81. −2531   83. All numbers less than 7   85. Positive   87. Positive

Margin Exercises, Section 2.3, pp. 97–99

1. −10   2. 3   3. −5   4. −2   5. −11   6. 4
7. −2   8. 3 − 10 = 3 + (−10); three minus ten is three plus negative ten.   9. 13 − 5 = 13 + (−5); thirteen minus five is thirteen plus negative five.
10. −12 − (−9) = −12 + 9; negative twelve minus negative nine is negative twelve plus nine.
11. −12 − 10 = −12 + (−10); negative twelve minus ten is negative twelve plus negative ten.
12. −14 − (−14) = −14 + 14; negative fourteen minus negative fourteen is negative fourteen plus fourteen.
13. −6   14. −16   15. 5   16. 3   17. −6   18. 13
19. −9   20. 17
21.

National League	HRs hit	HRs allowed	Diff.
Atlanta	215	117	98
St. Louis	223	151	72
Chicago	212	180	32
San Diego	167	139	28
Los Angeles	159	135	24
Houston	166	147	19
Colorado	183	174	9
Montreal	147	156	−9
San Francisco	161	171	−10
New York	136	152	−16
Arizona	159	188	−29
Cincinnati	138	170	−32
Milwaukee	152	188	−36
Pittsburgh	107	147	−40
Philadelphia	126	188	−62
Florida	114	182	−68

22. 50°C

Exercise Set 2.3, p. 101

1. −3   **3.** −8   **5.** −4   **7.** 0   **9.** −5   **11.** −7
13. −4   **15.** 0   **17.** 0   **19.** 14   **21.** 11   **23.** −14
25. 5   **27.** −7   **29.** −1   **31.** 18   **33.** −10   **35.** −3
37. −21   **39.** 5   **41.** −8   **43.** 12   **45.** −23
47. −68   **49.** −73   **51.** 116   **53.** 0   **55.** 55
57. 19   **59.** −62   **61.** −139   **63.** 6   **65.** 107
67. 219   **69.** −7 min   **71.** $470   **73.** −3220 ft
75. 64   **76.** 1   **77.** 8   **78.** 288 oz   **79.** 35   **80.** 3
81. ◆   **83.** ◆   **85.** 83,443   **87.** False; $0 - 3 \neq 3$
89. True   **91.** False; $3 - 3 = 0$, but $3 \neq -3$
93. (a) −2; (b) yes

Margin Exercises, Section 2.4, pp. 105–108

1. 20; 10; 0; −10; −20; −30   **2.** −18   **3.** −100
4. −9   **5.** −10; 0; 10; 20; 30   **6.** 12   **7.** 32   **8.** 7
9. 0   **10.** 0   **11.** 120   **12.** −120   **13.** 6   **14.** −8
15. 81   **16.** −1   **17.** −25   **18.** Negative eight squared; the opposite of eight squared

Calculator Spotlight, p. 108

1. 148,035,889   **2.** −1,419,857   **3.** −1,124,864
4. 1,048,576   **5.** −531,441   **6.** −117,649
7. −7776   **8.** −19,683

Exercise Set 2.4, p. 109

1. −21   **3.** −18   **5.** −48   **7.** −30   **9.** 10   **11.** 18
13. 42   **15.** 30   **17.** −120   **19.** 300   **21.** 72
23. −340   **25.** 0   **27.** 0   **29.** −96   **31.** 420
33. −70   **35.** 30   **37.** 0   **39.** −294   **41.** 25
43. −125   **45.** 10,000   **47.** −16   **49.** −243   **51.** 1
53. −729   **55.** −125   **57.** The opposite of seven to the fourth power   **59.** Negative nine to the sixth power   **61.** 532,500   **62.** 60,000,000   **63.** 80
64. 2550   **65.** 40 sq ft   **66.** 240   **67.** ◆
69. ◆   **71.** −8   **73.** 25   **75.** 294   **77.** −85,525,504
79. −32 m   **81.** (a) m and n must have different signs.
(b) At least one of m and n must be zero. (c) m and n must have the same sign.

Margin Exercises, Section 2.5, pp. 111–112

1. −2   **2.** 5   **3.** −3   **4.** 0   **5.** −6   **6.** −5
7. Undefined   **8.** 0   **9.** Undefined   **10.** 68   **11.** 3
12. −15

Calculator Spotlight, p. 112

1. −4   **2.** −2   **3.** 787

Exercise Set 2.5, p. 113

1. −7   **3.** −14   **5.** −9   **7.** 4   **9.** −9   **11.** 2
13. −43   **15.** −8   **17.** Undefined   **19.** −8   **21.** −23
23. 0   **25.** −19   **27.** −41   **29.** −7   **31.** 20
33. −334   **35.** 23   **37.** 8   **39.** 12   **41.** −10
43. −86   **45.** −9   **47.** 0   **49.** 10   **51.** −25
53. −7988   **55.** −3000   **57.** 60   **59.** 1   **61.** 2
63. 7   **65.** 2   **67.** 0   **69.** 28 sq in.   **70.** 42
71. 14 gal   **72.** 17 gal   **73.** 150 calories   **74.** 96 g
75. ◆   **77.** ◆   **79.** 0   **81.** −159   **83.** Negative
85. Negative

Margin Exercises, Section 2.6, pp. 115–116

1. 64   **2.** 28   **3.** −60   **4.** $-\frac{7}{x}; \frac{7}{-x}$ **5.** $\frac{-m}{n}; \frac{m}{-n}$
6. $-\frac{r}{4}; \frac{-r}{4}$ **7.** −7; −7; −7   **8.** 50   **9.** 48; 48
10. 81; 81   **11.** 9; −9   **12.** 4; −4   **13.** 32; −32

Exercise Set 2.6, p. 117

1. 14¢   **3.** −3   **5.** 1   **7.** 2   **9.** 13   **11.** 14 ft
13. 14 ft   **15.** 21   **17.** 21   **19.** 400 ft   **21.** −52
23. 10   **25.** −26   **27.** 36   **29.** $\frac{-3}{a}; \frac{3}{-a}$
31. $\frac{n}{-b}; -\frac{n}{b}$ **33.** $\frac{-9}{p}; -\frac{9}{p}$ **35.** $\frac{14}{-w}; -\frac{14}{w}$
37. −5; −5; −5   **39.** −27; −27; −27   **41.** 441; −147
43. 20; 20   **45.** 216; −216   **47.** 1; 1   **49.** 128; −128
51. −20   **53.** 370   **55.** 21   **57.** 100
59. Twenty-three million, forty-three thousand, nine hundred twenty-one   **60.** 901
61. $5280 - 2480 = 2800$ **62.** 994   **63.** 17 in.
64. 5 in.   **65.** ◆   **67.** ◆   **69.** −8454   **71.** 2
73. 20   **75.** False; $2^3 = 8$, but $-2^3 = -8$ **77.** True

Margin Exercises, Section 2.7, pp. 121–126

1.

	$1 \cdot x$	x
$x = 3$	3	3
$x = -6$	−6	−6
$x = 0$	0	0

2.

	$-1 \cdot x$	$-x$
$x = 2$	−2	−2
$x = -6$	6	6
$x = 0$	0	0

3. $4a + 4b$ **4.** $5x + 5y + 5z$ **5.** $3x - 3y$
6. $2a - 2b + 2c$ **7.** $3x - 15$ **8.** $5x - 5y + 20$
9. $-2x + 6$ **10.** $-5x + 10y - 20z$ **11.** $5x$; $-4y$; 3
12. $-4y$; $-2x$; $\frac{x}{y}$ **13.** $9a^3$ and a^3; $4ab$ and $3ab$
14. $3xy$ and $-4xy$ **15.** $11a$ **16.** $7x^2 - 6$
17. $5m - n^2 - 4$ **18.** 26 cm   **19.** 45 mm
20. 12 cm   **21.** 24 ft   **22.** 40 km   **23.** 24 ft

Exercise Set 2.7, p. 127

1. $5a + 5b$ **3.** $4x + 4$ **5.** $2b + 10$ **7.** $7 - 7t$
9. $30x + 12$ **11.** $8x + 56 + 48y$ **13.** $-7y + 14$
15. $45x + 54y - 72$ **17.** $-4x + 12y + 8z$
19. $8a - 24b + 8c$ **21.** $4x - 12y - 28z$
23. $20a - 25b + 5c - 10d$ **25.** $19a$ **27.** $9a$
29. $11x + 6z$ **31.** $-13a + 62$ **33.** $-4 + 4t + 6y$
35. $-7x$ **37.** $-16y$ **39.** $-1 + 17a - 12b$
41. $6x + 3y$ **43.** $7x + y$ **45.** $6a + 4b - 2$
47. $x^3 + 9x$ **49.** $2a^2 + 8a^3 + 5$ **51.** $7xy + 6y^2 - 1$
53. $4a^2b - 3ab^2 + 2ab$ **55.** $-4x^4 + 6y^4 + 8x^4y^4$
57. 17 mm **59.** 18 m **61.** 20 in. **63.** 300 yd
65. 124 yd **67.** 52 ft **69.** 56 in. **71.** 300 cm
73. 64 ft **75.** 17 **76.** 210 **77.** 3174 **78.** Three million, five hundred thirty-four thousand, five hundred twelve **79.** 709 R 4 **80.** 811 **81.** ◆
83. ◆ **85.** $59 \boxed{\times} 17 \boxed{+} 59 \boxed{\times} 8 = 1475$
87. $7x + 1$ **89.** $-29 - 3a$ **91.** $-10 - x - 27y$
93. $29.75

Margin Exercises, Section 2.8, pp. 132–134

1. 24 **2.** -3 **3.** 25 **4.** -14 **5.** 6 **6.** -8
7. -9 **8.** -12 **9.** 26 **10.** -50 **11.** -4

Exercise Set 2.8, p. 135

1. 12 **3.** -3 **5.** 8 **7.** -12 **9.** 23 **11.** -12
13. 17 **15.** -14 **17.** 8 **19.** -4 **21.** 13 **23.** 8
25. -27 **27.** -83 **29.** 7 **31.** 475 **33.** 5
35. -15 **37.** 35 **39.** 45 **41.** -7 **43.** -9
45. -190 **47.** -81 **49.** 29 **50.** 7 **51.** 8 **52.** 27
53. 16 **54.** 26 **55.** ◆ **57.** ◆ **59.** 19 **61.** 3
63. $-\frac{4913}{3375}$ **65.** 45 **67.** 7 **69.** -10

Summary and Review: Chapter 2, p. 137

1. $527; -53$ **2.** $>$ **3.** $<$ **4.** $>$ **5.** 39 **6.** 12
7. 0 **8.** 53 **9.** 29 **10.** -9 **11.** -11 **12.** 6
13. -24 **14.** -9 **15.** 23 **16.** -1 **17.** 7
18. -4 **19.** 12 **20.** 92 **21.** -84 **22.** -40
23. -3 **24.** -5 **25.** 0 **26.** -25 **27.** -20
28. 7 **29.** $-4; -4; -4$ **30.** $20x + 36$
31. $6a - 12b + 15$ **32.** $18a$ **33.** $6x$
34. $-3m + 6$ **35.** 36 in. **36.** 100 cm **37.** -8
38. -9 **39.** -6 **40.** Three hundred eighty-six thousand, four hundred fifty-one
41. $7300 - 2740 = 4560$ **42.** $2500 - 1700 = 800$
43. 136 **44.** 25,485 **45.** 21,286
46. ◆ The notation "$-x$" means "the opposite of x." If x is a negative number, then $-x$ is a positive number. For example, if $x = -2$, then $-x = 2$.
47. ◆ The expressions $(a - b)^2$ and $(b - a)^2$ are equivalent for all choices of a and b because $a - b$ and $b - a$ are opposites. When opposites are raised to an even power, the results are the same.
48. 662,582 **49.** $-88,174$ **50.** -240

Test: Chapter 2, p. 139

1. [2.1a] $-542; 307$ **2.** [2.1b] $>$ **3.** [2.1c] 429
4. [2.1d] -19 **5.** [2.2a] -11 **6.** [2.2a] -21
7. [2.2a] 9 **8.** [2.3a] -12 **9.** [2.3a] -15
10. [2.3a] -24 **11.** [2.3a] 19 **12.** [2.3a] 24
13. [2.4b] -64 **14.** [2.4a] -130 **15.** [2.4a] 0
16. [2.5a] 8 **17.** [2.5a] -8 **18.** [2.5b] -1
19. [2.5b] 25 **20.** [2.3b] 13°F **21.** [2.6a] -3
22. [2.7a] $14x + 21y - 7$ **23.** [2.7b] $4x - 17$
24. [2.8b] 5 **25.** [2.8a] -12 **26.** [1.1c] Two million, three hundred eight thousand, four hundred fifty-one **27.** [1.4b] $3200 - 1920 = 1280$
28. [1.4b] $9200 - 2900 = 6300$ **29.** [1.8a] 21
30. [1.5b] 5648 **31.** [1.5b] 20,536 **32.** [2.7c] 66 ft
33. [2.7a, b] $35x - 7$ **34.** [2.7a, b] $-24x - 57$

Cumulative Review: Chapters 1–2, p. 141

1. [1.1d] 584,017,800 **2.** [1.1c] Five million, three hundred eighty thousand, six hundred twenty-one
3. [1.2b] 17,797 **4.** [1.2b] 8857 **5.** [1.3d] 4946
6. [1.3d] 1425 **7.** [1.5b] 16,767 **8.** [1.5b] 8,266,500
9. [2.4a] -248 **10.** [2.4a] 72 **11.** [1.6c] 241 R 1
12. [1.6c] 62 **13.** [2.5a] 0 **14.** [2.5a] -5
15. [1.4a] 428,000 **16.** [1.4a] 5300
17. [1.4b] $749,600 + 301,400 = 1,051,000$
18. [1.5c] $700 \times 500 = 350,000$ **19.** [2.1b] $<$
20. [2.1c] 279 **21.** [1.9c] 36 **22.** [1.9d] 2
23. [2.5b] -86 **24.** [2.4b] 125 **25.** [2.6a] 3
26. [2.6a] 28 **27.** [2.7a] $-2x - 10$
28. [2.7a] $18x - 12y + 24$ **29.** [2.2a] -26
30. [2.3a] -31 **31.** [2.3a] -15 **32.** [2.3a] 13
33. [1.7b] 27 **34.** [2.8b] -3 **35.** [2.8a] -3
36. [1.7b] 24 **37.** [1.8a] 78,425 sq ft **38.** [1.8a] 13,931
39. [1.8a] $75 **40.** [1.8a] Westside Appliance
41. [2.7b] $23x - 14$ **42.** [1.8a] 5 cases, 3 six-packs, 4 loose cans **43.** [1.9d], [2.7b] $4a$

Chapter 3

Pretest: Chapter 3, p. 144

1. [3.1b] Yes **2.** [3.1b] Yes **3.** [3.2b] Prime
4. [3.2c] $2 \cdot 2 \cdot 2 \cdot 5 \cdot 7$ **5.** [3.3c] 1 **6.** [3.3c] $7x$
7. [3.3c] 0 **8.** [3.5b] $-\frac{1}{4}$ **9.** [3.5b] $\frac{2}{7}$ **10.** [3.5a] $\frac{12}{28}$
11. [3.6a] $\frac{6}{5}$ **12.** [3.6a] -20 **13.** [3.6a] $\frac{5a}{4}$
14. [3.7a] $\frac{8}{7}$ **15.** [3.7a] $\frac{1}{11}$ **16.** [3.7b] 24
17. [3.7b] $-\frac{3}{4}$ **18.** [3.7b] $\frac{2}{9x}$ **19.** [3.8a] $-\frac{2}{7}$
20. [3.8a] 30 **21.** [3.8a] $\frac{5}{8}$ **22.** [3.6b] $54

51. For all negative values of x **52.** For all values of x less than -2

23. [3.7c] $\frac{1}{24}$ m **24.** [3.6b] 26 ft²

Margin Exercises, Section 3.1, pp. 145–149

1. $5 = 1 \cdot 5$; $45 = 9 \cdot 5$; $100 = 20 \cdot 5$ **2.** $10 = 1 \cdot 10$; $60 = 6 \cdot 10$; $110 = 11 \cdot 10$ **3.** 5, 10, 15, 20, 25, 30, 35, 40, 45, 50 **4.** Yes **5.** Yes **6.** No **7.** Yes **8.** No **9.** Yes **10.** No **11.** Yes **12.** No **13.** Yes **14.** No **15.** Yes **16.** No **17.** No **18.** Yes **19.** No **20.** Yes **21.** No **22.** Yes **23.** No **24.** Yes **25.** No **26.** Yes **27.** Yes **28.** No **29.** No **30.** Yes

Calculator Spotlight, p. 146

1. Yes **2.** No **3.** No **4.** Yes

Exercise Set 3.1, p. 151

1. 6, 12, 18, 24, 30, 36, 42, 48, 54, 60 **3.** 20, 40, 60, 80, 100, 120, 140, 160, 180, 200 **5.** 3, 6, 9, 12, 15, 18, 21, 24, 27, 30 **7.** 13, 26, 39, 52, 65, 78, 91, 104, 117, 130 **9.** 10, 20, 30, 40, 50, 60, 70, 80, 90, 100 **11.** 9, 18, 27, 36, 45, 54, 63, 72, 81, 90 **13.** No **15.** Yes **17.** No **19.** No **21.** No **23.** 46, 224, 300, 36, 45,270, 4444, 256, 8064, 21,568 **25.** 300, 45,270 **27.** 300, 36, 45,270, 8064 **29.** 324, 42, 501, 3009, 75, 2001, 402, 111,111, 1005 **31.** 55,555, 200, 75, 2345, 35, 1005 **33.** 324 **35.** 53 **36.** 5 **37.** −8 **38.** −24 **39.** $680 **40.** 42 **41.** ◆ **43.** ◆ **45.** 99,969 **47.** 30 **49.** 60 **51.** 121

Margin Exercises, Section 3.2, pp. 153–156

1. 1, 2, 3, 6 **2.** 1, 2, 4, 8 **3.** 1, 2, 5, 10 **4.** 1, 2, 4, 8, 16, 32 **5.** 13, 19, 41 are prime; 4, 6, 8 are composite; 1 is neither **6.** $2 \cdot 3$ **7.** $2 \cdot 2 \cdot 3$ **8.** $3 \cdot 3 \cdot 5$ **9.** $2 \cdot 7 \cdot 7$ **10.** $2 \cdot 3 \cdot 3 \cdot 7$ **11.** $2 \cdot 2 \cdot 2 \cdot 2 \cdot 3 \cdot 3$

Exercise Set 3.2, p. 157

1. 1, 2, 3, 6, 9, 18 **3.** 1, 2, 3, 6, 9, 18, 27, 54 **5.** 1, 2, 4 **7.** 1, 7 **9.** 1 **11.** 1, 2, 7, 14, 49, 98 **13.** 1, 2, 3, 6, 7, 14, 21, 42 **15.** 1, 5, 7, 11, 35, 55, 77, 385 **17.** 1, 2, 3, 4, 6, 9, 12, 18, 36 **19.** 1, 3, 5, 9, 15, 25, 45, 75, 225 **21.** Prime **23.** Composite **25.** Composite **27.** Prime **29.** Neither **31.** Composite **33.** Prime **35.** Prime **37.** $2 \cdot 2 \cdot 2 \cdot 2$ **39.** $2 \cdot 7$ **41.** $2 \cdot 11$ **43.** $5 \cdot 5$ **45.** $2 \cdot 31$ **47.** $2 \cdot 2 \cdot 5 \cdot 7$ **49.** $2 \cdot 2 \cdot 5 \cdot 5$ **51.** $5 \cdot 7$ **53.** $2 \cdot 3 \cdot 13$ **55.** $7 \cdot 11$ **57.** $2 \cdot 2 \cdot 2 \cdot 2 \cdot 7$ **59.** $2 \cdot 2 \cdot 3 \cdot 5 \cdot 5$ **61.** −26 **62.** 256 **63.** 8 **64.** −23 **65.** 0 **66.** 1 **67.** ◆ **69.** ◆ **71.** $53 \cdot 53 \cdot 73 \cdot 139$ **73.** $2 \cdot 2 \cdot 2 \cdot 3 \cdot 3 \cdot 5 \cdot 7$ **75.** $2 \cdot 3 \cdot 3 \cdot 3 \cdot 37$ **77.** Answers may vary. One arrangement is a three-dimensional rectangular array consisting of 2 tiers of 12 objects each, where each tier consists of a rectangular array of 4 rows with 3 objects each.

Margin Exercises, Section 3.3, pp. 159–162

1. 5, numerator; 7, denominator **2.** $5a$, numerator; $7b$, denominator **3.** −22, numerator; 3, denominator **4.** $\$\frac{1}{2}$ **5.** $\frac{1}{3}$ mi **6.** $\frac{1}{3}$ gal **7.** $\frac{1}{6}$ hr **8.** $\$\frac{5}{8}$ **9.** $\frac{2}{3}$ mi **10.** $\frac{3}{4}$ gal **11.** $\frac{4}{6}$ yr **12.** $\frac{4}{3}$ mi **13.** $\$\frac{5}{5}$ **14.** $\frac{5}{4}$ mi **15.** $\frac{7}{4}$ gal **16.** $\frac{4}{7}$ **17.** $\frac{0}{3}$ **18.** $\frac{8}{6}$; $\frac{4}{6}$ **19.** 1 **20.** 1 **21.** 1 **22.** 1 **23.** 1 **24.** 1 **25.** 0 **26.** 0 **27.** 0 **28.** 0 **29.** Undefined **30.** Undefined **31.** 8 **32.** −10 **33.** −346 **34.** 1

Exercise Set 3.3, p. 163

1. 3, numerator; 4, denominator **3.** 7, numerator; −9, denominator **5.** $2x$, numerator; $3z$, denominator **7.** $\$\frac{3}{4}$ **9.** $\frac{2}{8}$ mi **11.** $\frac{4}{3}$ L **13.** $\frac{3}{4}$ acre **15.** $\frac{8}{16}$ lb **17.** $\frac{5}{12}$ **19.** $\frac{1}{4}$ **21.** 0 **23.** 15 **25.** 1 **27.** 1 **29.** 0 **31.** 1 **33.** 1 **35.** −63 **37.** 0 **39.** Undefined **41.** $7n$ **43.** Undefined **45.** −210 **46.** −322 **47.** 0 **48.** 0 **49.** $13,772 **50.** 201 min **51.** ◆ **53.** ◆ **55.** $\frac{52}{365}$ **57.** $\frac{9}{39}$

Margin Exercises, Section 3.4, pp. 165–168

1. $\frac{2}{3}$ **2.** $\frac{5}{8}$ **3.** $\frac{10}{3}$ **4.** $-\frac{33}{8}$ or $\frac{-33}{8}$ **5.** $\frac{46}{5}$ **6.** $\frac{4x}{9}$ **7.** $\frac{15}{56}$ **8.** $\frac{32}{15}$ **9.** $\frac{3}{100}$ **10.** $-\frac{7a}{b}$ or $\frac{-7a}{b}$

11.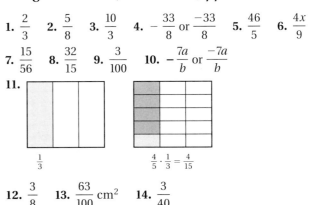

$\frac{1}{3}$ $\frac{4}{5} \cdot \frac{1}{3} = \frac{4}{15}$

12. $\frac{3}{8}$ **13.** $\frac{63}{100}$ cm² **14.** $\frac{3}{40}$

Exercise Set 3.4, p. 169

1. $\frac{3}{7}$ **3.** $\frac{-5}{6}$ or $-\frac{5}{6}$ **5.** $\frac{14}{3}$ **7.** $-\frac{7}{9}$ or $\frac{-7}{9}$ **9.** $\frac{2x}{5}$ **11.** $-\frac{6}{5}$ or $\frac{-6}{5}$ **13.** $\frac{3a}{4}$ **15.** $\frac{17m}{6}$ **17.** $\frac{6}{5}$ **19.** $\frac{2x}{7}$ **21.** $\frac{1}{10}$ **23.** $-\frac{1}{40}$ or $\frac{-1}{40}$ **25.** $\frac{2}{15}$ **27.** $\frac{2x}{5y}$ **29.** $\frac{9}{16}$ **31.** $\frac{14}{39}$ **33.** $-\frac{3}{50}$ or $\frac{-3}{50}$ **35.** $\frac{7a}{64}$ **37.** $\frac{1}{100y}$ **39.** $\frac{-182}{285}$ or $-\frac{182}{285}$ **41.** $\frac{12}{25}$ m² **43.** $\frac{1}{80}$ gal **45.** $\frac{6}{12}$ cup **47.** $\frac{2}{30}$ **49.** −4 **50.** 4 **51.** 76 **52.** 102 **53.** 6 hundred thousands **54.** 4 millions **55.** ◆ **57.** ◆ **59.** $-\frac{185,193}{226,981}$ or $\frac{-185,193}{226,981}$ **61.** $-\frac{3}{160}$ or $\frac{-3}{160}$ **63.** $\frac{4}{105}$

Margin Exercises, Section 3.5, pp. 171–173

1. $\frac{8}{16}$ 2. $\frac{3x}{5x}$ 3. $-\frac{52}{100}$ 4. $\frac{-16}{-6}$ 5. $\frac{12}{9}$ 6. $\frac{-18}{-24}$
7. $\frac{9x}{10x}$ 8. $\frac{9}{45}$ 9. $\frac{-56}{49}$ 10. $\frac{3}{7}$ 11. $\frac{-5}{6}$ 12. 5
13. $\frac{4}{3}$ 14. $-\frac{5}{3}$ 15. $\frac{7}{8}$ 16. $\frac{89}{78}$ 17. $\frac{-8}{7}$
18. $\frac{2}{100} = \frac{1}{50}$; $\frac{4}{100} = \frac{1}{25}$; $\frac{32}{100} = \frac{8}{25}$; $\frac{44}{100} = \frac{11}{25}$; $\frac{18}{100} = \frac{9}{50}$

Calculator Spotlight, p. 174

1. $\frac{14}{15}$ 2. $\frac{7}{8}$ 3. $\frac{138}{167}$ 4. $\frac{7}{25}$

Exercise Set 3.5, p. 175

1. $\frac{5}{10}$ 3. $\frac{-36}{-48}$ 5. $\frac{27}{30}$ 7. $\frac{11t}{5t}$ 9. $\frac{20}{48}$ 11. $-\frac{51}{54}$
13. $\frac{10}{-25}$ 15. $\frac{-42}{132}$ 17. $\frac{5x}{8x}$ 19. $\frac{7m}{11m}$ 21. $\frac{4ab}{9ab}$
23. $\frac{12b}{27b}$ 25. $\frac{1}{2}$ 27. $-\frac{2}{3}$ 29. $\frac{2}{5}$ 31. -3 33. $\frac{3}{4}$
35. $-\frac{12}{7}$ 37. $\frac{1}{3}$ 39. $\frac{-1}{3}$ 41. $\frac{7}{8}$ 43. $\frac{8}{9}$ 45. $\frac{3}{8}$
47. $\frac{3y}{2}$ 49. 3600 yd² 50. $928 51. -5 52. -18
53. 186 54. 2737 55. -5 56. 3520 57. ◆
59. ◆ 61. $\frac{11b}{13c}$ 63. $-\frac{11a}{17b}$ 65. $\frac{151}{139}$ 67. 69
69. (a) $1200; (b) $\frac{1}{12}$

Margin Exercises, Section 3.6, pp. 178–180

1. $\frac{7}{12}$ 2. $-\frac{1}{3}$ 3. 6 4. $\frac{15}{x}$ 5. 10 lb 6. 96 m²
7. $\frac{66}{5}$ cm² 8. 100 in²

Exercise Set 3.6, p. 181

1. $\frac{5}{8}$ 3. $-\frac{1}{8}$ 5. $\frac{3}{20}$ 7. $\frac{2}{9}$ 9. $-\frac{27}{10}$ 11. $\frac{7x}{9}$
13. 2 15. 5 17. -9 19. $3a$ 21. 1 23. 1
25. $\frac{40a}{3}$ 27. $-\frac{88}{3}$ 29. 1 31. 3 33. $\frac{119}{750}$
35. $-\frac{20}{187}$ 37. $-\frac{42}{275}$ 39. $-\frac{16}{5}$ 41. $-\frac{11}{40}$ 43. $\frac{5}{14}$
45. $42 47. 12 lb 49. 260 51. $\frac{1}{3}$ cup
53. $93,000 55. 160 mi 57. $6750 for food; $5400 for housing; $2700 for clothing; $3000 for savings; $6750 for taxes; $2400 for other expenses 59. 60 in²
61. $\frac{35}{4}$ mm² 63. $\frac{63}{8}$ m² 65. 92 mi² 67. 6800 ft²
69. 35 70. 8546 71. 477 72. 70 73. -111
74. -22 75. ◆ 77. ◆ 79. $\frac{219}{355}$ 81. 432 ft²
83. $\frac{1}{12}$ 85. $\frac{1}{168}$

Margin Exercises, Section 3.7, pp. 187–190

1. $\frac{5}{2}$ 2. $\frac{x}{-6}$ 3. $\frac{1}{9}$ 4. 5 5. $-\frac{10}{3}$ 6. $\frac{8}{7}$ 7. $-\frac{8}{3}$
8. $\frac{1}{10}$ 9. $\frac{100}{a}$ 10. -1 11. $\frac{14}{15}$ 12. 320 13. 12

Exercise Set 3.7, p. 191

1. $\frac{3}{7}$ 3. $\frac{1}{4}$ 5. 6 7. $-\frac{3}{10}$ 9. $\frac{21}{2}$ 11. $\frac{m}{-3n}$
13. $\frac{-15}{7}$ 15. $\frac{1}{7m}$ 17. $\frac{4}{5}$ 19. $-\frac{7}{10}$ 21. 4 23. -2
25. $\frac{1}{64}$ 27. $\frac{3}{7x}$ 29. -8 31. $35a$ 33. 1 35. $-\frac{2}{3}$
37. $\frac{99}{224}$ 39. $\frac{112a}{3}$ 41. 88 43. 9 days 45. 20
47. $\frac{3}{64}$ acre 49. $\frac{1}{10}$ m 51. 32 53. 32 55. 510
56. -292 57. 27; -27 58. 80; 80 59. 147, 147
60. 343; -343 61. ◆ 63. ◆ 65. $-\frac{15}{196}$ 67. $\frac{35}{33}$
69. $\frac{7}{5}$

Margin Exercises, Section 3.8, pp. 195–196

1. 15 2. -28 3. $-\frac{9}{32}$ 4. $-\frac{3}{4}$

Exercise Set 3.8, p. 197

1. 10 3. 9 5. -42 7. $\frac{2}{17}$ 9. $\frac{10}{21}$ 11. $-\frac{16}{21}$
13. $-\frac{2}{25}$ 15. $\frac{1}{6}$ 17. $-\frac{80}{3}$ 19. $-\frac{1}{6}$ 21. $-\frac{9}{13}$
23. $\frac{27}{31}$ 25. $\frac{5}{7}$ 27. $\frac{12}{5}$ 29. $-\frac{10}{7}$ 31. $\frac{24}{7}$ 33. 14
35. $\frac{10}{7}$ 37. 20 38. 6 39. $17x$ 40. $4a$
41. $7a + 3$ 42. $4x - 7$ 43. ◆ 45. ◆
47. $-\frac{432}{49}$ 49. 400 km; 160 km 51. $\frac{8}{21}$ mi

Summary and Review: Chapter 3, p. 199

1. No 2. No 3. Yes 4. $2 \cdot 5 \cdot 7$
5. $2 \cdot 2 \cdot 2 \cdot 3 \cdot 3$ 6. $2 \cdot 3 \cdot 5 \cdot 5$ 7. Prime
8. 9, numerator; 7, denominator 9. $\frac{3}{5}$ 10. $\frac{3}{8}$
11. 0 12. 1 13. 48 14. 1 15. $-\frac{2}{3}$ 16. $\frac{1}{4}$
17. -1 18. $\frac{3}{4}$ 19. $\frac{2}{5}$ 20. Undefined 21. $9n$
22. $\frac{1}{3}$ 23. $\frac{15}{21}$ 24. $\frac{-30}{55}$ 25. $\frac{9}{5}$ 26. $-\frac{1}{7}$ 27. 8
28. $\frac{5y}{3x}$ 29. $\frac{6}{35}$ 30. $\frac{4y}{9x}$ 31. $\frac{2}{3}$ 32. $-\frac{1}{14}$ 33. $\frac{1}{25}$

34. 1 **35.** $\frac{18}{5}$ **36.** $\frac{1}{4}$ **37.** 300 **38.** $\frac{1}{15}$ **39.** $4a$
40. -1 **41.** 220 mi **42.** $\frac{3}{8}$ cup **43.** $\frac{1}{6}$ mi **44.** 18
45. 42 m² **46.** $\frac{35}{2}$ ft² **47.** 240 **48.** $-\frac{3}{10}$ **49.** 24
50. 11 **51.** 14 **52.** 27 **53.** ◆ To simplify fractional notation, first factor the numerator and the denominator into prime numbers. Examine the factorizations for factors common to both the numerator and the denominator. Change the order of the factorizations, if necessary, so that the pairs of like factors are above and below each other. Factor the fraction, with each pair of like factors forming a factor equal to 1. Remove the factors equal to 1, and multiply the remaining factors in the numerator and in the denominator, if necessary. **54.** ◆ The fraction is equivalent to $\frac{2}{8}$, but it is not in simplest form. Fractional notation is not simplified until the numerator and the denominator contain no common factors. The notation $\frac{20}{80}$ simplifies to $\frac{1}{4}$.
55. $\frac{3}{17}$ **56.** $\frac{17}{6}$ **57.** 2, 8

Test: Chapter 3, p. 201

1. [3.1b] Yes **2.** [3.1b] No **3.** [3.2c] $2 \cdot 2 \cdot 3 \cdot 3$
4. [3.2c] $2 \cdot 2 \cdot 3 \cdot 5$ **5.** [3.2b] Composite
6. [3.3a] 4, numerator; 9, denominator **7.** [3.3b] $\frac{3}{4}$
8. [3.3c] 26 **9.** [3.3c] 1 **10.** [3.3c] 0 **11.** [3.5b] $\frac{-1}{3}$
12. [3.5b] $\frac{1}{9}$ **13.** [3.5b] $\frac{1}{14}$ **14.** [3.5a] $\frac{15}{40}$
15. [3.7a] $\frac{27}{x}$ **16.** [3.7a] $-\frac{1}{9}$ **17.** [3.4b] $\frac{35}{18}$
18. [3.7a] $\frac{28}{33}$ **19.** [3.4a] $\frac{3x}{8}$ **20.** [3.7b] $-\frac{98}{3}$
21. [3.6a] $\frac{6}{65}$ **22.** [3.7c] $\frac{3}{20}$ lb **23.** [3.6b] 125 lb
24. [3.8a] 64 **25.** [3.8a] $-\frac{7}{4}$ **26.** [3.6b] $\frac{91}{2}$ m²
27. [1.9c] 41 **28.** [1.7b], [2.8b] 101 **29.** [2.2a] -167
30. [2.4a] 63 **31.** [3.6b] $\frac{15}{8}$ tsp **32.** [3.6b] $\frac{7}{48}$ acre
33. [3.6a], [3.7b] $-\frac{7}{960}$ **34.** [3.8a] $\frac{7}{5}$

Cumulative Review: Chapters 1–3, p. 203

1. [1.1c] Two million, fifty-six thousand, seven hundred eighty-three **2.** [1.2b] 10,982 **3.** [2.2a] -43
4. [2.2a] -33 **5.** [1.3d] 2129 **6.** [2.3a] -23
7. [2.3a] -8 **8.** [1.5b] 16,905 **9.** [2.4a] -312
10. [3.6a] $-30x$ **11.** [3.6a] $\frac{7}{15}$ **12.** [1.6c] 235 R 3
13. [2.5a] -17 **14.** [3.7b] -28 **15.** [3.7b] $\frac{2}{3}$
16. [1.4a] 4510 **17.** [1.5c] $900 \times 500 = 450,000$
18. [2.1c] 879 **19.** [2.5b] 8 **20.** [3.2b] Composite
21. [2.6a] -21 **22.** [1.7b], [2.8b] 25
23. [1.7b], [2.8b] 9 **24.** [3.8a] -45 **25.** [1.8a] 8 mpg
26. [1.8a] 8 oz **27.** [2.7b] $2x - 4$ **28.** [2.7b] $3x + 7y$
29. [3.3c] 1 **30.** [3.3c] 59 **31.** [3.3c] 0
32. [3.5b] $-\frac{5}{27}$ **33.** [3.7a] $\frac{5}{2}$ **34.** [3.7a] $\frac{1}{17}$
35. [3.5a] $\frac{71}{70}$ **36.** [3.6b] \$36 **37.** [3.7c] 12
38. [3.6b] $\frac{3}{5}$ mi **39.** [2.6a], [3.6a], [3.7b] $-\frac{54}{169}$
40. [2.1c], [2.6a], [3.6a] $-\frac{9}{100}$ **41.** [3.6b] \$468

Chapter 4

Pretest: Chapter 4, p. 206

1. [4.1a] 75 **2.** [4.2c] $<$ **3.** [4.2a] $-\frac{1}{4}$ **4.** [4.2b] $\frac{11}{90}$
5. [4.3a] $\frac{9}{35}$ **6.** [4.5a] $\frac{43}{8}$ **7.** [4.5b] $3\frac{1}{4}$ **8.** [4.5c] $399\frac{1}{12}$
9. [4.6b] $6\frac{11}{30}$ **10.** [4.3b] $\frac{2}{9}$ **11.** [4.4a] -9
12. [4.7a] $18\frac{7}{11}$ **13.** [4.7a] $36\frac{5}{12}$ **14.** [4.7b] $-7\frac{5}{7}$
15. [4.7b] $1\frac{14}{39}$ **16.** [4.7c] $8\frac{26}{35}$ **17.** [4.6c] $38\frac{1}{4}$ lb
18. [4.7d] $4\frac{1}{4}$ cu ft **19.** [4.7d] $1577\frac{1}{2}$ mi
20. [4.6c] $6\frac{1}{4}$ cups

Margin Exercises, Section 4.1, pp. 207–210

1. 45 **2.** 40 **3.** 18 **4.** 24 **5.** 10 **6.** 200 **7.** 40
8. 360 **9.** 18 **10.** 24 **11.** 2520 **12.** xyz
13. $5a^3b$

Exercise Set 4.1, p. 211

1. 4 **3.** 50 **5.** 40 **7.** 54 **9.** 150 **11.** 120
13. 72 **15.** 420 **17.** 144 **19.** 288 **21.** 30
23. 105 **25.** 72 **27.** 60 **29.** 36 **31.** 900
33. 300 **35.** abc **37.** $9x^2$ **39.** $4x^3y$ **41.** Once every 60 yr **43.** 14 **44.** -27 **45.** 7935 **46.** $\frac{2}{3}$
47. $-\frac{8}{7}$ **48.** -167 **49.** ◆ **51.** ◆ **53.** 70,200
55. 2520 **57. (a)** Not the LCM because 8 is not a factor of $2 \cdot 2 \cdot 3 \cdot 3$; **(b)** not the LCM because 8 is not a factor of $2 \cdot 2 \cdot 3$; **(c)** not the LCM because neither 8 nor 12 is a factor of $2 \cdot 3 \cdot 3$; **(d)** the LCM because both 8 and 12 are factors of $2 \cdot 2 \cdot 2 \cdot 3$ and it is the smallest such number **59.** 27 and 2; 27 and 6; 27 and 18

Margin Exercises, Section 4.2, pp. 213–218

1. $\frac{4}{5}$ **2.** 1 **3.** $\frac{1}{2}$ **4.** $-\frac{3}{8}$ **5.** $-\frac{4}{x}$ **6.** $\frac{2}{5}a$
7. $\frac{7}{19} + \frac{10}{19}x$ **8.** $\frac{5}{6}$ **9.** $\frac{29}{24}$ **10.** $\frac{2}{9}$ **11.** $\frac{38}{5}$
12. $\frac{413}{1000}$ **13.** $\frac{157}{210}$ **14.** $<$ **15.** $>$ **16.** $>$ **17.** $<$
18. $<$ **19.** $<$ **20.** $\frac{11}{10}$ lb

Calculator Spotlight, p. 218

1. $\frac{5}{8}$ **2.** $\frac{67}{60}$ **3.** $\frac{52}{21}$ **4.** $\frac{201}{140}$ **5.** $\frac{253}{150}$ **6.** $\frac{792}{667}$

Exercise Set 4.2, p. 219

1. 1 **3.** $\frac{3}{4}$ **5.** $\frac{2}{5}$ **7.** $\frac{13}{a}$ **9.** $-\frac{2}{11}$ **11.** $\frac{7}{9}x$
13. $\frac{3}{8} + \frac{1}{2}t$ **15.** $-\frac{10}{x}$ **17.** $\frac{7}{24}$ **19.** $-\frac{1}{10}$ **21.** $\frac{19}{24}$
23. $\frac{83}{20}$ **25.** $\frac{5}{24}$ **27.** $\frac{37}{100}x$ **29.** $\frac{41}{60}$ **31.** $-\frac{9}{100}$ **33.** $\frac{67}{12}$
35. $-\frac{33}{7}t$ **37.** $-\frac{17}{24}$ **39.** $\frac{437}{1000}$ **41.** $\frac{5}{4}$ **43.** $\frac{239}{78}$
45. $\frac{59}{90}$ **47.** < **49.** < **51.** > **53.** > **55.** >
57. > **59.** $\frac{4}{15}, \frac{3}{10}, \frac{5}{12}$ **61.** $\frac{5}{6}$ lb **63.** $\frac{9}{8}$ mi **65.** $\frac{43}{48}$ qt
67. 690 kg; $\frac{14}{23}; \frac{5}{23}; \frac{4}{23}$; 1 **69.** $\frac{33}{20}$ mi **71.** $\frac{51}{32}$ in.
73. -13 **74.** 4 **75.** -8 **76.** -31 **77.** $\frac{10}{3}$
78. 42; 42 **79.** ◆ **81.** ◆ **83.** $\frac{22}{45} + \frac{17}{42}x$ **85.** <
87. $\frac{4}{15}$; $320

Margin Exercises, Section 4.3, pp. 223–226

1. $\frac{1}{2}$ **2.** $\frac{3}{5a}$ **3.** $-\frac{1}{2}$ **4.** $\frac{1}{12}$ **5.** $\frac{1}{6}$ **6.** $-\frac{3}{10}$ **7.** $-\frac{1}{6}$
8. $\frac{9}{112}$ **9.** $\frac{3}{10}x$ **10.** $\frac{3}{5}$ **11.** $\frac{1}{6}$ **12.** $-\frac{59}{40}$ **13.** $\frac{2}{15}$ cup

Exercise Set 4.3, p. 227

1. $\frac{2}{3}$ **3.** $-\frac{1}{4}$ **5.** $\frac{4}{a}$ **7.** $\frac{2}{t}$ **9.** $-\frac{4}{5a}$ **11.** $\frac{13}{16}$
13. $-\frac{1}{3}$ **15.** $\frac{7}{10}$ **17.** $-\frac{17}{60}$ **19.** $\frac{53}{100}$ **21.** $\frac{26}{75}$ **23.** $-\frac{21}{100}$
25. $\frac{13}{24}$ **27.** $\frac{1}{10}$ **29.** $\frac{23}{72}$ **31.** $\frac{1}{360}$ **33.** $\frac{2}{9}x$ **35.** $-\frac{3}{20}a$
37. $\frac{7}{9}$ **39.** $\frac{6}{11}$ **41.** $\frac{1}{9}$ **43.** $\frac{9}{8}$ **45.** $\frac{2}{15}$ **47.** $-\frac{7}{24}$
49. $\frac{1}{15}$ **51.** $-\frac{5}{4}$ **53.** $\frac{5}{12}$ hr **55.** $\frac{13}{24}$ mi **57.** $\frac{1}{32}$ in.
59. $\frac{3}{2}$ **60.** $\frac{4}{21}$ **61.** -21 **62.** -32 **63.** 6 lb
64. 9 cups **65.** ◆ **67.** ◆ **69.** $\frac{41}{3289}$ **71.** $-\frac{11}{10}$
73. $-\frac{64}{35}$ **75.** $\frac{31}{32}$ in. **77.** $\frac{3}{16}$

Margin Exercises, Section 4.4, pp. 231–234

1. Yes **2.** No **3.** 26 **4.** -15 **5.** -16 **6.** $-\frac{18}{7}$

Exercise Set 4.4, p. 235

1. 3 **3.** 5 **5.** 7 **7.** 12 **9.** -7 **11.** -5 **13.** 25
15. 10 **17.** 20 **19.** $\frac{5}{6}$ **21.** $\frac{21}{5}$ **23.** 6 **25.** $\frac{17}{4}$
27. $\frac{3}{4}$ **29.** The balance has decreased $150.
30. $1180 profit **31.** $\frac{5}{7m}$ **32.** $20n$ **33.** $3a + 3b$
34. $7m - 21$ **35.** ◆ **37.** ◆ **39.** 4 **41.** $\frac{436}{35}$
43. 2 cm

Margin Exercises, Section 4.5, pp. 237–240

1. $1\frac{2}{3}$ **2.** $8\frac{3}{4}$ **3.** $12\frac{2}{3}$ **4.** $\frac{22}{4}$ **5.** $\frac{61}{10}$ **6.** $\frac{29}{6}$ **7.** $\frac{37}{4}$
8. $\frac{62}{3}$ **9.** $-\frac{32}{5}$ **10.** $-\frac{59}{7}$ **11.** $2\frac{1}{3}$ **12.** $1\frac{1}{10}$ **13.** $18\frac{1}{3}$
14. $-2\frac{2}{5}$ **15.** $-11\frac{1}{6}$ **16.** $807\frac{2}{3}$ **17.** $55\frac{3}{4}$ qt

Calculator Spotlight, p. 240

1. $1476\frac{1}{6}$ **2.** $676\frac{4}{9}$ **3.** $800\frac{51}{56}$ **4.** $13{,}031\frac{1}{2}$
5. $51{,}626\frac{9}{11}$ **6.** $7330\frac{7}{32}$ **7.** $134\frac{1}{15}$ **8.** $2666\frac{130}{213}$
9. $3571\frac{51}{112}$ **10.** $12\frac{169}{454}$

Exercise Set 4.5, p. 241

1. $\frac{17}{5}$ **3.** $\frac{25}{4}$ **5.** $-\frac{161}{8}$ **7.** $\frac{51}{10}$ **9.** $\frac{103}{5}$ **11.** $-\frac{59}{6}$
13. $\frac{69}{10}$ **15.** $-\frac{51}{4}$ **17.** $\frac{57}{10}$ **19.** $-\frac{507}{100}$ **21.** $4\frac{2}{3}$
23. $-4\frac{1}{2}$ **25.** $5\frac{7}{10}$ **27.** $7\frac{4}{7}$ **29.** $7\frac{1}{2}$ **31.** $11\frac{1}{2}$
33. $-1\frac{1}{2}$ **35.** $4\frac{2}{3}$ **37.** $-55\frac{3}{4}$ **39.** $108\frac{5}{8}$
41. $906\frac{3}{7}$ **43.** $40\frac{4}{7}$ **45.** $-20\frac{2}{15}$ **47.** $-22\frac{3}{7}$ **49.** $2\frac{1}{5}$ g
51. $2\frac{3}{10}$ **53.** $\frac{8}{9}$ **54.** 18 **55.** $-\frac{5}{2}$ **56.** $\frac{1}{4}$
57. ◆ **59.** ◆ **61.** $297\frac{23}{349}$ **63.** $6\frac{5}{6}$ **65.** $52\frac{2}{7}$
67. Wire length: 3 ft, 11 in. Diameter: $4\frac{3}{4}$ in.

Margin Exercises, Section 4.6, pp. 243–248

1. $7\frac{2}{5}$ **2.** $12\frac{1}{10}$ **3.** $13\frac{7}{12}$ **4.** $1\frac{1}{2}$ **5.** $3\frac{1}{6}$ **6.** $3\frac{1}{3}$
7. $3\frac{2}{3}$ **8.** $12\frac{5}{6}t$ **9.** $2\frac{1}{4}x$ **10.** $14\frac{1}{30}x$ **11.** $7\frac{1}{12}$ yd
12. $\frac{3}{8}$ in. **13.** $-\frac{3}{4}$ **14.** $-3\frac{3}{4}$ **15.** $-1\frac{5}{6}$ **16.** $-13\frac{7}{30}$

Exercise Set 4.6, p. 249

1. $9\frac{1}{2}$ **3.** $2\frac{11}{12}$ **5.** $13\frac{7}{12}$ **7.** $12\frac{1}{10}$ **9.** $17\frac{5}{24}$ **11.** $21\frac{1}{2}$
13. $27\frac{7}{8}$ **15.** $1\frac{3}{5}$ **17.** $4\frac{1}{10}$ **19.** $21\frac{17}{24}$ **21.** $12\frac{1}{4}$
23. $15\frac{3}{8}$ **25.** $7\frac{5}{12}$ **27.** $11\frac{5}{18}$ **29.** $8\frac{13}{42}t$ **31.** $2\frac{1}{8}x$
33. $8\frac{31}{40}t$ **35.** $11\frac{34}{45}t$ **37.** $6\frac{1}{6}x$ **39.** $9\frac{31}{33}x$ **41.** $4\frac{1}{4}$ lb
43. $6\frac{7}{20}$ cm **45.** $19\frac{1}{16}$ ft **47.** 39 in. **49.** $17\frac{1}{8}$
51. $20\frac{1}{8}$ in. **53.** $3\frac{4}{5}$ hr **55.** $28\frac{3}{4}$ yd **57.** $7\frac{3}{8}$ ft
59. $1\frac{9}{16}$ in. **61.** $-\frac{4}{5}$ **63.** $-3\frac{1}{4}$ **65.** $-3\frac{13}{15}$ **67.** $-7\frac{3}{5}$
69. $-10\frac{29}{35}$ **71.** $-1\frac{8}{9}$ **73.** $\frac{1}{10}$ **74.** $-\frac{10}{13}$ **75.** $\frac{8}{11}$
76. $-\frac{7}{9}$ **77.** ◆ **79.** ◆ **81.** $1888\frac{1053}{2279}$
83. $5\frac{3}{4}$ ft **85.** $56\frac{8}{45}$

Margin Exercises, Section 4.7, pp. 253–258

1. 20 **2.** $1\frac{7}{8}$ **3.** $-12\frac{4}{5}$ **4.** $8\frac{1}{3}$ **5.** 16 **6.** $1\frac{7}{8}$
7. $-\frac{7}{10}$ **8.** $175\frac{1}{2}$ **9.** $159\frac{4}{5}$ **10.** $5\frac{3}{4}$ **11.** $227\frac{1}{2}$ mi
12. 20 mpg **13.** $240\frac{3}{4}$ ft^2 **14.** $9\frac{7}{8}$ in.

Calculator Spotlight, p. 258

1. $\frac{5}{8}$ **2.** $\frac{7}{10}$ **3.** $\frac{5}{7}$ **4.** $\frac{133}{68}$ **5.** $\frac{73}{150}$ **6.** $\frac{97}{116}$ **7.** $24\frac{3}{7}$
8. $\frac{115}{147}$ **9.** $13\frac{41}{63}$ **10.** $5\frac{59}{72}$ **11.** $3\frac{5}{8}$ **12.** $19\frac{5}{8}$

Exercise Set 4.7, p. 259

1. $28\frac{1}{3}$ **3.** $1\frac{2}{3}$ **5.** $-73\frac{1}{3}$ **7.** $16\frac{1}{3}$ **9.** $-10\frac{3}{25}$
11. $35\frac{91}{100}$ **13.** $7\frac{9}{13}$ **15.** $1\frac{1}{5}$ **17.** $3\frac{9}{16}$ **19.** $-1\frac{1}{8}$
21. $1\frac{8}{43}$ **23.** $-\frac{9}{40}$ **25.** $23\frac{2}{5}$ **27.** $15\frac{5}{7}$ **29.** $-28\frac{28}{45}$
31. $-1\frac{1}{3}$ **33.** $12\frac{1}{4}$ **35.** $8\frac{3}{20}$ **37.** 24 **39.** $13\frac{1}{3}$ tsp
41. $16\frac{1}{2}$ **43.** $343\frac{3}{4}$ lb **45.** $82\frac{1}{2}$ in. **47.** 68°
49. 69 kg **51.** $2\frac{41}{128}$ lb **53.** $18\frac{3}{5}$ days **55.** $76\frac{1}{4}$ ft^2
57. Yes; $\frac{7}{8}$ in. **59.** $-8x + 24$ **60.** 19°F **61.** -21
62. 198 **63.** 33 **64.** 0 **65.** ◆ **67.** ◆
69. $352\frac{44}{93}$ **71.** $18\frac{43}{48}$ **73.** $1\frac{5}{8}$ **75.** $-30\frac{5}{8}; -30\frac{5}{8}$

77. 6 ft, $6\frac{4}{5}$ in.   **79.** 5 chicken bouillon cubes; $3\frac{3}{4}$ cups hot water; $7\frac{1}{2}$ tbsp margarine; $7\frac{1}{2}$ tbsp flour; $6\frac{1}{4}$ cups diced cooked chicken; $2\frac{1}{2}$ cups cooked peas; $2\frac{1}{2}$ 4-oz cans sliced mushrooms, drained; $\frac{5}{6}$ cups sliced cooked carrots; $\frac{5}{8}$ cups chopped onions; 5 tbsp chopped pimiento; $2\frac{1}{2}$ tsp salt.

Summary and Review: Chapter 4, p. 265

1. 36   **2.** 90   **3.** 30   **4.** $\frac{7}{9}$ **5.** $\frac{7}{a}$ **6.** $-\frac{7}{15}$
7. $\frac{7}{16}$ **8.** $\frac{1}{3}$ **9.** $-\frac{1}{8}$ **10.** $\frac{5}{27}$ **11.** $\frac{11}{18}$ **12.** >
13. <   **14.** $\frac{19}{40}$ **15.** 4   **16.** $-\frac{5}{6}$ **17.** $\frac{12}{25}$ **18.** $\frac{15}{2}$
19. $\frac{67}{8}$ **20.** $-\frac{13}{3}$ **21.** $\frac{75}{7}$ **22.** $2\frac{1}{3}$ **23.** $-6\frac{3}{4}$
24. $12\frac{3}{5}$ **25.** $3\frac{1}{2}$ **26.** $-877\frac{1}{3}$ **27.** $81\frac{1}{3}$ **28.** $10\frac{2}{5}$
29. $11\frac{11}{15}$ **30.** -9 **31.** $1\frac{3}{4}$ **32.** $7\frac{7}{9}$ **33.** $4\frac{11}{15}$
34. $-5\frac{1}{8}$ **35.** $-14\frac{1}{4}$ **36.** $\frac{7}{9}x$ **37.** $3\frac{7}{40}a$ **38.** 16
39. $-3\frac{1}{2}$ **40.** $2\frac{21}{50}$ **41.** 6   **42.** 12   **43.** $-1\frac{7}{17}$
44. $\frac{1}{8}$ **45.** $\frac{9}{10}$ **46.** $13\frac{5}{7}$ **47.** $2\frac{8}{11}$ **48.** 15   **49.** $\$70\frac{3}{8}$
50. $\frac{3}{5}$ mi   **51.** $8\frac{3}{8}$ cups   **52.** $177\frac{3}{4}$ in^2 **53.** $50\frac{1}{4}$ in^2
54. $-\frac{6}{5}$ **55.** $-\frac{3}{2}$ **56.** $30 loss   **57.** $5a - 45$
58. ◆ The student multiplied the whole numbers and multiplied the fractions. The mixed numerals should be converted to fractional notation before multiplying.
59. ◆ Yes. We may need to find a common denominator before adding or subtracting. To find the least common denominator, we use the least common multiple of the denominators.
60. $3 \cdot 23 \cdot 41 \cdot 47 \cdot 59$, or 7,844,817   **61.** $\frac{600}{13}$
62. (a) 6; **(b)** 5; **(c)** 12; **(d)** 18; **(e)** -101; **(f)** -155; **(g)** -3; **(h)** -1 **63. (a)** 6; **(b)** 10; **(c)** 46; **(d)** 1; **(e)** -14; **(f)** -28; **(g)** -2; **(h)** -1

Test: Chapter 4, p. 269

1. [4.1a] 48   **2.** [4.2a] 3   **3.** [4.2b] $-\frac{5}{24}$ **4.** [4.3a] $\frac{2}{t}$
5. [4.3a] $\frac{1}{12}$ **6.** [4.3a] $-\frac{1}{12}$ **7.** [4.2c] >   **8.** [4.3b] $\frac{1}{4}$
9. [4.4a] $-\frac{12}{5}$ **10.** [4.4a] 18   **11.** [4.5a] $\frac{7}{2}$
12. [4.5a] $-\frac{79}{8}$ **13.** [4.5b] $-8\frac{2}{9}$ **14.** [4.5c] $162\frac{7}{11}$
15. [4.6a] $14\frac{1}{5}$ **16.** [4.6a] $14\frac{5}{12}$ **17.** [4.6b] $4\frac{7}{24}$
18. [4.6d] $8\frac{4}{7}$ **19.** [4.6d] $-5\frac{7}{10}$ **20.** [4.3a] $-\frac{1}{8}x$
21. [4.6b] $1\frac{54}{55}a$ **22.** [4.7a] 39   **23.** [4.7a] -18
24. [4.7b] 6   **25.** [4.7b] 2   **26.** [4.7c] $19\frac{3}{5}$
27. [4.7c] $28\frac{1}{20}$ **28.** [4.7d] $7\frac{1}{2}$ lb   **29.** [4.7d] 80
30. [4.5c] $148\frac{1}{2}$ lb   **31.** [4.3c] $\frac{13}{200}$ m   **32.** [2.7a] $9x - 54$
33. [3.7b] $\frac{8}{5}$ **34.** [3.6a] $\frac{10}{9}$ **35.** [2.3b] 1190 ft
36. [4.1a] $\frac{24}{25}$ min   **37.** [4.3c] Dolores; $\frac{17}{56}$ mi
38. [4.1a] **(a)** 24, 48, 72; **(b)** 24   **39.** [4.2a] **(a)** $\frac{1}{2}$; **(b)** $\frac{2}{3}$; **(c)** $\frac{3}{4}$; **(d)** $\frac{4}{5}$; **(e)** $\frac{9}{10}$

Cumulative Review: Chapters 1–4, p. 271

1. [1.1e] 5   **2.** [1.1a] 6 thousands + 7 tens + 5 ones
3. [1.1c] Twenty-nine thousand, five hundred
4. [1.2b] 623   **5.** [2.2a] -8 **6.** [4.2b] $\frac{5}{12}$ **7.** [4.6a] $8\frac{1}{4}$
8. [1.3d] 5124   **9.** [2.3a] 16   **10.** [4.3a] $-\frac{5}{12}$
11. [4.6b] $1\frac{1}{6}$ **12.** [1.5b] 5004   **13.** [2.4a] -145
14. [3.6a] $\frac{3}{2}$ **15.** [3.6a] -15 **16.** [4.7a] $7\frac{1}{3}$
17. [1.6c] 48 R 11   **18.** [1.6c] 56 R 11   **19.** [4.5c] $56\frac{11}{45}$
20. [3.7b] $-\frac{4}{7}$ **21.** [4.7b] $7\frac{1}{3}$ **22.** [1.4a] 38,500
23. [4.1a] 72   **24.** [3.1b] Yes   **25.** [3.2a] 1, 2, 4, 8, 16
26. [3.3b] $\frac{1}{6}$ **27.** [4.2c] >   **28.** [4.2c] <   **29.** [3.5b] $\frac{4}{5}$
30. [3.5b] -14 **31.** [4.5a] $\frac{37}{8}$ **32.** [4.5b] $-5\frac{2}{3}$
33. [1.7b], [2.8a] 93   **34.** [4.3b] $\frac{5}{9}$ **35.** [3.8a] $-\frac{12}{7}$
36. [4.4a] $\frac{2}{21}$ **37.** [2.6a] 4   **38.** [2.7a] $7b - 35$
39. [2.7a] $-3x + 6 - 3z$ **40.** [2.7b] $-6x - 9$
41. [1.8a] $235   **42.** [1.8a] $663   **43.** [1.8a] 297 ft^2
44. [1.8a] 31   **45.** [3.6b] $\frac{2}{5}$ tsp   **46.** [4.7d] 39 lb
47. [4.7d] 16   **48.** [4.2d] $\frac{19}{16}$ in.   **49.** [4.4a], [4.6a] $\frac{3}{7}$
50. [4.6c], [4.7d] 3780 m^2

Chapter 5

Pretest: Chapter 5, p. 274

1. [5.1a] Seventeen and three hundred sixty-nine thousandths   **2.** [5.1a] Six hundred twenty-five and $\frac{27}{100}$ dollars   **3.** [5.1b] $\frac{21}{100}$ **4.** [5.1b] $\frac{5408}{1000}$
5. [5.1c] 3.79   **6.** [5.1c] -0.0079 **7.** [5.1e] 21.0
8. [5.2a] 607.219   **9.** [5.2b] 91.732   **10.** [5.3a] 4.688
11. [5.2c] -252.9937 **12.** [5.4a] -30.4
13. [5.2d] $12.9a$ **14.** [5.2d] $-6.9x + 6.8$
15. [5.4b] -12.62 **16.** [5.6a] 7.6   **17.** [5.5c] 1.4
18. [5.5a] $4.\overline{142857}$ **19.** [5.5d] 1.7835
20. [5.5d] 4.34 m^2 **21.** [5.7a] 3.35   **22.** [5.7a] -84.26
23. [5.7a] 5.4   **24.** [5.7b] 6   **25.** [5.7b] 4.55
26. [5.8a] $285.95   **27.** [5.8a] 1081.6 mi
28. [5.8a] $89.70   **29.** [5.8a] $3397.71
30. [5.8a] 15.4 mpg

Margin Exercises, Section 5.1, pp. 276–280

1. Twenty and five tenths   **2.** Two and four thousand five hundred thirty-three ten-thousandths
3. Negative four hundred fifty-three and twenty-seven hundredths   **4.** Fifty-one thousand, seven hundred thirty-nine and eighty-two thousandths   **5.** Four thousand, two hundred seventeen and $\frac{56}{100}$ dollars
6. Thirteen and $\frac{98}{100}$ dollars   **7.** $\frac{896}{1000}$ **8.** $-\frac{3908}{100}$
9. $\frac{56,789}{10,000}$ **10.** $-\frac{37}{10}$ **11.** 7.43   **12.** 0.048
13. 6.7089   **14.** -0.9 **15.** -7.03 **16.** 23.047
17. 2.04   **18.** 0.06   **19.** 0.58   **20.** 1   **21.** 0.8989
22. 21.05   **23.** -34.008 **24.** -8.98 **25.** 2.8
26. 13.9   **27.** -234.4 **28.** 7.0   **29.** 0.64
30. -7.83 **31.** 34.70   **32.** -0.03 **33.** 0.943

34. −8.004 **35.** −43.112 **36.** 37.401 **37.** 7459.360
38. 7459.36 **39.** 7459.4 **40.** 7459 **41.** 7460
42. 7500 **43.** 7000

Exercise Set 5.1, p. 281

1. Four hundred eighty-one and twenty-seven hundredths **3.** One and five thousand five hundred ninety-nine ten-thousandths **5.** Thirty-four and eight hundred ninety-one thousandths **7.** Three hundred twenty-six and $\frac{48}{100}$ dollars **9.** Thirty-six and $\frac{72}{100}$ dollars **11.** $\frac{83}{10}$ **13.** $\frac{2036}{10}$ **15.** $-\frac{2703}{1000}$ **17.** $\frac{109}{10,000}$ **19.** $-\frac{6004}{1000}$ **21.** 0.8 **23.** −0.59 **25.** 3.798 **27.** 0.0078 **29.** −0.00018 **31.** 0.376193 **33.** 99.44 **35.** −8.431 **37.** 2.1739 **39.** 8.953073 **41.** 0.58 **43.** 0.91 **45.** −5.043 **47.** 235.07 **49.** $\frac{4}{100}$ **51.** −0.872 **53.** 0.1 **55.** −0.4 **57.** 3.0 **59.** −327.2 **61.** 0.89 **63.** −0.67 **65.** 1.00 **67.** −0.03 **69.** 0.325 **71.** 17.002 **73.** −20.202 **75.** 9.985 **77.** 809.5 **79.** 809.47 **81.** 0 **82.** $\frac{31}{45}$ **83.** $-\frac{2}{9}$ **84.** $\frac{2}{7}$ **85.** $\frac{29}{3}$ **86.** $-\frac{1}{14}$ **87.** ◆ **89.** ◆ **91.** 1979 **93.** 1998 **95.** 0.07070

Margin Exercises, Section 5.2, pp. 283–286

1. 10.917 **2.** 34.2079 **3.** 4.969 **4.** 3.5617 **5.** 9.40544 **6.** 912.67 **7.** 2514.773 **8.** 10.754 **9.** 0.339 **10.** 1.47 **11.** 0.24238 **12.** 7.36992 **13.** 1194.22 **14.** 4.9911 **15.** −1.96 **16.** 3.159 **17.** −8.55 **18.** −4.16 **19.** −7.44 **20.** 12.4 **21.** −2.7 **22.** $4.7x$ **23.** $1.7a$ **24.** $-2.7y + 5.4$

Exercise Set 5.2, p. 287

1. 334.37 **3.** 1576.215 **5.** 132.56 **7.** 64.413 **9.** 50.0248 **11.** 0.835 **13.** 771.967 **15.** 20.8649 **17.** 227.468 **19.** 2.1 **21.** 49.02 **23.** 2.4975 **25.** 85.921 **27.** 1.6666 **29.** 4.0622 **31.** 8.85 **33.** 3.37 **35.** 1.045 **37.** 3.703 **39.** 0.9092 **41.** 605.21 **43.** 0.994 **45.** 161.62 **47.** 44.001 **49.** 1.71 **51.** −2.5 **53.** −7.2 **55.** 3.379 **57.** −22.6 **59.** 2.5 **61.** −3.519 **63.** 9.601 **65.** 14.5 **67.** 3.8 **69.** −10.292 **71.** −8.8 **73.** $8.7x$ **75.** $4.86a$ **77.** $21.1t + 7.9$ **79.** $-5.311t$ **81.** $8.106y - 7.1$ **83.** $-0.9x + 3.1y$ **85.** $7.2 - 8.4t$ **87.** 0 **88.** $-\frac{1}{21}$ **89.** $-\frac{1}{10}$ **90.** $\frac{7}{3}$ **91.** 3 **92.** $-\frac{4}{9}$ **93.** ◆ **95.** ◆ **97.** $-18.183x - 0.02058y - 91.127z$ **99.** 345.8 **101.** $8744.16 should be $8744.17; $8764.65 should be $8723.68; $8848.65 should be $8808.68; $8801.05 should be $8760.08; $8533.09 should be $8492.13

Margin Exercises, Section 5.3, pp. 292–296

1. 625.66 **2.** 21.4863 **3.** 0.00943 **4.** −12.535374 **5.** 35.9 **6.** 0.7324 **7.** −0.058 **8.** 0.07236 **9.** 539.17 **10.** −6241.7 **11.** 64,700 **12.** 430,100 **13.** 4,300,000 **14.** $44,100,000,000 **15.** 1569¢ **16.** 17¢ **17.** $0.35 **18.** $5.77 **19.** 6.656 **20.** 8.125 sq cm **21.** 55.107

Calculator Spotlight, p. 296

2. (a) 8245.258176; (b) 1036.192585; (c) 11,814.1659

Exercise Set 5.3, p. 297

1. 47.6 **3.** 6.72 **5.** 0.252 **7.** 0.1032 **9.** 426.3 **11.** −783,686.852 **13.** −780 **15.** 7.918 **17.** 0.09768 **19.** −0.287 **21.** 43.68 **23.** 3.2472 **25.** 89.76 **27.** −322.07 **29.** 55.68 **31.** 3487.5 **33.** 0.1155 **35.** −9420 **37.** 0.00953 **39.** 2888¢ **41.** 66¢ **43.** $0.34 **45.** $34.45 **47.** $32,279,000,000 **49.** 1,030,000 **51.** 11,000 **53.** 26.025 **55.** (a) 44 ft; (b) 118.75 sq ft **57.** 1 **58.** $\frac{1}{15}$ **59.** $-\frac{1}{18}$ **60.** $\frac{2}{7}$ **61.** $\frac{3}{5}$ **62.** 0 **63.** ◆ **65.** ◆ **67.** 431.061084 **69.** 10^{15} **71.** $53.02

Margin Exercises, Section 5.4, pp. 299–304

1. 0.6 **2.** 1.5 **3.** 0.47 **4.** 0.32 **5.** −5.75 **6.** 0.25 **7.** (a) 375; (b) 15 **8.** 4.9 **9.** 12.8 **10.** 15.625 **11.** 12.78 **12.** 0.001278 **13.** 0.09847 **14.** −67.832 **15.** 0.2426 **16.** −7.4 **17.** 1.2825 billion

Calculator Spotlight, p. 300

1. 28 R 2 **2.** 116 R 3 **3.** 74 R 10 **4.** 415 R 3

Exercise Set 5.4, p. 305

1. 16.4 **3.** 23.78 **5.** 7.48 **7.** 7.2 **9.** −1.143 **11.** −0.9 **13.** 70 **15.** 40 **17.** −0.15 **19.** 48 **21.** 3.2 **23.** 0.625 **25.** 0.26 **27.** 2.34 **29.** −0.3045 **31.** −2.134567 **33.** 1023.7 **35.** −56,780 **37.** 9.7 **39.** −75,300 **41.** 230.01 **43.** 2107 **45.** −302.997 **47.** −178.1 **49.** 119.0176 **51.** −400.0108 **53.** 0.6725 **55.** 5.383 **57.** 10.5 **59.** 4.229 billion **61.** 59.49° **63.** $3\frac{1}{2}$ **64.** $-1\frac{1}{3}$ **65.** −8 **66.** $-\frac{29}{2}$, or $-14\frac{1}{2}$ **67.** 11 **68.** 15 **69.** ◆ **71.** ◆ **73.** 5.469156952 **75.** 10,000 **77.** 18.9 **79.** 590

Margin Exercises, Section 5.5, pp. 309–314

1. 0.4 **2.** 0.375 **3.** $0.1\overline{6}$ **4.** $0.\overline{6}$ **5.** $0.\overline{45}$ **6.** $-1.\overline{09}$ **7.** $0.\overline{714285}$ **8.** 0.7; 0.67; 0.667 **9.** 0.6; 0.61; 0.608 **10.** −7.3; −7.35; −7.349 **11.** 2.7; 2.69; 2.689 **12.** 0.8 **13.** −0.45 **14.** 0.035 **15.** 1.32 **16.** 0.72 **17.** 0.552 **18.** 4.225 ft²

Calculator Spotlight, p. 314

1. 123.150432 **2.** 52.59026102

Exercise Set 5.5, p. 315

1. 0.3125 **3.** 0.475 **5.** −0.2 **7.** 0.65 **9.** 0.425 **11.** 1.225 **13.** −0.52 **15.** 20.016 **17.** −0.25 **19.** 0.575 **21.** −0.625 **23.** 1.48 **25.** $0.5\overline{3}$ **27.** $0.\overline{3}$ **29.** $-1.\overline{3}$ **31.** $1.1\overline{6}$ **33.** $0.\overline{571428}$ **35.** $-0.91\overline{6}$ **37.** 0.5; 0.53; 0.533 **39.** 0.3; 0.33; 0.333

41. −1.3; −1.33; −1.333   **43.** 1.2; 1.17; 1.167
45. 0.6; 0.57; 0.571   **47.** −0.9; −0.92; −0.917
49. 0.7; 0.75; 0.747   **51.** −8.0; −7.97; −7.967
53. 9.485   **55.** −417.51$\overline{6}$ **57.** 0.09705   **59.** −1.5275
61. 24.375   **63.** 1.08 m²   **65.** 5.78 cm²   **67.** $3\frac{2}{5}$
68. $30\frac{7}{10}$ **69.** −95   **70.** −10   **71.** −7   **72.** 1
73. ◆   **75.** ◆   **77.** 0.$\overline{285714}$ **79.** 0.$\overline{571428}$
81. 0.$\overline{857142}$ **83.** 0.$\overline{01}$ **85.** 0.$\overline{0001}$ **87.** 6.16 cm²
89. 63.585 yd² or 63.61725124 yd²

Margin Exercises, Section 5.6, pp. 319–321

1. $540   **2.** $160   **3.** $1320   **4.** 4   **5.** 16   **6.** 18
7. 470   **8.** 0.07   **9.** 18   **10.** 125   **11.** (c)   **12.** (a)
13. (c)   **14.** (c)

Exercise Set 5.6, p. 323

1. $360   **3.** $50   **5.** $2700   **7.** 5   **9.** 1.6   **11.** 6
13. 60   **15.** 2.3   **17.** 180   **19.** (a)   **21.** (c)   **23.** (b)
25. (b)   **27.** 8000   **29.** 2 · 2 · 3 · 3 · 3
30. 2 · 2 · 2 · 2 · 5 · 5   **31.** 5 · 5 · 13   **32.** 2 · 3 · 3 · 37
33. $\frac{5}{16}$ **34.** $\frac{129}{251}$ **35.** $\frac{8}{9}$ **36.** $\frac{13}{25}$ **37.** ◆   **39.** ◆
41. No   **43.** Yes

Margin Exercises, Section 5.7, pp. 325–328

1. 4.2   **2.** 5.7   **3.** 2.6   **4.** −10.8   **5.** 3.6   **6.** −1.9
7. 3.5   **8.** 2.2   **9.** −4.5   **10.** 1.25

Exercise Set 5.7, p. 329

1. 5.4   **3.** 0.5239   **5.** −4.5   **7.** 10.3   **9.** 43.78
11. 7.7   **13.** −9.7   **15.** 5   **17.** 2.5   **19.** −2.8
21. −4.7   **23.** 1.7   **25.** 9   **27.** 1.25   **29.** −3.25
31. 30   **33.** 3.2   **35.** 4   **37.** 8.5   **39.** 2.9
41. 3.8575   **43.** −4.6   **45.** 7   **47.** −2.5   **49.** 1
50. $\frac{23}{18}$ **51.** $-\frac{29}{50}$ **52.** 0   **53.** −2   **54.** 8
55. ◆   **57.** ◆   **59.** −2.7   **61.** 1

Margin Exercises, Section 5.8, pp. 333–338

1. 8.4°   **2.** $37.28   **3.** $189.50   **4.** 28.6 mpg
5. 13.76 in²   **6.** 1.25 hr

Exercise Set 5.8, p. 339

1. $230.86   **3.** $26.50   **5.** $5.32   **7.** 8.17 mg
9. $368.75   **11.** 1.65 mg   **13.** $21,219.17
15. 4523.76 cm²   **17.** 93.9 km   **19.** $956.16
21. 78.1 cm   **23.** 28.5 cm   **25.** 2.31 cm
27. 20.2 mpg   **29.** $11.03   **31.** 331.74 ft²
33. 39.585 ft²   **35.** $505.75   **37.** 450.5 mi
39. 8.125 hr   **41.** 2.5 hr   **43.** 2152.56 yd²   **45.** 17
47. 0   **48.** $-\frac{1}{10}$ **49.** $-\frac{20}{33}$ **50.** −4   **51.** $6\frac{5}{6}$
52. 1   **53.** ◆   **55.** ◆   **57.** 3
59. 25 cm². We assume that the figures are nested squares formed by connecting the midpoints of consecutive sides of the next larger square.
61. $1.32

Summary and Review: Chapter 5, p. 345

1. Three and forty-seven hundredths
2. Five hundred ninety-seven and $\frac{25}{100}$ dollars
3. $\frac{9}{100}$ **4.** $-\frac{30,227}{10,000}$ **5.** −0.034   **6.** 27.91   **7.** 0.034
8. −0.19   **9.** 39.4   **10.** 39.43   **11.** 499.829
12. 29.148   **13.** 229.1   **14.** 685.0519   **15.** 57.3
16. 2.37   **17.** 12.96   **18.** −1.073   **19.** 24,680
20. 3.2   **21.** −1.6   **22.** 0.2763   **23.** 2.2x − 9.1y
24. −2.84a + 12.57   **25.** 925   **26.** 40.84   **27.** 11.3
28. 20   **29.** (c)   **30.** $15.49   **31.** 248.27   **32.** 2.6
33. 1.28   **34.** 3.25   **35.** −1.$\overline{16}$ **36.** 21.08   **37.** −3.2
38. −3   **39.** −7.5   **40.** 6.5   **41.** 11.16   **42.** 89.4 sec
43. 6365.1 bushels   **44.** $32.59   **45.** 24.36; 104.4
46. 8.4 mi   **47.** 1   **48.** $\frac{1}{18}$ **49.** $\frac{3}{10}$ **50.** $-\frac{1}{4}x - \frac{1}{4}y$
51. ◆ Each decimal place in the decimal notation corresponds to one zero in the denominator of the fractional notation. When the fractions are multiplied, the number of zeros in the denominator of the product is the sum of the number of zeros in each factor. So the number of decimal places in the product will be the sum of the number of decimal places in each factor.
52. ◆ The decimal points must be lined up before adding.   **53.** $\frac{-13}{15}, \frac{-17}{20}, -\frac{11}{13}, -\frac{15}{19}, \frac{-5}{7}, -\frac{2}{3}$ **54.** $35

Test: Chapter 5, p. 347

1. [5.1a] Six and four hundred one ten-thousandths
2. [5.1a] One thousand two hundred thirty-four and $\frac{78}{100}$ dollars   **3.** [5.1b] $-\frac{2}{10}$ **4.** [5.1b] $\frac{7308}{1000}$
5. [5.1c] 0.0049   **6.** [5.1c] −5.28   **7.** [5.1d] 0.162
8. [5.1d] −0.173   **9.** [5.1e] 9.5   **10.** [5.1e] 9.452
11. [5.2a] 405.219   **12.** [5.3a] 0.03   **13.** [5.3a] 0.21345
14. [5.2b] 44.746   **15.** [5.2a] 356.37   **16.** [5.2c] −2.2
17. [5.2b] 1.9946   **18.** [5.4a] 0.44   **19.** [5.4a] 30.4
20. [5.4a] −0.34682   **21.** [5.3b] 17,982¢
22. [5.2d] 9.8x − 3.9y − 4.6   **23.** [5.3c] 11.6
24. [5.4b] 7.6   **25.** [5.6a] 8 gal   **26.** [5.5b] 48.7
27. [5.5c] 1.6   **28.** [5.5c] 5.25   **29.** [5.5a] −0.4375
30. [5.5a] 1.$\overline{2}$ **31.** [5.5d] 9.72   **32.** [5.7a] −3.24
33. [5.7b] 10   **34.** [5.7b] 1.4   **35.** [5.8a] $1209.22
36. [5.8a] 33.46 sec   **37.** [5.8a] $119.70
38. [5.8a] 8.2 mi   **39.** [5.8a] 1.5 hr   **40.** [3.3c] 0
41. [4.2b] $\frac{3}{7}$ **42.** [4.3a] $-\frac{1}{30}$ **43.** [4.3a] $\frac{7}{10}x + \frac{1}{15}y$
44. [5.3a] (a) Always; (b) never; (c) sometimes; (d) sometimes

Cumulative Review: Chapters 1–5, p. 349

1. [1.1a] 1 ten thousand + 2 thousands + 7 hundreds + 5 tens + 8 ones   **2.** [5.1a] Eight hundred two and $\frac{53}{100}$ dollars   **3.** [5.1b] $\frac{1009}{100}$ **4.** [4.5a] $\frac{27}{8}$
5. [5.1c] −0.035   **6.** [3.2a] 1, 2, 3, 6, 11, 22, 33, 66
7. [3.2c] 2 · 3 · 11   **8.** [4.1a] 140   **9.** [1.4a] 7000
10. [5.1e] 6962.47   **11.** [4.6a] $6\frac{2}{9}$ **12.** [5.2a] 235.397
13. [1.2b] 5495   **14.** [4.2b] $-\frac{1}{30}$ **15.** [2.3a] 46
16. [5.2b] 8446.53   **17.** [4.3a] $\frac{1}{72}$ **18.** [4.6b] $3\frac{2}{5}$

19. [5.3a] 4.78   **20.** [3.6a] $-\frac{2}{7}$ **21.** [4.7a] $13\frac{7}{11}$
22. [3.6a] $\frac{3}{2}$ **23.** [4.7b] $1\frac{1}{2}$ **24.** [3.7b] $\frac{48}{35}$
25. [5.4a] $-43,795$ **26.** [5.4a] 20.6   **27.** [4.2c] >
28. [2.1b] >   **29.** [2.6a] -2 **30.** [2.7a] $4x - 4y + 12$
31. [2.7b] $7p - 5$ **32.** [2.7b] $14x - 11$ **33.** [5.7a] 0.78
34. [2.8b], [3.8b] -28 **35.** [5.7a] 8.62
36. [2.8a] 369,375   **37.** [4.3b] $\frac{1}{18}$ **38.** [3.8b] $\frac{1}{2}$
39. [5.7a] 3.8125   **40.** [4.4a] -15 **41.** [1.8a] 19,299
42. [3.7c] \$500   **43.** [1.8a] 86,400   **44.** [3.6b] \$2800
45. [5.8a] \$258.77   **46.** [4.6c] $6\frac{1}{2}$ lb   **47.** [3.6b] 88 ft²
48. [5.8a] 467.28 ft²   **49.** [4.7d] 144   **50.** [5.8a] \$1.58

Chapter 6

Pretest: Chapter 6, p. 352

1. [6.1a] **(a)** \$298; **(b)** \$172; **(c)** \$134
2. [6.2b]

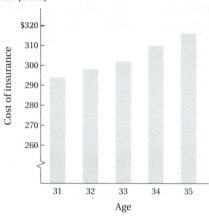

3. [6.2d]

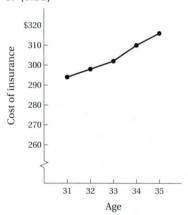

4. [6.2c] 260
5. [6.2c] 160 occurrences per 1000 people

6. [6.3a]

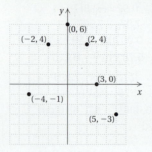

7. [6.3b] II   **8.** [6.3c] No
9. [6.4b]   **10.** [6.4b]

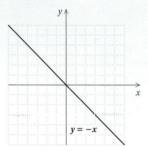   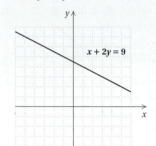

11. [6.4b]

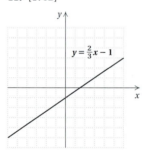

12. [6.5a, b, c] **(a)** 51; **(b)** 51.5; **(c)** None exists.
13. [6.5a, b, c] **(a)** 3; **(b)** 3; **(c)** 5, 1
14. [6.5a, b, c] **(a)** 12.75; **(b)** 17; **(c)** 4   **15.** [6.5a] 55 mph
16. [6.5a] 86   **17.** [6.6a] 50 per 1000 people
18. [6.6b] $\frac{1}{6}$

Margin Exercises, Section 6.1, pp. 353–356

1. Kellogg's Complete Bran Flakes   **2.** Kellogg's Special K   **3.** Kellogg's Special K, Wheaties
4. Ralston Rice Chex, Kellogg's Complete Bran Flakes, Honey Nut Cheerios   **5.** 510.66 mg   **6.** AT&T, LCI
7. GTE   **8.** 11.7¢/min   **9.** 14.7¢/min
10. 795 cups; answers may vary   **11.** 750 cups; answers may vary   **12.** 1650 cups; answers may vary
13.

Exercise Set 6.1, p. 357

1. 483,612,200 mi   **3.** Neptune   **5.** All   **7.** 11
9. 27,884.$\overline{1}$ mi   **11.** 1992   **13.** 59.29°, 59.58°
15. 59.50°, 59.56125°, 0.06125°   **17.** 1989 and 1990
19. 1.0 billion   **21.** 1999   **23.** 1650 and 1850
25. 2.0 billion   **27.** 1905   **29.** 8 hr   **31.** 4 hr/wk
33. 1991 and 1992, or 1992 and 1993, or 1995 and 1996
35.

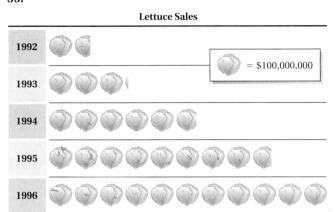

36. −2   **37.** −3   **38.** 1   **39.** 0.375   **40.** 0.06
41. 1.16   **42.** 0.8$\overline{3}$ **43.** ◆   **45.** ◆   **47.** 0.5$\overline{3}$ hr

Margin Exercises, Section 6.2, pp. 363–367

1. 16 g   **2.** Big Bacon Classic   **3.** Chicken Club, Big Bacon Classic, Single with Everything   **4.** 60
5. 85+   **6.** 60–64   **7.** Yes
8.

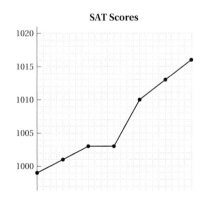

9. Month 7   **10.** Months 1 and 2, 4 and 5, 6 and 7, 11 and 12   **11.** Months 2, 5, 6, 7, 8, 9, 12
12.

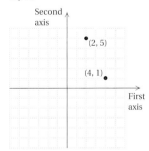

Exercise Set 6.2, p. 369

1. 190   **3.** 1 slice of chocolate cake with fudge frosting
5. 1 cup of premium chocolate ice cream
7. 120 calories   **9.** 920 calories   **11.** 28 lb
13. 920,000 hectares   **15.** Latin America
17. Africa   **19.** 880,000 hectares
21.

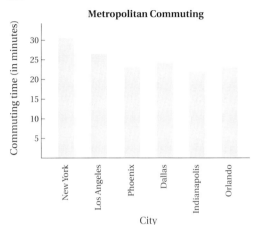

23. Indianapolis   **25.** 28.5 min
27. 580 calories   **29.** 1930 calories   **31.** 1998
33. About $0.55 million   **35.** 1994 and 1995
37.

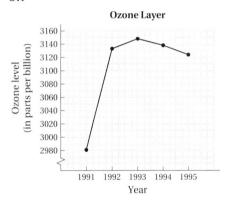

39. 1994 and 1995
41. 3135.75 parts per billion   **43.** 1995 and 1996
45. $48.35 million   **47.** 34   **48.** $\frac{9}{50}$ **49.** 72 fl oz
50. 32   **51.** 50   **52.** 18   **53.** ◆   **55.** ◆
57. 268 per 100,000

Margin Exercises, Section 6.3, pp. 375–378

1., 2.

3.–8.

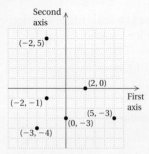

9. A: $(-5, 1)$; B: $(-3, 2)$; C: $(0, 4)$; D: $(3, 3)$; E: $(1, 0)$; F: $(0, -3)$; G: $(-5, -4)$ **10.** Both are negative numbers.
11. The first, or horizontal, coordinate is positive; the second, or vertical, coordinate is negative.   **12.** I
13. III   **14.** IV   **15.** II   **16.** No   **17.** Yes

Calculator Spotlight, p. 378

1. Yes   **2.** No   **3.** No   **4.** Yes   **5.** No   **6.** Yes
7. Yes   **8.** No

Exercise Set 6.3, p. 379

1.

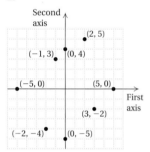

3.

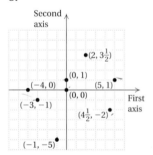

5. A: $(3, 3)$; B: $(0, -4)$; C: $(-5, 0)$; D: $(-1, -1)$; E: $(2, 0)$; F: $(-3, 5)$ **7.** A: $(5, 0)$; B: $(0, 5)$; C: $(-3, 4)$; D: $(2, -4)$; E: $(2, 3)$; F: $(-4, -2)$ **9.** II   **11.** IV   **13.** III   **15.** I
17. Positive; negative   **19.** III   **21.** IV; positive
23. Yes   **25.** No   **27.** Yes   **29.** No   **31.** Yes
33. No   **35.** 7   **36.** 9   **37.** 3   **38.** $\frac{30}{17}$ **39.** $\frac{61}{33}a$
40. $7x - 24$ **41.** ◆   **43.** ◆   **45.** No   **47.** III, IV
49. II, IV   **51.** $(5, 2)$; $(-7, 2)$; $(3, -8)$

53.

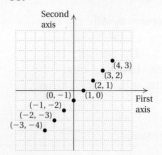

Answers may vary.
55. 65

Margin Exercises, Section 6.4, pp. 383–387

1. $(8, 5)$ **2.** $(1, 5)$; $(3, -5)$ **3.** $(1, 3)$, $(7, 0)$, $(5, 1)$
4. $(0, 7)$, $(2, 3)$, $(-2, 11)$
5. 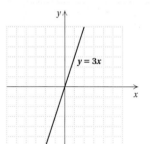   **6.**

7. 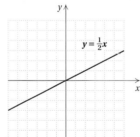   **8.**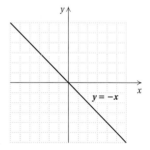

9. 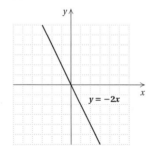   **10.**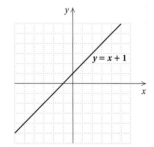

11. 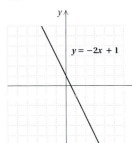   **12.** 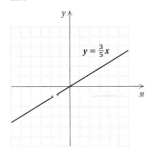   **45.** 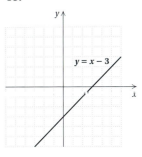   **47.**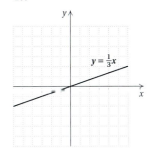

49. 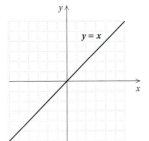   **51.**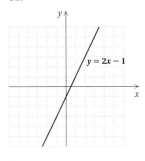

Calculator Spotlight, p. 388

1. $y = \frac{2}{3}x$

2. $y = x + 1$

53. 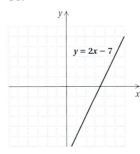   **55.**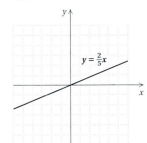

3. $y = -2x + 1$

4. $y = \frac{3}{5}x$

57.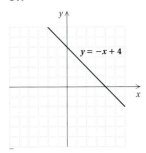

Exercise Set 6.4, p. 389

1. (8, 5)   **3.** (3, 1)   **5.** (5, 14)   **7.** (7, −2)   **9.** (1, 3)
11. (2, −1)   **13.** (−1, 8); (4, 3)   **15.** (7, 3); (10, 6)
17. (3, 3); (6, 1)   **19.** (3, −2); (1, 3)
21. (1, 4); (−2, −8)   **23.** $\left(0, \frac{3}{5}\right); \left(\frac{3}{2}, 0\right)$
25. (0, 12), (5, 7), (14, −2)   **27.** (0, 0), (1, 4), (2, 8)
29. (0, 13), (1, 10), (2, 7)   **31.** (0, −1), (2, 5), (−1, −4)
33. (0, 0), (1, −5), (−1, 5)   **35.** (0, −4), (4, 0), (1, −3)
37. (0, 6), (4, 0), $\left(1, \frac{9}{2}\right)$ **39.** (0, 2), (3, 3), (−3, 1)
41. 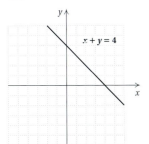   **43.**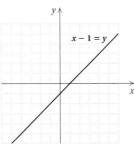

59. $1\frac{7}{8}$ cup   **60.** $-\frac{7}{11}$ **61.** 42   **62.** $-\frac{5}{8}$ **63.** 2.6
64. 2   **65.** ◆   **67.** ◆
69. (−3, 4.4), (−3.9, 5), (3, 0.4)

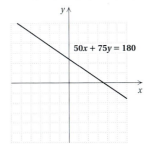

Chapter 6

A-17

71. (2, 4), (−3, −1), (−5, −3); answers may vary.
73. Answers may vary, but all should appear on this graph:

Margin Exercises, Section 6.5, pp. 393–396

1. 75 **2.** 54.9 **3.** 81 **4.** 19.4 **5.** 0.4 lb
6. 2.5 **7.** 94 **8.** 17 **9.** 17 **10.** 91 **11.** $1700
12. 67.5 **13.** 45 **14.** 34, 67 **15.** No mode exists
16. (a) 17 g; **(b)** 18 g; **(c)** 19 g2

Calculator Spotlight, p. 395

1. $203.\overline{3}$ **2.** The answers are the same.

Exercise Set 6.5, p. 397

1. Mean: 20; median: 18; mode: 29   **3.** Mean: 20; median: 20; modes: 5, 20   **5.** Mean: 5.2; median: 5.7; mode: 7.4   **7.** Mean: 137.5; median: 128.5; mode: 128
9. Mean: 21.2; median: 21; mode: 23   **11.** 33 mpg
13. 2.9   **15.** Mean: $9.19, median: $9.49; mode: $7.99
17. 90   **19.** 263 days   **21.** 196   **22.** 16   **23.** 1.96
24. $-\frac{10}{30}$ **25.** 2.5 hr   **26.** 6 hr   **27.** ◆   **29.** ◆
31. 171   **33.** 10   **35.** $3475

Margin Exercises, Section 6.6, pp. 399–402

1. 111 million   **2.** 94%   **3. (a)** $\frac{4}{25}$, or 0.16; **(b)** $\frac{6}{25}$, or 0.24; **(c)** $\frac{3}{5}$, or 0.6   **4. (a)** $\frac{1}{4}$, or 0.25; **(b)** $\frac{2}{13}$

Exercise Set 6.6, p. 403

1. 83   **3.** 3112 parts per billion   **5.** $148.8 billion
7. $27.30   **9.** $\frac{1}{6}$, or $0.1\overline{6}$ **11.** $\frac{1}{2}$, or 0.5   **13.** $\frac{1}{52}$
15. $\frac{2}{13}$ **17.** $\frac{3}{26}$ **19.** $\frac{4}{39}$ **21.** $\frac{34}{39}$ **23.** 3   **24.** $-\frac{5}{2}$
25. $-\frac{11}{5}$ **26.** 0.4   **27.** $1.1\overline{3}$ **28.** −0.625   **29.** ◆
31. ◆   **33.** $\frac{1}{2}$, or 0.5   **35.** $\frac{31}{366}$

Summary and Review: Chapter 6, p. 407

1. 179 lb   **2.** 118 lb   **3.** 5 ft, 3 in., medium frame
4. 5 ft, 11 in., medium frame
5.

6. 14,000   **7.** Los Angeles   **8.** Houston
9. 12,500   **10.** 2.5 million   **11.** Birds   **12.** Reptiles
13. 6 million   **14.** True   **15.** False   **16.** Under 20
17. Approximately 12
18. Approximately 13 per 100 drivers
19. Between 45 and 74   **20.** Approximately 11
21. Under 20   **22.** (−5, −1)   **23.** (−2, 5)   **24.** (3, 0)
25. (4, −2)
26.–28.

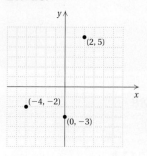

29. IV   **30.** III   **31.** I   **32.** No   **33.** Yes
34.   **35.**

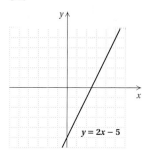 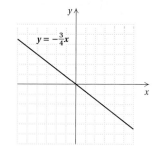

36.   **37.**

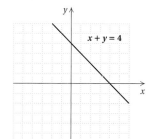 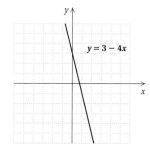

38. (a) 36; **(b)** 34; **(c)** 26   **39. (a)** 14; **(b)** 14; **(c)** 11, 17
40. (a) 322.5; **(b)** 375; **(c)** 470   **41. (a)** 1800; **(b)** 1900; **(c)** 700   **42. (a)** $17.50; **(b)** $17; **(c)** $17
43. (a) $37,500; **(b)** $27,500; **(c)** None exists.
44. $310.50, $307   **45.** $66\frac{1}{6}°$ **46.** 96
47. 28   **48.** $\frac{1}{52}$ **49.** $\frac{1}{2}$ **50.** −135   **51.** $\frac{1}{3}$ m²
52. −8   **53.** −1.45
54. ◆ It is not possible for the graph of a linear equation to pass through all four quadrants. Lines can be drawn through two quadrants, or three, but not four. Since the graph of a linear equation is a straight line, one will not pass through all four quadrants.
55. ◆ The attendance at four college football games was 21,500, 23,800, 24,400, and 27,000. What was the average attendance?

56. (1, 1.404255319), (0, 2.127659574), (2.941176471, 0);

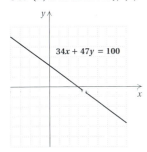

57. $11.52 **58.** 40.8 million **59.** 35.75 million
60. Yes **61.** Yes
62. **63.**

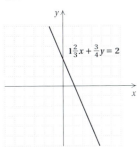

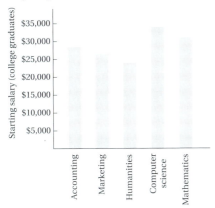

Test: Chapter 6, p. 411

1. [6.1a] Hiking with a 20-lb load **2.** [6.1a] Hiking with a 10-lb load
3. [6.2b]

4. [6.1b] 175 **5.** [6.1b] Alex Rodriguez
6. [6.1b] Aaron Boone **7.** [6.1b] Tony Gwynn and Tony Fernandez **8.** [6.2c] About $6.5 billion
9. [6.2c], [6.5a] About $5.3 billion
10. [6.2c] About $5.8 billion **11.** [6.2c] 1997
12. [6.2c], [6.5b] About $4.5 billion **13.** [6.6a] About $11.9 billion **14.** [6.3b] II **15.** [6.3b] III
16. [6.3a] (3, 4) **17.** [6.3a] (0, −4) **18.** [6.3a] (−4, 2)
19. [6.3c] Yes

20. [6.4b] **21.** [6.4b]

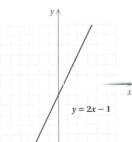

22. [6.4b]

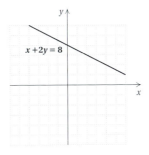

23. [6.5a] 50 **24.** [6.5a] 3 **25.** [6.5a] 15.5
26. [6.5b, c] Median: 50.5; mode: 54
27. [6.5b, c] Median: 3; no mode exists
28. [6.5b, c] Median: 17.5; modes: 17, 18
29. [6.5a] 58 km/h **30.** [6.5a] 76 **31.** [6.6b] $\frac{1}{4}$
32. [2.5a] −20 **33.** [3.5b] $\frac{5}{3}$ **34.** [3.4c] $\frac{1}{4}$ lb
35. [5.7a] −1.62
36. [6.4b] **37.** [6.4b]

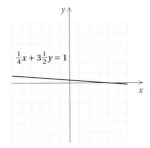 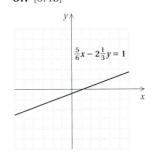

38. [6.3a] 56

Cumulative Review: Chapters 1–6, p. 415

1. [1.1a] 3 thousands + 6 hundreds + 7 tens + 1 one
2. [1.8a] 115 mi **3.** [1.1d] 8,000,000,000
4. [1.8a], [2.7c] Perimeter: 22 cm; area: 28 cm²
5. [1.8a] 216 g **6.** [1.9a] 7^4 **7.** [2.1a] −9, 4
8. [2.1b] > **9.** [4.2c] > **10.** [5.1d] < **11.** [2.1d] 5
12. [2.1d] 17 **13.** [2.6a] −2 **14.** [2.7b] $-2x + y$
15. [3.2a] 1, 2, 3, 4, 6, 9, 12, 18, 36 **16.** [3.1b] Yes
17. [2.6a] $\frac{7}{3}x$; $-\frac{7}{x}$ **18.** [2.7a] $10a - 15b + 5$
19. [3.3b] $\frac{3}{7}$ **20.** [3.5a] $\frac{10}{35}$ **21.** [1.3d] 29
22. [1.5b] 476 **23.** [2.5a] −9 **24.** [2.2a] −115
25. [4.2a] $\frac{5}{7}$ **26.** [3.7b] $\frac{5}{21}$ **27.** [4.3a] $\frac{13}{18}$ **28.** [4.2b] $\frac{1}{6}$
29. [3.6a] 1 **30.** [4.6a] $9\frac{1}{8}$ **31.** [4.6b] $2\frac{5}{12}x$

32. [4.7a] $13\frac{1}{5}$ **33.** [5.2a] 83.28 **34.** [5.4a] 62.345
35. [5.3a] −42.282 **36.** [3.3c] 1 **37.** [3.3c] $4x$
38. [3.3c] 0 **39.** [4.3b] $-\frac{16}{15}$ **40.** [4.4a] 32
41. [5.7b] $-\frac{17}{4}$ **42.** [6.3b] II
43. [6.4b]

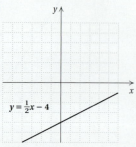

44. [6.5a] 36 **45.** [6.5b] 10.5 **46.** [6.5c] 49
47. [6.5a] $19\frac{2}{3}$ km/h **48.** [1.9c], [3.6a], [4.6b], [4.7a] $\frac{9}{32}$
49. [4.5b], [4.6a] $7\frac{47}{1000}$ **50.** [6.3a] (−2, −1), (−2, 7), (6, 7), (6, −1)

Chapter 7

Pretest: Chapter 7, p. 418

1. [7.1a] $\frac{31}{54}$ **2.** [7.1b] $\frac{3}{7}$ **3.** [7.2a] $24\frac{mi}{gal}$
4. [7.2b] $0.0656\frac{dollars}{oz}$ **5.** [7.2b] Cleary's
6. [7.3a] Yes **7.** [7.3b] 22.5 **8.** [7.3b] 0.75
9. [7.4a] 12 min **10.** [7.4a] Approximately 394 mi
11. [7.5b] $x = 3\frac{3}{4}$, $y = 2\frac{1}{2}$, $z = 3\frac{3}{4}$ **12.** [7.5a] 60 ft

Margin Exercises, Section 7.1, pp. 419–420

1. $\frac{5}{11}$ **2.** $\frac{57.3}{86.1}$ **3.** $\frac{6\frac{3}{4}}{7\frac{2}{5}}$ **4.** $\frac{7410}{28,500}$ **5.** $\frac{21.1}{182.5}$ **6.** $\frac{5}{9}$ **7.** $\frac{9}{5}$
8. $\frac{73}{27}$ **9.** $\frac{3}{4}$ **10.** 18 is to 27 as 2 is to 3.
11. 3.6 is to 12 as 3 is to 10.
12. 1.2 is to 1.5 as 4 is to 5.

Exercise Set 7.1, p. 421

1. $\frac{4}{5}$ **3.** $\frac{178}{572}$ **5.** $\frac{0.4}{12}$ **7.** $\frac{3.8}{7.4}$ **9.** $\frac{56.78}{98.35}$ **11.** $\frac{8\frac{3}{4}}{9\frac{5}{6}}$
13. $\frac{1}{4}$; $\frac{3}{1}$ **15.** $\frac{4}{1}$ **17.** $\frac{478}{213}$, $\frac{213}{478}$ **19.** $\frac{2}{3}$ **21.** $\frac{3}{5}$ **23.** $\frac{12}{25}$
25. $\frac{7}{9}$ **27.** $\frac{2}{3}$ **29.** $\frac{14}{25}$ **31.** $\frac{1}{2}$ **33.** $\frac{3}{4}$ **35.** $\frac{51}{49}$ **37.** $\frac{32}{101}$
39. < **40.** < **41.** > **42.** < **43.** $6\frac{7}{20}$ cm
44. $17\frac{11}{20}$ cm **45.** ◆ **47.** ◆ **49.** 1.011399197 to 1
51. $\frac{1}{3}$; $\frac{3}{2}$

Margin Exercises, Section 7.2, pp. 423–424

1. 5 mi/hr **2.** 12 mi/hr **3.** 0.3 mi/hr
4. 1100 ft/sec **5.** 4 ft/sec **6.** 14.5 ft/sec
7. 250 ft/sec **8.** 2 gal/day **9.** 20.64¢/oz
10. Container B

Exercise Set 7.2, p. 425

1. 40 km/h **3.** 11 m/sec **5.** 152 yd/day
7. 25 mi/hr; 0.04 hr/mi **9.** 57.5¢/min
11. 0.623 gal/ft^2 **13.** 1100 ft/sec **15.** 124 km/h
17. 560 mi/hr **19.** $19.50/yd **21.** 20.59¢/oz
23. $4.34/lb **25.** Tico's **27.** Dell's **29.** Shine
31. Big Net **33.** Four 12-oz bottles **35.** B
37. Bert's **39.** 1.7 million **40.** $25\frac{1}{2}$ **41.** 109.608
42. 67,819 **43.** IV **44.** II **45.** ◆ **47.** ◆
49. For about a month, the unit price decreased by 0.8¢/oz. Then it changed to the same unit cost as the 16-oz bottle. **51.** 16-in. pizza
53. About 7.27 at-bats per home run.

Margin Exercises, Section 7.3, pp. 430–432

1. Yes **2.** No **3.** No **4.** Yes **5.** No **6.** 14
7. $11\frac{1}{4}$ **8.** 10.5 **9.** 9 **10.** 10.8

Exercise Set 7.3, p. 433

1. No **3.** Yes **5.** Yes **7.** No **9.** 45 **11.** 12
13. 10 **15.** 20 **17.** 5 **19.** 18 **21.** 22 **23.** 28
25. $9\frac{1}{3}$ or $\frac{28}{3}$ **27.** $2\frac{8}{9}$ or $\frac{26}{9}$ **29.** 0.06 **31.** 5
33. 1 **35.** 1 **37.** 0 **39.** $\frac{51}{16}$, or $3\frac{3}{16}$ **41.** 12.5725
43. $\frac{1748}{249}$, or $7\frac{5}{249}$ **45.** III **46.** II **47.** IV **48.** I
49. −52 **50.** −19.75 **51.** 290.5 **52.** $1523.\overline{1}$
53. ◆ **55.** ◆ **57.** Approximately 66.85 **59.** −10.5

Margin Exercises, Section 7.4, pp. 435–436

1. 1120 mi **2.** 16 gal **3.** 8 **4.** $42\frac{3}{4}$ in. or less

Exercise Set 7.4, p. 437

1. 9.75 gal **3.** 175 **5.** 2975 ft^2 **7.** 450 **9.** 60
11. 13,500 mi **13.** 120 lb **15.** 64 cans **17.** 650 mi
19. 7
21.

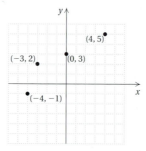

22. −79.13 **23.** −2.3 **24.** −7.3 **25.** −7.2
26. −1.8 **27.** ◆ **29.** ◆ **31.** $2990 **33.** $140,000

Margin Exercises, Section 7.5, pp. 439–441

1. 15 **2.** 24.75 ft **3.** 34.9 ft **4.** 21 cm **5.** 29 ft

Calculator Spotlight 7.5, p. 442

1. 27.5625 **2.** 25.6 **3.** 15.140625 **4.** 41.6875
5. 40.03952941 **6.** 39.74857143 **7.** 2.615217391
8. 119

Exercise Set 7.5, p. 443

1. 25 **3.** $\frac{4}{3}$, or $1\frac{1}{3}$ **5.** $x = \frac{27}{4}$, or $6\frac{3}{4}$; $y = 9$
7. $x = 7.5$; $y = 7.2$ **9.** 1.25 m **11.** 36 ft
13. 7 ft **15.** 100 ft **17.** 4 **19.** $10\frac{1}{2}$
21. $x = 6$; $y = 5.25$; $z = 3$ **23.** $x = 5\frac{1}{3}$, or $5.\overline{3}$; $y = 4\frac{2}{3}$, or $4.\overline{6}$; $z = 5\frac{1}{3}$, or $5.\overline{3}$ **25.** 20 ft **27.** $59.81
28. 9.63 **29.** 679.4020 **30.** 2.74568 **31.** 27,456.8
32. 0.549136 **33.** ◆ **35.** ◆ **37.** 19.25 ft
39. $x \approx 0.35$, $y \approx 0.4$ **41.** 86.68 cm^2

Summary and Review: Chapter 7, p. 447

1. $\frac{37}{85}$ **2.** $\frac{9}{10}$ **3.** $\frac{9}{11}$ **4.** $\frac{20}{57}$ km/min; 2.85 min/km
5. 3 min/mi **6.** $0.27/oz **7.** BiteFine **8.** No
9. 32 **10.** $\frac{1}{40}$ **11.** 7 **12.** 24 **13.** $6.65 **14.** 351
15. 6 **16.** 832 km **17.** 2,433,600 kg **18.** 9906
19. $\frac{14}{3}$ **20.** $x = \frac{56}{5}$, $y = \frac{63}{5}$ **21.** 40 ft
22. $x = 3$, $y = 9$, $z = 7\frac{1}{2}$ **23.** $1630.40
24. III, II, I, IV **25.** −10,672.74 **26.** 45.5
27. ◆ It took Leslie 4 gal of gasoline to drive 92 mi. How many gallons would she need to go 368 mi?
28. ◆ In terms of cost, a low faculty-to-student ratio is less expensive. In terms of quality education and student satisfaction, a high faculty-to-student ratio is better. To ensure a good education, the student-to-faculty ratio should be low.
29. $\frac{1827}{12,704} \approx 0.14$ **30.** $\frac{3121}{12,754} \approx 0.24$ **31.** $35.53/person
32. $100.75/person **33.** 28 gal **34.** 105 min

Test: Chapter 7, p. 449

1. [7.1a] $\frac{83}{94}$ **2.** [7.1a] $\frac{0.34}{124}$ **3.** [7.1b] $\frac{32}{15}$
4. [7.2a] $\frac{5}{8}$ ft/sec **5.** [7.2a] $1\frac{2}{3}$ min/km
6. [7.2b] $0.35/oz **7.** [7.3a] No **8.** [7.3b] 12
9. [7.3b] 360 **10.** [7.4a] 1512 km **11.** [7.4a] 4.8 min
12. [7.4a] 525 mi **13.** [7.5a] 66 m **14.** [7.5a] $x = 8$, $y = 8.8$ **15.** [7.5b] $x = 8$, $y = 8$, $z = 12$
16. [5.8a] 25.8 million lb **17.** [6.3b] II, IV, I, III
18. [5.3a] 173.736 **19.** [5.4a] −0.9944
20. [7.3b] $\frac{24}{13}$ **21.** [7.3b] $\frac{41}{10}$ **22.** [7.4a] 5888
23. [7.4a] $47.91 **24.** [7.5a] 10.5 ft

Cumulative Review: Chapters 1–7, p. 451

1. [5.2a] 643.502 **2.** [4.6a] $6\frac{3}{4}$ **3.** [4.2b] $\frac{7}{20}$
4. [5.2b] 50.501 **5.** [2.3a] −17 **6.** [4.3a] $\frac{7}{60}$
7. [5.3a] 222.076 **8.** [2.4a] −645 **9.** [4.7a] 3
10. [5.4a] 43 **11.** [2.5a] −51 **12.** [3.7b] $\frac{3}{2}$
13. [1.1a] 3 ten thousands + 7 tens + 4 ones
14. [5.1a] One hundred twenty and seven hundredths **15.** [5.1d] 0.7 **16.** [5.1d] −0.799
17. [3.2c] $2 \cdot 2 \cdot 2 \cdot 2 \cdot 3 \cdot 3$ **18.** [4.1a] 220
19. [3.3b] $\frac{5}{8}$ **20.** [3.5b] $\frac{5}{8}$ **21.** [5.5d] 5.718
22. [5.4b] −25.56 **23.** [6.5a] 48.75 **24.** [7.3a] Yes

25. [6.4b]

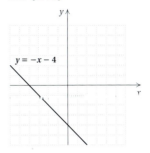

26. [2.6a] 5 **27.** [7.3b] $30\frac{6}{25}$ **28.** [2.8b], [3.8a] $-\frac{423}{16}$, or $-26\frac{7}{16}$ **29.** [4.4a] −4 **30.** [5.7b] 3
31. [5.7a] 33.34 **32.** [3.8a] $\frac{8}{9}$ **33.** [7.4a] 42.2025 mi
34. [7.4a] 7 min **35.** [5.8a] 976.9 mi
36. [1.8a] 414,000 **37.** [3.6b] 10 ft^2 **38.** [3.7c] 12
39. [4.6c] $2\frac{1}{4}$ cups **40.** [5.8a] 132 **41.** [7.2a] 60 mph
42. [7.2b] The 12-oz bag **43.** [7.4a] No; the money will be gone after 24 weeks. Hans will need $400 more. **44.** [7.4a] **(a)** CD player: $200; receiver–amplifier: $600; speakers: $400; **(b)** CD player: $500; receiver–amplifier: $1500; speakers: $1000

Chapter 8

Pretest: Chapter 8, p. 454

1. [8.1b] 0.87 **2.** [8.1b] 53.7% **3.** [8.1c] 75%
4. [8.1c] $\frac{37}{100}$ **5.** [8.3a, b] $a = 60\% \cdot 75$; 45
6. [8.2a, b] $\frac{n}{100} = \frac{35}{50}$; 70% **7.** [8.4a] 90 lb
8. [8.4b] 20% **9.** [8.5a] $17.16; $303.16
10. [8.5b] $5152
11. [8.5c] $112.50 discount; $337.50 sale price
12. [8.6a] $99.60 **13.** [8.6a] $20 **14.** [8.6b] $6242.40

Margin Exercises, Section 8.1, pp. 455–460

1.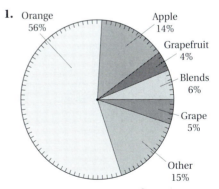

2. $\frac{70}{100}$; $70 \times \frac{1}{100}$; 70×0.01 **3.** $\frac{23.4}{100}$; $23.4 \times \frac{1}{100}$; 23.4×0.01 **4.** $\frac{100}{100}$; $100 \times \frac{1}{100}$; 100×0.01 **5.** 0.34
6. 0.789 **7.** 1.045 **8.** 0.042 **9.** 24% **10.** 347%
11. 100% **12.** 40% **13.** 25% **14.** 25% **15.** 87.5%
16. $66\frac{2}{3}\%$ **17.** 187.5% **18.** 57% **19.** 76% **20.** $\frac{9}{5}$
21. $\frac{13}{400}$ **22.** $\frac{2}{3}$

23.

$\frac{1}{5}$	$\frac{5}{6}$	$\frac{3}{8}$
0.2	0.83$\overline{3}$	0.375
20%	83$\frac{1}{3}$%	37$\frac{1}{2}$%

Calculator Spotlight, p. 459

1. 52% **2.** 38.46% **3.** 107.69% **4.** 171.43%
5. 59.62% **6.** 28.31%

Exercise Set 8.1, p. 461

1.

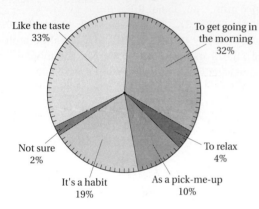

Reasons for Drinking Coffee
Like the taste 33%
To get going in the morning 32%
Not sure 2%
To relax 4%
It's a habit 19%
As a pick-me-up 10%

3. $\frac{90}{100}$; $90 \times \frac{1}{100}$; 90×0.01 **5.** $\frac{12.5}{100}$; $12.5 \times \frac{1}{100}$; 12.5×0.01 **7.** 0.12 **9.** 0.347 **11.** 0.5901
13. 0.1 **15.** 0.01 **17.** 3 **19.** 0.006 **21.** 0.0023
23. 1.0524 **25.** 0.4 **27.** 0.025 **29.** 0.622 **31.** 47%
33. 3% **35.** 870% **37.** 33.4% **39.** 75% **41.** 40%
43. 89.25% **45.** 0.0104% **47.** 24% **49.** 32.6%
51. 41% **53.** 5% **55.** 20% **57.** 28% **59.** 15%
61. 62.5%, or 62$\frac{1}{2}$% **63.** 80% **65.** 66.$\overline{6}$%, or 66$\frac{2}{3}$%
67. 58% **69.** 36% **71.** $\frac{17}{20}$ **73.** $\frac{5}{8}$ **75.** $\frac{1}{3}$ **77.** $\frac{1}{6}$
79. $\frac{29}{400}$ **81.** $\frac{1}{125}$ **83.** 21% **85.** 24% **87.** $\frac{11}{20}$
89. $\frac{19}{50}$ **91.** $\frac{11}{100}$ **93.** $\frac{33}{125}$ **95.** $\frac{457}{1000}$

97.

Fractional Notation	Decimal Notation	Percent Notation
$\frac{1}{8}$	0.125	12$\frac{1}{2}$%, or 12.5%
$\frac{1}{6}$	0.1$\overline{6}$	16$\frac{2}{3}$%, or 16.$\overline{6}$%
$\frac{1}{5}$	0.2	20%
$\frac{1}{4}$	0.25	25%
$\frac{1}{3}$	0.$\overline{3}$	33$\frac{1}{3}$%, or 33.$\overline{3}$%
$\frac{3}{8}$	0.375	37$\frac{1}{2}$%, or 37.5%
$\frac{2}{5}$	0.4	40%
$\frac{1}{2}$	0.5	50%

99.

Fractional Notation	Decimal Notation	Percent Notation
$\frac{1}{2}$	0.5	50%
$\frac{1}{3}$	0.$\overline{3}$	33$\frac{1}{3}$%, or 33.$\overline{3}$%
$\frac{1}{4}$	0.25	25%
$\frac{1}{6}$	0.1$\overline{6}$	16$\frac{2}{3}$%, or 16.$\overline{6}$%
$\frac{1}{8}$	0.125	12$\frac{1}{2}$%, or 12.5%
$\frac{3}{4}$	0.75	75%
$\frac{5}{6}$	0.8$\overline{3}$	83$\frac{1}{3}$%, or 83.$\overline{3}$%
$\frac{3}{8}$	0.375	37$\frac{1}{2}$%, or 37.5%

101. 33$\frac{1}{3}$ **102.** $-37\frac{1}{2}$ **103.** 400 **104.** 18.75
105. 23.125 **106.** 25.5 **107.** ◈ **109.** ◈
111. 5.$\overline{405}$% **113.** 329.$\overline{38479}$% **115.** 0.0158$\overline{3}$
117. 1.061$\overline{6}$

Margin Exercises, Section 8.2, pp. 468–470

1. $\frac{12}{100} = \frac{a}{50}$ **2.** $\frac{40}{100} = \frac{a}{60}$ **3.** $\frac{130}{100} = \frac{a}{72}$ **4.** $\frac{20}{100} = \frac{45}{b}$
5. $\frac{120}{100} = \frac{60}{b}$ **6.** $\frac{N}{100} = \frac{16}{40}$ **7.** $\frac{N}{100} = \frac{10.5}{84}$
8. $225 **9.** 35.2 **10.** 6 **11.** 50 **12.** 30%
13. 12.5%

Exercise Set 8.2, p. 471

1. $\frac{37}{100} = \frac{a}{74}$ **3.** $\frac{N}{100} = \frac{4.3}{5.9}$ **5.** $\frac{25}{100} = \frac{14}{b}$
7. $\frac{9}{10} = \frac{37}{b}$ **9.** $\frac{70}{100} = \frac{a}{660}$ **11.** 40 **13.** 2.88
15. 25% **17.** 102% **19.** 25% **21.** 93.75% **23.** $72
25. 90 **27.** 88 **29.** 20 **31.** 25 **33.** $780.20
35. 7.9

37.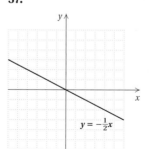

$y = -\frac{1}{2}x$

38.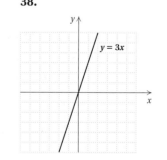

$y = 3x$

Answers

A-22

39. 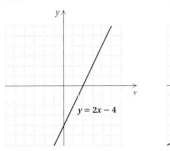 **40.**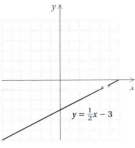

41. $\frac{43}{48}$ qt **42.** $\frac{1}{8}$ T **43.** ◆ **45.** ◆
47. 12,500 (can vary); 12,448

Margin Exercises, Section 8.3, pp. 473–476

1. 12% · 50 = a **2.** a = 40% · 60 **3.** 45 = 20% · t
4. 120% · y = 60 **5.** 16 = n · 40 **6.** n · 84 = 10.5
7. 6 **8.** $35.20 **9.** 225 **10.** $50 **11.** 40%
12. 12.5%

Exercise Set 8.3, p. 477

1. y = 32% · 78 **3.** 89 = a · 99 **5.** 13 = 25% · y
7. 234.6 **9.** 45 **11.** $18 **13.** 1.9 **15.** 78%
17. 200% **19.** 50% **21.** 125% **23.** 40 **25.** $40
27. 88 **29.** 20 **31.** 6.25 **33.** $846.60 **35.** 41.1
37. $\frac{623}{1000}$ **38.** $\frac{19}{10}$ **39.** $\frac{237}{100}$ **40.** 0.009 **41.** 0.39
42. 5.7 **43.** ◆ **45.** ◆
47. 62.5% (can vary); 64.5% **49.** $1875

Margin Exercises, Section 8.4, pp. 479–482

1. 16% **2.** 5 lb **3.** 4% **4.** $19,075 **5.** 9%

Exercise Set 8.4, p. 483

1. 20; 100 **3.** 402 **5.** 32.5%; 67.5%
7. 20.4 mL; 659.6 mL **9.** 25%
11. 166; 156; 146; 140; 122 **13.** 8% **15.** 20%
17. $30,030 **19.** $12,600
21. 6.096 billion; 6.194 billion; 6.293 billion
23. $36,400 **25.** $17.25; $39.10; $56.35
27. (a) 4.1%; (b) 31.0% **29.** 35.9% **31.** 2.$\overline{27}$
32. 0.44 **33.** 3.375 **34.** 4.$\overline{7}$ **35.** 0.92 **36.** 0.8$\overline{3}$
37. 0.4375 **38.** 2.317 **39.** 3.4809 **40.** 0.675
41. ◆ **43.** ◆ **45.** About 5 ft, 6 in. **47.** 83$\frac{1}{3}$%
49. 19%

Margin Exercises, Section 8.5, pp. 487–492

1. $40.14; $709.09 **2.** $1.03; $15.78 **3.** 6% **4.** $420
5. $5628 **6.** 25% **7.** $1675 **8.** $33.60; $106.40
9. 20%

Calculator Spotlight, p. 492

1. 50 **2.** 7001.88 **3.** 83,931.456 **4.** 36,458.03724

Exercise Set 8.5, p. 493

1. $29.30; $615.30 **3.** $16.56; $281.56 **5.** 4% **7.** 5%
9. $5000 **11.** $800 **13.** $719.86 **15.** 5.6%
17. $903 **19.** 50% **21.** $980 **23.** $5880 **25.** 12%
27. $420 **29.** $30; $270 **31.** $2.55; $14.45
33. $125; $112.50 **35.** 40%; $360 **37.** $387; 30.4%
39. $460; about 18% **41.** 18 **42.** $\frac{22}{7}$
43. 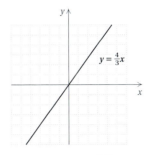 **44.**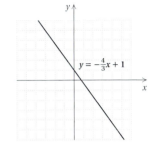

45. 0.$\overline{5}$ **46.** 2.$\overline{09}$ **47.** −0.91$\overline{6}$ **48.** −1.$\overline{857142}$
49. ◆ **51.** ◆ **53.** $91,112.50 **55.** $17,700
57. He bought the plaques for 166\frac{2}{3}$ + $250, or 416\frac{2}{3}$, and sold them for $400, so he lost money.

Margin Exercises, Section 8.6, pp. 497–500

1. $602 **2.** $451.50 **3.** $27.62; $4827.62
4. $2464.20 **5.** $8103.38

Calculator Spotlight, p. 500

1. $16,357.18 **2.** $12,764.72

Exercise Set 8.6, p. 501

1. $26 **3.** $124 **5.** $150.50 **7.** (a) $147.95;
(b) $10,147.95 **9.** (a) $128.22; (b) $6628.22
11. (a) $46.03; (b) $5646.03 **13.** $484 **15.** $236.75
17. $4284.90 **19.** $2604.52 **21.** $4101.01
23. $1324.58 **25.** 4.5 **26.** 8$\frac{3}{4}$ **27.** 8 **28.** 87.5
29. −3$\frac{13}{17}$ **30.** 3$\frac{5}{11}$ **31.** $\frac{18}{17}$ **32.** $\frac{209}{10}$ **33.** ◆
35. ◆ **37.** $33,981.98 **39.** 9.38%
41. $7883.24

Summary and Review: Chapter 8, p. 505

1. 48.3% **2.** 36% **3.** 37.5%, or 37$\frac{1}{2}$% **4.** 33.$\overline{3}$%,
or 33$\frac{1}{3}$% **5.** 0.735 **6.** 0.065 **7.** $\frac{6}{25}$ **8.** $\frac{63}{1000}$
9. 30.6 = x × 90; 34% **10.** 63 = 84% × n; 75
11. y = 38$\frac{1}{2}$% × 168; 64.68 **12.** $\frac{24}{100} = \frac{16.8}{b}$; 70
13. $\frac{42}{30} = \frac{N}{100}$; 140% **14.** $\frac{10.5}{100} = \frac{a}{84}$; 8.82 **15.** $598
16. 12% **17.** 82,400 **18.** 20% **19.** 168 **20.** $14.40
21. 5% **22.** 11% **23.** $42; $308 **24.** $42.70;
$262.30 **25.** $2940 **26.** $36 **27.** (a) $394.52;
(b) $24,394.52 **28.** $121 **29.** $7727.26

30. $9504.80 **31.** Approximately 25% **32.** $\frac{56}{3}$
33. 42
34. **35.**

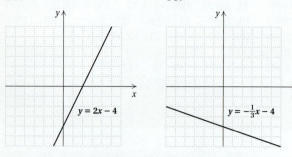

36. $3.\overline{6}$ **37.** $1.\overline{571428}$ **38.** $3\frac{2}{3}$ **39.** $17\frac{2}{7}$
40. ◆ No; the 10% discount was based on the original price rather than on the sale price.
41. ◆ A 40% discount is better. When successive discounts are taken, each is based on the previous discounted price, not the original price. A 20% discount followed by a 22% discount is the same as a 37.6% discount off the original price.
42. 19.5% **43.** $66\frac{2}{3}$% **44.** $168 **45.** 7

Test: Chapter 8, p. 507

1. [8.1b] 0.89 **2.** [8.1b] 67.4% **3.** [8.1c] 137.5%
4. [8.1c] $\frac{13}{20}$ **5.** [8.3a, b] $a = 40\% \cdot 55$; 22
6. [8.2a, b] $\frac{N}{100} = \frac{65}{80}$; 81.25% **7.** [8.4a] 28.75 lb
8. [8.4b] 140% **9.** [8.5a] $16.20; $340.20
10. [8.5b] $630 **11.** [8.5c] $40; $160 **12.** [8.6a] $8.52
13. [8.6a] $5356 **14.** [8.6b] $1102.50
15. [8.6b] $10,226.69 **16.** [8.5c] $131.95; 52.8%
17. [6.4b]

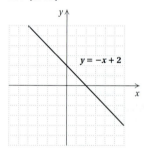

18. [7.3b] 16 **19.** [5.1c] $1.41\overline{6}$ **20.** [4.5b] $3\frac{21}{44}$
21. [8.5b] $117,800 **22.** [8.5b], [8.6b] $2546.16
23. (a) [3.6a], [4.2d] #1 pays $\frac{1}{25}$, #2 pays $\frac{9}{100}$, #3 pays $\frac{47}{300}$, #4 pays $\frac{77}{300}$, #5 pays $\frac{137}{300}$; (b) [8.1c], [8.4a] 4%, 9%, $15\frac{2}{3}$%, $25\frac{2}{3}$%, $45\frac{2}{3}$%; (c) [8.1c], [8.4a] 87%

Cumulative Review: Chapters 1–8, p. 507

1. [5.1b] $\frac{91}{1000}$ **2.** [5.5a] $2.1\overline{6}$ **3.** [8.1b] 0.03
4. [8.1c] 112.5% **5.** [7.1a] $\frac{5}{0.5}$ **6.** [7.2a] $23\frac{1}{3}$ km/h

7. [4.2c] < **8.** [5.1d] > **9.** [5.6a] 296,200
10. [1.4b] 50,000 **11.** [1.9c] 13
12. [2.7b] $-2x - 14$ **13.** [4.6a] $3\frac{1}{30}$
14. [5.2c] -14.2 **15.** [1.2b] 515,150
16. [5.2b] 0.02 **17.** [4.6b] $\frac{2}{3}$ **18.** [4.3a] $-\frac{110}{63}$
19. [3.6a] $\frac{1}{6}$ **20.** [2.4a] -384
21. [5.3a] 1.38036 **22.** [4.7b] $1\frac{1}{2}$ **23.** [5.4a] 12.25
24. [1.6c] 123 R 5 **25.** [2.8b], [3.8a] 95
26. [5.7a] 8.13 **27.** [3.8a] -9 **28.** [4.3b] $\frac{1}{12}$
29. [5.7b] $-\frac{16}{9}$ **30.** [7.3b] $8\frac{8}{21}$ **31.** [6.3b] III
32. [6.4b]

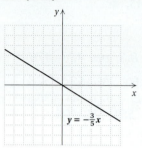

33. [6.5a] 33.2 **34.** [6.5b] 12 **35.** [2.7c] 60 in.
36. [1.5a] 3200 yd² **37.** [7.2b] $15\frac{¢}{\text{oz}}$ **38.** [7.4a] 608 km
39. [8.4b] 43.2% **40.** [8.4a] 1,033,710 mi²
41. [4.7d] 5 **42.** [4.2d] $\frac{3}{2}$ km
43. [8.4b] 12.5% increase **44.** [4.7d], [6.5a] 33.6 mpg
45. [6.2a] 1906, 1912, 1920, 1924, 1928 **46.** [6.2a] 1977

Chapter 9

Pretest: Chapter 9, p. 512

1. [9.1a] 108 **2.** [9.1a] $\frac{2}{3}$ **3.** [9.1b] 7320
4. [9.1b] 0.75 **5.** [9.3a] 432 **6.** [9.3a] 12
7. [9.3b] 420 **8.** [9.6b] 896 **9.** [9.7a] 14,000
10. [9.7a] 3 **11.** [9.7b] 0.423 **12.** [9.7b] 400,000
13. [9.2a] 130 ft² **14.** [9.2a] 4 m²
15. [9.2b] 235.5 in² **16.** [9.2b] 18.8 m
17. [9.2b] 220 cm **18.** [9.2b] 314 cm²
19. [9.6b] 0.045 L **20.** [9.5c] $c = 20$
21. [9.5c] $b = \sqrt{45}$; $b \approx 6.708$ **22.** [9.6a] 160 cm³
23. [9.6a] 1100 ft³ **24.** [9.6a] 38,808 yd³
25. [9.4c] 48° **26.** [9.4c] 155° **27.** [9.7c] 25°C
28. [9.7c] 98.6°F

Margin Exercises, Section 9.1, pp. 513–518

1. 2 **2.** 3 **3.** $1\frac{1}{2}$ **4.** $3\frac{1}{4}$ **5.** 288
6. 43.5 **7.** 240,768 **8.** 6 **9.** 8 **10.** 54
11. $11\frac{2}{3}$, or $11.\overline{6}$ **12.** 5 **13.** 31,680 **14.** 23,000
15. 400 **16.** 178 **17.** 9040 **18.** 7.814 **19.** 781.4
20. 8.72 **21.** 8,900,000 **22.** 6.78 **23.** 97.4
24. 0.1 **25.** 8.451 **26.** 90.909 **27.** 804.5
28. 1479.843

Exercise Set 9.1, p. 519

1. 12 **3.** $\frac{1}{12}$ **5.** 5280 **7.** 144 **9.** 7 **11.** $1\frac{1}{2}$
13. 26,400 **15.** 4 **17.** $6\frac{1}{3}$ **19.** 52,800 **21.** $2\frac{1}{2}$
23. 10 **25.** 110 **27.** 2 **29.** 300 **31.** 1
33. (a) 1000; (b) 0.001 **35.** (a) 10; (b) 0.1
37. (a) 0.01; (b) 100 **39.** 6700 **41.** 0.98
43. 8.921 **45.** 0.05666 **47.** 566,600 **49.** 4.77
51. 688 **53.** 0.1 **55.** 100,000 **57.** 142 **59.** 0.82
61. 450 **63.** 0.000024 **65.** 0.688 **67.** 230
69. 6.21 **71.** 35.56 **73.** 104.585 **75.** 100
77. 70.866 **79.** 32.727 **81.** $-\frac{3}{2}$ **82.** -2
83. $101.50 **84.** $44.50 **85.** 47% **86.** 35%
87. ◆ **89.** ◆ **91.** 393,700 **93.** 22.7 mph
95. > **97.** < **99.** >

Margin Exercises, Section 9.2, pp. 524–528

1. 43.8 cm² **2.** 12.375 km² **3.** 100 m²
4. 717.5 cm² **5.** 9″ **6.** 5 ft **7.** 62.8 m **8.** 15.7 m
9. 220 cm **10.** 34.296 yd **11.** $78\frac{4}{7}$ km²
12. 339.6 cm² **13.** A 12-ft-diameter flower bed is 13.04 ft² larger.

Calculator Spotlight, p. 528

1. Answers will vary. **2.** 1417.99 in.; 160,005.91 in²
3. 1729.27 in² **4.** 125,663.71 ft²

Exercise Set 9.2, p. 529

1. 32 cm² **3.** 104 ft² **5.** 64 m² **7.** 8.05 cm²
9. 144 mi² **11.** $55\frac{1}{8}$ ft² **13.** 49 m² **15.** 108 cm²
17. 68 yd² **19.** 14 cm **21.** $1\frac{1}{2}$ in. **23.** 16 ft
25. 0.7 cm **27.** 44 cm **29.** $4\frac{5}{7}$ in. **31.** 100.48 ft
33. 4.396 cm **35.** 154 cm² **37.** $1\frac{43}{56}$ in²
39. 803.84 ft² **41.** 1.5386 cm²
43. 2 cm; 6.28 cm; 3.14 cm² **45.** 151,976 mi²
47. 7.85 cm; 4.90625 cm² **49.** 15 in.
51. 439.7784 yd **53.** 45.68 ft **55.** 45.7 yd
57. 100.48 m² **59.** 6.9972 cm² **61.** $\frac{37}{400}$
62. $\frac{7}{8}$ **63.** 137.5% **64.** $66.\overline{6}$%, or $66\frac{2}{3}$%
65. 125% **66.** 160% **67.** ◆ **69.** ◆
71. 1267 cm² **73.** 360,000 **75.** Circumference

Margin Exercises, Section 9.3, pp. 535–536

1. 144 **2.** 1440 **3.** 63 **4.** 2.5 **5.** 3200
6. 1,000,000 **7.** 100 **8.** 28,800 **9.** 0.043
10. 0.678

Exercise Set 9.3, p. 537

1. 36 **3.** 1008 **5.** 3 **7.** 198 **9.** 396 **11.** 12,800
13. $7\frac{2}{3}$ **15.** 5 **17.** $\frac{1}{144}$ **19.** $\frac{1}{640}$ **21.** 17,000,000
23. 63,100 **25.** 23.456 **27.** 0.0349 **29.** 2500
31. 0.4728 **33.** 2 ft² **35.** $4\frac{1}{3}$ ft² **37.** 21 cm²
39. $17.50 **40.** $6.75 **41.** $8.78 **42.** $35.51
43. ◆ **45.** ◆ **47.** 10.89 **49.** 1.65 **51.** 1852.6 m²
53. $135.33

Margin Exercises, Section 9.4, pp. 541–544

1. Angle *GHI*, Angle *IHG*, Angle *H*, ∠*GHI*, ∠*IHG*, ∠*H*
2. Angle *PQR*, Angle *RQP*, Angle *Q*, ∠*PQR*, ∠*RQP*, ∠*Q*
3. 126° **4.** 33°
5. Times of Engagement of Married Couples

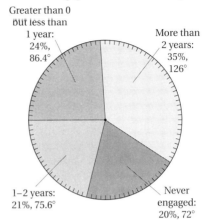

6. Right **7.** Acute **8.** Obtuse **9.** Straight
10. ∠1 and ∠2; ∠1 and ∠4; ∠2 and ∠3; ∠3 and ∠4
11. 45° **12.** 72° **13.** 5°
14. ∠1 and ∠2; ∠1 and ∠4; ∠2 and ∠3; ∠3 and ∠4
15. 142° **16.** 23° **17.** 90°

Exercise Set 9.4, p. 545

1. Angle *GHJ*, Angle *JHG*, Angle *H*, ∠*GHJ*, ∠*JHG*, ∠*H*
3. 10° **5.** 180° **7.** 90°
9. Sporting Goods Purchases

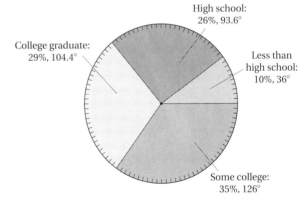

11. Obtuse **13.** Acute **15.** Straight **17.** Right
19. Acute **21.** Obtuse **23.** 79° **25.** 23° **27.** 32°
29. 61° **31.** 177° **33.** 41° **35.** 95° **37.** 78°
39. 0.561 **40.** 67.34% **41.** 2.5 **42.** 112.5%
43. −5.9 **44.** 5.5 **45.** ◆ **47.** ◆
49. $m\angle 1 = 55.93°$; $m\angle 4 = 55.93°$; $m\angle 5 = 42.17°$; $m\angle 6 = 81.9°$

Margin Exercises, Section 9.5, pp. 547–550

1. 81 **2.** 100 **3.** 121 **4.** 144 **5.** 169 **6.** 196
7. 225 **8.** 256 **9.** −10, 10 **10.** −9, 9 **11.** −7, 7
12. −14, 14 **13.** 7 **14.** 4 **15.** 11 **16.** 10 **17.** 9

18. 8 **19.** 15 **20.** 13 **21.** 1 **22.** 0 **23.** 2.236
24. 8.832 **25.** 12.961 **26.** $c = \sqrt{41}$; $c \approx 6.403$
27. $a = \sqrt{75}$; $a \approx 8.660$ **28.** $b = \sqrt{120}$; $b \approx 10.954$
29. $a = \sqrt{175}$; $a \approx 13.229$ **30.** $\sqrt{424}$ ft ≈ 20.6 ft

Calculator Spotlight, p. 548

1. 6.6 **2.** 9.7 **3.** 19.8 **4.** 17.3 **5.** 24.9
6. 24.5 **7.** 121.2 **8.** 115.6 **9.** 16.2 **10.** 85.4

Exercise Set 9.5, p. 551

1. −4, 4 **3.** −11, 11 **5.** −13, 13 **7.** −80, 80 **9.** 7
11. 9 **13.** 15 **15.** 25 **17.** 20 **19.** 100 **21.** 6.928
23. 2.828 **25.** 1.732 **27.** 3.464 **29.** 4.359
31. 10.488 **33.** $c = \sqrt{34}$; $c \approx 5.831$ **35.** $c = \sqrt{98}$;
$c \approx 9.899$ **37.** $a = 5$ **39.** $b = 8$ **41.** $c = 26$
43. $b = 12$ **45.** $b = \sqrt{1023}$; $b \approx 31.984$ **47.** $c = 5$
49. $\sqrt{208}$ ft ≈ 14.4 ft **51.** $\sqrt{16,200}$ ft ≈ 127.3 ft
53. $\sqrt{500}$ ft ≈ 22.4 ft
55. $\sqrt{211,200,000}$ ft $\approx 14,532.7$ ft **57.** $468
58. 12% **59.** 187,200 **60.** 40% **61.** About 324
62. $24.30 **63.** ◆ **65.** ◆
67. The areas are the same.
69. Length: 36.6 in.; width: 20.6 in.

Margin Exercises, Section 9.6, pp. 555–560

1. 12 cm³ **2.** 20 ft³ **3.** 128 ft³ **4.** 785 ft³
5. 67,914 m³ **6.** 904.32 ft³ **7.** $\frac{5632}{2625}$ in³ **8.** 20
9. 32 **10.** mL **11.** mL **12.** L **13.** L **14.** 970
15. 8.99 **16.** $1.08 **17.** $83.7\overline{3}$ mm³

Calculator Spotlight, p. 560

1. 523,598,775.6 m³ **2.** 105.89 cm³

Exercise Set 9.6, p. 561

1. 250 cm³ **3.** 135 in³ **5.** 75 m³ **7.** $357\frac{1}{2}$ yd³
9. 4082 ft³ **11.** 376.8 cm³ **13.** 41,580,000 yd³
15. $4,186,666.\overline{6}$ in³ **17.** Approximately 124.725 m³
19. $1437\frac{1}{3}$ km³ **21.** 1000; 1000 **23.** 87,000
25. 0.049 **27.** 27,300 **29.** 40 **31.** 320
33. 3 **35.** 32 **37.** 48 **39.** $1\frac{7}{8}$ **41.** 500 mL
43. 16 oz **45.** 56.52 in³ **47.** 4747.68 cm³
49. 646.74 cm³ **51.** 5832 yd³ **53.** 61,600 cm³; 61.6 L
55. $24 **56.** $175 **57.** $10.68 **58.** $\frac{7}{3}$ **59.** −1
60. 448 km **61.** ◆ **63.** ◆ **65.** 9.424779 L
67. 57,480 in³ **69.** 0.331 m³

Margin Exercises, Section 9.7, pp. 565–568

1. 80 **2.** 4.32 **3.** 32,000 **4.** kg **5.** kg **6.** mg
7. g **8.** t **9.** 6200 **10.** 0.0793 **11.** 77 **12.** 234.4
13. −4°C **14.** 85°F **15.** −25°C **16.** 50°F
17. 176°F **18.** 95°F **19.** 35°C **20.** 45°C

Exercise Set 9.7, p. 569

1. 16 **3.** 3 **5.** 48 **7.** 7000 **9.** 2.4 **11.** 4.5
13. 1000 **15.** 0.001 **17.** 0.01 **19.** 1000 **21.** 10
23. 725,000 **25.** 6.345 **27.** 0.000897 **29.** 7320
31. 6.78 **33.** 6.9 **35.** 800,000 **37.** 1000
39. 0.0034 **41.** 80°C **43.** 60°C **45.** 20°C
47. −10°C **49.** 190°F **51.** 140°F **53.** 10°F
55. 40°F **57.** 77°F **59.** 104°F **61.** 5432°F
63. 30°C **65.** 55°C **67.** 37°C **69.** 0.43%
70. 231% **71.** 10
72.

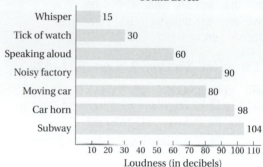

73. $\frac{22}{5}$ **74.** 2.5 **75.** ◆ **77.** ◆ **79.** 0.0022
81. (a) 200; (b) 3 **83.** 60 g **85.** 8 mL
87. About 5.1 g/cm³

Summary and Review: Chapter 9, p. 573

1. $3\frac{1}{3}$ **2.** 30 **3.** 0.07 **4.** 0.009 **5.** 400,000 **6.** $1\frac{1}{6}$
7. 144 **8.** 0.004 **9.** $13\frac{1}{2}$ **10.** 256 **11.** 0.06
12. 400 **13.** 1600 **14.** 200 **15.** 4700 **16.** 0.04
17. 45 **18.** 700,000 **19.** 7 **20.** 0.057 **21.** 31.4 m
22. $\frac{14}{11}$ in. **23.** 24 m **24.** 18 in² **25.** 29.82 ft²
26. 60 cm² **27.** 35 mm² **28.** 154 ft² **29.** 314 cm²
30. 26.28 ft² **31.** 49° **32.** 136° **33.** $c = \sqrt{850}$;
$c \approx 29.155$ **34.** $b = \sqrt{84}$; $b \approx 9.165$ **35.** $c = \sqrt{89}$ ft;
$c \approx 9.434$ ft **36.** $a = \sqrt{76}$ cm; $a \approx 8.718$ cm
37. 93.6 m³ **38.** 193.2 cm³ **39.** 28,260 ft³
40. $33.49\overline{3}$ cm³ **41.** 942 cm³ **42.** 95°F
43. 20°C **44.** 15.6 L **45.** $39.04 **46.** 47%
47. 0.567 **48.** −4.5 **49.** 24 mi **50.** 2.5
51. ◆ A square is a parallelogram because it is a four-sided figure with two pairs of parallel sides.
52. ◆ Since 1 m is slightly more than 1 yd, it follows that 1 m³ is larger than 1 yd³. Since 1 yd³ = 27 ft³, we see that 1 m³ is larger than 27 ft³.
53. ◆ Since 1 kg is about 2.2 lb and 32 oz is 32/16, or 2 lb, 1 kg weighs more than 32 oz. **54.** 961.625 cm²
55. 104.875 **56.** 34.375% **57.** $101.25

Test: Chapter 9, p. 577

1. [9.1a] 108 **2.** [9.1b] 2.8 **3.** [9.3a] 18
4. [9.1b] 5000 **5.** [9.1b] 0.87 **6.** [9.3b] 0.00452
7. [9.6b] 3.08 **8.** [9.7b] 3800 **9.** [9.6b] 1280
10. [9.6b] 240 **11.** [9.7a] 64 **12.** [9.7a] 8220

13. [9.2b] 8 cm   **14.** [9.2b] 50.24 m²   **15.** [9.2b] 88 ft
16. [9.2a] 25 cm²   **17.** [9.2b] 33.87 m²
18. [9.2a] 18 ft²   **19.** [9.2a] 13 in²
20. [9.4c] 115°   **21.** [9.5c] $c = \sqrt{2}$; $c \approx 1.414$
22. [9.5c] $b = \sqrt{51}$; $b \approx 7.141$ **23.** [9.5a] 11
24. [9.6a] 192 cm³   **25.** [9.6a] 628 ft³
26. [9.6a] $4186.\overline{6}$ yd³   **27.** [9.6a] 30 m³
28. [9.6c] 3000 mL   **29.** [9.7c] 30°C   **30.** [9.7c] 113°F
31. [8.6a] $440   **32.** [8.1b] 93%   **33.** [8.1b] 0.932
34. [5.7b] 4   **35.** [7.4a] $13.41
36. [5.7b] 2.5   **37.** [9.4c] 67.5°   **38.** [9.6c] $188.40

Cumulative Review: Chapters 1–9, p. 579

1. [4.6a] $10\frac{1}{6}$ **2.** [5.5d] 49.2   **3.** [5.2b] 87.52
4. [2.5a] −1234   **5.** [1.9d] 2   **6.** [1.9c] 1565
7. [5.1b] $-\frac{623}{100}$ **8.** [8.1c] $\frac{21}{10}$ **9.** [4.2c] <
10. [4.2c] >   **11.** [9.7a] $\frac{3}{8}$ **12.** [9.7c] 59°F
13. [9.6b] 87   **14.** [9.1a] 90   **15.** [9.3a] 27
16. [9.1b] 0.17   **17.** [2.7c], [3.6b] 52 m; 120 m²
18. [2.7b] $9a - 16$
19. [6.4b]

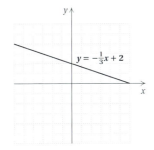

20. [6.2c] 27.5 lb   **21.** [6.2c] 1994   **22.** [7.3b] 14.4
23. [5.7b] 1   **24.** [2.8b], [3.8a] $-\frac{53}{3}$ **25.** [4.3b] $\frac{1}{8}$
26. [1.8a] 7   **27.** [1.8a] 528 million   **28.** [6.5a] 58.6
29. [8.6a] $24   **30.** [9.5c] 17 m   **31.** [8.5a] 6%
32. [4.6c] $2\frac{1}{8}$ yd   **33.** [5.8a] $16.99
34. [7.2b] The 8-qt box   **35.** [3.6b] $\frac{7}{20}$ km
36. [9.6c] 272 ft³   **37.** [2.7c] No; the sum of the perimeters can be measured as 14 ft or 16 ft, or both of which exceed 108 in.

Chapter 10

Pretest: Chapter 10, p. 582

1. [10.1a] $12x^2 - 2x - 10$
2. [10.1b] $-(6a^3b^2 - 7a^2b - 5)$, $-6a^3b^2 + 7a^2b + 5$
3. [10.1c] $-4x^2 - 17x + 21$ **4.** [10.2a] 57   **5.** [10.4a] 1
6. [10.1d] 48   **7.** [10.2a] $-18a^5b^3$
8. [10.2b] $35x^5 - 20x^3 + 5x^2$ **9.** [10.3a] $a^2 - 3a - 28$
10. [10.3b] $2a^5 - 2a^4 - a^3 + 4a^2 - 3a$
11. [10.2c] $3a(7a^2 + 10a - 4)$ **12.** [10.2c] $5xy(3x - 4y^2)$
13. [10.4b] $\frac{1}{7^2}$; $\frac{1}{49}$ **14.** [10.4b] $\frac{5}{x^3}$ **15.** [10.4b] $\frac{b^5}{a^8c}$
16. [10.4b] $\left(\frac{2}{5}\right)^2$; $\frac{4}{25}$ **17.** [10.4b] x^{-9} **18.** [10.4b] $\frac{1}{a^{11}}$
19. [10.4b] $\frac{-24x^9}{y^5}$

Margin Exercises, Section 10.1, pp. 584–586

1. $9a^2 + 3a - 1$ **2.** $7x^2y + 3x^2 + 4x + 4$
3. $2a^3 + 2a^2 - 9a + 17$ **4.** $-(12x^4 - 3x^2 + 4x)$;
$-12x^4 + 3x^2 - 4x$ **5.** $-(-4x^4 + 3x^2 - 4x)$;
$4x^4 - 3x^2 + 4x$ **6.** $-\left(-13x^6 + 2x^4 - 3x^2 + x - \frac{5}{13}\right)$;
$13x^6 - 2x^4 + 3x^2 - x + \frac{5}{13}$
7. $-(-8a^3b + 5ab^2 - 2ab)$; $8a^3b - 5ab^2 + 2ab$
8. $-4x^3 + 6x - 3$ **9.** $-5x^3y - 3x^2y^2 + 7xy^3$
10. $-14x^{10} + \frac{1}{2}x^5 - 5x^3 + x^2 - 3x$ **11.** $2x^3 + 2x + 8$
12. $x^2 - 6x - 2$ **13.** $2x^3 + 3x^2 - 4xy - 2$
14. 41; 41   **15.** 132   **16.** 360 ft   **17.** 260 ft

Exercise Set 10.1, p. 587

1. $-2x + 10$ **3.** $x^2 - 8x + 2$ **5.** $2x^2$
7. $6t^4 + 4t^3 + 5t^2 - t$ **9.** $9 + 12x^2$
11. $9x^8 + 8x^7 - 3x^4 + 2x^2 - 2x + 5$
13. $14t^4 + 4t^3 - t^2 + 4t - 7$
15. $-5x^4y^3 + 2x^3y^3 + 4x^3y^2 - 4xy^2 - 5xy$
17. $13a^3b^2 + 4a^2b^2 - 4a^2b + 6ab^2$
19. $-0.6a^2bc + 17.5abc^3 - 5.2abc$ **21.** $-(-5x)$; $5x$
23. $-(-x^2 + 10x - 2)$; $x^2 - 10x + 2$
25. $-(-12x^4 - 3x^3 + 3)$; $-12x^4 + 3x^3 - 3$ **27.** $-3x + 7$
29. $-4x^2 + 3x - 2$ **31.** $4x^4 - 6x^2 - \frac{3}{4}x + 8$
33. $7x - 1$ **35.** $4t^2 + 6t + 6$ **37.** $-x^2 - 7x + 5$
39. $5a^2 + 5a - 16$ **41.** $6x^4 + 3x^3 - 4x^2 + 3x - 4$
43. $4.6x^3 + 9.2x^2 - 3.8x - 23$ **45.** $\frac{3}{4}x^3 - \frac{1}{2}x$
47. $2x^3y^3 + 8x^2y^2 + 2x^2y + 4xy$ **49.** −18   **51.** 19
53. −12   **55.** 2   **57.** 4   **59.** 11   **61.** About 449
63. 1024 ft   **65.** $18,750   **67.** $155,000
69. $\frac{7}{10}$ serving per pound   **70.** $3560   **71.** $67.50
72. 26 m²   **73.** 1256 cm²   **74.** $395 per week
75. ◆   **77.** ◆   **79.** 209 million   **81.** $5x^2 - 9x - 1$
83. $2 + x + 2x^2 + 4x^3$ **85.** $8t^4 + 9t^3 - 2t^3 + 2t^2 - 2t^2 + t - 4t - 3 + 7 = 8t^4 + 7t^3 - 3t + 4$

Margin Exercises, Section 10.2, pp. 591–595

1. $18a^2$ **2.** $-14x^2$ **3.** $48a^2$ **4.** $-5m^2$ **5.** $42ab$
6. a^9 **7.** $8x^{13}$ **8.** $35m^{11}$ **9.** $15a^7b^{12}$ **10.** 7; −19
11. $12x^2 + 20x$ **12.** $6a^3 - 15a^2 + 21a$
13. $8a^5b^2 + 20a^3b^6$ **14.** $6(z - 2)$ **15.** $3(x - 2y + 3)$
16. $2(8a - 18b + 21)$ **17.** $-4(3x - 8y + 4z)$
18. $5a(a^2 + 1)$ **19.** $7x(2x^2 - x + 3)$
20. $3ab(3a - 2b)$

Exercise Set 10.2, p. 597

1. $45a^2$ **3.** $-60x^2$ **5.** $28x^8$ **7.** $-0.02x^{10}$
9. $35x^6y^{12}$ **11.** $12a^8b^8c^2$ **13.** $-24x^{11}$
15. $-3x^2 + 15x$ **17.** $-3x^2 + 3x$ **19.** $x^5 + x^2$
21. $6x^3 - 18x^2 + 3x$ **23.** $12x^3y + 8xy^2$
25. $12a^7b^3 - 9a^4b^3$ **27.** $2(x + 3)$ **29.** $7(a - 3)$
31. $7(2x + 3y)$ **33.** $9(a - 3b + 9)$ **35.** $6(4 - m)$
37. $-8(2 + x - 5y)$ **39.** $3x(x^4 + 1)$ **41.** $a^2(a - 8)$

43. $2x(4x^2 - 3x + 1)$ **45.** $6a^4b^2(2b + 3a)$ **47.** 210 ft
48. 14 mpg   **49.** 17.5 mpg   **50.** $418.95   **51.** 62.5%
52. 37.68 cm   **53.** ◆   **55.** ◆
57. $19a^{437}(37 + 23a^{266})$
59.

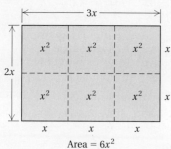

Area = $6x^2$

Margin Exercises, Section 10.3, pp. 599–600

1. $x^2 + 13x + 40$ **2.** $x^2 + x - 20$ **3.** $5x^2 - 17x - 12$
4. $6x^2 - 19x + 15$ **5.** $x^2 + 8x + 15$
6. $x^2 - 11x + 24$ **7.** $x^4 + 3x^3 + x^2 + 15x - 20$
8. $6y^5 - 20y^3 + 15y^2 + 14y - 35$
9. $3x^3 + 13x^2 - 6x + 20$

Exercise Set 10.3, p. 601

1. $x^2 + 9x + 14$ **3.** $x^2 + 3x - 10$ **5.** $x^2 + 8x + 15$
7. $x^2 - 10x + 21$ **9.** $x^2 - 36$ **11.** $18 + 12x + 2x^2$
13. $9x^2 - 24x + 16$ **15.** $x^2 - \frac{21}{10}x - 1$ **17.** $x^3 - 1$
19. $4x^3 + 14x^2 + 8x + 1$
21. $3y^4 - 6y^3 - 7y^2 + 18y - 6$ **23.** $x^6 + 2x^5 - x^3$
25. $6t^4 + t^3 - 16t^2 - 7t + 4$ **27.** $x^9 - x^5 + 2x^3 - x$
29. 912 m²   **30.** $123,000   **31.** $133\frac{1}{3}$%, or $133.\overline{3}$%
32. 26   **33.** 68.5%   **34.** 154 ft²   **35.** ◆   **37.** ◆
39. $2x^2 - 3x + 22$
41. (a) $4m + 20$; (b) $m^2 + 10m + 25$
43. $V = 4x^3 - 48x^2 + 144x$; $S = -4x^2 + 144$
45. $81x^2 + 72x + 16$

Margin Exercises, Section 10.4, pp. 603–606

1. 1   **2.** 6   **3.** 1   **4.** -3 **5.** 1   **6.** $\frac{1}{4^3}$; $\frac{1}{64}$
7. $\frac{1}{5^2}$; $\frac{1}{25}$ **8.** $\frac{1}{2^4}$; $\frac{1}{16}$ **9.** $\frac{1}{(-2)^3}$; $\frac{1}{-8}$ **10.** $\left(\frac{3}{5}\right)^2$; $\frac{9}{25}$
11. 9^{-2} **12.** $7x^{-4}$ **13.** m^3n^5 **14.** abc **15.** a^4b^6
16. x^7yz^4 **17.** $\frac{a^4c^3}{b^7}$ **18.** 5^2, or 25   **19.** $\frac{1}{x^7}$
20. $20x^9y^6$ **21.** $\frac{b^3}{a^7}$

Calculator Spotlight, p. 606

1. 4.21399177; 4.21399177   **2.** 4.768371582; 4.768371582
3. -0.2097152; -0.2097152
4. -0.0484002582; -0.0484002582 **5.** 2.0736; 2.0736
6. 0.4932701843; 0.4932701843

Exercise Set 10.4, p. 607

1. 1   **3.** 1   **5.** -19.57 **7.** 1   **9.** 1   **11.** 1
13. 17   **15.** 1   **17.** 7   **19.** 5   **21.** 2   **23.** $\frac{1}{3^2}$; $\frac{1}{9}$
25. $\frac{1}{10^4}$; $\frac{1}{10,000}$ **27.** $\frac{1}{a^3}$ **29.** $\frac{1}{(-5)^2}$; $\frac{1}{25}$ **31.** $\frac{3}{x^7}$
33. xy^4 **35.** a^3b^4 **37.** $\frac{-7}{a^9}$ **39.** $\frac{25}{4}$
41. $\frac{a^3}{125}$ **43.** 4^{-3} **45.** $9x^{-3}$ **47.** $\frac{1}{x}$ **49.** 1
51. $\frac{1}{x^{13}}$ **53.** $6a^3b^2$ **55.** $-\frac{6y^6}{x^2}$ **57.** $\frac{6c^3}{a^6b^4}$
59. 50 km/h   **60.** 60%   **61.** 87.5%   **62.** 6%
63. 31.4 cm   **64.** 22.5 mpg   **65.** ◆   **67.** ◆
69. $\frac{1}{5}$, or 0.2; $\frac{1}{5}$, or 0.2   **71.** y^{5x} **73.** a^{4t}

Summary and Review: Chapter 10, p. 609

1. $2x - 4$ **2.** $5x^4 - 4x^3 - x - 2$
3. $8a^5 + 12a^3 - a^2 + 4a + 9$
4. $5a^3b^3 + 12a^2b^3 - 7$ **5.** $-(12x^3 - 4x^2 + 9x - 3)$; $-12x^3 + 4x^2 - 9x + 3$ **6.** -72 **7.** 1   **8.** -42
9. $30x^7$ **10.** $18x^4 - 12x^2 - 3x$ **11.** $14a^7b^4 + 10a^6b^4$
12. $x^2 + 2x - 63$ **13.** $10x^2 - 11x + 3$
14. $a^4 + 2a^3 - 2a^2 - 2a + 1$ **15.** $5x(9x^2 - 2)$
16. $7(a - 5b - 7ac)$ **17.** $3x^2y(2x - 3y^4)$ **18.** $\frac{1}{12^2}$; $\frac{1}{144}$
19. $\frac{8}{a^7}$ **20.** $\frac{z^6}{x^3y^5}$ **21.** $\left(\frac{5}{4}\right)^2$; $\frac{25}{16}$ **22.** x^{-7} **23.** $\frac{1}{x^{17}}$
24. $\frac{15x^9}{y^9}$ **25.** $\frac{2}{15}$ slice per minute; $\frac{15}{2}$ minutes per slice
26. 30% defective; 70% not defective   **27.** $189
28. 346.5 in²   **29.** ◆ The expression x^{-2} is equivalent to $1/x^2$. Since x^2 means $x \cdot x$ and a number times itself is never negative, x^2 can never be negative. Thus, $1/x^2$ and x^{-2} can never represent a negative number.
30. ◆ The student is probably adding the coefficients and multiplying the exponents, instead of multiplying the coefficients and adding the exponents.
31. $1,158,057x^{17} + 1,226,352x^{12} - 213,129x^7$
32. $-40x^8$ **33.** $13a^2b^5c^5(3ab^2c - 10c^3 + 4a^2b)$
34. $w^5x^2y^3z^3(x^4yz^2 - w^2xy^4 + wy^2z^3 - wx^5z)$

Test: Chapter 10, p. 611

1. [10.1a] $17a^3 - 5a^2 - a + 7$
2. [10.1b] $-(-9a^4 + 7b^2 - ab + 3)$; $9a^4 - 7b^2 + ab - 3$ **3.** [10.1c] $4x^4 - x^2 - 13$
4. [10.2a] 193   **5.** [10.4a] 1   **6.** [10.1d] 12.4 m
7. [10.2a] $-10x^6y^8$ **8.** [10.2b] $35a^3 - 20a^2 + 15a$
9. [10.3a] $x^2 + x - 42$ **10.** [10.3b] $2a^3 - 5a^2 + a + 2$
11. [10.2c] $5x^2(7x^4 - 5x + 3)$
12. [10.2c] $3(2ab - 3bc + 4ac)$ **13.** [10.4b] $\frac{1}{5^3}$; $\frac{1}{125}$
14. [10.4b] $\frac{5b^2}{a^3}$ **15.** [10.4b] $\left(\frac{5}{3}\right)^3$; $\frac{125}{27}$

16. [10.4b] $\frac{1}{x^{16}}$ 17. [10.4b] $\frac{-6a^3}{b^3}$
18. [7.2a] $\frac{5}{3}$ nails per minute; $\frac{3}{5}$ minute per nail
19. [8.4a] 40% satisfied; 60% not satisfied
20. [8.5a] $8500   21. [9.2b] 94.2 cm
22. [10.1d] 2.92 L
23. [10.2c], [10.3a], [10.4b] $\frac{3a^4}{a^4-6a^2+9}$

Final Examination, p. 613

1. [1.1e] 6   2. [1.1a] 8 thousands + 4 hundreds + 9 ones   3. [1.2b] 1018   4. [1.2b] 21,085
5. [4.2b] $\frac{9}{26}$ 6. [4.6a] $8\frac{7}{9}$ 7. [2.2a] 24
8. [2.2a] -762 9. [5.2c] -44.364
10. [10.1a] $10x^5 - x^4 + 2x^3 - 3x^2 + 2$
11. [1.3d] 243   12. [2.7b] $-19x$ 13. [4.3a] $-\frac{19}{40}$
14. [4.6b] $2\frac{17}{24}$ 15. [5.2b] 19.9973
16. [10.1c] $2x^3 + 5x^2 + 7x$ 17. [10.1c] $-4a^2b + 7ab$
18. [1.5b] 4752   19. [2.4a] $-74,337$ 20. [4.7b] $4\frac{7}{12}$
21. [3.6a] $-\frac{6}{5}$ 22. [3.6a] 10   23. [5.3a] 259.084
24. [2.7a] $24x - 15$ 25. [10.2a] $27a^8b^3$
26. [10.2b] $21x^5 - 14x^3 + 56x^2$
27. [10.3a] $x^2 - 5x - 14$
28. [10.3b] $a^3 - 2a^2 - 11a + 12$ 29. [1.6c] 573
30. [1.6c] 56 R 10   31. [3.7b] $-\frac{3}{2}$ 32. [4.7b] $\frac{7}{90}$
33. [5.4a] 39   34. [4.5c] $56\frac{5}{17}$ 35. [1.9c] 75
36. [1.9c], [2.1c] -2 37. [1.9a] 17^4 38. [1.4a] 68,000
39. [5.5b] 21.84   40. [3.1b] Yes   41. [3.2a] 1, 3, 5, 15
42. [4.1a] 105   43. [3.5b] $\frac{7}{10}$ 44. [3.5b] $-\frac{58}{3}$
45. [4.5b] $-3\frac{3}{5}$ 46. [2.1b] >   47. [4.2c] <
48. [5.1d] 1.001   49. [2.6a] 29   50. [10.2c] $5(8 - t)$
51. [10.2c] $3a(6a^2 - 5a + 2)$ 52. [3.3b] $\frac{3}{5}$
53. [5.1c] 0.037   54. [5.5a] -0.52 55. [5.5a] $0.\overline{8}$
56. [8.1b] 0.07   57. [5.1b] $\frac{671}{100}$ 58. [4.5a] $-\frac{29}{4}$
59. [8.1c] $\frac{2}{5}$ 60. [8.1c] 85%   61. [8.1b] 150%
62. [5.6a] 13.6   63. [2.8a] 555   64. [5.7a] 64
65. [3.8a] $\frac{5}{4}$ 66. [7.3b] $\frac{153}{2}$ 67. [4.4a] 5
68. [5.7b] $-\frac{2}{5}$ 69. [6.5a] $23.75   70. [1.8a] 65 min
71. [4.6c] $25\frac{3}{4}$ 72. [5.8a] 485.9 mi
73. [1.8a] $24,000   74. [1.8a] $595   75. [3.4c] $\frac{3}{10}$ km
76. [5.8a] $84.96   77. [7.4a] 13 gal   78. [7.2b] $17\frac{¢}{oz}$
79. [8.6a] $240   80. [8.5b] 7%   81. [8.4b] 30,160
82. [5.8a] 220 mi   83. [1.9b] 324   84. [10.4a] 1
85. [9.5a] 11   86. [10.4b] $\frac{1}{4^3}$; $\frac{1}{64}$ 87. [10.4b] $\left(\frac{4}{5}\right)^2$; $\frac{16}{25}$
88. [6.2c] 85+; 8 fatalities per 100 million vehicle miles traveled   89. [6.2c] 1 fatality per 100 million vehicle miles traveled   90. [6.5a, b, c] Mean: $28\frac{2}{7}$; median: 26; mode: 20   91. [6.6b] $\frac{3}{20}$
92. [6.3a]   93. [6.4b]

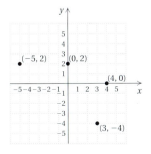
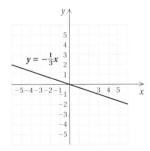

94. [7.5a] $x = 16, y = 14$ 95. [9.1a] 12   96. [9.1b] 428
97. [9.1b] 3700   98. [9.7b] 20   99. [9.7a] 160
100. [9.7b] 0.08   101. [9.6b] 8.19   102. [9.6b] 5
103. [2.7c] 88 in.   104. [9.2a] 61.6 cm²
105. [3.6b] 25 in²   106. [9.2a] 128.65 yd²
107. [5.8a] 25.75 m²   108. [9.2b] Diameter: 20.8 in.; circumference: 65.312 in.; area: 339.6224 in²
109. [9.6a] 52.9 m³   110. [9.6a] 803.84 ft³
111. [9.6a] $267.94\overline{6}$ mi³   112. [9.5c] $\sqrt{85}$ ft; 9.220 ft

Developmental Units

Margin Exercises, Section A, pp. 620–624

1. 9   2. 7   3. 14   4. 16   5. 16   6. 16   7. 8
8. 8   9. 7   10. 46   11. 13   12. 58
13.

+	1	2	3	4	5
1	2	3	4	5	6
2	3	4	5	6	7
3	4	5	6	7	8
4	5	6	7	8	9
5	6	7	8	9	10

14.

+	6	5	7	4	9
7	13	12	14	11	16
9	15	14	16	13	18
5	11	10	12	9	14
8	14	13	15	12	17
4	10	9	11	8	13

15. 16   16. 9   17. 11   18. 15   19. 59   20. 549
21. 9979   22. 5496   23. 56   24. 85   25. 829
26. 1026   27. 12,698   28. 13,661   29. 71,328

Exercise Set A, p. 625

1. 17   2. 15   3. 13   4. 14   5. 12   6. 11   7. 17
8. 16   9. 12   10. 10   11. 10   12. 11   13. 7
14. 7   15. 11   16. 0   17. 3   18. 18   19. 14
20. 10   21. 4   22. 14   23. 11   24. 15   25. 16
26. 9   27. 13   28. 14   29. 11   30. 7   31. 13
32. 14   33. 12   34. 6   35. 10   36. 12   37. 10
38. 8   39. 2   40. 9   41. 13   42. 8   43. 10
44. 9   45. 10   46. 6   47. 13   48. 9   49. 8
50. 11   51. 12   52. 13   53. 10   54. 12   55. 17
56. 13   57. 10   58. 16   59. 16   60. 15   61. 39
62. 89   63. 87   64. 999   65. 900   66. 868
67. 999   68. 848   69. 877   70. 17,680   71. 10,873
72. 4699   73. 10,867   74. 9895   75. 3998
76. 18,222   77. 16,889   78. 64,489   79. 99,999
80. 77,777   81. 46   82. 26   83. 55   84. 101
85. 1643   86. 1412   87. 846   88. 628   89. 1204

90. 607 **91.** 10,000 **92.** 1010 **93.** 1110 **94.** 1227 **95.** 1111 **96.** 1717 **97.** 10,138 **98.** 6554 **99.** 6111 **100.** 8427 **101.** 9890 **102.** 11,612 **103.** 11,125 **104.** 15,543 **105.** 16,774 **106.** 68,675 **107.** 34,437 **108.** 166,444 **109.** 101,315 **110.** 49,449

Margin Exercises, Section S, pp. 627–632

1. 4 **2.** 7 **3.** 8 **4.** 3 **5.** 12 − 8 = 4; 12 − 4 = 8 **6.** 13 − 6 = 7; 13 − 7 = 6 **7.** 8 **8.** 7 **9.** 9 **10.** 4 **11.** 14 **12.** 20 **13.** 241 **14.** 2025 **15.** 17 **16.** 36 **17.** 454 **18.** 250 **19.** 376 **20.** 245 **21.** 2557 **22.** 3674 **23.** 8 **24.** 6 **25.** 30 **26.** 90 **27.** 500 **28.** 900 **29.** 538 **30.** 677 **31.** 42 **32.** 224 **33.** 138 **34.** 44 **35.** 2768 **36.** 1197 **37.** 1246 **38.** 4728

Exercise Set S, p. 633

1. 7 **2.** 0 **3.** 0 **4.** 5 **5.** 3 **6.** 8 **7.** 8 **8.** 6 **9.** 7 **10.** 3 **11.** 7 **12.** 9 **13.** 6 **14.** 6 **15.** 2 **16.** 4 **17.** 3 **18.** 2 **19.** 0 **20.** 3 **21.** 1 **22.** 1 **23.** 7 **24.** 0 **25.** 4 **26.** 4 **27.** 5 **28.** 4 **29.** 4 **30.** 5 **31.** 9 **32.** 9 **33.** 7 **34.** 9 **35.** 6 **36.** 6 **37.** 2 **38.** 6 **39.** 7 **40.** 8 **41.** 33 **42.** 21 **43.** 247 **44.** 193 **45.** 500 **46.** 654 **47.** 202 **48.** 617 **49.** 288 **50.** 1220 **51.** 2231 **52.** 4126 **53.** 3764 **54.** 1691 **55.** 1226 **56.** 99,998 **57.** 10 **58.** 10,101 **59.** 11,043 **60.** 11,111 **61.** 65 **62.** 29 **63.** 8 **64.** 9 **65.** 308 **66.** 126 **67.** 617 **68.** 214 **69.** 89 **70.** 4402 **71.** 3555 **72.** 5889 **73.** 2387 **74.** 3649 **75.** 3832 **76.** 8144 **77.** 7750 **78.** 10,445 **79.** 33,793 **80.** 281 **81.** 455 **82.** 571 **83.** 6148 **84.** 2200 **85.** 2113 **86.** 3748 **87.** 5206 **88.** 1459 **89.** 305 **90.** 4455

Margin Exercises, Section M, pp. 635–640

1. 56 **2.** 36 **3.** 28 **4.** 42 **5.** 0 **6.** 0 **7.** 8 **8.** 23

9.

×	2	3	4	5
2	4	6	8	10
3	6	9	12	15
4	8	12	16	20
5	10	15	20	25
6	12	18	24	30

10.

×	6	7	8	9
5	30	35	40	45
6	36	42	48	54
7	42	49	56	63
8	48	56	64	72
9	54	63	72	81

11. 70 **12.** 450 **13.** 2730 **14.** 100 **15.** 1000 **16.** 560 **17.** 360 **18.** 700 **19.** 2300 **20.** 72,300 **21.** 10,000 **22.** 100,000 **23.** 5600 **24.** 1600 **25.** 9000 **26.** 852,000 **27.** 10,000 **28.** 12,000 **29.** 72,000 **30.** 54,000 **31.** 5600 **32.** 56,000 **33.** 18,000 **34.** 28 **35.** 116 **36.** 148 **37.** 4938 **38.** 6740 **39.** 116 **40.** 148 **41.** 4938 **42.** 6740 **43.** 5968 **44.** 59,680 **45.** 596,800

Exercise Set M, p. 641

1. 12 **2.** 0 **3.** 7 **4.** 0 **5.** 10 **6.** 30 **7.** 10 **8.** 63 **9.** 54 **10.** 12 **11.** 0 **12.** 72 **13.** 8 **14.** 0 **15.** 28 **16.** 24 **17.** 45 **18.** 18 **19.** 0 **20.** 35 **21.** 45 **22.** 40 **23.** 0 **24.** 16 **25.** 25 **26.** 81 **27.** 1 **28.** 0 **29.** 4 **30.** 36 **31.** 8 **32.** 0 **33.** 27 **34.** 18 **35.** 0 **36.** 10 **37.** 48 **38.** 54 **39.** 0 **40.** 72 **41.** 15 **42.** 8 **43.** 9 **44.** 2 **45.** 32 **46.** 6 **47.** 15 **48.** 6 **49.** 8 **50.** 20 **51.** 20 **52.** 16 **53.** 10 **54.** 0 **55.** 80 **56.** 70 **57.** 160 **58.** 210 **59.** 450 **60.** 780 **61.** 560 **62.** 360 **63.** 800 **64.** 300 **65.** 900 **66.** 1000 **67.** 345,700 **68.** 1200 **69.** 4900 **70.** 4000 **71.** 10,000 **72.** 7000 **73.** 9000 **74.** 2000 **75.** 457,000 **76.** 6,769,000 **77.** 18,000 **78.** 20,000 **79.** 48,000 **80.** 16,000 **81.** 6000 **82.** 1,000,000 **83.** 1200 **84.** 200 **85.** 4000 **86.** 2500 **87.** 12,000 **88.** 6000 **89.** 63,000 **90.** 120,000 **91.** 800,000 **92.** 120,000 **93.** 16,000,000 **94.** 80,000 **95.** 147 **96.** 444 **97.** 2965 **98.** 4872 **99.** 6293 **100.** 3460 **101.** 16,236 **102.** 13,508 **103.** 87,554 **104.** 195,384 **105.** 3480 **106.** 2790 **107.** 3360 **108.** 7020 **109.** 20,760 **110.** 10,680 **111.** 358,800 **112.** 109,800 **113.** 583,800 **114.** 299,700 **115.** 11,346,000 **116.** 23,390,000 **117.** 61,092,000 **118.** 73,032,000

Margin Exercises, Section D, pp. 643–650

1. 4 **2.** 8 **3.** 9 **4.** 3 **5.** 12 ÷ 2 = 6; 12 ÷ 6 = 2 **6.** 42 ÷ 6 = 7; 42 ÷ 7 = 6 **7.** 7 **8.** 9 **9.** 8 **10.** 9 **11.** 6 **12.** 13 **13.** 1 **14.** 2 **15.** Not defined **16.** 0 **17.** Not defined **18.** Not defined **19.** 10 **20.** Not defined **21.** 0 **22.** 1 **23.** 1 **24.** 1 **25.** 1 **26.** 17 **27.** 1 **28.** 75 R 4 **29.** 23 R 11 **30.** 969 R 6 **31.** 47 R 4 **32.** 96 R 1 **33.** 1263 R 5 **34.** 324 **35.** 417 R 5 **36.** 37 R 8 **37.** 23 R 17

Exercise Set D, p. 651

1. 3 **2.** 8 **3.** 4 **4.** 1 **5.** 32 **6.** 9 **7.** 7 **8.** 5 **9.** 37 **10.** 5 **11.** 9 **12.** 4 **13.** 6 **14.** 9 **15.** 5 **16.** 8 **17.** 9 **18.** 6 **19.** 3 **20.** 2 **21.** 9 **22.** 2 **23.** 3 **24.** 7 **25.** 8 **26.** 4 **27.** 7 **28.** 2 **29.** 3 **30.** 6 **31.** 1 **32.** 4 **33.** 6 **34.** 3 **35.** 7 **36.** 9 **37.** 0 **38.** Not defined **39.** Not defined **40.** 7 **41.** 8 **42.** 7 **43.** 1 **44.** 5 **45.** 0 **46.** 0 **47.** 3 **48.** 1 **49.** 4 **50.** 7 **51.** 1 **52.** 6 **53.** 1 **54.** 5 **55.** 2 **56.** 8 **57.** 0 **58.** 0 **59.** 8 **60.** 3 **61.** 4 **62.** Not defined **63.** 4 **64.** 1 **65.** 69 R 1 **66.** 199 R 1 **67.** 92 R 1 **68.** 138 R 3 **69.** 1723 R 4 **70.** 2925 **71.** 864 R 1 **72.** 522 R 3 **73.** 527 R 2 **74.** 2897 **75.** 43 R 15 **76.** 32 R 27 **77.** 47 R 12 **78.** 42 R 9 **79.** 91 R 22 **80.** 52 R 13 **81.** 29 **82.** 55 R 2 **83.** 228 R 22 **84.** 20 R 379 **85.** 21 **86.** 118 **87.** 91 R 1 **88.** 321 R 2 **89.** 964 R 3 **90.** 679 R 5 **91.** 1328 R 2 **92.** 27 R 8 **93.** 23 **94.** 321 R 18 **95.** 224 R 10 **96.** 27 R 60 **97.** 22 R 85 **98.** 49 R 46 **99.** 383 R 91 **100.** 74 R 440

Index

A

Absolute value, 87
Accidents, daily, 589
Acre, 535
Acute angle, 543
Addends, 9, 620
  missing, 20
Addition
  addends, 9, 620
  associative law of, 12
  basic, 620
  and carrying, 11, 283, 623, 624
  and combining, 9
  commutative law of, 12, 621
  with decimal notation, 283, 285
  estimating sums, 29, 319
  of exponents, 592, 609
  using fractional notation, 213, 214
  of integers, 93, 94
  using mixed numerals, 243
  on the number line, 93
  of polynomials, 583
  and regrouping, 11
  related subtraction sentence, 20, 627
  repeated, 33, 635
  sum, 9, 620
    estimating, 29, 319
  symbol, 620
  of three numbers, 12, 621
  of whole numbers, 11, 620–624
  of zero, 620
Addition principle, 131, 231, 265
Additive identity, 94, 620
Additive inverse, 88, 94, 584. *See also* Opposite.
Algebraic expressions, 115. *See also* Expressions, Polynomials.
  factoring, 594
American system of measures
  area, 535
  capacity, 557
  changing units, 513, 535, 558, 565
    to metric, 518

  length, 513
  temperature, 567
  weight, 565
Angle, 541
  acute, 543
  complementary, 543
  degrees, 541
  measure, 541
  naming, 541
  obtuse, 543
  rays, 541
  right, 543
  sides, 541
  straight, 543
  supplementary, 544
  unit, 541
  vertex, 541
Annual compounding of interest, 498
Applied problems, 1, 2, 9, 10, 15, 19–21, 23, 24, 34, 35, 39, 42–44, 51, 54, 61–72, 76, 78–86, 91, 99, 102–104, 110, 114, 120, 126, 129, 130, 137–140, 142–144, 152, 164, 167, 168, 170, 176, 178, 180, 183–186, 190, 193, 194, 198, 200, 202, 204–206, 212, 217, 221, 222, 226, 229, 230, 236, 240, 242, 245, 246, 250–252, 256–258, 260–264, 266–268, 270, 272–274, 282, 290, 296, 304, 307, 308, 314, 317, 319, 320, 323, 324, 333–348, 350–367, 369–374, 392–408, 410–412, 414–428, 435–438, 440, 441, 443–450, 452–455, 461–464, 472, 478–512, 522, 528, 532, 534, 540, 542, 545, 550, 553, 554, 556, 559, 560, 563, 571, 572, 575–581, 586, 589, 590, 598, 602, 608, 610–612, 616–618. *See also* Solving problems; Index of Applications.

Approximately equal to ($\approx$), 30
Approximating, *see* Estimating
Approximating square roots, 547
Area
  American units of, 535
  of a circle, 318, 337, 527, 528, 573
  metric units of, 536
  of a parallelogram, 523, 528, 573
  of a rectangle, 36, 168, 199
  of a square, 73, 337
  of a trapezoid, 524, 528, 573
  of a triangle, 179, 199
Array, rectangular
  in division, 43, 643
  in multiplication, 33, 635
Associative law
  of addition, 12
  of multiplication, 38
Augustus, 456
Average, 76, 240, 393. *See also* Mean.
  bowling, 398
  grade point, 394
Axes, 375
Axis, 375

B

Bar graphs, 363
Base(s)
  in exponential notation, 73
  of a parallelogram, 523
  of a trapezoid, 524
Basic addition, 620
Basic division, 643
Basic multiplication, 635
Basic subtraction, 627
Billions period, 5
Binomial, 593
  and multiplication, 599
Board foot, 578
Borrowing, 22, 244, 284, 629–631
Bowling average, 398
Braces, 76
Brackets, 76

British-American system of measures, *see* American system

C

Calculator spotlight, 48, 75, 86, 108, 112, 146, 174, 218, 240, 296, 300, 314, 326, 378, 388, 395, 442, 459, 492, 500, 528, 548, 606
Canceling, 174, 178, 189, 514
Capacity, 557–559
Carrying, 11, 283, 623, 624
Cartesian coordinate system, 377
cc, 559
Celsius temperature, 567, 568, 573
Center point of statistics, 393
Centigram, 566
Centimeter, 515
   cubic, 558, 559
Cents, converting from/to dollars, 295
Changing the sign, 89, 584
Changing units, *see* American system of measures; Metric system of measures
Checking
   division, 46, 299, 648
   solutions of equations, 132, 377, 378
   solutions of applied problems, 61
   subtraction, 22, 631
Check-writing, 276
Circle, 525
   area, 318, 337, 527, 528, 573
   circumference, 526, 528, 573
   diameter, 337, 525, 528, 573
   radius, 337, 525, 573
Circle graph, 160, 455, 542
Circular cylinder, 556, 573
Circumference of a circle, 526, 528, 573
Coefficient, 196
Collecting like terms, 124. *See also* Combining like terms.
Combining like terms, 124, 224, 245, 286
   in equation solving, 327
Commission, 489, 505
Common denominators, 214
Common factor, 594
Common multiple, least, 207
Commutative law
   of addition, 12, 620
   of multiplication, 37
Complement of an angle, 543
Complementary angles, 543
Composite numbers, 154

Compound interest, 498–500, 505
Constant, 115
Converting
   American to metric measures, 518
   area units, 535, 536
   capacity units, 558, 559
   cents to dollars, 295
   decimal notation to fractional notation, 276, 277
   decimal notation to percent notation, 457
   dollars to cents, 295
   expanded notation to standard notation, 4
   fractional notation to decimal notation, 277, 278, 309, 312
   fractional notation to mixed numerals, 238, 239
   fractional notation to percent notation, 458, 459
   length units, 513, 516, 517
   mass units, 566
   mentally, in metric system, 517, 536, 566
   metric to American measures, 518
   mixed numerals to decimal notation, 277
   mixed numerals to fractional notation, 237, 238
   percent notation to decimal notation, 456
   percent notation to fractional notation, 460
   standard notation to expanded notation, 3
   standard notation to word names, 4, 5
   temperature units, 567, 568, 572, 573
   weight units, 565
   word names to standard notation, 5
Coordinate system, 375, 377
Coordinates, 375
Cross-products, 429
Cube, unit, 555
Cubic centimeter, 558, 559
Cup, 557
Cylinder, circular, 556, 573

D

Decigram, 566
Decimal notation, 275
   addition with, 283, 285
   and combining like terms, 286
   converting
      from/to fractional notation, 276–278, 309, 312

      from mixed numerals, 277, 278
      from/to percent notation, 456, 457
   division with, 299–303
   and evaluating expressions, 296
   and fractional notation together in calculations, 313
   and money, 276, 295
   multiplication with, 291–295
   and order, 279
   order of operations with, 303
   percent notation as, 455
   place-value, 275
   repeating, 309, 310
      rounding, 311
   rounding, 280, 311
      truncating, 282
   subtraction with, 284, 285
   terminating, 309
   word names for, 275
Decimeter, 515
Decrease, percent of, 480–482
Degree measure, 541
Dekagram, 566
Dekameter, 515
Denominators, 159
   common, 214
   least common, 214
   like, 213, 223
   of zero, 162, 199
Depreciation, 345, 370
Descartes, René, 377
Descending order, 584
Diameter, 337, 525–528, 573
Difference, 19
   estimating, 30, 319
Digits, 5
Discount, 490, 505
Distance
   and absolute value, 87
   of a timed fall, 589
Distributive law, 35, 122, 137
   and combining like terms, 124
   and factoring, 594
Dividend, 43
Divisibility, 145
   by 2, 147
   by 3, 147
   by 4, 149
   by 5, 149
   by 6, 147
   by 7, 149
   by 8, 149
   by 9, 148
   by 10, 149
   and calculators, 146
   and simplifying fractional notation, 173
Division
   basic, 643

checking, 46, 299, 648
with decimal notation, 299–303
definition, 111
dividend, 43
divisor, 43
    rounding the divisor, 48
by estimating multiples, 47, 48, 648
estimating quotients, 319
with fractional notation, 187, 643
by guess, multiply, and subtract, 647
of integers, 111
with mixed numerals, 254
of a number by itself, 646, 647
by one, 162, 199, 645
partial quotient, 647
by a power of 10, 303
quotient, 43
    expressed as a mixed numeral, 240
and reciprocals, 188, 254
and rectangular arrays, 43, 643
related multiplication sentence, 44
with remainders, 46
as repeated subtraction, 45
undefined, 111, 646
of whole numbers, 45–48, 643–650
by zero, 111, 645, 646
of zero by a nonzero number, 111, 646
zeros in quotients, 49
Division principle, 133
Divisor, 43
Dollars, converting from/to cents, 295

E

Effective yield, 504
Eight, divisibility by, 149
English system of measures, *see* American system
Equality of fractions, test for, 429
Equation, 30, 55. *See also* Solving equations.
    equivalent, 56, 131, 325
    graph of, 384
    linear, 384
    Pythagorean, 548, 573
    solution of, 55
    solving, 55–58, 131, 133, 195, 225, 231, 325–328
Equivalent
    equations, 56, 131, 325
    expressions, 121, 131
    numbers, 171

Estimating, 29. *See also* Rounding.
    in checking, 61–64, 297, 334, 335, 474
    differences, 30, 319
    in division process, 47, 48, 648
    products, 38, 319
    quotients, 319
    sums, 29, 319
Evaluating
    algebraic expressions, 88, 115, 255, 296
    exponential notation, 73, 603
    polynomials, 586
Even number, 147
Expanded notation, 3
    converting to standard notation, 4
Exponent(s), 73
    adding, 592, 609
    on a calculator, 75, 108
    negative integers as, 604, 609
    one as, 593
    zero as, 603
Exponential notation, 73
    and least common multiples, 210
    and multiplying, 592, 609
Expressions
    algebraic, 115
    equivalent, 121, 131
    evaluating, 88, 115, 255, 296
    simplifying, 74
    terms, 123
    value of, 115
Extrapolation, 400

F

Factor, 33, 153, 635
    common, 594
Factor tree, 156
Factoring algebraic expressions, 594
Factorization
    and LCMs, 208
    of natural numbers, 153
    prime, 155
Factorization method for finding LCM, 208, 209
Factors, *see* Factor
Fahrenheit temperature, 567, 568, 573
Fall, distance of, 589
Familiarize, 61
First coordinate, 375
Five, divisibility by, 149
FOIL, 599
Foot, 513
Four, divisibility by, 149
Fraction, *see* Fractions
Fraction key on a calculator, 174, 218

Fractional notation, 159. *See also* Fractions.
    addition, 213, 214
    converting
        from/to decimal notation, 276–278, 309, 312
        from/to mixed numerals, 237–239
        from/to percent notation, 458–460
    and decimal notation together in calculations, 313
    denominator, 159
        of zero, 162, 199
    in division, 187, 643
    for integers, 161, 162, 199
    multiplication using, 165
    numerator, 159
    for one, 161, 199
    order, 216
    percent notation as, 455
    for ratios, 419
    simplest, 172
    simplifying, 172
    subtraction, 223
    for zero, 162, 199
Fractions, 159. *See also* Fractional notation.
    cross-products, 429
    equality test for, 429
    improper, 239
    multiplication by an integer, 165

G

GPA, 394
Gallon, 557
Grade point average, 394
Gram, 565, 566
Graph
    bar, 363
    circle, 160
    line, 366
    of a linear equation, 384, 388
    picture, 355
Graphing calculator, 388
Greater than (>), 30, 87
Grid, 375
Grouping symbols on a calculator, 112
Guess, multiply, and subtract, for division, 647

H

Hectogram, 566
Hectometer, 515
Height of a parallelogram, 523

Home-run differential, 99
Hundred thousandths, 275
Hundredths, 275
Hypotenuse, 548

I

Identity
   additive, 94, 620
   multiplicative, 636
Improper fraction, 239
Inch, 513
Increase, percent of, 480–482
Inequality, 30
Inequality symbols, 30
Integers, 85
   addition of, 93, 94
   division of, 111
   fractional notation, 162, 199
   multiplication of, 105–108
   powers of, 107
   subtraction of, 97
Interest
   compound, 498–500, 505
   computing on a calculator, 500
   effective yield, 504
   rate, 497
   simple, 497, 505
Interpolation, 399
Inverse, additive, 88, 94, 584. *See also* Opposite.

K

Kelvin temperature, 572
Key on a pictograph, 355
Kilogram, 566
Kilometer, 515

L

Laws
   associative, 12, 38
   commutative, 12, 37, 620
   distributive, 35, 122, 137
LCM, *see* Least common multiple
Least common denominators, 214
   and addition, 214
   and subtraction, 223
Least common multiple (LCM), 207
   of denominators, 207
   methods for finding, 207–209
Legs of a right triangle, 548
Leibniz, Gottfried Wilhelm von, 33
Length
   American units, 513
   metric units, 515

Less than (<), 30, 87
Like denominators, 213, 223
Like terms, 124
   combining, 124, 224, 245, 286
Line graph, 366
Linear equation, 384
   graph of, 384
   solution, 383
Linear measure, *see* American system of measures; Metric system of measures.
Liter, 558

M

Marked price, 490
Mass, 565. *See also* Weight.
Math study skills, 6, 14, 50, 90, 100, 150, 322, 368, 596
Mean, 393
   on a calculator, 395
Measures, *see* American system of measures; Angle; Metric system of measures
Median, 395
Meter, 515
Metric system of measures
   area, 536
   capacity, 558, 559
   changing units, 516, 517, 536, 559, 566
      to American system, 518
      mentally, 517, 536, 566
   length, 515
   mass, 566
   temperature, 567
Metric ton, 566
Mile, 513
Milligram, 566
Milliliter, 559
Millimeter, 515
Millions period, 5
Minuend, 19
Missing addend, 20
Mixed numerals, 237
   and addition, 243
   and combining like terms, 245
   converting to decimal notation, 277, 278
   converting from/to fractional notation, 237–239
   and division, 254
   and evaluating expressions, 255
   and multiplication, 253
   negative, 247
   as quotients, 240
   and subtraction, 244
Mode, 396

Money and decimal notation, 276, 295
Monomial, 583
   and multiplication, 591
Multiples, 145
   least common, 207
   of 10, 100, 1000, multiplying by, 640
Multiples method for finding LCM, 207
Multiplication
   associative law of, 38
   basic, 635
   of binomials, 599
   commutative law of, 37
   with decimal notation, 291–295
   estimating products, 38, 319
   using exponential notation, 73, 107, 592, 609
   factors, 33, 635
   of a fraction by an integer, 165
   using fractional notation, 165
   of integers, 105–108
      and fraction, 165
   using mixed numerals, 253
   of monomials, 591
   by multiples of 10, 100, and 1000; 293, 640
   by negative one, 107
   by one, 171, 214, 224, 514, 516, 518, 535, 558, 559, 565, 636
   and parentheses, 33
   of polynomials, 591–594, 599, 600
   by a power of 10, 293, 637, 638
   product, 33, 635
      estimating, 38, 319
   as a rectangular array, 33
   related division sentence, 45, 644
   as repeated addition, 33, 635
   and simplifying, 177
   of whole numbers, 35, 635–640
   by zero, 106, 137, 635
   zeros in, 37
Multiplication principle, 195, 199, 231, 265
Multiplicative identity, 636
Multiplying, *see* Multiplication
Multistep problems, 67

N

Naming an angle, 541
Natural numbers, 3
   composite, 154
   divisibility, 145–149
   factorization, 153
   least common multiple, 207
   multiples, 145
   prime, 154

Negative integers, 85
   as exponents, 604, 609
Negative mixed numerals, 247
Negative numbers on a calculator, 86
Nine, divisibility by, 148
Notation
   decimal, 275
   expanded, 3
   exponential, 73
   fractional, 159
   percent, 455
   ratio, 419
   standard, for numbers, 3
Number line, 27
   addition on, 93
   order on, 87
Numbers
   composite, 154
   digits, 5
   equivalent, 171
   even, 147
   expanded notation, 3
   factoring, 153
   integers, 85
   natural, 3
   negative, on a calculator, 86
   periods, 4
   place-value, 4, 5, 275
   prime, 154
   rational, 275
   standard notation, 3
   whole, 3
   word names for, 4, 275
Numerals, mixed, 237. *See also*
   Mixed numerals.
Numerator, 159

O

Obtuse angle, 543
One
   division by, 162, 199, 645
   as an exponent, 593
   fractional notation for, 161, 199
   multiplying by, 171, 214, 224, 514, 516, 518, 535, 558, 559, 565, 636
Ones period, 5
Operations, order of, 74, 112
Opposite, 85, 88
   of a polynomial, 584
   and subtracting, 97, 585
Order
   of addition, *see* Commutative law of addition
   in decimal notation, 279
   descending, 584
   in fractional notation, 216
   of multiplication, *see* Commutative law of multiplication

   on the number line, 87
   of operations, 74, 112, 303
      and calculators, 75
   of whole numbers, 30
Ordered pair, 375
   as the solution of an equation, 377
Origin, 375
Original price, 490, 505
Ounce, 557, 565

P

Parallelogram, 382, 523
   area, 523, 528, 573
Parentheses
   in multiplication, 33
   within parentheses, 76
Partial quotient, 647
Percent, *see* Percent notation
Percent key on a calculator, 492
Percent of decrease, 480–482
Percent of increase, 480–482
Percent notation, 455
   converting
      from/to decimal notation, 456, 457
      from/to fractional notation, 458–460
   solving problems involving, 467–470, 473–476
Percent symbol, 456
Perimeter, 10, 125
   of a rectangle, 125, 137
   of a square, 126, 137
Periods in word names, 4
Pi (π), 337, 526, 534
   on a calculator, 314, 528
Picture graphs, 355
Pictographs, 355
Pie chart, 160, 455, 542
Pint, 557
Pixels, 39
Place-value, 4, 5, 275
Plane, 375
Plotting points, 375
Points, 375
Polygon, 125
Polynomials, 583
   addition of, 583
   additive inverse, 584
   applications, 581, 586, 590, 611, 612
   binomial, 593
   descending order, 584
   evaluating, 586
   factoring, 594
   monomials, 583
   multiplication of, 591–594, 599, 600
   opposite of, 584
   subtraction, 585

   terms, 583
   trinomial, 600
Positive integers, 85
Pound, 565
Power, 73
   of an integer, 107
Predictions, 399
Price
   marked, 490
   original, 490, 505
   purchase, 487
   sale, 490, 505
   total, 487
   unit, 424
Primary principle of probability, 402
Prime factorization, 155
   in finding LCMs, 208
Prime number, 154
Prime, relatively, 158
Principal, 497
Principles
   addition, 131, 231, 265
   multiplication, 195, 199, 231, 265
Probability, 401
   primary principle of, 402
Problem solving, 61. *See also* Applied problems, Index of Applications.
Product rule, 592, 609
Products, 33, 635
   cross, 429
   estimating, 38, 319
Proportional, 429
Proportions, 429
   and geometric figures, 441
   and similar triangles, 439
   solving, 430–432, 442
   used in solving percent problems, 467–470
Protractor, 541
Purchase price, 487
Pythagoras, 550
Pythagorean equation, 548, 573
Pythagorean theorem, 548

Q

Quadrants, 377
Quart, 557
Quotient, 43
   estimating, 319
   as a mixed numeral, 240
   partial, 647
   zeros in, 49

R

Radical sign ($\sqrt{\ }$), 547
Radius, 337, 525, 573

Rate
  commission, 489, 505
  of discount, 490, 505
  interest, 497
  ratio as, 423
  sales tax, 487
  unit, 424
Ratio, 419
  percent as, 455
  and proportion, 429
  as a rate, 423
  simplifying, 420
Rational numbers, 275
Rays, 541
Reciprocal, 187
  and division, 188, 254
  of zero, 187
Rectangle, 125
  area of, 36, 168, 199
  perimeter of, 125, 137
Rectangular array
  in division, 43, 643
  in multiplication, 33, 635
Rectangular solid, volume of, 555, 556, 573
Reflection of a number, 88
Regrouping
  in addition, 11
  in multiplication, 38
Related sentences
  addition, subtraction, 20, 627
  multiplication, division, 44, 45, 644
Relatively prime, 158
Remainder, 46
Removing a factor equal to one, 172
  and canceling, 174, 178, 189
Repeated addition, 33, 635
Repeated subtraction, 45
Repeating decimals, 309, 310
  rounding, 311
Representative line, 400
Right angle, 543
Right triangle, 548
Rounding
  in checking, 61–64, 299, 334, 335, 474
  of decimal notation, 280
    repeating, 311
  the divisor, 48
  and estimating, 29
  by truncating, 28, 282
  of whole numbers, 27

S

Salary, 489
Sale price, 490
Sales tax, 490, 505

Scale
  on a bar graph, 365
  on a line graph, 366
Second coordinate, 375
Segment, unit, 513
Semiannually compounding, interest, 499
Sentences, see Related sentences
Seven, divisibility by, 149
Sides of an angle, 541
Signs of numbers, changing, 89
Similar geometric figures, 441. See also Similar triangles.
Similar terms, 124. See also Like terms.
Similar triangles, 439
Simple interest, 497, 505
Simplest form of a number, 172
Simplifying
  expressions, 74
  fractional notation, 172
    after multiplying, 177
  for ratios, 420
Six, divisibility by, 147
Solution of an equation, 55, 377. See also Solving equations.
Solving equations, 55
  using addition principle, 131, 225, 325
  checking solutions, 132, 377, 378
  and combining like terms, 327
  by dividing on both sides, 57, 133
  using division principle, 133
  using multiplication principle, 195, 325
  using the principles together, 231, 326
  by subtracting on both sides, 56
  by trial, 55
Solving problems, 61. See also Applied problems, Index of Applications.
  multistep problems, 67
Solving proportions, 430–432
  on a calculator, 442
Speed, 423
Sphere, 557, 573
Square, 126
  area, 73, 337
  perimeter, 126, 137
Square of a number, 547
Square roots, 547
  on a calculator, 548
Standard notation for numbers, 3
State the answer, 61
Statistic, 393
Straight angle, 543
Study skills, 6, 14, 50, 90, 100, 150, 322, 368, 596
Substituting for a variable, 115

Subtraction
  basic, 627
  and borrowing, 22, 244, 284, 629–631
  checking, 22, 631
  with decimal notation, 284, 285
  definition, 20, 97
  difference, 19
    estimating, 30, 319
  estimating differences, 30, 319
  using fractional notation, 223
  as "how much more?", 20, 628
  of integers, 97
  minuend, 19
  using mixed numerals, 244
  and opposites, 97
  of polynomials, 585
  related addition sentence, 20
  repeated, 45
  subtrahend, 19
  as "take away," 19, 627
  of whole numbers, 21, 627–632
  zeros in, 631, 632
Subtrahend, 19
Sum, 9, 620
  estimating, 29, 319
Supplement of an angle, 544
Supplementary angles, 544
Symbol
  for addition (+), 620
  approximately equal to (≈), 30
  greater than (>), 30
  less than (<), 30
  for multiplication (× and ·), 33
  percent (%), 456
  radical (√), 547
Systems of measure, see American system of measures; Metric system of measures.

T

Table of data, 353
Tax, sales, 487
Temperature, 567, 568, 572, 573
Ten, divisibility by, 149
Ten thousandths, 275
Tenths, 275
Terminating decimal, 309
Terms, 123, 583
  like, or similar, 124
Test for equality of fractions, 429
Theorem, Pythagorean, 548
Thousands period, 5
Thousandths, 275
Three, divisibility by, 147
Ton
  American, 565
  metric, 566

Total price, 487
Trace, 388
Translate to an equation, 61, 473
Translating to proportions, 467
Trapezoid, 524
   area, 524, 528, 573
Tree, factor, 156
Triangle(s)
   area, 179, 199
   right, 548
   similar, 439
Trillions period, 5
Trinomial, 600
Truncating, 28, 282
Two, divisibility by, 147

U

Undefined
   division, 111, 646
   fractional notation, 162, 199
Unit angle, 541
Unit cube, 555
Unit price, 424
Unit rate, 424
Unit segment, 513

V

Value of an expression, 115
Variable, 55, 115
   substituting for, 115
Vertex of an angle, 541
Volume
   of a circular cylinder, 556, 573
   of a rectangular solid, 555, 556, 573
   of a sphere, 557, 573

W

Weight, 565. *See also* Mass.
Whole numbers, 3
   addition, 11, 620–624
   division, 45–48, 643–650
   expanded notation, 3
   multiplication, 36, 635–640
   order, 30
   as remainders on a calculator, 300
   rounding, 27
   standard notation, 3
   subtraction, 21, 627–632
   word names for, 4

Widmann, Johann, 620
Window, 388
Word names
   for decimal notation, 275
   for whole numbers, 4

Y

Yard, 513
Yield, effective, 504

Z

Zero
   addition of, 620
   denominator of, 162, 199
   divided by a nonzero number, 111, 646
   division by, 111, 645, 646
   as an exponent, 603
   fractional notation for, 162, 199
   in multiplication, 37
   multiplication by, 106, 137, 635
   in quotients, 49
   reciprocal, 187
   in subtraction, 631, 632

Index of Applications

Biology, Health, and Life Sciences

Allergy shots, 339, 512
Anatomy, human, 480
Asthma inhaler actuations, 511, 572
Birth weight, 205, 258, 262
Body temperature, 333, 339
Body weight, ideal, 407, 410
Bones in hands and feet, 68
Botany, 532
Breast cancer rates, 364, 374
Caloric expenditure, 371, 374, 411, 435
Cholesterol and heart disease, 352
Coffee bean production, 437
Colds from kissing, 483
Drunk driving, 485
Exercise and fitness, 229, 230, 260, 270, 346, 371
Fetal acoustic stimulation, 485
Heart rate, maximum, 403, 484
Heights, 250, 398, 486
Lung capacity, 612
Medicine
 amount in bloodstream, 590
 capsule volume, 560
 dosage, 435, 563, 572, 575, 578
Muscles, weight of, 454, 507
Nursing, 339
Nutrition, 242, 353, 363, 369
Organ transplants, 350
Pregnancy, length of, 398
Sodium consumption, 260
Sugaring, 437
Trampoline injuries, 484
Waist-to-hip ratio, 436
Weightlifting, 261
Weight loss, 67, 68, 72, 79, 103, 484

Business and Economics

Advertising budget, 483
Automobile value, 484
Bicycle
 production, 400
 rental, 343
 sales, 69
Car expense, 419
Car rental, 342, 344, 616
Cellular phone sales, 581, 590
Checking account balance, 62, 70, 79, 110, 236, 274, 290, 350, 448
Commission, 454, 489, 490, 494, 496, 505, 506, 507, 508, 590, 602, 616
Compound interest, 454, 499, 500, 502, 503, 504, 506, 507, 508
Concert revenue, 356
Cost, total, 589
Culinary arts, 339
Diet cola packaging, 65, 66
Dunkin Donuts sales, 6
Electric rates, 343
Effective yield, 504
Farming, 532, 563
FedEx rates, 404
Household expenses, 419
Interest, *see* Compound interest, Simple interest
IRS mileage allowance, 334
Junk mail, 479
Lettuce sales, 361
Loan payments, 70, 82, 273, 334, 335, 339, 350
Lottery sales, 545
Mailing lists, 183
Manufacturing, 186, 262, 307
Merchandising, 563
Military expenditures, 193
Minivan sales, 61
Motion picture expenses, 373, 403
Movie attendance, 304
Movie revenue, 304, 324
National debt, 522
New home sales, 366
Nike revenue, 412
Overtime pay, 342, 343
Packaging, 193, 262, 420
Phone bills, 343
Popcorn sales, 556
Profit and loss, 236, 268
Publishing, 437
Quality control, 437, 447, 610
Real-estate values, 438
Revenue, total, 589
Sales tax, 454, 487, 488, 493, 494, 505, 507, 580, 590, 598, 608, 610, 612
Service calls, 343, 348
Simple interest, 454, 497, 501, 502, 506, 507, 522, 540, 564, 576, 578, 580, 616
Sports marketing, 534
Starting salaries, college graduates, 411
Stock prices, 104, 245, 251, 267, 616
Swimsuit testing, 51
Taxes, 340
Taxi fares, 346
Teachers, new jobs for, 62
Truck rental, 338
Wages, 481, 484
Word processing, 262

Construction and Engineering

Blueprints, 440
Carpentry, 251, 342, 485, 540
Floor coverings, 540
Home furnishings, 257, 260, 267
Ladder height, 550
Land cost, 274
Landscaping, 338, 342, 534
Lot size, 262
Lumber, 576, 578
Masonry, 222, 532
Office space, 272
Painting, 186, 251, 437, 448, 616
Plumbing, 250
Scale models and drawings, 417, 440, 441

Surface area of a building, 185
Surveying, 553
Town planning, 193
Waterproofing, 437
Window area, 314

Consumer Applications

Budget, 80, 184, 452, 505
Calculator purchase, 63
Caloric content, 23, 114
Charities, 240
Coffee drinking, 461
Cooling costs, 485
Coupons, 346, 350, 486
Credit card spending, 102, 404
Depreciation, 482, 485
Discount, 454, 491, 495, 496, 505, 506, 507, 508
Electric bills, 298, 308
Estimating costs, 319, 320, 323
Ethnic foods, popularity, 464
Fat content, 114
Fruit juice sales, 455
Gasoline mileage, 66, 67, 114, 204, 256, 274, 336, 341, 397, 418, 437, 510, 598, 608
Gasoline price, 559
Laptop computer cost, 64
Life insurance, 352
Lottery prize, 448
Meals tax, 554
Phone rates, 354, 362
Price negotiations, 398
Recordings purchased, 464
Sale price, 454, 484, 491, 495, 505, 506, 507
Shopping, 250
Stereo purchase, 438, 452
Unit prices, 418, 424, 426–428, 447, 449, 452, 510, 580, 616
Vacation spending, 461
VCR cost, 64
Woodworking, 226

Environment

Conservation, 563
Deforestation, 370
Drinking water purity, 483
Energy consumption, 464, 480, 481
Food waste, 393
Garbage produced, 447
Global warming, 282, 307, 358
Municipal waste, 170
Oceanography, 563
Ozone layer, 373, 403
Paper recycling, 479
Paper waste, 453
Recyclables, 478
Temperature extremes, 99
Water supplies, 542

Geometry

Area
   of circular objects, 337, 342, 344, 528, 532, 602, 610
   of rectangular objects, 2, 42, 65, 71, 72, 78, 81, 110, 114, 129, 144, 168, 170, 176, 262, 272, 274, 296, 340, 342, 343, 350, 509, 577, 602
   surface, see Surface area
   of triangular objects, 144, 180, 185, 186, 314, 317, 350, 452
Circumference, 532, 598, 612
Diameter, 532
Perimeter
   of rectangular objects, 69, 71, 84, 130 296, 598, 618
   of square objects, 126, 130, 509, 586
Radius, 532
Similar geometric figures, 441, 444–446, 448
   triangles, 440, 443, 444, 446, 448, 449
Surface area, 82, 534, 602
Volume, 556, 557, 560, 576, 578, 580, 602

Miscellaneous

Art supplies, 193
Baking, 217, 221, 222, 267, 472
Bartending, 222
Book size, 250
Candy colors, 405
Cassette tape music, 256
Class size, 437
Coin toss, 401, 406
Cooking, 143, 168, 170, 178, 183, 200, 202, 226, 230, 264, 270, 272, 392, 414
Cord of wood, 556
Elvis impersonators, 70
Fast food, drive-through, 419
Furniture cleaner, 229
Gardening, 193
Grade point average, 351, 352, 394, 395, 397, 398, 410, 414
Grading, 437
Hard-drive space, 62
Interstate highway speed, 23
Knitting, 194
Lincoln-head pennies, 80
Meal planning, 193, 204, 261, 428, 437
Package shipping, 580
Paper quantity, 70
Pixels, 39, 51
Playing cards, 402, 405, 410
Quiz scores, average, 82
Reading rate, 70
Rolling a die, 401, 404, 406
SAT scores, 367
Sewing, 170, 194
Sociology, 183
Study time and test scores, 400, 403
Ticket purchases, 436
Weight of LAV, 70
Writing supplies, 250, 270

Physical Science

Aeronautics, 260
Beaker contents, 51, 80, 415
Cloud volume, 560
Distance traveled, 255
   by falling object, 589
Height of a thrown object, 611
Light distance, 8
Metallurgy, 437, 563
Planet orbits, 212
Planets, 357
Pluto, 8, 563
Rate of travel, average, 352
Reaction temperature, 572
Snow to water, 437
Sound levels, 571
Speed, 423, 426
   of light, 72, 425
   of sound, 425
Temperatures, 261, 263
Test tube contents, 70, 190
Tire revolutions, 534
Water, weight of, 261, 340

Sports, Hobbies, and Entertainment

Aaron, Hank, home runs, 398
At-bats to home-run ratio, 423, 428
Athletics, 586
Audio recording, 361
Baseball diamond, 553, 586
Basketball scoring, 393, 397
Blackjack counting system, 104
Boating, 314
Bowlers, left-handed, 483
Bowling
   averages, 398
   ball, 557
Cheers episodes, 70

Dean, James, 81
Digital-camera screen size, 482
Dining out, 333, 344
Fishing, 259
Football, 170, 246
Gardening, 486
Golfing, 393
Hockey goals, 446
Home-run differential, 99
Knives, 246
Lotteries, 340
Major-league baseball players
   average salary, 372
   number of hits, 412
   payroll, 422
Marathon qualifying times, 510
Monopoly, 7
Moviegoers' ages, 483
Movie releases, 367, 400
NBA gross revenue, 18
NBA salaries, 7
NBA tall men, 76, 264
Nintendo, 324
Pet ownership, 408, 410, 580
Sailing, 185
Soccer field
   dimensions, 343
   goalie's position, 450

Softball diamond, 553, 586
Sporting goods purchases, 545
Strike zone, 486
Television ratings, 308
Television screen dimensions, 420, 554
Tipping, 485
Track and field, 522, 532, 572, 576
TV news magazines programs 360, 362
TV usage, 483
Vertical leaps, 262
Video recording, 261
Young, Cy, earned runs, 438

Statistics and Demographics

Accident incidence, 408, 410, 589
Area of Greenland, 8
Banana consumption, 570
Beverage consumption, 339, 346, 355, 419
Brides, median age, 340
Canyonlands, 51
Centenarians, 365
Desserts sold, 479
Driver fatality rate, 617

Election turnouts, 464
Engagements, length of, 455, 464, 542
Kangaroos, 23
Lake area, 69
Land area of the United States, 506
Lefties, 437, 617
Map drawing and scaling, 71, 183, 418, 438, 449
Oceanography, 7, 103
Offshore oil, 103
Panama Canal length, 340
Population, 2, 7, 307, 359, 454 464, 484, 485, 505, 507, 554, 616
Police officers, number of, 407
River length, 69
San Francisco International passengers, 79
Temperature, 397, 410
Urban planning, 534
Waterfall, average height, 76
White House living area, 540

Using a Scientific Calculator

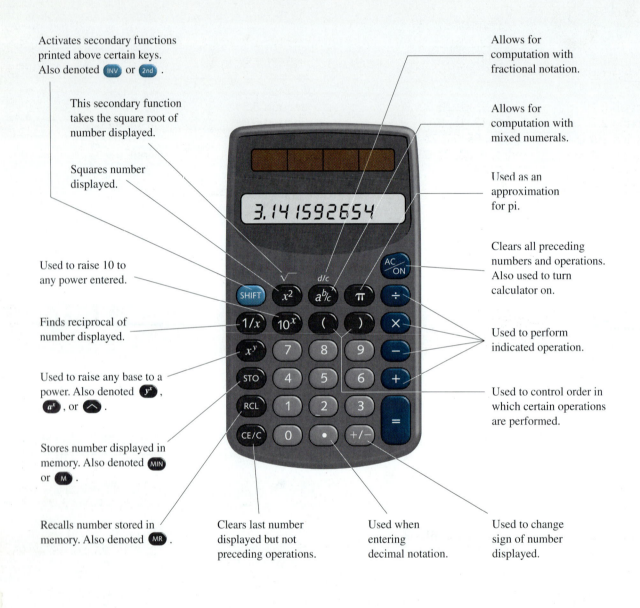

Activates secondary functions printed above certain keys. Also denoted INV or 2nd.

This secondary function takes the square root of number displayed.

Squares number displayed.

Used to raise 10 to any power entered.

Finds reciprocal of number displayed.

Used to raise any base to a power. Also denoted y^x, a^x, or ∧.

Stores number displayed in memory. Also denoted MIN or M.

Recalls number stored in memory. Also denoted MR.

Clears last number displayed but not preceding operations.

Used when entering decimal notation.

Allows for computation with fractional notation.

Allows for computation with mixed numerals.

Used as an approximation for pi.

Clears all preceding numbers and operations. Also used to turn calculator on.

Used to perform indicated operation.

Used to control order in which certain operations are performed.

Used to change sign of number displayed.